MATHEMATICS
Its Contents, Methods and Meaning

THREE VOLUMES BOUND AS ONE

Edited by

A. D. Aleksandrov,
A. N. Kolmogorov,
M. A. Lavrent'ev

Translation edited by
S. H. Gould

Bibliographical Note

This Dover edition, first published in 1999, is an unabridged reprint in one volume of the second edition of *Mathematics: Its Content, Methods and Meaning,* published in three volumes by The M.I.T. Press, Cambridge, MA, in 1969.

Library of Congress Cataloging-in-Publication Data

Matematika, ee soderzhanie metody i znachenie. English.
  Mathematics, its content, methods, and meaning / edited by A.D. Aleksandrov, A.N. Kolmogorov, M.A. Lavrent'ev ; translation edited by S.H. Gould.
     p.   cm.
  Previously published: Cambridge, Mass., M.I.T. Press, 1964, c1963.
  Includes index.
  ISBN 0-486-40916-3 (pbk.)
  1. Mathematics. I. Aleksandrov, A. D. (Aleksandr Danilovich), 1912– . II. Kolmogorov, A. N. (Andrei Nikolaevich), 1903– . III. Lavrent'ev, Mikhail Alekseevich, 1900– . IV. Title.
QA36.M2913   1999
510—dc21
                                                                99-33023
                                                                CIP

Manufactured in the United States of America
Dover Publications, Inc., 31 East 2nd Street, Mineola, N.Y. 11501

PREFACE TO
THE RUSSIAN EDITION

Mathematics, which originated in antiquity in the needs of daily life, has developed into an immense system of widely varied disciplines. Like the other sciences, it reflects the laws of the material world around us and serves as a powerful instrument for our knowledge and mastery of nature. But the high level of abstraction peculiar to mathematics means that its newer branches are relatively inaccessible to nonspecialists. This abstract character of mathematics gave birth even in antiquity to idealistic notions about its independence of the material world.

In preparing the present volume, the authors have kept in mind the goal of acquainting a sufficiently wide circle of the Soviet intelligentsia with the various mathematical disciplines, their content and methods, the foundations on which they are based, and the paths along which they have developed.

As a minimum of necessary mathematical knowledge on the part of the reader, we have assumed only secondary-school mathematics, but the volumes differ from one another with respect to the accessibility of the material contained in them. Readers wishing to acquaint themselves for the first time with the elements of higher mathematics may profitably read the first few chapters, but for a complete understanding of the subsequent parts it will be necessary to have made some study of corresponding textbooks. The book as a whole will be understood in a fundamental way only by readers who already have some acquaintance with the applications of mathematical analysis; that is to say, with the differential and integral calculus. For such readers, namely teachers of mathematics and instructors in engineering and the natural sciences, it will be particularly important to read those chapters which introduce the newer branches of mathematics.

iii

Naturally it has not been possible, within the limits of one book, to exhaust all the riches of even the most fundamental results of mathematical research; a certain freedom in the choice of material has been inevitable here. But along general lines, the present book will give an idea of the present state of mathematics, its origins, and its probable future development. For this reason the book is also intended to some extent for persons already acquainted with most of the factual material in it. It may perhaps help to remove a certain narrowness of outlook occasionally to be found in some of our younger mathematicians.

The separate chapters of the book are written by various authors, whose names are given in the Contents. But as a whole the book is the result of collaboration. Its general plan, the choice of material, the successive versions of individual chapters, were all submitted to general discussion, and improvements were made on the basis of a lively exchange of opinions. Mathematicians from several cities in the Soviet Union were given an opportunity, in the form of organized discussion, to make many valuable remarks concerning the original version of the text. Their opinions and suggestions were taken into account by the authors.

The authors of some of the chapters also took a direct share in preparing the final version of other chapters: The introductory part of Chapter II was written essentially by B. N. Delone, while D. K. Faddeev played an active role in the preparation of Chapter IV and Chapter XX.

A share in the work was also taken by several persons other than the authors of the individual chapters: §4 of Chapter XIV was written by L. V. Kantorovič, §6 of Chapter VI by O. A. Ladyženskaja, §5 of Chapter 10 by A. G. Postnikov; work was done on the text of Chapter V by O. A. Oleĭnik and on Chapter XI by Ju. V. Prohorov.

Certain sections of Chapters I, II, VII, and XVII were written by V. A. Zalgaller. The editing of the final text was done by V. A. Zalgaller and V. S. Videnskiĭ with the cooperation of T. V. Rogozkinaja and A. P. Leonovaja.

The greater part of the illustrations were prepared by E. P. Sen'kin.

Moscow
1956   Editorial Board

FOREWORD BY THE EDITOR OF THE TRANSLATION

Mathematics, in view of its abstractness, offers greater difficulty to the expositor than any other science. Yet its rapidly increasing role in modern life creates both a need and a desire for good exposition.

In recent years many popular books about mathematics have appeared in the English language, and some of them have enjoyed an immense sale. But for the most part they have contained little serious mathematical instruction, and many of them have neglected the twentieth century, the undisputed "golden age" of mathematics. Although they are admirable in many other ways, they have not yet undertaken the ultimate task of mathematical exposition, namely the large-scale organization of modern mathematics in such a way that the reader is constantly delighted by the obvious economizing of his own time and effort. Anyone who reads through some of the chapters in the present book will realize how well this task has been carried out by the Soviet authors, in the systematic collaboration they have described in their preface.

Such a book, written for "a wide circle of the intelligentsia," must also discuss the general cultural importance of mathematics and its continuous development from the earliest beginnings of history down to the present day. To form an opinion of the book from this point of view the reader need only glance through the first chapter in Part 1 and the introduction to certain other chapters; for example, Analysis, or Analytic Geometry.

In translating the passages on the history and cultural significance of mathematical ideas, the translators have naturally been aware of even greater difficulties than are usually associated with the translation of scientific texts. As organizer of the group, I express my profound gratitude to the other two translators, Tamas Bartha and Kurt Hirsch, for their skillful cooperation.

v

The present translation, which was originally published by the American Mathematical Society, will now enjoy a more general distribution in its new format. In thus making the book more widely available the Society has been influenced by various expressions of opinion from American mathematicians. For example, ". . . the book will contribute materially to a better understanding by the public of what mathematicians are up to. . . . It will be useful to many mathematicians, physicists and chemists, as well as to laymen. . . . Whether a physicist wishes to know what a Lie algebra is and how it is related to a Lie group, or an undergraduate would like to begin the study of homology, or a crystallographer is interested in Fedorov groups, or an engineer in probability, or any scientist in computing machines, he will find here a connected, lucid account."

In its first edition this translation has been widely read by mathematicians and students of mathematics. We now look forward to its wider usefulness in the general English-speaking world.

August, 1964

S. H. GOULD
Editor of Translations
American Mathematical Society
Providence, Rhode Island

CONTENTS

VOLUME ONE, PART 1

PART 2

CONTENTS

VOLUME TWO, PART 3

PART 4

CONTENTS

VOLUME THREE, PART 5

PART 6

CHAPTER XVIII   TOPOLOGY   193

P. S. Aleksandrov

CHAPTER XIX   FUNCTIONAL ANALYSIS   227

I. M. Gel'fand

CHAPTER XX   GROUPS AND OTHER ALGEBRAIC
                  SYSTEMS   263

A. I. Mal'cev

VOLUME ONE

PART 1

CHAPTER I

A GENERAL VIEW
OF MATHEMATICS

An adequate presentation of any science cannot consist of detailed information alone, however extensive. It must also provide a proper view of the essential nature of the science as a whole. The purpose of the present chapter is to give a general picture of the essential nature of mathematics. For this purpose there is no great need to introduce any of the details of recent mathematical theories, since elementary mathematics and the history of the science already provide a sufficient foundation for general conclusions.

§1. The Characteristic Features of Mathematics

1. Abstractions, proofs, applications. With even a superficial knowledge of mathematics, it is easy to recognize certain characteristic features: its abstractness, its precision, its logical rigor, the indisputable character of its conclusions, and finally, the exceptionally broad range of its applications.

The abstractness of mathematics is easy to see. We operate with abstract numbers without worrying about how to relate them in each case to concrete objects. In school we study the abstract multiplication table, that is, a table for multiplying one abstract number by another, not a number of boys by a number of apples, or a number of apples by the price of an apple.

Similarly in geometry we consider, for example, straight lines and not stretched threads, the concept of a geometric line being obtained by abstraction from all other properties, excepting only extension in one direction. More generally, the concept of a geometric figure is the result of abstraction from all the properties of actual objects except their spatial form and dimensions.

Abstractions of this sort are characteristic for the whole of mathematics. The concept of a whole number and of a geometric figure are only two of the earliest and most elementary of its concepts. They have been followed by a mass of others, too numerous to describe, extending to such abstractions as complex numbers, functions, integrals, differentials, functionals, *n*-dimensional, and even infinite-dimensional spaces, and so forth. These abstractions, piled up as it were on one another, have reached such a degree of generalization that they apparently lose all connection with daily life and the "ordinary mortal" understands nothing about them beyond the mere fact that "all this is incomprehensible."

In reality, of course, the case is not so at all. Although the concept of *n*-dimensional space is no doubt extremely abstract, yet it does have a completely real content, which is not very difficult to understand. In the present book it will be our task to emphasize and clarify the concrete content of such abstract concepts as those mentioned earlier, so that the reader may convince himself that they are all connected with actual life, both in their origin and in their applications.

But abstraction is not the exclusive property of mathematics; it is characteristic of every science, even of all mental activity in general. Consequently, the abstractness of mathematical concepts does not in itself give a complete description of the peculiar character of mathematics.

The abstractions of mathematics are distinguished by three features. In the first place, they deal above all else with quantitative relations and spatial forms, abstracting them from all other properties of objects. Second, they occur in a sequence of increasing degrees of abstraction, going very much further in this direction than the abstractions of other sciences. We will illustrate these two features in detail later, using as examples the fundamental notions of number and figure. Finally, and this is obvious, mathematics as such moves almost wholly in the field of abstract concepts and their interrelations. While the natural scientist turns constantly to experiment for proof of his assertions, the mathematician employs only argument and computation.

It is true that mathematicians also make constant use, to assist them in the discovery of their theorems and methods, of models and physical analogues, and they have recourse to various completely concrete examples. These examples serve as the actual source of the theory and as a means of discovering its theorems, but no theorem definitely belongs to mathematics until it has been rigorously proved by a logical argument. If a geometer, reporting a newly discovered theorem, were to demonstrate it by means of models and to confine himself to such a demonstration, no mathematician would admit that the theorem had been proved. The demand for a proof of a theorem is well known in high school geometry, but it pervades the whole of mathematics. We could measure the angles at the base of a thousand isosceles triangles with extreme accuracy, but such a procedure would never provide us with a mathematical proof of the theorem that the angles at the base of an isosceles triangle are equal. Mathematics demands that this result be deduced from the fundamental concepts of geometry, which at the present time, in view of the fact that geometry is nowadays developed on a rigorous basis, are precisely formulated in the axioms. And so it is in every case. To prove a theorem means for the mathematician to deduce it by a logical argument from the fundamental properties of the concepts occurring in that theorem. In this way, not only the concepts but also the methods of mathematics are abstract and theoretical.

The results of mathematics are distinguished by a high degree of logical rigor, and a mathematical argument is conducted with such scrupulousness as to make it incontestable and completely convincing to anyone who understands it. The scrupulousness and cogency of mathematical proofs are already well known in a high school course. Mathematical truths are in fact the prototype of the completely incontestable. Not for nothing do people say "as clear as two and two are four." Here the relation "two and two are four" is introduced as the very image of the irrefutable and incontestable.

But the rigor of mathematics is not absolute; it is in a process of continual development; the principles of mathematics have not congealed once and for all but have a life of their own and may even be the subject of scientific quarrels.

In the final analysis the vitality of mathematics arises from the fact that its concepts and results, for all their abstractness, originate, as we shall see, in the actual world and find widely varied application in the other sciences, in engineering, and in all the practical affairs of daily life; to realize this is the most important prerequisite for understanding mathematics.

The exceptional breadth of its applications is another characteristic feature of mathematics.

In the first place we make constant use, almost every hour, in industry and in private and social life, of the most varied concepts and results of mathematics, without thinking about them at all; for example, we use arithmetic to compute our expenses or geometry to calculate the floor area of an apartment. Of course, the rules here are very simple, but we should remember that in some period of antiquity they represented the most advanced mathematical achievements of the age.

Second, modern technology would be impossible without mathematics. There is probably not a single technical process which can be carried through without more or less complicated calculations; and mathematics plays a very important role in the development of new branches of technology.

Finally, it is true that every science, to a greater or lesser degree, makes essential use of mathematics. The "exact sciences," mechanics, astronomy, physics, and to a great extent chemistry, express their laws, as every schoolboy knows, by means of formulas and make extensive use of mathematical apparatus in developing their theories. The progress of these sciences would have been completely impossible without mathematics. For this reason the requirements of mechanics, astronomy, and physics have always exercised a direct and decisive influence on the development of mathematics.

In other sciences mathematics plays a smaller role, but here too it finds important applications. Of course, in the study of such complicated phenomena as occur in biology and sociology, the mathematical method cannot play the same role as, let us say, in physics. In all cases, but especially where the phenomena are most complicated, we must bear in mind, if we are not to lose our way in meaningless play with formulas, that the application of mathematics is significant only if the concrete phenomena have already been made the subject of a profound theory. In one way or another, mathematics is applied in almost every science, from mechanics to political economy.

Let us recall some particularly brilliant applications of mathematics in the exact sciences and in technology.

The planet Neptune, one of the most distant in the Solar System, was discovered in the year 1846 on the basis of mathematical calculations. By analyzing certain irregularities in the motion of Uranus, the astronomers Adams and Leverrier came to the conclusion that these irregularities were caused by the gravitational attraction of another planet. Leverrier calculated on the basis of the laws of mechanics exactly where this planet must be, and an observer to whom he communicated his results caught sight of it in his telescope in the exact position indicated by Leverrier. This discovery was a triumph not only for mechanics and astronomy, and in particular for the system of Copernicus, but also for the powers of mathematical calculation.

Another example, no less impressive, was the discovery of electromagnetic waves. The English physicist Maxwell, by generalizing the laws of electromagnetic phenomena as established by experiment, was able to express these laws in the form of equations. From these equations he deduced, by purely mathematical methods, that electromagnetic waves could exist and that they must be propagated with the speed of light. On the basis of this result, he proposed the electromagnetic theory of light, which was later developed and deepened in every direction. Moreover, Maxwell's results led to the search for electromagnetic waves of purely electrical origin, arising for example from an oscillating charge. These waves were actually discovered by Hertz. Shortly afterwards, A. S. Popov, by discovering means for exciting, transmitting, and receiving electromagnetic oscillations made them available for a wide range of applications and thereby laid the foundations for the whole technology of radio. In the discovery of radio, now the common possession of everyone, an important role was played by the results of a purely mathematical deduction.

So from observation, as for example of the deflection of a magnetic needle by an electric current, science proceeds to generalization, to a theory of the phenomena, and to formulation of laws and to mathematical expression of them. From these laws come new deductions, and finally, the theory is embodied in practice, which in its turn provides powerful new impulses for the development of the theory.

It is particularly remarkable that even the most abstract constructions of mathematics, arising within that science itself, without any immediate motivation from the natural sciences or from technology, nevertheless have fruitful applications. For example, imaginary numbers first came to light in algebra, and for a long time their significance in the actual world remained uncomprehended, a circumstance indicated by their very name. But when about 1800 a geometrical interpretation (see Chapter IV, §2) was given to them, imaginary numbers became firmly established in mathematics, giving rise to the extensive theory of functions of a complex variable, i.e., of a variable of the form $x + y \sqrt{-1}$. This theory of "imaginary" functions of an "imaginary" variable proved itself to be far from imaginary, but rather a very practical means of solving technological problems. Thus, the fundamental results of N. E. Jukovski concerning the lift on the wing of an airplane are proved by means of this theory. The same theory is useful, for example, in the solution of problems concerning the oozing of water under a dam, problems whose importance is obvious during the present period of construction of huge hydroelectric stations.

Another example, equally impressive, is provided by non-Euclidean geometry,* which arose from the efforts, extending for 2000 years from the time of Euclid, to prove the parallel axiom, a problem of purely

---

* Here we merely point out this example without further explanation, for which the reader may turn to Chapter XVII.

mathematical interest. N. I. Lobačevskiǐ himself, the founder of the new geometry, was careful to label his geometry "imaginary," since he could not see any meaning for it in the actual world, although he was confident that such a meaning would eventually be found. The results of his geometry appeared to the majority of mathematicians to be not only "imaginary" but even unimaginable and absurd. Nevertheless, his ideas laid the foundation for a new development of geometry, namely the creation of theories of various non-Euclidean spaces; and these ideas subsequently became the basis of the general theory of relativity, in which the mathematical apparatus consists of a form of non-Euclidean geometry of four-dimensional space. Thus the abstract constructions of mathematics, which at the very least seemed incomprehensible, proved themselves a powerful instrument for the development of one of the most important theories of physics. Similarly, in the present-day theory of atomic phenomena, in the so-called quantum mechanics, essential use is made of many extremely abstract mathematical concepts and theories, as for example the concept of infinite-dimensional space.

There is no need to give any further examples, since we have already shown with sufficient emphasis that mathematics finds widespread application in everyday life and in technology and science; in the exact sciences and in the great problems of technology, applications are found even for those theories which arise within mathematics itself. This is one of the characteristic peculiarities of mathematics, along with its abstractness and the rigor and conclusiveness of its results.

2. The essential nature of mathematics. In discussing these special features of mathematics we have been far from explaining its essence; rather we have merely pointed out its external marks. Our task now is to explain the essential nature of these characteristic features. For this purpose it will be necessary to answer, at the very least, the following questions:

What do these abstract mathematical concepts reflect? In other words, what is the actual subject matter of mathematics?

Why do the abstract results of mathematics appear so convincing, and its initial concepts so obvious? In other words, on what foundation do the methods of mathematics rest?

Why, in spite of all its abstractness, does mathematics find such wide application and does not turn out to be merely idle play with abstractions? In other words, how is the significance of mathematics to be explained?

Finally, what forces lead to the further development of mathematics, allowing it to unite abstractness with breadth of application? What is the basis for its continuing growth?

In answering these questions we will form a general picture of the content of mathematics, of its methods, and of its significance and its development; that is, we will understand its essence.

Idealists and metaphysicists not only fall into confusion in their attempts to answer these basic questions but they go so far as to distort mathematics completely, turning it literally inside out. Thus, seeing the extreme abstractness and cogency of mathematical results, the idealist imagines that mathematics issues from pure thought.

In reality, mathematics offers not the slightest support for idealism or metaphysics. We will convince ourselves of this as we attempt, in general outline, to answer the listed questions about the essence of mathematics. For a preliminary clarification of these questions, it is sufficient to examine the foundations of arithmetic and elementary geometry, to which we now turn.

§2. Arithmetic

1. The concept of a whole number. The concept of number (for the time being, we speak only of whole positive numbers), though it is so familiar to us today, was worked out very slowly. This can be seen from the way in which counting has been done by various races who until recent times have remained at a relatively primitive level of social life. Among some of them, there were no names for numbers higher than two or three; among others, counting went further but ended after a few numbers, after which they simply said "many" or "countless." A stock of clearly distinguished names for numbers was only gradually accumulated among the various peoples.

At first these peoples had no concept of what a number is, although they could in their own fashion make judgments about the size of one or another collection of objects met with in their daily life. We must conclude that a number was directly perceived by them as an inseparable property of a collection of objects, a property which they did not, however, clearly distinguish. We are so accustomed to counting that we can hardly imagine this state of affairs, but it is possible to understand it.*

At the next higher level a number already appears as a property of a

---

* In fact, every collection of objects, whether it be a flock of sheep or a pile of firewood, exists and is immediately perceived in all its concreteness and complexity. The distinguishing in it of separate properties and relationships is the result of conscious analysis. Primitive thought does not yet make this analysis, but considers the object only as a whole. Similarly, a man who has not studied music perceives a musical composition without distinguishing in it the details of melody, tonality, and so forth, while at the same time a musician easily analyzes even a complicated symphony.

collection of objects, but it is not yet distinguished from the collection as an "abstract number," as a number in general, not connected with concrete objects. This is obvious from the names of numbers among certain peoples, as "hand" for five and "wholeman" for twenty. Here five is to be understood not abstractly but simply in the sense of "as many as the fingers on a hand," twenty is "as many as the fingers and toes on a man" and so forth. In a completely analogous way, certain peoples had no concept of "black," "hard," or "circular." In order to say that an object is black, they compared it with a crow for example, and to say that there were five objects, they directly compared these objects with a hand. In this way it also came about that various names for numbers were used for various kinds of objects; some numbers for counting people, others for counting boats, and so forth, up to as many as ten different kinds of numbers. Here we do not have abstract numbers, but merely a sort of "appellation," referring only to a definite kind of objects. Among other peoples there were in general no separate names for numbers, as for example, no word for "three," although they could say "three men" or "in three places," and so forth.

Similarly among ourselves, we quite readily say that this or that object is black but much more rarely speak about "blackness" in itself, which is a more abstract concept.*

The number of objects in a given collection is a property of the collection, but the number itself, as such, the "abstract number," is a property abstracted from the concrete collection and considered simply in itself, like "blackness" or "hardness." Just as blackness is the property common to all objects of the color of coal, so the number "five" is the common property of all collections containing as many objects as there are fingers on a hand. In this case the equality of the two numbers is established by simple comparison: We take an object from the collection, bend one finger over, and count in this way up to the end of the collection. More generally, by pairing off the objects of two collections, it is possible, without making any use of numbers at all, to establish whether or not the collections contain the same number of objects. For example, if guests are taking their places at the table they can easily, without any counting, make it clear to the hostess that she has forgotten one setting, since one guest will be without a setting.

---

* In the formation of concepts about properties of objects, such as color or the numerosity of a collection, it is possible to distinguish three steps, which we must not, of course, try to separate too sharply from one another. At the first step the property is defined by direct comparison of objects: like a crow, as many as on a hand. At the second, an adjective appears: a black stone or (the numerical adjective being quite analogous) five trees. At the third step the property is abstracted from the objects and may appear "as such"; for example "blackness," or the abstract number "five."

In this way it is possible to give the following definition of a number: Each separate number like "two," "five," and so forth, is that property of collections of objects which is common to all collections whose objects can be put into one-to-one correspondence with one another and which is different for those collections for which such a correspondence is impossible. In order to discover this property and to distinguish it clearly, that is, in order to form the concept of a definite number and to give it a name "six," "ten," and so forth, it was necessary to compare many collections of objects. For countless generations people repeated the same operation millions of times and in that way discovered numbers and the relations among them.

2. Relations among the whole numbers. Operations with numbers arose in their turn as a reflection of relations among concrete objects. This is observable even in the names of numbers. For example, among certain American Indians the number "twenty-six" is pronounced as "above two tens I place a six," which is clearly a reflection of a concrete method of counting objects. Addition of numbers corresponds to placing together or uniting two or more collections, and it is equally easy to see the concrete meaning of subtraction, multiplication, and division. Multiplication in particular arose to a great extent, it seems clear, from the habit of counting off equal collections: that is, by twos, by threes, and so forth.

In the process of counting, men not only discovered and assimilated the relations among the separate numbers, as for example that two and three are five, but also they gradually established certain general laws. By practical experience, it was discovered that a sum does not depend on the order of the summands and that the result of counting a given set of objects does not depend on the order in which the counting takes place, a fact which is reflected in the essential identity of the "ordinal" and "cardinal" numbers: first, second, third, and one, two, three. In this way the numbers appeared not as separate and independent, but as interrelated with one another.

Some numbers are expressed in terms of others in their very names and in the way they are written. Thus, "twenty" denotes "two (times) ten"; in French, eighty is "four-twenties" (quatre-vingt), ninety is "four-twenties-ten"; and the Roman numerals VIII, IX denote that $8 = 5 + 3$, $9 = 10 - 1$.

In general, there arose not just the separate numbers but a system of numbers with mutual relations and rules.

The subject matter of arithmetic is exactly this, the system of numbers with its mutual relations and rules.* The separate abstract number by itself does not have tangible properties, and in general there is very little to be said about it. If we ask ourselves, for example, about the properties of the number six, we note that $6 = 5 + 1, 6 = 3 \cdot 2$, that 6 is a factor of 30 and so forth. But here the number 6 is always connected with other numbers; in fact, the properties of a given number consist precisely of its relations with other numbers.† Consequently, it is clear that every arithmetical operation determines a connection or relation among numbers. Thus the subject matter of arithmetic is relations among numbers. But these relations are the abstract images of actual quantitative relations among collections of objects; so we may say that arithmetic is the science of actual quantitative relations considered abstractly, that is, purely as relations. Arithmetic, as we see, did not arise from pure thought, as the idealists represent, but is the reflection of definite properties of real things; it arose from the long practical experience of many generations.

3. Symbols for the numbers. As social life became more extensive and complicated, it posed broader problems. Not only was it necessary to take note of the number of objects in a set and to tell others about it, a necessity which had already led to formulation of the concept of number and to names for the numbers, but it became essential to learn to count increasingly larger collections, of animals in a herd, of objects for exchange, of days before a fixed date, and so forth, and to communicate the results of the count to others. This situation absolutely demanded improvement in the names and also in the symbols for numbers.

The introduction of symbols for the numbers, which apparently occured as soon as writing began, played a great role in the development of arithmetic. Moreover, it was the first step toward mathematical signs and formulas in general. The second step, consisting of the introduction of signs for arithmetical operations and of a literal designation for the unknown (x), was taken considerably later.

The concept of number, like every other abstract concept, has no immediate image; it cannot be exhibited but can only be conceived in the

---

* The word "arithmetic," meaning the "art of calculation," is derived from the Greek adjective "arithmetic" formed from the noun "arithmos," meaning "number." The adjective modifies a noun "techne" (art, technique), which is here understood.

† This is understandable from the most general considerations. An arbitrary abstraction, removed from its concrete basis (just as a number is abstracted from a concrete collection of objects), has no sense "in itself"; it exists only in its relations with other concepts. These relations are already implicit in any statement about the abstraction, in the most incomplete definition of it. Without them the abstraction lacks content and significance, i.e., it simply does not exist. The content of the concept of an abstract number lies in the rules, in the mutual relations of the system of numbers.

mind. But thought is formulated in language, so that without a name there can be no concept. The symbol is also a name, except that it is not oral but written and presents itself to the mind in the form of a visible image. For example, if I say "seven," what do you picture to yourself? Probably not a set of seven objects of one kind or another, but rather the symbol "7," which forms a sort of tangible framework for the abstract number "seven." Moreover, a number 18273 is considerably harder to pronounce than to write and cannot be pictured with any accuracy in the form of a set of objects. In this way it came about, though only after some lapse of time, that the symbols gave rise to the conception of numbers so large that they could never have been discovered by direct observation or by enumeration. With the appearance of government, it was necessary to collect taxes, to assemble and outfit an army, and so forth, all of which required operations with very large numbers.

Thus the importance of symbols for the numbers consists, in the first place, in their providing a simple embodiment of the concept of an abstract number.* This is the role of mathematical designations in general: They provide an embodiment of abstract mathematical concepts. Thus $+$ denotes addition, x denotes an unknown number, a an arbitrary given number, and so forth. In the second place the symbols for numbers provide a particularly simple means of carrying out operations on them. Everyone knows how much easier it is "to calculate on paper" than "in one's head." Mathematical signs and formulas have this advantage in general: They allow us to replace a part of our arguments with calculations, with something that is almost mechanical. Moreover, if a calculation is written down, it already possesses a definite authenticity; everything is visible, everything can be checked, and everything is defined by exact rules. As examples one might mention addition by individual columns or any algebraic transformation such as "taking over to the other side of the equation with change of sign." From what has been said, it is clear that without suitable symbols for the numbers arithmetic could not have made much progress. Even more is it true that contemporary mathematics would be impossible without its special signs and formulas.

It is obvious that the extremely convenient method of writing numbers that is in use today could not have been worked out all at once. From ancient times there appeared among various peoples, from the very

---

* It is worth remarking that the concept of number, which was worked out with such difficulty in a long period of time, is mastered nowadays by a child with relative ease. Why? The first reason is, of course, that the child hears and sees adults constantly making use of numbers, and they even teach him to do the same. But a second reason, and this is the one to which we wish to draw special attention, is that the child already has at hand words and symbols for the numbers. He first learns these external symbols for number and only later masters the meaning of them.

Table

	Slavic		Chinese			
	Cyrillic	Glagolitic	Ancient	Commercial	Scientific	Greek
0				○	○	
1	$\tilde{\Delta}$	✛	一	|	$\hat{\text{I}}$	$\bar{\alpha}$
2	$\tilde{\text{в}}$	Ⰱ	二	||	||	$\bar{\beta}$
3	$\tilde{\Gamma}$	Ⰲ	三	|||	|||	$\bar{\gamma}$
4	$\tilde{A}$	Ⰳ	四	X	||||	$\bar{\delta}$
5	$\tilde{\epsilon}$	Ⰴ	五	⅄	|||||	$\bar{\epsilon}$
6	$\tilde{s}$	Ⰵ	六	⅃	Т	$\bar{s}$
7	$\tilde{z}$	Ⰶ	七	⊥	Π	$\bar{\zeta}$
8	$\tilde{H}$	Ⰷ	八	⊥	Ⅲ	$\bar{\eta}$
9	$\tilde{\theta}$	Ⰸ	九	⅄	ⅢⅠ	$\bar{\theta}$
10	$\tilde{\text{I}}$	Ⰹ	十	+	Ⅰ○	$\bar{\iota}$
20	$\tilde{K}$	Ⰺ	$\stackrel{=}{+}$	$\stackrel{\|\|}{+}$	Ⅱ○	$\bar{\kappa}$
30	$\tilde{\Lambda}$	Ⰻ	$\stackrel{\equiv}{+}$	$\stackrel{\|\|\|}{+}$	Ⅲ○	$\bar{\lambda}$
100	$\tilde{\rho}$	Ⰼ	百	ⅅ	Ⅰ○○	$\bar{\rho}$
1000	$\diagup\tilde{\Delta}$	Ⰽ	千	千	Ⅰ○○○	$\diagup\bar{\alpha}$

1

	Arabic	Georgian	Egyptian Hieroglyphic	Egyptian Hieratic	Roman	Mayan
0						⬭
1	ī	ᔕ	I	I	I	•
2	ᴗ	𝟠	II	u	II	••
3	ح	𝟔	III	ui	III	•••
4	৬	𝒬	IIII	ши	IV	••••
5	•	𝒪	|||| (III/II)	𝟏	V	—
6	𝟗	𝟥	|||| (III/III)	𝟤	VI	⎯•⎯
7	ز	𝟠	|||| (IIII/III)	⌒	VII	⎯••⎯
8	ح	𝟪	|||| (IIII/IIII)	⇒	VIII	⎯•••⎯
9	𝖻	𝔬	|||| (III/III/III)	𝟥	IX	⎯••••⎯
10	ᔕ	⌒	∩	⋀	X	⩵
20	ᔕ	𝔧	∩∩	𝝀	XX	
30	ᔍ	𝔐	∩∩∩	X	XXX	
100	ᴅ̈	𝔥	ᕴ	⟋	C	
1000	غ	𝔞	𝔭	𝔰	M	

beginnings of their culture, various symbols for the numbers, which were very unlike our contemporary ones not only in their general appearance but also in the principles on which they were chosen. For example, the decimal system was not used everywhere, and among the ancient Babylonians there was a system that was partly decimal and partly sexagesimal. Table 1 gives some of the symbols for numbers among various peoples. In particular, we see that the ancient Greeks, and later also the Russians, made use of letters to designate numbers. Our contemporary "Arabic" symbols and, more generally, our method of forming the numbers, were brought from India to Europe by the Arabs in the 10th century and became firmly rooted there in the course of the next few centuries.

The first peculiarity of our system is that it is a decimal system. But this is not a matter of great importance, since it would have been quite possible to use, for example, a duodecimal system by introducing special symbols for ten and eleven. The most important peculiarity of our system of designating numbers is that it is "positional"; that is, that one and the same number has a different significance depending upon its position. For example, in 372 the number 3 denotes the number of hundreds and 7 the numbers of tens. This method of writing is not only concise and simple but makes calculations very easy. The Roman numerals were in this respect much less convenient, the same number 372 being written in the form CCCLXXII; it is a very laborious task to multiply together two large numbers written in Roman numerals.

Positional writing of numbers demands that in one way or another we take note of any category of numbers that has been omitted, since if we do not do this, we will confuse, for example, thirty-one with three-hundred-and-one. In the position of the omitted category we must place a zero, thereby distinguishing 301 and 31. In a rudimentary form, zero already appears in the late Babylonian cuneiform writings, but its systematic introduction was an achievement of the Indians:* It allowed them to proceed to a completely positional system of writing just as we have it today.

But in this way zero also became a number and entered into the system of numbers. By itself zero is nothing; in the Sanskrit language of ancient India, it is called exactly that: "empty" (çūnga); but in connection with other numbers, zero acquires content, and well-known properties; for example, an arbitrary number plus zero is the same number, or when an arbitrary number is multiplied by zero it becomes zero.

---

* The first Indian manuscript in which zero appears comes from the end of the 9th century; in it the number 270 is written exactly as we would write it today. But it is probable that zero was introduced in India still earlier, in the 6th century.

4. The theory of numbers as a branch of pure mathematics. Let us return to the arithmetic of the ancients. The oldest texts that have been preserved from Babylon and Egypt go back to the second millennium B.C. These and later texts contain various arithmetical problems with their solutions, among them certain ones that today belong to algebra, such as the solution of quadratic and even cubic equations or progressions; all this being presented, of course, in the form of concrete problems and numerical examples. Among the Babylonians we also find certain tables of squares, cubes, and reciprocals. It is to be supposed that they were already beginning to form mathematical interests which were not immediately connected with practical problems.

In any case arithmetic was well developed in ancient Babylon and Egypt. However, it was not yet a mathematical theory of numbers but rather a collection of solutions for various problems and of rules of calculation. It is exactly in this way that arithmetic is taught up to the present time in our elementary schools and is understood by everyone who is not especially interested in mathematics. This is perfectly legitimate, but arithmetic in this form is still not a mathematical theory. There are no general theorems about numbers.

The transition to theoretical arithmetic proceeded gradually.

As was pointed out, the existence of symbols allows us to operate with numbers so large that it is impossible to visualize them as collections of objects or to arrive at them by the process of counting in succession from the number one. Among primitive tribes special numbers were worked out up to 3, 10, 100 and so forth, but after these came the indefinite "many." In contrast to this situation the use of symbols for numbers enabled the Chinese, the Babylonians, and the Egyptians to proceed to tens of thousands and even to millions. It was at this stage that the possibility was noticed of indefinitely extending the series of numbers, although we do not know how soon this possibility was clearly perceived. Even Archimedes (287–212 B.C.) in his remarkable essay "The Sand Reckoner" took the trouble to describe a method for naming a number greater than the number of grains of sand sufficient to fill up the "sphere of the fixed stars." So the possibility of naming and writing such a number still required at his time a detailed explanation.

By the 3rd century B.C., the Greeks had clearly recognized two important ideas: first, that the sequence of numbers could be indefinitely extended and second, that it was not only possible to operate with arbitrarily given numbers but to discuss numbers in general, to formulate and prove general theorems about them. This idea represents the generalization of an immense amount of earlier experience with concrete numbers, from which arose the rules and methods for *general* reasoning about numbers. A transition took place to a higher level of abstraction: from separate given (though abstract) numbers to number in general, to any possible number.

From the simple process of counting objects one by one, we pass to the unbounded process of formation of numbers by adding one to the number already formed. The sequence of numbers is regarded as being indefinitely continuable, and with it there enters into mathematics the notion of infinity. Of course, we cannot in fact, by the process of adding one, proceed arbitrarily far along the sequence of numbers: Who could reach as far as a million-million, which is almost forty times the number of seconds in a thousand years? But that is not the point; the process of adding ones, the process of forming arbitrary large collections of objects is in principle unlimited, so that the possibility exists of continuing the sequence of numbers beyond all limits. The fact that in actual practice counting is limited is not relevant; an abstraction is made from it. It is with this indefinitely prolonged sequence that general theorems about numbers have to deal.

General theorems about any property of an arbitrary number already contain in implicit form infinitely many assertions about the properties of separate numbers and are therefore qualitatively richer than any particular assertions that could be verified for specific numbers. It is for this reason that general theorems must be proved by general arguments proceeding from the fundamental rule for the formation of the sequence of numbers. Here we perceive a profound peculiarity of mathematics: Mathematics takes as its subject not only given quantitative relationships but all possible quantitative relationships and therefore infinity.

In the famous "Elements" of Euclid, written in the 3rd century B.C., we already find general theorems about whole numbers, in particular, the theorem that there exist arbitrarily large prime numbers.*

Thus arithmetic is transformed into the theory of numbers. It is already removed from particular concrete problems to the region of abstract concepts and arguments. It has become a part of "pure" mathematics. More precisely, this was the moment of the birth of pure mathematics itself with the characteristic features discussed in our first section. We must, of course, take note of the fact that pure mathematics was born simultaneously from arithmetic and geometry and that there were already to be found in the general rules of arithmetic some of the rudiments of algebra, a subject which was separated from arithmetic at a later stage. But we will discuss this later.

It remains now to summarize our conclusions up to this point, since we

---

* We recall that a prime number is defined as a positive integer greater than unity which is divisible without remainder only by the number itself and by unity.

have now traced out, though in very hurried fashion, the process whereby theoretical arithmetic arose from the concept of number.

5. The essential nature of arithmetic. Since the birth of theoretical arithmetic is part of the birth of mathematics, we may reasonably expect that our conclusions about arithmetic will throw light on our earlier questions concerning mathematics in general. Let us recall these questions, particularly in their application to arithmetic.

1. How did the abstract concepts of arithmetic arise and what do they reflect in the actual world?

This question is answered by the earlier remarks about the birth of arithmetic. Its concepts correspond to the quantitative relations of collections of objects. These concepts arose by way of abstraction, as a result of the analysis and generalization of an immense amount of practical experience. They arose gradually; first came numbers connected with concrete objects, then abstract numbers, and finally the concept of number in general, of any possible number. Each of these concepts was made possible by a combination of practical experience and preceding abstract concepts. This, by the way, is one of the fundamental laws of formation of mathematical concepts: They are brought into being by a series of successive abstractions and generalizations, each resting on a combination of experience with preceding abstract concepts. The history of the concepts of arithmetic shows how mistaken is the idealistic view that they arose from "pure thought," from "innate intuition," from "contemplation of a priori forms," or the like.

2. Why are the conclusions of arithmetic so convincing and unalterable?

History answers this question too for us. We see that the conclusions of arithmetic have been worked out slowly and gradually; they reflect experience accumulated in the course of unimaginably many generations and have in this way fixed themselves firmly in the mind of man. They have also fixed themselves in language: in the names for the numbers, in their symbols, in the constant repetition of the same operations with numbers, in their constant application to daily life. It is in this way that they have gained clarity and certainty. The methods of logical reasoning also have the same source. What is essential here is not only the fact that they can be repeated at will but their soundness and perspicuity, which they possess in common with the relations among things in the actual world, relations which are reflected in the concepts of arithmetic and in the rules for logical deduction.

This is the reason why the results of arithmetic are so convincing; its conclusions flow logically from its basic concepts, and both of them, the methods of logic and the concepts of arithmetic, were worked out and firmly fixed in our consciousness by three thousand years of practical experience, on the basis of objective uniformities in the world around us.

3. Why does arithmetic have such wide application in spite of the abstractness of its concepts?

The answer is simple. The concepts and conclusions of arithmetic, which generalize an enormous amount of experience, reflect in abstract form those relationships in the actual world that are met with constantly and everywhere. It is possible to count the objects in a room, the stars, people, atoms, and so forth. Arithmetic considers certain of their general properties, in abstraction from everything particular and concrete, and it is precisely because it considers only these general properties that its conclusions are applicable to so many cases. The possibility of wide application is guaranteed by the very abstractness of arithmetic, although it is important here that this abstraction is not an empty one but is derived from long practical experience. The same is true for all mathematics, and for any abstract concept or theory. The possibilities for application of a theory depend on the breadth of the original material which it generalizes.

At the same time every abstract concept, in particular the concept of number, is limited in its significance as a result of its very abstractness. In the first place, when applied to any concrete object it reflects only one aspect of the object and therefore gives only an incomplete picture of it. How often it happens, for example, that the mere numerical facts say very little about the essence of the matter. In the second place, abstract concepts cannot be applied everywhere without certain limiting conditions; it is impossible to apply arithmetic to concrete problems without first convincing ourselves that their application makes some sense in the particular case. If we speak of addition, for example, and merely unite the objects in thought, then naturally no progress has been made with the objects themselves. But if we apply addition to the actual uniting of the objects, if we in fact put the objects together, for example by throwing them into a pile or setting them on a table, in this case there takes place not merely abstract addition but also an actual process. This process does not consist merely of the arithmetical addition, and in general it may even be impossible to carry it out. For example, the object thrown into a pile may break; wild animals, if placed together, may tear one another apart; the materials put together may enter into a chemical reaction: a liter of water and a liter of alcohol when poured together produced not 2, but 1.9 liters of the mixture as a result of partial solution of the liquids; and so forth.

If other examples are needed they are easy to produce.

To put it briefly, truth is concrete; and it is particularly important to remember this fact with respect to mathematics, exactly because of its abstractness.

4. Finally, the last question we raised had to do with the forces that led to the development of mathematics.

For arithmetic the answer to this question also is clear from its history. We saw how people in the actual world learned to count and to work out the concept of number, and how practical life, by posing more difficult problems, necessitated symbols for the numbers. In a word, the forces that led to the development of arithmetic were the practical needs of social life. These practical needs and the abstract thought arising from them exercise on each other a constant interaction. The abstract concepts provide in themselves a valuable tool for practical life and are constantly improved by their very application. Abstraction from all nonessentials uncovers the kernel of the matter and guarantees success in those cases where a decisive role is played by the properties and relations picked out and preserved by the abstraction; namely, in the case of arithmetic, by the quantitative relations.

Moreover, abstract reflection often goes farther than the immediate demands of a practical problem. Thus the concept of such large numbers as a million or a billion arose on the basis of practical calculations but arose earlier than the practical need to make use of them. There are many such examples in the history of science; it is enough to recall the imaginary numbers mentioned earlier. This is just a particular case of a phenomenon known to everyone, namely the interaction of experience and abstract thought, of practice and theory.

§3. Geometry

1. The concept of a geometric figure. The history of the origin of geometry is essentially similar to that of arithmetic. The earliest geometric concepts and information also go back to prehistoric times and also result from practical activity.

Early man took over geometric forms from nature. The circle and the crescent of the moon, the smooth surface of a lake, the straightness of a ray of light or of a well-proportioned tree existed long before man himself and presented themselves constantly to his observation. But in nature itself our eyes seldom meet with really straight lines, with precise triangles or squares, and it is clear that the chief reason why men gradually worked out a conception of these figures is that their observation of nature was an active one, in the sense that, to meet their practical needs, they manufactured objects more and more regular in shape. They built dwellings, cut stones, enclosed plots of land, stretched bowstrings in their bows, modeled their clay pottery, brought it to perfection and correspondingly formed the notion that a pot is *curved,* but a stretched bowstring is *straight.* In short, they first gave form to their material and only then recognized form as that which is impressed on material and can therefore be considered in itself, as an abstraction from material. By recognizing the form of bodies, man was able to improve his handiwork and thereby to work out still more precisely the abstract notion of form. Thus practical activity served as a basis for the abstract concepts of geometry. It was necessary to manufacture thousands of objects with straight edges, to stretch thousands of threads, to draw upon the ground a large number of straight lines, before men could form a clear notion of the straight line in general, as that quality which is common to all these particular cases. Nowadays we learn early in life to draw a straight line, since we are surrounded by objects with straight edges that are the result of manufacture, and it is only for this reason that in our childhood we already form a clear notion of the straight line. In exactly the same way the notion of geometric magnitudes, of length, area, and volume, arose from practical activity. People measured lengths, determined distances, estimated by eye the area of surfaces and the volumes of bodies, all for their practical purposes. It was in this way that the simplest general laws were discovered, the first geometric relations: for example, that the area of a rectangle is equal to the product of the lengths of its sides. It is useful for a farmer to be aware of such a relation, in order that he may estimate the area he has sowed and consequently the harvest he may expect.

So we see that geometry took its rise from practical activity and from the problems of daily life. On this question the ancient Greek scholar, Eudemus of Rhodes, wrote as follows: "Geometry was discovered by the Egyptians as a result of their measurement of land. This measurement was necessary for them because of the inundations of the Nile, which constantly washed away their boundaries.* There is nothing remarkable in the fact that this science, like the others, arose from the practical needs of men. All knowledge that arises from imperfect circumstances tends to perfect itself. It arises from sense impressions but gradually becomes an object of our contemplation and finally enters the realm of the intellect."

Of course, the measurement of land was not the only problem that led the ancients toward geometry. From the fragmentary texts that have survived, it is possible to form some idea of various problems of the ancient Egyptians and Babylonians and of their methods for solving them. One of the oldest Egyptian texts goes back to 1700 B.C. This is a manual

---

* What is meant here is the boundaries between shares of land. Let us note, parenthetically, that *geometry* means land-measurement (in ancient Greek "ge" is land, and "metron" is measure).

of instruction for "secretaries" (royal officers), written by a certain Ahmes. It contains a collection of problems on calculating the capacity of containers and warehouses, the area of shares of land, the dimensions of earthworks, and so forth.

The Egyptians and Babylonians were able to determine the simplest areas and volumes, they knew with considerable exactness the ratio of the circumference to the diameter of a circle, and perhaps they were even able to calculate the surface area of a sphere; in a word, they already possessed a considerable store of geometrical knowledge. But so far as we can tell, they were still not in possession of geometry as a theoretical science with theorems and proofs. Like the arithmetic of the time, geometry was basically a collection of rules deduced from experience. Moreover, geometry was in general not distinguished from arithmetic. Geometric problems were at the same time problems for calculation in arithmetic.

In the 7th century B.C., geometry passed from Egypt to Greece, where it was further developed by the great materialist philosophers, Thales, Democritus, and others. A considerable contribution to geometry was also made by the successors of Pythagoras, the founders of an idealistic religiophilosophical school.

The development of geometry took the direction of compiling new facts and clarifying their relations with one another. These relations were gradually transformed into logical deductions of certain propositions of geometry from certain others. This had two results: first, the concept of a geometrical theorem and its proof; and second, the clarification of those fundamental propositions from which the others may be deduced, namely, the axioms.

In this way geometry gradually developed into a mathematical theory.

It is well known that systematic expositions of geometry appeared in Greece as far back as the 5th century B.C., but they have not been preserved, for the obvious reason that they were all supplanted by the "Elements" of Euclid (3rd century B.C.). In this work, geometry was presented as such a well-formed system that nothing essential was added to its foundations until the time of N. I. Lobačevskiĭ, more than two thousand years later. The well-known school text of Kiselev, like school books over the whole world, represented in its older editions, nothing but a popular reworking of Euclid. Very few other books in the world have had such a long life as the "Elements" of Euclid, this perfect creation of Greek genius. Of course, mathematics continued to advance, and our understanding of the foundations of geometry has been considerably deepened; nevertheless the "Elements" of Euclid became, and to a great extent remains, the model of a book on pure mathematics. Bringing together the accomplishments of his predecessors, Euclid presented the mathematics of his time as an independent theoretical science; that is, he presented it essentially as it is understood today.

2. The essential nature of geometry.   The history of geometry leads to the same conclusions as that of arithmetic. We see that geometry arose from practical life and that its transformation to a mathematical theory required an immense period of time.

Geometry operates with "geometric bodies" and figures; it studies their mutual relations from the point of view of magnitude and position. But a geometric body is nothing other than an actual body considered solely from the point of view of its spatial form,* in abstraction from all its other properties such as density, color, or weight. A geometric figure is a still more general concept, since in this case it is possible to abstract from spatial extension also; thus a surface has only two dimensions, a line, only one dimension, and a point, none at all. A point is the abstract concept of the end of a line, of a position defined to the limit of precision so that it no longer has any parts. It is in this way that all these concepts are defined by Euclid.

Thus geometry has as its object the spatial forms and relations of actual bodies, removed from their other properties and considered from the purely abstract point of view. It is just this high level of abstraction that distinguishes geometry from the other sciences that also investigate the spatial forms and relations of bodies. In astronomy for example, the mutual positions of bodies are studied, but they are the actual bodies of the sky; in geodesy it is the form of the earth that is studied, in crystallography, the form of crystals, and so forth. In all these other sciences, the form and the position of concrete bodies are studied in their dependence on other properties of the bodies.

This abstraction necessarily leads to the purely theoretical method of geometry; it is no longer possible to set up experiments with breadthless straight lines, with "pure forms." The only possibility is to make use of logical argument, deriving some conclusions from others. A geometrical theorem must be proved by reasoning, otherwise it does not belong to geometry; it does not deal with "pure forms."

The self-evidence of the basic concepts of geometry, the methods of reasoning and the certainty of their conclusions, all have the same source as in arithmetic. The properties of geometric concepts, like the concepts themselves, have been abstracted from the world around us. It was necessary for people to draw innumerable straight lines before they could take it as an axiom that through every two points it is possible to draw a

---

* By form we mean also dimensions.

straight line; they had to move various bodies about and apply them to one another on countless occasions before they could generalize their experience to the notion of superposition of geometric figures and make use of this notion for the proof of theorems, as is done in the well-known theorems about congruence of triangles.

Finally, we must emphasize the generality of geometry. The volume of a sphere is equal to $4/3\pi R^3$ quite independently of whether we are speaking of a spherical vessel, of a steel sphere, of a star, or of a drop of water. Geometry can abstract what is common to all bodies, because every actual body does have more or less definite form, dimensions, and position with respect to other bodies. So it is no cause for wonder that geometry finds application almost as widely as arithmetic. Workmen measuring the dimensions of a building or reading a blueprint, an artillery man determining the distance to his target, a farmer measuring the area of his field, an engineer estimating the volume of earthworks, all these people make use of the elements of geometry. The pilot, the astronomer, the surveyor, the engineer, the physicist, all have need of the precise conclusions of geometry.

A clear example of the abstract-geometrical solution of an important problem in physics is provided by the investigations of the well-known crystallographer and geometer, E. S. Fedorov. The problem he set himself of finding all the possible forms of symmetry for crystals is one of the most fundamental in theoretical crystallography. To solve this problem, Fedorov made an abstraction from all the physical properties of a crystal, considering it only as a regular system of geometric bodies "in place of a system of concrete atoms." Thus the problem became one of finding all the forms of symmetry which could possibly exist in a system of geometric bodies. This purely geometrical problem was completely solved by Fedorov, who found all the possible forms of symmetry, 230 in number. His solution proved to be an important contribution to geometry and was the source of many geometric investigations.

In this example, as in the whole history of geometry, we detect the prime moving force in the development of geometry. It is the mutual influence of practical life and abstract thought. The problem of discovering possible symmetries originated in physical observation of crystals but was transformed into an abstract problem and so gave rise to a new mathematical theory, the theory of regular systems, or of the so-called Fedorov groups.* Subsequently this theory not only found brilliant confirmation in the practical observation of crystals but also served as a general guide in the development of crystallography, giving rise to new investigations, both in experimental physics and in pure mathematics.

---

* Compare Chapter XX.

§4. Arithmetic and Geometry

1. The origin of fractions in the interrelation of arithmetic and geometry.
Up to now we have considered arithmetic and geometry apart from each
other. Their mutual relation, and consequently the more general interrelation of all mathematical theories, has so far escaped our attention.
Nevertheless this relation has exceptionally great significance. The interaction of mathematical theories leads to advances in mathematics itself
and also uncovers a rich treasure of mutual relations in the actual world
reflected by the these theories.

Arithmetic and geometry are not only applied to each other but they
also serve thereby as sources for further general ideas, methods, and
theories. In the final analysis, arithmetic and geometry are the two roots
from which has grown the whole of mathematics. Their mutual influence
goes back to the time when both of them had just come into being. Even
the simple measurement of a line represents a union of geometry and
arithmetic. To measure the length of an object we *apply* to it a certain
unit of length and *calculate* how many times it is possible to do this; the
first operation (application) is geometric, the second (calculation) is
arithmetical. Everyone who counts off his steps along a road is already
uniting these two operations.

In general, the measurement of any magnitude combines calculation
with some specific operation which is characteristic of this sort of
magnitude. It is sufficient to mention measurement of a liquid in a graduated container or measurement of an interval of time by counting the
number of strokes of a pendulum.

But in the process of measurement it turns out, generally speaking, that
the chosen unit is not contained in the measured magnitude an integral
number of times, so that a simple calculation of the number of units is not
sufficient. It becomes necessary to divide up the unit of measurement in
order to express the magnitude more accurately by parts of the unit;
that is, no longer by whole numbers but by fractions. It was in this way
that fractions actually arose, as is shown by an analysis of historical and
other data. They arose from the division and comparison of continuous
magnitudes; in other words, from measurement. The first magnitudes to
be measured were geometric, namely lengths, areas of fields, and volumes
of liquids or friable materials, so that in the earliest appearance of fractions
we see the mutual action of arithmetic and geometry. This interaction
leads to the appearance of an important new concept, namely of fractions,
as an extension of the concept of number from whole numbers to fractional
numbers (or as the mathematicians say, to rational numbers, expressing
the ratio of whole numbers). Fractions did not arise, and could not arise,

from the division of whole numbers, since only whole objects are counted by whole numbers. Three men, three arrows, and so forth, all these make sense, but two-thirds of a man and even two-thirds of an arrow are senseless concepts; even three separate thirds of an arrow will not kill a deer, for this it is necessary to have a *whole* arrow.

2. Incommensurable magnitudes. In the development of the concept of number, arising from the mutual action of arithmetic and geometry, the appearance of fractions was only the first step. The next was the discovery of incommensurable intervals. Let us recall that intervals are called incommensurable if no interval exists which can be applied to each of them a whole number of times or, in other words, if their ratio cannot be expressed by an ordinary fraction; that is, by a ratio of whole numbers.

At first people simply did not think about the question whether every interval can be expressed by a fraction. If in dividing up or measuring an interval they came upon very small parts, they merely discarded them; in practice, it made no sense to speak of infinite precision of measurement. Democritus even advanced the notion that geometrical figures consist of atoms of a particular kind. This notion, which to our view seems quite strange, proved very fruitful in the determination of areas and volumes. An area was calculated as the sum of rows consisting of atoms, and a volume as the sum of atomic layers. It was in this way, for example, that Democritus found the volume of a cone. A reader who understands the integral calculus will note that this method already forms the prototype of the determination of areas and volumes by the methods of the integral calculus. Moreover, in returning in thought to the times of Democritus, one must attempt to free oneself of the customary notions of today, which have become firmly fixed in our minds by the development of mathematics. At the time of Democritus, geometrical figures were not yet separated from actual ones to the same extent as is now the case. Since Democritus considered actual bodies as consisting of atoms, he naturally also regarded geometrical figures in the same light.

But the notion that intervals consist of atoms comes into contradiction with the theorem of Pythagoras, since it follows from this theorem that incommensurable intervals exist. For example, the diagonal of a square is incommensurable with its side; in other words, the ratio of the two cannot be expressed as the ratio of whole numbers.

We shall prove that the side and the diagonal of a square are in fact incommensurable. If a is the side and b is the diagonal of a square, then according to the theorem of Pythagoras $b^2 = a^2 + a^2 = 2a^2$ and therefore

$$\left(\frac{b}{a}\right)^2 = 2.$$

But there is no fraction such that its square is equal to 2. In fact, if we suppose that there is, let p and q be whole numbers for which

$$\left(\frac{p}{q}\right)^2 = 2,$$

where we may assume that p and q have no common factor, since otherwise we could simplify the fraction. But if $(p/q)^2 = 2$, then $p^2 = 2q^2$, and therefore p^2 is divisible by 2. In this case p^2 is also divisible by 4, since it is the square of an even number. So $p^2 = 4q_1$; that is, $2q^2 = 4q_1$, and $q^2 = 2q_1$. From this it follows that q must also be divisible by 2. But this contradicts the supposition that p and q have no common factor. This contradiction proves that the ratio b/a cannot be expressed by a rational number. The diagonal and the side of a square are incommensurable.

This discovery made a great impression on the Greek scientists. Nowadays, when we are accustomed to irrational numbers and calculate freely with square roots, the existence of incommensurable intervals does not disturb us. But in the 5th century B.C., the discovery of such intervals had a completely different aspect for the Greeks. Since they did not have the concept of an irrational number and never wrote a symbol like $\sqrt{2}$, the previous result indicated that the ratio of the diagonal and the side of the square was not represented by any number at all.

In the existence of incommensurable intervals the Greeks discovered a profound paradox inherent in the concept of continuity, one of the expressions of the dialectical contradiction comprised in continuity and motion. Many important Greek philosophers considered this contradiction; particularly well-known among them, because of his paradoxes, is Zeno the Eleatic.

The Greeks founded a theory of ratios of intervals, or of magnitudes in general, which takes into consideration the existence of incommensurable intervals;* it is expounded in the "Elements" of Euclid, and in simplified form is explained today in high school courses in geometry. But to recognize that the ratio of one interval to another (if the second interval is taken as the unit of length, this ratio is simply the length of the first interval) may also be considered as a number, whereby the very concept of number is generalized, to this idea the Greeks were not able to rise: The concept of an irrational number simply did not originate among them.† This step was taken at a later period by the mathematicians

---

* This theory is ascribed to the Greek scientist Eudoxus, who lived in the 4th century B.C.

† As a result of the fact that the theory of the measurement of magnitudes did not become part of arithmetic but passed over into geometry, mathematics among the

of the East; and in general, a mathematically rigorous definition of a real number, not depending immediately on geometry, was given only recently: in the seventies of the last century.* The passage of such an immense period of time after the founding of the theory of ratios shows how difficult it is to discover abstract concepts and give them exact formulation.

3. The real number. In describing the concept of a real number, Newton in his "General Arithmetic" wrote: "by number we mean not so much a collection of units as an abstract ratio of a certain quantity to another quantity taken as the unit." This number (ratio) may be integral, rational, or if the given magnitude is incommensurable with the unit, irrational.

A real number in its original sense is therefore nothing but the ratio of one magnitude to another taken as a unit; in particular cases this is a ratio of intervals, but it may also be a ratio of areas, weights, and so forth.

Consequently, a real number is a ratio of magnitudes in general, considered in abstraction from their concrete nature.

Just as *abstract* whole numbers are of mathematical interest only in their relations with one another, so *abstract* real numbers have content and become an object of mathematical attention only in relation with one another in the system of real numbers.

In the theory of real numbers, just as in arithmetic, it is first necessary to define operations on numbers: addition, subtraction, multiplication, division, and also the relations expressed by such words as "greater than" or "less than." These operations and relations reflect actual connections among the various magnitudes; for example, addition reflects the placing together of intervals. A beginning on operations with abstract real numbers was made in the Middle Ages by the mathematicians of the East. Later came the gradual discovery of the most important property of the system

---

Greeks was engulfed by geometry. Such questions, for example, as the solution of quadratic equations, which today we treat in an algebraic way, they stated and solved geometrically. The "Elements" of Euclid contain a considerable number of such questions, which obviously represented for contemporary mathematicians a summary of the foundations not only of geometry in our sense but of mathematics in general. This domination by geometry continued up to the time when Descartes, on the contrary, subjected geometry to algebra. Traces of the long domination by geometry are preserved, for example, in such names as "square" and "cube" for the second and third powers: "*a* cubed" is a cube with side *a*.

* We are speaking here not of a descriptive definition, but of a definition which serves as the immediate basis for proofs of theorems about the properties of real numbers. It is natural that such definitions should arise at a later period, when the development of mathematics, and in particular of the infinitesimal analysis, required a suitable definition of the real number represented by "the variable *x*." This definition was given in various forms in the seventies of the last century by the German mathematicians Weierstrass, Dedekind, and Cantor.

of real numbers, its continuity. The system of real numbers is the abstract image of all the possible values of a continuously varying magnitude.

In this way, as in the similar case of whole numbers, the arithmetic of real numbers deals with the actual quantitative relations of continuous magnitudes, which it studies in their general form, in complete abstraction from all concrete properties. It is precisely because real numbers deal with what is common to all continuous magnitudes that they have such wide application: The values of various magnitudes, a length, a weight, the strength of an electric current, energy and so forth, are expressed by numbers, and the interdependence or relations among these entities are mirrored as relations among their numerical values.

To show how the general concept of real numbers can serve as the basis of a mathematical theory, we must give their mathematical definition in a formal way. This may be done by various methods, but perhaps the most natural is to proceed from the very process of measurement of magnitudes which actually did lead in practical life to this generalization of the concept of number. We will speak about the length of intervals, but the reader will readily perceive that we could argue in exactly the same way about any other magnitudes which permit indefinite subdivision.

Let us suppose that we wish to measure the interval AB by means of the interval CD taken as a unit (figure 1).

FIG. 1.

We apply the interval CD to AB, beginning for example with the point A, as long as CD goes into AB. Suppose this is n_0 times. If there still remains from the interval AB a remainder PB, then we divide the interval CD into ten parts and measure the remainder with these tenths. Suppose that n_1 of the tenths go into the remainder. If after this there is still a remainder, we divide our measure into ten parts again; that is, we divide CD into a hundred parts, and repeat the same operation, and so forth. Either the process of measurement comes to an end, or it continues. In either case we reach the result that in the interval AB the whole interval CD is contained n_0 times, the tenths are contained n_1 times, the hundredths n_2 times and so forth. In a word, we derive the ratio of AB to CD with increasing accuracy: up to tenths, to hundredths, and so forth. So the ratio itself is represented by a decimal fraction with n_0 units, n_1 tenths and so forth

$$\frac{AB}{CD} = n_0 \cdot n_1 n_2 n_3 \cdots .$$

This decimal fraction may be infinite, corresponding to the possibility of indefinite increase in the precision of measurement.

Thus the ratio of two intervals, or of two magnitudes in general, is always representable by a decimal fraction, finite or infinite. But in the decimal fraction there is no longer any trace of the concrete magnitude itself; it represents exactly the abstract ratio, the real number. Thus a real number may be formally defined if we wish, as a finite or infinite decimal fraction.*

Our definition will be complete if we say what we mean by the operations of addition and so forth for decimal fractions. This is done in such a way that the operations defined on decimal fractions correspond to the operations on the magnitudes themselves. Thus, when intervals are put together their lengths are added; that is, the length of the interval $AB + BC$ is equal to the sum of the length AB and BC. In defining the operations on real numbers, there is a difficulty that these numbers are represented in general by *infinite* decimal fractions, while the well-known rules for these operations refer to finite decimal fractions. A rigorous definition of the operations for infinite decimals may be made in the following way. Suppose, for example, that we must add the two numbers a and b. We take the corresponding decimal fractions up to a given decimal place, say the millionth, and add them. We thus obtain the sum $a + b$ with corresponding accuracy, up to two millionths, since the errors in a and b may be added together. So we are able to define the sum of two numbers *with an arbitrary degree of accuracy*, and in that sense their sum is completely defined, although at each stage of the calculation it is known only with a a certain accuracy. But this corresponds to the essential nature of the case, since each of the magnitudes a and b is also measured only with a certain accuracy, and the exact value of each of the corresponding infinite fractions is obtained as the result of an indefinitely extended increase in accuracy. The relations "greater than" and "less than" may then be defined by means of addition: a is greater than b if there exists a magnitude c such that $a = b + c$, where we are speaking, of course, of positive numbers.

The continuity of the sequence of real numbers finds expression in the fact that if the numbers $a_1, a_2, \cdots$ increase and $b_1, b_2, \cdots$ diminish but

---

* Fractions with the periodic digit nine are not considered here, they are identical with the corresponding fraction without nines according to the well-known rule, which is clear from the example: $0.139999 \cdots = 0.140000 \cdots$.

always remain greater than the a_i , then between the one series of numbers and the other there is always a number c. This may be visualized on a straight line if its points are put into correspondence with the numbers (figure 2) according to the well-known rule.

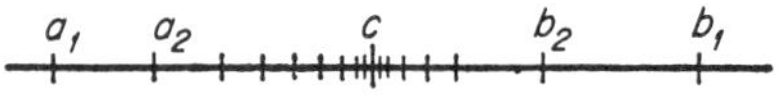

FIG. 2.

Here it is clearly seen that the presence of the number c and of the point corresponding to it signify the absence of a break in the series of numbers, which is what is meant by their continuity.

4. The conflict of opposites: concrete and abstract.   Already in the example of the interaction of arithmetic and geometry we can see that the development of mathematics is a process of conflict among the many contrasting elements: the concrete and the abstract, the particular and the general, the formal and the material, the finite and the infinite, the discrete and the continuous, and so forth. Let us try, for example, to trace the contrast between concrete and abstract in the formation of the concept of a real number. As we have seen, the real number reflects an infinitely improvable process of measurement or, in slightly different terms, an absolutely accurate determination of a magnitude. This corresponds to the fact that in geometry we consider ideally precise forms and dimensions of bodies, abstracting altogether from the mobility of concrete objects and from a certain indefiniteness in their actual forms and dimensions; for example, the interval measured (figure 1) was a completely ideal one.

But ideally precise geometric forms and absolutely precise values for magnitudes represent abstractions. No concrete object has absolutely precise form nor can any concrete magnitude be measured with absolute accuracy, since it does not even *have* an absolutely accurate value. The length of a line segment, for example, has no sense if one tries to make it precise beyond the limits of atomic dimensions. In every case when one passes beyond well-known limits of quantitative accuracy, there appears a qualitative change in the magnitude, and in general it loses its original meaning. For example, the pressure of a gas cannot be made precise beyond the limits of the impact of a single molecule; electric charge ceases to be continuous when one tries to make it precise beyond the charge on an electron and so forth. In view of the absence in nature of objects of ideally precise form, the assertion that the ratio of the diagonal of a square to the side is equal to the $\sqrt{2}$ not only cannot be deduced with absolute accuracy from immediate measurement but does not even have any absolutely accurate meaning for an actual concrete square.

The conclusion that the diagonal and the side of a square are incommensurable comes, as we have seen, from the theorem of Pythagoras. This is a *theoretical* conclusion based on a development of the data of experience; it is a result of the application of logic to the original premises of geometry, which are taken from experience.

In this way the concept of incommensurable intervals, and all the more of real numbers, is not a simple immediate reflection of the facts of experience but goes beyond them. This is quite understandable. The real number does not reflect any given concrete magnitude but rather magnitude in general, in abstraction from all concreteness; in other words, it reflects what is *common* to particular concrete magnitudes. What is common to all of them consists in particular in this, that the value of the magnitude can be determined more and more precisely; and if we abstract from concrete magnitudes, then the limit of this possible increase in precision, which depends on the concrete nature of the magnitude, becomes indefinite and disappears.

In this way a *mathematical* theory of magnitudes, since it considers magnitudes in abstraction from their individual nature, must inevitably consider the possibility of unlimited accuracy for the value of the magnitude and *must* thereby lead to the concept of a real number. At the same time, since it reflects only what is common to various magnitudes, mathematics takes no account of the peculiarity of each individual magnitude.

Since mathematics selects only general properties for consideration, it operates with its clearly defined abstractions quite independently of the actual limits of their applicability, as must happen precisely because these limits are different in different particular cases. These limits depend on the concrete properties of the phenomena under consideration and on the qualitative changes that take place in them. So in making an application of mathematics, it is necessary to verify the actual applicability of the theory in question. To consider matter as continuous and to describe its properties by continuous magnitudes is permissible only if we may abstract from its atomic structure, and this is possible only under well-known conditions.

Nevertheless, the real numbers represent a trustworthy and powerful instrument for the mathematical investigation of actual continuous magnitudes and processes. Their theory is based on practice, on an immense field of applications in physics, technology, and chemistry. Consequently, practice shows that the concept of the real number correctly reflects the general properties of magnitudes. But this correctness is not without limits; it is not possible to consider the theory of real numbers as something absolute, allowing an unlimited abstract development in complete separation from reality. The very concept of the real number is continuing to develop and is in fact still far from being complete.

5. The conflict of opposites: discrete and continuous. The role of another of the mentioned contrasts, the contrast between the discrete and the continuous, may also be illustrated by the development of the concept of number. We have already seen that fractions arose from the division of continuous magnitudes.

On this theme of division there is a humorous question which is extraordinarily instructive. Grandmother has bought three potatoes and must divide them equally between two grandsons. How is she to do it? The answer is: make mashed potatoes.

The joke reveals the very essence of the matter. Separate objects are indivisible in the sense that, when divided, the object almost always ceases to be what it was before, as is clear from the example of "thirds of a man" or "thirds of an arrow." On the other hand, continuous and homogenous magnitudes or objects may easily be divided and put together again without losing their essential character. Mashed potatoes offer an excellent example of a homogeneous object, which in itself is not separated into parts but may nevertheless be divided in practice into as small parts as desired. Lengths, areas, and volumes have the same property. Although they are continuous in their very essence and are not actually divided into parts, nevertheless they offer the possibility of being divided without limit.

Here we encounter two contrasting kinds of objects: on the one hand, the indivisible, separate, discrete objects; and on the other, the objects which are completely divisible and yet are not divided into parts but are continuous. Of course, these contrasting characteristics are always united, since there are no absolutely indivisible and no completely continuous objects. Yet these aspects of the objects have an actual existence, and it often happens that one aspect is decisive in one case and the other in another.

In abstracting forms from their content, mathematics by this very act sharply divides these forms into two classes, the discrete and the continuous.

The mathematical model of a separate object is the unit, and the mathematical model of a collection of discrete objects is a sum of units, which is, so to speak, the image of pure discreteness, purified of all other qualities. On the other hand, the fundamental, original mathematical model of continuity is the geometric figure; in the simplest case, the straight line.

We have before us therefore two contrasts, discreteness and continuity, and their abstract mathematical images: the whole number and the geometric extension. Measurement consists of the union of these contrasts:

The continuous is measured by separate units. But the inseparable units are not enough; we must introduce fractional parts of the original unit. In this way the fractional numbers arise and the concept of numbers develops precisely as a result of the union of the mentioned contrasts.

Then, on a more abstract level, appeared the concept of incommensurable intervals, and, as a result, the real number as an abstract image of unlimited increase in accuracy in the determination of a magnitude. This concept was not formed immediately, and the long path of its development led through many a conflict between these same two contrasting elements, the discrete and the continuous.

In the first place, Democritus represented figures as consisting of atoms and in this way reduced the continuous to the discrete. But the discovery of incommensurable intervals led to the abandonment of such a representation. After this discovery continuous magnitudes were no longer thought of as consisting of separate elements, atoms or points, and they were not represented by numbers, since numbers other than the whole numbers and the fractions were not known at that time.

The contrast between the continuous and the discrete appeared in mathematics again with renewed force in the 17th century, when the foundations of the differential and integral calculus were being laid. Here it was the infinitesimal that was under discussion. In some accounts the infinitesimal was thought of as a real, "actually" infinitesimal, "indivisible" particle of the continuous magnitude, like the atoms of Democritus, except that now the number of these particles was considered to be infinitely great. Calculation of areas and volumes, or in other words integration, was thought of as summation of an infinite number of these infinitely small particles.

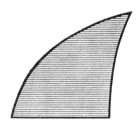

Fig. 3.

An area, for example, was understood as "the sum of the lines from which it is formed" (figure 3). Consequently, the continuous was again reduced to the discrete, but now in a more complicated way, on a higher level. But this point of view also proved unsatisfactory, and, as a counterweight to it, there appeared, on the basis of Newton's work, the notion of *continuous* variables, of the infinitesimal as a *continuous* variable decreasing without limit. This conception finally carried the day at the beginning of the 19th century, when the rigorous theory of limits was founded. An interval was now thought of as consisting not of points or "indivisibles," but as an extension, as a continuous medium, where it was only possible to fix separate points, separate values of a continuous magnitude. Mathematicians then spoke of "extension." In the union of the discrete and the continuous, it was again the continuous that dominated.

But the development of analysis demanded further precision in the theory of variable magnitudes and above all in the general definition of a real number as an arbitrary possible value of a variable magnitude. In the seventies of the last century there arose a theory of real numbers which represents an interval as a set of points, and correspondingly the range of variation of a variable as a set of real numbers. The continuous again consisted of separate discrete points and the properties of continuity were again expressed in the structure of the set of points that formed it. This conception led to immense progress in mathematics and became dominant. But again profound difficulties were discovered in it, and these led to attempts to return on a new level to the notion of pure continuity. Other attempts were made to change the concept of an interval as a set of points. New points of view appeared for the concepts of number, variable, and function. The development of the theory is continuing, and we must await its further progress.

6. Further results of the interaction of arithmetic and geometry. The interaction of geometry and arithmetic played a role elsewhere than in the formation of the concept of a real number. The same interaction of geometry with arithmetic, or more accurately with algebra, also showed itself in the formation of negative and complex numbers, that is of numbers of the form $a + b \sqrt{-1}$. Negative numbers are represented by points of the straight line to the left of the point representing zero. It was exactly this geometric representation which gave imaginary numbers a firm place in mathematics; up to that time they had not been understood. New concepts of magnitude appeared: for example, vectors, which are represented by directed line segments; and tensors, which are still more general magnitudes; in these again algebra is united with geometry.

The union of various mathematical theories has always played a great and sometimes decisive role in the development of mathematics. We shall see this further on in the rise of analytic geometry, differential and integral calculus, the theory of functions of a complex variable, the recent so-called functional analysis, and other theories. Even in the theory of numbers itself, that is in the study of whole numbers, methods are applied with great success which depend on continuity (namely on the infinitesimal analysis) and on geometry. These methods have given rise to extensive chapters in the theory of numbers, the "analytic theory of numbers," and the "geometry of numbers."

From a certain well-known point of view, it is possible to regard the foundations of mathematics as the union of concepts arising from geometry and arithmetic; that is to say, of the general concepts of continuity and of algebraic operations (as generalizations of arithmetic operations). But we will not be able to speak here of these difficult theories. The aim of the present chapter has been to give an impression of the general interaction of concepts, of the union and the conflict between contrasting ideas in mathematics, as illustrated by the interaction of arithmetic and geometry in the development of the concept of number.

§5. The Age of Elementary Mathematics

1. The four periods of mathematics.   The development of mathematics cannot be reduced to the simple accumulation of new theorems but includes essential qualitative changes. These qualitative changes take place, however, not in a process of destruction or abolition of already existing theories but in their being deepened and generalized, so as to form more general theories, for which the way has been prepared by preceding developments.

From the most general point of view, we may distinguish in the history of mathematics four fundamental, qualitatively distinct periods. Of course, it is not possible to draw exact boundary lines between these stages, since the essential traits of each period appeared more or less gradually, but the distinctions among the stages and the passages from one to another are completely clear.

The first stage (or period) is the period of the rise of mathematics as an independent and purely theoretical science. It begins in the most ancient times and extends to the 5th century B.C., or perhaps earlier, when the Greeks laid the foundations of "pure" mathematics with its logical connection between theorems and proofs (in that century there appeared, in particular, systematic expositions of geometry like the "Elements" of Hippocrates of Chios). This first stage was the period of the formation of arithmetic and geometry, in the form considered earlier. At this time mathematics consisted of a collection of separate rules deduced from experience and immediately connected with practical life. These rules did not yet form a logically unified system, since the theoretical character of mathematics with its logical proof of theorems was formed very slowly, as material for it was accumulated. Arithmetic and geometry were not separated but were closely interwoven with each other.

The second period may be characterized as the period of elementary mathematics, of the mathematics of constant magnitudes; its simple fundamental results now form the content of a high school course. This period extended for almost 2000 years and ended in the 17th century with the rise of "higher" mathematics. It is with this period that we will be concerned in greater detail in the present section. The following sections will be devoted to the third and fourth periods, namely to the founding and development of analysis and to the period of contemporary mathematics.

2. Mathematics in Greece. The period of elementary mathematics may in its turn be divided into two parts, distinguished by their basic content: the period of the development of geometry (up to the 2nd century A.D.) and the period of the predominance of algebra (from the 2nd to the 17th century). With respect to historical conditions it is divided into three parts, which may be called "Greek," "Eastern," and "European Renaissance." The Greek period coincides in time with the general flowering of Greek culture, beginning with the 7th century B.C., reaching its culmination in the 3rd century B.C. at the time of the great geometers of antiquity, Euclid, Archimedes, and Apollonius, and ending in the 6th century A.D. Mathematics, and especially geometry, enjoyed a wonderful development in Greece. We know the names and the results of numerous Greek mathematicians, although only a few genuine works have come down to us. It is to be remarked that Rome gave nothing to mathematics though it reached its zenith in the 1st century A.D. at a time when the science of Greece, which had been conquered by Rome, was still flourishing.

The Greeks not only developed and systematized elementary geometry to the extent to which it is given in the "Elements" of Euclid and is now taught in our secondary schools, but achieved considerably higher results. They studied the conic sections: ellipse, hyperbola, parabola; they proved certain theorems relating to the elements of what is called projective geometry; guided by the needs of astronomy, they worked out spherical geometry (in the 1st century A.D.) and also the elements of trigonometry, and calculated the first tables of sines (Hipparchus, 2nd century B.C. and Claudius Ptolemy, 2nd century A.D.);* they determined the areas and volumes of a number

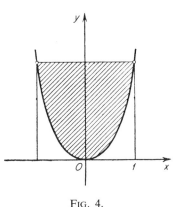

Fig. 4.

---

* Ptolemy is widely known as the author of a system in which the Earth is considered as the center of the universe and the motion of the heavenly bodies is described as proceeding around it. This system was supplanted by the Copernician system.

of complicated figures; for example, Archimedes found the area of the segment of a parabola by proving that it is 2/3 of the area of the rectangle containing it (figure 4). The Greeks were also acquainted with the theorem that of all bodies with a given surface area the sphere has the greatest volume, but their proof has not been preserved and was probably not complete. Such a proof is quite difficult and was first discovered in the 19th century, by means of the integral calculus.

In arithmetic and in the elements of algebra, the Greeks also made considerable progress. As was mentioned earlier, they laid the foundation for the theory of numbers. Here belong, for example, their investigations on prime numbers (the theorem of Euclid on the existence of an infinite number of prime numbers and the "sieve" of Eratosthenes for finding prime numbers) and the solution of equations in whole numbers (Diophantus about 246–330 A.D.).

We have already said that the Greeks discovered irrational magnitudes but considered them geometrically, as line segments. So the problems that today we deal with algebraically were treated geometrically by the Greeks. It was in this way that they solved quadratic equations and transformed irrational expressions. For example, the equation that we today write in the form $x^2 + ax = b^2$, they stated as follows: Find a segment x such that if to the square constructed on it we add a rectangle constructed on the same segment and on the given segment a, we obtain a rectangle equal in area to a given square. This dominance of geometry lasted a long time after the Greeks. They were also acquainted with (geometric) methods for extracting square roots and cube roots and with the properties of arithmetic and geometric progressions.

In this way the Greeks were already in possession of much of the material of contemporary elementary algebra but not, however, of the following essential elements: negative numbers and zero, irrational numbers abstracted entirely from geometry, and finally a well-developed system of literal symbols. It is true that Diophantus made use of literal symbols for the unknown quantity and its powers and also of special symbols for addition, subtraction, and equality, but his algebraic equations were still written with concrete numerical coefficients.

In geometry the Greeks attained what we now call "higher" mathematics. Archimedes made use of integral calculus for the calculation of areas and volumes and Apollonius used analytic geometry in his investigations on conic sections. Apollonius actually gives the equations of these curves*

---

* He gives the "equations" of conic sections referred to a vertex. For example "the equation" of the parabola $y^2 = 2px$ is formulated thus: The square on the side y is equal in area to the rectangle with sides $2p$ and x. Of course, in place of the symbols p, x, y he uses the corresponding line segments.

but expresses them in geometric language. In these equations there does not yet appear the general notion of an arbitrary constant or of a variable magnitude; and the necessary means of expressing such concepts, namely the literal symbols of algebra, appear only at a later age; they alone could convert such investigations into a source of new theories, which would be truly a part of higher mathematics. The founders of these new theories were guided, a thousand years later, by the legacy of the Greek scientists; in fact, the "Geometry" of Descartes (1637), which laid the foundation for analytic geometry, begins with a selection of problems left by the Greeks.

Such is the general rule. The old theories, by giving rise to new and profound problems, outgrow themselves, as it were, and demand for further progress new forms and new ideas. But these forms and ideas may demand new historical conditions for their birth. In ancient society the conditions necessary for the passage to higher mathematics did not and could not exist; they came on the scene with the development of the natural sciences in modern times, a development which in its turn was conditioned in the 16th and 17th centuries by the new demands of techno- logy and of manufacturing and was connected in this way with the birth and development of capitalism.

The Greeks practically exhausted the possibilities of elementary mathe- matics, which is the explanation of the fact that the brilliant progress of geometry dried up at the beginning of our era and was replaced by trigonometry and algebra in the works of Ptolemy, Diophantus, and others. In fact, one may consider the works of Diophantus as the beginning of the period in which algebra played the leading role. But the society of the ancients, already verging to its decline, was no longer able to advance science in this new direction.

It should be noted that, a few centuries earlier, arithmetic had already reached a high level in China. The Chinese scientists of the 2nd and 1st centuries B.C. described the rules for arithmetical solution of a system of three equations of the first degree. It is here for the first time in history that negative coefficients are made use of and the rules for operating with negative quantities are formulated. But the solutions themselves were sought only in the form of positive numbers, just as later in the works of Diophantus. These Chinese books also include a method for the extraction of square roots and cube roots.

3. The Middle East. With the end of Greek science a period of scientific stagnation began in Europe, the center of mathematical develop- ment being shifted to India, Central Asia, and the Arabic countries.*

---

* To give some orientation in the dates we list here the times of some of the out-

For a period of about a thousand years, from the 5th to the 15th century, mathematics developed chiefly in connection with the demands of computation, particularly in astronomy, since the mathematicians of the East were for the most part also astronomers. It is true that they added nothing of importance to Greek geometry; in this field they only preserved for later times the results of the Greeks. But the Indian, Arabic, and Central Asian mathematicians achieved immense successes in the fields of arithmetic and algebra.*

As has been mentioned in §2, the Indians invented our present system of numeration. They also introduced negative numbers, comparing the contrast between positive and negative numbers with the contrast between property and debt or between the two directions on a straight line. Finally, they began to operate with irrational magnitudes exactly as with rational, without representing them geometrically, in contrast to the Greeks. They also had special symbols for the algebraic operations, including extraction of roots. For the very reason that the Indian and Central Asian scholars were no longer embarrassed by the difference between the irrational and rational magnitudes, they were able to overcome the "dominance" of geometry, which was characteristic of Greek mathematics, and to open up paths for the development of contemporary algebra, free of the heavy geometric framework into which it had been forced by the Greeks.

The great poet and mathematician, Omar Khayyam (about 1048–1122), and also the Azerbaijanian mathematician, Nasireddin Tusi (1201–1274), clearly showed that every ratio of magnitudes, whether commensurable or incommensurable, may be called a number; in their works we find the same general definition of number, both rational and irrational, as was introduced above in Newton's formulation, in §4. The magnitude of these achievements becomes particularly clear when we recall that complete recognition of negative and irrational numbers was attained by European mathematicians only very slowly, even after the beginning of the Renaissance of mathematics in Europe. For example, the celebrated French mathematician Viète (1540–1603), to whom algebra owes a great deal, avoided negative numbers, and in England protests against them lasted even into the 18th century. These numbers were considered absurd, since they were less than zero, that is "less than nothing at all." Nowadays they

standing mathematicians of the East. From India: Aryabhata, born about 476 A.D.; Brahmagupta, about 598–660; Bhaskara, 12th century; from Kharizm: Al-Kharizmi, 9th century; Al-Biruni, 973–1048; from Azerbaijan: Nasireddin Tusi, 1201–74; from Samarkand: Gyaseddin Jamschid, 15th century.

* One should keep in mind that it is wrong to associate the development of mathematics in this period chiefly with the Arabs. The term "Arabic" mathematics came into use chiefly because most of the scholars of the East wrote in the Arabic language, which had been spread abroad by the Arab conquests.

have become familiar, if only in the form of negative temperature; everyone reads the newspapers and understands what is mean by "the temperature in Moscow is $-8°$."

The word "algebra" comes from the name of a treatise of the mathematician and astronomer Mahommed ibn Musa al-Kharizmi (Mahommed, son of Musa, native of Kharizm), who lived in the 9th century. His treatise on algebra was called Al-jebr w'al-muqabala, which means "transposition and removal." By transposition (al-jebr) is understood the transfer of negative terms to the other side of an equation, and by removal (al-muqabala), cancellation of equal terms on both sides.

The Arabic word "al-jebr" became in Latin transcription "algebra" and the word al-muqabala was discarded, which accounts for the modern term "algebra."*

The origin of this term corresponds very well to the actual content of the science itself. Algebra is basically the doctrine of arithmetical operations considered formally from a general point of view, with abstraction from the concrete numbers. Its problems bring to the fore the formal rules for transformation of expressions and solution of equations. Al-Kharizmi placed on the title page of his book the actual names of two most general formal rules, expressing in this way the true spirit of algebra.

Subsequently, Omar Khayyam defined algebra as the science of solving equations. This definition retained its significance up to the end of the 19th century, when algebra, along with the theory of equations, struck out in new directions, essentially changing its character but not changing its spirit of generality as the science of formal operations.

The mathematicians of Central Asia found methods for calculation, both exact and approximate, of the roots of a number of equations; they discovered the general formula for the "binomial of Newton," although they expressed it in words; they greatly advanced and systematized the science of trigonometry, and calculated very accurate tables of sines. These tables were computed, for astronomical purposes, by the mathematician Gyaseddin (about 1427) who was working with the famous Uzbek astronomer Ulug Begh; Gyaseddin also invented decimal fractions 150 years before they were reinvented in Europe.

To sum up, in the course of the Middle Ages in India and in Central Asia the present decimal system of numeration (including fractions) was almost completely built up, as were also elementary algebra and trigonometry. During the same period the achievements of Chinese science began to make their way into the neighboring countries; about the 6th

---

* It is to be noted also that the mathematical term "algorithm," denoting a method or set of rules for computation, comes from the name of the same al-Kharizmi.

century B.C. the Chinese already had methods for the solution of the simplest indeterminate equations, for approximate calculations in geometry, and for the first steps in approximate solution of equations of the third degree. Essentially the only parts of our present high school course in algebra that were not known before the 16th century were logarithms and imaginary numbers. However, there did not yet exist a system of literal symbols: The content of algebra had outdistanced its form. Yet the form was indispensable: The abstraction from concrete numbers and the formulation of general rules demanded a corresponding method of expression; it was essential to have some way of denoting *arbitrary* numbers and operations on them. The algebraic symbolism is the necessary form corresponding to the content of algebra. Just as in remote antiquity it had been necessary, in order to operate with whole numbers, that symbols should be invented for them, so now, to operate with arbitrary numbers and to give general rules for their use, it was necessary to work out corresponding symbols. This task, begun at the time of the Greeks, was not brought to completion until the 17th century, when the present system of symbols was finally set up in the works of Descartes and others.

4. Renaissance Europe. At the time of the Renaissance the Europeans became acquainted with Greek mathematics by way of the Arabic translations. The books of Euclid, Ptolemy, and Al-Kharizmi were translated in the 12th century from Arabic into Latin, the common scientific language of Western Europe, and at the same time, the earlier system of calculation, as derived from the Greeks and Romans, was gradually replaced by the present-day Indian method, which was borrowed by the Europeans from the Arabs.

It was only in the 16th century that European science finally surpassed the achievements of its predecessors. Thus the Italians, Tartaglia and Ferrari, solved the general cubic equation, and later, the general equation of the fourth degree (see Chapter IV). Let us note that although these results are not taught in school, they belong, with respect to the methods employed in them, to elementary algebra. To higher algebra we must however refer the general theory of equations.

During the same period imaginary numbers began for the first time to be used; at first this was done in a purely formal manner, without logical foundation, which came considerably later at the beginning of the 19th century. Our present-day algebraic symbols were also worked out; in particular, literal symbols were used by Viète in 1591 not only for unknown quantities but also for given ones.

Many mathematicians took a share in this development of algebra. At the same time decimal fractions appeared in Europe; they were invented by the Dutch scholar Stevin, who wrote about them in 1585.

Finally, Napier in Great Britain invented logarithms as an aid in astronomical calculations and wrote about them in 1614; Briggs calculated the first decimal tables of logarithms, which were published in 1624.*

At the same time there appeared in Europe the "theory of combinations" and the general formula for the" binomial of Newton";† the progressions being already known, and in this way the structure of elementary algebra was completed. Therewith came to an end, at the beginning of the 17th century, the whole period of the mathematics of constant magnitudes, of elementary mathematics as it is now taught, with a few additions, in our schools. Arithmetic, elementary geometry, trigonometry, and elementary algebra were now essentially complete. There followed a transition to higher mathematics, to the mathematics of variable magnitudes.

It is not to be thought, however, that the development of elementary mathematics ceased at this time; for example, new results were discovered and are being constantly discovered today in elementary geometry. Furthermore, it is precisely because of the subsequent development of higher mathematics that we now understand more clearly the essential nature of elementary mathematics itself. But the leading role in mathematics was now taken over by the concepts of variable magnitude, function, and limit. The problems, that led from elementary mathematics to higher mathematics are nowadays clarified and solved by the concepts and methods of higher mathematics (occasionally they are not solvable at all by elementary methods), and there are other problems which may be stated in terms of elementary mathematics but which serve even today as a source of more general results and even of entire theories. Examples are provided by the earlier mentioned theory of regular systems of figures or by problems of the theory of numbers which are elementary in their formulation but far from elementary in the methods by which they are solved. For further details the reader may consult Chapter X.

---

* It is interesting to note that Napier did not define logarithms as they are defined nowadays, when we say that in the formula $x = a^y$ the number y is the logarithm of x to the base a. This definition of logarithms appeared later. Napier's definition was related to the concepts of a variable magnitude and an infinitesimal and amounted to saying that the logarithm of x is a function $y = f(x)$ whose rate of growth is inversely proportional to x; that is, $y' = c/x$ (see Chapter II). In this way the basis of the definition was essentially a differential equation, defining the logarithm, although differentials had not yet been invented.

† The formula bears the name of Newton not because he was the first to discover it but because he generalized it from integral exponents to arbitrary fractional and irrational exponents.

§6. Mathematics of Variable Magnitudes

1. Variable and function. In the 16th century the investigation of motion was the central problem of physics. The physical sciences were led to this problem, and to the study of various others involving interdependence of variable magnitudes, by the demands of practical life and by the whole development of science itself.

As a reflection of the general properties of change, there arose in mathematics the concepts of a variable magnitude and a function, and it was this cardinal extension of the subject matter of mathematics that determined the transition to a new stage, to the mathematics of variable magnitudes.

The law of motion of a body in a given trajectory, for example along a straight line, is defined by the manner in which the distance covered by the body increases with time.

Thus Galileo (1564–1642) discovered the law of falling bodies by establishing that the distance fallen increases proportionally to the square of the time. This fact is expressed in the well-known formula

$$s = \frac{gt^2}{2}, \tag{1}$$

where g is approximately equal to 9.81 m/sec².

In general, the law of motion expresses the distance covered in the time t. Here the time t and the distance s are respectively the "independent" and the "dependent" variable, and the fact that to each time t there corresponds a definite distance s is what is meant by saying that the distance s is a function of the time t.

The mathematical concepts of variable and function are the abstract generalization of concrete variables (such as time, distance, velocity, angle of rotation, and area of surface traced out) and of the interdependences among them (the distance depends on the time and so forth). Just as the concept of a real number is the abstract image of the actual value of an arbitrary magnitude, so a "variable" is the abstract image of a varying magnitude, which assumes various values during the process under consideration. A mathematical variable x is "something" or, more accurately, "anything" that may take on various numerical values. This is the meaning of a variable in general; in particular, we may understand by it the time, the distance, or any other variable magnitude.

In exactly the same way, a function is the abstract image of the dependence of one magnitude on another. The assertion that y is a function of x means in mathematics only that to each possible value of x there corresponds a definite value of y. This correspondence between the values of y and the values of x is called a function. For example, according to the law of falling bodies, the distance covered corresponds to the time of fall by formula (1). The distance is a function of the time. Let us look at some other examples.

The energy of a falling body is expressed by its mass and its velocity according to the formula

$$E = \frac{mv^2}{2}. \tag{2}$$

For a given body the energy is a function of the velocity v.

By a familiar law the quantity of heat generated in a conductor in unit time by the passage of an electric current is expressed by the formula

$$Q = \frac{RI^2}{2}, \tag{3}$$

where I is the magnitude of the current and R is the resistance of the conductor. For a given resistance there corresponds to every current I a definite amount of heat Q, generated in unit time. That is, Q is a function of I.

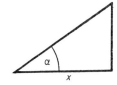

The area of a right-angled triangle S with a given acute angle α and corresponding side x (see figure 5) is expressed by the formula

$$S = \tfrac{1}{2} x^2 \tan \alpha. \tag{4}$$

FIG. 5

For a given angle α the area is a function of the side x.

All these formulas (1)–(4) may be united in the one

$$y = \tfrac{1}{2} ax^2. \tag{5}$$

This general formula represents a transition from the concrete variable magnitudes t, s, E, Q, v and so forth to the general variables x and y, and from the concrete dependences (1), (2), (3), (4) to their general form (5). Mechanics and the theory of electricity have to do with concrete formulas (1), (2), (3), interrelating concrete magnitudes, but the mathematical theory of functions deals with the general formula (5), without associating this formula with any concrete magnitudes.

The next degree of abstraction from the concrete consists in our examining not a given dependence of y on x, like $y = \tfrac{1}{2} ax^2$, $y = \sin x$, $y = \log x$ and so forth, but the general dependence of y on x expressed in the abstract formula

$$y = f(x).$$

This formula states that the magnitude y is in general some function of x; that is, to each value assumed by x there corresponds, in some fashion or another, a definite value y. The subject matter of mathematics thus consists not only of certain given functions ($y = \frac{1}{2} ax^2$, $y = \sin x$, and so forth), but of *arbitrary* (more accurately, more or less arbitrary) functions. These degrees of abstraction, first from concrete magnitudes and then from concrete functions, are analogous to the degrees of abstraction observed in the formation of the concept of a whole number: First, abstraction from concrete collections of objects led to the concept of whole numbers (1, 3, 12, and so forth), and then a further abstraction led to the concept of an arbitrary whole number in general. This generalization is the result of a profound interraction between analysis and synthesis: analysis of separate interrelations and synthesis, in the form of new concepts, of their common features.

The branch of mathematics devoted to the study of functions is called analysis, or often, infinitesimal analysis, since one of the most important elements in the study of functions is the concept of the infinitesimal (the meaning of this concept and its significance are explained in Chapter II).

Since a function is the abstract image of a dependence of one magnitude on another, we may say that analysis takes as its subject matter dependences between variable magnitudes, not between one concrete magnitude and another but between variables in general, in abstraction from their content. An abstraction of this sort guarantees great breadth of application, since one formula or one theorem contains an infinite number of possible concrete cases. An example of this is given already by our simple formulas (1)–(5). So the complete analogy of analysis with arithmetic and algebra becomes evident. They all originate in definite practical problems and give a general abstract expression to concrete relationships in the actual world.

2. Analytic geometry and analysis. Thus the new period of mathematics, beginning in the 17th century, may be defined as the period of the birth and development of analysis. (This is the third of the three important periods mentioned earlier.) It is to be understood, of course, that no theory arises as a result of the mere formation of new concepts, that analysis could not result from the mere existence of the concepts of variable and function. For the founding of a theory, and all the more of a complete branch of science like mathematical analysis, it is necessary that the new concepts become active, so to speak, that among them there be discovered new relationships, and that they permit the solution of new problems.

But more than that, new concepts can originate and develop, and become more general and precise, only on the basis of the very problems they enable us to solve, only through those theorems of which they form a part. The concepts of variable and function did not arise in complete form in the mind of Galileo, Descartes, Newton, or anybody else. They occurred to many mathematicians (for example Napier in connection with logarithms) and gradually assumed a more or less clear, but still by no means final, form with Newton and Leibnitz, being made still more precise and general in the subsequent development of analysis. Their present-day definition was laid down only in the 19th century, but even it is not *absolutely* rigorous or *altogether* final. The development of the concept of a function is continuing even at the present time.

Mathematical analysis was based on material furnished by the new science of mechanics, and on problems of geometry and algebra. The first definite step toward the mathematics of variable magnitudes was the appearance in 1637 of the "geometry" of Descartes, where the foundations were laid for the so-called analytic geometry. The basic ideas of Descartes are as follows.

Suppose we are given, for example, the equation

$$x^2 + y^2 = a^2. \tag{6}$$

In algebra x and y were understood as unknowns, and since the given equation does not allow us to determine them, it did not present any essential interest for algebra. But Descartes did not consider x and y as unknowns, to be found from the equation, but as *variables*; so that the given equation expresses the interdependence of two variables. Such an equation may be written in general form, by taking all its terms to the left-hand side, thus:

$$F(x, y) = 0.$$

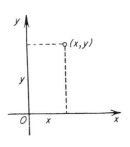

FIG. 6.

Further, Descartes introduced into the plane the coordinates x, y which are now called Cartesian (figure 6). In this way, to each pair of values x and y there corresponds a point, and conversely to each point there corresponds a pair of coordinates x, y. Consequently, the equation $F(x, y) = 0$ determines the geometric locus of those points on the plane whose coordinates satisfy the equation. In general, this will be a curve. For example, equation (6) determines the circumference of a circle of radius a with center at the origin. In fact, as is obvious from figure 7, by the theorem of Pythagoras, $x^2 + y^2$ is the square of the distance from the origin O to the point M with coordinates x and y. So equation (6) represents the geometric locus of those points whose distance from the origin is equal to a, which is the circumference of a circle.

Conversely, a geometric locus of points, given by a geometric condition, may also be given by an equation expressing the same condition in the language of algebra by means of coordinates. For example, the geometric condition defining the circumference of a circle, namely that it is a geometric locus of points equidistant from a given point, may be expressed in algebraic language by equation (6).

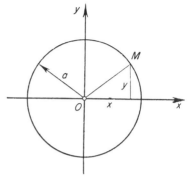

FIG. 7.

Thus the general problem and the general method of analytic geometry are as follows: We represent a given equation in two variables by a curve on the plane, and from the algebraic properties of the equation we investigate the geometric properties of the corresponding curve; and conversely, from the geometric properties of the curve we find the equation, and then from the algebraic properties of the equation we investigate the geometric properties of the curve. In this way geometric problems may be reduced to algebraic, and so finally to computation.

The content of analytic geometry will be discussed in detail in Chapter III. We now wish to direct attention to the fact that, as is evident from our short explanation, it originated in a union of geometry, algebra, and the general idea of a variable magnitude. The main geometric content of the early beginnings of analytic geometry was the theory of conic sections, ellipse, hyperbola, and parabola. This theory, as we have pointed out, was developed by the ancient Greeks; the results of Apollonius already contained in geometric form the equations of the conic sections. The union of this geometric content with algebraic form, developed after the time of the Greeks, and with the general idea of a variable magnitude, arising from the study of motion, produced analytic geometry.

Among the Greeks the conic sections were a subject of purely mathematical interest, but by the time of Descartes they were of practical importance for astronomy, mechanics, and technology. Kepler (1571–1630) discovered that the planets move around the sun in ellipses, and Galileo established the fact that a body thrown in the air, whether it is a stone or a cannonball, moves along a parabola (to the first approximation, if we may neglect air resistance). As a result, the calculation of various magnitudes referring to the conic sections became an urgent necessity, and it was the method of Descartes that solved this problem. So the way was prepared for his method by the preceding development of mathematics, and the method itself was brought into existence by the insistent demands of science and technology.

3. Differential and integral calculus. The next decisive step in the mathematics of variable magnitudes was taken by Newton and Leibnitz during the second half of the 17th century, in the founding of the differential and integral calculus. This was the actual beginning of analysis, since the subject matter of this calculus is the properties of functions themselves, as distinct from the subject matter of analytic geometry, which is geometric figures. In fact Newton and Leibnitz only brought to completion an immense amount of preparatory work, shared by many mathematicians and going back to the methods for determining areas and volumes worked out by the ancient Greeks.

Here we shall not explain the fundamental concepts of differential and integral calculus and of the theories of analysis that followed them, since this will be done in the special chapters devoted to these theories. We wish only to draw attention to the sources of the calculus, which were mainly the new problems of mechanics and the old problems of geometry, the latter consisting of drawing a tangent to a given curve and of determining areas and volumes. These geometric problems had already been studied by the ancients (it is sufficient to mention Archimedes), and also by Kepler, Cavalieri, and others at the beginning of the 17th century. But the decisive event was the discovery of the remarkable relation between these two types of problems and the formulation of a general method for solving them; this was the achievement of Newton and Leibnitz.

This relation, allowing us to connect the problems of mechanics with those of geometry, was discovered because of the possibility, arising from the method of coordinates, of making a graphical representation of the dependence of one variable on another, or in other words of a function. With the help of this graphical representation, it is easy for us to formulate the earlier mentioned relation, between the problems of mechanics and geometry, which was the source of the differential and integral calculus, and consequently to describe the general content of these two types of calculus.

The differential calculus is basically a method for finding the velocity of motion when we know the distance covered at any given time. This problem is solved by "differentiation." It turns out that the problem is completely equivalent to that of drawing a tangent to the curve representing the dependence of distance on time. The velocity at the moment t

is equal to the slope of the tangent to the curve at the point corresponding to t (figure 8).

The integral calculus is basically a method of finding the distance covered when the velocity is known, or more generally of finding the total result of the action of a variable magnitude. This problem is obviously

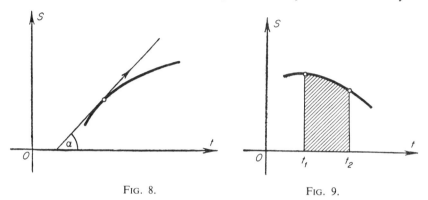

FIG. 8.   FIG. 9.

the converse of the problem of the differential calculus (the problem of finding the velocity); it is solved by "integration." It turns out that the problem of integration is completely equivalent to that of finding the area under the curve representing the dependence of the velocity on time. The distance covered in the interval of time from the moment t_1 to the moment t_2 is equal to the area under the curve between the straight lines corresponding on the graph to the values t_1 and t_2 (figure 9).

By abstracting from the mechanical formulation of the problems of the calculus and by dealing with functions rather than with dependence of distance or velocity on time, we obtain the general concept of the problems of differential and integral calculus in abstract form.

Fundamental to the calculus, as to the whole subsequent development of analysis, is the concept of a limit, which was formulated somewhat later than the other fundamental concepts of variable and function. In the early days of analysis the role later played by the limit was taken by the somewhat nebulous concept of an infinitesimal. The methods for actual calculation of velocity, given the distance covered (namely, differentiation), and of distance, given the velocity (integration), were founded on a union of algebra with the concept of limit. Analysis originated in the application of these concepts and methods to the aforementioned problems of mechanics and geometry (and also to certain other problems; for example, problems of maxima and minima). The science of analysis was in turn absolutely necessary for the development of mechanics, in the formulation of whose laws its concepts had already appeared in latent form. For example, the second law of Newton, as formulated by Newton himself, states that "the change in momentum is proportional to the acting force" (more precisely: The rate of change of momentum is proportional to the force). Consequently, if we wish to make any use of this law, we must be able to define the rate of change of a variable, that is, to differentiate. (If we state the law in the form that the acceleration is proportional to the force, the problem remains the same, because acceleration is proportional to rate of change of momentum.) Also, it is perfectly clear that in order to state the law governing a motion when the force is variable (in other words, the motion proceeds with a variable acceleration), we must be able to solve the inverse problem of finding a magnitude given its rate of change; in other words, we must be able to integrate. So one might say that Newton was simply *compelled* to invent differentiation and integration in order to develop the science of mechanics.

4. Other branches of analysis.   Along with the differential and integral calculus, other branches of analysis arose: The theory of series (see Chapter II, §14), the theory of differential equations (Chapters V and VI), and the application of analysis to geometry, which later became a special branch of geometry, called differential geometry and dealing with the general theory of curves and surfaces (Chapter VII). All these theories were brought to life by the problems of mechanics, physics, and technology.

The theory of differential equations, the most important branch of analysis, has to do with equations in which the unknown is no longer a magnitude but a function, or in other words a law governing the dependence of one magnitude on another or on several others. It is easy to understand how such equations arose. In mechanics we seek to determine the whole law of motion of a body under given conditions and not just one value of the velocity or of the distance covered. In the mechanics of fluids it is necessary to find the distribution of velocity over the whole mass of fluid in motion, or in other words to find the dependence of the velocity on all three space coordinates and on time. Analogously, in the theory of electricity and magnetism we must find the tension in the field throughout all space; that is, the dependence of this tension on the same three space coordinates, and similarly in other cases.

Problems of this sort arose continually in the various branches of mechanics, including hydrodynamics and the theory of elasticity, in acoustics, in the theory of electricity and magnetism, and in the theory of heat. From the very moment of its birth, analysis remained in close contact with mechanics and with physics in general, its most important achievements being invariably connected with the solution of problems posed by the exact sciences. Beginning with Newton, the greatest analysts, D. Bernoulli (1700–1782), L. Euler (1707–1783), J. Lagrange (1736–1813), H. Poincaré (1854–1912), M. V. Ostrogradskiĭ (1801–1861) and A. M. Lyapunov (1857–1918), as well as many others who laid new foundations in analysis, started as a rule from the urgent problems of contemporary physics.

In this way new theories arose: In direct connection with mechanics, Euler and Lagrange founded a new branch of analysis, called the calculus of variations (see Chapter VIII), and at the end of the 19th century Poincaré and Lyapunov, starting again from the problems of mechanics, founded the so-called qualitative theory of differential equations (see Chapter V, §7).

In the 19th century analysis was enriched by an important new branch, the theory of functions of a complex variable (see Chapter IX). The rudiments of it are to be found in the works of Euler and certain other mathematicians, but its transformation into a well-formed theory took place in the middle of the 19th century and was carried out to a great extent by the French mathematician Cauchy (1789–1857). This theory rapidly underwent an imposing development with numerous significant results that allowed mathematicians to penetrate more deeply into many of the laws of analysis and found important applications in problems of mathematics itself, and of physics and technology.

Analysis developed rapidly; not only did it form the center and the most important part of mathematics but it also penetrated into the older regions: algebra, geometry, and even the theory of numbers. Algebra began to be thought of as basically the doctrine of functions expressed in the form of polynomials of one or several variables.* Analytic and differential geometry began to dominate the field of geometry. As far back as Euler, methods of analysis were introduced into the theory of numbers and formed in this way the beginning of the so-called analytic theory of numbers, which contains some of the most profound achievements of the science of whole numbers.

Through the influence of analysis, with its concepts of variable, function, and limit, the whole of mathematics was penetrated by the idea of motion and change, and therefore of dialectic. In exactly the same way, basically through analysis, mathematics was affected by the exact sciences and

---

* Polynomials are functions of the form $y = a_0x^n + a_1x^{n-1} + \cdots + a_n$. The fundamental problem of the algebra of the period, namely the solution of the equation $a_0x^n + a_1x^{n-1} + \cdots + a_n = 0$, simply means the search for values of x for which the function $y = a_0x^n + a_1x^{n-1} + \cdots + a_n$ is equal to zero. The very existence of a solution, of a root of the equation, which is called the fundamental theorem of algebra, is proved by means of analysis (see Chapter IV, §3).

technology and in turn played a role in their development, since it was the means of giving exact expression to their laws and of solving their problems. Just as among the Greeks mathematics was basically geometry, one may say that after Newton it was basically analysis. Of course, analysis did not completely absorb the whole of mathematics; in geometry, in the theory of numbers, and in algebra the problems and methods characteristic of these sciences were everywhere continued. Thus in the 17th century there arose, along with analytic geometry, another branch of geometry, namely projective geometry, in which purely geometric methods played a dominant role. It originated chiefly in problems of the representation of objects on a plane (projection), and as a result it is particularly useful in descriptive geometry.

At the same time there was developed an important new branch of mathematics, the theory of probability, which takes as its subject matter the uniformities observable in large masses of phenomena, such as a long series of rifle shots or tosses of a coin. In the succeeding period it acquired a special importance in physics and technology and its development was conditioned by the problems which came to it from those branches of science. The characteristic feature of this theory is that it deals with the laws of "random events," providing mathematical methods for investigation of the irregularities that necessarily appear in random events. The basic features of the theory of probability will be explained in Chapter XI.

5. Applications of analysis. Analysis in all its branches provided physics and technology with powerful methods for the solution of problems of many different kinds. We have already mentioned the earliest of these: to find the rate of change of a magnitude when we know how the magnitude itself depends on time; to find the area of curvilinear figures and the volumes of solids; and to find the total result of some process or another or the total action of a variable magnitude. Thus, the integral calculus allows us to determine the work done by an expanding gas as the pressure changes according to a well-known law; the same integral calculus allows us to compute, for example, the tension of an electric field with an arbitrarily given system of charges, basing our work on the law of Coulomb which determines the tension of a field resulting from a point charge, and so forth.

Further, analysis provided a method for finding the maximum and the minimum values of a magnitude under given conditions. Thus, with the help of analysis it is easy to determine the shape of a cylindrical cistern which for a given volume will have the smallest surface and consequently will require the smallest outlay of material. It turns out that the cistern will have this property if its height is equal to the diameter of its base (figure 10). Analysis allows us to determine the shape of the curve along which a body must roll in order to fall in the shortest time from one given point to another (this curve is the so-called cycloid; figure 11).

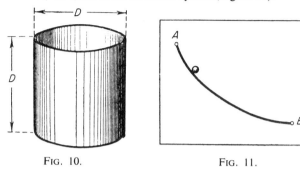

FIG. 10.   FIG. 11.

For the solution of these and other problems the reader may turn to Chapters II and VIII.

Analysis, or more precisely the theory of differential equations, allows us not merely to find separate values for variable magnitudes but also to determine unknown functions; that is, to find laws of dependence of certain variables on others. Thus we have the possibility, on the basis of the general laws of electricity, of computing how the current varies with time in a circuit with arbitrary resistance, capacitance, and self-induction. We can determine laws for the distribution of velocities throughout the whole mass of a fluid under given conditions. We can deduce general laws for the vibration of strings and membranes, and for the propagation of vibrations in various media; here we are referring to sound waves, electromagnetic waves, or elastic vibrations propagated through the Earth by earthquakes or explosions. Parenthetically, we may remark that new methods are thereby provided for searching for useful minerals and for carrying out investigations far below the surface of the Earth. Individual problems of this sort will be found in Chapters V and VI.

Finally, analysis not only provides us with methods for solving special problems; it also gives us general methods for mathematical formulation of the quantitative laws of the exact sciences. As was mentioned, earlier, the general laws of mechanics could not be formulated mathematically without recourse to the concepts of analysis, and without such a formulation we would not be able to solve the problems of mechanics. In exactly the same way the general laws for heat conduction, diffusion through porous materials, propagation of vibrations, the course of chemical reactions, the basic laws of electromagnetism, and many other laws simply could not be given a mathematical formulation without the concepts of analysis. It is only as the result of such a formulation that these laws can be applied to the most varied concrete cases, providing a basis for exact mathematical conclusions in the special problems of heat conduction, vibrations, chemical solution, electromagnetic fields, and other problems of mechanics, astronomy, and all the numerous branches of physics, chemistry, heat engineering, power, machine construction, electrical engineering, and so forth.

6. Critical examination of the foundations of analysis. Just as in the history of geometry among Greeks the rigorous and systematic presentation given by Euclid brought to completion a long previous development, so in the development of analysis there arose the necessity of placing it upon a firmer basis than had been provided by the first creators of its powerful methods: Newton, Euler, Lagrange, and others. As the analysis founded by them grew more extensive, it began, on the one hand, to deal with more profound and difficult problems, and on the other, to require from its very extent a more systematic and carefully reasoned basis. The growth of the theory necessitated a systematization and critical analysis of its foundations. To put a theory on a firm foundation requires examination of its entire development and should by no means be considered as a starting point for the theory itself, since without the theory we would simply have no idea of what it is that we need to provide with a foundation. By the way, certain contemporary formalists forget this fact when they consider it advisable to found and develop a theory starting from axioms that have not been selected on the basis of any analysis of the actual material which they are supposed to summarize. But the axioms themselves require a justification of their content; they only sum up other material and provide a foundation for the logical construction of a theory.*

The necessary period of criticism, systematization, and laying of foundations occurred in analysis at the beginning of the last century. Through the efforts of a number of eminent scientists this important and difficult work was brought to a successful completion. In particular, precise definitions were given for the basic concepts of real number, variable, function, limit, and continuity.

However, as we have already had occasion to mention, none of these definitions may be considered as absolutely rigorous or final. The development of these concepts is continuing. Euclid and all the mathematicians in the course of 2,000 years after him no doubt considered his "Elements"

---

* This double role of the axioms is sometimes lost from view even in works of a methodological character, which thereby attribute to the construction of axioms a significance which does not at all belong to it, namely that of the total construction of a theory.

as the practical limit of logical rigor. But to a contemporary view the Euclidean foundations of geometry seem quite superficial. This historical example shows that we ought not to flatter ourselves with any idea of "absolute" or "final" rigor in contemporary mathematics. In a science that is not yet dead and mummified, there is not and cannot be anything perfect. But we can say with confidence that the foundations of analysis as they exist at present correspond in a quite satisfactory way to the contemporary problems of science and the contemporary conception of logical precision; and second, that the continued deepening of these concepts and the discussions that are now taking place about them give us no cause, and will not give us cause, simply to reject them; these discussions will lead us to a new, more precise, and more profound understanding, the results of which it is still difficult to estimate.

Although the establishment of the basic principles of a theory forms a summary of its development, it does not represent the end of the theory; on the contrary, it is conducive to further development. This is exactly what happened in analysis. In connection with the deepening of its foundations there arose a new mathematical theory, created by the German mathematician Cantor in the seventies of the last century, namely the general theory of infinite sets of arbitrary abstract objects, whether numbers, points, functions or any other "elements". On the basis of these ideas there grew up a new chapter in analysis, the so-called theory of functions of a real variable, whose concepts, along with those of the foundations of analysis and the theory of sets, are explained in Chapter XV. At the same time the general ideas of the theory of sets penetrated every branch of mathematics. But this "set-theoretical point of view" is inseparably connected with a new stage in the development of mathematics, which we will now consider briefly.

§7. Contemporary Mathematics

1. The more advanced character of present-day mathematics. To the four stages of the develoment of mathematics mentioned in §5 there naturally correspond stages in our mathematical education, the material learned at each stage of our study consisting, to a fair degree of approximation, of the basic content of the corresponding period in the history of mathematics.

The basic results of arithmetic and geometry, obtained in the first period of the development of mathematics, form the subject of primary education and are known to us all. For example, when we determine the quantity of material necessary to carry out a certain job, let us say to cover a floor, we are already making use of these first results of mathematics. The most important achievements of the second period, the period of elementary mathematics, are taught in the high schools. The basic results of the third period, the foundations of analysis, the theory of differential equations, higher algebra, and so forth, form the mathematical instruction of an engineer; they are studied in all the schools of higher education, except those devoted purely to the humanities. In this way the basic ideas and results of the mathematics of that period are widely known, use being made of them to some extent by almost every engineer and scientist.

On the other hand, the ideas and results of the present-day period of mathematics are studied almost exclusively in graduate departments of mathematics and physics. Beside mathematical specialists, they are used by researchers in the fields of mechanics and physics, and in a number of the newer branches of technology. Of course, this does not at all mean that they have no practical application, but since they represent the most recent results of science, they are naturally more complicated. Consequently, as we now pass to a general description of the latest stage in the development of mathematics we can no longer consider that everything which we mention briefly will be altogether clear. We will try to present in a few lines the most general character of the new branches of mathematics; their content will be explained in greater detail in the corresponding chapters of the book.

If the present section seems overly difficult it may be passed over at first reading and taken up again after study of the special chapters.

2. Geometry. The beginning of the present-day development of mathematics is characterized by profound changes in all its basic fields: algebra, geometry, and analysis. This change may perhaps be followed most clearly in the field of geometry. In the year 1826 Lobačevskiĭ, and almost simultaneously with him the Hungarian mathematician Janos Bolyai, developed the new non-Euclidean geometry. The ideas of Lobačevskiĭ were far from being immediately understood by all mathematicians. They were too bold and unexpected. But from this moment there began a fundamental new development of geometry; the very conception of what is meant by geometry was changed. Its subject matter and the range of its applications were rapidly extended. The most important step, after Lobačesvskiĭ, in this direction was taken in 1854 by the celebrated German mathematician, Riemann. He clearly formulated the general idea that an unlimited number of "spaces" could be investigated by geometry, and at the same time he indicated their possible significance in the real world. In the new development of geometry two features were characteristic.

In the first place the earlier geometry studied only the *spatial* forms and relations of the material world, and then only to the extent in which they appear in the framework of Euclidean geometry, but now the subject matter of geometry began to include also many *other* forms and relations of the actual world, provided only they were similar to the spatial ones and therefore allowed the use of geometric methods. The term "space" thereby took on in mathematics a new meaning, broader and at the same time more special. Simultaneously, the methods of geometry became much richer and more varied. In their turn they provide us with more complete means for learning about the physical world around us, the world from which geometry in its original form was abstracted.

In the second place, even in Euclidean geometry important progress was made: In it were studied the properties of incomparably more complicated figures, even including arbitrary sets of points. Also a fundamentally new attitude appeared toward the properties of the figures under investigation. Separate groups of properties were distinguished, which could be investigated in abstraction from others, and this very abstraction *within* geometry gave rise to many characteristic branches of the subject, which essentially became independent "geometries." The development of geometry in all these directions is being continued and more and more new "spaces" and their "geometries" are being studied: the space of Lobačesvskiǐ, projective space, Euclidean and other spaces of various dimensions, in particular four-dimensional space, Riemann spaces, Finsler spaces, topological spaces, and so forth. These theories find important application in mathematics itself, outside of geometry, and also in physics and mechanics; particularly noteworthy are their applications in the theory of relativity of contemporary physics, which is a theory of space, time, and gravitation. From what has been said it is clear that we are dealing here with a qualitative change in geometry.

The ideas of contemporary geometry and some of the elements of the theory of various spaces investigated in it will be explained in Chapters XVII and XVIII.

3. Algebra. Algebra too underwent a qualitative change. In the first half of the 19th century new theories arose, which led to changes in its character, and to an extension of its subject matter and its range of application.

In its original form, as pointed out in §5, algebra dealt with mathematical operations on numbers considered from a formal point of view, in abstraction from given concrete numbers. This abstraction found expression in the fact that in algebra magnitudes are denoted by letters, on which calculations are carried out according to well-known formal rules.

Contemporary algebra retains this basis but widens it in a very extensive way. It now considers "magnitudes" of a much more general nature than numbers, and studies operations on these "magnitudes" which are to some extent analogous in their formal properties to the ordinary operations of arithmetic: addition, subtraction, multiplication, and divison. A very simple example is offered by vector magnitudes, which may be added by the well-known parallelogram rule. But the generalization carried out in contemporary algebra is such that even the very term "magnitude" often loses its meaning and one speaks more generally of "elements" on which it is possible to perform operations similar to the usual algebraic ones. For example, two motions carried out one after the other are evidently equivalent to a certain single motion, which is the sum of the two; two algebraic transformations of a formula may be equivalent to a single transformation that produces the same result, and so forth; and so it is possible to speak of a characteristic "addition" of motions or transformations. All this and much else is studied in a general abstract form in contemporary algebra.

The new algebraic theories in this direction arose in the first half of the 19th century in the investigations of a number of mathematicians, among whom we should particularly mention the French mathematician Galois (1811–1832). The concepts, methods, and results of contemporary algebra find important applications in analysis, geometry, physics, and crystallography. In particular, the theory mentioned at the end of §3 concerning the symmetry of crystals, which was developed by E. S. Fedorov, is based on a union of geometry with one of the new algebraic theories, the so-called theory of groups.

As we see, we are dealing here with a fundamental, qualitative generalization of the subject matter of algebra with a change in the very concept of what algebra is. The ideas of contemporary algebra and the basic elements of some of its theories will be explained in Chapter XX and XVI.

4. Analysis.   Analysis in all its branches also made profound progress. In the first place, as was already mentioned in the preceding section, its foundations were made more precise; in particular, its basic concepts were given exact and general definitions: such concepts as function, limit, integral and finally, the basic concept of a variable magnitude (a rigorous definition was given for the real number). A beginning of the process of putting analysis on a more precise foundation was made by the Czech mathematician Bolzano (1781–1848), the French mathematician Cauchy (1789–1857), and a number of others. This greater precision was gained at the same time as the new developments in algebra and geometry were being made; it was brought to completion in its present well-known form in the eighties of the 19th century by the German mathematicians Weierstrass, Dedekind, and Cantor. As was mentioned at the end of §6, Cantor also laid the foundation for the theory of transfinite sets, which plays such a large role in the development of the newer ideas in mathematics.

The increase in precision in the concepts of variable and function in connection with the theory of sets laid the foundation for a further development of analysis. A transition was made to the study of more general functions; and in the same direction the apparatus of analysis, namely the integral and differential calculus, was also generalized. Thus, on the threshold of the present century, there arose the new branch of analysis already mentioned in §6, the so-called theory of functions of a real variable. The development of this theory is chiefly connected with the French mathematicians, Borel and Lebesgue and others, and with N. N. Luzin (1883–1950) and his school. In general, the newer branches of analysis are called modern analysis in contradistinction to the earlier so-called classical analysis.

Other new theories arose in analysis. Thus a special branch was formed by the theory of approximation of functions, which studies questions of the best approximate representation of general functions by various "simple" functions, above all by polynomials, that is by functions of the form

$$a_0 x^n + a_1 x^{n-1} + \cdots + a_{n-1} x + a_n .$$

The theory of approximation of functions has great importance, if only for the reason that it lays down general foundations for the practical calculation of functions, for the approximate replacement of complicated functions by simpler ones. The rudiments of this theory go back to the very beginnings of analysis. Its modern direction was given to it by the great Russian mathematician P. L. Čebyšev (1821–1894). This direction was later developed into the so-called constructive theory of functions, chiefly in the works of Soviet mathematicians, particularly S. N. Bernšteĭn (born 1880), to whom belong the most important results in this field. Chapter XII deals with approximation of functions.

We spoke earlier about the development of the theory of functions of a complex variable. We must still mention the so-called qualitative theory of differential equations, originating in the works of Poincaré (1854–1912) and A. M. Lyapunov (1857–1918), about which some ideas will be given in Chapter V, and also the theory of integral equations. These theories have great practical importance in mechanics, physics, and technology. Thus, the qualitative theory of differential equations provides solutions of problems concerning stability of motion, and the action of mechanisms or of vibrating electric systems and the like. Stability of a process means in the most general sense that if small changes are made in the initial data or in the conditions of the motion, then the motion itself during the whole of its course will change only slightly. The technical significance of questions of this sort hardly needs to be emphasized.

5. Functional analysis. On the ground prepared by the development of analysis and mathematical physics, along with the new ideas of geometry and algebra, there has grown up an extensive new division of mathematics, the so-called functional analysis, which plays an exceptionally important role in modern mathematics. Many mathematicians shared in creating it; let us mention, for example, the greatest German mathematician of recent times, Hilbert (1862–1943), the Hungarian mathematician Riesz (1880–1956) and the Polish mathematician Banach (1892–1945). The separate Chapter XIX is devoted to functional analysis.

The essence of this new branch of mathematics consists briefly in the following. In classical analysis the variable is a magnitude, or "number," but in functional analysis the function itself is regarded as the variable. The properties of the given function are determined here not in themselves but in relation to other functions. What is under study is not a separate function but a whole collection of functions characterized by one property or another; for example, the collection of all continuous functions. Such a collection of functions forms the so-called functional space. This procedure corresponds, for example, to the fact that we may consider the collection of all curves on a surface or of all possible motions of a given mechanical system, thereby defining the properties of the separate curves or motions in their relation to other curves or motions.

The transition from the investigation of separate functions to a *variable* function is similar to the transition from unknown numbers x, y to variables x, y; that is, it is similar to the idea of Descartes mentioned in a preceding paragraph. On the basis of this idea Descartes produced his well-known union of algebra and geometry, of an equation and a curve, which is one of the most important elements in the rise of analysis. Similarly, the union of the concept of a variable function with the ideas of contemporary algebra and geometry produced the new functional analysis. Just as analysis was necessary for the development of the mechanics of the time, so functional analysis provided new methods for the solution of present-day problems of mathematical physics and produced the mathematical apparatus for the new quantum mechanics of the atom. History repeats itself as usual, but in a new way, on a higher plane. As we have said, functional analysis unites the basic ideas and methods of analysis, of modern algebra, and of geometry and in its turn exercises an influence on the development of these branches of mathematics. The problems arising in classical analysis now find new, more general solutions, often almost at a single step, by means of functional analysis. Here, as at a focus, are gathered together, in a very productive way, the most general and abstract ideas of modern mathematics.

From this short sketch, from this mere enumeration of the new directions of analysis(the theory of functions of a real variable, theory of approximation of functions, qualitative theory of differential equations, theory of integral equations, and functional analysis) it may be seen that we are dealing here in fact with an essentially new stage in the development of analysis.

6. Computational mathematics and mathematical logic. At all periods the technical level of the means of computation has had an essential influence on mathematical methods. But the equipment for carrying out calculations which has been at our disposal up until most recent times has been very limited. The simplest devices, such as the abacus, tables of logarithms and the logarithmic sliderule, the calculating machine, and finally more complicated calculators and the automatic calculating machine, these were the basic implements for computation existing up to the forties of the 20th century. These implements made it possible to carry out more or less quickly the separate operations of addition, multiplication, and so forth. But to carry through to final numerical result the practical problems that arise nowadays requires a colossal number of such operations, following one another in a complicated program that sometimes depends on results obtained during the course of the calculation. The solution of such problems proved to be practically impossible or completely valueless on account of the length of the process of solution. But in the last ten years a radical change has taken place in the whole science of computation. Modern calculating machines, constructed on new principles, allow us to make computations with exceptionally great speed and at the same time to carry out complicated chains of calculations automatically, according to extremely flexible programs arranged in advance. Some of the questions connected with the construction and significance of modern calculating machines will be discussed in Chapter XIV.

The new techniques not only enable us to carry out investigations that were formerly quite impracticable but also lead us to change our estimate of the value of many well-known mathematical results. For example, they have given a special stimulus to the development of approximative methods; that is, methods which allow us, by a chain of elementary operations, to reach a desired numerical result with sufficiently great accuracy. The mathematical methods themselves must now be estimated from the point of view of their suitability for corresponding machines.

In close connection with the development of calculating techniques is the subject of mathematical logic. It was developed primarily as a result of intrinsic difficulties arising in mathematics itself, its subject matter being the analysis of mathematical proof. It is itself a branch of mathematics, and includes those branches of general logic that can be objectively formulated and developed by the mathematical method.

Although on the one hand mathematical logic thus goes back to the very sources and foundations of mathematics, it is closely connected, on the other hand, with the most modern questions of computational technique. Naturally, for example, a proof that leads to the setting up of a definite preassigned process, permitting us to approach a desired result with an arbitrary degree of accuracy, is essentially different from more abstract proofs on the existence of the given result.

There also arises here a characteristic range of questions concerning the degree of generality possible in problems that can be dealt with by a method which is completely defined in advance at every step. Profound results have been reached along these lines in mathematical logic, results that are extremely important from a general epistemological point of view.

It would not be an exaggeration to say that with the development of the new computational techniques and the achievements of mathematical logic a new period has begun in modern mathematics, characterized by the fact that its subject matter is not only the study of one object or another but also all the ways and means by which such an object can be defined; not only certain problems, but also all possible methods of solving them.

To what has been said it is only necessary to add that also in the older branches of mathematics, the theory of numbers, Euclidean geometry, classical algebra and analysis, and the theory of probability, rapid development has continued throughout the whole period of modern mathematics so that these fields have been enriched by many new fundamental ideas and results; let us mention, for example, the results attained in the theory of numbers and in the geometry of everyday space by the Russian and Soviet mathematicians P. L. Čebyšev, E. S. Fedorov, I. M. Vinogradov, and others. The development on a wide front of the theory of probability has been connected with the extraordinarily important regularities observable in statistical physics and in contemporary problems of technology.

7. Characteristic features of modern mathematics.   What are the most general characteristics of modern mathematics as a whole, distinguishing it from the earlier development of geometry, algebra, and analysis?

First of all is the immense extension of the subject matter of mathematics and of its applications. Such an extension of subject matter and range of application represents an enormous quantitative and qualitative growth, brought about by the appearance of powerful new theories and methods that allow us to solve problems completely inaccessible up to now. This extension of the subject matter of mathematics is characterized by the fact that contemporary mathematics conscientiously sets itself the task of studying all possible types of quantitative relationships and spatial forms.

A second characteristic feature of modern mathematics is the formation of general concepts on a new and higher level of abstraction. It is precisely this feature that guarantees preservation of the unity of mathematics, in spite of its immense growth in widely differing branches. Even in parts of mathematics that are extremely far from one another similarities of structure are brought to light by the general concepts and theories of the present day. They guarantee that contemporary mathematical methods will have great generality and breath of application; in particular, they produce a profound interpenetration of the fundamental branches of mathematics: geometry, algebra, and analysis.

As one of the characteristic features of modern mathematics, we must also mention the obvious dominance of the set-theoretical point of view. Of course, this point of view owes its significance to the fact that it summarizes in a certain sense the rich content of all the preceding developments of mathematics. Finally, one of the most characteristic features of modern mathematics is the profound analysis of its foundations, of the mutual influence of its concepts, of the structure of its separate theories, and of the methods of mathematical proof. Without such an analysis of foundations it would not be possible to improve or develop any further the principles and theories that have led to the present generalizations.

The characteristic feature of modern mathematics may be said to be that its subject matter consists not only of given quantitative relations and forms but of all possible ones. In geometry, we speak not only of spatial relations and forms but of all possible forms similar to spatial ones. In algebra, we speak of various abstract systems of objects with all possible laws of operation on them. In analysis, not only magnitudes are considered as variables but the very functions themselves. In a functional space all the functions of a given type (all the possible interdependences among the variables) are brought together. Summing up, it is possible to say that while elementary mathematics deals with constant magnitudes, and the next period with variable magnitudes, *contemporary mathematics is the mathematics of all possible (in general, variable) quantitative relations and interdependences among magnitudes.* This definition is, of course, incomplete, but it does emphasize the characteristic feature of modern mathematics which distinguishes it from the mathematics of preceding ages.*

Suggested Reading

Preliminary remark. The original Russian text of *Mathematics: its content, methods, and meaning* contains a list of recommended books at the end of each of its twenty chapters. In the present translation these books have been retained only if they have been translated into English. In compensation, the lists given here contain many other, readily available, works in the English language.

Books dealing with mathematics in general

E. T. Bell, *The development of mathematics*, 2d ed., McGraw-Hill, New York, 1945.

R. Courant and H. Robbins, *What is mathematics?* Oxford University Press, New York, 1941.

H. Eves and C. V. Newsom, *An introduction to the foundations and fundamental concepts of mathematics*, Rinehart, New York, 1958.

G. H. Hardy, *A mathematician's apology*, Macmillan, New York, 1940.

R. L. Wilder, *Introduction to the foundations of mathematics*, Wiley, New York, 1952.

Books of a historical character

R. C. Archibald, *Outline of the history of mathematics*, 5th ed., Mathematical Association of America, Oberlin, Ohio, 1941.

F. Cajori, *History of mathematics*, 2d ed., Macmillan, New York, 1919.

Euclid, The thirteen books of Euclid's *Elements* translated with an introduction and commentary by T. L. Heath, 2d ed., 3 vols., Dover, New York, 1956.

O. E. Neugebauer, *The exact sciences in antiquity*, Princeton University Press, Princeton, N. J., 1952.

D. E. Smith, *History of mathematics*, Vol. I. *General survey of the history of elementary mathematics*, Vol. II. *Special topics of elementary mathematics*, Dover, New York, 1958.

D. J. Struik, *A concise history of mathematics*, Dover, New York, 1948.

B. L. van der Waerden, *Science awakening*, P. Noordhoff, Groningen, 1954.

---

* This section is followed in the original Russian text by two sections entitled "The essential nature of mathematics" and "The laws of the development of mathematics." These sections are omitted in the present translation in view of the fact that they discuss in more detail, and in the more general philosophical setting of dialectical materialism, points of view already stated with great clarity in the preceding sections.

CHAPTER II

ANALYSIS

§1. Introduction

The rise at the end of the Middle Ages of new conditions of manufacture in Europe, namely the birth of capitalism, which at this time was replacing the feudal system, was accompanied by important geographical discoveries and explorations. In 1492, relying on the idea that the earth is spherical, Columbus discovered the New World. The discovery by Columbus greatly extended the boundaries of the known world and produced a revolution in the minds of men. The end of the 15th century and the beginning of the 16th saw the creative activity of the great artist-humanists Leonardo da Vinci, Raphael, and Michelangelo, which gave new meaning to art. In 1543 Copernicus published his work "On the revolution of the heavenly bodies," which completely changed the face of astronomy; in 1609 appeared the "New astronomy" of Kepler, containing his first and second laws for the motion of the planets around the sun, and in 1618 his book "Harmony of the world," containing the third law. Galileo, on the basis of his study of the works of Archimedes and his own bold experiments, laid the foundations for the new mechanics, an indispensable science for the newly arising technology. In 1609 Galileo directed his recently constructed telescope, though still small and imperfect, toward the night sky; the first glance in a telescope was enough to destroy the ideal celestial spheres of Aristotle and the dogma of the perfect form of celestial bodies. The surface of the moon was seen to be covered with mountains and pitted with craters. Venus displayed phases like the Moon, Jupiter was surrounded by four satellites and provided a miniature visual model of the solar system. The Milky Way fell apart into separate stars, and for the first time men felt the staggeringly immense distance of the stars. No other scientific discovery has ever made such an impression on the civilized world.*

The further development of navigation, and consequently of astronomy, and also the new development of technology and mechanics necessitated the study of many new mathematical problems. The novelty of these problems consisted chiefly in the fact that they required mathematical study of the laws of motion in a broad sense of the word.

The state of rest and motionlessness is unknown in nature. The whole of nature, from the smallest particles up to the most massive bodies, is in a state of eternal creation and annihilation, in a perpetual flux, in unceasing motion and change. In the final analysis, every natural science studies some aspect of this motion. Mathematical analysis is that branch of mathematics that provides methods for the quantitative investigation of various processes of change, motion, and dependence of one magnitude on another. So it naturally arose in a period when the development of mechanics and astronomy, brought to life by questions of technology and navigation, had already produced a considerable accumulation of observations, measurements, and hypotheses and was leading science straight toward quantitative investigation of the simplest forms of motion.

The name "infinitesimal analysis" says nothing about the subject matter under discussion but emphasizes the method. We are dealing here with the special mathematical method of infinitesimals, or in its modern form, of limits. We now give some typical examples of arguments which make use of the method of limits and in one of the later sections we will define the necessary concepts.

Example 1. As was established experimentally by Galileo, the distance s covered in the time t by a body falling freely in a vacuum is expressed by the formula

$$s = \frac{gt^2}{2} \tag{1}$$

(g is a constant equal to 9.81 m/sec².)† What is the velocity of the falling body at each point in its path?

Let the body be passing through the point A at the time t and consider what happens in the short interval of time of length Δt; that is, in the time from t to $t + \Delta t$. The distance covered will be increased by a certain

---

* This section is based on the beautiful essay of Academician S. I. Vavilov "Galileo" (Great Soviet Encyclopedia, Volume 10, 1952).

† Nowadays formula (1) is deduced from the general laws of mechanics, but historically it was just this formula which, after being established experimentally by Galileo, served as a part of the accumulation of experience that was subsequently generalized by those laws.

increment Δs. The original distance is $s_1 = gt^2/2$; the increased distance is

$$s_2 = \frac{g(t + \Delta t)^2}{2} = \frac{gt^2}{2} + \frac{g}{2}(2t\Delta t + \Delta t^2).$$

From this we find the increment

$$\Delta s = s_2 - s_1 = \frac{g}{2}(2t\Delta t + \Delta t^2).$$

This represents the distance covered in the time from t to $t + \Delta t$. To find the average velocity over the section of the path Δs, we divide Δs by Δt:

$$v_{\mathrm{av}} = \frac{\Delta s}{\Delta t} = gt + \frac{g}{2}\Delta t.$$

Letting Δt approach zero we obtain an average velocity which approaches as close as we like to the true velocity at the point A. On the other hand, we see that the second summand on the right-hand side of the equation becomes vanishingly small with decreasing Δt, so that the average v_{av} approaches the value gt, a fact which it is convenient to write as follows:

$$v = \lim_{\Delta t \to 0} v_{\mathrm{av}} = \lim_{\Delta t \to 0} \frac{\Delta s}{\Delta t}$$

$$= \lim_{\Delta t \to 0}\left(gt + \frac{g}{2}\Delta t\right) = gt.$$

Consequently, gt is the true velocity at the time t.

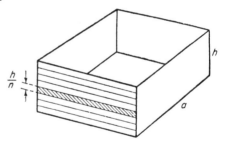

Fig. 1.

Example 2. A reservoir with a square base of side a and vertical walls of height h is full to the top with water (figure 1). With what force is the water acting on one of the walls of the reservoir?

We divide the surface of the wall into n horizontal strips of height h/n. The pressure exerted at each point of the vessel is equal, by a well-known law, to the weight of the column of water lying above it. So at the lower edge of each of the strips the pressure, expressed in suitable units, will be equal respectively to

$$\frac{h}{n}, \frac{2h}{n}, \frac{3h}{n}, \dots, \frac{(n-1)h}{n}, h.$$

We obtain an approximate expression for the desired force P, if we assume that the pressure is constant over each strip. Thus the approximate value of P is equal to

$$P \approx \frac{ah}{n} \cdot \frac{h}{n} + \frac{ah}{n} \cdot \frac{2h}{n} + \cdots + \frac{ah}{n} \frac{(n-1)h}{n} + \frac{ah}{n} h$$

$$= \frac{ah^2}{n^2}(1 + 2 + \cdots + n) = \frac{ah^2}{n^2} \cdot \frac{n(n+1)}{2} = \frac{ah^2}{2}\left(1 + \frac{1}{n}\right).$$

To find the true value of the force, we divide the side into narrower and narrower strips, increasing n without limit. With increasing n the magnitude $1/n$ in the above formula will become smaller and smaller and in the limit we obtain the exact formula

$$P = \frac{ah^2}{2}.$$

The idea of the method of limits is simple and amounts to the following. In order to determine the exact value of a certain magnitude, we first determine not the magnitude itself but some approximation to it. However, we make not one approximation but a whole series of them, each more accurate than the last. Then from examination of this chain of approximations, that is from examination of the process of approximation itself, we uniquely determine the exact value of the magnitude. By this method, which is in essence a profoundly dialectical one, we obtain a fixed constant as the result of a process or motion.

The mathematical method of limits was evolved as the result of the persistent labor of many generations on problems that could not be solved by the simple methods of arithmetic, algebra, and elementary geometry.

What were the problems whose solution led to the fundamental concepts of analysis, and what were the methods of solution that were set up for these problems? Let us examine some of them.

The mathematicians of the 17th century gradually discovered that a large number of problems arising from various kinds of motion with consequent dependence of certain variables on others, and also from geometric problems which had not yielded to former methods, could be reduced to two types. Simple examples of problems of the first type are: find the velocity at any time of a given nonuniform motion (or more generally, find the rate of change of a given magnitude), and draw a tangent to a given curve. These problems (our first example is one of them) led to a branch of analysis that received the name "differential calculus." The simplest examples of the second type of problem are:

find the area of a curvilinear figure (the problem of quadrature), or the distance traversed in a nonuniform motion, or more generally the total effect of the action of a continuously changing magnitude (compare the second of our two examples). This group of problems led to another branch of analysis, the "integral calculus." Thus two fundamental problems were singled out: the problem of tangents and the problem of quadratures.

In this chapter we will describe in detail the underlying ideas of the solution of these two problems. Particularly important here is the theorem of Newton and Leibnitz to the effect that the problem of quadratures is the inverse, in a well-known sense, of the problem of tangents. For solving the problem of tangents, and problems that can be reduced to it, there was worked out a suitable algorithm, a completely general method leading directly to the solution, namely the method of derivatives or of differentiation.

The history of the creation and development of analysis and of the role played in its growth by the analytic geometry of Descartes has already been described in Chapter I. We see that in the second half of the 17th century and the first half of the 18th a complete change took place in the whole of mathematics. To the divisions that already existed, arithmetic, elementary geometry, and the rudiments of algebra and trigonometry, were added such general methods as analytic geometry, differential and integral calculus, and the theory of the simplest differential equations. It was now possible to solve problems whose solutions up to now had been quite inaccessible.

It turned out that if the law for the formation of a given curve is not too complicated, then it is always possible to construct a tangent to it at an arbitrary point; it is only necessary to calculate, with the help of the rules of differential calculus, the so-called derivative, which in most cases requires a very short time. Up till then it had been possible to draw tangents only to the circle and to one or two other curves, and no one had suspected the existence of a general solution of the problem.

If we know the distance traversed by a moving point up to any desired instant of time, then by the same method we can at once find the velocity of the point at a given moment, and also its acceleration. Conversely, from the acceleration it is possible to find the velocity and the distance, by making use of the inverse of differentiation, namely integration. As a result, it was not very difficult, for example, to prove from the Newtonian laws of motion and the law of universal gravitation that the planets must move around the sun in ellipses according to the laws of Kepler.

Of the greatest importance in practical life is the problem of the greatest and least values of a magnitude, the so-called problem of maxima and minima. Let us take an example: From a log of wood with circular cross section of given radius we wish to cut a beam of rectangular cross section such that it will offer the greatest resistance to bending. What should be the ratio of the sides? A short argument on the stiffness of beams of rectangular cross section (applying simple concepts from the integral calculus), followed by the solving of a maximum problem (which involves calculating a derivative) provides the answer that the greatest stiffness is produced for a rectangular cross section whose height is in the ratio to its base of $\sqrt{2}:1$. The problems of maxima and minima are solved as simply as those of drawing tangents.

At various points of a curved line, if it is not a straight line or a circle, the curvature is in general different. How can we calculate the radius of a circle with the same curvature as the given line at the given point, the so-called radius of curvature of the curve at the point? It turns out that this is equally simple; it is only necessary to apply the operation of differentiation twice. The radius of curvature plays a great role in many questions of mechanics.

Before the invention of the new methods of calculation, it had been possible to find the area only of polygons, of the circle, of a sector or a segment of the circle, and of two or three other figures. In addition, Archimedes had already invented a way to calculate the area of a segment of a parabola. The extremely ingenious method which he used in this problem was based on special properties of the parabola and consequently gave rise to the idea that every new problem in the calculation of area would very likely require its own methods of investigation, even more ingenious and difficult than those of Archimedes. So mathematicians were greatly pleased when it turned out that the theorem of Newton and Leibnitz, to the effect that the inversion of the problem of tangents would solve the problem of quadrature, at once provided a method of calculating the areas bounded by curves of widely different kinds. It became clear that a general method exists, which is suitable for an infinite number of the most different figures. The same remark is true for the calculation of volumes, surfaces, the lengths of curves, the mass of inhomogeneous bodies, and so forth.

The new method accomplished even more in mechanics. It seemed that there was no problem in mechanics that the new calculations would not clarify and solve.

Not long before, Pascal had explained the increase in the size of the Torricelli vacuum with increasing altitude as a consequence of the decrease in atmospheric pressure. But exactly what is the law governing this decrease? The question is answered immediately by the investigation of a simple differential equation.

It is well known to sailors that they should take two or three turns of the mooring cable around the capstan if one man is to be able to keep a large vessel at its mooring. Why is this? It turned out that from a mathematical point of view the problem is almost completely identical with the preceding one and can be solved at once.

Thus, after the creation of analysis, there followed a period of tempestuous development of its applications to the most varied branches of technology and natural science. Since it is founded on abstraction from the special features of particular problems, mathematical analysis reflects the actual deep-lying properties of the material world; and this is the reason why it provides the means for investigation of such a wide range of practical questions. The mechanical motion of solid bodies, the motion of liquids and gases of their particular particles, their laws of flow in the mass, the conduction of heat and electricity, the course of chemical reactions, all these phenomena are studied in the corresponding sciences by means of mathematical analysis.

At the same time as its applications were being extended, the subject of analysis itself was being immeasurably enriched by the creation and development of various new branches, such as the theory of series, applications of geometry to analysis, and the theory of differential equations.

Among mathematicians of the 18th century, there was a widespread opinion that any problem of the natural sciences, provided only that one could find a correct mathematical description of it, could be solved by means of analytic geometry and the differential and integral calculus.

Mathematicians proceeded gradually to more complicated problems of natural science and technology, which demanded further development of their methods. For the solution of such problems it became necessary to create further branches of mathematics: the calculus of variations, the theory of functions of a complex variable, field theory, integral equations, and functional analysis. But all these new methods of calculation were essentially immediate extensions and generalizations of the remarkable methods discovered in the 17th century. The greatest mathematicians of the 18th century, David Bernoulli (1700–1782), Leonard Euler (1707–1783) and Lagrange (1736–1813), who blazed new paths in science, constantly took as their starting point the fundamental problems of the exact sciences. This energetic development of analysis was continued into the 19th century by such famous mathematicians as Gauss (1777–1855), Cauchy (1789–1857), M. V. Ostrogradskiĭ (1801–1861), P. L. Čebyšev (1821–1894), Riemann (1826–1866), Abel (1802–1829), Weierstrass (1815–1897), all of whom made truly remarkable contributions to the development of mathematical analysis.

The Russian mathematical genius, N. I. Lobačevskiĭ, had an influence on the development of certain questions of mathematical analysis, and we should also mention the leading mathematicians who were active at the turn of the 20th century: A. A. Markov (1856–1922), A. M. Lyapunov (1857–1918), H. Poincaré (1854–1912), F. Klein (1849–1925), D. Hilbert (1862–1943).

The second half of the 19th century witnessed a profound critical examination and clarification of the foundations of analysis. The various powerful methods that had accumulated were now put on a uniform systematic basis, corresponding to the advanced level of mathematical rigor. All these methods are the means by which, along with arithmetic, algebra, geometry and trigonometry, we give a mathematical interpretation to the world around us, describe the course of actual events, and solve the important practical problems connected with them.

Analytic geometry, differential and integral calculus, and the theory of differential equations are studied at all technical institutes, so that these branches of mathematics are known to millions of citizens; the elements of these sciences are also taught at many technical schools; there is also some question of their being introduced into the secondary schools.

In most recent times the general use of rapid calculating machines has introduced a new era in mathematics. These machines, in conjunction with the branches of mathematics just mentioned, open up strange new possibilities for mankind.

At the present time, analysis and the branches arising from it represent a widely diversified mathematical science, consisting of several broad independent disciplines closely connected with one another; each of these disciplines is being developed and perfected.

More than ever before, a significant role is being played in analysis by the requirements of daily life, by problems connected with the imposing development of technology. Of great importance are the aerodynamical problems of hypersonic velocities, which are being solved with constant success. The most difficult problems of mathematical physics have now reached the stage where they can be solved in practical numerical form. In contemporary physics such theories as quantum mechanics (which studies the problems peculiar to the microcosm of the atom) not only require the most advanced branches of contemporary mathematical analysis for solving their problems but could not even describe their fundamental concepts without the use of analysis.

The purpose of the present chapter is to give a popular presentation, suitable for a reader acquainted only with elementary mathematics, of the growth and the simplest applications of such basic concepts of analysis as function, limit, derivative, and integral. Since the various special branches of analysis will be dealt with in other chapters of the book, the present chapter has a more elementary character and a reader who has already studied a usual first course in analysis may omit it without harm to his understanding of the rest of the book.

§2. Function

The concept of a function. The various objects or phenomena that we observe in nature are organically connected with one another; they are interdependent. The simplest relations of this sort have long been known to mankind and information about them has been accumulated and formulated as physical laws. These laws indicate that the various magnitudes characterizing a given phenomenon are so closely related to one another that some of them are completely determined by the values of others. For example, the length of the sides of a rectangle completely determines its area, the volume of a given amount of gas at a given temperature is determined by the pressure, and the elongation of a given metallic rod is determined by its temperature. It was uniformities of this sort that served as the origin of the concept of *function*.

Consider an algebraic formula which, corresponding to each value of the literal magnitudes occurring in it, allows us to find the value of the magnitude expressed by the formula; the basic idea here is that of a function. Let us consider some examples of functions expressed by such formulas.

1. Let us suppose that at the beginning of a certain period of time a material point was at rest and that subsequently it began to fall as the result of gravity. Then the distance s traced out by the point up to time t is expressed by the formula

$$s = \frac{gt^2}{2}, \tag{1}$$

where g is the acceleration of gravity.

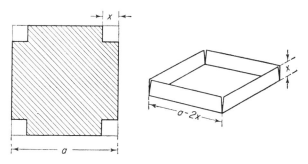

Fig. 2.

2. From a square of side a we construct an open rectangular box of height x (figure 2). The volume V of the box is calculated from the formula

$$V = x(a - 2x)^2. \tag{2}$$

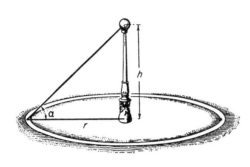

Formula (2) allows us, for every height x under the obvious restriction $0 \leqslant x \leqslant a/2$, to find the volume of the box.

3. Let a pillar (figure 3) be erected at the center of a circular skating rink with a light at height h. The illumination T at the edge of the circle may be expressed by the formula

FIG. 3.

$$T = \frac{A \sin \alpha}{h^2 + r^2}, \tag{3}$$

where r is the radius of the circle, $\tan \alpha = h/r$, and A is a certain magnitude characterizing the power of the light. If we know the height h we can calculate T from formula (3).

4. The roots of the quadratic equation

$$x^2 + px - 1 = 0 \tag{4}$$

are given by the formula

$$x = -\frac{p}{2} \pm \sqrt{1 + \frac{p^2}{4}}. \tag{5}$$

The characteristic feature of a formula in general, and of the examples just given in particular, is that the formula enables us, for any given value of one of the variables (the time t, the height x of the box, the height h of the pillar, the coefficient p of the quadratic equation), which is called the independent variable, to calculate the value of the other variable (the distance s, the volume V, the illumination T, the root x of the equation), which is called a dependent variable or a function of the first variable.

Each of the formulas introduced provides an example of a function: the distance s traced by the point is a function to the time t; the volume

V of the box is a function of height x; the illumination T of the edge of the rink is a function of the height h of the pillar; the two roots of the quadratic equation (4) are functions of the coefficient p.

It should be remarked that in some cases the independent variable may assume any desired numerical value, as in example 4 where the coefficient p of the quadratic equation (4) may be an arbitrary number. In other cases the independent variable may take an arbitrary value from some set (or collection) of numbers determined in advance; as in example 2, where the volume of the box is a function of its height x, which can take any value from the set of numbers x satisfying the inequality $0 < x < a/2$. Similarly, in example 3 the illumination T at the edge of the rink is a function of the height h of the pillar, which theoretically can take any value satisfying the inequality $h > 0$, but in practice h must satisfy the inequalities $0 < h \leqslant H$, where the magnitude H is determined by the technical facilities at the disposal of the administration of the rink.

Let us introduce other examples of this kind. The formula

$$y = \sqrt{1 - x^2}$$

determines a real function (expressing a relationship between the real numbers x and y) only for those values of x which satisfy the inequalities $-1 \leqslant x \leqslant +1$, and the formula $y = \log (1 - x^2)$ only for those x which satisfy the inequalities $-1 < x < 1$.

So it is necessary to take account of the fact that actual functions may not be defined for all numerical values of the independent variable but only for those values which belong to a certain set, which most often fills out an interval on the x-axis, with or without the end points.

We are now in the position to give the definition of a function accepted in present-day mathematics.

The (dependent) magnitude y is a function of the (independent) magnitude x if there exists a rule whereby to each value of x belonging to a certain set of numbers there corresponds a definite value of y.

The set of values x appearing in this definition is called the *domain of the function.*

Every new concept gives rise to a new symbolism. The transition from arithmetic to algebra was made possible by the construction of formulas which were valid for arbitrary numbers, and the search for general solutions gave rise to the literal symbolism of algebra.

The problem of analysis is the study of functions, that is of the dependence of one variable on another. Consequently, just as in algebra a transition took place from concrete numbers to arbitrary numbers, denoted by letters, so in analysis there was the corresponding transition from concrete formulas to arbitrary formulas. The phrase "y is a function of x" is conventionally written as

$$y = f(x).$$

Just as in algebra different letters are used for different numbers, so in analysis different notations are used for different types of dependence, that is for different functions: thus we write $y = F(x)$, $y = \phi(x)$, $\cdots$

Graphs of functions. One of the most fruitful and brilliant ideas of the second half of the 17th century was the idea of the connection between the concept of a function and the geometric representation of a line. This connection can be realized, for example, by means of a rectangular Cartesian system of coordinates, with which the reader is certainly familiar in a general way from his secondary school mathematics.

Let us set up on the plane a rectangular Cartesian system of coordinates. This means that on the plane we choose two mutually perpendicular lines (the axis of abscissas and the axis of ordinates), on each of which we fix a positive direction. Then to each point M of the plane we may assign two numbers (x, y), which are its coordinates, expressing in the given system of measurement the distance, taken with the proper sign,* of the point M from the axis of ordinates and the axis of abscissas respectively.

With such a system of coordinates we may represent functions graphically in the form of certain lines. Suppose we are given a function

$$y = f(x). \tag{6}$$

This means, as we know, that for every value of x belonging to the domain of definition of the given function, it is possible to determine by some means, for example by calculation, a corresponding value y. Let us give to x all possible numerical values, for each x determine y according to our rule (6), and construct on the plane the point with coordinates x and y. In this way, for every point M' on the x-axis (figure 4) there will correspond

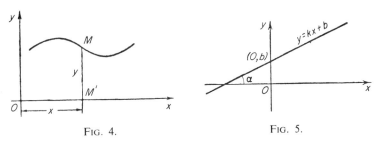

FIG. 4.   FIG. 5.

---

* The number x is the abscissa and y is the ordinate of the point M.

a point M with coordinates x and $y = f(x)$. The set of all points M forms a certain line, which we call the graph of the function $y = f(x)$.

Thus, *the graph of the function $f(x)$* is the geometric locus of the points whose coordinates satisfy equation (6).

In school we became acquainted with the graphs of the simplest functions. Thus the reader probably knows that the function $y = kx + b$, where k and b are constants, is the graph (figure 5) of a straight line forming the angle α with the positive direction of the x-axis, where $\tan \alpha = k$, and intersecting the y-axis at the point $(0, b)$. This function is called a *linear function*.

Linear functions occur very frequently in the applications. Let us recall that many physical laws are represented, with considerable accuracy, by linear functions. For example, the length l of a body may be considered with good approximation as a linear function of its temperature

$$l = l_0 + \alpha l_0 t,$$

where α is the coefficient of linear expansion, and l_0 is the length of the body for $t = 0$. If x is the time and y is the distance covered by a moving point, then the linear function $y = kx + b$ obviously expresses the fact that the point is moving with uniform velocity k; and the number b denotes the distance, at time $x_0 = 0$, of the moving point from the fixed zero-point from which we measure our distances. Linear functions are extremely useful because of their simplicity and because it is possible to consider nonuniform changes as being approximately linear, even if only for small intervals.

But in many cases it is necessary to make use of nonlinear functional dependence. Let us recall for example the law of Boyle-Mariotte

$$v = \frac{c}{p},$$

where the magnitudes p and v are inversely proportional. The graph of such a relation represents a hyperbola (figure 6).

The physical law of Boyle-Mariotte corresponds actually to the case that p and v are positive; it represents a branch of the hyperbola lying in the first quadrant.

The general class of oscillatory processes includes periodic motions, which are usually described by the familiar trigonometric functions. For example, if we extend a hanging spring from its position of equilibrium, then, so long as we stay within the elastic limits of the spring, the point A will perform vertical oscillations which are quite accurately expressed by the law

$$x = a \cos(pt + \alpha),$$

where x is the displacement of the point A from its position of equilibrium, t is the time, and the numbers a, p and α are certain constants determined by the material, the dimensions, and the initial extension of the spring.

It should be kept in mind that a function may be defined in various domains by various formulas, determined by the circumstances of the

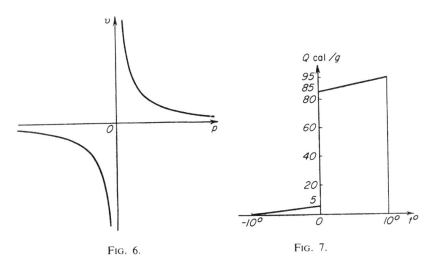

FIG. 6.   FIG. 7.

case. For example, the relation $Q = f(t)$ between the temperature t of a gram of water (or ice) and the quantity of heat Q in it, as t varies between $-10°$ and $+ 10°$, is a completely determined function which it is difficult to express in a single formula,* but it is easy to represent this function by two formulas. Since the specific heat of ice is equal to 0.5 and that of water is equal to 1, this function, if we agree that $Q = 0$ at $-10°$, is represented by the formula

$$Q = 0.5t + 5,$$

as t varies in the interval $-10° \leqslant t < 0°$ and by another formula

$$Q = t + 85,$$

as t varies in the interval $0° < t \leqslant 10°$. For $t = 0$ this function is indefinite or multiple-valued; for convenience, we may agree that at $t = 0$ it takes some well-defined value, for example $f(0) = 45$. The graph of the function $Q = f(t)$ is given in figure 7.

---

* This does not mean that such an expression is impossible. In Chapter XII we will show how to obtain a single formula.

We have introduced many examples of functions given by formulas. The possibility of representing a function by means of formulas is extremely important from the mathematical point of view, since such formulas provide very favorable conditions for investigating the properties of the functions by mathematical methods.

But one must not think that a formula is the only method of defining a function. There are many other methods; for example, the graph of the function, which gives a visual geometric picture of it. The following example gives a good illustration of another method.

To record variation of the temperature of the air during the course of 24 hours, meteorological stations make use of an instrument called the thermograph. A thermograph consists of a drum rotated about its axis by a clockwork mechanism, and of a curved brass framework that is extremely sensitive to changes of temperature. As a result, a pen fastened to the framework by a system of levers rises with rising temperature; and conversely, a fall in the temperature lowers the pen. On the drum is wound a ribbon of graph paper, on which the pen draws a continuous line, forming the graph of the function $T = f(t)$, which expresses the interdependence of the time and the temperature of the air. From this graph we may determine, without calculation, the value of the temperature at any moment of time t.

This example shows that a graph in itself determines a function independently of whether the function is given by a formula or not.

Incidentally, we shall return to this question (see Chapter XII) and shall prove the following important assertion: Every continuous graph can be represented by a formula, or, as it is still customary to say, by an analytic expression. This statement is also true for many discontinuous graphs.*

We remark that the truth of this statement, which is of great theoretical importance, was completely realized in mathematics only in the middle of the past century. Up to that time mathematicians understood by the term "function" only an analytic expression (formula). But they were under the mistaken impression that many discontinuous graphs did not correspond to any analytic expression, since they assumed that if a function was given by a formula, then its graph must possess certain particularly desirable properties in comparison with the other graphs.

But in the 19th century, it was discovered that every continuous graph may be represented by a more or less complicated formula. Thus the exceptional role of the analytic expression as a means of definition of

---

* Of course, the above statement will be completely clear to the reader only after we have given a precise definition of exactly what is meant in mathematics by the term "formula" and "analytic expression."

functions was weakened and there came into existence the new, more flexible definition given above for the concept of a function. By this definition a variable y is called a function of a variable x if there exists a rule whereby to every value of x in the domain of definition of the function there corresponds a completely determined value y, independent of the way in which this rule is given: by a formula, a graph, a table or in any other way.

We may remark here that in the mathematical literature the above definition of a function is often associated with the name of Dirichlet, but it is worth emphasizing that this definition was given simultaneously and independently by N. I. Lobačevskiĭ. Finally we suggest as an exercise that the reader sketch the graphs of the functions x^3, $\sqrt{x}$, $\sin x$, $\sin 2x$, $\sin (x + \pi/4)$, $\ln x$, $\ln(1 + x)$, $|x - 3|$, $(x + |x|)/2$.

We should also note that the graph of a function which for all values of x satisfies the relation

$$f(-x) = f(x)$$

is symmetric with respect to the y-axis and in the case

$$f(-x) = -f(x)$$

the graph is symmetric with respect to the origin of coordinates. Consider also how to obtain the graph of a function $f(a + x)$, when a is a constant, from the graph of $f(x)$. Finally, consider how, using the graphs of the functions $f(x)$ and $\phi(x)$, it is possible to find the values of the composite function $y = f[\phi(x)]$.

§3. Limits

In §1 it was stated that modern mathematical analysis uses a special method, which was worked out in the course of many centuries and serves now as its basic instrument. We are speaking here of the method of infinitesimals, or, as is essentially the same, of limits. We shall try to give some idea of these concepts. For this purpose we consider the following example.

We wish to calculate the area bounded by the parabola with equation $y = x^2$, by the x-axis and by the straight line $x = 1$ (figure 8). Elementary mathematics will not furnish us with a means for solving this problem. But here is how we may proceed.

We divide the interval $[0, 1]$ along the x-axis into n equal parts at the points

$$0, \frac{1}{n}, \frac{2}{n}, \cdots, \frac{n-1}{n}, 1$$

and on each of these parts construct the rectangle whose left side extends up to the parabola. As a result we obtain the system of rectangles shaded in figure 8, the sum S_n of whose areas is given by

$$S_n = 0 \cdot \frac{1}{n} + \left(\frac{1}{n}\right)^2 \frac{1}{n} + \left(\frac{2}{n}\right)^2 \frac{1}{n} + \cdots + \left(\frac{n-1}{n}\right)^2 \frac{1}{n}$$

$$= \frac{1^2 + 2^2 + \cdots + (n-1)^2}{n^3} = \frac{(n-1)\,n(2n-1)}{6n^3}. *$$

Let us express S_n in the following form:

$$S_n = \frac{1}{3} + \left(\frac{1}{6n^2} - \frac{1}{2n}\right) = \frac{1}{3} + \alpha_n. \tag{7}$$

The quantity α_n, which depends on n, is admittedly rather unwieldy in appearance, but it possesses a certain remarkable property: If n is increased beyond all bounds, then α_n approaches 0. This property may also be expressed as follows: If we are given an arbitrary positive number ϵ, then it is possible to choose an integer N sufficiently large that for all n greater than N the number α_n will be less than the given ϵ in absolute value.[†]

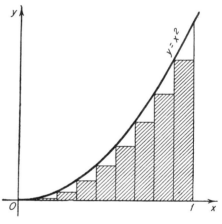

Fig. 8.

---

* If in the obvious equalities $(k+1)^3 - k^3 = 3k^2 + 3k + 1$, for the different values $k = 1, 2, \cdots, n-1$, we add the left and right sides separately, we obtain the equation

$$n^3 - 1 = 3\sigma_n + \frac{3(n-1)n}{2} + n - 1$$

where $\sigma_n = 1^2 + 2^2 + \cdots + (n-1)^2$. Solving this equation for σ_n, we get

$$\sigma_n = \frac{(n-1)n(2n-1)}{6}.$$

[†] For example, if $\epsilon = 0.001$, we may take $N = 500$. In fact, since

$$\frac{1}{6n^2} < \frac{1}{2n}$$

for positive integral n, therefore

$$|\alpha_n| = \left|\frac{1}{6n^2} - \frac{1}{2n}\right| = \frac{1}{2n} - \frac{1}{6n^2} < \frac{1}{2n} < 0.001$$

The magnitude α_n is an example of an infinitesimal in the sense in which that word is used in modern mathematics.

In figure 8 we see that if we increase the number n beyond all bounds, the sum S_n of the areas of the shaded rectangles will approach the desired area of the curvilinear figure. On the other hand, equation (7), in view of the fact that α_n approaches zero as n increases beyond all bounds, shows that the sum S_n at the same time approaches 1/3. From this it follows that the desired area S of the figure is equal to 1/3, and we have solved our problem.

So the method under discussion amounts to this, that in order to find a certain magnitude S we introduce another magnitude S_n, a *variable* magnitude which approaches S through particular values S_1, S_2, S_3, $\cdots$, which depend according to some law on the natural numbers $n = 1, 2, \cdots$. Then, from the fact that the variable S_n may be represented as the sum of a constant $\frac{1}{3}$ and an infinitesimal α_n, we conclude that S_n approaches $\frac{1}{3}$ and so $S = \frac{1}{3}$. In the language of the modern theory of limits we may say that for increasing n the variable magnitude S_n approaches a limit, which is equal to $\frac{1}{3}$.

Now let us give a precise definition of the concepts introduced here.

If a variable magnitude $\alpha_n (n = 1, 2, \cdots)$ has the property that for every arbitrarily small positive number ϵ it is possible to choose an integer N so large that for all $n > N$ we have $|\alpha_n| < \epsilon$, then we say that α_n is an *infinitesimal* and we write

$$\lim_{n\to\infty} \alpha_n = 0 \text{ or } \alpha_n \to 0.$$

On the other hand, if a variable x_n may be represented as a sum

$$x_n = a + \alpha_n,$$

where a is constant and α_n is an infinitesimal, then we say that the variable x_n, for n increasing beyond all bounds, approaches the number a and we write

$$\lim x_n = a \text{ or } x_n \to a.$$

The number a is called the *limit* of x_n. In particular the limit of an infinitesimal is obviously zero.

---

for arbitrary $n > 500$. In the same way it would be possible to assign arbitrarily small values ϵ, for example:

$$\epsilon_1 = 0.0001, \qquad \epsilon_2 = 0.00001, \cdots,$$

and for each of them to choose, as above, appropriate values $N = N_1, N_2, \cdots$.

Let us consider the following examples of variable magnitudes

$$x_n = \frac{1}{n}, y_n = -\frac{1}{n^2}, z_n = \frac{(-1)^n}{n}, u_n = \frac{n-1}{n} = 1 - \frac{1}{n};$$

$$v_n = (-1)^n \ (n = 1, 2, \cdots).$$

It is clear that x_n, y_n, and z_n are infinitesimals, the first of them approaching zero through decreasing values, the second through increasing negative values, while the third takes on values which oscillate around zero. Further, $u_n \to 1$, while v_n does not have a limit at all, since with increasing n it does not approach any constant number but continually oscillates, taking on the values 1 and -1.

Another important concept in analysis is that of an *infinitely large* magnitude, which is defined as a variable x_n $(n = 1, 2, \cdots)$, with the property that after choice of an arbitrarily large positive number M it is possible to find a number N such that for all $n > N$

$$| x_n | > M.$$

The fact that the magnitude x_n is infinitely large is written thus

$$\lim x_n = \infty \text{ or } x_n \to \infty.$$

Such a magnitude x_n is said to approach infinity. If it is positive (negative) from some value on, this fact is expressed thus: $x_n \to + \infty (x_n \to - \infty)$. For example, for $n = 1, 2, \cdots$

$$\lim n^2 = + \infty, \lim (-n^3) = - \infty;$$

$$\lim \log \frac{1}{n} = - \infty, \lim \tan \left(\frac{\pi}{2} + \frac{1}{n}\right) = - \infty.$$

It is easy to see that if a magnitude α_n is infinitely large, then $\beta_n = 1/\alpha_n$ is infinitely small, and conversely.

Two variable magnitudes x_n and y_n may be added, subtracted, multiplied, and divided the one by the other so as to produce new magnitudes that are in general also variable: namely their sum $x_n + y_n$, their difference $x_n - y_n$, their product $x_n y_n$, and their quotient x_n/y_n. Correspondingly their particular values will be

$$x_1 \pm y_1, x_2 \pm y_2, x_3 \pm y_3, \cdots$$

$$x_1 y_1, x_2 y_2, x_3 y_3, \cdots$$

$$\frac{x_1}{y_1}, \frac{x_2}{y_2}, \frac{x_3}{y_3}, \cdots.$$

It is also possible to prove, as is fairly evident, that if the variables x_n and y_n approach finite limits, then their sum, difference, product, and quotient also approach limits which are correspondingly equal to the sum, difference, product, and quotient of these limits. This fact may be expressed thus:

$$\lim (x_n \pm y_n) = \lim x_n \pm \lim y_n; \lim (x_n y_n) = \lim x_n \lim y_n;$$

$$\lim \frac{x_n}{y_n} = \frac{\lim x_n}{\lim y_n}.$$

However, in the case of the quotient it is necessary to assume that the limit of the denominator ($\lim y_n$) is not equal to zero. If $\lim y_n = 0$ and $\lim x_n \neq 0$, then the ratio of x_n to y_n will not have a finite limit but will approach infinity.

Especially interesting, and at the same time important, is the case when the numerator and the denominator simultaneously approach zero. Here it is impossible to state in advance whether the ratio x_n/y_n will approach a limit, and if it does, what that limit will be, since the answer to this question depends entirely on the character of the approach of x_n and y_n to zero. For example, if

$$x_n = \frac{1}{n}, y_n = \frac{1}{n^2}, z_n = \frac{(-1)^n}{n} \ (n = 1, 2, \cdots),$$

then

$$\frac{y_n}{x_n} = \frac{1}{n} \to 0, \frac{x_n}{y_n} = n \to \infty.$$

On the other hand, the magnitude

$$\frac{x_n}{z_n} = (-1)^n$$

evidently does not approach any limit.

Thus the case when the numerator and the denominator of the fraction both approach zero cannot be dealt with in advance by general theorems, and for each particular fraction of this kind it is necessary to make a special investigation.

We shall see later that the fundamental problem of the differential calculus, which may be considered as the problem of determining the velocity of a nonuniform motion at a given moment, reduces to determining the limit of the ratio of two infinitesimal magnitudes, namely the increase of the distance covered and the increase in the time.

So far we have considered variables x_n which take on a sequence of numerical values $x_1, x_2, x_3, \cdots, x_n, \cdots$, while the index n runs through the sequence of natural numbers $n = 1, 2, 3, \cdots$. But it is also possible to consider the case that n varies continuously, like the time for example, and here also to determine the limit of the variable x_n. The properties of such limits are completely analogous to those formulated earlier for discrete (that is, discontinuous) variables. We also note that there is no special significance in the fact that n increases beyond all bounds. It is equally possible to consider the case that, while varying continuously, n approaches a given value n_0.

As an example let us investigate the variation in the magnitude of $(\sin x)/x$ as x approaches zero. Table 1 shows the values of this magnitude for certain values of x:

Table 1

x	$\dfrac{\sin x}{x}$
0.50	0.9589 ...
0.10	0.9983 ...
0.05	0.9996 ...
...	...

(it is assumed that the values of x are given in radian measure).

It is obvious that as x approaches zero the magnitude $(\sin x)/x$ approaches 1, but of course we must still give a rigorous proof of this fact. The proof may be obtained, for example, from the following inequality, which is valid for all nonzero angles in the first quadrant:

$$\sin x < x < \tan x.$$

If we divide both sides of this inequality by $\sin x$, we obtain

$$1 < \frac{x}{\sin x} < \frac{1}{\cos x},$$

from which follows

$$\cos x < \frac{\sin x}{x} < 1.$$

But as x decreases to zero $\cos x$ approaches 1, so that the magnitude $(\sin x)/x$, being contained in the interval between $\cos x$ and 1, also approaches 1, that is

$$\lim_{x \to 0} \frac{\sin x}{x} = 1.$$

We shall have occasion below to make use of this fact.

Our equation has been proved for the case that x approaches zero through positive values. But by changing the proof in an obvious way, it is possible to obtain the same result when x approaches zero through negative values.

Let us now discuss for a moment the following question. A variable magnitude may or may not have a limit and the question arises whether it is possible to give a criterion for determining the existence of a limit for a variable. We will confine ourselves to an important and sufficiently general case, for which such a criterion can be given. Let us suppose that the variable magnitude x_n increases or at least does not decrease; that is, it satisfies the inequalities

$$x_1 \leqslant x_2 \leqslant x_3 \leqslant \cdots,$$

and let us also suppose we have determined that none of its values exceeds a certain fixed number M; that is, $x_n \leqslant M$ ($n = 1, 2, \cdots$). If we mark the values of x_n and the number M on the x-axis, we see that the variable point x_n moves along the axis to the right but constantly remains to the left of the point M. It is rather obvious that the variable point x_n must inevitably approach a certain limit point a, situated to the left of M or at most coinciding with M.

So, in the case under consideration, the limit

$$\lim x_n = a$$

of our variable exists.

The above argument has an intuitive character but we may consider it as a proof. In a course in modern analysis a complete proof of this fact is given on the basis of the theory of real numbers.

As an example let us consider the variable

$$u_n = \left(1 + \frac{1}{n}\right)^n \ (n = 1, 2, 3, \cdots).$$

The first few values are $u_1 = 2$, $u_2 = 2.25$, $u_3 \approx 2.37$, $u_4 \approx 2.44$, $\cdots$, which are seen to increase. From the binomial theorem of Newton it is possible to prove that this increase holds for arbitrary n. Moreover, it is also easy to prove that for all n the inequality $u_n < 3$ is valid. Consequently, our variable must have a limit which is not greater than 3. We shall see that this limit plays a very important role in mathematical physics and in a certain sense is the most natural base for logarithms of numbers.

It is customary to denote this limit by the letter e. It is equal to

$$e = \lim_{n \to \infty} \left(1 + \frac{1}{n}\right)^n = 2.718281828459045 \cdots .$$

A more detailed analysis shows that the number e is not rational.*

It is also possible to show that the limit under consideration exists and is equal to e not only when $n \to +\infty$ but also when $n \to -\infty$. In both cases n may also take on noninteger values.

Let us mention an important application to physics of the concept of a limit. It consists of the remarkable fact that only by using the concept of a limit (passage to the limit) is it possible for us to give a complete definition of many of the concrete magnitudes encountered in physics.

Let us also consider for the moment the following geometric example. In elementary geometry the figures considered first are those bounded by straight line segments. But later there arises the more difficult task of finding the length of the circumference of a circle with given radius.

If we analyze the difficulties connected with the solution of this problem, we find that they reduce to the following.

We must give an answer to the question, what is meant by the length of the circumference; that is, we must give a precise definition of this length. It is essential that the definition should be expressible in terms of the lengths of straight-line segments and also that it should provide us with the possibility of effectively calculating the length of the circumference.

It is understood, of course, that the result of this calculation should be in agreement with practical experience. For example, if we consider a circumference consisting of an actual thread, then, if we cut the thread and stretch it out, we must obtain a segment whose length, within the limits of accuracy of measurement, coincides with our computed length.

As is known from elementary geometry, the solution of this problem reduces to the following definition. The length of a circumference is defined to be the limit approached by the perimeter of a regular† polygon inscribed in it as the number of sides of the polygon increases beyond all bounds. Thus the solution of the problem is based essentially on the concept of a limit.

The length of an arbitrary smooth curve is defined in the same way.

---

* In this connection we should remark that addition, subtraction, multiplication, and division (excluding division by zero) of rational numbers, that is numbers of the form p/q where p and q are integers, leads to rational numbers. But this is not necessarily the case for the operation of taking a limit. The limit of a sequence of rational numbers may be an irrational number.

† It is not important that the polygon should be regular. The only essential feature is that the greatest side of the variable inscribed polygon should approach zero.

In the paragraphs just following, we will meet with a number of examples of geometric and physical magnitudes that can be defined only with the concept of a limit.

The concepts of limit and infinitesimal were given a definitive formulation at the beginning of the last century. The definitions introduced here are connected with the name of Cauchy, before whose time mathematicians operated with concepts that were less clear. The present-day concepts of a limit, of an infinitesimal as a variable magnitude, and of a real number, resulted from the development of mathematical analysis and were at the same time the means of stating and clarifying its many achievements.

§4. Continuous Functions

Continuous functions form the basic class of functions for the operations of mathematical analysis. The general idea of a continuous function may be obtained from the fact that its graph is continuous; that is, its curve may be drawn without lifting the pencil from the paper.

A continuous function gives the mathematical expression of a situation often encountered in practical life, namely that to a small increase in an independent variable there corresponds a small increase in the dependent variable, or function. Excellent examples of a continuous function are given by the various rules governing the motion of bodies $s = f(t)$, expressing the dependence of the distance s on the time t. Since the time and the distance are continuous, a law of motion of the body $s = f(t)$ sets up between them a definite continuous relation, characterized by the fact that to a small increase in the time corresponds a small increase in the distance.

Mankind arrived at the abstraction of continuity by observing the surrounding so-called dense media, namely solids, liquids, and gases; for example, metals, water, and air. In actual fact, as is well known now, every physical medium represents the accumulation of a large number of separate particles in motion. But these particles and the distances between them are so small in comparison with the dimensions of the media in which the phenomena of microscopic physics take place that many of these phenomena may be studied with sufficient accuracy if we consider the medium as being approximately without interstices, that is as continuously distributed over the occupied space. It is on such an assumption that many of the physical sciences are based, for example, hydrodynamics, aerodynamics, and the theory of elasticity. The mathematical concept of continuity naturally plays a large role in these sciences, and in many others as well.

Let us consider an arbitrary function $y = f(x)$ and some specific value of the independent variable x_0. If our function reflects a continuous process, then to values x which differ only slightly from x_0 will correspond values of the function $f(x)$ differing only slightly from the value $f(x_0)$ at the point x_0. Thus if the increment $x - x_0$ of the independent variable is small, then the corresponding increment $f(x) - f(x_0)$ of the function will also be small. In other words if the increment of the independent variable $x - x_0$ approaches zero, then the increment $f(x) - f(x_0)$ of the function must also approach zero, a fact which may be expressed in the following way:

$$\lim_{x - x_0 \to 0} [f(x) - f(x_0)] = 0. \tag{8}$$

This relation constitutes the mathematical definition of continuity of the function at the point x_0; namely, the function $f(x)$ is said to be *continuous at the point x_0*, if equality (8) holds.

Finally, we give the following definition. A function is said to be *continuous in a given interval*, if it is continuous at every point x_0 of this interval; that is, if at every such point equality (8) is fulfilled.

Thus, in order to introduce a mathematical definition of the property of a function reflected in the fact that its graph is continuous (in the everyday sense of this word), it was necessary first to define local continuity (continuity at the point x_0) and then on this basis to define continuity of the function in the whole interval.

This definition, first introduced at the beginning of the last century by Cauchy, is now generally adopted in contemporary mathematical analysis. The test of many concrete examples has shown that it corresponds very well to the practical notion we have of a continuous function, for instance, as represented by its continuous graph.

As examples of continuous functions, the reader may consider the elementary functions well known to him from school mathematics x^n, $\sin x$, $\cos x$, a^x, $\log x$, arc $\sin x$, arc $\cos x$. All these functions are continuous in the intervals for which they are defined.

If continuous functions are added, subtracted, multiplied, or divided (except for division by zero), the result is also a continuous function. But in the case of division the continuity is usually destroyed for those values x_0 for which the function in the denominator vanishes. The result of the division in that case is a function which is *discontinuous* at the point x_0.

The function $y = 1/x$ may serve as an example of a function which is discontinuous at the point $x = 0$. Other discontinuous functions are represented by the graphs in figure 9.

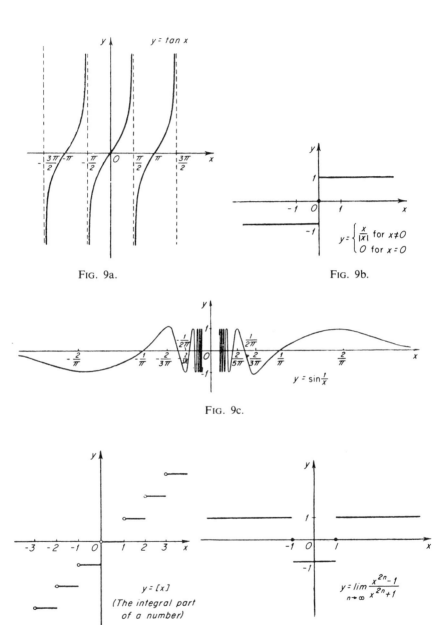

Fig. 9a.

Fig. 9b.

Fig. 9c.

Fig. 9d.

Fig. 9e.

We recommend that the reader examine these graphs carefully. He will notice that the breaks in the functions are of different kinds: In some cases a limit $f(x)$ exists as x approaches the point x_0 where the function suffers a discontinuity, but this limit is different from $f(x_0)$. In other cases, as in figure 9c, the limit simply does not exist. It may also happen that as x approaches x_0 from one side $f(x) - f(x_0) \to 0$, but as $x \to x_0$ from the other side, $f(x) - f(x_0)$ does not approach zero. In this case, of course, the function has a discontinuity, but we may say that at such a point it is "continuous from one side." All these cases are represented in the graphs of figure 9.

As an exercise we recommend to the reader to consider the question, what value must be given to the functions

$$\frac{\sin x}{x}, \frac{1 - \cos x}{x^2}, \frac{x^3 - 1}{x - 1}, \frac{\tan x}{x}$$

at those points where they are not defined (that is, at the points where the denominator is equal to zero), in order that they may be continuous at these points. Also, is it possible to find such numbers for the functions

$$\tan x, \frac{1}{x - 1}, \frac{x - 2}{(x^2 - 4)} \ ?$$

These discontinuous functions in mathematics represent the numerous jumplike processes to be met with in nature. In the case of a sudden blow, for example, the value of the velocity of a body changes in such a jumplike fashion. Many qualitative transitions take place with such jumps. In §2 we introduced the function $Q = f(t)$, expressing the way in which the quantity of heat in a given quantity of water (or ice) depends on the temperature. In the neighborhood of the melting point of ice the quantity of heat $Q = f(t)$ changes in a jumplike fashion with changing t.

Functions with isolated discontinuities are encountered quite often in analysis, along with the continuous functions. But as an example of a more complicated function, where the number of discontinuities is infinite, let us consider the so-called Riemann function, which is equal to zero at all irrational points and equal to $1/q$ at rational points of the form $x = p/q$ (where p/q is a fraction in its lowest terms). This function is discontinuous at all rational points and continuous at irrational points. By altering it slightly we may easily obtain an example of a function which is discontinuous at all points.* Let us remark by the way that even for such complicated functions modern analysis has discovered many in-

---

* It is sufficient to set the function equal to unity at the irrational points.

teresting laws, which are investigated in one of the independent branches of analysis, the theory of functions of a real variable. This theory has developed with extraordinary rapidity during the past 50 years.

§5. Derivative

The next fundamental concept of analysis is the concept of *derivative*. Let us consider two problems from which it arose historically.

Velocity.   At the beginning of the present chapter we defined the velocity of a freely falling body. To do so we made use of a passage to the limit from the average velocity over short distances to the velocity at the given point and the given time. The same procedure may be used to define the instantaneous velocity for an arbitrary nonuniform motion. In fact, let the function

$$s = f(t) \tag{9}$$

express the dependence of the distance s covered by the material point in the time t. To find the velocity at the moment $t = t_0$, let us consider the interval of time from t_0 to $t_0 + h$ ($h \neq 0$). During this time the point will cover the distance

$$\Delta s = f(t_0 + h) - f(t_0) .$$

The average velocity v_{av} over this part of the path will depend on h

$$v_{av} = \frac{\Delta s}{h} = \frac{1}{h} [f(t_0 + h) - f(t_0)],$$

and will represent the actual velocity at the point t_0 with greater and greater accuracy as h becomes smaller. It follows that the true velocity at the time t_0 is equal to the limit

$$v = \lim_{h \to 0} \frac{f(t_0 + h) - f(t_0)}{h}$$

of the ratio of the increase in the distance to the increase in the time, as the latter approaches zero without ever being actually equal to zero. In order to calculate the velocity for different forms of motion, we must discover how to find this limit for various functions $f(t)$.

Tangent.   We are led to investigate a precisely analogous limit by another problem, this time a geometric one, namely the problem of drawing a tangent to an arbitrary plane curve.

Let the curve C be the graph of a function $y = f(x)$, and let A be the point on the curve C with abscissa x_0 (figure 10). Which straight line shall we call the tangent to C at the point A? In elementary geometry this question does not arise. The only curve studied there, namely the circumference of a circle, allows us to define the tangent as a straight line which has only one point in common with the curve. But for other curves such a definition will clearly not correspond to our intuitive picture of "tangency." Thus, of the two straight lines L and M in figure 11, the first is obviously not tangent to the curve drawn there (a sinusoidal curve), although it has only one point in common with it; while the second straight line has many points in common with the curve, and yet it is tangent to the curve at each of these points.

To define the tangent, let us consider on the curve C (figure 10) another point A', distinct from A, with abscissa $x_0 + h$. Let us draw the secant AA' and denote the angle which it forms with the x-axis by β. We now allow the point A' to approach A along the curve C. If the secant AA' correspondingly approaches a limiting position, then the straight line T which has this limiting position is called the *tangent* at the point A. Evidently the angle α formed by the straight line T with the x-axis, must be equal to the limiting value of the variable angle β.

The value of $\tan \beta$ is easily determined from the triangle ABA' (figure 10):

$$\tan \beta = \frac{BA'}{AB} = \frac{f(x_0 + h) - f(x_0)}{h}.$$

For the limiting position we must have

$$\tan \alpha = \lim_{A' \to A} \tan \beta = \lim_{h \to 0} \frac{f(x_0 + h) - f(x_0)}{h},$$

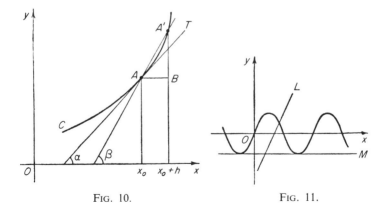

FIG. 10.   FIG. 11.

that is, the trigonometric tangent of the angle of inclination of the tangent line is equal to the limit of the ratio of the increase in the function $f(x)$ at the point x_0 to the corresponding increase in the independent variable, as the latter approaches zero without ever being actually equal to zero.

Let us give still another example leading to the calculation of an analogous limit. Let us suppose that a variable electric current is flowing through a conductor. Let us assume that we know the function $Q = f(t)$ expressing the quantity of electricity that has passed through a fixed cross section of the conductor up to time t. In the period from t_0 to $t_0 + h$, there will flow through this cross section a quantity of electricity ΔQ equal to $f(t_0 + h) - f(t_0)$. The average value of the current will therefore be equal to

$$I_{\text{av}} = \frac{\Delta Q}{h} = \frac{f(t_0 + h) - f(t_0)}{h}.$$

The limit of this ratio as $h \to 0$ will give us the value of the current at the time t_0

$$I = \lim_{h \to 0} \frac{f(t_0 + h) - f(t_0)}{h}.$$

All the three problems discussed, in spite of the fact that they refer to different branches of science, namely mechanics, geometry, and the theory of electricity, have led to one and the same mathematical operation to be performed on a given function, namely to find the limit of the ratio of the increase of the function to the corresponding increase h of the independent variable as $h \to 0$. The number of such widely different problems could be increased at will, and their solution would lead to the same operation. To it we are led, for example, by the question of the rate of a chemical reaction, or of the density of a nonhomogeneous mass and so forth. In view of the exceptional role played by this operation on functions, it has received a special name, differentiation, and the result of the operation is called the *derivative* of the function.

Thus, the *derivative of the function* $y = f(x)$, or more precisely, the *value of the derivative at the given point* x is the limit* approached by the ratio of the increase $f(x + h) - f(x)$ of the function to the increase h of the independent variable, as the latter approaches zero. We often write $h = \Delta x$, and $f(x + \Delta x) - f(x) = \Delta y$, in which case the definition of the derivative is written in the concise form:

$$\lim_{\Delta x \to 0} \frac{\Delta y}{\Delta x}.$$

---

* It is understood that we are speaking here of the case where the limit in question actually exists. If this limit does not exist, then we say that at the point x the function does not have a derivative.

The value of the derivative obviously depends on the point x at which it is found. Thus the derivative of a function $y = f(x)$ is itself a function of x. It is customary to denote the derivative thus

$$f'(x) = \lim_{h \to 0} \frac{f(x + h) - f(x)}{h} = \lim_{\Delta x \to 0} \frac{\Delta y}{\Delta x}.$$

Certain other notations are also customary for the derivative:

$$\frac{df(x)}{dx}, \text{ or } \frac{dy}{dx}, \text{ or } y', \text{ or } y'_x.$$

We should also remark that the notation $\frac{dy}{dx}$ looks like a fraction, although it is read as a single symbol for the derivative. In the following sections the numerator and the denominator of this "fraction" will take on independent meaning, in such a way that their ratio will coincide with the derivative so that this manner of writing is completely justified.

The results of these examples may now be formulated as follows.

The velocity of a point for which the distance s is a given function of the time $s = f(t)$ is equal to the derivative of this function

$$v = s' = f'(t).$$

More concisely, the velocity is the derivative of the distance with respect to time.

The trigonometric tangent of the angle of inclination of the tangent line to the curve $y = f(x)$ at the point with abscissa x is equal to the derivative of the function $f(x)$ at this point:

$$\tan \alpha = y' = f'(x).$$

The strength of the current I at the time t, if $Q = f(t)$ is the quantity of electricity which up to time t has passed through a cross section of the conductor, is equal to the derivative

$$I = Q' = f'(t).$$

Let us make the following remark. The velocity of a nonuniform motion at a given time is a purely physical concept, arising from practical experience. Mankind arrived at it as the result of numerous observations on different concrete motions. The study of nonuniform motion of a body on different parts of its path, the comparison of different motions of this sort taking place simultaneously, and in particular the study of the phenomena of collisions of bodies, all represented an accumulation of practical experience that led to the setting up of the physical concept of the velocity of a nonuniform motion at a given time. But the exact definition of velocity necessarily depended upon the method of defining its numerical value, and to define this value was possible only with the concept of the derivative.

In mechanics the velocity of a body moving according to the rule $s = f(t)$ at the time t is defined as the derivative of the function $f(t)$ for this value of t.

The discussion at the beginning of the present section has shown, on the one hand, the advantages of introducing the operation of finding the derivative, and on the other has given a reasonable justification for the above formulated definition of the velocity at any given moment.

Thus, when we raised the question of finding the velocity of a point in nonuniform motion we had, properly speaking, only an empirical notion of its value but no exact definition. But now, as a result of our analysis, we have reached an exact definition of the value of the velocity at a given moment, namely the derivative of the distance with respect to the time. This result is extremely important from a practical point of view, since our empirical knowledge of the velocity has been greatly enriched by the fact that we can now make an exact numerical calculation.

What has just been said refers equally well, of course, to the strength of a current and to many other concepts expressing the rate of some process, physical, chemical, and so forth.

This situation may serve as an example for numerous others of a similar nature, where practical experience has led to the formation of a concept relating to the external world (velocity, work, density, area, and so forth) and then mathematics has enabled us to define this concept precisely, whereupon we can make use of the concept in practical calculations.

We have already noted at the beginning of the chapter that the concept of a derivative arose chiefly as the result of many centuries of effort directed towards the solving of two problems: drawing a tangent to a curve and finding the velocity of a nonuniform motion. These problems, and also the calculation of areas discussed later, interested mathematicians in ancient times. But until the 16th century the statement and the method of solution for each problem of this sort bore an extremely special character. The accumulation of all this extensive material was reduced to a theoretically complete system in the 17th century in the work of Newton and Leibnitz. An important contribution to the foundations of present-day analysis was also made by Euler.

But it must be said that Newton and Leibnitz and their contemporaries provided very little logical basis for their great mathematical discoveries;

in their methods of reasoning and in the concepts with which they operated there was much that is unclear from our point of view. Even at that time the mathematicians themselves were quite conscious of this, as is shown by the embittered discussions to be found in their correspondence with one another. However, these mathematicians of the 17th and 18th centuries carried on their purely mathematical activities in very close association with the research of other investigators, in the various branches of natural science (physics, mechanics, chemistry, technology). The statement of a mathematical problem usually arose from practical needs or from a wish to understand some phenomenon of nature, and as soon as the problem was solved, the solution was submitted in one way or another to a practical test. Consequently, in spite of a certain lack of logical basis, mathematics was able to advance in extremely useful directions.

Examples for the calculation of derivatives. The definition of the derivative as the limit

$$f'(x) = \lim_{h \to 0} \frac{f(x + h) - f(x)}{h}$$

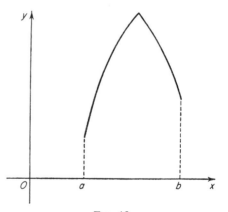

allows us to calculate the derivative of any given concrete function.

Of course, it must be admitted that cases are possible where the function at one point or another or even at many points simply does not have a derivative; in other words, the ratio

$$\frac{f(x + h) - f(x)}{h}$$

FIG. 12.

as $h \to 0$ does not approach a finite limit. This case obviously occurs at every point of discontinuity of the function $f(x)$, since here the ratio

$$\frac{f(x + h) - f(x)}{h} \tag{10}$$

has a numerator which does not approach zero while the denominator decreases without bound. The derivative may also fail to exist at a point where the function is continuous. A simple example is given by any point where the graph of the function forms an angle (figure 12). At such a point the curve of the graph has no definite tangent, and consequently the function has no derivative. Often at such points the expression (10) approaches different values, depending on whether h approaches zero from the right or from the left, so that if h approaches zero in an arbitrary manner, the ratio (10) simply has no limit. An example of a more complicated function without a derivative is given by

$$y = \begin{cases} x \sin \dfrac{1}{x} & \text{for } x \neq 0, \\ 0 & \text{for } x = 0. \end{cases}$$

The graph of this function is drawn in figure 13. At the point $x = 0$ it has no derivative because, as is evident from the graph, the secant OA does not approach any definite position even when $A \to 0$ from one side. In fact, the secant OA oscillates endlessly back and forth between the straight line OM and the straight line OL. The corresponding ratio (10) in this case has no limit, even if h preserves the same sign as it approaches zero.

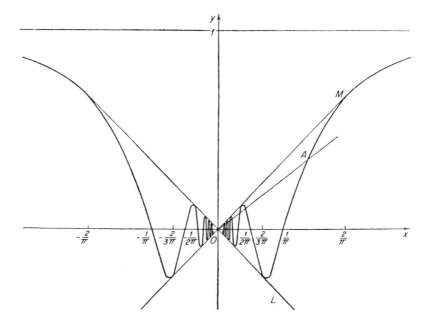

FIG. 13.

Let us remark finally that it is possible to define, in a purely analytic way by means of a formula, a continuous function which does not have a derivative at any point. An example of such a function was first given by the outstanding German mathematician of the last century, Weierstrass.

Consequently the class of differentiable functions is considerably narrower than that of continuous functions.

Let us pass now to the actual calculation of the derivatives of the simplest functions.

1. $y = c$, where c is a constant. A constant may be considered as a special case of a function that remains equal to the same number for arbitrary x. Its graph is a straight line parallel to the x-axis at a distance equal to c. This straight line forms with the x-axis an angle $\alpha = 0$, and obviously the derivative of a constant is identically equal to zero: $y' = (c)' = 0$. From the point of view of mechanics, this equation means that the velocity of a fixed point is equal to zero.

2. $y = x^2$

$$\frac{f(x + h) - f(x)}{h} = \frac{(x + h)^2 - x^2}{h} = 2x + h.$$

As $h \to 0$ we obtain* in the limit $2x$; consequently

$$y' = (x^2)' = 2x.$$

3. $y = x^n$ (n a positive integer).

$$\frac{f(x + h) - f(x)}{h} = \frac{(x + h)^n - x^n}{h}$$

$$= \frac{x^n + nx^{n-1}h + \frac{n(n - 1)}{2!} x^{n-2}h^2 + \cdots + h^n - x^n}{h}$$

$$= nx^{n-1} + \frac{n(n - 1)}{2!} x^{n-2}h + \cdots + h^{n-1}.$$

Every addend on the right side, beginning with the second, approaches zero as $h \to 0$; consequently

$$y' = (x^n)' = nx^{n-1}.$$

This formula remains true for arbitrary n positive or negative, fractional

---

* We always assume here that $h \neq 0$.

or even irrational, although the proof must then be different. We will make use of this fact without proving it. Thus for example

$$(\sqrt{x})' = (x^{\frac{1}{2}})' = \frac{1}{2}x^{-\frac{1}{2}} = \frac{1}{2\sqrt{x}}, (x > 0);$$

$$(\sqrt[3]{x})' = (x^{\frac{1}{3}})' = \frac{1}{3}x^{-\frac{2}{3}} = \frac{1}{3\sqrt[3]{x^2}}, (x \neq 0);$$

$$\left(\frac{1}{x}\right)' = (x^{-1})' = -1 \cdot x^{-2} = -\frac{1}{x^2}, (x \neq 0);$$

$$(x^{\pi})' = \pi x^{\pi-1}, \qquad\qquad (x > 0).$$

4. $y = \sin x$.

$$\frac{\sin(x+h) - \sin x}{h} = \frac{2\sin h/2 \cos(x+h/2)}{h} = \frac{\sin h/2}{h/2} \cdot \cos\left(x + \frac{h}{2}\right).$$

As explained earlier, the first fraction approaches unity as $h \to 0$, and $\cos(x + h/2)$ obviously approaches $\cos x$. Thus the derivative of the sine is equal to the cosine

$$y' = (\sin x)' = \cos x.$$

We suggest to the reader that by the same sort of argument he prove that

$$(\cos x)' = -\sin x.$$

5. Earlier (Chapter II, §3) we have already noted the existence of the limit

$$\lim_{n\to\infty}\left(1 + \frac{1}{n}\right)^n = e = 2.71828\cdots.$$

We also remarked that for the calculation of this limit no essential role is played by the fact that n took on only positive integral values. It is important only that the infinitesimal $1/n$, which is being added to unity, and the exponent n, which is increasing beyond all bounds, should be reciprocal to each other.

Making use of this assertion, we can easily find the derivative of the logarithm $y = \log_a x$

$$\frac{\log_a(x+h) - \log_a x}{h} = \frac{1}{h}\log_a\frac{x+h}{x} = \frac{1}{x}\log_a\left(1 + \frac{h}{x}\right)^{x/h}.$$

The continuity of the logarithm allows us to replace the quantity under the log sign by its limit, which is equal to e; thus

$$\lim_{h \to 0} \left(1 + \frac{h}{x}\right)^{x/h} = e$$

(in this case the role of $n \to \infty$ is played by the increasing quantity x/h). As a result, we obtain the rule for differentiating a logarithm

$$(\log_a x)' = \frac{1}{x} \log_a e.$$

This rule becomes particularly simple if as the base of our logarithms we choose the number e. Logarithms taken to this base are called *natural logarithms* and are denoted by ln x. We may write

$$(\log_e x)' = \frac{1}{x}$$

or again

$$(\ln x)' = \frac{1}{x}.$$

§6. Rules for Differentiation

From the examples given earlier it may appear that the calculation of the derivative of every new function demands the invention of new methods. This is not the case. The development of analysis was made possible to no small extent by the discovery of a simple unified method for finding the derivative of an arbitrary "elementary" function (that is, a function which may be expressed by a formula consisting of a finite combination of the fundamental algebraic operations, the trigonometric functions, the operation of raising to a power, and the taking of logarithms). At the basis of this method are the so-called *rules of differentiation*. They consist of a number of theorems that allow us to reduce more complicated problems to simpler ones.

We will explain here the rules of differentiation and will try to be very brief in deducing them. If the reader wishes to form merely a general idea of analysis, he may omit the present section, remembering only that there exists a means of actually finding the derivative of any elementary function. In this case it will be necessary, of course, for him to take on faith some of the calculations in our later examples.

Derivative of a sum. Assume that y is given as a function of x by the expression

$$y = \phi(x) + \psi(x),$$

where $u = \phi(x)$ and $v = \psi(x)$ are known functions of x. We assume moreover that we can find the derivatives of the functions u and v. How then are we to find the derivative of the function y? The answer is simple

$$y' = (u + v)' = u' + v'. \tag{11}$$

In fact, let us give x an increment Δx; then u, v, and y will each receive an increment Δu, Δv, and Δy, connected by the equation

$$\Delta y = \Delta u + \Delta v.$$

Thus*

$$\frac{\Delta y}{\Delta x} = \frac{\Delta u}{\Delta x} + \frac{\Delta v}{\Delta x},$$

and after the passage to the limit for $\Delta x \to 0$ we at once get formula (11), if, of course, the functions u and v have derivatives.

Analogously we may derive the formula for differentiating the difference of two functions

$$(u - v)' = u' - v'. \tag{12}$$

Derivative of a product. The rule for the differentiation of a product is somewhat more complicated. The derivative of the product of two functions, each of which has a derivative, exists, and is equal to the sum of the product of the first function by the derivative of the second and the product of the second by the derivative of the first; that is

$$(uv)' = uv' + vu'. \tag{13}$$

In fact, let us give x an increment Δx. Then the functions u, v and $y = uv$ will receive the increments Δu, Δv, Δy, satisfying the relation

$$\Delta y = (u + \Delta u)(v + \Delta v) - uv = u\,\Delta v + v\,\Delta u + \Delta u\,\Delta v,$$

from which

$$\frac{\Delta y}{\Delta x} = u\frac{\Delta v}{\Delta x} + v\frac{\Delta u}{\Delta x} + \Delta u\frac{\Delta v}{\Delta x}.$$

After passage to the limit for $\Delta x \to 0$ the first two summands on the right side produce the right side of formula (13) while the third summand vanishes.† Consequently, in the limit we obtain the rule (13).

---

* Here Δx is never equal to zero.

† The final summand here approaches zero for $\Delta x \to 0$, since $\Delta v/\Delta x$ approaches a finite number, equal to the derivative v', which was assumed from the beginning to exist, and $\Delta u \to 0$, since the function u, assumed to have a derivative, is continuous.

In the particular case $v - c$ = constant, we have

$$(cu)' = cu' + uc' = cu', \tag{14}$$

since the derivative of a constant is equal to zero.

Derivative of a quotient. Let $y = u/v$, where u and v have a derivative for a given x, with $v \neq 0$ for that value of x. Obviously

$$\Delta y = \frac{u + \Delta u}{v + \Delta v} - \frac{u}{v} = \frac{v \Delta u - u \Delta v}{(v + \Delta v) v},$$

from which

$$\frac{\Delta y}{\Delta x} = \frac{v \dfrac{\Delta y}{\Delta x} - u \dfrac{\Delta v}{\Delta x}}{(v + \Delta v) v} \rightarrow \frac{vu' - uv'}{v^2} \ (\Delta x \rightarrow 0).$$

Here we have again made use of the fact that for a function v which has a derivative we necessarily have $\Delta v \rightarrow 0$, when $\Delta x \rightarrow 0$. Thus

$$\left(\frac{u}{v}\right)' = \frac{vu' - uv'}{v^2}. \tag{15}$$

Let us give some examples of the application of these rules

$$(2x^3 - 5)' = 2(x^3)' - (5)' = 2 \cdot 3x^2 - 0 = 6x^2;$$
$$(x^2 \sin x)' = x^2(\sin x)' + (x^2)' \sin x = x^2 \cos x + 2x \sin x;$$
$$(\tan x)' = \left(\frac{\sin x}{\cos x}\right)' = \frac{\cos x(\sin x)' - \sin x(\cos x)'}{\cos^2 x}$$
$$= \frac{\cos x \cdot \cos x - \sin x(-\sin x)}{\cos^2 x} = \frac{1}{\cos^2 x} = \sec^2 x.$$

We recommend to the reader to prove for himself the formula

$$(\cot x)' = -\csc^2 x.$$

Derivative of the inverse function. Let us consider a function $y = f(x)$, which is continuous and increasing (decreasing) on the interval $[a, b]$. By increasing (decreasing) we mean that to a greater value of x in the interval $[a, b]$ corresponds a greater (smaller) value of y (figure 14).

Let $c = f(a)$ and $d = f(b)$. In figure 14 it is evident that for each value of y from the interval $[c, d]$ (or $[d, c]$, respectively) there corresponds exactly one value of x from the interval $[a, b]$ such that $y = f(x)$. Thus on the interval $[c, d]$ (or $[d, c]$) we have a completely determined function $x = \phi(y)$, which is called the *inverse function* of $y = f(x)$. In figure 14 it is clear that the function $\phi(y)$ is continuous, a fact which is proved in modern analysis by strictly analytical methods. Now let Δx and Δy

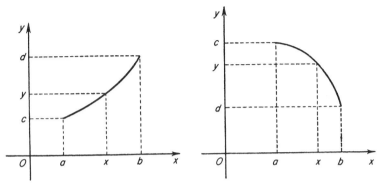

FIG. 14.

correspond respectively to the increments in x and y. It is evident that

$$\frac{\Delta y}{\Delta x} = \frac{1}{\Delta x/\Delta y}, \text{ if } \Delta y \neq 0.$$

In the limit this gives us a simple relation between derivatives of the direct and inverse functions

$$y'_x = \frac{1}{x'_y}. \tag{16}$$

Let us make use of this relation to find the derivative of the function $y = a^x$. The inverse function is $x = \log_a y$, which we are already able to differentiate, and so we may write

$$(a^x)'_x = \frac{1}{(\log_a y)'_y} = \frac{1}{1/y \,(\log_a e)} = y \log_e a = a^x \ln a. \tag{17}$$

In particular $(e^x)' = e^x$.

As another example let us take $y = \arcsin x$. The inverse function is $x = \sin y$. Thus

$$(\arcsin x)'_x = \frac{1}{(\sin y)'_y} = \frac{1}{\cos y} = \frac{1}{\sqrt{1 - (\sin y)^2}} = \frac{1}{\sqrt{1 - x^2}}.$$

Table of derivatives. Let us tabulate the derivatives of the simplest elementary functions (Table 2).

Table 2

y	y'	y	y'	y	y'
c	0	$\ln x$	$\dfrac{1}{x}$	$\tan x$	$\sec^2 x$
x^a	ax^{a-1}	$\log_a x$	$\dfrac{1}{x}\log_a e$	arc sin x	$\dfrac{1}{\sqrt{1-x^2}}$
e^z	e^z	$\sin x$	$\cos x$	arc cos x	$-\dfrac{1}{\sqrt{1-x^2}}$
a^z	$a^z \ln a$	$\cos x$	$-\sin x$	arc tan x	$\dfrac{1}{1+x^2}$

These formulas have been calculated and explained earlier, with the exception of the last two which the reader may, if he wishes, easily derive for himself by using the rule for differentiation of an inverse function.

Calculation of the derivative of a function of a function. It remains to consider the last and most difficult rule for differentiation. The reader in possession of this rule and of a set of tables may with perfect right consider that he is able to differentiate any elementary function.

In order to apply the rule we are about to give, it is necessary to be completely clear about how the function we wish to differentiate is constructed; that is, which operations must we perform on the independent variable x, and in which order, to produce the value of the dependent variable y.

For example, to calculate the function

$$y = \sin x^2,$$

it is necessary first of all to raise x to the second power and then to take the sine of the magnitude so obtained, a procedure which may be described in the following way: $y = \sin u$, where $u = x^2$.

On the other hand, in order to calculate the function

$$y = \sin^2 x,$$

it is necessary first of all to find the sine of x, and then to raise the value so found to the second power, a procedure which may be written thus: $y = u^2$, where $u = \sin x$.

Here are some examples:

1. $y = (3x + 4)^3$, $y = u^3$, $u = 3x + 4$.
2. $y = \sqrt{1 - x^2}$, $y = u^{\frac{1}{2}}$, $u = 1 - x^2$.
3. $y = e^{kx}$; $y = e^u$, $u = kx$.

In more complicated cases we have a chain of simple relations, which may have several links. For example,

4. $y = \cos^3 x^2$; $y = u^3$; $u = \cos v$; $v = x^2$.

If y is a function of the variable u

$$y = f(u), \tag{18}$$

and u in its turn is a function of the variable x

$$u = \phi(x), \tag{19}$$

then y, being a function of u, is also a certain function of x, which may be denoted as follows

$$y = F(x) = f[\phi(x)]. \tag{20}$$

By considering more complicated cases we may form, for example, the function

$$y = \Phi(x) = f\{\phi[\psi(x)]\},$$

which is equivalent to the equations

$$y = f(u), \quad u = \phi(v), \quad v = \psi(x),$$

and we could form still longer chains.

We now show how to calculate the derivative of the function $F(x)$ defined by equation (20) if we know the derivative of $f(u)$ with respect to u and the derivative of $\phi(x)$ with respect to x.

Let us give to x the increment Δx; then by (19) u will receive a certain increment Δu and by (18) y will receive an increment Δy. Thus we may write

$$\frac{\Delta y}{\Delta x} = \frac{\Delta y}{\Delta u} \cdot \frac{\Delta u}{\Delta x}.$$

Now let Δx approach zero. Then $\Delta u/\Delta x \rightarrow u'_x$. Furthermore, from the continuity of u, the increase $\Delta u \rightarrow 0$, and therefore $\Delta y/\Delta u \rightarrow y'_u$ (the existence of the derivatives y'_u and u'_x was assumed).

Thus we have proved the important formula for the derivative of a function of a function*

$$y'_x = y'_u u'_x. \tag{21}$$

Let us calculate, from formula (21) and the fundamental table of derivatives given, the derivatives of the functions we have been considering:

1. $y = (3x + 4)^3 = u^3, y'_x = (u^3)'_u (3x + 4)'_x = 3u^2 \cdot 3 = 9(3x + 4)^2.$

2. $y = \sqrt{1 - x^2} = u^{\frac{1}{2}}, y'_x = (u^{\frac{1}{2}})'_x (1 - x^2)'_x = \frac{1}{2} u^{-\frac{1}{2}} (-2x)$

$$= -\frac{x}{\sqrt{1 - x^2}}.$$

3. $y = e^{kx} = e^u, y'_x = (e^u)'_u \cdot u'_x = e^u \cdot k = ke^{kx}.$

If $y = f(u)$, $u = \phi(v)$, $v = \psi(x)$, then

$$y'_x = y'_u \cdot u_x = y'_u(u'_v \cdot v'_x) = y'_u \cdot u'_v \cdot v'_x.$$

It is clear how to generalize this formula for the case of an arbitrary (finite) number of functions in the chain. For example,

4. $y = \cos^3 x^2; y'_x = (u^3)'_u (\cos v)'_v \cdot (x^2)'_x = 3u^2(-\sin v) \cdot 2x$

$$= -6x \cos^2 x^2 \sin x^2.$$

In our explanation of how to calculate the derivative of a function of a function, we have introduced intermediate variables $u, v, \cdots$. But in fact, after a little practice one may dispense with them, simply keeping in mind the functions they denote.

The elementary functions. To close the present section let us remark that the functions whose derivatives were listed in tabular form (Table 2) may be used to define the so-called elementary functions. These *elementary functions* are defined as those functions that may be obtained from the preceding simple functions by the four arithmetical operations and the operation of taking a function of a function, each of these operations being performed a finite number of times.

For example, the polynomial $x^2 - 2x^2 + 3x - 5$ is an elementary function since it is obtained by arithmetic operations from a number of functions of the form x^k. The function $\ln \sqrt{1 - x^2}$ is also elementary,

---

* In deducing this formula we have tacitly assumed that, as Δx approaches zero, Δu is never equal to zero. But the formula remains true even when this assumption does not hold.

since it is obtained from the polynomial $u = 1 - x^2$ by the operation $v = u^{1/2}$, and subsequently the operation $\ln v$.

The rules for differentiation discussed earlier are sufficient to obtain the derivative of any elementary function, as soon as we know the derivatives of the simplest elementary functions.

§7. Maximum and Minimum; Investigation of the Graphs of Functions

One of the simplest and most important applications of the derivative is in the theory of maxima and minima. Let us suppose that on a certain interval $a \leqslant x \leqslant b$ we are given a function $y = f(x)$ which is not only continuous but also has a derivative at every point. Our ability to calculate the derivative enables us to form a clear picture of the graph of the function. On an interval on which the derivative is always positive the tangent to the graph will be directed upward. On such an interval the

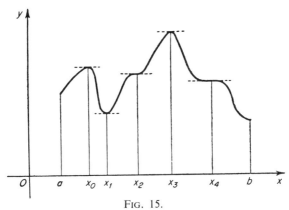

Fig. 15.

function will increase; that is, to a greater value of x will correspond a greater value of $f(x)$. On the other hand, on an interval where the derivative is always negative, the function will decrease; the graph will run downward.

Maximum and minimum. In figure 15 we have drawn the graph of a function $y = f(x)$ defined on the interval (a, b). Of a special interest are the points of this graph whose abcissas are x_0, x_1, x_3.

At the point x_0 the function $f(x)$ is said to have a local maximum; by this we mean that at this point $f(x)$ is greater than at neighboring points; more precisely $f(x_0) \geqslant f(x)$ for every x in a certain interval around the point x_0.

A local minimum is defined analogously.

For our function a local maximum occurs at the points x_0 and x_3, and a local minimum at the point x_1.

At every maximum or minimum point, if it is inside the interval $[a, b]$, i.e., if it does not coincide with one of the end points a or b, the derivative must be equal to zero.

This last statement, a very important one, follows immediately from the definition of the derivative as the limit of the ratio $\Delta y/\Delta x$. In fact, if we move a short distance from the maximum point, then $\Delta y \leqslant 0$. Thus for positive Δx the ratio $\Delta y/\Delta x$ is nonpositive, and for negative Δx the ratio $\Delta y/\Delta x$ is nonnegative. The limit of this ratio, which exists by hypothesis, can therefore be neither positive nor negative and there remains only the possibility that it is zero. By inspection of the diagram it is seen that this means that at maximum or minimum points (it is customary to leave out the word "local," although it is understood) the tangent to the graph is horizontal. In figure 15 we should remark that at the points x_2 and x_4 also the tangent is horizontal, just as it is at the points x_0, x_1, x_3, although at these points the function has neither maximum nor minimum. In general, there may be more points at which the derivative of the function is equal to zero (stationary points) than there are maximum or minimum points.

Determination of the greatest and least values of a function. In numerous technical questions it is necessary to find the point x at which a given function $f(x)$ attains its greatest or its least value on a given interval.

In case we are interested in the greatest value, we must find x_0 on the interval $[a, b]$ for which among all x on $[a, b]$ the inequality $f(x_0) \geqslant f(x)$ is fulfilled.

But now the fundamental question arises, whether in general there exists such a point. By the methods of modern analysis it is possible to prove the following existence theorem: If the function $f(x)$ is continuous on a finite interval, then there exists at least one point on the interval for which the function attains its maximum (minimum) value on the interval $[a, b]$.

From what has been said already, it follows that these maximum or minimum points must be sought among the "stationary" points. This fact is the basis for the following well-known method for finding maxima and minima.

First we find the derivative of $f(x)$ and then solve the equation obtained by setting it equal to zero

$$f'(x) = 0.$$

If x_1, x_2, $\cdots$, x_n are the roots of this equation, we then compare the numbers $f(x_1)$, $f(x_2)$, $\cdots$, $f(x_n)$ with one another. Of course, it is necessary to take into account that the maximum or minimum of the function may be found not within the interval but at the end (as is the case with the minimum in figure 15) or at a point where the function has no derivative (as in figure 12). Thus to the points x_1, x_2, $\cdots$, x_n we must add the ends a and b of the interval and also those points, if they exist, at which there is no derivative. It only remains to compare the values of the function at all these points and to choose among them the greatest or the least.

With respect to the stated existence theorem, it is important to add that this theorem ceases, in general, to hold in the case that the function $f(x)$ is continuous only on the interval (a, b); that is, on the set of points x satisfying the inequalities $a < x < b$. We leave it to the reader to consider the fact that the function $1/x$ has neither a maximum nor a minimum on the interval $(0, 1)$.

Let us look at some examples.

From a square piece of tin of side a it is required to make a rectangular open box of maximum volume. If from the corners of the original square we take away squares of side x (see §2, example 2) we get a box with the volume

$$V = x(a - 2x)^2.$$

Our problem then becomes to find the value of x for which the function $V(x)$ attains its greatest value on the interval $0 \leqslant x \leqslant a/2$. In accordance with the rule, we find the derivative and set it equal to zero

$$V'(x) = (a - 2x)^2 - 4x(a - 2x) = 0.$$

Solving this equation, we find the two roots

$$x_1 = \frac{a}{2}, x_2 = \frac{a}{6}.$$

To these we adjoin the left end of the interval (the right end is identical with x_1) and compare the values of the function at these points

$$V(0) = 0; V\left(\frac{a}{6}\right) = \frac{2}{27} a^3, V\left(\frac{a}{2}\right) = 0.$$

Thus the box will have the greatest volume, equal to $2/27 \, a^3$, for the height $x = a/6$.

As a second example, let us examine the problem of the lamp at the skating rink (see §2, example 3). At what height h should we place the lamp in order that the edge of the rink may receive the greatest illumination?

For formula (3) §2, our problem reduces to determining the value of h for $T = A \sin \alpha / h^2 + r^2$ takes on its greatest value. Instead of h it is more convenient here to find the angle α (figure 3, Chapter I). We have

$$h = r \tan \alpha,$$

so that

$$T = \frac{A}{r^2} \frac{\sin \alpha}{1 + \tan^2 \alpha} = \frac{A}{r^2} \sin \alpha \cos^2 \alpha.$$

Then it is required to find the maximum of the function $T(\alpha)$ among those values of α which satisfy the inequality $0 < \alpha < \pi/2$. To do this, we find the derivative and set it equal to zero

$$T'(\alpha) = \frac{A}{r^2} (\cos^3 \alpha - 2 \sin^2 \alpha \cos \alpha) = 0.$$

This equation splits into two

$$\cos \alpha = 0, \cos^2 \alpha - 2 \sin^2 \alpha = 0.$$

The first equation has the root $\alpha = \pi/2$, which coincides with the end of the interval $(0, \pi/2)$. The second equation may be put in the form

$$\tan^2 \alpha = \tfrac{1}{2}.$$

But since $0 < \alpha < \pi/2$, we have the result $\alpha \approx 35°15'$. So this is the value for which the function $T(\alpha)$ attains its maximum (at the ends of the interval, $T = 0$). The desired height h is thus equal to

$$h = r \tan \alpha = \frac{r}{\sqrt{2}} \approx 0.7r.$$

For best illumination of the edge of the rink the lamp should be placed at a height equal to about 0.7 times the radius.

But now let us suppose that the facilities at our disposal do not allow us to raise the lamp to a height greater than a certain H. Then the angle α may vary not from 0 to $\pi/2$ but only within the narrower limits $0 < \alpha \leqslant \arctan (H/r)$. For example, let $r = 12$ meters and $H = 9$ meters. In this case, it is in fact possible to raise the lamp to the height $h = r/\sqrt{2}$, which amounts to somewhat more than 8 meters, so that this is what we ought to do. But if H is less than 8 meters (for example, if we have at our disposal only a pole of length 6 meters), then it turns out that the derivative of the function $T(\alpha)$ in the interval $[0, \arctan (H/r)]$ is nowhere equal to zero. In this case the maximum is attained at the end of the interval, and the lamp should be raised to the greatest possible height $H = 6$ meters.

Up to now we have considered a function on a finite interval. If the interval is infinite in length, then even a continuous function may fail to attain its greatest or least value but may, for example, continue to grow or to decrease as x approaches infinity.

Thus the functions $y = kx + b$ (see figure 5, Chapter I), $y = $ arc tan x (figure 16a), $y = \ln x$ (figure 16b) nowhere attain either a

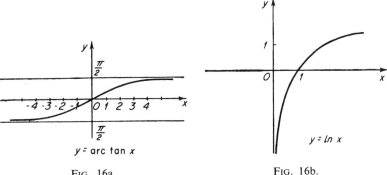

$y = $ arc tan x

FIG. 16a.

$y = \ln x$

FIG. 16b.

maximum or a minimum. The function $y = e^{-x^2}$ (figure 16c) attains its maximum at the point $x = 1$, but nowhere attains a minimum. As for the function $y = x/(1 + x^2)$ (figure 16d), it reaches its minimum at the point $x = -1$ and its maximum at the point $x = 1$.

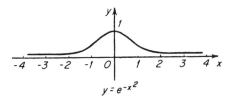

$y = e^{-x^2}$

FIG. 16c.

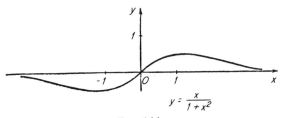

$y = \dfrac{x}{1 + x^2}$

FIG. 16d.

In the case of an interval of infinite length the investigation may be reduced to the ordinary rules. It is only necessary to consider in place of $f(a)$ and $f(b)$ the limits

$$A = \lim_{x \to -\infty} f(x), \quad B = \lim_{x \to +\infty} f(x).$$

Derivatives of higher orders. We have just seen how, for closer study of the graph of a function, we must examine the changes in its derivative $f'(x)$. This derivative is a function of x, so that we may in turn find its derivative.

The derivative of the derivative is called the second derivative and is denoted by

$$[y']' = y'' \quad \text{or} \quad [f'(x)]' = f''(x).$$

Analogously, we may calculate the third derivative

$$[y'']' = y''' \quad \text{or} \quad [f''(x)]' = f'''(x)$$

and more generally the nth derivative or, as it is also called, the derivative of nth order

$$y^{(n)} = f^{(n)}(x).$$

Of course, it must be kept in mind that, for a certain value of x (or even for all values of x) this sequence may break off at the derivative of some order, say the kth; it may happen that $f^{(k)}(x)$ exists but not $f^{(k+1)}(x)$. Derivatives of arbitrary order will appear later in §9 in connection with the Taylor formula. For the moment we confine ourselves to the second derivative.

Significance of the second derivative; convexity and concavity. The second derivative has a simple significance in mechanics. Let $s = f(t)$ be a law of motion along a straight line; then s' is the velocity and s'' is the "velocity of the change in the velocity" or more simply the "acceleration" of the point at time t. For example, for a falling body under the force of gravity

$$s = \frac{gt^2}{2} + v_0 t + s_0,$$

$$s' = gt + v_0,$$

$$s'' = g,$$

that is, the acceleration of falling bodies is constant.

The second derivative also has a simple geometric meaning. Just as the sign of the first derivative determines whether the function is increasing or decreasing, so the sign of the second derivative determines the side toward which the graph of the function will be curved.

Suppose, for example, that on a given interval the second derivative is everywhere positive. Then the first derivative increases and therefore

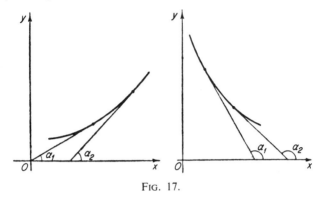

FIG. 17.

$f'(x) = \tan \alpha$ increases and the angle α of inclination of the tangent line itself increases (figure 17). Thus as we move along the curve it keeps turning constantly to the same side, namely upward, and is thus, as they say, "convex downward."

On the other hand, in a part of a curve where the second derivative is constantly negative (figure 18) the graph of the function is "convex upward."*

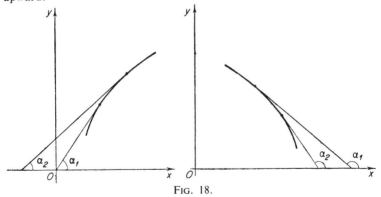

FIG. 18.

---

* Strictly defined, the "convexity upward" is that property of the curve that consists of its lying above (more precisely "not below") the chord joining any two of its points; analogously, for "convexity downward" (which is also simply called "concavity"), the curve does not lie above its chords.

Criteria for maxima and minima; study of the graphs of curves. If throughout the whole interval over which x varies the curve is convex upward and if at a certain point x_0 of this interval the derivative is equal to zero, then at this point the function necessarily attains its maximum; and its minimum in the case of convexity downward. This simple consideration often allows us, after finding a point at which the derivative is equal to zero, to decide thereupon whether at this point the function has a local maximum or minimum.*

Example 1. Let us study the appearance of the graph of the function

$$f(x) = \frac{x^3}{3} - \frac{5x^2}{2} + 6x - 2.$$

We take its first derivative and set it equal to zero,

$$f'(x) = x^2 - 5x + 6 = 0.$$

The roots of the equation obtained in this way are $x_1 = 2$, $x_2 = 3$. The corresponding values of the function are

$$f(2) = 2\tfrac{2}{3}, \; f(3) = 2\tfrac{1}{2}.$$

We then mark these two points on the diagram. Along with these we may also mark the point with coordinates $x = 0$ and $y = f(0) = -2$ where the graph intersects the y-axis. The second derivative is $f''(x) = 2x - 5$. This reduces to zero for $x = \tfrac{5}{2}$, so that

$$f''(x) > 0 \text{ for } x > \tfrac{5}{2},$$

$$f''(x) < 0 \text{ for } x < \tfrac{5}{2}.$$

The point

$$x = \frac{5}{2}, y = f\left(\frac{5}{2}\right) = 2\tfrac{7}{12}$$

is a *point of inflection* of the graph. To the left of this point the curve is convex upward, and to the right it is convex downward.

It is now evident that the point $x = 2$ is a maximum point and the point $x = 3$ is a minimum point for the function.

---

* In more complicated cases, where the second derivative itself changes sign, the problem of determining the character of the stationary point is solved by means of the Taylor formula (§9).

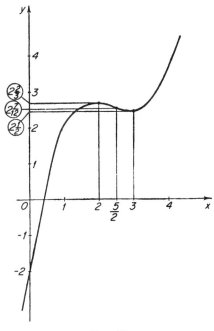

FIG. 19.

On the basis of these results we conclude that the graph of the function $y = f(x)$ has the appearance sketched in figure 19. To the right of the point $(0, -2)$ the curve rises with increasing x, is convex upwards, and attains its maximum at the point $(2, 2\frac{2}{3})$, after which it begins to fall. At the point $(2\frac{1}{2}, 2\frac{7}{12})$, where $f''(x) = 0$, the convexity changes to concavity. Then at the point $(3, 2\frac{1}{2})$ the function attains its minimum and from there on rises to infinity. The final statement comes from the fact that the first term of the function, the one containing the highest (third) power of x, approaches infinity faster than the second and third terms. For the same reason the graph of the function approaches $-\infty$ as x assumes numerically larger negative values.

Example 2. We shall prove the inequality $e^x \geqslant 1 + x$ for arbitrary x. For this purpose we consider the function $f(x) = e^x - x - 1$. Its first derivative is $f'(x) = e^x - 1$, which reduces to zero only for $x = 0$. The second derivative $f''(x) = e^x > 0$ for all x. Consequently the graph of the function $f(x)$ is convex downward. The number $f(0) = 0$ is a minimum for the function and $e^x - x - 1 \geqslant 0$ for all x.

The study of graphs has many different purposes. They often show very clearly, for example, the number of real roots of a given equation. Thus, in order to demonstrate that the equation

$$xe^x = 2$$

has a single real root, we may study the graphs of the functions $y = e^x$ and $y = 2/x$ (as sketched in figure 20). It is easy to see that these graphs intersect at only one point, so that the equation $e^x = 2/x$ has exactly one root.

The methods of analysis are extensively applied to questions of approximate calculation of the roots of an equation. On this subject, see Chapter IV, §5.

§8. Increment and Differential of a Function

The differential of a function.   Let us consider a function $y = f(x)$ that has a derivative. The increment of this function

$$\varDelta y = f(x + \varDelta x) - f(x),$$

corresponding to the increment $\varDelta x$, has the property that the ratio $\varDelta y / \varDelta x$, as $\varDelta x \to 0$, approaches a finite limit, equal to the derivative

$$\frac{\varDelta y}{\varDelta x} \to f'(x).$$

This fact may be written as an equality

$$\frac{\varDelta y}{\varDelta x} = f'(x) + \alpha,$$

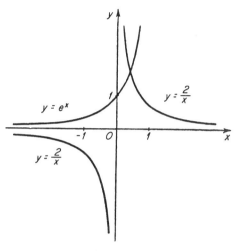

FIG. 20.

where the value of α depends on $\varDelta x$ in such a way that as $\varDelta x \to 0$, α also approaches zero. Thus the increment of a function may be represented in the form

$$\varDelta y = f'(x)\, \varDelta x + \alpha \varDelta x,$$

where $\alpha \to 0$, if $\varDelta x \to 0$.

The first summand on the right side of this equality depends on $\varDelta x$ in a very simple way, namely it is proportional to $\varDelta x$. It is called the *differential* of the function, at the point x, corresponding to the given increment $\varDelta x$, and is denoted by

$$dy = f'(x)\, \varDelta x.$$

The second summand has the characteristic property that, as $\varDelta x \to 0$, it approaches zero more rapidly than $\varDelta x$, as a result of the presence of the factor α. It is therefore said to be an infinitesimal of higher order than Δx and, in case $f'(x) \neq 0$, it is also of higher order than the first summand.

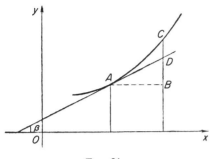

FIG. 21.

By this we mean that for sufficiently small Δx the second summand is small in itself and its ratio to Δx is also arbitrarily small.

The decomposition of Δy into two summands, of which the first (the principal part) depends linearly on Δx and the second is negligible for small Δx, may be illustrated by figure 21. The segment $BC = \Delta y$, where $BC = BD + DC$, $BD = \tan \beta \cdot \Delta x = f'(x)\,\Delta x = dy$, and DC is an infinitesimal of higher order than Δx.

In practical problems the differential is often used as an approximate value for the increment in the function. For example, suppose we have the problem of determining the volume of the walls of a closed cubical box whose interior dimensions are $10 \times 10 \times 10$ cm and the thickness of whose walls is 0.05 cm. If great accuracy is not required, we may argue as follows. The volume of all the walls of the box represents the increment Δy of the function $y = x^3$ for $x = 10$ and $\Delta x = 0.1$. So we find approximately

$$\Delta y \approx dy = (x^3)'\,\Delta x = 3x^2 \Delta x = 3 \cdot 10^2 \cdot 0.1 = 30 \text{ cm}^3.$$

For symmetry in the notation it is customary to denote the increment of the independent variable by dx and to call it also a differential. With this notation the differential of the function may be written thus:

$$dy = f'(x)\,dx.$$

Then the derivative is the ratio $f'(x) = dy/dx$ of the differential of the function to the differential of the independent variable.

The differential of a function originated historically in the concept of an "indivisible." This concept, which from a modern point of view was never very clearly defined, was in its time, in the 18th century, a fundamental one in mathematical analysis. The ideas concerning it have undergone essential changes in the course of several centuries. The indivisible, and later the differential of a function, were represented as actual infinitesimals, as something in the nature of an extremely small constant magnitude, which however was not zero. The definition given in this section is the one accepted in present-day analysis. According to this definition the differential is a finite magnitude for each increment Δx and is at the same time proportional to Δx. The other fundamental property of the differential, the character of its difference from Δy, may be recognized only in motion, so to speak: if we consider an increment Δx which is approaching zero (which is infinitesimal), then the difference between dy and Δy will be arbitrarily small even in comparison with Δx.

This substitution of the differential in place of small increments of the function forms the basis of most of the applications of infinitesimal analysis to the study of nature. The reader will see this in a particularly clear way in the case of differential equations, dealt with in this book in Chapters V and VI.

Thus, in order to determine the function that represents a given physical process, we try first of all to set up an equation that connects this function in some definite way with its derivatives of various orders. The method of obtaining such an equation, which is called a differential equation, often amounts to replacing increments of the desired functions by their corresponding differentials.

As an example let us solve the following problem. In a rectangular system of coordinates $Oxyz$, we consider the surface obtained by rotation of the parabola whose equation (in the Oyz plane) is $z = y^2$. This surface is called a paraboloid of revolution (figure 22). Let v denote the volume of the body bounded by the paraboloid and the plane parallel to the Oxy plane at a distance z from it. It is evident that v is a function of z ($z > 0$).

To determine the function v, we attempt to find its differential dv.

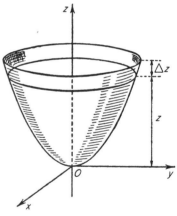

Fig. 22.

The increment Δv of the function v at the point z is equal to the volume bounded by the paraboloid and by two planes parallel to the Oxy plane at distances z and $z + \Delta z$ from it.

It is easy to see that the magnitude of Δv is greater than the volume of the circular cylinder of radius $\sqrt{z}$ and height Δz but less than that of the circular cylinder with radius $\sqrt{z + \Delta z}$ and height Δz.

Thus
$$\pi z \, \varDelta z < \varDelta v < \pi(z + \varDelta z) \, \varDelta z$$
and so
$$\varDelta v = \pi(z + \theta \, \varDelta z) \, \varDelta z = \pi z \, \varDelta z + \pi \theta \, \varDelta z^2 \,,$$

where θ is some number depending on $\varDelta z$ and satisfying the inequality $0 < \theta < 1$.

So we have succeeded in representing the increment $\varDelta v$ in the form of a sum, the first summand of which is proportional to $\varDelta z$, while the second is an infinitesimal of higher order than $\varDelta z$ (as $\varDelta z \to 0$). It follows that the first summand is the differential of the function v

$$dv = \pi z \, \varDelta z,$$
or
$$dv = \pi z \, dz,$$

since $\varDelta z = dz$ for the independent variable z.

The equation so obtained relates the differentials dv and dz (of the variables v and z) to each other and thus is called a differential equation.

If we take into account that

$$\frac{dv}{dz} = v',$$

where v' is the derivative of v with respect to the variable z, our differential equation may also be written in the form

$$v' = \pi z.$$

To solve this very simple differential equation we must find a function of z whose derivative is equal to πz. Problems of this sort are treated in a general way in §§10 and 11, but for the moment we urge the reader to verify that a solution of our equation is given by $v = \pi z^2/2 + C$, where for C we may choose an arbitrary number.* In our case the volume of the body is obviously zero for $z = 0$ (see figure 22), so that $C = 0$. Thus our function is given by $v = \pi z^2/2$.

The mean value theorem and examples of its application. The differential expresses the approximate value of the increment of the function in terms of the increment of the independent variable and of the derivative at the initial point. So for the increment from $x = a$ to $x = b$, we have

$$f(b) - f(a) \approx f'(a)(b - a).$$

---

* This formula gives all the solutions.

It is possible to obtain an exact equation of this sort if we replace the derivative $f'(a)$ at the initial point by the derivative at some intermediate point, suitably chosen in the interval (a, b). More precisely: *If $y = f(x)$ is a function which is differentiable on the interval $a \leqslant x \leqslant b$, then there exists a point ξ, strictly within this interval, such that the following exact equality holds*

$$f(b) - f(a) = f'(\xi)(b - a). \tag{22}$$

The geometric interpretation of this "mean-value theorem" (also called Lagrange's formula or the finite-difference formula) is extraordinarily simple. Let A, B be the points on the graph of the function $f(x)$ which correspond to $x = a$ and $x = b$, and let us join A and B by the chord AB (figure 23). Now let us move the straight line AB, keeping it constantly

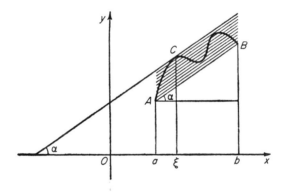

Fɪɢ. 23.

parallel to itself, up or down. At the moment when this straight line cuts the graph for the last time, it will be tangent to the graph at a certain point C. At this point (let the corresponding abscissa be $x = \xi$), the tangent line will form the same angle of inclination α as the chord AB. But for the chord we have

$$\tan \alpha = \frac{f(b) - f(a)}{b - a}.$$

On the other hand at the point C

$$\tan \alpha = f'(\xi).$$

This equation

$$\frac{f(b) - f(a)}{b - a} = f'(\xi)$$

is exactly the mean-value theorem.*

Formula (22) has the peculiar feature that the point ξ appearing in it is unknown to us; we know only that it lies "somewhere in the interval (a, b)." But in spite of this indeterminacy, the formula has great theoretical significance and is part of the proof of many theorems in analysis. The immediate practical importance of this formula is also very great, since it enables us to estimate the increase in a function when we know the limits between which its derivative can vary. For example,

$$| \sin b - \sin a | = | \cos \xi | (b - a) \leqslant b - a.$$

Here a, b and ξ are angles, expressed in radian measure; ξ is some value between a and b; ξ itself is unknown, but we know that $| \cos \xi | \leqslant 1$.

From formula (22) it is clear that a function whose derivative is everywhere equal to zero must be a constant; at no part of the interval can it receive an increment different from zero. Analogously, the reader will easily prove that a function whose derivative is everywhere positive must everywhere increase, and if its derivative is negative, the function must decrease. We give here without proof one of the many generalizations of the mean-value theorem.

For arbitrary functions $\phi(x)$ and $\psi(x)$ differentiable in the interval $[a, b]$, provided only that $\psi'(x) \neq 0$ in (a, b), the following equation[†] holds

$$\frac{\phi(b) - \phi(a)}{\psi(b) - \psi(a)} = \frac{\phi'(\xi)}{\psi'(\xi)} , \qquad (23)$$

where ξ is some point in the interval (a, b).[‡]

From this theorem we can derive a general method for calculating the limits of an expression like

$$\lim_{x \to 0} \frac{\phi(x)}{\psi(x)} , \qquad (24)$$

---

* Of course these arguments only give a geometric interpretation of the theorem and by no means form a rigorous proof.

† Formula (23) can be derived by a simple application of the mean-value theorem to the function

$$f(x) = \phi(x) - \frac{\phi(b) - \phi(a)}{\psi(b) - \psi(a)} \psi(x).$$

‡ By the symbols $[a, b]$ and (a, b) we denote the sets of values of x satisfying the inequalities $a \leqslant x \leqslant b$ and $a < x < b$ respectively.

if $\phi(0) = \psi(0) = 0$. From formula (23) we have

$$\frac{\phi(x)}{\psi(x)} = \frac{\phi(x) - \phi(0)}{\psi(x) - \psi(0)} = \frac{\phi'(\xi)}{\psi'(\xi)},$$

where ξ is between 0 and x, and therefore $\xi \to 0$ together with x. This allows us to calculate the limit

$$\lim_{x \to 0} \frac{\phi'(x)}{\psi'(x)},$$

instead of the limit (24), which is in many cases very much easier.*

Example. Let us find the $\lim\limits_{x \to 0} \dfrac{x - \sin x}{x^3}$. By making use of the rule three times, we have successively

$$\lim_{x \to 0} \frac{x - \sin x}{x^3} = \lim_{x \to 0} \frac{1 - \cos x}{3x^2} = \lim_{x \to 0} \frac{\sin x}{6x} = \lim_{x \to 0} \frac{\cos x}{6} = \frac{1}{6}.$$

§9. Taylor's Formula

The function

$$p(x) = a_0 + a_1 x + a_2 x^2 + \cdots + a_n x^n,$$

where the coefficients a_k are constants, is called a polynomial of degree n. In particular, $y = ax + b$ is a polynomial of the first degree and $y = ax^2 + bx + c$ is a polynomial of the second degree. Polynomials may be considered as the simplest of all functions. In order to calculate their value for a given x, we require only the operations of addition, subtraction, and multiplication; not even division is needed. Polynomials are continuous for all x and have derivatives of arbitrary order. Also, the derivative of a polynomial is again a polynomial, of degree lower by one, and the derivatives of order $n + 1$ and higher of a polynomial of degree n are equal to zero.

If to the polynomials we adjoin functions of the form

$$y = \frac{a_0 + a_1 x + \cdots + a_n x^n}{b_0 + b_1 x + \cdots + b_m x^m},$$

---

* The same rule is valid for finding the limit of a fractional expression in which the numerator and the denominator both approach infinity. This method, which is very convenient for finding such limits (or, as we say, for the removal of indeterminacies), will be used, for example, in §3 of Chapter XII.

for the calculation of which we also need division, and also the functions $\sqrt{x}$ and $\sqrt[3]{x}$ and, finally, arithmetical combinations of these functions, we obtain essentially all the functions whose values can be calculated by methods learned in the secondary school.

While we were still in school, we formed some notion of a number of other functions, like

$$\sqrt[5]{x},\ \log x,\ \sin x,\ \arctan x,\ \cdots.$$

But though we became acquainted with the most important properties of these functions, we found no answer in elementary mathematics to the question: How can we calculate them? What sort of operations, for example, is it necessary to perform on x in order to obtain $\log x$ or $\sin x$? The answer to this question is given by methods that have been worked out in analysis. Let us examine one of these methods.

Taylor's formula. On an interval containing the point a, let there be given a function $f(x)$ with derivatives of every order. The polynomial of first degree

$$p_1(x) = f(a) + f'(a)\,(x - a)$$

has the same value as $f(x)$ at the point $x = a$ and also, as is easily verified, has the same derivative as $f(x)$ as this point. Its graph is a straight line, which is tangent to the graph of $f(x)$ to the point a. It is possible to choose a polynomial of the second degree, namely

$$p_2(x) = f(a) + f'(a)(x - a) + \frac{f''(a)}{2}(x - a)^2,$$

which at the point of $x = a$ has with $f(x)$ a common value and a common first and second derivative. Its graph at the point a will follow that of $f(x)$ even more closely. It is natural to expect that if we construct a polynomial which at $x = a$ has the same first n derivatives as $f(x)$ at the same point, then this polynomial will be a still better approximation to $f(x)$ at points x near a. Thus we obtain the following approximate equality, which is Taylor's formula

$$f(x) \approx f(a) + f'(a)(x - a) + \frac{f''(a)}{2!}(x - a)^2 + \cdots + \frac{f^{(n)}(a)}{n!}(x - a)^n. \quad (25)$$

The right side of this formula is a polynomial of degree n in $(x - a)$. For each x the value of this polynomial can be calculated if we know the values of $f(a),\ f'(a),\ \cdots,\ f^{(n)}(a)$.

For functions which have an $(n + 1)$th derivative, the right side of this formula, as is easy to show, differs from the left side by a small quantity which approaches zero more rapidly than $(x - a)^n$. Moreover, it is the only possible polynomial of degree n that differs from $f(x)$, for x close to a, by a quantity that approaches zero, as $x \to a$, more rapidly than $(x - a)^n$. If $f(x)$ itself is an algebraic polynomial of degree n, then the approximate equality (25) becomes an exact one.

Finally, and this is particularly important, we can give a simple expression for the difference between the right side of formula (25) and the actual value of $f(x)$. To make the approximate equality (25) exact, we must add to the right side a further term, called the "remainder term"

$$f(x) = f(a) + f'(a)(x - a) + \cdots + \frac{f^{(n)}(a)}{n!}(x - a)^n + \frac{f^{(n+1)}(\xi)}{(n + 1)!}(x - a)^{n+1} \tag{26}$$

This final supplementary term*

$$R_{n+1}(x) = \frac{f^{(n+1)}(\xi)}{(n + 1)!}(x - a)^{n+1}$$

has the peculiarity that the derivative appearing in it is to be calculated in each case not at the point a but at a suitably chosen point ξ, which is unknown but lies somewhere in the interval between a and x.

The proof of equality (26) is rather cumbersome but quite simple in essence. We shall give here a somewhat artificial version of the proof, which has the merit of being concise.

In order to find out by how much the left side in the approximate formula (25) differs from the right, let us consider the ratio of the difference between the two sides in equality (25) to the quantity $-(x - a)^{n+1}$

$$\frac{f(x) - \left[f(a) + f'(a)(x - a) + \cdots + \frac{f^{(n)}(a)}{n!}(x - a)^n \right]}{-(x - a)^{n+1}}. \tag{27}$$

We also introduce the function

$$\phi(u) = f(u) + f'(u)(x - u) + \cdots + \frac{f^{(n)}(u)}{n!}(x - u)^n$$

of a variable u, taking x to be fixed (constant). Then the numerator in (27) will represent the increase of this function as we pass from $u = a$ to $u = x$, and the denominator will be the increase over the same interval of the function

$$\psi(u) = (x - u)^{n+1}.$$

---

* This is only one of the possible forms for the remainder term $R_{n+1}(x)$.

We now make use of the generalized mean-value theorem quoted earlier

$$\frac{\phi(x) - \phi(a)}{\psi(x) - \psi(a)} = \frac{\phi'(\xi)}{\psi'(\xi)}.$$

Differentiating the functions $\phi(u)$ and $\psi(u)$ with respect to u (it must be recalled that the value of x has been fixed) we find that

$$\frac{\phi'(\xi)}{\psi'(\xi)} = -\frac{f^{(n+1)}(\xi)}{(n+1)!}.$$

The equality of this last expression with the original quantity (27) gives Taylor's formula in the form (26).

   In the form (26) Taylor's formula not only provides a means of approximate calculation of $f(x)$ but also allows us to estimate the error. Let us consider the simple example

$$y = \sin x.$$

The values of the function $\sin x$ and of its derivatives of arbitrary order are known for $x = 0$. Let us make use of these values to write Taylor's formula for $\sin x$, choosing $a = 0$ and limiting ourselves to the case $n = 4$. We find successively

$$f(x) = \sin x, \qquad f'(x) = \cos x, \qquad f''(x) = -\sin x,$$

$$f'''(x) = -\cos x, \qquad f^{IV}(x) = \sin x, \qquad f^{V}(x) = \cos x;$$

$$f(0) = 0 \qquad\qquad f'(0) = 1, \qquad\qquad f''(0) = 0,$$

$$f'''(0) = -1. \qquad\quad f^{IV}(0) = 0, \qquad\quad f^{V}(\xi) = \cos \xi.$$

Therefore

$$\sin x = x - \frac{x^3}{6} + R_5, \quad \text{where} \quad R_5 = \frac{x^5}{120} \cos \xi.$$

Although the exact value R_5 is unknown, still we can easily estimate it from the fact that $|\cos \xi| \leqslant 1$. For all values of x between 0 and $\pi/4$ we have

$$|R_5| = \left| \frac{x^5}{120} \cos \xi \right| < \frac{1}{120} \left(\frac{\pi}{4} \right)^5 < \frac{1}{400}.$$

Consequently, on the interval $[0, \pi/4]$ the function $\sin x$ may be considered, with accuracy up to $\frac{1}{400}$, as equal to the polynomial of third degree

$$\sin x = x - \frac{1}{6} x^3.$$

If we were to take more terms in Taylor's expansion for sin x, we would obtain a polynomial of higher degree which would approximate sin x still more closely.

The tables for trigonometric and other functions are calculated by similar methods.

The laws of nature, as a rule, can be expressed with good approximation by functions that may be differentiated as often as we like and that in their turn may be approximated by polynomials, the degree of the polynomial being determined by the accuracy desired.

Taylor's series. If in formula (25) we take a larger and larger number of terms, then the difference between the right side and $f(x)$, expressed by the remainder term $R_{n+1}(x)$, may tend to zero. Of course this will not always occur: neither for all functions nor for all values of x. But there exists a broad class of functions (the so-called *analytic* functions) for which the remainder term $R_{n+1}(x)$ does in fact approach zero as $n \to \infty$, at least for all values of x within a certain interval around the point a. For these functions the Taylor formula allows us to calculate $f(x)$ with any desired degree of accuracy. Let us examine such functions more closely.

If $R_{n+1}(x) \to 0$ as $n \to \infty$, then from (26) it follows that

$$f(x) = \lim_{n \to \infty} \left[f(a) + f'(a)(x - a) + \cdots + \frac{f^{(n)}(a)}{n!}(x - a)^n \right]$$

In this case we say that $f(x)$ has been expanded in a convergent infinite series

$$f(x) = f(a) + f'(a)(x - a) + \frac{f''(a)}{2!}(x - a)^2 + \cdots,$$

in increasing powers of $(x - a)$. This series is called a *Taylor series*, and $f(x)$ is said to be the sum of the series. Let us consider some examples (with $a = 0$):

1. $(1 + x)^n = 1 + nx + \frac{n(n - 1)}{2!} x^2 + \frac{n(n - 1)(n - 2)}{3!} x^3 + \cdots$

   (valid for $|x| < 1$ and for arbitrary real n).

2. $\quad \sin x = x - \frac{x^3}{3!} + \frac{x^5}{5!} - \frac{x^7}{7!} + \cdots$ (valid for all x).

3. $\cos x = 1 - \dfrac{x^2}{2!} + \dfrac{x^4}{4!} - \dfrac{x^6}{6!} + \cdots$ (valid for all x).

4. $e^x = 1 + x + \dfrac{x^2}{2!} + \dfrac{x^3}{3!} + \cdots$ (valid for all x).

5. $\arctan x = x - \dfrac{x^3}{3} + \dfrac{x^5}{5} - \cdots$ (valid for $|x| \leqslant 1$).

The first of these examples is the famous binomial theorem of Newton, which was obtained by Newton for all n but completely proved in his time only for integral n. This example served as a model for the establishment of the general Taylor formula. The last two formulas allow us, for $x = 1$, to calculate with arbitrarily good approximation the numbers e and π.

The Taylor formula, which opens up the way for most of the calculations in applied analysis, is extremely important from the practical point of view.

Many of the laws of nature, physical and chemical processes, the motion of bodies, and the like, are expressed with great accuracy by functions which may be expanded in a Taylor series. The theory of such functions can be formulated in a clearer and more complete way if we consider them as functions of a complex variable (see Chapter IX).

The idea of approximating a function by polynomials or of representing it as the sum of an infinite number of simpler functions underwent far-reaching developments in analysis, where it now forms an independent branch, the theory of approximation of functions (see Chapter XII).

§10. Integral

From Chapter I and from §1 of the present chapter the reader already knows that the concept of the integral, and more generally of the integral calculus, had its historical origin in the need for solving concrete problems, a characteristic example of which is the calculation of the area of a curvilinear figure. The present section is devoted to these questions. In it we will also discuss the aforementioned connection between the problems of the differential and the integral calculus, which was not fully cleared up until the 18th century.

Area.   Let us suppose that a curve above the x-axis forms the graph of the function $y = f(x)$. We attempt to find the area S of the segment bounded by the line $y = f(x)$, by the x-axis and by the straight lines drawn through the points $x = a$ and $x = b$ parallel to the y-axis.

To solve this problem we proceed as follows. We divide the interval [a, b] into n parts, not necessarily equal. We denote the length of the first part by Δx_1, of the second by Δx_2, and so forth up to the final part Δx_n. In each segment we choose points ξ_1, ξ_2, $\cdots$, ξ_n and set up the sum

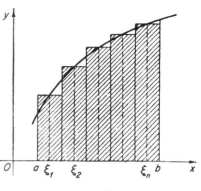

$$S_n = f(\xi_1) \, \Delta x_1 + f(\xi_1) \, \Delta x_2 + \cdots + f(\xi_n) \, \Delta x_n. \quad (28)$$

The magnitude S_n is obviously equal to the sum of the areas of the rectangles shaded in figure 24.

FIG. 24.

The finer we make the subdivision of the segment [a, b], the closer S_n will be to the area S. If we carry out a sequence of such constructions, dividing the interval [a, b] into successively smaller and smaller parts, then the sums S_n will approach S.

The possibility of dividing [a, b] into unequal parts makes it necessary for us to define what we mean by "successively smaller" subdivisions. We assume not only that n increases beyond all bounds but also that the length of the greatest Δx_i in the nth subdivision approaches zero. Thus

$$S = \lim_{\max \Delta x_i \to 0} [f(\xi_1) \, \Delta x_1 + f(\xi_2) \, \Delta x_2 + \cdots + f(\xi_n) \, \Delta x_n]$$

$$= \lim_{\max \Delta x_i \to 0} \sum_{i=1}^{n} f(\xi_i) \, \Delta x_i. \quad (29)$$

The calculation of the desired area has in this way been reduced to finding the limit (29).

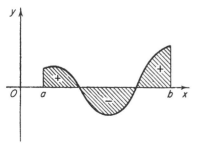

FIG. 25.

We note that when we first set up the problem, we had only an empirical idea of what we mean by the area of our curvilinear figure, but we had no precise definition. But now we have obtained an exact definition of the concept of area: It is the limit (29). We now have not only an intuitive notion of area but also a mathematical definition, on the basis of which we can calculate the area numerically (compare the remarks at the end of §3, concerning velocity and the length of a circumference).

We have assumed that $f(x) \geqslant 0$. If $f(x)$ changes sign, then in figure 25, the limit (29) will give us the algebraic sum of the areas of the segments lying between the curve $y = f(x)$ and the x-axis, where the segments above the x-axis are taken with a plus sign and those below with a minus sign.

Definite integral.   The need to calculate the limit (29) arises in many other problems. For example, suppose that a point is moving along a straight line with variable velocity $v = f(t)$. How are we to determine the distance s covered by the point in the time from $t = a$ to $t = b$?

Let us assume that the function $f(t)$ is continuous; that is, in small intervals of time the velocity changes only slightly. We divide the interval $[a, b]$ into n parts, of length $\Delta t_1, \Delta t_2, \cdots, \Delta t_n$. To calculate an approximate value for the distance covered in each interval Δt_i, we will suppose that the velocity in this period of time is constant, equal throughout to its actual value at some intermediate point ξ_1. The whole distance covered will then be expressed approximately by the sum

$$s_n = \sum_{i=1}^{n} f(\xi_i) \, \Delta t_i \, ,$$

and the exact value of the distance s covered in the time from a to b, will be the limit of such sums for finer and finer subdivisions; that is, it will be the limit (29)

$$s = \lim_{\max \, \Delta t_i \to 0} \sum_{i=1}^{n} f(\xi_i) \, \Delta t_i \, .$$

It would be easy to give many examples of practical problems leading to the calculation of such a limit. We will discuss some of them later, but for the moment the examples already given will sufficiently indicate the importance of this idea. The limit (29) is called the *definite integral* of the function $f(x)$ taken over the interval $[a, b]$, and it is denoted by

$$\int_a^b f(x) \, dx.$$

The expression $f(x) \, dx$ is called the integrand, a and b are the limits of integration; a is the lower limit, b is the upper limit.

The connection between differential and integral calculus. As an example of the direct calculation of a definite integral, we may take example 2, §1. We may now say that the problem considered there reduces to calculation of the definite integral

$$\int_0^h ax\, dx.$$

Another example was considered in §3, where we solved the problem of finding the area bounded by the parabola $y = x^2$. Here the problem reduces to calculation of the integral

$$\int_0^1 x^2\, dx.$$

We were able to calculate both these integrals directly, because we have simple formulas for the sum of the first n natural numbers and for the sum of their squares. But for an arbitrary function $f(x)$, we are far from being able to add up the sum (28) (that is, to express the result in a simple formula) if the points ξ_i and the increments Δx_i are given to suit some particular problem. Moreover, even when such a summation is possible, there is no general method for carrying it out; various methods, each of a quite special character, must be used in the various cases.

So we are confronted by the problem of finding a general method for the calculation of definite integrals. Historically this question interested mathematicians for a long period of time, since there were many practical aspects involved in a general method for finding the area of curvilinear figures, the volume of bodies bounded by a curved surface, and so forth.

We have already noted that Archimedes was able to calculate the area of a segment and of certain other figures. The number of special problems that could be solved, involving areas, volumes, centers of gravity of solids, and so forth, gradually increased, but progress in finding a general method was at first extremely slow. The general method could not be discovered until sufficient theoretical and computational material had been accumulated through the demands of practical life. The work of gathering and generalizing this material proceeded very gradually until the end of the Middle Ages; and its subsequent energetic development was a direct consequence of the rapid growth in the productive powers of Europe resulting from the breakup of the former (feudal) methods of manufacturing and the creation of new ones (capitalistic).

The accumulation of facts connected with definite integrals proceeded alongside of the corresponding investigations of problems related to the derivative of a function. The reader already knows from §1 that this immense preparatory labor was crowned with success in the 17th century by the work of Newton and Leibnitz. It is in this sense that Newton and Leibnitz are the creators of the differential and integral calculus.

One of the fundamental contributions of Newton and Leibnitz consists of the fact that they finally cleared up the profound connection between differential and integral calculus, which provides us, in particular, with a general method of calculating definite integrals for an extremely wide class of functions.

To explain this connection, we turn to an example from mechanics.

We suppose that a material point is moving along a straight line with velocity $v = f(t)$, where t is the time. We already know that the distance σ covered by our point in the time between $t = t_1$ and $t = t_2$ is given by the definite integral

$$\sigma = \int_{t_1}^{t_2} f(t)\,dt.$$

Now let us assume that the law of motion of the point is known to us; that is, we know the function $s = F(t)$ expressing the dependence on the time t of the distance s calculated from some initial point A on the straight line. The distance σ covered in the interval of time $[t_1,\, t_2]$ is obviously equal to the difference

$$\sigma = F(t_2) - F(t_1).$$

In this way we are led by physical considerations to the equality

$$\int_{t_1}^{t_2} f(t)\,dt = F(t_2) - F(t_1),$$

which expresses the connection between the law of motion of our point and its velocity.

From a mathematical point of view the function $F(t)$, as we already know from §5, may be defined as a function whose derivative for all values of t in the given interval is equal to $f(t)$, that is

$$F'(t) = f(t).$$

Such a function is called a *primitive* for $f(t)$.

We must keep in mind that if the function $f(t)$ has at least one primitive, then along with this one it will have an infinite number of others; for if $F(t)$ is a primitive for $f(t)$, then $F(t) + C$, where C is an arbitrary constant, is also a primitive. Moreover, in this way we exhaust the whole set of primitives for $f(t)$, since if $F_1(t)$ and $F_2(t)$ are primitives for the same function $f(t)$, then their difference $\phi(t) = F_1(t) - F_2(t)$ has a derivative

$\phi'(t)$ that is equal to zero at every point in a given interval so that $\phi(t)$ is a constant.*

From a physical point of view the various values of the constant C determine laws of motion which differ from one another only in the fact that they correspond to all possible choices for the initial point of the motion.

We are thus led to the result that for an extremely wide class of functions $f(x)$, including all cases where the function $f(x)$ may be considered as the velocity of a point at the time x, we have the following equality†

$$\int_a^b f(x)\,dx = F(b) - F(a), \tag{30}$$

where $F(x)$ is an arbitrary primitive for $f(x)$.

This equality is the famous *formula of Newton and Leibnitz*, which reduces the problem of calculating the definite integral of a function to finding a primitive for the function and in this way forms a link between the differential and the integral calculus.

Many particular problems that were studied by the greatest mathematicians are automatically solved by this formula, stating that the definite integral of the function $f(x)$ on the interval $[a, b]$ is equal to the difference between the values of any primitive at the left and right ends of the interval.‡ It is customary to write the difference (30) thus:

$$F(x)\,\Big|_a^b = F(b) - F(a).$$

Example 1. The equality

$$\left(\frac{x^3}{3}\right)' = x^2$$

shows that the function $x^3/3$ is a primitive for the function x^2. Thus, by the formula of Newton and Leibnitz,

$$\int_0^a x^2\,dx = \frac{x^3}{3}\,\Big|_0^a = \frac{a^3}{3} - \frac{0}{3} = \frac{a^3}{3}.$$

---

* By the mean value theorem
$$\phi(t) - \phi(t_0) = \phi'(v)(t - t_0) = 0,$$
when v lies between t and t_0. Thus $\phi(t) = \phi(t_0) = \text{const}$ for all t.

† It is possible to prove mathematically, without recourse to examples from mechanics, that if the function $f(x)$ is continuous (and even if it is discontinuous but Lebesgue-summable; see Chapter XV) on the interval $[a, b]$, then there exists a primitive $F(x)$ satisfying equality (30).

‡ This formula has been generalized in various ways (see for example §13, the formula of Ostrogradskiĭ).

Example 2. Let c and c' be two electric charges, on a straight line at distance r from each other. The attraction F between them is directed along this straight line and is equal to

$$F = \frac{a}{r^2}$$

($a = kcc'$, where k is a constant). The work W done by this force, when the charge c remains fixed but c' moves along the interval $[R_1, R_2]$, may be calculated by dividing the interval $[R_1, R_2]$ into parts Δr_i. On each of these parts we may consider the force to be approximately constant, so that the work done on each part is equal to $a/r_i^2 \, \Delta r_i$. Making the parts smaller and smaller, we see that the work W is equal to the integral

$$W = \lim_{n \to \infty} \sum_{i=1}^{n} \frac{a}{r_i^2} \, \Delta r_i = \int_{R_1}^{R_2} \frac{a}{r^2} \, dr.$$

The value of this integral can be calculated at once, if we recall that

$$\frac{a}{r^2} = \left(-\frac{a}{r}\right)',$$

so that

$$W = -\frac{a}{r}\Big|_{R_1}^{R_2} = a\left(\frac{1}{R_1} - \frac{1}{R_2}\right).$$

In particular, the work done by a force F as the charge c', initially at a distance R_1 from c, moves out to infinity, is equal to

$$W = \lim_{R_2 \to \infty} a\left(\frac{1}{R_1} - \frac{1}{R_2}\right) = \frac{a}{R_1}.$$

From the arguments given above for the formula of Newton and Leibnitz, it is clear that this formula gives mathematical expression to an actual tie existing in the objective world. It is a beautiful and important example of how mathematics gives expression to objective laws. We should remark that in his mathematical investigations, Newton always took a physical point of view. His work on the foundations of differential and integral calculus cannot be separated from his work on the foundations of mechanics.

The concepts of mathematical analysis, such as the derivative or the integral, as they presented themselves to Newton and his contemporaries, had not yet completely "broken away" from their physical and geometric origins, such as velocity and area. In fact, they were half mathematical in character and half physical. The conditions existing at that time were not yet suitable for producing a purely mathematical definition of these concepts. Consequently, the investigator could handle them correctly in complicated situations only if he remained in close contact with the practical aspects of his problem even during the intermediate (mathematical) stages of his argument.

From this point of view the creative work of Newton was different in character from that of Leibnitz.* Newton was guided at all stages by a physical way of looking at the problem. But the investigations of Leibnitz do not have such an immediate connection with physics, a fact that in the absence of clear-cut mathematical definitions sometimes led him to mistaken conclusions. On the other hand, the most characteristic feature of the creative activity of Leibnitz was his striving for generality, his efforts to find the most general methods for the problems of mathematical analysis.

The greatest merit of Leibnitz was his creation of a mathematical symbolism expressing the essence of the matter. The notations for such fundamental concepts of mathematical analysis as the differential dx, the second differential d^2x, the integral $\int y\,dx$, and the derivative d/dx were proposed by Leibnitz. The fact that these notations are still used shows how well they were chosen.

One advantage of a well-chosen symbolism is that it makes our proofs and calculations shorter and easier; also, it sometimes protects us against mistaken conclusions. Leibnitz, who was well aware of this, paid especial attention in all his work to the choice of notation.

The evolution of the concepts of mathematical analysis (derivative, integral, and so forth) continued, of course, after Newton and Leibnitz and is still continuing in our day; but there is one stage in this evolution that should be mentioned especially. It took place at the beginning of the last century and is related particularly to the work of Cauchy.

Cauchy gave a clear-cut formal definition of the concept of a limit and used it as the basis for his definitions of continuity, derivative, differential, and integral. These definitions have been introduced at the corresponding places in the present chapter. They are widely used in present-day analysis.

The great importance of these achievements lies in the fact that it is now possible to operate in a purely formal way not only in arithmetic,

---

* The discoveries of Newton and Leibnitz were made independently.

algebra, and elementary geometry, but also in this new and very extensive branch of mathematics, in mathematical analysis, and to obtain correct results in so doing.

Regarding practical application of the results of mathematical analysis, it is now possible to say: If the original data are verified in the actual world, then the results of our mathematical arguments will also be verified there. If we are properly assured of the accuracy of the original data, then there is no need to make a practical check of the correctness of the mathematical results; it is sufficient to check only the correctness of the formal arguments.

This statement naturally requires the following limitation. In mathematical arguments the original data, which we take from the actual world, are true only up to a certain accuracy. This means that at every step of our mathematical argument the results obtained will contain certain errors, which may accumulate as the number of steps in the argument increases.*

Returning now to the definite integral, let us consider a question of fundamental importance. For what functions $f(x)$, defined on the interval $[a, b]$, is it possible to guarantee the existence of the definite integral $\int_a^b f(x)\, dx$, namely a number to which the sum $\Sigma_1^n f(\xi_i)\, \Delta x_i$ tends as limit as max $\Delta x_i \to 0$? It must be kept in view that this number is to be the same for all subdivisions of the interval $[a, b]$ and all choices of the points ξ_i.

Functions for which the definite integral, namely the limit (29), exists are said to be *integrable* on the interval $[a, b]$. Investigations carried out in the last century show that all continuous functions are integrable.

But there are also discontinuous functions which are integrable. Among them, for example, are those functions which are bounded and either increasing or decreasing on the interval $[a, b]$.

The function that is equal to zero at the rational points in $[a, b]$ and equal to unity at the irrational points, may serve as an example of a nonintegrable function, since for an arbitrary subdivision the integral sum s_n will be equal to zero or unity, depending on whether we choose the points ξ_i as rational numbers or irrational.

Let us note that in many cases the formula of Newton and Leibnitz provides an answer to the practical question of calculating a definite integral. But here arises the problem of finding a primitive for a given

---

* For example, it follows formally from $a = b$ and $b = c$ that $a = c$. But in practice this relation appears as follows: From the facts that $a = b$ is known with accuracy up to ϵ and $b = c$ is known with the same accuracy, it follows that $a = c$ is known with accuracy up to 2ϵ.

function; that is, of finding a function that has the given function for its derivative. We now proceed to discuss this problem. Let us note by the way that the problem of finding a primitive has great importance in other branches of mathematics also, particularly in the solution of differential equations.

§11. Indefinite Integrals; the Technique of Integration

An arbitrary primitive of a given function $f(x)$ is usually called an *indefinite integral* of $f(x)$ and is written in the form

$$\int f(x)\, dx.$$

In this way, if $F(x)$ is a completely determined primitive of $f(x)$, then the indefinite integral of $f(x)$ is given by

$$\int f(x)\, dx = F(x) + C, \tag{31}$$

where C is an arbitrary constant.

Let us also note that if the function $f(x)$ is given on the interval $[a, b]$ and, if $F(x)$ is a primitive for $f(x)$ and x is a point in the interval $[a, b]$, then by the formula of Newton and Leibnitz we may write

$$F(x) = F(a) + \int_a^x f(t)\, dt.$$

Here the integral on the right side differs from the primitive $F(x)$ only by the constant $F(a)$. In such a case this integral, if we consider it as a function of its upper limit x (for variable x), is a completely determined primitive of $f(x)$. Consequently, an indefinite integral of $f(x)$ may also be written as follows:

$$\int f(x)\, dx = \int_a^x f(t)\, dt + C,$$

where C is an arbitrary constant.

Let us set up a fundamental table of indefinite integrals, which can be obtained directly from the corresponding table of derivatives (see §6):

$$\int x^a \, dx = \frac{x^{a+1}}{a+1} + C(a \neq -1),$$

$$\int \frac{dx}{x} = \ln |x| + C,^*$$

$$\int a^x \, dx = \frac{a^x}{\ln a} + C,$$

$$\int e^x \, dx = e^x + C,$$

$$\int \sin x \, dx = -\cos x + C,$$

$$\int \cos x \, dx = \sin x + C,$$ (32)

$$\int \sec^2 x \, dx = \tan x + C,$$

$$\int \frac{dx}{\sqrt{1-x^2}} = \text{arc sin } x + C$$

$$= -\text{arc cos } x + C_1 \left(C_1 - C = \frac{\pi}{2} \right),$$

$$\int \frac{dx}{1+x^2} = \text{arc tan } x + C.$$

The general properties of indefinite integrals may also be deduced from the corresponding properties of derivatives. For example, from the rule for the differentiation of a sum we obtain the formula

$$\int [f(x) \pm \phi(x)] \, dx = \int f(x) \, dx \pm \int \phi(x) \, dx + C,$$

and from the corresponding rule expressing the fact that a constant factor k may be taken outside the sign of differentiation we get

$$\int kf(x) \, dx = k \int f(x) \, dx + C.$$

For example,

$$\int \left(3x^2 + 2x - \frac{3}{\sqrt{x}} + \frac{4}{x} - 1 \right) dx$$

$$= 3\frac{x^3}{3} + \frac{2x^2}{2} - 3\frac{x^{-1/2+1}}{-\frac{1}{2}+1} + 4 \ln |x| - x + C.$$

---

* For $x > 0$, $(\ln |x|)' = (\ln x)' = 1/x$; for $x < 0$, $(\ln |x|)' = [\ln(-x)]' = 1/-x(-1) = 1/x$.

There are a number of methods for calculating indefinite integrals. Let us consider one of them, namely the *method of substitution* or change of variable, which is based on the following equality

$$\int f(x)\, dx = \int f[\phi(t)]\, \phi'(t)\, dt + C, \tag{33}$$

where $x = \phi(t)$ is a differentiable function. The relation (33) is to be understood in the sense that if in the function

$$F(x) = \int f(x)\, dx,$$

on the left side of equality (33), we set $x = \phi(t)$, we thereby obtain a function $F[\phi(t)]$ whose derivative with respect to t is equal to the expression under the sign of integration on the right side of equality (33). This fact follows immediately from the theorem on the derivative of a function of a function.

Let us give some examples of this method of substitution

$$\int e^{kx}\, dx = \int e^t \frac{1}{k}\, dt = \frac{1}{k}\int e^t\, dt = \frac{1}{k} e^t + C = \frac{e^{kx}}{k} + C$$

(substitution of $kx = t$, from which $k\, dx = dt$).

$$\int \frac{d\cdot}{\sqrt{c^{2}}} \frac{}{x^2} = -\int dt = -t + C = -\sqrt{a^2 - x^2} + C$$

$\left(\text{substitution of } t = \sqrt{a^2 - x^2}, \text{ from which } dt = -\dfrac{x\, dx}{\sqrt{a^2 - x^2}}\right).$

$$\int \sqrt{a^2 - x^2}\, dx = \int \sqrt{a^2 - a^2 \sin^2 u}\; a \cos u\, du = a^2 \int \cos^2 u\, du$$

$$= a^2 \int \frac{1 + \cos 2u}{2}\, du = \frac{a^2}{2}\left(u + \frac{\sin 2u}{2}\right) + C$$

$$= \frac{a^2}{2}(u + \sin u \cos u) + C$$

$$= \frac{a^2}{2}\left(\arcsin \frac{x}{a} + \frac{x}{a^2} \sqrt{a^2 - x^2}\right) + C$$

(substitution of $x = a \sin u$).

As can be seen from these examples, the method of substitution or change of variables greatly extends the class of elementary functions that we are able to integrate; that is, for which we can find primitives that are themselves elementary functions. But it must be noted that from the point of view of actually calculating the result, we are in a much worse position, generally speaking, with respect to integration than for differentiation.

From §6 we know that the derivative of an arbitrary elementary function is itself an elementary function, which we may effectively calculate by making use of the rules of differentiation. But the converse statement is in general untrue, since there exist elementary functions whose indefinite integrals are not elementary functions. Examples are e^{-x^2}, $1/(\ln x)$, $(\sin x)/x$ and so forth. To obtain integrals of these functions we must make use of approximative methods and also introduce new functions which can not be reduced to elementary ones. We can not spend more time here on this question but must simply note that even in elementary mathematics it is possible to find many examples in which a direct operation can be carried out on a certain class of numbers, while the inverse operation can not be carried out on the same class; thus, a square of an arbitrary rational number is again a rational number, but the square root of a rational number is by no means always rational. Analogously, differentiation of elementary functions produces a function that is again elementary, but integration may lead us outside the class of elementary functions.

Some of the integrals that cannot be expressed in terms of elementary functions have great importance in mathematics and its applications. An example is

$$\int_0^x e^{-t^2}\, dt,$$

which plays a very important role in the theory of probability (see Chapter XI). Other examples are the integrals

$$\int_0^\phi \frac{d\theta}{\sqrt{1 - k^2 \sin^2 \theta}} \quad \text{and} \quad \int_0^\phi \sqrt{1 - k^2 \sin^2 \theta}\, d\theta \; (k^2 < 1),$$

which are called *elliptic integrals* of the first and second kind respectively. We are led to the calculation of these integrals by a large number of problems in physics (see Chapter V, §1, example 3). Detailed tables of these integrals for various values of the arguments x and ϕ have been calculated by approximate methods but with great accuracy.

It must be emphasized that the proof of the very fact that a given elementary function cannot be integrated in terms of elementary functions is in each case quite difficult. Such questions occupied the attention of outstanding mathematicians in the last century and have played an important role in the development of analysis. Fundamental results were obtained here by Čebyšev, who gave a complete answer to the question of expressing in terms of elementary functions the integrals of the form

$$\int x^m(a + bx^s)^p \, dx,$$

where m, s, and p are rational numbers. Up to his time three relations, obtained by Newton, were known for the exponents m, s, and p, which implied the integrability of this integral in terms of elementary functions. Čebyšev proved that in all other cases the integral cannot be expressed in terms of elementary functions.

We introduce here another method of integration, namely integration by parts. It is based on the formula we already know

$$(uv)' = uv' + u'v,$$

for the derivative of the product of the functions u and v. This formula may also be written

$$uv' = (uv)' - u'v.$$

Let us now integrate the left and right sides, keeping in mind that

$$\int (uv)' \, dx = uv + C.$$

We now finally obtain the equality

$$\int uv' \, dx = uv - \int u'v \, dx,$$

which is also called the *formula of integration by parts*. We have not written the constant C since we may consider that it is included in one of the indefinite integrals occurring in this equation.

Let us introduce some applications of this formula. Suppose we have to calculate $\int xe^x \, dx$. Here we will take $u = x$ and $v' = e^x$, and thus $u' = 1$, $v = e^x$, and consequently

$$\int xe^x \, dx = xe^x - \int 1 \cdot e^x \, dx = xe^x - e^x + C.$$

In the integral $\int \ln x \, dx$ it is convenient to take $u = \ln x$, $v' = 1$, so that $u' = 1/x$, $v = x$ and

$$\int \ln x \, dx = x \ln x - \int dx = x \ln x - x + C.$$

In the following characteristic example it is necessary to integrate twice by parts and then to find the desired integral from the equations so obtained:

$$\int e^x \sin x \, dx = e^x \sin x - \int e^x \cos x \, dx$$

$$= e^x \sin x - e^x \cos x - \int e^x \sin x \, dx,$$

from which

$$\int e^x \sin x \, dx = \frac{e^x}{2} (\sin x - \cos x) + C.$$

We end this section here; from it the reader will have obtained only a superficial idea of the theory of integration. We have not given any attention to many different methods in this theory. In particular we have not touched here on the very interesting question of the integration of rational fractions, a theory in which an important contribution was made by the well-known mathematician and mechanician, Ostrogradskiĭ.

§12. Functions of Several Variables

Up to now we have spoken only of functions of one variable, but in practice it is often necessary to deal also with functions depending on two, three, or in general many variables. For example, the area of a rectangle is a function

$$S = xy$$

of its base x and its height y. The volume of a rectangular parallelepiped is a function

$$v = xyz$$

of its three dimensions. The distance between two points A and B is a function

$$r = \sqrt{(x_1 - x_2)^2 + (y_1 - y_2)^2 + (z_1 - z_2)^2}$$

of the six coordinates of these points. The well-known formula

$$pv = RT$$

expresses the dependence of the volume v of a definite amount of gas on the pressure p and absolute temperature T.

Functions of several variables, like functions of one variable, are in many cases defined only on a certain region of values of the variables themselves. For example, the function

$$u = \ln(1 - x^2 - y^2 - z^2) \tag{34}$$

is defined only for values of x, y and z that satisfy the condition

$$x^2 + y^2 + z^2 < 1. \tag{35}$$

(For other x, y, z its values are not real numbers.) The set of points of space whose coordinates satisfy the inequality (35) obviously fills up a sphere of unit radius with its center at the origin of coordinates. The points on the boundary are not included in this sphere; the surface of the sphere has been so to speak "peeled off." Such a sphere is said to be open. The function (34) is defined only for such sets of three numbers (x, y, z) as are coordinates of points in the open sphere G. It is customary to state this fact concisely by saying that the function (34) is defined on the sphere G.

Let us give another example. The temperature of a nonuniformly heated body V is a function of the coordinates x, y, z of the points of the body. This function is not defined for all sets of three numbers x, y, z but only for such sets as are coordinates of points of the body V.

Finally, as a third example, let us consider the function

$$u = \phi(x) + \phi(y) + \phi(z),$$

where ϕ is a function of one variable defined on the interval $[0, 1]$. Obviously the function u is defined only for sets of three numbers (x, y, z) which are coordinates of points in the cube:

$$0 \leqslant x \leqslant 1, \ 0 \leqslant y \leqslant 1, \ 0 \leqslant z \leqslant 1.$$

We now give a formal definition of a function of three variables. Suppose that we are given a set E of triples of numbers (x, y, z) (points of space). If to each of these triples of numbers (points) of E there corresponds a definite number u in accordance with some law, then u is said to be a function of x, y, z (of the point), defined on the set of triples of numbers (on the points) E, a fact which is written thus:

$$u = F(x, y, z).$$

In place of F we may also write other letters: f, ϕ, ψ.

In practice the set E will usually be a set of points, filling out some geometrical body or surface: sphere, cube, annulus, and so forth, and then we simply say that the function is defined on this body or surface. Functions of two, four, and so forth, variables are defined analogously.

Implicit definition of a function. Let us note that functions of two variables may serve, under certain circumstances, as a useful means for the definition of functions of one variable. Given a function $F(x, y)$ of two variables let us set up the equation

$$F(x, y) = 0. \tag{36}$$

In general, this equation will define a certain set of points (x, y) of the surface on which our function is equal to zero. Such sets of points usually represent curves that may be considered as the graphs of one or several one-valued functions $y = \phi(x)$ or $x = \psi(y)$ of one variable. In such a case these one-valued functions are said to be defined implicitly by the equation (36). For example, the equation

$$x^2 + y^2 - r^2 = 0$$

gives an implicit definition of two functions of one variable

$$y = + \sqrt{r^2 - x^2} \quad \text{and} \quad y = - \sqrt{r^2 - x^2}.$$

But it is necessary to keep in mind that an equation of the form (36) may fail to define any function at all. For example, the equation

$$x^2 + y^2 + 1 = 0$$

obviously does not define any real function, since no pair of real numbers satisfies it.

Geometric representation. Functions of two variables may always be visualized as surfaces by means of a system of space coordinates. Thus the function

$$z = f(x, y) \tag{37}$$

is represented in a three-dimensional rectangular coordinate system by a surface, which is the geometric locus of points M whose coordinates x, y, z satisfy equation (37) (figure 26).

There is another, extremely useful method, of representing the function (37), which has found wide application in practice. Let us choose a sequence of numbers z_1, z_2, $\cdots$, and then draw on one and the same plane Oxy the curves

$$z_1 = f(x, y), \; z_2 = f(x, y),$$

which are the so-called level lines of the function $f(x, y)$. From a set of level lines, if they correspond to values of z that are sufficiently close to one another, it is possible to form a very good image of the variation of the function $f(x, y)$, just as from the level lines of a topographical map one may judge the variation in altitude of the locality.

Figure 27 shows a map of the level lines of the function $z = x^2 + y^2$, the diagram at the right indicating how the function is built up from its level lines. In Chapter III, figure 50, a similar map is drawn for the level lines of the function $z = xy$.

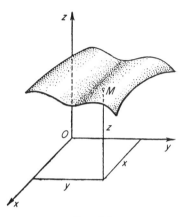

Fig. 26.

Partial derivatives and differential. Let us make some remarks about the differentiation of the functions of several variables. As an example we take the arbitrary function

$$z = f(x, y)$$

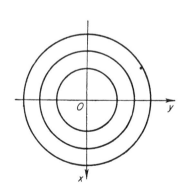
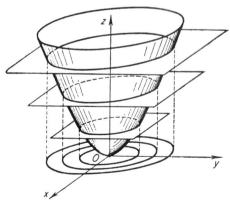

Fig. 27.

of two variables. If we fix the value of y, that is if we consider it as not varying, then our function of two variables becomes a function of the one variable x. The derivative of this function with respect to x, if it exists, is called the *partial derivative* with respect to x and is denoted thus:

$$\frac{\partial z}{\partial x}, \quad \text{or} \quad \frac{\partial f}{\partial x}, \quad \text{or} \quad f_x'(x, y).$$

The last of these three notations indicates clearly that the partial derivative with respect to x is in general a function of x and y. The partial derivative with respect to y is defined similarly.

Geometrically the function $f(x, y)$ represents a surface in a rectangular three-dimensional system of coordinates. The corresponding function of x for fixed y represents a plane curve (figure 28) obtained from the intersection of the surface with a plane parallel to the plane Oxz and at a distance y from it. The partial derivative $\partial z/\partial x$ is obviously equal to the trigonometric tangent of the angle between the tangent to the curve at the point (x, y) and the positive direction of the x-axis.

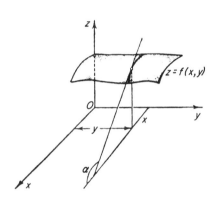

FIG. 28.

More generally, if we consider a function $z = f(x_1, x_2, \cdots, x_n)$ of the n variables $x_1, x_2, \cdots, x_n$, the partial derivative $\partial z/\partial x_i$ is defined as the derivative of this function with respect to x_i, calculated for fixed values of the other variables:

$$x_1, x_2, \ldots, x_{i-1}, x_{i+1}, \ldots, x_n.$$

We may say that the partial derivative of a function with respect to the variable x_i is the rate of change of this function in the direction of the change in x_i. It would also be possible to define a derivative in an arbitrary assigned direction, not necessarily coinciding with any of the coordinate axis, but we will not take the time to do this.

Examples.

1. $z = \dfrac{x}{y}, \dfrac{\partial z}{\partial x} = \dfrac{1}{y}, \dfrac{\partial z}{\partial y} = -\dfrac{x}{y^2}.$

2. $u = \dfrac{1}{\sqrt{x^2 + y^2 + z^2}}$,

$$\frac{\partial u}{\partial x} = -\frac{1}{x^2 + y^2 + z^2} \cdot \frac{2x}{2\sqrt{x^2 + y^2 + z^2}} = -\frac{x}{(x^2 + y^2 + z^2)^{3/2}}.$$

It is sometimes necessary to form the partial derivatives of these partial derivatives; that is; the so-called partial derivatives of second order. For functions of two variables there are four of them

$$\frac{\partial^2 u}{\partial x^2}, \frac{\partial^2 u}{\partial x\,\partial y}, \frac{\partial^2 u}{\partial y\,\partial x}, \frac{\partial^2 u}{\partial y^2}.$$

However, if these derivatives are continuous, then it is not hard to prove that the second and third of these four (the so-called mixed derivatives) coincide:

$$\frac{\partial^2 u}{\partial x\,\partial y} = \frac{\partial^2 u}{\partial y\,\partial x}.$$

For example, in the case of first function considered,

$$\frac{\partial^2 z}{\partial x^2} = 0,\ \frac{\partial^2 z}{\partial x\,\partial y} = -\frac{1}{y^2},\ \frac{\partial^2 z}{\partial y\,\partial x} = -\frac{1}{y^2},\ \frac{\partial^2 z}{\partial y^2} = \frac{2x}{y^3},$$

the two mixed derivatives are seen to coincide.

For functions of several variables, just as was done for functions of one variable, we may introduce the concept of a differential.

For definiteness let us consider a function

$$z = f(x, y)$$

of two variables. If it has continuous partial derivatives, we can prove that its increment

$$\Delta z = f(x + \Delta x, y + \Delta y) - f(x, y),$$

corresponding to the increments Δx and Δy of its arguments, may be put in the form

$$\Delta z = \frac{\partial f}{\partial x}\Delta x + \frac{\partial f}{\partial y}\Delta y + \alpha\sqrt{\Delta x^2 + \Delta y^2},$$

where $\partial f/\partial x$ and $\partial f/\partial y$ are the partial derivatives of the function at the point (x, y) and the magnitude α depends on Δx and Δy in such a way that $\alpha \to 0$ as $\Delta x \to 0$ and $\Delta y \to 0$.

The sum of the first two components

$$dz = \frac{\partial f}{\partial x} \Delta x + \frac{\partial f}{\partial y} \Delta y$$

is linearly dependent* on Δx and Δy and is called the *differential of the function*. The third summand, because of the presence of the factor α, tending to zero with Δx and Δy, is an infinitesimal of higher order than the magnitude

$$\rho = \sqrt{\Delta x^2 + \Delta y^2},$$

describing the change in x and y.

Let us give an application of the concept of differential. The period of oscillation of a pendulum is calculated from the formula

$$T = 2\pi \sqrt{\frac{l}{g}} \, ,$$

where l is its length and g is the acceleration of gravity. Let us suppose that l and g are known with errors respectively equal to Δl and Δg. Then the error in the calculation of T will be equal to the increment ΔT corresponding to the increments of the arguments Δl and Δg. Replacing ΔT approximately by dT, we will have

$$\Delta T \approx dT = \pi \left(\frac{\Delta l}{\sqrt{lg}} - \frac{\sqrt{l} \, \Delta g}{\sqrt{g^3}} \right).$$

The signs of Δl and Δg are unknown, but we may obviously estimate ΔT by the inequality

$$| \Delta T | \leqslant \pi \left(\frac{| \Delta l |}{\sqrt{lg}} + \sqrt{\frac{l}{g^3}} \, | \Delta g | \right),$$

from which after division by T we get

$$\frac{| \Delta T |}{T} < \left(\frac{| \Delta l |}{l} + \frac{| \Delta g |}{g} \right).$$

Thus we may consider in practice that the relative error for T is equal to the sum of the relative errors for l and g.

---

* In general a function $Ax + By + C$, where A, B, C are constants, is called a linear function of x and y. If $C = 0$, it is called a homogeneous linear function. Here we omit the word "homogeneous."

For symmetry of notation, the increments of the independent variables Δx and Δy are usually denoted by the symbols dx and dy and are also called differentials. With this notation the differential of the function $u = f(x, y, z)$ may be written thus:

$$du = \frac{\partial f}{\partial x} dx + \frac{\partial f}{\partial y} dy + \frac{\partial f}{\partial z} dz.$$

Partial derivatives play a large role whenever we have to do with functions of several variables, as happens in many of the applications of analysis to technology and physics. We shall be dealing in Chapter VI with the problem of reconstructing a function from the properties of its partial derivatives.

In the following paragraphs, we give some simple examples of applications of partial derivatives in analysis.

Differentiation of implicit functions. Suppose we wish to find the derivative of y, where y is a function of x defined implicitly by the relation

$$F(x, y) = 0 \tag{38}$$

between these variables. If x and y satisfy the relation (38) and we give x the increment Δx, then y will receive an increment Δy such that $x + \Delta x$ and $y + \Delta y$ again satisfy (38). Consequently*

$$F(x + \Delta x, y + \Delta y) - F(x, y) = \frac{\partial F}{\partial x} \Delta x + \frac{\partial F}{\partial y} \Delta y + \alpha \sqrt{\Delta x^2 + \Delta y^2} = 0.$$

Thus, provided $\partial F/\partial y \neq 0$, it follows that

$$\lim_{\Delta x \to 0} \frac{\Delta y}{\Delta x} = y'_x = -\frac{\dfrac{\partial F}{\partial x}}{\dfrac{\partial F}{\partial y}}.$$

In this way we have obtained a method for finding the derivative of an implicit function y without first solving the equation (38) for y.

Maximum and minimum problems. If a function, let us say of two variables $z = f(x, y)$, attains its maximum at the point (x_0, y_0), that is if $f(x_0, y_0) \geqslant f(x, y)$ for all points (x, y) close to (x_0, y_0), then this point must also be the point of maximum altitude for any line formed by the

---

* We assume that $F(x, y)$ has continuous derivatives with respect to x and y.

intersection of the surface $z = f(x, y)$ with a plane parallel to Oxz or Oyz. So at such a point we must have

$$f'_x(x, y) = 0, f'_y(x, y) = 0. \tag{39}$$

The same equations must also hold for a point of local minimum. Consequently, the greatest or least values of the function are to be sought first of all at points where the conditions (39) are satisfied, but we must also not forget about points on the boundary of the domain of definition of the function and points where the function fails to have a derivative, if such points exist.

To establish whether a point (x, y) satisfying (39) is actually a maximum or minimum point, use is frequently made of various indirect arguments. For example, if for any reason it is clear that the function is differentiable and attains its minimum inside the region and that there is only one point where the conditions (39) are fulfilled, then obviously the minimum must be attained at this point.

For example, let it be required to make a rectangular tin box (without lid) with assigned volume V, using the smallest possible amount of material. If the sides of the base of this box are denoted by x and y, then its height h will be equal to V/xy, and consequently the surface S will be given by the function

$$S = xy + \frac{V}{xy}(2x + 2y) = xy + 2V\left(\frac{1}{x} + \frac{1}{y}\right) \tag{40}$$

of x and y. Since x and y by the terms of the problem must be positive, the question has been reduced to finding the minimum of the function $S(x, y)$ for all possible points (x, y) in the first quadrant of the plane (x, y), which we will denote by the letter G.

If the minimum is attained at some point of the region G, then the partial derivatives must be equal to zero

$$\frac{\partial S}{\partial x} = y - \frac{2V}{x^2} = 0,$$

$$\frac{\partial S}{\partial y} = x - \frac{2V}{y^2} = 0,$$

that is $yx^2 = 2V$, $xy^2 = 2V$, from which we find as the dimensions of the box:

$$x = y = \sqrt[3]{2V} \quad \text{and} \quad h = \sqrt[3]{\frac{V}{4}}. \tag{41}$$

We have solved the problem but have not altogether proved that our solution is correct. A rigorous mathematician will say to us: "You have supposed from the very beginning that under the given conditions the box with minimum surface actually exists and, proceeding from this assumption, you have found its dimensions. So you have really obtained only the following result: If there exists a point (x, y) in G for which the function S attains its minimum, then the coordinates of this point must necessarily be determined by the equation (41). But now you must show that the minimum of S does exist for some point in G and then I will admit the correctness of your result." This remark is a very reasonable one, since, for example, our function S, as we shall soon see, does not possess any maximum in the region G. But let us show how it is possible to convince ourselves that in the given case the function actually does attain its minimum at a certain point (x, y) of the region G.

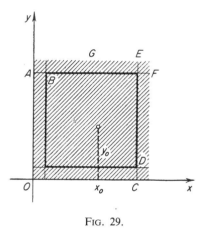

Fig. 29.

The fundamental theorem on which we shall base our argument is one that is proved in analysis with complete rigor; it amounts to the following. If the function f of one or several variables is everywhere continuous in a certain finite region H which is bounded and includes its boundary, then there always exists in H at least one point at which the function attains its minimum (maximum). With this theorem we can easily complete our analysis of the problem.

Let us consider an arbitrary point (x_0, y_0) of the region G; at this point let $S(x_0, y_0) = N$. Let us also choose a number R satisfying the two inequalities $R > N$, $2VR > N$ and construct a square Ω_R with side R^2, as in figure 29, where $AB = CD = 1/R$.

We now give a lower bound for the values of our function $S(x, y)$ at points of the region G lying outside the square Ω_R. If the point of the region G has abscissa $x < 1/R$, then

$$S(x, y) = xy + 2V\left(\frac{1}{x} + \frac{1}{y}\right) > 2V\frac{1}{x} > 2VR > N.$$

Analogously, if the point of the region G has its ordinate $y < 1/R$, then also $S > N$. Also, if the point of the region G has its abscissa $x > 1/R$

and if it lies above the straight line AF or has its ordinate $y > 1/R$ and lies to the right of the straight line CE, then

$$S(x, y) > xy > \frac{1}{R} R^2 = R > N.$$

Thus, for all points (x, y) of the region G lying outside the square Ω_k, the inequality $S(x, y) > N$ holds, and since $S(x_0, y_0) = N$, the point (x_0, y_0) must belong to the square and consequently the minimum of our function on G is equal to its minimum on the square.

But the function $S(x, y)$ is continuous in this square and on its boundary, so that by the theorem stated earlier there exists in the square a point (x, y) where our function assumes its minimum for points in the square and consequently for the entire region G. Thus the existence of a minimum has been proved.

This argument may serve as an example of the way that it is possible to discuss the existence of a maximum or a minimum for a function defined on an unbounded domain.

The Taylor formula. Like functions of one variable, functions of several variables may be represented by a Taylor formula. For example, an expansion of the function

$$u = f(x, y)$$

in the neighborhood of the point (x_0, y_0) has the following form, if we confine ourselves to the first and second powers of $x - x_0$ and $y - y_0$:

$$f(x, y) = f(x_0, y_0) + [f_x'(x_0, y_0)(x - x_0) + f_y'(x_0, y_0)(y - y_0)]$$

$$+ \frac{1}{2!} [f_{xx}''(x_0, y_0)(x - x_0)^2 + 2f_{xy}''(x_0, y_0)(x - x_0)(y - y_0)$$

$$+ f_{yy}''(x_0, y_0)(y - y_0)^2] + R_3.$$

If the function $f(x, y)$ has continuous partial derivatives of the second order, the remainder term here will approach zero faster than

$$r^2 = (x - x_0)^2 + (y - y_0)^2,$$

that is, faster than the square of the distance between the points (x, y) and (x_0, y_0), as $r \to 0$. The Taylor formula provides a widely used method of defining and approximately calculating the values of various functions.

Let us note that with the help of this formula we can also answer the question asked earlier, whether a given function actually has a maximum or minimum at a point where $\partial f/\partial x = \partial f/\partial y = 0$. In fact, if these conditions are satisfied at the point (x_0, y_0), then for points (x, y) close to (x_0, y_0), the value of the function will, by the Taylor formula, differ from $f(x_0, y_0)$ by the amount

$$f(x, y) - f(x_0, y_0)$$

$$= \frac{1}{2!} [A(x - x_0)^2 + 2B(x - x_0)(y - y_0) + C(y - y_0)^2] + R_3, \qquad (42)$$

where A, B, and C denote respectively the second partial derivatives $f''_{xx}, f''_{xy}, f''_{yy}$ at the point (x_0, y_0).

If it turns out that the function

$$\Phi(x, y) = A(x - x_0)^2 + 2B(x - x_0)(y - y_0) + C(y - y_0)^2$$

is positive for arbitrary values of $(x - x_0)$ and $(y - y_0)$ not both equal to zero, then the right side of equation (42) will also be positive for small values of $(x - x_0)$ and $(y - y_0)$, since for sufficiently small $(x - x_0)$ and $(y - y_0)$ the quantity R_3 is known to be less in absolute value than $\frac{1}{2} \Phi(x, y)$. Thus it will follow that at the point (x_0, y_0) the function f attains its minimum. On the other hand, if the function $\Phi(x, y)$ is negative for arbitrary $(x - x_0)$ and $(y - y_0)$ the right side of (42) will be negative for $(x - x_0)$ and $(y - y_0)$, so that at the point (x_0, y_0) the function will have a maximum. In more complicated cases it is necessary to consider the succeeding terms in the Taylor formula.

Problems concerning the maximum or the minimum of functions of three or more variables may be treated in a completely analogous fashion. As an exercise the reader may prove that if given masses

$$m_1, m_2, \cdots, m_n$$

are arranged in space at given points

$$P_1(x_1, y_1, z_1), P_2(x_2, y_2, z_2), \cdots, P_n(x_n, y_n, z_n),$$

the moment (of inertia) M of this system of masses about the point $P(x, y, z)$, defined as the sum of the products of the masses and the squares of their distances from the point P,

$$M(x, y, z) = \sum_{i=1}^{n} m_i[(x - x_i)^2 + (y - y_i)^2 + (z - z_i)^2],$$

will be a minimum if the point P is at the so-called center of gravity of the system, with the coordinates

$$x = \frac{\sum_{i=1}^n m_i x_i}{\sum_{i=1}^n m_i}, \, y = \frac{\sum_{i=1}^n m_i y_i}{\sum_{i=1}^n m_i}, \, z = \frac{\sum_{i=1}^n m_i z_i}{\sum_{i=1}^n m_i}.$$

Maxima and minima with subsidiary conditions. For functions of several variables we may set up various problems concerning maximum and minimum. Let us illustrate with a simple example. Suppose that among all rectangles inscribed in a circle of radius R, we wish to find the one with greatest area. The area of a rectangle is equal to the product xy of its sides, where x and y are positive numbers connected in this case by the relation $x^2 + y^2 = (2R)^2$, as is clear from figure 30. Thus we are required to find the maximum of the function $f(x, y) = xy$ for all x and y satisfying the relation $x^2 + y^2 = 4R^2$.

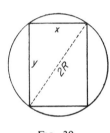

Problems of this sort, where it is necessary to find the maximum (or minimum) of a function $f(x, y)$ for those values only of x and y that satisfy a certain relation that $\phi(x, y) = 0$ are very common in practice.

Of course, it would be possible to solve the equation $\phi(x, y) = 0$ for y, to substitute the solution into the function $f(x, y)$ and in this way to seek the ordinary maximum for a function of one variable x. But this method is usually complicated and sometimes impossible.

FIG. 30.

For the solution of such problems in analysis, a much more convenient procedure called the method of Lagrange multipliers, has been worked out. The idea behind it is extremely simple. Let us consider the function

$$F(x, y) = f(x, y) + \lambda\phi(x, y),$$

where λ is an arbitrary positive number. Obviously, for x, y satisfying the condition $\phi(x, y) = 0$, the values of $F(x, y)$ coincide with those of $f(x, y)$.

For function $F(x, y)$ let us seek a maximum without conditions of any kind on x and y. At the maximum point the conditions $\partial F/\partial x = \partial F/\partial y = 0$* must hold; in other words

$$\frac{\partial f}{\partial x} + \lambda\frac{\partial \phi}{\partial x} = 0; \tag{43}$$

$$\frac{\partial f}{\partial y} + \lambda\frac{\partial \phi}{\partial y} = 0. \tag{44}$$

---

* We are speaking here, of course, of a maximum attained in the domain of definition of the function $F(x, y)$. The functions $f(x, y)$ and $\phi(x, y)$ are assumed to be differentiable.

The values of x and y at the maximum point for $F(x, y)$, being a solution of the system (43) and (44), depend on the coefficient λ in these equations. Let us now suppose that we have succeeded in choosing the number λ in such a way that the coordinates of the maximum point satisfy the condition

$$\phi(x, y) = 0. \tag{45}$$

Then this point will be an exact local maximum for the original problem.

In fact, we may consider the problem geometrically as follows. The function $f(x, y)$ is defined on a certain region G (figure 31). The condition $\phi(x, y) = 0$ will ordinarily be satisfied by the points of some curve Γ. We are required to find the greatest value of x and y on points of the line Γ. If $F(x, y)$ attains its maximum on the curve Γ, then $F(x, y)$ does not increase for small shifts in an arbitrary direction from this point, and in particular for shifts along the curve Γ. But for shifts along Γ, the values of $F(x, y)$, coincide

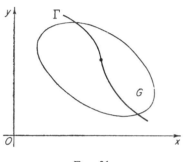

FIG. 31.

with those of $f(x, y)$ which means that for small shifts along the curve the function $f(x, y)$ does not increase, or in other words it has the local maximum at the point.

These arguments indicate a simple method of solving the problem. We solve equations (43), (44), (45) for the unknowns x, y, and λ, obtaining one or more solutions

$$(x_1, y_1, \lambda_1), \ (x_2, y_2, \lambda_2), \ \cdots. \tag{46}$$

To the points $(x_1, y_1), (x_2, y_2), \cdots$ so determined we adjoin those points of the boundary of G where the curve Γ leaves the region G. Then from all these points we choose that one at which $f(x, y)$ takes on its greatest (or smallest) value.

Of course, the arguments here are far from proving the correctness of the method. In fact, we have not yet even proved that the points of local maximum for $f(x, y)$ on the curve Γ can be obtained as maximum points for the function $F(x, y)$ for some value of λ. However, it is possible to prove, as is done in the textbooks in analysis, that every point (x_0, y_0) where $f(x, y)$ has a local maximum on the curve will be obtained by the method indicated, provided only that at this point the partial derivatives $\phi'_x(x_0, y_0)$ and $\phi'_y(x_0, y_0)$ are not both equal to zero.*

Let us use the method of Lagrange to solve the problem at the beginning of the present section. In this case $f(x, y) = xy$; $\phi(x, y) = x^2 + y^2 - 4R^2$. We set up the equations (43), (44), (45)

$$y + 2\lambda x = 0,$$

$$x + 2\lambda y = 0,$$

$$x^2 + y^2 = 4R^2,$$

for which, taking into account that x and y are positive, we find the unique solution

$$x = y = R\sqrt{2}\left(\lambda = -\frac{1}{2}\right).$$

For these values of x and y, which are equal to one another so that the inscribed rectangle is a square, the area is in fact a maximum.

The method of Lagrange may be extended to deal with functions of three or more variables. There may be any number of subsidiary conditions (smaller than the number of variables) of the type of condition (45), and we will introduce the corresponding number of auxiliary multipliers.

Let us give some examples of problems involving maxima or minima with subsidiary conditions.

Example 1.   For what height h and radius r will an open cylindrical tank of given volume V require the least amount of sheet metal for its manufacture; that is, the area of its sides and circular base will be a minimum?

The problem obviously reduces to finding the minimum of the function of the variables r and h

$$f(r, h) = 2\pi rh + \pi r^2$$

under the condition $\pi r^2 h = V$, which may be written in the form

$$\phi(r, h) = \pi r^2 h - V = 0.$$

---

* In the course in higher mathematics of V. I. Smirnov, the reader will find a simple example where this particular feature of the situation would lead to the loss of a solution if we apply the method of Lagrange mechanically and do not consider, in addition to the points mentioned above, a point where not only (45) holds but also:

$$\phi'_x(x_0, y_0) = 0, \qquad \phi'_y(x_0, y_0) = 0$$

Example 2. A moving point is required to pass from A to B (figure 32). On the path AM it moves with the velocity of v_1, and on MB with the velocity v_2. Where should the point M be placed on the line DD' so that the entire path from A to B may be covered as quickly as possible?

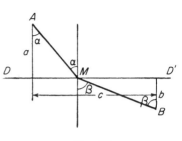

Let us take as unknowns the angles α and β marked in figure 32. The lengths a and b of the perpendiculars from the points A and B to the straight line DD' and the distance c between them are known. The time

Fig. 32.

required for covering the entire path is represented, as can easily be seen, by the formula

$$f(\alpha, \beta) = \frac{a}{v_1 \cos \alpha} + \frac{b}{v_2 \cos \beta}.$$

It is required to find the minimum of this expression, taking into account the fact that α and β are connected by the relation

$$a \tan \alpha + b \tan \beta = c.$$

The reader may solve these examples by the Lagrange method. In the second example he will find that the best position for M is given by the condition

$$\frac{\sin \alpha}{\sin \beta} = \frac{v_1}{v_2}.$$

This is the well-known law for the refraction of light. Consequently, a ray of light will be refracted in its passage from one medium to another in such a way that the time for its passage from a point in one medium to a point in the other is a minimum. Conclusions of this sort are interesting not only for computational purposes but also from a general philosophical point of view; they have inspired researchers in the exact sciences to penetrate further and further into the profound and general laws of nature.

Finally let us note that the multipliers λ, introduced in the solution of problems by the method of Lagrange, are not merely auxiliary numbers. In each case they are closely connected with the essential nature of the particular problem and have a concrete interpretation.

§13. Generalizations of the Concept of Integral

In §10 we defined the definite integral of the function $f(x)$ on the interval $[a, b]$ as the limit of the sum

$$\sum_{i=1}^{n} f(\xi_i)\, \Delta x_i$$

when the length of the greatest segment Δx_i in the subdivision of $[a, b]$ approaches zero. In spite of the fact that the class of functions $f(x)$ for which this limit actually exists (the class of integrable functions) is a very wide one, and in particular includes all continuous and even many discontinuous functions, this class of functions has a serious shortcoming. If we add, subtract, or multiply, or under certain conditions divide the values of two integrable functions $f(x)$ and $\phi(x)$, we obtain functions which, as may easily be proved, are again integrable. For $f(x)/\phi(x)$ this will be true in all cases in which $1/\phi(x)$ remains bounded on $[a, b]$. But if a function is obtained as a result of a limiting process from a sequence of approximating integrable functions $f_1(x), f_2(x), f_3(x), \cdots$ such that for all values of x in the interval $[a, b]$

$$f(x) = \lim_{n \to \infty} f_n(x),$$

then the limit function $f(x)$ is not necessarily integrable.

In many cases this and other circumstances give rise to considerable complication, since the process of passing to a limit is widely used.

A way out of the difficulty was discovered by making further generalizations of the concept of an integral. The most important of these is the integral of Lebesgue, with which the reader will become acquainted in Chapter XV on the theory of functions of a real variable. But here we will confine ourselves to generalizations of the integral in other directions, which are also of the greatest importance in practice.

Multiple integrals.   We have already studied the process of integration for functions of one variable defined on a one-dimensional region, namely an interval. But the analogous process may be extended to functions of two, three, or more variables, defined on corresponding regions.

For example, let us consider a surface

$$z = f(x, y)$$

defined in a rectangular system of coordinates, and on the plane Oxy let there be given a region G bounded by a closed curve Γ. It is required to find the volume bounded by the surface, by the plane Oxy and by the cylindrical surface passing through the curve Γ with generators parallel to the Oz axis (figure 33). To solve this problem we divide the plane region G into subregions by a network of straight lines parallel to the axes Ox and Oy and denote by

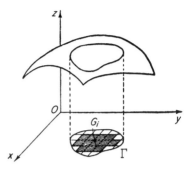

FIG. 33.

$$G_1, G_2, \cdots, G_n$$

those subregions which consist of complete rectangles. If the net is sufficiently fine, then practically the whole of the region G will be covered by the enumerated rectangles. In each of them we choose at will a point

$$(\xi_1, \eta_1), (\xi_2, \eta_2), \cdots, (\xi_n, \eta_n)$$

and, assuming for simplicity that G_i denotes not only the rectangle but also its area, we set up the sum

$$S_n = f(\xi_1, \eta_1)\, G_1 + f(\xi_2, \eta_2)\, G_2 + \cdots + f(\xi_n, \eta_n)\, G_n = \sum_{i=1}^{n} f(\xi_i, \eta_i)\, G_i. \tag{47}$$

It is clear that, if the surface is continuous and the net is sufficiently fine, this sum may be brought as near as we like to the desired volume V. We will obtain the desired volume exactly if we take the limit of the sum (47) for finer and finer subdivisions (that is, for subdivisions such that the greatest of the diagonals of our rectangles approaches zero)

$$\lim_{\max d(G_i) \to 0} \sum_{i=1}^{n} f(\xi_i, \eta_i)\, G_i = V. \tag{48}$$

From the point of view of analysis it is therefore necessary, in order to determine the volume V, to carry out a certain mathematical operation on the function $f(x, y)$ and its domain of definition G, an operation indicated by the left side of equality (48). This operation is called the integration of the function f over the region G, and its result is the integral of f over G. It is customary to denote this result in the following way:

$$\iint_{G} f(x, y)\, dx\, dy = \lim_{\max d(G_i) \to 0} \sum_{i=1}^{n} f(\xi_i, \eta_i)\, G_i. \tag{49}$$

Similarly, we may define the integral of a function of three variables over a three-dimensional region G, representing a certain body in space. Again we divide the region G into parts, this time by planes parallel to the coordinate planes. Among these parts we choose the ones which represent complete parallelepipeds and enumerate them

$$G_1, G_2, \cdots, G_n.$$

In each of these we choose an arbitrary point

$$(\xi_1, \eta_1, \zeta_1), (\xi_2, \eta_2, \zeta_2), \cdots, (\xi_n, \eta_n, \zeta_n)$$

and set up the sum

$$S = \sum_{i=1}^{n} f(\xi_i, \eta_i, \zeta_i) G_i, \qquad (50)$$

where G_i denotes the volume of the parallelepiped G_i. Finally we define the integral of $f(x, y, z)$ over the region G as the limit

$$\lim_{\max d(G_i) \to 0} \sum_{i=1}^{n} f(\xi_i, \eta_i, \zeta_i) G_i = \iiint_G f(x, y, z)\, dx\, dy\, dz, \qquad (51)$$

to which the sum (50) tends when the greatest diagonal $d(G_i)$ approaches zero.

Let us consider an example. We imagine the region G is filled with a nonhomogeneous mass whose density at each point in G is given by a known function $\rho(x, y, z)$. The density $\rho(x, y, z)$ of the mass at the point (x, y, z) is defined as the limit approached by the ratio of the mass of an arbitrary small region containing the point (x, y, z) to the volume of this region as its diameter approaches zero.* To determine the mass of the body G it is natural to proceed as follows. We divide the region G into parts by planes parallel to the coordinate planes and enumerate the complete parallelepipeds formed in this way

$$G_1, G_2, \cdots, G_n.$$

Assuming that the dividing planes are sufficiently close to one another, we will make only a small error if we neglect the irregular regions of the body and define the mass of each of the regular regions G_i (the complete parallelepipeds) as the product

$$\rho(\xi_i, \eta_i, \zeta_i) G_i,$$

---

* The diameter of a region is defined as the least upper bound of the distance between two points of the region.

where (ξ_i, η_i, ζ_i) is an arbitrary point G_i. As a result the approximate value of the mass M will be expressed by the sum

$$S_n = \sum_{i=1}^{n} \rho(\xi_i, \eta_i, \zeta_i) \, G_i,$$

and its exact value will clearly be the limit of this sum as the greatest diagonal G_i approaches zero; that is

$$M = \iiint\limits_{G} \rho(x, y, z) \, dx \, dy \, dz = \lim_{\max d(G_i) \to 0} \sum_{i=1}^{n} \rho(\xi_i, \eta_i, \zeta_i) \, G_i.$$

The integrals (49) and (51) are called double and triple integrals respectively.

Let us examine a problem which leads to a double integral. We imagine that water is flowing over a plane surface. Also, on this surface the underground water is seeping through (or soaking back into the ground) with an intensity $f(x, y)$ which is different at different points. We consider a region G bounded by a closed contour (figure 34) and assume that at every point of G we know the intensity $f(x, y)$, namely the amount of underground water seeping through per minute per cm^2 of surface; we will have $f(x, y) > 0$ where

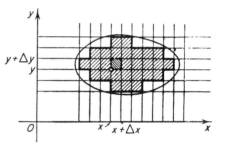

FIG. 34.

the water is seeping through and $f(x, y) < 0$ where it is soaking into the ground. How much water will accumulate on the surface G per minute?

If we divide G into small parts, consider the rate of seepage as approximately constant in each part and then pass to the limit for finer and finer subdivisions, we will obtain an expression for the whole amount of accumulated water in the form of an integral

$$\iint\limits_{G} f(x, y) \, dx \, dy.$$

Double (two-fold) integrals were first introduced by Euler. Multiple integrals form an instrument which is used everyday in calculations and investigations of the most varied kind.

It would also be possible to show, though we will not do it here, that calculation of multiple integrals may be reduced, as a rule, to iterated calculation of ordinary one-dimensional integrals.

Contour and surface integrals.   Finally, we must mention that still other generalizations of the integral are possible. For example, the problem of defining the work done by a variable force applied to a material point, as the latter moves along a given curve, naturally leads to a so-called curvilinear integral, and the problem of finding the general charge on a surface on which electricity is continuously distributed with a given surface density leads to another new concept, an integral over a curved surface.

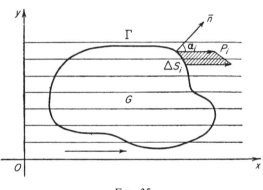

For example, suppose that a liquid is flowing through space (figure 35) and that the velocity of a particle of the liquid at the point (x, y) is given by a function $P(x, y)$, not depending on z. If we wish to determine the amount of liquid flowing per minute through the contour Γ,* we may

FIG. 35.

reason in the following way. Let us divide Γ up into segments Δs_i. The amount of water flowing through one segment Δs_i is approximately equal to the column of liquid shaded in figure 35; this column may be considered as the amount of liquid forcing its way per minute through that segment of the contour. But the area of the shaded parallelogram is equal to

$$P_i(x, y) \cdot \Delta s_i \cdot \cos \alpha_i ,$$

where α_i is the angle between the direction $\bar{x}$ of the x-axis and the outward normal of the surface bounded by the contour Γ; this normal is the perpendicular $\bar{n}$ to the tangent, which we may consider as defining the direction of the segment Δs_i. By summing up the areas of such parallelograms and passing to the limit for finer and finer subdivisions

---

* More precisely, through a cylindrical surface with the contour for its base and with height equal to unity.

of the contour Γ, we determine the amount of water flowing per minute through the contour Γ; it is denoted thus:

$$\int_\Gamma P(x, y) \cos (\bar{n}, \bar{x}) \, ds$$

and is called a curvilinear integral. If the flow is not everywhere parallel, then its velocity at each point (x, y) will have a component $P(x, y)$ along the x-axis and a component $Q(x, y)$ along the y-axis. In this case we can show by an analogous argument that the quantity of water flowing through the contour will be equal to

$$\int_\Gamma [P(x, y) \cos (\bar{n}, \bar{x}) + Q(x, y) \cos (\bar{n}, \bar{y})] \, ds. *$$

When we speak of an integral over a curved surface G for a function $f(M)$ of its points $M(x, y, z)$, we mean the limit of sums of the form

$$\lim \sum_{i=1}^{n} f(M_i) \, \Delta \sigma_i = \iint_G f(x, y, z) \, d\sigma$$

for finer and finer subdivisions of the region G into segments whose areas are equal to $\Delta \sigma_i$.

General methods exist for transforming multiple, curvilinear, and surface integrals into other forms and for calculating their values, either exactly or approximately.

Formula of Ostrogradskiĭ. Several important and very general formulas relating an integral over a volume to an integral over its surface (and also an integral over a surface, curved or plane, to an integral around its boundary) were discovered in the middle of the past century by Ostrogradskiĭ.

We shall not try to give here a proof of the general formula of Ostrogradskiĭ, which has very wide application, but will merely illustrate it by an example of its simplest particular case.

Let us imagine, as we did before, that over a plane surface there is a horizontal flow of water that is also soaking into the ground or seeping out again from it. We mark off a region G, bounded by a curve Γ, and

---

* Since for small displacements along the curve the differential of the coordinate y is equal to $\cos(\bar{n}, \bar{x}) \, ds$ and the differential dx is equal to $-\cos(\bar{n}, \bar{y}) \, ds$, this latter integral is often written in the form

$$\int_\Gamma [P(x, y) \, dy - Q(x, y) \, dx].$$

assume that for each point of the region we know the components $P(x, y)$ and $Q(x, y)$ of the velocity of the water in the direction of the x-axis and of the y-axis respectively.

Let us calculate the rate at which the water is seeping from the ground at a point with coordinates (x, y). For this purpose we consider a small rectangle with sides Δx and Δy situated at the point (x, y).

As a result of the velocity $P(x, y)$ through the left vertical edge of this rectangle, there will flow approximately $P(x, y)\Delta y$ units of water per minute into the rectangle, and through the right side in the same time will flow out approximately $P(x + \Delta x, y)\Delta y$ units. In general, the net amount of water leaving a square unit of surface as a result of the flow through its left and right vertical sides will be approximately

$$\frac{[P(x + \Delta x, y) - P(x, y)]\,\Delta y}{\Delta x\,\Delta y}\,.$$

If we let Δx approach zero, we obtain in the limit

$$\frac{\partial P}{\partial x}\,.$$

Correspondingly, the net rate of flow of water per unit area in the direction of the y-axis will be given by

$$\frac{\partial Q}{\partial y}\,.$$

This means that the intensity of the seepage of ground water at the point with coordinates (x, y) will be equal to

$$\frac{\partial P}{\partial x} + \frac{\partial Q}{\partial y}\,.$$

But in general, as we saw earlier, the quantity of water coming out from the ground will be given by the double integral of the function expressing the intensity of the seepage of ground water at each point, namely

$$\iint_G \left(\frac{\partial P}{\partial x} + \frac{\partial Q}{\partial y}\right) dx\, dy. \tag{52}$$

But, since the water is incompressible, this entire quantity must flow out during the same time through the boundaries of the contour Γ. The quantity of water flowing out through the contour Γ is expressed, as we saw earlier, by the curvilinear integral over Γ

$$\int_\Gamma [P(x, y)\cos(\bar{n}, \bar{x}) + Q(x, y)\cos(\bar{n}, \bar{y})]\, ds. \tag{53}$$

The equality of the magnitudes (52) and (53) expresses the formula of Ostrogradskiĭ in its simplest two-dimensional case

$$\iint_{G} \left(\frac{\partial P}{\partial x} + \frac{\partial Q}{\partial y} \right) dx\, dy = \int_{\Gamma} [P(x, y) \cos(\bar{n}, \bar{x}) + Q(x, y) \cos(\bar{n}, \bar{y})]\, ds.$$

We have merely explained the meaning of this formula by a physical example, but it can be proved mathematically.

In this way the mathematical theorem of Ostrogradskiĭ reflects a widespread phenomenon in the external world, which in our example we interpreted in a readily visualized way as preservation of the volume of an incompressible fluid.

Ostrogradskiĭ established a considerably more general formula expressing the connection between an integral over a multidimensional volume and an integral over its surface. In particular, for a three-dimensional body G, bounded by the surface Γ, his formula is

$$\iiint_{G} \left(\frac{\partial P}{\partial x} + \frac{\partial Q}{\partial y} + \frac{\partial R}{\partial z} \right) dx\, dy\, dz$$
$$= \iint_{\Gamma} [P \cos(\bar{n}, \bar{x}) + Q \cos(\bar{n}, \bar{y}) + R \cos(\bar{n}, \bar{z})]\, d\sigma,$$

where $d\sigma$ is the element of surface.

It is interesting to note that the fundamental formula of the integral calculus

$$\int_{a}^{b} f(x)\, dx = F(b) - F(a) \tag{54}$$

may be considered as a one-dimensional case of the formula of Ostrogradskiĭ. The equation (54) connects the integral over an interval with the "integral" over its "null-dimensional" boundary, consisting of the two end points.

Formula (54) may be illustrated by the following analogy. Let us imagine that in a straight pipe with constant cross section $s = 1$ water is flowing with velocity $F(x)$, which is different for different cross sections (figure 36). Through the porous walls of the pipe, water is seeping into it

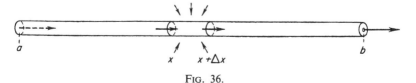

FIG. 36.

(or out of it) at a rate which is also different for different cross sections.

If we consider a segment of the pipe from x to $x + \Delta x$, the quantity of water seeping into it in unit time must be compensated by the difference $F(x + \Delta x) - F(x)$ between the quantity flowing out of this segment and the quantity flowing into it along the pipe. So the quantity seeping into the segment is equal to the difference $F(x + \Delta x) - F(x)$, and consequently the rate of seepage per unit length of pipe (the ratio of the seepage over an infinitesimal segment to the length of the segment) will be equal to

$$f(x) = \lim_{\Delta x \to 0} \frac{F(x + \Delta x) - F(x)}{\Delta x} = F'(x).$$

More generally, the quantity of water seeping into the pipe over the whole section $[a, b]$ must be equal to the amount lost by flow through the ends of the pipe. But the amount seeping through the walls is equal to $\int_a^b f(x)\,dx$ and the amount lost by flow through the ends is $F(b) - F(a)$. The equality of these two magnitudes produces formula (54).

§14. Series

Concept of a series. A series in mathematics is an expression of the form

$$u_0 + u_1 + u_2 + \cdots.$$

The numbers u_k are called the terms of the series. There is an infinite number of them, and they are arranged in a definite order, so that to each natural number $k = 0, 1, 2, \cdots$ there corresponds a definite value u_k.

The reader must keep in mind that we have not said whether it is possible to calculate a value for such expressions or, in case it is possible, how to do it. The presence of a plus sign between the terms u_k in our expression seems to indicate that in some way all the terms should be added. But there are infinitely many of them and addition of numbers is defined only for a finite number of terms.

Let us denote by S_n the sum of the first n terms of the series; we will call it the *nth partial sum*. As a result we obtain a sequence of numbers

$$S_1 = u_0,$$
$$S_2 = u_0 + u_1,$$
$$\cdots \cdots \cdots \cdots \cdots \cdots \cdots$$
$$S_n = u_0 + u_1 + \cdots + u_{n-1},$$
$$\cdots \cdots \cdots \cdots \cdots \cdots \cdots$$

and we may speak of a variable quantity S_n, where $n = 1, 2, \cdots$.

The series is said to be *convergent* if, as $n \to \infty$, the variable S_n approaches a definite finite limit

$$\lim_{n\to\infty} S_n = S.$$

This limit is called the *sum of the series*, and in this case we write

$$S = u_0 + u_1 + u_2 + \cdots.$$

But if, as $n \to \infty$, the limit S_n does not exist, then the series is said to be *divergent* and in this case there is no sense in speaking of its sum.* But if all the u_n have the same sign, then it is customary to say that the sum of the series is equal to infinity with the corresponding sign.

As an example, let us consider the series

$$1 + x + x^2 + \cdots,$$

whose terms form a geometric progression with common ratio x.

The sum of the first n terms is equal to

$$S_n(x) = \frac{1 - x^n}{1 - x} \ (x \neq 1); \tag{55}$$

if $|x| < 1$ this sum has a limit

$$\lim_{n\to\infty} S_n(x) = \frac{1}{1 - x},$$

and so for $|x| < 1$ we may write

$$\frac{1}{1 - x} = 1 + x + x^2 + \cdots.$$

If $|x| > 1$, then obviously

$$\lim_{n\to\infty} S_n(x) = \infty,$$

and the series diverges. The same situation holds also for $x = 1$, as may be seen immediately without use of formula (55), which for $x = 1$ has no meaning. Finally, if $x = -1$ the partial sums take the values $+1$ and 0 alternately, so that this series also is divergent.

---

* Let us note that it is also possible to give generalized definitions of the sum of a series, by virtue of which it is possible to assign to certain divergent series a more or less natural concept of "generalized sum." Such series are said to be summable. Operations with generalized sums of divergent series are sometimes useful.

To each series there corresponds a definite sequence of values of its partial sums $S_1, S_2, S_3, \cdots$ such that the convergence of the series depends on the fact that the sums approach a limit. Conversely, to an arbitrary sequence of numbers $S_1, S_2, S_3, \cdots$ corresponds a series

$$S_1 + (S_2 - S_1) + (S_3 - S_2) + \cdots,$$

the partial sums of which will be the numbers of the sequence. Thus the theory of variables ranging over a sequence may be reduced to the theory of the corresponding series, and conversely. Yet each of these theories has independent significance. In some cases it is more convenient to study the variable directly and in others to consider the equivalent series.

Let us note that series have long served as an important method of representing various entities (above all, functions) and of calculating their value. Of course, the views of mathematicians concerning series have changed with the passage of time, corresponding to the changes in their ideas about infinitesimals. The above clear-cut definition of convergence and divergence of a series was formulated at the beginning of the last century at the same time as the closely associated concept of a limit.

If the series converges, then its general term approaches zero with increasing n, since

$$\lim_{n \to \infty} u_n = \lim_{n \to \infty} (S_{n+1} - S_n) = S - S = 0.$$

From examples given in the following paragraphs, it will be clear that the converse statement is in general false. But the criterion is still a useful one, since it provides a *necessary* condition for the convergence of a series. For example, the divergence of a geometric progression with common ratio $x > 1$ follows immediately from the fact its general term does not approach zero.

If the series consists of positive terms, then its partial sum S_n increases with increasing n and only two cases can exist: Either the variable S_n becomes and remains greater than any preassigned number A for sufficiently large n, in which case $\lim_{n \to \infty} S_n = \infty$, so that the series diverges; or else there exists a number A such that for all n the value of S_n does not exceed A; but then the variable S_n necessarily approaches a definite finite limit not greater than A and the series is convergent.

Convergence of a series. The question whether a given series converges or diverges may often be settled by comparing it with another series. Here it is customary to make use of the following criterion.

If we are given two series

$$u_0 + u_1 + u_2 + \cdots,$$
$$v_0 + v_1 + v_2 + \cdots$$

with positive terms such that for all values of n, beginning with a certain one, we have the inequality

$$u_n \leqslant v_n,$$

then the convergence of the second series implies the convergence of the first, and the divergence of the first implies the divergence of the second.

For example, let us consider the so-called harmonic series

$$1 + \frac{1}{2} + \frac{1}{3} + \frac{1}{4} + \frac{1}{5} + \frac{1}{6} + \frac{1}{7} + \frac{1}{8} + \cdots.$$

Its terms are correspondingly not less than the terms of the series

$$1 + \underline{\frac{1}{2}} + \underline{\frac{1}{4} + \frac{1}{4}} + \underline{\frac{1}{8} + \frac{1}{8} + \frac{1}{8} + \frac{1}{8}} + \underbrace{\frac{1}{16} + \cdots + \frac{1}{16}}_{8 \text{ times}} + \cdots,$$

in which the sum of the underlined terms in each case is equal to $\frac{1}{2}$.

It is clear that the sum S_n of the second series approaches infinity with increasing n, and consequently that the harmonic series diverges.

The series

$$1 + \frac{1}{2^\alpha} + \frac{1}{3^\alpha} + \frac{1}{4^\alpha} + \cdots, \tag{56}$$

where α is a positive number less than unity, also obviously diverges, since for arbitrary n

$$\frac{1}{n^\alpha} > \frac{1}{n} \, (0 < \alpha < 1).$$

On the other hand, it is possible to prove that series (56) for $\alpha > 1$ is convergent. We will prove this here only for the case $\alpha \geqslant 2$; for this purpose we consider the series

$$\left(1 - \frac{1}{2}\right) + \left(\frac{1}{2} - \frac{1}{3}\right) + \cdots + \left(\frac{1}{n-1} - \frac{1}{n}\right) + \cdots$$

with positive terms. It converges to unity as its sum, since its partial sums S_n are equal to

$$S_n = 1 - \frac{1}{n+1} \to 1 \, (n \to \infty).$$

On the other hand, the general term of this series satisfies the inequality

$$\frac{1}{n-1} - \frac{1}{n} = \frac{1}{(n-1)\,n} > \frac{1}{n^2},$$

from which it follows that the series

$$1 + \frac{1}{2^2} + \frac{1}{3^2} + \frac{1}{4^2} + \cdots$$

converges. All the more then will the series (56) converge with $\alpha > 2$.

Let us give here without proof another useful criterion for convergence and divergence of series with positive terms, the so-called criterion of d'Alembert.

Let us suppose that, as n approaches infinity, the ratio $(u_n + 1)/u_n$ has a limit q. Then for $q < 1$ the sequence will certainly converge, while for $q > 1$ it will diverge. But for $q = 1$ the question of its convergence remains open.

The sum of a finite number of summands does not change if we permute the summands. But in general this is no longer true for infinite series. There exist convergent series for which it is possible to permute the terms in such a way as to change their sum and even to turn them into divergent series. Series with unstable sums of this sort fail to possess one of the fundamental properties of ordinary sums, permutability of the summands. So it is important to distinguish those series which preserve this property. It turns out that they are the so-called absolutely convergent series. The series

$$u_0 + u_1 + u_2 + u_3 + \cdots$$

is said to be *absolutely convergent* if the series

$$|\,u_0\,| + |\,u_1\,| + |\,u_2\,| + |\,u_3\,| + \cdots$$

of absolute values of its terms is also convergent. It is possible to prove that an absolutely convergent series is always convergent; in other words, that its partial sums S_n approach a finite limit. It is obvious that every convergent series with terms of one sign is absolutely convergent.

The series

$$\frac{\sin x}{1^2} + \frac{\sin 2x}{2^2} + \frac{\sin 3x}{3^2} + \cdots$$

is an example of an absolutely convergent series, since the terms of the series

$$\left|\,\frac{\sin x}{1^2}\,\right| + \left|\,\frac{\sin 2x}{2^2}\,\right| + \left|\,\frac{\sin 3x}{3^2}\,\right| + \cdots$$

arc not greater than the corresponding terms of the convergent series

$$1 + \frac{1}{2^2} + \frac{1}{3^2} + \cdots.$$

An example of a series which is convergent, but not absolutely convergent, is the following

$$1 - \frac{1}{2} + \frac{1}{3} - \frac{1}{4} + \cdots$$

as the reader may prove for himself.

Series of functions; uniformly convergent series. In analysis we often have to do with series whose terms are functions of x. In the preceding paragraphs we have already had examples of this sort, for instance, the series $1 + x + x^2 + x^3 + \cdots$. For some values of x this series converges, but for others it diverges. Particularly important in applications are series of functions convergent for all values of x belonging to a certain interval, which may in particular be the whole of the real axis or the positive half of it and so forth. Then the necessity arises for differentiating such series term by term, integrating them, deciding whether their sum is continuous, and so forth. For the familiar case of the sum of a finite number of terms, there are simple general rules. We know that the derivative of a sum of differentiable functions is equal to the sum of their derivatives, the integral of a sum of continuous functions is the sum of their integrals, and a sum of continuous functions is itself a continuous function: All this holds for the sum of a finite number of terms.

But for infinite series these simple rules are in general no longer true. We could give many examples of convergent series of functions for which the rules of termwise integration and differentiation are false. In the same way a series of continuous functions may turn out to have a discontinuous sum. On the other hand many infinite series behave like finite sums with respect to these rules.

Profound investigations of this question have shown that these rules may still be applied if the infinite series in question are not only convergent at each separate point of the interval of definition (the domain over which x varies) but if they are *uniformly* convergent over the whole interval. In this way there was crystallized in mathematical analysis, in the middle of the 19th century, the important concept of the uniform convergence of a series.

Let us consider the series

$$S(x) = u_0(x) + u_1(x) + u_2(x) + \cdots,$$

whose terms are functions defined on the interval $[a, b]$. We suppose that for each separate value of x in the interval this series converges to a certain sum $S(x)$. The sum of the first n terms of the series

$$S_n(x) = u_0(x) + u_1(x) + \cdots + u_{n-1}(x)$$

is also a certain function of x, defined on $[a, b]$.

We now introduce a magnitude η_n, which is equal to the least upper bound of the values* $|S(x) - S_n(x)|$, as x varies on the interval $[a, b]$. This magnitude is written as follows†

$$\eta_n = \sup_{a \leqslant x \leqslant b} |S_n(x) - S(x)|.$$

In case the quantity $S(x) - S_n(x)$ attains its maximum value, which will certainly occur for example, when $S(x)$ and $S_n(x)$ are continuous, then η_n is simply the maximum of $|S(x) - S_n(x)|$ on $[a, b]$.

From the assumed convergence of our series, we have for every individual value of x in the interval $[a, b]$

$$\lim_{n \to \infty} |S(x) - S_n(x)| = 0.$$

But the magnitude η_n may approach zero or it may not. If η_n approaches zero as $n \to \infty$, then the series is said to be *uniformly convergent*, and in the opposite case nonuniformly convergent. In the same sense it is possible to speak of the uniform or nonuniform convergence of a sequence of functions $S_n(x)$ without necessarily interpreting them as partial sums of a series.

Example 1.   The series of functions

$$\frac{1}{x + 1} - \frac{1}{(x + 1)(x + 2)} - \frac{1}{(x + 2)(x + 3)} - \cdots,$$

which we take to be defined only for nonnegative values of x, namely on the half line $[0, \infty)$, may be written in the form

$$\frac{1}{x + 1} + \left(\frac{1}{x + 2} - \frac{1}{x + 1}\right) + \left(\frac{1}{x + 3} - \frac{1}{x + 2}\right) + \cdots,$$

from which we see that its partial sums are equal to

$$S_n(x) = \frac{1}{x + n}$$

and

$$\lim_{n \to \infty} S_n(x) = 0.$$

---

* See Chapter XV.

† sup is an abbreviation for the Latin word *supremum* (highest).

Thus the series is convergent for all nonnegative x and has the sum $S(x) = 0$. Furthermore,

$$\eta_n = \sup_{0 \leqslant x < \infty} |S_n(x) - S(x)| = \sup_{0 \leqslant x < \infty} \frac{1}{x+n} = \frac{1}{n} \to 0 \ (n \to \infty),$$

so that the series is uniformly convergent to zero on the half axis $[0, \infty)$. Figure 37 shows the graphs of some of the partial sums $S_n(x)$.

Example 2. The series

$$x + x(x - 1) + x^2(x - 1) + \cdots$$

may be written in the form

$$x + (x^2 - x) + (x^3 - x^2) + \cdots,$$

from which

$$S_n(x) = x^n,$$

and therefore

$$\lim_{n \to \infty} S_n(x) = \begin{cases} 0, \text{ if } 0 \leqslant x < 1; \\ 1, \text{ if } x = 1. \end{cases}$$

Thus the sum of the series is discontinuous on the interval $[0, 1]$ with a discontinuity at the point $x = 1$. The quantity $|S_n(x) - S(x)|$ is less than unity for every x in $[0, 1]$ but for x close to $x = 1$ it is arbitrarily close to unity. So,

$$\eta_n = \sup_{0 \leqslant x \leqslant 1} |S_n(x) - S(x)| = 1$$

for all $n = 1, 2, \cdots$. Thus the series is nonuniformly convergent on the interval $[0, 1]$. Figure 38 shows some of the graphs of the function $S_n(x)$. The graph of the sum of the series consists of the segment $0 \leqslant x < 1$ of the x-axis omitting the right end point and of the point $(1, 1)$.

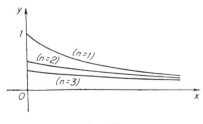

FIG. 37.

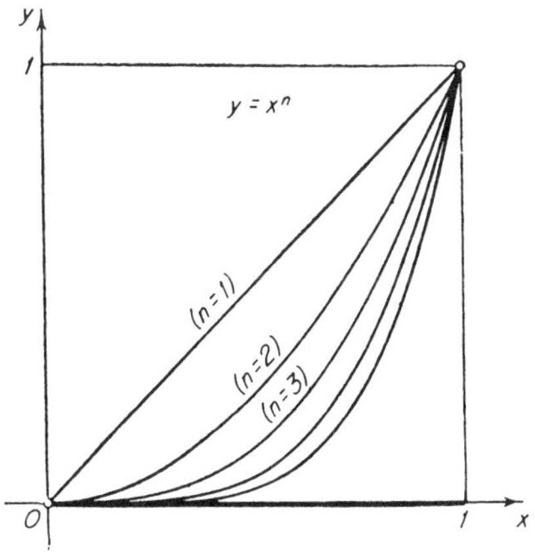

FIG. 38.

This example shows that the sum of a nonuniformly convergent series of continuous functions may in fact be a discontinuous function.

On the other hand, if we consider the series on the interval $0 \leqslant x \leqslant q$ with $q < 1$, then

$$\eta_n = \sup_{0 \leqslant x \leqslant q} |S_n(x) - S(x)| = \max_{0 \leqslant x \leqslant q} x^n = q^n \underset{n \to \infty}{\to} 0,$$

so that on this interval the series converges uniformly and its sum is continuous. The fact that the sum of a uniformly convergent series of continuous functions is itself a continuous function is a general rule, as was pointed out earlier, which can be rigorously proved.

Example 3. The sum of the first n terms of the series $S_n(x)$ has the graph represented by the heavy broken line in figure 39. Obviously, for all n we have $S_n(0) = 0$, but if $0 < x \leqslant 1$, then for all $n \geqslant 1/x$, we will have $S_n(x) = 0$, and consequently for arbitrary x in the interval $[0, 1]$,

$$S(x) = \lim_{n \to \infty} S_n(x) = 0.$$

On the other hand,

$$\eta_n = \sup_{0 \leqslant x \leqslant 1} |S_n(x) - S(x)| = \sup |S_n(x)| = n^2.$$

So the quantity η_n does not approach zero but even approaches infinity. We now note that the series corresponding to this sequence $S_n(x)$ cannot be integrated term by term on the interval $[0, 1]$, since

$$\int_0^1 S(x)\,dx = 0, \int_0^1 S_n(x)\,dx = \frac{1}{2}n^2\frac{1}{n} = \frac{n}{2},$$

so that the series

$$\int_0^1 S_1(x)\,dx + \int_0^1 [S_2(x) - S_1(x)]\,dx + \int_0^1 [S_3(x) - S_2(x)]\,dx + \cdots$$

reduces to the divergent series

$$\frac{1}{2} + \left(\frac{2}{2} - \frac{1}{2}\right) + \left(\frac{3}{2} - \frac{2}{2}\right) + \left(\frac{4}{2} - \frac{3}{2}\right) + \cdots.$$

Let us state without proof the fundamental properties of uniformly convergent series:

1. The sum of a series of continuous functions which is uniformly convergent on the interval $[a, b]$ is a continuous function on this interval.

2. If the series of continuous functions

$$S(x) = u_0(x) + u_1(x) + u_2(x) + \cdots \tag{57}$$

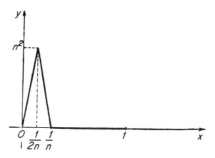

FIG. 39.

converges uniformly on the interval $[a, b]$, then it may be integrated term by term on this interval; that is, for all x_1, x_2 in $[a, b]$ we have the equality

$$\int_{x_1}^{x_2} S(t)\, dt = \int_{x_1}^{x_2} u_0(t)\, dt + \int_{x_1}^{x_2} u_1(t)\, dt + \cdots.$$

3. If on the interval $[a, b]$ the series (57) converges and the functions $u_k(x)$ have continuous derivatives, then the equality

$$S'(x) = u_0'(x) + u_1'(x) + u_2'(x) + \cdots, \tag{58}$$

obtained by termwise differentiation of (57) will be valid on the interval $[a, b]$ if the series on the right in (58) converges uniformly.

Power series. In §9, a function $f(x)$ defined on an interval $[a, b]$ was called analytic, if on this interval it has derivatives of arbitrary order and if in a sufficiently small neighborhood of any point x_0 of the interval $[a, b]$ it may be expanded in a convergent Taylor series

$$f(x) = f(x_0) + \frac{f'(x_0)}{1}(x - x_0) + \frac{f''(x_0)}{2!}(x - x_0)^2 + \cdots. \tag{59}$$

If we introduce the notation

$$a_n = \frac{f^{(n)}(x_0)}{n!},$$

this series may be written in the following form

$$f(x) = a_0 + a_1(x - x_0) + a_2(x - x_0)^2 + \cdots. \tag{60}$$

A series of this sort, where the numbers a_1, a_2, $\cdots$ are constants independent of x, is called a *power series*.

As an example let us consider the power series

$$1 + x + x^2 + x^3 + \cdots, \qquad (61)$$

whose terms form a geometric progression.

We know that for all values of x in the interval $-1 < x < 1$ this series converges and its sum is equal to

$$S(x) = \frac{1}{1-x}.$$

For other values of x the series diverges.

It is also easy to see that the difference between the sum of the series and the sum of its first n terms is given by the formula

$$S(x) - S_n(x) = \frac{x^n}{1-x}, \qquad (62)$$

and if $-q \leqslant x \leqslant q$, where q is a positive number less than unity, then

$$\eta_n = \max |S(x) - S_n(x)| = \frac{q^n}{1-q}.$$

From this it is clear that η_n approaches zero with increasing n so that the series is uniformly convergent on the interval $-q \leqslant x \leqslant q$, for all positive values of $q < 1$.

It is easy to verify that the function

$$S(x) = \frac{1}{1-x}$$

has a derivative of nth order, which is equal to

$$S^{(n)}(x) = \frac{n!}{(1-x)^{n+1}},$$

from which

$$S^{(n)}(0) = n!$$

and the sum of the first n terms of the Taylor series for the function $S(x)$ exactly coincides for $x_0 = 0$ with the sum of the first n terms of the series (59). Moreover, we know that the remainder term of the formula, given by the equality (62), approaches zero with increasing n, for all x

on thc interval $-1 < x < 1$. Thus we have shown that the series (61) is the Taylor series of its sum $S(x)$.

Let us note one further fact. From the interval of convergence $-1 < x < 1$ of our series, let us choose an arbitrary point x_0. It is easy to see that for all x sufficiently close to x_0, namely for all x satisfying the inequality

$$\frac{|x - x_0|}{1 - x_0} < 1,$$

we have the equality

$$
\begin{aligned}
S(x) &= \frac{1}{1 - x} = \frac{1}{1 - x_0} \frac{1}{\left(1 - \dfrac{x - x_0}{1 - x_0}\right)} \\
&= \frac{1}{1 - x_0}\left[1 + \frac{x - x_0}{1 - x_0} + \left(\frac{x - x_0}{1 - x_0}\right)^2 + \cdots\right] \\
&= \frac{1}{1 - x_0} + \frac{x - x_0}{(1 - x_0)^2} + \frac{(x - x_0)^2}{(1 - x_0)^3} + \cdots. \tag{63}
\end{aligned}
$$

The reader may prove without difficulty that

$$\frac{S^{(n)}(x_0)}{n!} = \frac{1}{(1 - x_0)^{n+1}}.$$

Consequently, series (63) is the Taylor series of its sum $S(x)$ and converges to it in a sufficiently small neighborhood of any point x_0 belonging to the interval of convergence of (61). Since the point x_0 is arbitrary, this means that the function $S(x)$ is analytic on the interval.

All these facts that we have observed for the particular power series (61) hold for arbitrary power series.* Namely, for every power series of the form (60) where the constants a_k are chosen by any given law, there exists a certain nonnegative number R (which may also be infinite), called the *radius of convergence of the series* (60), with the following properties:

1. For all values of x from the interval $x_0 - R < x < x_0 + R$, which is called its interval of convergence, the series converges and its sum $S(x)$ is an analytic function of x in its interval. Here the convergence is uniform for every interval $[a, b]$ lying completely within the interval of convergence. The series itself is the Taylor series of its sum.

2. At the end points of the interval of convergence, the series may converge or diverge, depending on its individual character. But it will certainly diverge outside the closed interval $x_0 - R \leqslant x \leqslant x_0 + R$.

---

* For more detailed information on this point see Chapter IX.

We suggest to the reader that he consider the power series

$$1 + \frac{x}{1} + \frac{x^2}{2!} + \frac{x^3}{3!} + \cdots,$$

$$1 + x + 2!x^2 + 3!x^3 + \cdots,$$

$$1 + x + \frac{x^2}{2} + \frac{x^3}{3} + \cdots$$

and convince himself that their radii of convergence are respectively infinity, zero, and unity.

By the definition given earlier every analytic function may be expanded, in a sufficiently small neighborhood of an arbitrary point where it is defined, into a power series which converges to the function. Conversely, from what has been said it follows that the sum of every power series whose radius of convergence is not zero is an analytic function in its interval of convergence.

So we see that power series are organically connected with analytic functions. We could even say that on their interval of convergence power series are the natural means of representing analytic functions, and consequently they are also the natural means of approximating analytic functions by algebraic polynomials.*

For example, from the fact that the function $1/(1 - x)$ may be expanded in the power series

$$\frac{1}{1 - x} = 1 + x + x^2 + x^3 + \cdots,$$

which is convergent on the interval $-1 < x < 1$, it follows that the power series is uniformly convergent on an arbitrary interval $-a \leqslant x \leqslant a$ with $a < 1$, and this implies the possibility of approximating the function on the whole interval $[-a, a]$ by the partial sums of the series with any preassigned degree of accuracy.

Let us suppose that we are required to approximate the function $1/(1 - x)$ by polynomials on the interval $[-\frac{1}{2}, \frac{1}{2}]$ with an accuracy of 0.01. We note that for all x in this interval we have the inequality

$$\left| \frac{1}{1 - x} - 1 - x - \cdots - x^n \right| = \left| x^{n+1} + x^{n+2} + \cdots \right|$$

$$\leqslant |x|^{n+1} + |x|^{n+2} + \cdots \leqslant \frac{1}{2^{n+1}} + \frac{1}{2^{n+2}} + \cdots = \frac{1}{2^n},$$

---

* Approximations going beyond the limits of the interval of convergence of a power series require other methods. (See Chapter XII.)

and since $2^6 = 64$, and $2^7 = 128$, the desired polynomial, approximating the function on the whole interval $[-\frac{1}{2}, \frac{1}{2}]$ with an accuracy of 0.01, will have the form

$$\frac{1}{1-x} \approx 1 + x + x^2 + \cdots + x^7.$$

Let us note one further extremely valuable property of power series: They may be differentiated termwise everywhere in the interval of convergence. This property finds extremely wide application in the solution of various problems in mathematics.

For example, let it be required to find the solution of the differential equation $y' = y$ under the auxiliary condition $y(0) = 1$. We will seek the solution in the form of a power series,

$$y = a_0 + a_1 x + a_2 x^2 + \cdots.$$

Because of the auxiliary condition, we must set $a_0 = 1$. Assuming that this series converges, we may differentiate it termwise; as a result we obtain

$$y' = a_1 + 2a_2 x + 3a_3 x^2 + \cdots.$$

If we substitute these two series into the differential equation and equate coefficients for each of the powers of x, we obtain

$$a_k = \frac{1}{k!} \ (k = 1, 2, \cdots)$$

and the desired solution has the form

$$y = 1 + \frac{x}{1} + \frac{x^2}{2!} + \frac{x^3}{3!} + \cdots.$$

It is well known that this series converges for all values of x and that its sum is equal to $y = e^x$.

In this case we have obtained a series whose sum is a well-known elementary function. But this does not always happen; it may turn out that a convergent power series so obtained has a sum that is not an elementary function. An example is the series

$$y_p(x) = x^p \left[1 - \frac{x^2}{2(2p+2)} + \frac{x^4}{2 \cdot 4(2p+2)(2p+4)} - \cdots \right],$$

obtained as a solution of Bessel's differential equation, which is of great importance in applications. In this way power series may serve to define functions.

Suggested Reading

R. Courant, *Differential and integral calculus*, 2 vols., Interscience, New York, 1938.

A. Dresden, *Introduction to the calculus*, Henry Holt, New York, 1940.

F. Klein, *Elementary mathematics from an advanced standpoint. Arithmetic—algebra—analysis*, Dover, New York, 1953.

K. Knopp, *Infinite sequences and series*, Dover, New York, 1956.

PART 2

CHAPTER III

ANALYTIC GEOMETRY

§1. Introduction

In the first half of the 17th century a completely new branch of mathematics arose, the so-called analytic geometry, establishing a connection between curves in a plane and algebraic equations in two unknowns.

A quite rare event thereby happened in mathematics: In one or two decades there appeared a great, entirely new branch of mathematics based on a very simple concept, which until then had not received proper attention. The appearance of analytic geometry in the first half of the 17th century was not accidental. The transition in Europe to the new capitalistic methods of manufacture required the advance of a whole series of sciences. A short time before, contemporary mechanics was being created by Galileo and other scientists, experimental data were being accumulated in all regions of natural science, the means of observation were being perfected, and instead of absolute scholastic theories new ones were being created. In astronomy, among the foremost scientists the teachings of Copernicus had finally triumphed. The rapid development of long-range navigation insistently called for knowledge of astronomy and the elements of mechanics.

The art of warfare also required mechanics. Ellipses and parabolas, whose geometric properties as conic sections were already well known in detail to the ancient Greeks almost 2000 years earlier, ceased to be only part of geometry, as they were to the Greeks. After Kepler had discovered that the planets revolve around the sun in ellipses, and Galileo that a stone thrown into the air traces out a parabola, it was necessary to calculate these ellipses and to find the parabolas along which bullets fly from a gun; it was necessary to discover the law by which the atmospheric pressure, discovered by Pascal, decreases with the height; it was necessary actually to calculate the volumes of various bodies, and so forth.

All these questions almost simultaneously called to life three entirely new mathematical sciences: analytic geometry, differential calculus, and integral calculus, including the solution of the simplest differential equations.

These three new fields qualitatively changed the face of the whole of mathematics. They made it possible to solve problems never even dreamed of before.

In the first half of the 17th century, i.e., at the beginning of the 1600's, a group of the most outstanding mathematicians was already close to the idea of analytic geometry, but there were two of them, in particular, who understood clearly the possibility of creating a new branch of mathematics. These were Pierre Fermat, a counsellor of the parliament of the French city of Toulouse and a world-famous mathematician, and the famous French philosopher René Descartes. Descartes is credited with being the chief creator of analytic geometry. He was the one who, as a philosopher, raised the question of its complete generality. Descartes published the great philosophical treatise "Discourse on the method of rightly conducting the reason and seeking the truth in the sciences, with applications: dioptrics, meteorology and geometry."

The last part of this work, entitled "Geometry" and published in 1637, contains a sufficiently complete, although somewhat confusing, presentation of the mathematical theory that since then has been called analytic geometry.

§2. Descartes' Two Fundamental Concepts

Descartes wished to create a method that could equally well be applied to the solution of all problems of geometry, that is, which would provide a general method for their solution. Descartes' theory is based on two concepts: the concept of coordinates and the concept of representing by the coordinate method any algebraic equation with two unknowns in the form of a curve in the plane.

The concept of coordinates.   By the *coordinates of a point* in the plane Descartes means the abscissa and ordinate of this point, i.e., the numerical values x and y of its distances (with corresponding signs) to two mutually perpendicular straight lines (coordinate axes) chosen in this plane (see Chapter II). The point of intersection of the coordinate axes, i.e., the point having coordinates (0, 0) is called the *origin*.

With the introduction of coordinates Descartes constructed, so to speak, an "arithmetization" of the plane. Instead of determining any point geometrically, it is sufficient to give a pair of numbers x, y and conversely (figure 1).

The notion of comparison of equations with two unknowns with curves in the plane. Descartes' second concept is the following. Up to the time of Descartes, where an algebraic equation in two unknowns $F(x, y) = 0$ was given, it was said that the problem was indeterminate, since from the equation it was impossible to determine these unknowns; any value could be assigned to one of them, for example to x, and substituted in the equation; the result was an equation with only one unknown y, for which, in general, the equation could be solved. Then this *arbitrarily chosen x* together with the *so-obtained y* would satisfy the given equation. Consequently, such an "indeterminate" equation was not considered interesting.

Descartes looked at the matter differently. He proposed that in an equation with two unknowns x be regarded as the abscissa of a point and the corresponding y as its ordinate. Then if we vary the unknown x, to every value of x the corresponding y is computed from the equation, so that we obtain, in general, a set of points which form a curve (figure 2).*

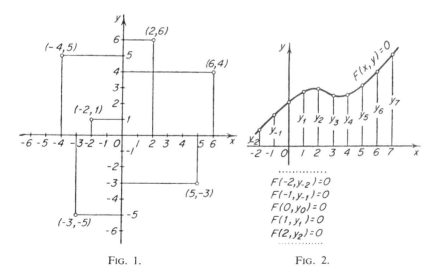

FIG. 1. FIG. 2.

---

* Sometimes, the equation is not satisfied by any point (x, y) with real coordinates, sometimes by one or a few such points. In this case we say that the curve is imaginary or reduces to points (see §7).

Thus, to each algebraic equation with two variables, $F(x, y) = 0$, corresponds a completely determined curve of the plane, namely a curve representing the totality of all those points of the plane whose coordinates satisfy the equation $F(x, y) = 0$.

This observation of Descartes *opened up an entire new science.*

The basic problems solved by analytic geometry and the definition of analytic geometry.   Analytic geometry provides the possibility: (1) of solving construction problems by computation (see for example, the division of a segment in a given ratio, see §3); (2) of finding the equation of curves defined by a geometric property (for example, of an ellipse defined by the condition that the sum of distances to two given points is constant, see §7); (3) of proving new geometric theorems algebraically (see, for example, the derivation of Newton's theory of diameters, §6); (4) conversely, of representing an algebraic equation geometrically, to clarify its algebraic properties (see, for example, the solution of third- and fourth-degree equations from the intersection of a parabola with a circle, §5).

Thus, analytic geometry is that part of mathematics which, applying the coordinate method, investigates geometric objects by algebraic means.

§3. Elementary Problems

The coordinates of a point that divide a segment in a given ratio.   Given the coordinates (x_1, y_1) and (x_2, y_2) of two points M_1 and M_2, let us find the coordinates (x, y) of the point M dividing the segment M_1M_2 in the ratio m to n (figure 3). From the similarity of the shaded triangles we obtain:

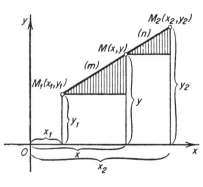

FIG. 3.

$$\frac{x - x_1}{x_2 - x} = \frac{m}{n} \text{ , from which}$$

$$x = \frac{nx_1 + mx_2}{m + n} \text{ ,}$$

$$\frac{y - y_1}{y_2 - y} = \frac{m}{n} \text{ , from which}$$

$$y = \frac{ny_1 + my_2}{m + n} \text{ .}$$

Distance between two points. Let us find the distance between the points M_1 and M_2, whose coordinates are (x_1, y_1) and

(x_2, y_2) respectively. From the shaded right triangle (figure 4), we obtain by the theorem of Pythagoras

$$d = \sqrt{(x_2 - x_1)^2 + (y_2 - y_1)^2}.$$

The area of a triangle. Let us find the area S of the triangle $M_1 M_2 M_3$ (figure 5) if the coordinates of its vertices are respectively (x_1, y_1), (x_2, y_2), (x_3, y_3). Considering the area of the triangle as the sum of the areas of trapezoids with bases y_1, y_3 and y_3, y_2 minus the area of the trapezoid with bases y_1, y_2 and writing the product $-(y_1 + y_2)(x_2 - x_1)$ in the form $(y_1 + y_2)(x_1 - x_2)$, we obtain

$$S = \tfrac{1}{2}\,[(y_1 + y_2)(x_1 - x_2) + (y_2 + y_3)(x_2 - x_3) + (y_3 + y_1)(x_3 - x_1)].$$

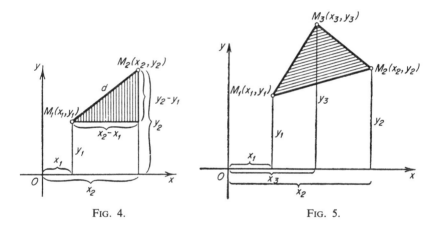

FIG. 4.   FIG. 5.

In these problems it only remains to verify that the derived formulas remain valid without any change in those cases when one or more coordinates or their differences are negative. Such verification easily follows.

Determination of the points of intersection of two curves. Relying on the second fundamental idea that the equation $F(x, y) = 0$ represents a curve, it is particularly simple to find the points of intersection of two curves. In order to find the coordinates of the points of intersection of two curves, it is obviously necessary to solve simultaneously the equations that represent them. The pair of numbers x, y obtained from the ordinary solution of these two equations will determine the point whose coordinates satisfy both of the equations, i.e., the point that lies on the first as well as on the second curve, and this is the point of their intersection.

The solution of geometric problems by the tools of analytic geometry, as we see, is very convenient for practical purposes, especially because every solution is at once obtained in the convenient form of numbers. *Such a geometry, such a science, was exactly what was lacking at that time.*

§4. Discussion of Curves Represented by First- and Second-Degree Equations

First degree equation.   Making use of his second idea, Descartes first of all examined what curves correspond to an equation of the first-degree,

$$Ax + By + C = 0, \tag{1}$$

i.e., to an equation where A, B, C are numerical coefficients with A and B not both zero. It turned out that in the plane a straight line always corresponds to such an equation.

We shall prove that equation (1) always represents a straight line, and conversely, that to every line in the plane there corresponds a completely determined equation of the form (1). In fact, let us suppose, for example, that $B \neq 0$; then equation (1) can be solved for y

$$y = kx + l,$$

where $k = -\dfrac{A}{B}$; $l = -\dfrac{C}{B}$.

We examine first the equation $y = kx$. It obviously represents a straight line passing through the origin and making an angle ϕ with the x-axis whose tangent $\tan \phi$ is k (figure 6). Indeed, the equation can be

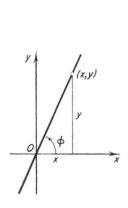

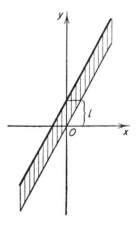

FIG. 6.   FIG. 7

written as $y/x - k$, so that the coordinates of every point (x, y) on the straight line satisfy the equation, and the coordinates of no point $(x, \bar{y})$ not lying on the straight line satisfy the equation, since for such a point y/x will be either greater than or smaller than k. In addition, if $\tan \phi > 0$, then for this line either both x and y are positive or both negative, and if $\tan \phi < 0$ their signs are opposite.

Thus, the equation $y = kx$ represents a straight line passing through the origin O, and consequently the equation $y = kx + l$ also represents a line, namely the one which is obtained from the previous line by the parallel translation such that the ordinate of each of its points is increased by l (figure 7).

The earlier derived formulas of the coordinates of a point dividing a segment in a given ratio, the distance between two given points, and the area of a triangle as well as the information about the equation of a straight line already enable us to solve a large number of problems.

The equation of a straight line passing through one or two given points.
Let M_1 be the point with coordinates x_1, y_1 and let k be a given number. The equation $y = kx + l$ represents a straight line making with the Ox-axis an angle whose tangent is equal to k and intersecting the Oy-axis at a distance l from O. Let us choose l such that this line goes through the point (x_1, y_1). For this, the coordinates of the point M_1 must satisfy the equation, i.e., we must have $y_1 = kx_1 + l$, from which $l = y_1 - kx_1$.

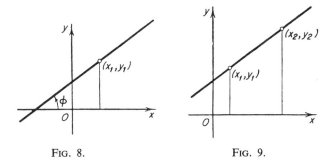

FIG. 8.   FIG. 9.

Substituting this value for l, we obtain the equation of the line that passes through the given point (x_1, y_1) and makes with the Ox-axis an angle whose tangent is equal to k (figure 8). This equation is $y = kx + y_1 - kx_1$ or

$$y - y_1 = k(x - x_1).$$

Example. Let the angle between the line and the Ox-axis be equal to 45°, and let the point M have coordinates $(3, 7)$; then the equation of the corresponding line (since $\tan 45° = 1$) will be: $y - 7 = 1 \cdot (x - 3)$ or $x - y + 4 = 0$.

If we require that the line passing through the point (x_1, y_1) also go through the point (x_2, y_2), it follows that the condition $y_2 - y_1 = k(x_2 - x_1)$ must be imposed on k. Finding k from this and substituting it in the previous equation, we obtain the equation of the line passing through two given points (figure 9):

$$\frac{x - x_1}{x_2 - x_1} = \frac{y - y_1}{y_2 - y_1} .$$

Descartes' result concerning second-degree equations. Descartes also investigated the question as to what kinds of curves in the plane are represented by the second-degree equation with two variables whose general form is

$$Ax^2 + Bxy + Cy^2 + Dx + Ey + F = 0,$$

and showed that such an equation, generally speaking, represents an ellipse, a hyperbola, or a parabola; i.e., curves very well known to the mathematicians of antiquity.

These are Descartes' most important achievements. However, his book was far from being restricted to these topics; he also investigated the equations of a number of interesting geometric loci, examined certain theorems on transformation of algebraic equations, mentioned without proof his famous *law of signs* for the number of positive roots of an equation whose roots are all real (see Chapter IV, §4) and, finally, presented a remarkable method for determining the real roots of third- and forth-degree equations from the intersection of the parabola $y = x^2$ with circles.

§5. Descartes' Method of Solving Third- and Fourth-Degree Algebraic Equations

Transformation of third- and fourth-degree equations to an equation of the fourth-degree not involving the x^3-term. We will show that the solution of an arbitrary third- or fourth-degree equation can be reduced to the solution of an equation of the form

$$x^4 + px^2 + qx + r = 0. \tag{2}$$

Let the given third-degree equation be $z^3 + az^2 + bz + c = 0$. Substituting $z = x - a/3$, we obtain

$$(x - a/3)^3 + a(x - a/3)^2 + b(x - a/3) + c = 0.$$

The x^2-terms in the expansion of the parentheses will cancel out, so that we get an equation of the form $x^3 + px + q = 0$. Multiplying this equation by x, we bring it to the form (2) with $r = 0$, which also admits a root $x_4 = 0$.

An equation of the fourth-degree $z^4 + az^3 + bz^2 + cz + d = 0$ can be reduced to the form (2) by the substitution $z = x - a/4$. Hence, the solution of all third- and fourth-degree equations can be reduced to the solution of an equation of the form (2).

The solution of third- and fourth-degree equations by the intersection of a circle with the parabola $y = x^2$. Let us first derive the equation of a circle with center (a, b) and radius R. If (x, y) is any of its points, then the square of its distance to the point (a, b) is equal to $(x - a)^2 + (y - b)^2$ (see §3). Thus, the equation of the circle in question is

$$(x - a)^2 + (y - b)^2 = R^2.$$

Now we try to find the points of intersection of this circle with the parabola $y = x^2$. In order to do this, by virtue of what was said in §3, it is necessary to solve simultaneously the equation of this circle

$$x^2 + y^2 - 2ax - 2by + a^2 + b^2 - R^2 = 0$$

and the equation of the parabola

$$y = x^2.$$

Substituting y from the second equation into the first, we obtain a fourth-degree equation in x:

$$x^2 + x^4 - 2ax - 2bx^2 + a^2 + b^2 - R^2 = 0$$

or

$$x^4 + (1 - 2b) x^2 - 2ax + a^2 + b^2 - R^2 = 0.$$

If we choose a, b and R^2 such that

$$1 - 2b = p, \ -2a = q, a^2 + b^2 - R^2 = r,$$

then exactly equation (2) is obtained. For this purpose we have to take

$$a = -\frac{q}{2}, b = \frac{1-p}{2}, R^2 = \frac{q^2}{4} + \frac{(1-p)^2}{4} - r. \tag{3}$$

In the last formula (3), generally speaking, R^2 may turn out to be negative. However, in the case when equation (2) has even one real root x_1, the following equality holds

$$x_1^4 + (1 - 2b) x_1^2 - 2ax_1 + a^2 + b^2 - R^2 = 0. \qquad (4)$$

Denoting x_1^2 by y_1, equation (4) can be rewritten as

$$x_1^2 + y_1^2 - 2ax_1 - 2by_1 + a^2 + b^2 - R^2 = 0$$

or as

$$(x_1 - a)^2 + (y_1 - b)^2 = R^2.$$

Hence, in the case when equation (2) has a real root, the number $R^2 = [(1 - p)^2 + q^2]/4 - r$ is *positive*, the equation

$$(x - a)^2 + (y - b)^2 = R^2$$

is the equation of a circle, and all real roots of equation (2) are the abscissas of points of intersection of the parabola $y = x^2$ with this circle. (In case $r = 0$, $R^2 = a^2 + b^2$, this circle passes through the origin.)

Thus, if the coefficients p, q, r of equation (2) are given, and it is necessary to find a, b and R^2 by formulas (3), then if $R^2 < 0$, equation (2) is known to have no real roots. But, if $R^2 \geqslant 0$ then the abscissas of the points of intersection of the circle with center (a, b) and radius R with the parabola $y = x^2$ (drawn once and for all) give all the real roots of equation (2); and also in case $R^2 < 0$, the resulting circle cannot intersect the parabola and equation (2) does not have real roots.

Example. Let the given fourth-degree equation be:

$$x^4 - 4x^2 + x + \frac{5}{2} = 0.$$

Then we have

$$a = -\frac{1}{2}, b = \frac{5}{2} = 2\frac{1}{2},$$

$$R = \sqrt{\frac{1}{4} + \frac{25}{4} - \frac{5}{2}} = 2.$$

Figure 10 shows the corresponding circle and the roots x_1, x_2, x_3, x_4 of the given equation.

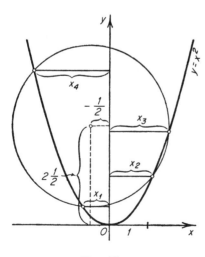

FIG. 10.

The 1st, 2nd, 3rd and 4th sections above contain, in an abbreviated and somewhat more modern form, the essential content of Descartes' book.

From Descartes' time up to the present, analytic geometry has undergone an immense development that has been very fruitful for many different parts of mathematics. We will attempt in the following sections of this chapter to trace the most important stages of this development.

First of all, it is necessary to say that the inventors of the infinitesimal analysis were already in possession of Descartes' method. Whether it was a question of tangents or normals (perpendiculars to the tangents at the point of tangency) to curves, or of maxima or minima of functions considered geometrically, or of the radius of curvature of a curve at a given point, etc., the equation of the curve was considered first, by the method of Descartes, and then the equations of the normal, the tangent, and so forth, were found. Thus infinitesimal analysis, namely the differential and integral calculus, would have been inconceivable without the preliminary development of analytic geometry.

§6. Newton's General Theory of Diameters

The first mathematician to take a further great step forward in analytic geometry itself was Newton. In 1704 he examined the theory of third-order curves, i.e., curves which are represented by third-degree algebraic equations in two unknowns. At the same time he found, among other things, an elegant general theorem about "diameters," which correspond to secants in a given direction. He proved the following.

Let an nth-order curve be given, i.e., a curve which is represented by an nth-degree algebraic equation in two unknowns; then an arbitrary straight line intersecting it has in general n common points with it. Let M be the point of the secant that is the "center of gravity" of these points of its intersec-

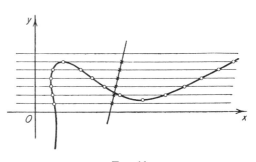

Fɪɢ. 11.

tion with the given nth-order curve, i.e., the center of gravity of a set of n equal point masses situated at these points. It turns out that if we take all possible sets of mutually parallel secants and for each of them consider these centers of mass M, then for any given set of parallel secants all the points M lie on a straight line. Newton called this line the "diameter" of the nth-order curve corresponding to the given direction of the secants. Since the proof of this theorem is quite easy with the help of analytic geometry, we will give it here.

Let an nth-order curve be given and some set of mutually parallel secants of the curve. Choose the coordinate axes so that these secants are parallel to the Ox-axis (figure 11). Then their equations will have the form $y = l$, where the constant l will be different for different secants. Let $F(x, y) = 0$ be the equation that represents the nth-order curve with respect to these coordinate axes. It is easy to show that under a transformation from one rectangular coordinate system to another, although the equation of the curve changes, its order does not change (this will be shown in §8). Therefore $F(x, y)$ will also be an nth-degree polynomial. To determine the abscissas of the points of intersection of the curve with the secant $y = l$, it is necessary to solve the simultaneous equations $F(x, y) = 0$ and $y = l$; as a result, in general, an nth-degree equation in x is obtained

$$F(x, l) = 0, \qquad\qquad (5)$$

from which we find the abscissas $x_1, x_2, \cdots, x_n$. The abscissa x_c of the center of gravity of the n points of intersection is equal, by the very definition of center of gravity, to

$$x_c = \frac{x_1 + x_2 + \cdots + x_n}{n} .$$

But, as is known from the theory of algebraic equations, the sum $x_1 + x_2 + \cdots + x_n$ of the roots of an equation is equal to the coefficient of the $(n-1)$th power of the unknown x, taken with the opposite sign, divided by the coefficient of the nth power of x. But because the sum of the exponents of x and y in every term of $F(x, y)$ is equal to or less than n, the term in x^n does not contain y at all but has the form Ax^n, where A is a constant; and if the terms in x^{n-1} contain y, they do so to no higher than the first power; i.e., they have the form $x^{n-1}(By + C)$. Consequently, the coefficient of x^n is A and that of x^{n-1} is $Bl + C$, and we have for any given l

$$x_c = -\frac{Bl + C}{nA} .$$

But the secant is parallel to the Ox-axis so that for all of its points $y = l$, and hence the ordinate y of the center of gravity of the points of its intersection with the given nth-order curve is also equal to l; thus finally we obtain $nAx_c + By_c + C = 0$, i.e., the coordinates x_c, y_c of the centers of gravity for all these secants satisfy a first-degree equation, and consequently lie on a straight line.

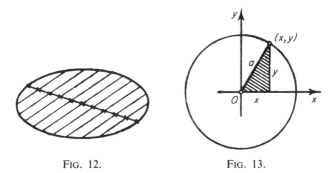

FIG. 12.   FIG. 13.

The case when $F(x, y)$ does not contain x^n can be investigated analogously.

In case the curve is of the 2nd order ($n = 2$) the center of gravity of two points is simply the midpoint between them, so that the locus of midpoints of parallel chords of a second-order curve is a straight line (figure 12), a result that for the ellipse, as well as for the hyperbola and the parabola, was already well known to the ancients. But this was proved by them, even though only for these partial cases, with quite difficult geometric arguments, and here a new general theorem, unknown to the ancients, is proved in an entirely simple way.

Such examples reveal the power of analytic geometry.

§7. Ellipse, Hyperbola, and Parabola

In this and the following sections, we consider second-order curves. Before investigating the general second-degree equation, it is useful to examine some of its simplest forms.

The equation of a circle with center at the origin. First of all, we consider the equation

$$x^2 + y^2 = a^2.$$

It evidently represents a circle with center at the origin and radius a, as follows from the theorem of Pythagoras applied to the shaded right triangle (figure 13), since whatever point (x, y) of this circle is taken, its x and y coordinates satisfy this equation, and conversely, if the coordinates x, y of a point satisfy the equation, then the point belongs to the circle; i.e., the circle is the set of all those points of the plane that satisfy the equation.

The equation of an ellipse and its focal property. Let two points F_1 and F_2 be given, the distance between which is equal to $2c$. We will find the equation of the locus of all points M of the plane; the sum of whose distances to the points F_1 and F_2 is equal to a constant $2a$ (where, of course, a is greater than c). Such a curve is called an *ellipse* and the points F_1 and F_2 are its *foci*.

Let us choose a rectangular coordinate system such that the points F_1 and F_2 lie on the Ox-axis and the origin is halfway between them. Then the coordinates of the points F_1 and F_2 will be $(c, 0)$ and $(- c, 0)$. Let us take an arbitrary point M with coordinates (x, y), belonging to the locus in question, and let us write that the sum of its distances to the points F_1 and F_2 is equal to $2a$,

$$\sqrt{(x - c)^2 + (y - 0)^2} + \sqrt{(x + c)^2 + (y - 0)^2} = 2a. \tag{6}$$

This equation is satisfied by the coordinates (x, y) of any point of the locus under consideration. Obviously the converse is also true, namely that any point whose coordinates satisfy equation (6) belongs to this locus. Equation (6) is therefore the equation of the locus. It remains to simplify it.

Raising both sides to the second power, we obtain

$$x^2 - 2cx + c^2 + y^2 + 2\sqrt{(x^2 - 2cx + c^2 + y^2)(x^2 + 2cx + c^2 + y^2)} \\ + x^2 + 2cx + c^2 + y^2 = 4a^2,$$

or after simplification

$$x^2 + y^2 + c^2 - 2a^2 = -\sqrt{(x^2 + y^2 + c^2)^2 - 4c^2x^2}.$$

Squaring again both sides, we obtain

$$(x^2 + y^2 + c^2)^2 - 4a^2(x^2 + y^2 + c^2) + 4a^4 = (x^2 + y^2 + c^2)^2 - 4c^2x^2$$

or after simplification

$$(a^2 - c^2)x^2 + a^2y^2 = (a^2 - c^2)a^2.$$

Let us set $a^2 - c^2 = b^2$ (as may be done since $a > c$); then we obtain $b^2x^2 + a^2y^2 = a^2b^2$, and dividing by a^2b^2 we have

$$\frac{x^2}{a^2} + \frac{y^2}{b^2} = 1. \tag{7}$$

The coordinates (x, y) of any point M of the locus thus satisfy equation (7).

It can be shown on the other hand that if the coordinates of a point satisfy equation (7) then they also satisfy equation (6). Consequently, equation (7) is the equation of this locus, i.e., the equation of the ellipse (figure 14).

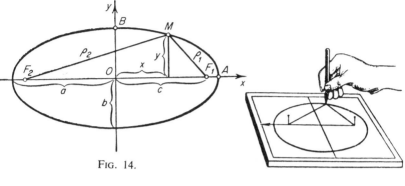

FIG. 14.

FIG. 15.

This argument is a classical example of finding the equation of a curve given by some of its geometrical properties.

The well-known method of tracing an ellipse by means of a thread (figure 15) is based on the property of the ellipse that the sum of the distances of any of its points to two given points is a constant.

Remark. In order to determine an ellipse, we could have taken, instead of the focal property considered here, any other geometric property characteristic of it, for example, that the ellipse is the result of a "uniform contraction" of a circle toward one of its diameters or any other property.

Substituting $y = 0$ in equation (7) of the ellipse, we obtain $x = \pm a$, i.e., a is the length of the segment OA (see figure 14), which is called the *major semiaxis* of the ellipse. Analogously, substituting $x = 0$, we obtain $y = \pm b$, i.e., b is the length of the segment OB, which is called the *minor semiaxis* of the ellipse.

The number $\epsilon = c/a$ is called the *eccentricity* of the ellipse, so that, since $c = \sqrt{a^2 - b^2} < a$, the eccentricity of an ellipse is less than 1. In the case of a circle, $c = 0$ and consequently $\epsilon = 0$; both foci are at one point, the center of the circle (since $OF_1 = OF_2 = 0$), but the previous method of drawing the curve with a thread is still valid.

Laws of planetary motion. In studying Tycho Brahe's long-continued observations on the motion of the planet Mars, Kepler discovered that the planets revolve around the Sun in ellipses such that the Sun occupies one focus of the ellipse (the other focus remains unoccupied and plays no part in the motion of a planet around the Sun) (figure 16). Kepler also observed that the focal radius ρ in equal times sweeps out sectors of equal area,* and Newton showed that the necessity of such a motion follows mathematically from the law of inertia (proportionality of acceleration to force) and the law of universal gravitation.

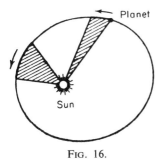

Fig. 16.

The ellipse of inertia. As an example of the application of the ellipse in a technical problem, we consider the so-called ellipse of inertia of a plate.

Let the plate be of uniform thickness and homogeneous material, for example a zinc plate of arbitrary shape. We rotate it around an axis in its plane. A body in rectilinear motion has, as is well known, an inertia with respect to this rectilinear motion that is proportional to its mass (independently of the shape of the body and the distribution of the mass). Similarly, a body rotating around an axis, for instance a flywheel, has inertia with respect to this rotation. But in the case of rotation, the inertia is not only proportional to the mass of the rotating body but

Fig. 17a.   Fig. 17b.

---

* The eccentricities of planetary orbits are not very large, so that the orbits of planets are almost circles.

also depends on the distribution of the mass of the body with respect to the axis of rotation, since the inertia with respect to rotation is greater if the mass is farther from the axis. For example, it is very easy to bring a stick at once into fast rotation around its longitudinal axis (figure 17a). But if we try to bring it at once to fast rotation around an axis perpendicular to its length, even if the axis passes through its midpoint, we will find that unless this stick is very light, we must exert considerable effort (figure 17b).

It is possible to show that the inertia of a body with respect to rotation about an axis, the so-called *moment of inertia* of the body relative to the axis, is equal to $\Sigma r_i^2 m_i$ (where by $\Sigma r_i^2 m_i$ we mean the sum $r_1^2 m_1 + r_2^2 m_2 + \cdots + r_n^2 m_n$ and think of the body as decomposed into very small elements, with m_i as the mass of the ith element and r_i the distance of the ith element from the axis of rotation, the summation being taken over all elements of the body).

Let us return to our plate. Let O (figure 18) be a point of this plate. We consider the moments of inertia J_u of the plate relative to an axis

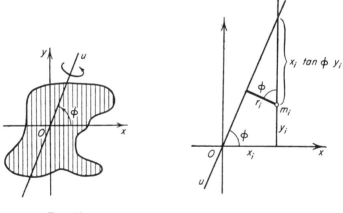

Fig. 18.   Fig. 19.

of rotation u passing through O and lying in the plane of the plate. For this purpose we take the point O as the origin of a Cartesian coordinate system and choose arbitrary axes Ox and Oy in the plane of the plate; then we will characterize the axis of rotation u by the angle ϕ which it makes with the Ox-axis. It is easy to see (figure 19) that

$$r_i = |(x_i \tan \phi - y_i) \cos \phi| = |x_i \sin \phi - y_i \cos \phi|.$$

Hence

$$\Sigma \, r_i^2 m_i = \Sigma \, (x_i^2 \sin^2 \phi - 2x_i y_i \sin \phi \cos \phi + y_i^2 \cos^2 \phi) \, m_i$$

$$= \sin^2 \phi \, \Sigma \, x_i^2 m_i - 2 \sin \phi \cos \phi \, \Sigma \, x_i y_i m_i + \cos^2 \phi \, \Sigma \, y_i^2 m_i \, .$$

The quantities $\sin^2 \phi$, $2 \sin \phi \cos \phi$, and $\cos^2 \phi$ are taken outside the summation sign, since they are constant for a given axis u. We now write

$$\Sigma \, x_i^2 m_i = A, \quad - \Sigma \, x_i y_i m_i = B, \quad \Sigma \, y_i^2 m_i = C.$$

The quantities A, B, and C do not depend on the choice of the axis u, but only on the shape of the plate, the distribution of its mass, and the fixed choice of the coordinate axes Ox and Oy. Consequently,

$$J_u = A \sin^2 \phi + 2B \sin \phi \cos \phi + C \cos^2 \phi.$$

We consider all possible axes u in the plane of the plate passing through the point O and lay off on each of these axes from the point O a length equal to ρ, the inverse of the square root of the moment of inertia J_u of the plate relative to that axis, i.e., $\rho = 1/\sqrt{J_u}$. Then we obtain

$$\frac{1}{\rho^2} = A \sin^2 \phi + 2B \sin \phi \cos \phi + C \cos^2 \phi.$$

But

$$x = \rho \cos \phi, \quad y = \rho \sin \phi,$$

so that the equation of this locus has the following form:

$$Cx^2 + 2Bxy + Ay^2 = 1.$$

A second-order curve is obtained that is evidently finite and closed; i.e., it is an ellipse (figure 20), since all other second-order curves, as we will later show, are either infinite or reduce to one point.

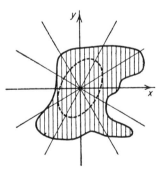

FIG. 20.

The following remarkable result is obtained: Whatever may be the form and size of a plate and the distribution of its mass, the magnitude of its moment of inertia (more precisely, of the quantity ρ inversely proportional to the square root of the moment of inertia) with respect to the various axes lying in the plane of the plate and passing through the given point O, is characterized by a certain ellipse. This ellipse is called the ellipse of inertia of the plate relative to the point O. If the point O is the center of gravity of the plate, then the ellipse is called its central ellipse of inertia.

The ellipse of inertia plays a great role in mechanics; in particular, it has an important application in the strength of materials. In the theory of strength of materials, it is proved that the resistance to bending of a beam with given cross section is proportional to the moment of inertia of its cross section relative to the axis through the center of gravity of the cross section and perpendicular to the direction of the bending force. Let us clarify this by an example. We assume that a bridge across a stream consists of a board that sags under the weight of a pedestrian passing over it. If the same board (no thicker than before) is placed "on its edge," it scarcely bends at all, i.e., a board placed on its edge is, so to speak, stronger. This follows from the fact that the moment of inertia of the cross section of the board (it has the shape of an elongated rectangle that we may think of as evenly covered with mass) is greater relative to the axis perpendicular to its long side than relative to the axis parallel to its long side. If we set the board not exactly flat nor on edge but obliquely, or even if we do not take a board at all but a rod of arbitrary cross section, for example a rail, the resistance to bending will still be proportional to the moment of inertia of its cross section relative to the corresponding axis. The resistance of a beam to bending is therefore characterized by the ellipse of inertia of its cross section.

For an ordinary rectangular beam this ellipse will have the form shown in figure 21. The rigidity of such a beam under a load in the direction of the Oz-axis is proportional to bh^3.

Steel beams often have a ∫ -shaped cross section; for such beams the cross section and the ellipse of inertia are represented in figure 22. The

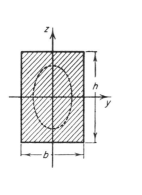

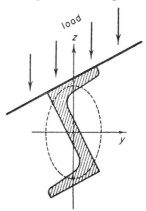

FIG. 21.   FIG. 22.

greatest resistance to bending is in the z direction. When they are used, for example as roof rafters under a load of snow and their own weights, they work directly against bending in a direction close to this most advantageous direction.

The hyperbola and its focal property.   Now we consider the equation

$$\frac{x^2}{a^2} - \frac{y^2}{b^2} = 1,$$

representing a curve which is called a *hyperbola*. If we denote by c a number such that $c^2 = a^2 + b^2$, then it is possible to show that a hyperbola is the locus of all points the difference of whose distances to the points F_1 and F_2 on the Ox-axis with abscissas c and $-c$ is a constant: $\rho_2 - \rho_1 = 2a$ (figure 23). The points F_1 and F_2 are called the *foci*.

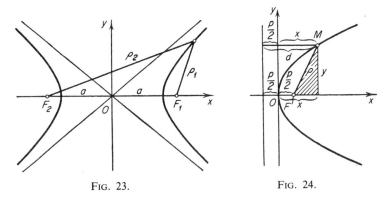

FIG. 23.   FIG. 24.

The parabola and its directrix.   Finally, we consider the equation

$$y^2 = 2px$$

and call the corresponding curve a *parabola*. The point F lying on the Ox-axis with abscissa $p/2$ is called the *focus* of the parabola, and the straight line $y = -p/2$, parallel to the Oy-axis, is its *directrix*. Let M be any point of the parabola (figure 24), ρ the length of its focal radius MF, and d the length of the perpendicular dropped from it to the directrix. Let us compute ρ and d for the point M. From the shaded triangle we obtain $\rho^2 = (x - p/2)^2 + y^2$. As long as the point M lies on the parabola, we have $y^2 = 2px$, hence

$$\rho^2 = \left(x - \frac{p}{2}\right)^2 + 2px = \left(x + \frac{p}{2}\right)^2.$$

But directly from the figure it is clear that $d = x + p/2$. Therefore $\rho^2 = d^2$, i.e., $\rho = d$. The inverse argument shows that if for a given point we have $\rho = d$, then the point lies on the parabola. Thus a parabola is the locus of points equidistant from a given point F (called the focus) and a given straight line d (called the directrix).

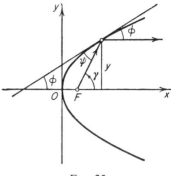

FIG. 25.

The property of the tangent to a parabola. Let us examine an important property of the tangent to a parabola and its application in optics.
Since for a parabola $y^2 = 2px$ we have $2y\,dy = 2p\,dx$, it follows that the derivative, or the slope of the tangent, is equal to $dy/dx = \tan\phi = p/y$ (figure 25).

On the other hand, it follows directly from the figure that

$$\tan\gamma = \frac{y}{x - p/2}\,.$$

But

$$\tan 2\phi = \frac{2\,p/y}{1 - p^2/y^2} = \frac{2py}{y^2 - p^2} = \frac{2py}{2px - p^2} = \frac{y}{x - p/2},$$

i.e., $\gamma = 2\phi$, and since $\gamma = \phi + \psi$, therefore $\psi = \phi$. Consequently, by virtue of the law (angle of incidence is equal to angle of reflection) a beam of light, starting from the focus F and reflected by an element of the parabola (whose direction coincides with the direction of the tangent) is reflected parallel to the Ox-axis, i.e., parallel to the axis of symmetry of the parabola.

On this property of the parabola is based the construction of reflecting telescopes, as invented by Newton. If we manufacture a concave mirror whose surface is a so-called paraboloid of revolution, i.e., a surface obtained by the rotation of a parabola around its axis of symmetry, then all the light rays originating from any point of a heavenly body lying strictly in the direction of the "axis" of the mirror are collected by the mirror (figure 26) at one point, namely its focus. The rays originating from some other point of the heavenly body, being not exactly parallel to the axis of the mirror, are collected almost at one point in the neighborhood of the focus. Thus, in the so-called focal plane through the focus of the mirror and perpendicular to its axis, the inverse image of the star is obtained; the farther away this image is from the focus, the more diffuse it will be, since it is only the rays exactly parallel to the axis of the mirror that are collected by the mirror at one point. The image so obtained can be viewed in a special microscope, the so-called

FIG. 26.   FIG. 27.

eye piece of the telescope, either directly or, in order not to cut off the light from the star with one's own head, after reflection in a small plane mirror, attached to the telescope near the focus (somewhat nearer than the focus to the concave mirror) at an angle of 45°.

The searchlight (figure 27) is based on the same property of the parabola. In it, conversely, a strong source of light is placed at the focus of a paraboloidal mirror, so that its rays are reflected from the mirror in a beam parallel to its axis. Automobile headlights are similarly constructed (figure 28).

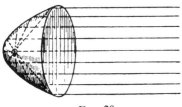
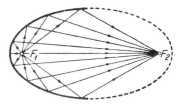

FIG. 28.   FIG. 29.

In the case of an ellipse, as it is easy to show, the rays issuing from one of its foci F_1 and reflected by the ellipse are collected at the other focus F_2 (figure 29), and in the hyperbola the rays originating from one of its foci F_1 are reflected by it as if they originated from the other focus F_2 (figure 30).

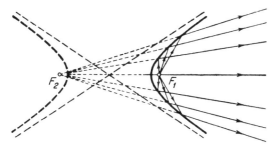

FIG. 30.

The directrices of the ellipse and the hyperbola. Like the parabola, the ellipse and the hyperbola have directrices, in this case two apiece. If we consider a focus and the directrix "on the same side with it," then for all points M of the ellipse we have $\rho/d = \epsilon$, where the constant ϵ is the eccentricity, which for an ellipse is always smaller than 1; and for all points of the corresponding branch of the hyperbola, we also have $\rho/d = \epsilon$, where ϵ is again the eccentricity, which for a hyperbola is always greater than 1.

Thus the ellipse, the parabola and one branch of the hyperbola are the loci of all those points in the plane for which the ratio of their distance ρ from the focus to their distance d from the directrix is constant (figures 31 and 32). For the ellipse this constant is smaller than unity, for the parabola it is equal to unity, and for the hyperbola it is greater than unity. In

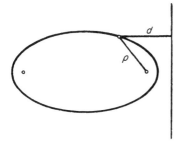

FIG. 31.

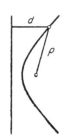

FIG. 32.

this sense the parabola is the "limiting" or "transition" case from the ellipse to the hyperbola.

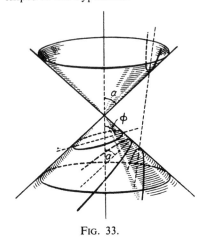

FIG. 33.

Conic sections. The ancient Greeks had already investigated in detail the curves obtained by intersecting a straight circular cone by a plane. If the intersecting plane makes with the axis of the cone an angle ϕ of 90°, i.e., is perpendicular to it, then the section obtained is a circle. It is easy to show that if the angle ϕ is smaller than 90°, but greater than the angle α which the generators of the cone make with its axis, then an ellipse is obtained. If ϕ is equal to α, a parabola results and if ϕ is smaller than α, then we obtain a hyperbola as the section (figure 33).

The parabola as the graph of quadratic proportion and the hyperbola as the graph of inverse proportion. We recall that the graph of quadratic proportion

$$y = kx^2$$

is a parabola (figure 34) and that the graph of inverse proportion

$$y = \frac{k}{x} \quad \text{or} \quad xy = k$$

is a hyperbola (figure 35), as we will easily prove later. A hyperbola was defined earlier as the curve represented by the equation

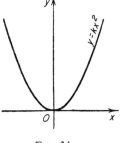

FIG. 34.

$$\frac{x^2}{a^2} - \frac{y^2}{b^2} = 1.$$

In the special case $a = b$ the so-called *rectangular hyperbola* plays the same role among hyperbolas as the circle plays among ellipses. In this case, if we rotate the coordinate axes by 45° (figure 36) the equation in the new coordinates (x', y') will have the form

$$x'y' = k.$$

We have now considered three important second-order curves: the ellipse, the hyperbola, and the parabola, and for their definitions we have taken the so-called canonical equations

$$\frac{x^2}{a^2} + \frac{y^2}{b^2} = 1, \frac{x^2}{a^2} - \frac{y^2}{b^2} = 1 \quad \text{and} \quad y^2 = 2px,$$

by which they are represented.

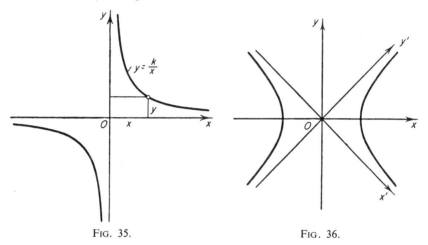

FIG. 35.   FIG. 36.

We now pass to the study of the general second-degree equation in two unknowns, namely to the question what kinds of curves are represented by this equation.

§8. The Reduction of the General Second-Degree Equation to Canonical Form

The first consistent presentation of analytic geometry by Euler. A significant step in the development of analytic geometry was the appearence in 1748 of the book "Introduction to analysis" in the second volume of which, among other things related to the theory of functions and other branches of analysis, for the first time a presentation was given of analytic geometry in the plane with a detailed investigation of second-order curves, very close to the one given in contemporary textbooks of analytic geometry, and also with an investigation of higher order curves. This was the first exposition of analytic geometry in the contemporary sense of the word.

The notion of reducing an equation to canonical form.   A second-degree equation*

$$Ax^2 + 2Bxy + Cy^2 + 2Dx + 2Ey + F = 0$$

contains six terms, not three or only two as in the above canonical equations of the ellipse, hyperbola, and parabola. This is not because such an equation represents a more complicated curve but because the system of coordinates is possibly not suited to it. It turns out that if we select a suitable Cartesian coordinate system, then a second-degree equation with two variables always can be reduced to one of the following canonical forms:

1. $\dfrac{x^2}{a^2} + \dfrac{y^2}{b^2} - 1 = 0.$ Ellipse

2. $\dfrac{x^2}{a^2} + \dfrac{y^2}{b^2} + 1 = 0.$ Imaginary ellipse

3. $\dfrac{x^2}{a^2} + \dfrac{y^2}{b^2} = 0.$ Point (a pair of imaginary lines intersecting in a real point)

4. $\dfrac{x^2}{a^2} - \dfrac{y^2}{b^2} - 1 = 0.$ Hyperbola

5. $\dfrac{x^2}{a^2} - \dfrac{y^2}{b^2} = 0.$ A pair of intersecting lines

6. $y^2 - 2px = 0.$ Parabola

7. $x^2 - a^2 = 0.$ A pair of parallel lines

8. $x^2 + a^2 = 0.$ A pair of imaginary parallel lines

9. $x^2 = 0.$ A pair of coincident straight lines

where a, b, p, are not equal to zero.

Equations 1, 4, and 6 of the enumerated canonical forms are already well known to us; these are the canonical equations of the ellipse, hyperbola, and parabola. Two of them are not satisfied by any points, namely equations 2 and 8. Indeed, the square of a real number is always positive or zero, so that on the left-hand side of equation 2 the sum of the terms $x^2/a^2 + y^2/b^2$ is never negative, and since the term $+1$ also appears, the

* The coefficients of xy, x, y will be denoted not by B, D, E but by $2B$, $2D$, $2E$ for simplicity of subsequent formulas.

left-hand side cannot be equal to zero; analogously in equation 8, the number x^2 is not negative, and a^2 is positive. From these considerations, it follows that only $(x = 0, y = 0)$ satisfies equation 3, i.e., one point, the origin. Equation 5 can be written as $(x/a - y/b)(x/a + y/b) = 0$, from which we see that it is satisfied by those points and only those points for which one of the first-degree expressions $x/a - y/b$ or $x/a + y/b$ is equal to zero; so the curve it represents is this pair of intersecting lines. Equation 7 analogously gives $(x - a)(x + a) = 0$; i.e., the corresponding curve is a pair of parallel lines $x = a$ and $x = -a$. Finally, curve 9 is a special limiting case of curve 7, when $a = 0$; i.e., it is a pair of coincident lines.

Formulas of coordinate transformations. In order to obtain the indicated important result about the possible types of second-order curves, it is necessary first to deduce the formulas by which the rectangular coordinates of points vary under a change of the coordinate system.

Let x, y be the coordinates of a point M relative to the axes Oxy. Let us translate these axes parallel to themselves to the position $O'x'y'$ and let the coordinates of the new origin O' relative to the old axes

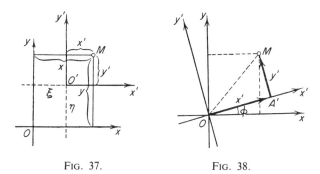

Fig. 37.   Fig. 38.

be ξ and η. It is evident (figure 37) that the new coordinates x', y' of the point M are connected with its old coordinates x, y by the formulas

$$x = x' + \xi,$$

$$y = y' + \eta,$$

which are the formulas of the so-called parallel translation of axes. If we rotate the original axes Oxy about the origin counterclockwise by an angle ϕ then, as is easy to see (figure 38), if we project the polygonal line $OA'M$ composed of the new coordinate segments x', y' on the Ox-axis and the Oy-axis, respectively, we obtain

$$x = x' \cos \phi - y' \sin \phi,$$
$$y = x' \sin \phi + y' \cos \phi,$$

which are the formulas for transformation of coordinates under rotation of a rectangular coordinate system.

If we are given an equation $F(x, y) = 0$ of a curve relative to the axes Oxy and we wish to write the transformed equation of the same curve, i.e., relative to the new axes $O'x'y'$, then we must replace x and y in the equation $F(x, y) = 0$ by their expressions in terms of x' and y', given by the formulas of the transformation. For example, under parallel translation of the axes, we obtain the transformed equation

$$F(x' + \xi, y' + \eta) = 0,$$

and under rotation of the axes the equation

$$F(x' \cos \phi - y' \sin \phi, x' \sin \phi + y' \cos \phi) = 0.$$

We note that under a transformation to new axes the degree of an equation does not change. Indeed, the degree cannot increase, since the transformation formulas are of the first-degree. But the degree cannot decrease either, since then the inverse coordinate transformation would increase it (and it is also of the first degree).

The reduction of a general second-degree equation to one of the 9 canonical forms. We now show that given any second-degree equation in two unknowns we can always first rotate the axes and then translate them parallel to themselves in such a way that the transformed equation for the final axes will have one of the forms 1, 2, ⋯, 9.

Indeed, let the given second-degree equation have the form

$$Ax^2 + 2Bxy + Cy^2 + 2Dx + 2Ey + F = 0. \tag{8}$$

Let us rotate the axes through some angle ϕ, which we select in the following way. Replacing x and y in equation (8) by their expressions in terms of the new coordinates (according to the formulas for rotation), we find, after collecting similar terms, that the coefficient $2B'$ in the transformed equation

$$A'x'^2 + 2B'x'y' + C'y'^2 + 2D'x' + 2E'y' + F' = 0$$

is equal to

$$2B' = -2A \sin \phi \cos \phi + 2B(\cos^2 \phi - \sin^2 \phi) + 2C \sin \phi \cos \phi$$
$$= 2B \cos 2\phi - (A - C) \sin 2\phi.$$

Setting it equal to zero, we obtain $2B \cos 2\phi = (A - C) \sin 2\phi$, from which

$$\cot 2\phi = \frac{A - C}{2B}.$$

Since the cotangent varies from $-\infty$ to $+\infty$, we can always find an angle ϕ for which this equality is satisfied. By rotating the axes through this angle, we find that for the rotated axes $Ox'y'$ the equation of our curve, represented for the initial axes by equation (8), has the form

$$A'x'^2 + C'y'^2 + 2D'x' + 2E'y' + F = 0, \tag{9}$$

i.e., that it does not contain the term with the product of the coordinates (F remains unchanged, since the formulas of rotation do not contain constant terms).

Now we translate the already rotated axes $Ox'y'$ parallel to themselves to the position $O''x''y''$, and let the coordinates of the new origin O'' relative to the axes $Ox'y'$ be ξ', η'. The equation of our curve for these final axes will be

$$A'(x'' + \xi')^2 + C'(y'' + \eta')^2 + 2D'(x'' + \xi') + 2E'(y'' + \eta') + F = 0. \tag{10}$$

We now show that we can always select ξ' and η' (i.e., we can translate the axes $Ox'y'$ parallel to themselves) in such a way that the final equation for the axes $O''x''y''$ has one of the canonical forms 1, 2, $\cdots$, 9.

Removing all parentheses in equation (10) and collecting similar terms, we obtain

$$A'x''^2 + C'y''^2 + 2(A'\xi' + D')x'' + 2(C'\eta' + E')y'' + F' = 0, \tag{10'}$$

where we have denoted by F' the sum of all constant terms; its value does not interest us at the moment.

We consider three possible cases.

I. A' and C' both not equal to zero. In this case, taking $\xi' = -D'/A'$, $\eta' = -E'/C'$, we eliminate the terms with the first powers of x'' and y'' and obtain an equation of the form

$$A'x''^2 + C'y''^2 + F' = 0. \tag{I}$$

II. $A' \neq 0$, $C' = 0$, but $E' \neq 0$. Letting $\xi' = -D'/A'$, $\eta' = 0$, i.e., $y'' = y'$, we obtain the equation

$$A'x''^2 + 2E'y' + F' = 0,$$

or

$$A'x''^2 + 2E'\left(y' + \frac{F'}{2E'}\right) = 0.$$

Then making a parallel translation along the Oy'-axis by an amount $\eta'' = -F'/2E'$, we find that $y' = y'' - F'/2E'$, i.e., $y' + F'/2E' = y''$ so that we obtain the equation

$$A'x''^2 + 2E'y'' = 0. \tag{II}$$

If we have $A' = 0$, $C' \neq 0$, $D' \neq 0$, we can simply interchange the roles of x and y and obtain the same result.

III. $A' \neq 0$, $C' = 0$, $E' = 0$. Taking again $\xi' = -D'/A'$, $\eta' = 0$, we obtain the equation

$$A'x''^2 + F' = 0. \tag{III}$$

If we have $A' = 0$, $C' \neq 0$, $D' = 0$, we can again interchange the roles of x and y.

We have now considered all the possibilities, in view of the fact that A' and C' cannot simultaneously be zero, since then the degree of the equation would be reduced, and we have seen that under our coordinate transformations this degree does not change.

Thus, with the appropriate choice of rectangular coordinates every second-degree equation can be brought to one of the three so-called "reduced" equations (I), (II), (III).

Let the equation have the form (I) (in this case A' and C' are not equal to zero). If $F' \neq 0$, then writing equation (I) as

$$\frac{x''^2}{-F'/A'} + \frac{y''^2}{-F'/C'} - 1 = 0,$$

we arrive, depending on the signs of A', C', F', at one of the equations 1, 2, or 4. If the denominator of x''^2 is negative and that of y''^2 is positive, then we must also interchange the axes $O''x''$ and $O''y''$.

If $F' = 0$, then equation (I) can be written in the form

$$\frac{x''^2}{1/A'} + \frac{y''^2}{1/C'} = 0,$$

and we arrive at equations 3 or 5.

If the equation has the form (II) (in this case A' and E' are not both zero), then we can write it as

$$x''^2 + \frac{2E'}{A'}y'' = 0,$$

and denoting $-E'/A'$ by p and interchanging the names of the axes $O''x''$ and $O''y''$ we obtain equation 6.

Finally, if we have an equation of form (III) (where $A' \neq 0$), it can be rewritten as $x''^2 + F'/A' = 0$ and one of the equations 7, 8, or 9 is obtained.

This important theorem on the possibility of reducing every 2nd-degree equation to one of the 9 canonical forms was already examined in detail by Euler. The arguments in Euler's book differ only in form from the ones just given.

§9. The Representation of Forces, Velocities, and Accelerations by Triples of Numbers; Theory of Vectors

Following Euler an important step was taken by Lagrange. In his "Analytic mechanics," published in 1788, Lagrange arithmetized forces, velocities and accelerations in the same way as Descartes arithmetized points. This idea that Lagrange developed in his book subsequently took the form of the so-called theory of vectors and proved to be an important help in physics, mechanics, and technology.

Rectangular coordinates in space. We remark, first of all, that neither Descartes nor Newton developed analytic geometry in space. This was done later on, in the first half of the 18th century, by Laguerre and Clairaut. In order to specify a point M in space they selected three mutually perpendicular axes Ox, Oy, and Oz and considered (figure 39) the numerical values of the distances of the point M from the planes Oyz, Oxz, and Oxy, taken with the corresponding signs, the so-called *abscissa* x, *ordinate* y, and *altitude* z of the point M.

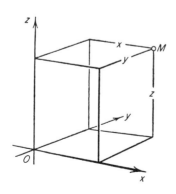

Fig. 39.

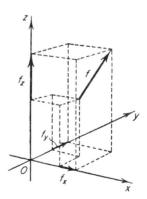

Fig. 40.

Arithmetization of forces, velocities, and accelerations, introduced by Lagrange. We consider (figure 40) a force f which can be represented in conventional units by a segment with an arrow, having a specific length and direction. Lagrange points out that this force f can be decomposed into three components f_x, f_y, and f_z in the direction of the corresponding axes Ox, Oy, and Oz; these components, as directed segments on the axes, can be given simply by numbers, positive or negative depending on whether the component is directed in the positive or the opposite direction of the axis. Thus, we can consider, for example, the force (2, 3, 4) or the force (1, —2, 5), etc. In the composition of forces according to the parallelogram law, as can easily be shown (it will be shown later), their corresponding components have to be added. For example, the sum of the given forces is the force

$$(2 + 1, 3 - 2, 4 + 5) = (3, 1, 9).$$

The same can be done for velocities and accelerations. In every problem of mechanics, all the equations connecting forces, velocities, and accelerations can also be written as equations connecting their components, i.e., connecting simply numbers; then the mechanical equation will necessarily be written in the form of three equations: first for the x's, the second for the y's, and the third for the z's.

But it was only after a hundred years from the time of Lagrange that mathematicians and physicists, particularly under the influence of the developing theory of electricity, began on a wide scale to consider the general theory of such segments, having a definite length and direction. Such segments were called vectors.

The theory of vectors has a great significance in mechanics, physics, and technology, and its algebraic side, the so-called algebra of vectors (in contrast to vector analysis) appears at once as an essential constituent part of analytic geometry.

Algebra of vectors. Any directed segment (whether it represents a force, a velocity, an acceleration, or some other entity) i.e., a segment having a given length and a definite direction, is called a *vector*. Two vectors are said to be *equal*, if they have the same length and the same direction; i.e., in the very concept of "vector" only its length and its direction are taken into account. Vectors can be added. Let the vectors **a**, **b**, ⋯ , **d** be given. We lay out the vector **a** from some point, then from its end point we draw the vector **b**, etc. We obtain a so-called vector polygon **ab** ⋯ **d** (figure 41). The vector **m** whose initial point coincides with the initial point of the first vector **a** of this polygon, and whose end point coincides with the end point of the last vector **d**, is called the *sum* of these vectors

$$\mathbf{m} = \mathbf{a} + \mathbf{b} + \cdots + \mathbf{d}. \qquad (11)$$

It is easy to show that the vector **m** does not depend on the order in which the summands **a**, **b**, $\cdots$, **d** are taken.

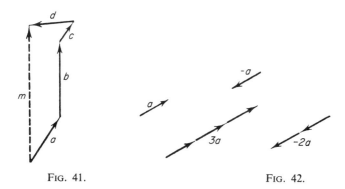

FIG. 41.　　　　　　　　　　　FIG. 42.

The vector equal in length to the vector **a** but opposite in direction is called its *inverse* vector and is denoted by —**a**.

Subtraction of the vector **a** is defined as addition of its inverse vector.

In vector calculus ordinary real numbers are customarily called scalars. Let a vector **a** (figure 42) and a scalar λ be given, then by the product of the vector **a** with the scalar (number) λ, i.e., $\lambda\mathbf{a}$, is meant the vector whose length is equal to the product of the length |**a**| of the vector **a**

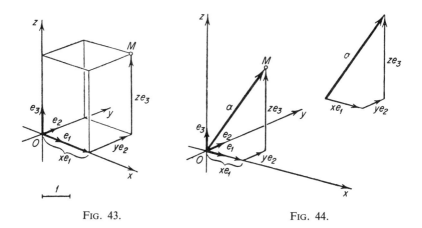

FIG. 43.　　　　　　　　　　　FIG. 44.

and the absolute value $|\lambda|$ of the number λ, and whose direction is the same as that of **a** if $\lambda > 0$ and the opposite if $\lambda < 0$.

Let us consider a system of rectangular Cartesian coordinates $Oxyz$ and the vectors $\mathbf{e}_1, \mathbf{e}_2, \mathbf{e}_3$ having length equal to unity and directions coinciding with the positive directions of the axes Ox, Oy, Oz, respectively. It is obvious that any given point M (figure 43) of space can be reached from the origin O by traversing a certain number of "times" (an integral, fractional or irrational, positive or negative "number of times") the vector $\mathbf{e}_1$, then so many "times" the vector $\mathbf{e}_2$, and finally so many "times" the vector $\mathbf{e}_3$. It is clear that the numbers x, y, z showing how many "times" it is necessary to traverse the vectors $\mathbf{e}_1, \mathbf{e}_2, \mathbf{e}_3$, are simply the Cartesian coordinates of the point M.

Let a vector **a** be given; if we cause a point to move from the initial point of **a** to its end point and decompose this motion into motions parallel to the axes Ox, Oy, and Oz, and if it is hereby necessary to shift the point through a distance $x\mathbf{e}_1$ parallel to the Ox-axis, through $y\mathbf{e}_2$ parallel to the Oy-axis and through $z\mathbf{e}_3$ parallel to the Oz-axis, then

$$\mathbf{a} = x\mathbf{e}_1 + y\mathbf{e}_2 + z\mathbf{e}_3 . \tag{12}$$

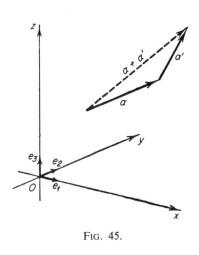

FIG. 45.

The numbers x, y, z are called the *coordinates* of the vector **a**. These are obviously just the coordinates of the end point M of this vector, if its initial point lies at the origin O of the coordinate system (figure 44). From this it follows at once that in adding vectors their corresponding coordinates are to be added, and in subtraction they are to be subtracted. If the first vector "carries" us along the Ox-axis by a distance $x\mathbf{e}_1$, and the second by $x'\mathbf{e}_1$, then clearly their sum "carries" us along the Ox-axis by a distance $(x + x')\mathbf{e}_1$, etc. (figure 45). It also follows at once that in multiplication of a vector by a number, its coordinates are multiplied by the number.

Scalar product and its properties. If we are given two vectors **a** and **b**, then the number equal to the product of their lengths by the cosine of the angle between them $|\mathbf{a}||\mathbf{b}|\cos\phi$ is called their *scalar product* and is denoted by **ab** or (**ab**). Let x, y, z be the coordinates of the vector **a**

and $\bar{x}$, $\bar{y}$, $\bar{z}$ the coordinates of the vector **b**; then the scalar product is equal to

$$\mathbf{ab} = x\bar{x} + y\bar{y} + z\bar{z}, \tag{13}$$

i.e., to the sum of the products of their corresponding coordinates.

This important result can be proved as follows. First we make the following remarks:

(1) If we multiply one of the vectors of a scalar product, for example **a**, by a number λ, this is obviously the same as multiplying their scalar product by the same number, i.e.,

$$(\lambda\mathbf{a})\mathbf{b} = \lambda(\mathbf{ab}).$$

(2) The scalar product is distributive, i.e., if $\mathbf{a} = \mathbf{a}_1 + \mathbf{a}_2$, then $\mathbf{ab} = \mathbf{a}_1\mathbf{b} + \mathbf{a}_2\mathbf{b}$.

In fact, the left-hand side of this equality is equal to the product of the length of the vector **b** by the numerical value of the projection of the vector **a** on the axis of the vector **b** (figure 46), and the right-hand side is equal to the product of the length of **b** by the sum of the numerical values

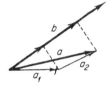

FIG. 46.

of the projections of the vectors $\mathbf{a}_1$ and $\mathbf{a}_2$ on the axis of **b**. But proj $\mathbf{a} = $ proj $\mathbf{a}_1 + $ proj $\mathbf{a}_2$, which proves the equality.

Now we consider two vectors **a** and **b** whose decompositions in terms of the vectors $\mathbf{e}_1$, $\mathbf{e}_2$, $\mathbf{e}_3$ are $\mathbf{a} = x\mathbf{e}_1 + y\mathbf{e}_2 + z\mathbf{e}_3$, $\mathbf{b} = \bar{x}\mathbf{e}_1 + \bar{y}\mathbf{e}_2 + \bar{z}\mathbf{e}_3$, so that

$$\mathbf{ab} = (x\mathbf{e}_1 + y\mathbf{e}_2 + z\mathbf{e}_3)(\bar{x}\mathbf{e}_1 + \bar{y}\mathbf{e}_2 + \bar{z}\mathbf{e}_3).$$

By the distributivity (2) of the scalar product, the sums of vectors in parentheses can be multiplied as polynomials, and by (1) the scalar factors in each of the terms can be taken outside the parentheses, so that

$$\mathbf{ab} = x\bar{x}\mathbf{e}_1\mathbf{e}_1 + x\bar{y}\mathbf{e}_1\mathbf{e}_2 + x\bar{z}\mathbf{e}_1\mathbf{e}_3 + y\bar{x}\mathbf{e}_2\mathbf{e}_1 + y\bar{y}\mathbf{e}_2\mathbf{e}_2 + y\bar{z}\mathbf{e}_2\mathbf{e}_3$$
$$+ z\bar{x}\mathbf{e}_3\mathbf{e}_1 + z\bar{y}\mathbf{e}_3\mathbf{e}_2 + z\bar{z}\mathbf{e}_3\mathbf{e}_3.$$

But

$$|\mathbf{e}_1| = |\mathbf{e}_2| = |\mathbf{e}_3| = 1, \quad \cos 0° = 1 \text{ and } \cos 90° = 0.$$

Consequently,

$$\mathbf{e}_1\mathbf{e}_1 = 1, \quad \mathbf{e}_1\mathbf{e}_2 = 0, \quad \mathbf{e}_1\mathbf{e}_3 = 0,$$
$$\mathbf{e}_2\mathbf{e}_1 = 0, \quad \mathbf{e}_2\mathbf{e}_2 = 1, \quad \mathbf{e}_2\mathbf{e}_3 = 0,$$
$$\mathbf{e}_3\mathbf{e}_1 = 0, \quad \mathbf{e}_3\mathbf{e}_2 = 0, \quad \mathbf{e}_3\mathbf{e}_3 = 1.$$

Thus,

$$\mathbf{ab} = x\bar{x} + y\bar{y} + z\bar{z}. \tag{14}$$

We remark, in particular, that if the vectors **a** and **b** are mutually perpendicular, then $\phi = 90°$ and $\cos\phi = 0$. Therefore the equality

$$x\bar{x} + y\bar{y} + z\bar{z} = 0 \tag{15}$$

serves as an easily verifiable condition of perpendicularity of the vectors **a** and **b**.

Angle between two directions. Let us consider a direction characterized by its angles α, β, γ with the coordinate axes. We draw the line

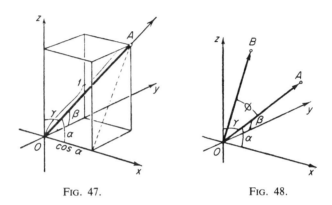

FIG. 47.   FIG. 48.

in this direction through the origin of the coordinate system and mark off on it from the origin a segment OA of unit length (figure 47). In this case the coordinates of the point A, i.e., the coordinates of the vector $\overrightarrow{OA}$ are exactly $\cos\alpha$, $\cos\beta$, and $\cos\gamma$. If we have a second direction given by the angles $\bar{\alpha}$, $\bar{\beta}$, $\bar{\gamma}$, then the analogous vector $\overrightarrow{OB}$ for this second direction has coordinates $\cos\bar{\alpha}$, $\cos\bar{\beta}$, $\cos\bar{\gamma}$ (figure 48). Let ϕ be the angle between these vectors; then their scalar product is equal to $1\cdot1\cos\phi$ from which we find

$$\cos\phi = \cos\alpha\cos\bar{\alpha} + \cos\beta\cos\bar{\beta} + \cos\gamma\cos\bar{\gamma}. \tag{16}$$

This is the very important formula for the cosine of the angle between two directions.

§10. Analytic Geometry in Space; Equations of a Surface in Space and Equations of a Curve

If an equation $z = f(x, y)$ is given and if x and y are regarded as the abscissa and ordinate and z the altitude of a point, then this equation itself represents some surface P, which can be obtained by erecting perpendiculars of length z at the points (x, y) of the Oxy-plane. The locus of the end points of these perpendiculars gives the surface P represented by this equation. If the equation connecting x, y, and z is not already solved with respect to z, then it can be solved for z and after that we can construct the surface P. In general, in analytic geometry the totality of all those points of space whose coordinates x, y, z satisfy a given equation (figure 49) in three variables x, y, z is said to be the surface represented by the equation.

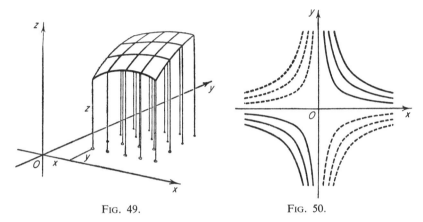

FIG. 49.   FIG. 50.

A function of two variables $f(x, y)$, as was pointed out already in Chapter II, can represent not only a surface P, but also its system of level curves, i.e., curves in the Oxy-plane on each of which the function $f(x, y)$ has a constant value. This system of curves is clearly nothing else than the topographical map of the surface P on the Oxy-plane.

Example. The equation $xy = z$ gives, for instance, the level curves: $\cdots$, $xy = -3$, $xy = -2$, $xy = -1$, $xy = 0$, $xy = 1$; $xy = 2$, $xy = 3$, $\cdots$. All of them are hyperbolas (figure 50) except $xy = 0$, which represents the two coordinate axes. What is obtained is clearly a saddlelike surface (figure 51) (the so-called hyperbolic paraboloid).

In order to define a curve in space, we can give the equations of any two surfaces P and Q which intersect along the curve. For example, the system

$$xy = z,$$
$$x^2 + y^2 = 1$$

gives a space curve (figure 52). The equation $xy = z$ determines the earlier hyperbolic paraboloid, and the equation $x^2 + y^2 = 1$ determines a circular cylinder of unit radius, whose axis is the Oz-axis. The system of equations consequently defines the curve of intersection of the paraboloid with the cylinder, which is represented in figure 52.

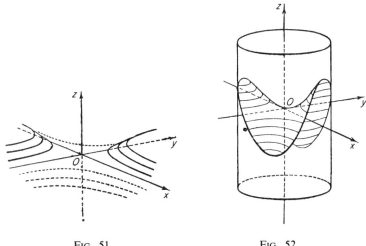

FIG. 51.   FIG. 52.

If in this system one of the unknowns, say x, is chosen arbitrarily, and then the system is solved with respect to y and z, we will obtain the coordinates x, y, z of the various points of the curve.

Equation of a plane and equations of a straight line. It can be shown that every equation of the first degree with three variables

$$Ax + By + Cz + D = 0$$

represents a plane, and conversely. By what has already been said, it is clear that a line can be given by a system of two such equations:

$$A_1x + B_1y + C_1z + D_1 = 0,$$
$$A_2x + B_2y + C_2z + D_2 = 0,$$

i.e., as the curve of intersection of two planes.

The general second-degree equation in three variables and its 17 canonical forms. A second-degree equation in three variables

$$A_1x^2 + A_2y^2 + A_3z^2 + 2B_1yz + 2B_2xz + 2B_3xy$$
$$+ 2C_1x + 2C_2y + 2C_3z + D = 0, \quad (17)$$

contains 10 terms. Analogously to what was done earlier for an equation with two variables, it can be shown that by a suitable rotation of the given coordinate system about the origin, equation (17) can be reduced to the form

$$A_1'x'^2 + A_2'y'^2 + A_3'z'^2 + 2C_1'x' + 2C_2'y' + 2C_3'z' + D = 0, \qquad (18)$$

i.e., so as to eliminate the terms with products of the variables. However, the proof here of the possibility of such a simplification of the equation is considerably more difficult than in the case of the plane. The difficulty of the proof arises from the fact that in the plane a rotation about a point is given by one angle ϕ, which we selected suitably, while in space the rotation of a body about a fixed point is given by three independent angles (Euler angles) ϕ, θ, ψ and in a quite complicated way. So the equation must be cleared the terms with products of variables in a roundabout way (see Chapter XVI on the theory of reduction by orthogonal transformations of a quadratic form to a sum of squares). Then, as in the case of the plane, a parallel translation of the axes is made and the equation is simplified, after which equation (18) finally assumes one of the following canonical forms:

1. $\dfrac{x^2}{a^2} + \dfrac{y^2}{b^2} + \dfrac{z^2}{c^2} - 1 = 0$ Ellipsoid

2. $\dfrac{x^2}{a^2} + \dfrac{y^2}{b^2} + \dfrac{z^2}{c^2} + 1 = 0$ Imaginary ellipsoid

3. $\dfrac{x^2}{a^2} + \dfrac{y^2}{b^2} - \dfrac{z^2}{c^2} - 1 = 0$ Hyperboloid of one sheet

4. $\dfrac{x^2}{a^2} + \dfrac{y^2}{b^2} - \dfrac{z^2}{c^2} + 1 = 0$ Hyperboloid of two sheets

5. $\dfrac{x^2}{a^2} + \dfrac{y^2}{b^2} - \dfrac{z^2}{c^2} = 0$ Second-order cone

6. $\dfrac{x^2}{a^2} + \dfrac{y^2}{b^2} + \dfrac{z^2}{c^2} = 0$ Imaginary second-order cone

7. $\dfrac{x^2}{a^2} + \dfrac{y^2}{b^2} - 2cz = 0$ Elliptic paraboloid

8. $\dfrac{x^2}{a^2} - \dfrac{y^2}{b^2} - 2cz = 0$ Hyperbolic paraboloid

9. $\dfrac{x^2}{a^2} + \dfrac{y^2}{b^2} - 1 = 0$ Elliptic cylinder

10. $\dfrac{x^2}{a^2} + \dfrac{y^2}{b^2} + 1 = 0$ Imaginary elliptic cylinder

11. $\dfrac{x^2}{a^2} + \dfrac{y^2}{b^2} = 0$ A pair of intersecting
                                                                 imaginary planes

12. $\dfrac{x^2}{a^2} - \dfrac{y^2}{b^2} - 1 = 0$ Hyperbolic cylinder

13. $\dfrac{x^2}{a^2} - \dfrac{y^2}{b^2} = 0$ A pair of intersecting planes

14. $y^2 - 2px = 0$ Parabolic cylinder

15. $x^2 - a^2 = 0$ A pair of parallel planes

16. $x^2 + a^2 = 0$ A pair of imaginary parallel
                                               planes

17. $x^2 = 0$ A pair of coincident planes.

The last nine canonical equations 9-17 do not contain terms in z and represent exactly the canonical equations of second-order curves in the Oxy-plane. In space these equations represent cylinders, whose directrices are the corresponding second-order curves in the Oxy-plane and whose generators are parallel to the Oz-axis. Indeed, if one of these equations is satisfied by a point with coordinates $(x_1, y_1, 0)$, then it will also be satisfied by any point with coordinates (x_1, y_1, z) whatever z may be, since there are in any case no terms with z in the equation.

Among the equations 1-8 as can easily be seen, equation 2 is not satisfied by any point with real x, y, z and equation 6 is satisfied only by one such point $(0, 0, 0)$, i.e., the origin. It remains, therefore, to study only the six equations 1, 3, 4, 5, 7, 8.

Ellipsoid. Let us compare the surfaces represented by the equations $x^2/a^2 + y^2/b^2 + z^2/c^2 - 1 = 0$ and $x^2 + y^2 + z^2 - 1 = 0$. The second of these is obviously the equation of a sphere C with center at the origin and with unit radius, since $x^2 + y^2 + z^2$ is the square of the distance from the point (x, y, z) to the origin O. If (x, y, z) is a point lying on the sphere, i.e., satisfying the second equation, then (ax, by, cz) is a point whose coordinates satisfy the first equation. The surface represented by the first equation is thus obtained from the sphere C if all abscissas x of points of the sphere are replaced by ax, y by by, and z by cz, i.e., if the

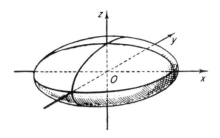

FIG. 53.

sphere C is uniformly stretched from the Oyz-, Oxz-, and Oxy-planes with coefficients of expansion a, b and c, respectively. This surface is called an ellipsoid (figure 53).

Hyperboloids and the second-order cone. Let us consider equations 3, 4, and 5, i.e., the equation of the form

$$\frac{x^2}{a^2} + \frac{y^2}{b^2} - \frac{z^2}{c^2} = \delta, \tag{19}$$

where $\delta = 1, -1$ or 0. Let us compare it with the equation

$$\frac{x^2}{a^2} + \frac{y^2}{a^2} - \frac{z^2}{c^2} = \delta, \tag{20}$$

in which the denominator of y^2 is also a^2 and not b^2, as in equation (19). As before, we observe that surface (19) is obtained from surface (20) by expansion from the Oxz-plane with coefficient b/a.

Let us now see what surface is represented by (20). We take a plane $z = h$ perpendicular to the Oz-axis and examine its intersection with the surface (20). Substituting $z = h$ in equation (20), we obtain

$$x^2 + y^2 = a^2 \left(\delta + \frac{h^2}{c^2} \right).$$

If $\delta + h^2/c^2$ is positive, then this equation together with $z = h$ gives a circle, lying in the plane $z = h$ with center on the Oz-axis. If $\delta + h^2/c^2$ is negative, which can be the case only with $\delta = -1$ and h^2 small, then the plane $z = h$ does not intersect surface (20) at all, since the sum of squares $x^2 + y^2$ cannot be a negative number.

The whole surface (20) thus consists of circles lying in planes perpendicular to the Oz-axis and having their centers on the Oz-axis. But in this case the surface (20) is a surface of revolution about the Oz-axis.

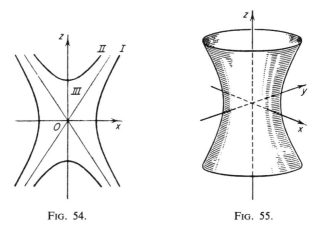

FIG. 54.   FIG. 55.

If we intersect it with a plane passing through the Oz-axis, we obtain its "meridian," i.e., a curve, lying in a plane passing through the axis, by the revolution of which the surface is generated.

If we intersect the surface (20) with the coordinate plane Oxz, i.e., the plane $y = 0$ (figure 54), by substituting $y = 0$ in equation (20), we obtain the equation of the meridian $x^2/a^2 - z^2/c^2 = \delta$. In case $\delta = 1$

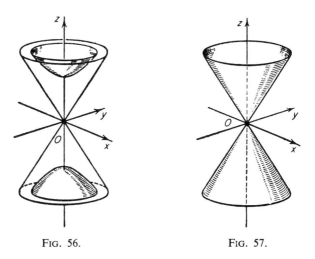

FIG. 56.   FIG. 57.

this is the hyperbola *I*, for δ = —1, it is the hyperbola *II*, and for δ = 0, the pair of intersecting lines *III*. By revolution around the *z*-axis these produce, respectively, a so-called hyperboloid of revolution of one sheet (figure 55), a hyperboloid of revolution of two sheets (figure 56) and a straight circular cone (figure 57).

The general hyperboloid of one sheet, hyperboloid of two sheets, and second-order cone 3, 4, and 5 are obtained from these surfaces of revolution by an expansion from the *Oxz*-plane with coefficient *b/a*.

Paraboloids. Only equations 7 and 8 remain. Let us compare the first of these $x^2/a^2 + y^2/b^2 = 2cz$ with the equation

$$\frac{x^2}{a^2} + \frac{y^2}{a^2} = 2cz,$$

which we investigate in the same way as before. It represents a surface obtained by revolving the parabola $x^2 = 2a^2cz$ about the *Oz*-axis,

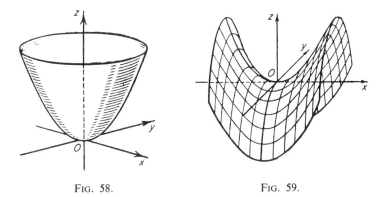

FIG. 58.   FIG. 59.

namely the so-called paraboloid of revolution (figure 58) discussed earlier in connection with parabolic mirrors. The general elliptic paraboloid 7 is obtained from the paraboloid of revolution by an expansion from the *Oxz*-plane.

The surface 8 has to be studied in a different way, namely by examining its intersections with planes $z = h$, which are hyperbolas. The contour map of the surface 8 is represented in figure 50; in a different position of the coordinate axes we considered this surface in figure 51. It is saddle-shaped, as illustrated in figure 59 and is called a hyperbolic paraboloid. Its intersections with planes parallel to the *Oxz*-plane turn out to be identical parabolas. The same result is obtained by intersections with planes parallel to the *Oyz*-plane.

Rectilinear generators of a hyperboloid of one sheet. It is a very curious and not at all obvious fact that the hyperboloid of one sheet and the hyperbolic paraboloid can be obtained, just like the cone and the cylinder, by the motion of a straight line. In case of the hyperboloid, it is sufficient to prove this fact for a hyperboloid of revolution of one sheet $x^2/a^2 + y^2/b^2 - z^2/c^2 = 1$, since the general hyperboloid of one sheet is obtained by a uniform expansion from the Oxz-plane and under such an expansion any straight line will go into a straight line. Let us

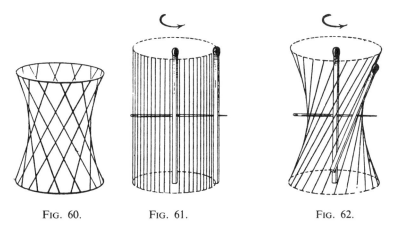

FIG. 60.   FIG. 61.   FIG. 62.

intersect the hyperboloid of revolution with the plane $y = a$ parallel to the Oxz-plane. Substituting $y = a$ we obtain

$$\frac{x^2}{a^2} + \frac{a^2}{a^2} - \frac{z^2}{c^2} = 1 \quad \text{or} \quad \frac{x^2}{a^2} - \frac{z^2}{c^2} = 0.$$

But this equation together with $y = a$ gives in the plane $y = a$ a pair of intersecting lines: $x/a - z/c = 0$ and $x/a + z/c = 0$.

Thus we have already discovered that there is a pair of intersecting lines lying on the hyperboloid. If now we revolve the hyperboloid about the Oz-axis, then each of these lines obviously traces out the entire hyperboloid (figure 60).

It is easy to show that: (1) two arbitrary straight lines of one and the same family of lines so obtained do not lie in the same plane (i.e., they are skew lines), (2) any line of one of these families intersects all the lines of the other family (except its opposite, which is parallel to it), and (3) three lines of one and the same family are not parallel to any one and the same plane.

With two matches and a needle it is easy to obtain a representation of the hyperboloid of revolution of one sheet. Let us puncture one of the matches through its middle by the needle, and on the sharp end point of the needle we pin the other match parallel to the first match. If we then revolve the whole apparatus about the first match as an axis, the second match will trace the surface of a cylinder (figure 61). But if the second match is not parallel to the first match, then during a revolution it will trace the surface of a hyperboloid of revolution of one sheet, as can easily be visualized if the rotation is rapid (figure 62).

Summary of the investigation of the second-degree equation. Although the general second-degree equation with three variables can represent essentially 17 different surfaces, it is not difficult to remember them. The last nine are cylinders over the nine possible second-order curves, while the first eight are divided into four pairs: two ellipsoids (real and imaginary), two hyperboloids (of one sheet and two sheets), two second-order cones (real and imaginary), and two paraboloids (elliptic and hyperbolic). All these surfaces play an essential role in mechanics, physics, and technology (ellipsoid of inertia, ellipsoid of elasticity, hyperboloid in the Lorentz transformation in physics, paraboloid of revolution for parabolic mirrors, etc.).

§11. Affine and Orthogonal Transformations

The next important step in the development of analytic geometry was the introduction into it, and into geometry in general, of the theory of transformations. Here it will be necessary to explain the matter in some detail.

"Contraction" of the plane toward a line. Let us consider one of the simplest transformations of the plane, namely uniform "contraction" toward a line with coefficient k. In the plane let there be given a line a and a positive coefficient k, for example, $k = 2/3$. All points of the line a are fixed, and every point M not lying on this line is sent into the point M' such that M' lies on the same side of the line as M on the perpendicular from M to a at a distance from a equal to $2/3$ of the distance from M to a. If the coefficient k, as here, is smaller than unity, then we have a proper contraction of the

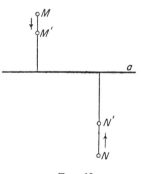

Fig. 63.

plane to the line; but if k is greater than unity, we have an expansion of the plane from the line, but for convenience we will in this and other cases talk about "contraction," except that the word "contraction" will be put in quotation marks.

The point or figure to be transformed is called the preimage and the one into which it is sent is its image. The point M', for example, is the image of the point M (figure 63).

We show that under a uniform "contraction" of a plane to a line, any line of the plane is transformed into a line. For let the plane be "contracted" to a line a lying in it with coefficient of "contraction" k. Let b be any line of the plane, O the point in which it intersects the line a, B another arbitrary point of b, and BA the perpendicular to the line a from the point B (figure 64). In the "contraction" the point B goes to

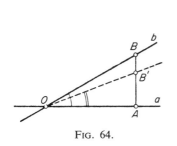

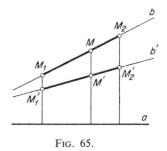

FIG. 64.   FIG. 65.

the point B' on this perpendicular such that $B'A = k \cdot BA$. Therefore, the tangent of the angle $B'OA$ will be equal to $AB'/OA = k \cdot AB/OA$, i.e., will be equal to k times the tangent of the angle which the line b makes with line a, i.e., for all points B' into which different points of the line b are transformed, it will be one and the same. All points B' consequently lie on one and the same line, passing through the point O and making with line a an angle with this tangent.

Under "contraction" parallel lines remain parallel. Indeed, if the tangents of the angles which lines b and c make with line a are the same, then the tangents of those angles which the images b' and c' make with a differ from them only by a factor k, i.e., they are still equal to each other, which means that the lines b' and c' are also parallel to each other.

Any rectilinear segment of the plane under "contraction" to a line is contracted (or expanded) uniformly (although to various degrees for segments of various directions). When we speak here of "uniform" contraction, we mean that the midpoint of the segment remains the midpoint, the third remains the third, etc., i.e., the segment shrinks uniformly along its full length. Indeed, in whatever ratio the point M

divides the segment M_1M_2, its image M' will divide $M'_1M'_2$, in the same ratio, since parallel lines (in this case perpendiculars to the line a) cut lines intersecting them (in this case b and b') in proportional parts (figure 65).

The ellipse as the result of "contraction" of a circle. We consider a circle with center at the origin and radius a. By the theorem of Pythagoras its equation is $x^2 + \bar{y}^2 = a^2$, where we have written $\bar{y}$ instead of y, since y will be needed later. Let us see what this circle is contracted into if we "contract" the plane to the Ox-axis with coefficient b/a (figure 66). After this "contraction" the x-values of all points remain the same, but the $\bar{y}$-values become equal to $y = \bar{y}(b/a)$, i.e., $\bar{y} = (a/b)\,y$. Substituting for $\bar{y}$ in the above equation of the circle, we will have:

$$x^2 + \frac{a^2}{b^2}y^2 = a^2 \quad \text{or} \quad \frac{x^2}{a^2} + \frac{y^2}{b^2} = 1$$

as the equation, in the same coordinate system, of the curve obtained from the given circle by contraction to the Ox-axis. As we see, we obtain an ellipse. Thus we have proved that an ellipse is the result of a "contraction" of a circle.

From the fact that an ellipse is a "contraction" of a circle, many properties of the ellipse follow directly. For example, the aforementioned property of diameters, namely that if parallel secants of an ellipse are given, then their midpoints lie on a straight line (see figure 12), can be shown in the following way. We perform the inverse expansion of the ellipse into the circle. Under this expansion parallel chords of the ellipse go into parallel chords of the circle, and their midpoints into the

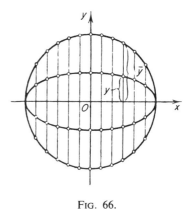

FIG. 66.

midpoints of these chords. But the midpoints of parallel chords of a circle lie on a diameter, i.e., on a straight line, and so that the midpoints of parallel chords of the ellipse also lie on a straight line. Namely, they lie on that line which is obtained from the diameter of the circle under the "contraction" which sends the circle into the ellipse.

Here is another application of the theory of "contraction." Since any vertical strip of the circle under its contraction to the Ox-axis does not change its width and its length is multiplied by b/a, the area of this strip after contraction is equal to its initial area multiplied by b/a, and since the area of the circle is equal to πa^2, the area of the corresponding ellipse is equal to $\pi a^2 (b/a) = \pi ab$.

Example of the solution of a more complicated problem. Let an ellipse be given and let it be required to find the triangle with smallest area circumscribed to this ellipse. We first solve the problem for a circle. We show that in the case of a circle, this is an equilateral triangle. Indeed, let the circumscribed triangle be nonequilateral; i.e., the smallest of its angles (denoted by B) is less than $60°$, and the largest $C > 60°$. If then, without varying the angle A, we move side BC into the position B_0C_0 (figure 67) by shifting the vertex B toward A until one of the angles B_0 or C_0 becomes equal to $60°$, we obtain a circumscribed triangle AB_0C_0

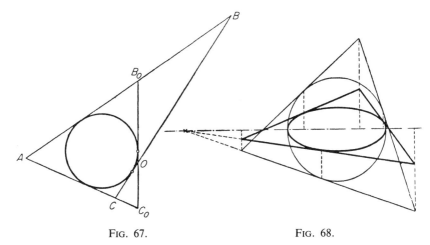

FIG. 67.   FIG. 68.

with smaller area, since here* $OC < OB$, $OC_0 \leqslant OB_0$ and therefore the discarded area OBB_0 is greater than the added one OCC_0. If the triangle so obtained is not equilateral, then by repeating the above procedure we reduce its area still further and arrive at an equilateral triangle. Hence, any nonequilateral triangle circumscribed to a given circle has a greater area than an equilateral one.

We now return to the ellipse. Let us make an expansion of it from the major axis, thereby converting it back into the circle from which it was obtained by "contraction." Under this expansion (figure 68): (1) all

---

* As can be easily shown.

triangles circumscribed to the ellipse are transformed into triangles circumscribing the resultant circle; (2) the areas of all figures, and in particular of these triangles are increased in one and the same ratio. From this we see that the triangles circumscribing the given ellipse with the smallest area will be those that are converted into equilateral triangles circumscribing the circle. There are infinitely many such triangles; each of them has its center of gravity at the center of the ellipse and the points of tangency are in the middle of its sides. Any of these triangles can easily be constructed (figure 68), starting from the aforementioned circle.

"Contractions" of the plane to a line are only a particular case of more general, so-called *affine* transformations of the plane.

General affine transformations. A pair of vectors e_1, e_2 starting from a common origin O and not lying on the same line will be called a co-ordinate "frame" of the plane. The coordinates of a point M of the plane relative to this frame Oe_1e_2 will then be numbers x, y such that in order to reach the point M from the origin O it is necessary to lay off from the point O x-times the vector e_1 and then y-times the vector e_2. This is a general Cartesian coordinate system of the plane. Analogously, a general Cartesian coordinate system can be introduced in space. The ordinary, so-called rectangular Cartesian coordinate system that we have made use of up to now corresponds to the particular case when the coordinate vectors e_1, e_2 are mutually perpendicular and their lengths are equal to the unit of measurement.

A general affine transformation of the plane is one under which a given net of equal parallelograms is transformed into another arbitrary net of equal parallelograms. More precisely, it is a transformation of the plane under which a given coordinate frame Oe_1e_2 is transformed into a certain other frame (generally speaking, with another "metric," i.e., with different lengths for the vectors e_1' and e_2' and a different angle between them) and an arbitrary point M is sent into the point M' having the same coordinates relative to the new frame as M had relative to the old (figure 69).

"Contraction" to the Ox-axis with coefficient k is a special case in which the rectangular frame $Oe_1'e_2'$ passes into the frame $Oe_1'ke_2'$.

It can easily be shown that under an affine transformation every straight line is sent into a straight line, parallel lines are mapped into parallel lines, and if a point divides a segment in a given ratio, then its image divides the image of this segment in the same ratio. Moreover, we can prove the remarkable theorem that any affine transformation of the plane can be obtained by performing a certain rigid motion of the plane onto itself, and then, in general, two "contractions" with different coefficients k_1 and k_2 to two mutually perpendicular lines.

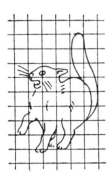

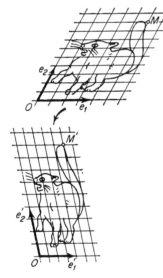

FIG. 69.

For the proof of this assertion, we consider all radii of some circle of the plane (figure 70). Let radius OA be the one which, after the transformation, turns out to be the shortest, and let it be mapped into $O'A'$. The perpendicular AB to OA is then transformed into $A'B'$, which must be perpendicular to $O'A'$, since if the perpendicular $O'C'$ were different from $O'A'$, then it would be the image of the oblique OC, and the image $O'D'$ of the radius OD would be a part of the perpendicular $O'C'$, i.e., shorter than the oblique $O'A'$, contrary to assumption.

The mutually perpendicular lines OA and AB are therefore mapped into mutually perpendicular lines $O'A'$ and $A'B'$. Consequently, the square net constructed on OA and AB is transformed into a net of equal rectangles (figure 71) and uniform "contractions" take place along the straight lines of this square net.

In a completely analogous way a general affine transformation of space can be defined as one under which a space coordinate frame $Oe_1e_2e_3$ is transformed into some other frame $O'e_1'e_2'e_3'$, generally speaking, with a different "metric," i.e., with unit segments of different lengths and with different angles between them, and a point M is sent into point M' having the same coordinates relative to the new frame as those of the point M relative to the old frame.

All the properties enumerated here also hold for affine transformations of space, except that in the last theorem there will be a rigid motion of space and then three "contractions" to three mutually perpendicular planes with certain coefficients k_1, k_2, k_3.

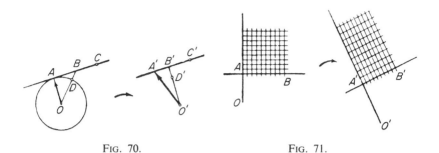

FIG. 70.   FIG. 71.

Applications of affine transformations. The most important applications of affine transformations are:

1. In the first place there is the application in geometry to solving problems concerning affine properties of figures, i.e., properties that are preserved under affine transformations. The theorem about the diameters of an ellipse and the problem of circumscribed triangles were examples. To solve such problems we make an affine transformation of the figure to some simpler one, for which we prove the desired property and then return to the original figure.

2. Second, there is the application in analytic geometry to the classification of second-order curves and surfaces. The main point is, as can be shown, that different ellipses are related to one another in the sense that one can be obtained from another by an affine transformation (the Latin word *affinis* means "related"). Also all hyperbolas are affine to one another, and so are all parabolas. But we cannot convert an ellipse into a parabola, or a hyperbola into a parabola, by an affine transformation, i.e., they are not affinely related to one another. It is natural to divide up all second-order curves into affine classes of curves, affinely related to one another. It turns out that the reduction of an equation to canonical form gives exactly this classification; i.e., there are nine affine classes of second-order curves. (We will not go into detail why imaginary ellipses and pairs of imaginary parallel lines belong to different affine classes. Properly speaking, neither in one case nor in the other are there any curves on the plane at all. The question here is really about algebraic properties of the equation itself.)

Similarly, the classification of second-order surfaces according to their canonical equations into 17 forms is the same as the affine classification.

Let us give a simple example of the application of the affine classification of second-order surfaces. We show that if we arbitrarily select in space three lines *a*, *b*, *c* such that (1) any two of them do not lie in the same plane (i.e., they are skew to each other) and (2) they are not all parallel to one and the same plane, then the set of all straight lines d of space, each of which simultaneously intersects all three given lines a, b, c (figure 72) constitutes the entire surface of a hyperboloid of one sheet.

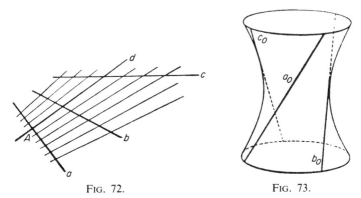

FIG. 72.   FIG. 73.

Let us explain more fully the set of lines d we are discussing here. Through an arbitrary point A of line a, we can pass a plane P containing the line b and a plane Q containing the line c. These planes P and Q intersect in a unique line d, which passes through the point A of line a and intersects lines b and c. Drawing all such lines d through arbitrary points of line a, we obtain the set of all those lines d of space each of which intersects all three given lines a, b, and c. This collection of lines determines a surface. We note that any given hyperboloid of one sheet can be obtained in this way, since we only need to take for the lines a, b, and c three distinct straight lines a_0, b_0, c_0 of one family (figure 73) and for the lines d all the straight lines of the other family. Conversely, let there be given three arbitrary pairwise skew lines of space a, b, c, not all parallel to one and the same plane. Then, as can be shown, these lines always form the three edges (without common points) of some parallelepiped (figure 74). After constructing such parallelepipeds for the given lines a, b, c and for three lines a_0, b_0, c_0 of one and the same family of an arbitrary hyperboloid of one sheet, we make an affine transformation of space that sends the parallelepiped a_0, b_0, c_0 into the parallelepiped a, b, c; obviously, this transformation maps the hyperboloid onto the surface in question. But according to the affine classification of second-order surfaces, the affine image of a hyperboloid of one sheet is again a hyperboloid of one sheet.

3. Third, there is the application to the theory of continuous transformations of continuous media, for example, in the theory of elasticity,

in the theory of electric or magnetic fields, etc. Very small elements of the given continuous medium transform "almost" affinely. We say, "in the small the transformation is linear" (we call a first-degree expression linear, and in the following section we see that in analytic geometry the formulas of affine transformations are of the first degree). This is evident in figure 75. On the lines of the large square net, their distortion or

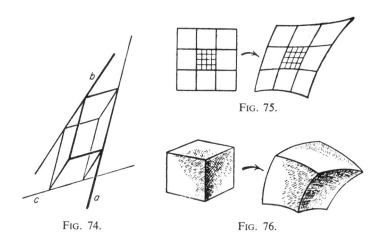

Fig. 75.

Fig. 74.   Fig. 76.

"fanning out" is clearly noticeable. But for a small piece of the very dense square net, all this shows itself very little, and the square net transforms "almost" into a net of equal parallelograms. A similar picture is obtained in space also (figure 76). By the fact that any affine transformation of space reduces to a motion and three mutually perpendicular "contractions," it follows that an element of a body under an elastic deformation first moves as a rigid body and then undergoes three mutually perpendicular "contractions."

Formulas of affine transformations. If the frame Oe_1e_2 is affinely transformed and $O'e_1'e_2'$ is its image, while the coordinates of the new origin O' relative to the old frame are ξ, η and the coordinates of the vectors e_1' and e_2' relative to the old frame are a_1, a_2 and b_1, b_2, respectively, then the formulas of the affine transformation, as can be easily seen from figure 77, are

$$x' = a_1x + b_1y + \xi,$$
$$y' = a_2x + b_2y + \eta$$

in the sense that if x, y are the coordinates of any point M relative to the old frame Oe_1e_2, then x', y' given by these formulas, are the coordinates relative to the same frame of the image M' of this point.

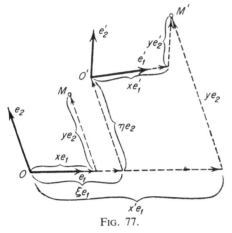

FIG. 77.

Indeed, let Oe_1e_2 be a frame before transformation, and $O'e_1'e_2'$ its image, while M is an arbitrary point of the plane and M' is its image. Then by the very definition of an affine transformation, if the coordinates of the point M relative to the frame Oe_1e_2 are x, y, then the coordinates of its image M' relative to the image $O'e_1'e_2'$ of this frame are exactly the same x, y.

Now consider a vector $\boldsymbol{m}'$ joining the origin O of the old frame to the image M' of the point M. Then $\boldsymbol{m}' = x'\boldsymbol{e}_1 + y'\boldsymbol{e}_2$. But this vector is equal to a certain vector sum

$$\boldsymbol{m}' = \xi\boldsymbol{e}_1 + \eta\boldsymbol{e}_2 + x\boldsymbol{e}_1' + y\boldsymbol{e}_2',$$

and the vectors $\boldsymbol{e}_1'$ and $\boldsymbol{e}_2'$ are

$$\boldsymbol{e}_1' = a_1\boldsymbol{e}_1 + a_2\boldsymbol{e}_2, \quad \boldsymbol{e}_2' = b_1\boldsymbol{e}_1 + b_2\boldsymbol{e}_2$$

so that

$$\boldsymbol{m}' = \xi\boldsymbol{e}_1 + \eta\boldsymbol{e}_2 + a_1x\boldsymbol{e}_1 + a_2x\boldsymbol{e}_2 + b_1y\boldsymbol{e}_1 + b_2y\boldsymbol{e}_2$$

or

$$\boldsymbol{m}' = (a_1x + b_1y + \xi)\boldsymbol{e}_1 + (a_2x + b_2y + \eta)\boldsymbol{e}_2.$$

Comparing this expression with the first expression for $\boldsymbol{m}'$ we obtain

$$\begin{aligned} x' &= a_1x + b_1y + \xi, \\ y' &= a_2x + b_2y + \eta. \end{aligned} \tag{21}$$

The determinant

$$\Delta = \begin{vmatrix} a_1 & b_1 \\ a_2 & b_2 \end{vmatrix} = a_1b_2 - a_2b_1,$$

as can be shown, is not zero and is equal to the ratio of area of the parallelogram constructed on the vectors of the new frame to the area

of the same parallelogram constructed on the vectors of the old frame.
Analogous formulas are obtained for space

$$\left.\begin{aligned}
x' &= a_1x + b_1y + c_1z + \xi, \\
y' &= a_2x + b_2y + c_2z + \eta, \\
z' &= a_3x + b_3y + c_3z + \zeta,
\end{aligned}\right\} \tag{22}$$

where $(\xi,\ \eta,\ \zeta)$ are the coordinates of the origin O' of the transformed
frame $O'e_1'e_2'e_3'$ and $(a_1, a_2, a_3), (b_1, b_2, b_3), (c_1, c_2, c_3)$ are the coordinates of its vectors e_1', e_2', e_3' relative to the old frame $Oe_1e_2e_3$.

The determinant*

$$\Delta = \begin{vmatrix} a_1 & b_1 & c_1 \\ a_2 & b_2 & c_2 \\ a_3 & b_3 & c_3 \end{vmatrix} = a_1b_2c_3 + a_2b_3c_1 + a_3b_1c_2 - a_1b_3c_2 - a_2b_1c_3 - a_3b_2c_1$$

is not zero and is equal to the ratio of the volume of the parallelepiped
formed by the vectors of the new frame to the volume of the parallelepiped
formed by the vectors of the old frame.

Orthogonal transformations. Rigid motions of the plane onto itself
or such motions plus a reflection about a line lying in the plane, are
called *orthogonal transformations of the plane*, and rigid motions of space,
or such motions plus a reflection of the space about one of its planes,
are called *orthogonal transformations of space*. It is clear that orthogonal
transformations are affine transformations under which the "metric" of
the frame does not change, since it only undergoes a rigid motion, or
else such a motion plus a reflection.

We will investigate orthogonal transformations by means of rectangular
coordinates, i.e., when the vectors of the original frame are mutually
perpendicular and have lengths equal to the unit of measurement. After
an orthogonal transformation the vectors of the frame remain mutually
perpendicular, i.e., their scalar product remains equal to zero and their
lengths remain equal to unity. Therefore (see formula (14), this chapter)
in the case of the plane, we have

$$a_1b_1 + a_2b_2 = 0, \quad a_1^2 + a_2^2 = 1, \quad b_1^2 + b_2^2 = 1, \tag{21'}$$

and in the case of space

$$\begin{aligned}
a_1b_1 + a_2b_2 + a_3b_3 &= 0, & a_1^2 + a_2^2 + a_3^2 &= 1, \\
a_1c_1 + a_2c_2 + a_3c_3 &= 0, & b_1^2 + b_2^2 + b_3^2 &= 1, \\
b_1c_1 + b_2c_2 + b_3c_3 &= 0, & c_1^2 + c_2^2 + c_3^2 &= 1.
\end{aligned} \tag{22'}$$

---

* On determinants see Chapter XVI.

Hence, if the initial frame is taken to be rectangular, then formulas (21) give an orthogonal transformation if and only if the conditions (21′) of orthogonality are fulfilled, and formulas (22) give an orthogonal transformation of space if the conditions (22′) of orthogonality are satisfied. It can be shown that if $\Delta > 0$, we have a rigid motion, and if $\Delta < 0$ a rigid motion plus a reflection.

§12. Theory of Invariants

The concept of invariant.* Invariants of a second-degree equation with two variables. In the second half of the last century still another important new concept was introduced, that of invariant.

Consider, for example, a second-degree polynomial in two variables

$$Ax^2 + 2Bxy + Cy^2 + 2Dx + 2Ey + F. \tag{23}$$

If we regard x, y as rectangular coordinates and make a transformation to new rectangular axes, then after replacing x, y in (23) by their expressions in terms of the new coordinates x', y', removing parentheses, and reducing similar terms, we obtain a new transformed polynomial with different coefficients

$$A'x'^2 + 2B'x'y' + C'y'^2 + 2D'x' + 2E'y' + F'. \tag{24}$$

It turns out that there exist expressions formed from the coefficients which under this transformation do not change their numerical value, although the coefficients themselves change. Such an expression in A', B', C', D', E', F' has exactly the same numerical value as when it is formed with the A, B, C, D, E, F.

Expressions of this kind are called *invariants* of the polynomial (23) with respect to the group of orthogonal transformations (i.e., relative to transformations from one set of rectangular coordinates x, y to any other rectangular coordinates x', y').

Invariants of this sort, as it turns out, are

$$I_1 = A + C,$$

$$I_2 = \begin{vmatrix} A & B \\ B & C \end{vmatrix} = AC - B^2,$$

$$I_3 = \begin{vmatrix} A & B & D \\ B & C & E \\ D & E & F \end{vmatrix} = ACF + 2BDE - AE^2 - CD^2 - FB^2,$$

---

* *Invarians* in Latin means "unchanged."

i.e.,

$$A + C = A' + C', \quad AC - B^2 = A'C' - B'^2,$$
$$ACF + 2BDE - AE^2 - CD^2 - FB^2$$
$$= A'C'F' + 2B'D'E' - A'E'^2 - C'D'^2 - F'B'^2.$$

It is possible to prove the important theorem that any orthogonal invariant of the polynomial (23) can be expressed in terms of these three basic invariants.

If we equate the polynomial (23) to zero, we obtain an equation of some second-order curve. Any quantity, connected with this curve but not with its location in the plane, will clearly not depend on what coordinates its equation is written in, and therefore, when expressed in terms of the coefficients, it will be an orthogonal invariant of the polynomial (23), and thus it will be expressible in terms of the three basic invariants. Moreover, since under multiplication of all six coefficients of the equation by any given number t (different from zero) the curve represented by the equation remains the same, an expression of any property of the curve in terms of the I_1, I_2, I_3 must certainly be such that if the A, B, C, D, E, F in it are multiplied by t, the number t cancels out. The expression in question must be, as they say, homogeneous of degree zero relative to A, B, C, D, E, F.

Let us verify this by an example. For instance, let the equation

$$Ax^2 + 2Bxy + Cy^2 + 2Dx + 2Ey + F = 0$$

represent an ellipse. Since the equation completely determines this ellipse, we can calculate from it (i.e., from its coefficients) all the basic quantities connected with the ellipse. For example, we can calculate its semiaxes a and b, i.e., we can express the semiaxes in terms of the coefficients. The expressions for these semiaxes will be invariants and therefore, expressible in terms of I_1, I_2, I_3. By reduction of the equation to canonical form and some subsequent calculation, the following rather complicated expressions for the semiaxes are obtained in terms of I_1, I_2, I_3:

$$\sqrt{\frac{2\,|I_3|}{|I_2|\,|I_1 \pm \sqrt{I_1^2 - 4I_2}|}},$$

which are homogeneous relative to A, B, C, D, E, F.

From this it is clear that the invariants I_1, I_2, I_3, themselves, being homogeneous but not of degree zero, do not have straightforward geometric meanings; they are algebraic entities.

It can be shown that the expression

$$K_1 = \begin{vmatrix} A & D \\ D & F \end{vmatrix} + \begin{vmatrix} C & E \\ E & F \end{vmatrix} = AF - D^2 + CF - E^2$$

can be varied by parallel translation but not by pure rotation of the given rectangular axes, and it is therefore called a *semi-invariant*.

As an example of an application of invariants and semi-invariants, we give Table 1, which if we calculate I_1, I_2, I_3 and K_1, allows us to determine directly from its equation the affine class of a second-order curve.

In Table 1 the necessary and sufficient conditions are given that an

Table 1

Criterion of the class	Name	Reduction equation	Canonical equation
$I_2 > 0$, $I_1 I_3 < 0$	ellipse		$\dfrac{x^2}{a^2} + \dfrac{y^2}{b^2} = 1$
$I_2 > 0$, $I_1 I_3 > 0$	imaginary ellipse		$\dfrac{x^2}{a^2} + \dfrac{y^2}{b^2} = -1$
$I_2 > 0$, $I_3 = 0$	point	$\lambda_1 x^2 + \lambda_2 y^2 + \dfrac{I_3}{I_2} = 0$	$\dfrac{x^2}{a^2} + \dfrac{y^2}{b^2} = 0$
$I_2 < 0$, $I_3 \neq 0$	hyperbola		$\dfrac{x^2}{a^2} - \dfrac{y^2}{b^2} = 1$
$I_2 < 0$, $I_3 = 0$	pair of intersecting lines		$\dfrac{x^2}{a^2} - \dfrac{y^2}{b^2} = 0$
$I_2 = 0$, $I_3 \neq 0$	parabola	$I_1 x^2 + 2\sqrt{-\dfrac{I_3}{I_1}}\, y = 0$	$x^2 = 2py$
$I_1 = 0, I_3 = 0$, $\quad K_1 < 0$	pair of parallel lines		$x^2 = a^2$
$I_2 = 0$, $I_3 = 0$, $\quad K_1 > 0$	pair of imaginary parallel lines	$I_1 x^2 + \dfrac{K_1}{I_1} = 0$	$x^2 = -a^2$
$I_2 = 0$, $I_3 = 0$, $\quad K_1 = 0$	pair of coincident lines		$x^2 = 0$

equation of a second-order curve be reducible to one or another of the nine canonical forms ($I_1 I_3$ designates the product of I_1 and I_3).

Consider, for example, the equation $x^2 - 6xy + 5y^2 - 2x + 4y + 3 = 0$. We have $A = 1$, $B = -3$, $C = 5$, $D = -1$, $E = 2$, $F = 3$, so that $I_1 = 6$, $I_2 = -4$, $I_3 = -9$. The conditions of the 4th line of the table are satisfied: $I_2 < 0$, $I_3 \neq 0$, i.e. this is a hyperbola. Its semiaxes are equal to

$$\sqrt{\frac{2 \cdot 9}{4 \cdot |6 \pm \sqrt{36 + 16}|}} \approx 0.57 \quad \text{and} \quad 1.93.$$

The coefficients of the reduced equation (I), (II) and (III) are given in terms of invariants and semi-invariants as follows:

$$\lambda_1 x''^2 + \lambda_2 y''^2 + \frac{I_3}{I_2} = 0, \tag{I}$$

$$I_1 x''^2 + 2\sqrt{-\frac{I_3}{I_1}}\, y'' = 0, \tag{II}$$

$$I_1 x''^2 + \frac{K_1}{I_1} = 0, \tag{III}$$

where λ and λ_2 are the roots of the so-called characteristic quadratic equation

$$\lambda^2 - I_1 \lambda + I_2 = 0.$$

Formulas (I-III) allow a quick calculation of the semiaxes a and b of an ellipse and a hyperbola, the parameter p of an ellipse and the distance $2a$ between parallel lines. The formulas for semiaxes were given earlier. The parameter p is equal to

$$p = \sqrt{-\frac{I_3}{I_1^3}} \quad \text{and the distance} \quad 2a = 2\sqrt{-\frac{K_1}{I_1^2}}.$$

A completely analogous theory of invariants and semi-invariants, with a corresponding table for the determination of the affine class and the formulas of the coefficients of reduced equations, can be given for second-order surfaces in three-dimensional space.

It should be pointed out that so far we have been discussing only those invariants that are considered in analytic geometry for curves and surfaces of the second order. The concept of invariant, however, has a far broader meaning.

By an invariant of some object under study, relative to certain of its transformations, we mean any quantity numerical, vectorial, etc. connected with this object that does not vary under these transformations. In the previous problem the object is a second-degree polynomial with two variables (i.e., more precisely, the set of its coefficients), and the transformations are those of the polynomial obtained by the transition from one rectangular coordinate system to another.

Another example: The object is a given mass of a given gas under a given temperature. The transformations are changes in volume or pressure of this mass of gas. The invariant, according to the Boyle-Mariotte law, is the product of the volume by the pressure. We can speak of lengths of segments in space or the size of angles as invariants of the group of motions of space, of ratios in which a point divides a segment, or of ratios of areas, as the invariants of the group of affine transformations of space, etc.

Various invariants are particularly important in physics.

§13. Projective Geometry

Perspective projections.   Artists began long ago to study the laws of perspectivity. This was necessary because a human being sees objects in perspective projection on the retina of the eye, in such a way that the

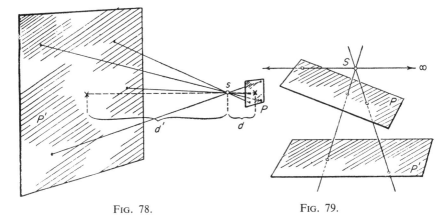

FIG. 78.   FIG. 79.

form and mutual location of objects are distorted in a characteristic manner. For example, telegraph poles in the distance look smaller and closer together, parallel tracks of a railway seem to converge, etc. We will not consider here space perspectivities, i.e., properties of perspective projections of objects in space onto a plane but only the properties of perspective projections of a plane onto a plane.

Let us consider a photograph (for example, one frame of a moving picture film) P, a screen P', and between them a lens S (figure 78). Then, if the photograph is transparent and is illuminated from behind (if it is nontransparent, let it be illuminated from the front, i.e., from the side where the lens is), then the illuminated points of the photograph radiate beams of light, which are collected by the lens in such a way that they appear again on the screen P' in the form of points. We will assume that this projection takes place as if the points of the photograph P were projected on the screen P' on straight lines passing through the optical center S of the lens.

The situation will be a very simple one if the planes P and P' are parallel. In this case we will obviously obtain on the plane P' an undistorted image of everything that is on the plane P. This image will be smaller or larger than the original depending on whether the ratio $d':d$, where d and d' are the distances from the center of the lens to the planes P and P' respectively, is smaller or larger than 1.

The situation will be considerably more difficult if the planes P and P' are not parallel (figure 79). In this case, under projection through the point S not only the size of the figure changes but also its form is distorted. Parallel lines under such projection may become convergent, the ratio in which a point divides a segment may change, etc. In general, some of the relations that remain invariant under arbitrary affine transformation may change here.

This sort of projection takes place, for example, in aerial photography. The airplane oscillates in flight and therefore the photographic apparatus (figure 80a) rigidly attached to it is, in general, not oriented altogether vertically but at the moment of exposure is usually in an oblique position, i.e., we obtain a distorted image of the locality (which we assume to be plane).

How are we to correct this image? For this it is necessary to study the properties of projection of a plane P onto another plane II (in general, the two planes are not parallel) by lines passing through a point S which is not on plane P nor on plane II. Such projections are called *perspective projections*.

We will prove later the following important theorem.

Theorem. *If we have two perspective projections of a plane P on plane* II *such that under both projections the points A, B, C, D of a quadruple of points of "general position" on the plane P (i.e., a quadruple in which no three of the points lie on one line), are projected into the same points A'*,

B', C', D' respectively of plane II, *then all points of plane P are also projected under both projections into the same points of plane* II.

In other words, the result of a perspective projection is completely determined if it is known into which points this projection sends the points of an arbitrary quadruple of points of general position in the figure to be projected.

This is the so-called uniqueness theorem of the theory of projective transformations or the fundamental theorem of plane perspectivity.

Application of the fundamental theorem of plane perspectivity in aerial photography. Let us show how this theorem provides a suitable method for correcting this image in photography.

If at the moment of aerial exposure, we imagine a horizontal screen II placed at a distance h below the center S of the lens (figure 80a), then the projection onto this screen through the center S of the image recorded on the photographic plate P will obviously not be distorted but will be similar to the horizontal locality with a scale $h:H$, where H is the height of the airplane at the moment of exposure. In order to correct the image received on the photograph P so as to convert it into an undistorted image, we treat it as follows. The developed photograph P is placed in a projecting apparatus resting on a special tripod on which, by means of adjustable screws, the apparatus can be moved closer to the screen II or farther from it and can be rotated in every way.

To the screen II (figure 80b) we attach a topographical map of the locality made by measurements on the surface of the Earth (not a detailed map, since the details of interest to us are to be provided by the aerial photograph). On this map attached to the screen II we select four points A', B', C', D' that can be found easily on the photograph also (for example, an intersection of roads, a corner of a house, etc.), and at the corresponding points A, B, C, D of the picture P we pierce the film with a needle. We then place a projection lamp behind the plate P in such a position that the picture is projected onto the screen II through a lens S of the supporting apparatus. By using the adjustable screws we arrange that the light beams from the pinholes fall on the corresponding points A', B', C', D' of the map attached to the screen. After this has been done, we replace the topographical map by a plateholder with a photographic plate and then, without changing the settings of the screws, we photograph the image projected on the screen II of the picture P taken from the airplane.

By the theorem stated previously, we thereby obtain a true (i.e., similar to the locality) and not a distorted map of the photographed region.

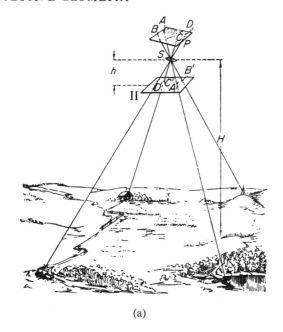

(a)

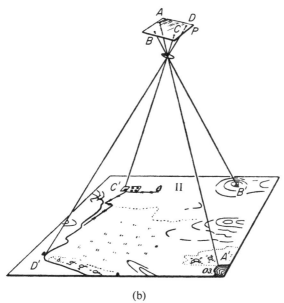

(b)

FIG. 80.

We now pass to the presentation of the theory necessary for proving the fundamental theorem.

The projective plane. The totality of *all* lines and planes of space passing through a given point S of the space is called the projecting bundle of lines and planes with center S. If this bundle is intersected by a plane P, not passing through the center, then to every point of the plane P will correspond a line of the bundle intersecting the plane P in this point, and to each line of the plane P will correspond that plane of the bundle which intersects the plane P along this line. However, we do not in this way establish a one-to-one mapping from the set of lines and planes of the bundle on the set of points and lines of the plane P. As a matter of fact, the lines and planes of the bundle which are parallel to the plane P do not in this sense correspond to any points or lines of the plane P, since they do not intersect it. Nevertheless, we *agree* to say that these lines of the bundle intersect the plane P but in its ideal (or infinitely distant) points, lying in the corresponding directions, and that such a plane of the bundle intersects the plane P along an ideal (or infinitely distant) line. The plane P, complemented by these ideal points and ideal line, is called a *complemented* or *projective plane*. We will denote it by P^*. The sets of lines and planes of the bundle S are then mapped one-to-one onto the sets of points (real and ideal) and lines (real and ideal) of this projective plane P^*.

Hence, we *agree* to say that a point (real or ideal) lies on a line (real or ideal) of the projective plane P^* if the corresponding line of the bundle lies in the corresponding plane of the bundle. From this point of view, any two lines of the projective plane intersect (in a real or ideal point), since any two planes of the bundle intersect along some line of the bundle. It follows from this, among other things, that the ideal line consists simply of the set of all ideal points.

In essence, the complementation of the plane by its ideal elements means that we use this plane as a cross section to study the bundle of all lines and planes passing through one point.

Projective mappings; the fundamental theorem. By a *projective mapping* we understand such a mapping of a projective plane P^* onto some other projective plane $P^{*\prime}$ (which can also coincide with the plane P^*, in which case we speak of a projective transformation of the plane P^*), which, first of all, is pointwise a one-to-one mapping, and second, is such that collinear sets of points of the plane P^* go into collinear sets of points of the plane $P^{*\prime}$, and conversely. (Here, by points and lines we always understand real as well as ideal points and lines.)

It is clear that two arbitrary perspective projections of one and the same plane P^* onto a plane II^* may be obtained from each other by projective transformations.

In fact, 1, their points (real or ideal) are in one-to-one correspondence with the points (real or ideal) of the protective plane P^* and consequently with each other, and 2, collinear points of the first projection correspond to collinear points of the plane P^* and consequently also of the second projection, and conversely. Therefore, the aforementioned theorem of the theory of perspectivities is a direct consequence of the following theorem about projective transformations: if under a projective transformation of the plane II^*, four of its points A, B, C, D, forming a quadruple of general position remain fixed, then all of its points remain fixed.

Let us outline the idea of the proof of this theorem by means of the so-called Möbius net.

We note that (1) if under a projective transformation two points remain fixed, then the line that passes through them is mapped into itself, and (2) if two lines are mapped into themselves, then the point of their intersection remains fixed. Therefore, from the fact that the points A, B, C, D of the plane II^* remain fixed, it follows in turn that also the points E, F, G, H, K, L, etc. remain fixed (figure 81). The con-

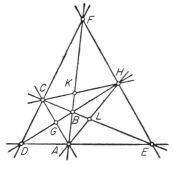

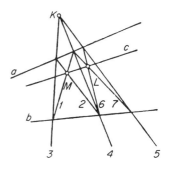

FIG. 81.   FIG. 82.

struction of such points can be continued by joining the points already obtained. This is the so-called Möbius net. By continuing its construction, we can find points as densely placed as we like. It can be shown that the set of these nodes everywhere densely covers the whole plane. Therefore, if we further assume the continuity of a projective transformation (which in fact, already follows from its definition, although the proof of this fact is not easy), the result is that if under a projective transformation of the plane II* the points A, B, C, D remain fixed, then all the points of the plane II* remain fixed.

Projective geometry. By *two-dimensional projective geometry* we mean the totality of theorems about those properties of figures in the projective plane, i.e., the ordinary plane complemented by ideal elements, which do not change under arbitrary projective transformations.

Here is an example of a problem of projective geometry. Given two lines a and b and a point M (figure 82), the problem is to construct the line c passing through the point M and through the point of intersection of lines a and b, not using this point of intersection (as may be necessary, if this point is very distant). If through the point M we draw the two secants 1 and 2 and then the lines 3 and 4 through the points of their intersection with lines a and b, we obtain the point K. Let us draw through it line 5 and secants 6 and 7; then it can be shown that the line c passing through point L of intersection of lines 6 and 7 and point M, is the desired line.

From the theory of conic sections, it follows (figure 83) that the ellipse, hyperbola and parabola are perspective projections of one another, and moreover all of them are perspective projections of the circle.

If we regard perspective projections as projective transformations of projective planes P^* and $P^{*\prime}$ one onto the other, then by superposing these planes we obtain the result that all ellipses, hyperbolas, and parabolas are projective transformations of the circle. The difference in them

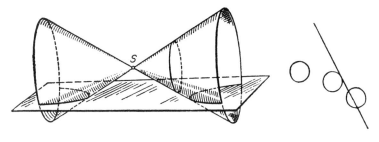

FIG. 83.   FIG. 84.

is that projective images of the circle under transformations in which a line not intersecting the circle is mapped into the infinitely distant line are ellipses; on the other hand, if a line tangent to the circle is mapped into the infinitely distant line, then a parabola is obtained, and if a secant, then a hyperbola (figure 84).

The notation of projective transformations in formulas. If on the plane P^* we take an ordinary Cartesian coordinate system, then, as can be shown, the formulas for projective transformations of the plane are as follows

$$x' = \frac{a_1x + b_1y + c_1}{a_3x + b_3y + c_3}, \quad y' = \frac{a_2x + b_2y + c_2}{a_3x + b_3y + c_3},$$

where the determinant

$$\begin{vmatrix} a_1 & b_1 & c_1 \\ a_2 & b_2 & c_2 \\ a_3 & b_3 & c_3 \end{vmatrix} \neq 0,$$

and conversely.

If for some point (x, y) the denominators are equal to zero, this means that its image (x', y') is an ideal (infinitely distant) point. The equation

$$a_3x + b_3y + c_3 = 0$$

represents the line which under the given projective transformation goes into the ideal (infinitely distant) line.

§14. Lorentz Transformations

The derivation of the formulas of the Lorentz transformation for motion on a straight line and in the plane from the condition of the constancy of the speed of light. At the very end of the 19th century a fundamental contradiction was discovered in physics. Michelson's well-known experiment, in which the speed of light (which is about 300,000 km/sec) was measured in the direction of motion of the Earth along its orbit around the Sun (the speed of the Earth is about 30 km/sec) and perpendicular to this direction, showed irrefutably that all moving bodies in nature, even if they are moving in a vacuum, are contracted in the direction of motion. The theory of this contraction was investigated in detail by the Dutch physicist, Lorentz. He showed that this contraction is greater as the speed of the moving body gets closer to the speed of light in a vacuum, and at a speed equal to the speed of light the contraction becomes infinite. Lorentz derived the formulas for this contraction. But shortly afterwards the physicist Einstein introduced into this problem a completely different point of view, to which Poincaré was already close. Einstein argued as follows. If we assume that for the propagation of light, as for ordinary motion of a material body, Galileo's law of composition of velocities is valid, then the speed of light is $c' = c + v$, where v is the speed of the observer moving toward the source of the light, and c is the speed of light for a stationary observer. From Michelson's experiment it follows that $c' = c$. The law $c' = c + v$ is based on the transformation

$$x' = x + v_x t,$$
$$t' = t, \tag{25}$$

connecting the coordinate x of a point relative to a coordinate system I with its coordinate x' relative to a coordinate system II which has its axes parallel to the axes of system I and which moves parallel to the Ox-axis with velocity v_x relative to system I. Clearly, these are the formulas, as Einstein says, that must be changed.

It can be shown, as was recently done, for example, by A. D. Aleksandrov, that from the equality of the speed of light in both coordinate systems x, y, z, t and x', y', z', t' it already follows that the formulas of transformation from coordinates x, y, z, t to coordinates x', y', z', t' are linear and homogeneous, i.e., have the form

$$\begin{aligned}
x' &= a_1 x + b_1 y + c_1 z + d_1 t, \\
y' &= a_2 x + b_2 y + c_2 z + d_2 t, \\
z' &= a_3 x + b_3 y + c_3 z + d_3 t, \\
t' &= a_4 x + b_4 y + c_4 z + d_4 t.
\end{aligned} \tag{26}$$

From other considerations one can show that their determinant* is equal to unity.

If a point in system I moves rectilinearly and uniformly in an arbitrary given direction with the speed of light c, then $x = v_x t$, $y = v_y t$, $z = v_z t$ and $v_x^2 + v_y^2 + v_z^2 = c^2$, from which

$$x^2 + y^2 + z^2 - c^2 t^2 = 0. \tag{27}$$

But according to Michelson's experiment this point in system II also necessarily moves with the same speed of light c, so that it is also necessary that

$$x'^2 + y'^2 + z'^2 - c^2 t'^2 = 0.$$

Consequently the formulas (26) are not just arbitrary transformations which are linear, homogeneous, and with determinant equal to 1, but must at the same time satisfy the condition that if the coordinates x, y, z, t are such that

$$x^2 + y^2 + z^2 - c^2 t^2 = 0,$$

then the transformed coordinates x', y', z', t' must also satisfy this equation. Such transformations (26) are called *Lorentz transformations*.

---

* See Chapter XVI.

Let us first consider the simplest case, when the point moves along the Ox-axis. In this case formulas (26) have the form

$$x' = a_1 x + d_1 t,$$
$$t' = a_2 x + d_2 t, \tag{26'}$$

and equation (27)

$$x^2 - c^2 t^2 = 0. \tag{27'}$$

Let us introduce the notation $ct = u$, when formulas (26') and equation (27') take the form

$$x' = a_1 x + \frac{d_1}{c} u,$$
$$u' = a_2 c x + \frac{d_2 c}{c} u \tag{26_1}$$

and

$$x^2 - u^2 = 0.$$

Let us find the explicit forms of formulas (26_1). Consider x and u as a Cartesian rectangular system in the plane, i.e., consider the problem geometrically; then we may regard formulas (26_1) as those of an affine transformation of the plane Oxu (whose determinant, as was shown is equal to 1). We will denote this transformation by L. If, as we assume, $x^2 - u^2 = 0$ implies $x'^2 + u'^2 = 0$, then this transformation translates the intersecting straight lines

$$x^2 - u^2 = 0$$

into themselves. The transformation L is therefore a combination of a contraction and expansion with identical coefficients τ along these lines.

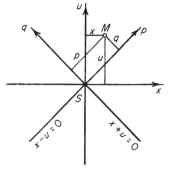

Fig. 85.

From figure 85 we obtain

$$x = \frac{p}{\sqrt{2}} - \frac{q}{\sqrt{2}},$$

$$u = \frac{p}{\sqrt{2}} + \frac{q}{\sqrt{2}}.$$

But after the transformation L, the numbers p and q will go into $p' = p/\tau$ and $q' = q\tau$, so that

$$x' \sqrt{2} = \frac{p}{\tau} - q\tau,$$

$$u' \sqrt{2} = \frac{p}{\tau} + q\tau.$$

Expressing p and q in terms of x and u from the first pair of equations, substituting into the second and simplifying, we obtain

$$x' = \frac{x - \dfrac{\tau^2 - 1}{\tau^2 + 1}\, ct}{\dfrac{2\tau}{\tau^2 + 1}}, \quad t' = \frac{t - \dfrac{1}{c}\dfrac{\tau^2 - 1}{\tau^2 + 1}\, x}{\dfrac{2\tau}{\tau^2 + 1}},$$

or, setting $(\tau^2 - 1)/(\tau^2 + 1)c = v$, we have

$$x' = \frac{x - vt}{\sqrt{1 - (v/c)^2}}, \quad t' = \frac{t - (vx/c^2)}{\sqrt{1 - (v/c)^2}},$$

which are the famous *Lorentz formulas*.

   In particular, if we take $x = 0$, i.e., if we consider the motion of the origin of coordinate system I, we obtain

$$x' = \frac{-vt}{\sqrt{1 - (v/c)^2}}, \quad t' = \frac{t}{\sqrt{1 - (v/c)^2}},$$

or $x' = -vt'$, from which obviously v is the speed of motion of coordinate system II relative to system I.

   Suppose, for example, that we are given two points on the Ox-axis with coordinates x_1 and x_2 relative to system I, so that the distance between them relative to system I is $r = |x_1 - x_2|$. Let us see what the distance between them is for an observer attached to system II. We have

$$x_1' = \frac{x_1 - vt}{\sqrt{1 - (v/c)^2}}, \quad x_2' = \frac{x_2 - vt}{\sqrt{1 - (v/c)^2}},$$

from which

$$r' = |x_1' - x_2'| = \frac{|x_1 - x_2|}{\sqrt{1 - (v/c)^2}}.$$

   The factor $\sqrt{1 - (v/c)^2}$ is exactly the coefficient of the Lorentz contraction. Since c is very large, this coefficient is very close to 1 for moderately large v, and therefore the contraction is not significant. But such elementary particles as electrons or positrons often move with velocities comparable to the speed of light, and therefore in studying their motion it is necessary to take this contraction into account, or, as they say, to consider the relativistic effect.

We pass now to the case next in complexity, namely, when the point moves in the Oxy-plane. For this case the transformations (26) will have the form

$$\begin{aligned} x' &= a_1 x + b_1 y + d_1 t, \\ y' &= a_2 x + b_2 y + d_2 t, \\ t' &= a_3 x + b_3 y + d_3 t, \end{aligned} \tag{26''}$$

where

$$\begin{vmatrix} a_1 & b_1 & d_1 \\ a_2 & b_2 & d_2 \\ a_3 & b_3 & d_3 \end{vmatrix} = 1,$$

and equation (27) will be

$$x^2 + y^2 - c^2 t^2 = 0. \tag{27''}$$

These are the Lorentz formulas for motion in the Oxy-plane.

Again we put $ct = u$. Then transformations (26'') can be rewritten as

$$x' = a_1 x + b_1 y + \frac{d_1}{c} u,$$

$$y' = a_2 x + b_2 y + \frac{d_2}{c} u, \tag{26_2}$$

$$u' = a_3 c x + b_3 c y + \frac{d_3 c}{c} u,$$

where the determinant will again be equal to one, and equation (27'') will assume the simpler form

$$x^2 + y^2 - u^2 = 0. \tag{27_2}$$

We will regard x, y, u as the Cartesian rectangular coordinates of a point in ordinary three-dimensional space and will consider formulas (26_2) as those of affine transformations of this space. Equation (27_2) represents a straight circular cone K with an angle of $90°$ at the vertex (figure 86).

From the point of view of this geometric interpretation (we call it geometric because here we regard $u = ct$ simply as a space coordinate) of a Lorentz transformation, the set of motions in the plane is identical with the set of all equi-affine (i.e., affine and volume-preserving) transformations of the space which map the cone K onto itself.

Let us consider some special Lorentz transformations.

1. It is clear that any simple rigid rotation about the axis of the cone K through an angle ω is an equi-affine transformation of space, mapping the cone K into itself, i.e., it is a special Lorentz transformation. We will denote it by ω.

2. Reflections of the space in an arbitrary plane π passing through the axis of the cone K are clearly also Lorentz transformations. We will denote them by π.

3. Finally, let us consider the following transformation (figure 87). Let v and w be any pair of opposite generators of the cone, and let P and Q be the planes tangent to the cone along these generators. These

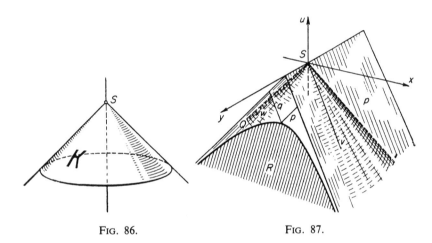

FIG. 86.   FIG. 87.

planes are mutually perpendicular. Let us make a contraction of the space to the plane P and an expansion of it with the same coefficient from the plane Q, or conversely. For example, we contract the space by a factor of three to the plane P and expand it also by a factor of three from the plane Q. Such a transformation of space is clearly also affine and preserves all volumes. We will denote it by L. We show that this transformation maps the cone K into itself. Because the cone K has the axis u as its axis of revolution, any figure can be rotated in such a way that the generators v and w lie, for example, in the plane Sxu. Therefore it is sufficient to carry out the proof for this case.

For the proof we intersect the cone K by an arbitrary plane R parallel to the Sxu plane. The equation of this plane is $y = b$, where b is a constant. Substituting this value in the equation of the cone K, we obtain

$$x^2 - u^2 = -b^2.$$

This is the equation of a hyperbola for which the lines of intersection of the plane R with the planes P and Q are exactly the asymptotes. But since for a point of such a hyperbola it is characteristic that the product of distances p and q to the asymptotes, i.e., to the planes P and Q, is constant, under transformation L all points of this hyperbola remain on the same hyperbola, and the hyperbola is mapped onto itself. But the whole surface of the cone K consists of such hyperbolas, and therefore under the transformation L of the space the cone K is sent into itself. This transformation L is therefore also a Lorentz transformation.

Since under affine transformations straight lines go into straight lines, and intersecting lines go into intersecting lines, therefore a bundle S of straight lines under any Lorentz transformation is mapped one-to-one onto itself. Moreover, under affine transformations of space all planes go into planes, so that under these transformations of the bundle S onto itself a projective transformation of the bundle is obtained. If we intersect this bundle by a plane II perpendicular to the axis of the cone K, which as a whole is not altered by the given Lorentz transformation of space, and extend this plane to the projective plane II* and then trace the points of intersection of the lines of the bundle S with the plane II*, we have the result that the Lorentz transformations of the bundle will simultaneously produce projective transformations Λ of the plane II* and these latter will transform the circle α, in which the plane II* intersects the interior part of the cone K, into itself. To analyze the properties of Lorentz transformations, it is easier to examine these projective transformations Λ of the circle α into itself.

Projective transformations of a circle into itself. A point, a ray or half line issuing from it, and one of the half planes cut off by the entire line will be called a "frame" of the plane II* (not to be confused with a coordinate frame, §11). We show (figure 88) that if we take two arbitrary frames M and M' containing interior points of the circle α, then by means of the transformations L, ω, π we can send one of these frames into the other. For this it is sufficient to make the transformations $\Lambda = L_1 \cdot \omega \cdot L_2^{-1}$ (or else $\Lambda = L_1 \cdot \omega \cdot \pi \cdot L_2^{-1}$). The transformation L_1 sends the first frame M to the center O of the circle α, the transformation ω rotates it as necessary, and finally the transformation L_2^{-1} brings it into coincidence with the second frame M'.

Let us show, in addition, that there is only one transformation Λ which translates a given frame M into a given frame M'. In order to do this we observe first that if there were two transformations Λ_1 and Λ_2 sending frame M into frame M', then the transformation $\Lambda = \Lambda_1\Lambda_2^{-1}$ would not be the identity transformation Λ and would send frame M into itself. Therefore, it is sufficient to show that if a transformation Λ sends frame M into itself, then it is the identity, i.e., leaves all points of the plane of circle α fixed.

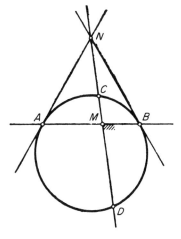

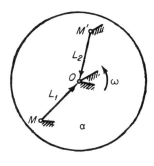

FIG. 88.   FIG. 89.

Let us show this. Suppose that the transformation Λ sends frame M into itself (figure 89). Then it maps the line AB of this frame into itself, but since it sends the circumference of the circle α into itself, it therefore leaves points A and B fixed, or else interchanges them. The latter, however, is impossible, since the half line of the frame is mapped into itself. Let us draw the tangents at points A and B to the circle α. They are mapped into themselves, since if such a tangent were mapped into a secant $A\bar{A}$, then the inverse transformation would send the different points A and $\bar{A}$ of the circle α into the one point A. But the Λ are projective transformations, and consequently one-to-one. Since under the transformations Λ these tangents go into themselves, therefore the point N of their intersection remains fixed, and consequently the line MN is mapped into itself. From the fact that the half line of the frame M is mapped into itself, we conclude as above that points C and D are not interchanged, but remain fixed. Hence, under the given projective transformation Λ of the projective plane II* four of its points A, B, C, D, no three of which lie on the same line, remain fixed. According to the uniqueness theorem of projective transformations this is the identity transformation.

Later, in §5 of Chapter XVII it will be shown that by using these properties of the Lorentz group, it is easy to construct a model of

Lobačevskiĭ's plane geometry, and if we consider Lorentz transformations for the general case of motion of a point in space, then we can do the same for Lobačevskiĭ's space geometry, and thereby prove its consistency.

We see that the theory of Lorentz transformations, projective geometry and the theory of perspectivity and non-Euclidean geometry are closely related to one another. It turns out that there is still another theory that is also closely related to them, namely, the so-called conformal transformations in the theory of functions of a complex variable, which solve such important problems of mathematical physics as the distribution of temperature in a heated plate, the flow of air around the wing of an airplane, the distribution of charge in a plane electrostatic field, the problems of elasticity in the plane, and many others.

Conclusion

Analytic geometry is an absolutely indispensable method for the investigation of other branches of mathematics, physics, and other natural sciences. Therefore it is studied not only at universities but in all technical higher institutions of learning, and also in some vocational schools. It is also a question of whether we should not include a fairly detailed treatment of the elements of analytic geometry in high school courses.

Various coordinates. The essential elements of the concept of analytic geometry, as we have seen, are the coordinate method and the investigation of equations connecting these coordinates. Besides Cartesian coordinates, other different ones can be considered. For example, in the plane, we can choose a point P (the so-called pole) and a ray originating from it (the polar axis) and determine the position of a point M by the length ρ of the polar radius from the pole to the point and the value ω of the angle made by this radius with the polar axis (figure 90).

In particular, the ellipse, hyperbola, or parabola, if for the pole we take

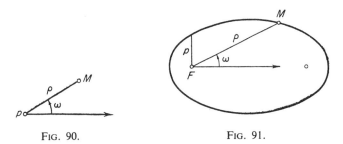

Fig. 90.   Fig. 91.

a focus, and for the polar axis the ray passing from the focus along the
axis of symmetry to the side opposite the nearer vertex (figure 91), have
one and the same equation

$$\rho = \frac{p}{1 - \epsilon \cos \omega},$$

where ϵ is the eccentricity of the curve, and p is its so-called parameter.
This equation is of a great importance in astronomy. For it was with
its help that the result was derived, from the law of inertia and the law
of universal gravitation, that the planets revolve about the Sun in ellipses.

The geographical coordinates, latitude and longitude, by which the
position of a point is given on a sphere, are well known.

Analogously, we can take a coordinate network on an arbitrary surface,
as is done in differential geometry (see Chapter VII), etc.

**Many-dimensional and infinite-dimensional analytic geometry; algebraic
geometry.** It would seem that in the 19th century analytic geometry underwent such an immense development, described earlier in a general way, and
produced so many ideas, that it would have necessarily exhausted itself,
but this is not so. In very recent times, two new, extensive branches of
mathematics have been rapidly developed and have extended the concepts
of analytic geometry, namely so-called *functional analysis* and general
algebraic geometry. It is true that both of these only halfway represent
a straightforward continuation of classical analytic geometry: Much of
functional analysis is analysis, and in algebraic geometry there is more
than a little of the theory of functions and of topology.

Let us explain what we mean. In the middle of the last century
mathematicians had already begun to consider four-dimensional and
general n-dimensional analytic geometry, i.e., to study those questions
of algebra that are straightforward generalizations of algebraic questions
of the kind involved in two- and three-dimensional analytic geometry,
to the case when there are four or n unknowns. At the very end of the
19th century a series of outstanding analysts came to the idea that for
the purposes of analysis and mathematical physics it is significant to
consider infinite-dimensional analytic geometry.

At first glance it may seem that n-dimensional or even four-dimensional
spaces seem like farfetched mathematical fictions, then the same can also
be said about an infinite-dimensional space. But it is not really so. The
arguments concerning an infinite-dimensional space are not at all difficult.
They now constitute an important branch of mathematics, functional
analysis (see Chapter XIX).

It is curious that infinite-dimensional analytic geometry has most important practical applications and plays a fundamental role in contemporary physics.

As to algebraic geometry, it is a more immediate continuation of ordinary analytic geometry, which is itself only a part of algebraic geometry. Algebraic geometry can be regarded as that part of mathematics which is occupied with curves, surfaces, and hypersurfaces, represented in Cartesian coordinates by algebraic equations of not only first and second degree, but also of higher degrees. It turns out that in these investigations it is advantageous to consider not only real but also complex coordinates, i.e., to consider everything in a so-called complex space. The most important results in this domain were obtained in the last century by Riemann. As a brilliant example of theorems about higher order curves, we point out a remarkably general result of I. G. Petrovskiĭ about the number of ovals into which an nth-order curve can be decomposed. Petrovskiĭ showed that if p is the number of such ovals which do not lie at all in other ovals, or lie in an even number of ovals, and m is the number of those ovals which lie in an odd number of ovals, and if we consider only curves whose component ovals neither intersect themselves nor each other (figure 92), then

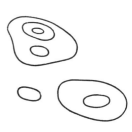

FIG. 92.

$$p - m \leqslant \frac{3n^2 - 6n}{8} + 1,$$

where n is the order of the curve, i.e., the degree of the equation by which the curve is represented.

This result is the more important as up to then almost nothing had been known about the general form of a higher order curve. It is no doubt one of the most important recent general theorems in analytic geometry.

Suggested Reading

A. A. Albert, *Solid analytic geometry*, McGraw-Hill, New York, 1949.

J. L. Coolidge, *A history of the conic sections and quadratic surfaces*, Oxford University Press, Oxford, 1945.

H. S. M. Coxeter, *The real projective plane*, 2nd ed., Cambridge University Press, New York, 1960.

L. P. Eisenhart, *Coordinate geometry*, Dover, New York, 1960.

A. Jaeger, *Introduction to analytic geometry and linear algebra*, Holt, Rinehart and Winston, New York, 1960.

D. J. Struik, *Lectures on analytic and projective geometry*, Addison-Wesley, Cambridge, Mass., 1953.

CHAPTER IV

ALGEBRA: THEORY
OF ALGEBRAIC EQUATIONS

§1. Introduction

The characteristic features of algebra are well known to everyone, since the elementary but fundamental information about it is already given in high school. Algebra is characterized, first of all, by its method, involving the use of letters, and expressions in letters, on which we perform operations according to definite laws. In elementary algebra the letters denote ordinary numbers, so that the laws of operations on expressions in letters are based on the general laws of operations on numbers. For example, the sum does not depend on the order of the summands, a fact which in algebra is written as: $a + b = b + a$; in multiplying the sum of two numbers, we can multiply each one of the numbers individually and then add the products so obtained: $(a + b)c = ac + bc$, etc.

If we trace the proof of an algebraic theorem, it is easy to see that it depends only on these laws for operations on numbers and not at all on what the letters represent.

The algebraic method, i.e., the method of calculations with letters, penetrates all of mathematics. In fact, a substantial part of the solution of a mathematical problem often turns out to be nothing but a more or less complicated algebraic computation. Besides, in mathematics we employ various symbolic calculations in which the letters no longer denote numbers but some other entities, where the laws for operations on these entities may be different from the laws of elementary algebra. For example, in geometry, mechanics, and physics we make use of vectors, and as is well known, the laws for operations on vectors are in part the same as for numbers and in part essentially different.

The significance of the algebraic method in modern mathematics and the range of its applications have greatly increased in recent decades.

First of all, the growing demands of technology force us to reduce to numerical results the solutions of difficult problems of mathematical analysis, and this usually proves to be feasible only after the algebraization of these problems, a process which in turn creates new and sometimes difficult problems in algebra itself.

Second, certain problems of analysis became clear and understandable only after they were attacked by algebraic methods based on a profound generalization (to the case of infinitely many unknowns) of the theory of systems of equations of the first degree.

Finally, the more advanced parts of algebra have found application in contemporary physics: In fact, the fundamental concepts of quantum mechanics are expressed in terms of complicated and nonelementary algebraic entities.

The basic features of the history of algebra are as follows.

First of all we must point out that our ideas regarding what algebra is and what its fundamental problem consists of have changed twice: once in the first half of the past century, and the second time at the beginning of our century. Thus, algebra has meant at different times three quite different things. In this respect the history of algebra differs from the history of the three famous branches of mathematics: analytic geometry, differential calculus, and integral calculus, which were forged into shape at the hands of their creators, Fermat, Descartes, Newton, Leibnitz, and others and were later rapidly developed and amplified, sometimes by the addition of great new sections, but were comparatively little changed in their fundamental character.

In ancient times any law that was discovered for the solution of a class of mathematical problems was recorded simply in words, since symbolic calculations had not yet been invented. The word "algebra" itself was created from the name of the important work of the Kharizmian scientist of the 9th century, Mohammed Al-Kharizmi (see Chapter I), in whose works the first general law for the solution of first- and second-degree equations was deduced. However, the introduction of the symbolic notation itself is usually associated with the name of Viète, who not only began to denote the unknowns by letters but also the given quantities. Descartes also did a great deal for the development of symbolic notation, and he too, of course, took the letters to mean ordinary numbers. It is at this moment that algebra really begins as the science of symbolic calculations, of transformations of formulas composed of letters, of algebraic equations, and so forth, in contrast to arithmetic, which always operates on concrete numbers. Only now did complicated mathematical concepts become perspicuous and accessible to investigation, since by taking a look at a formula in letters, it is in most cases possible for us to see its general arrangement or law of formation and to subject it to suitable transformations. At that time everything in mathematics which was neither geometry nor infinitesimal analysis was called algebra. This is the first, the so to say Viète point of view, concerning algebra. It was very clearly expressed in the well-known book "Introduction to Algebra" by a member of the Russian Academy of Sciences, the famous L. Euler, written in the 1760's, i.e., 200 years ago.

Euler defined algebra as the theory of calculations with quantities of various kinds. The first part of his book contains the theory of calculation with integral rational numbers, ordinary fractions, square and cube roots, the theory of logarithms, progressions, the theory of calculations with polynomials, and the theory of Newton's binomial series and its applications. The second part consists of the theory of first-degree equations and of systems of such equations, the theory of quadratic equations and of solutions of third- and fourth-degree equations by radicals, and also an extensive section on methods of solutions of various indeterminate equations in integers. For example, it was shown that Fermat's equation $x^3 + y^3 = z^3$ cannot be solved in integers x, y, z.

At the end of the 18th and the beginning of the 19th century, one of the problems of algebra gradually began to occupy the central place, namely the theory of solution of algebraic equations, in which the fundamental difficulty is the solution of an nth-degree algebraic equation with one unknown

$$x^n + a_1 x^{n-1} + a_2 x^{n-2} + \cdots + a_{n-1} x + a_n = 0.$$

This happened as a natural consequence of the importance of the problem for the whole of pure and applied mathematics, and also because of the difficulty and depth of the majority of the theorems connected with it.

The general formula for the solution of a quadratic equation,

$$x = -\frac{p}{2} \pm \sqrt{\frac{p^2}{4} - q},$$

was known to everybody. Italian algebraists of the 16th century found analogous, though more complicated, general rules for the solution of arbitrary third- and fourth-degree equations. Further investigations in this direction for higher degree equations, however, met with insurmountable difficulties. The greatest mathematicians of the 16th, 17th, 18th and the beginning of the 19th century (Tartaglia, Cardan, Descartes, Newton, d'Alembert, Tschirnhausen, Bézout, Lagrange, Gauss, Abel, Galois,

Lobačevskiĭ, Sturm, and others) created an impressive edfice of theorems and methods connected with this problem. The two-volume algebra of Serret (an epoch-making work of its time, since it presented for the first time the high point of the theory of algebraic equations, namely the theory of Galois), appeared in the middle of the 19th century, exactly 100 years after Euler's text; in it algebra was already defined as the theory of algebraic equations. This is the second point of view concerning what algebra is.

In the second half of the past century there occured, on the basis of the ideas of Galois about the theory of algebraic equations, a profound development of group theory* and the theory of algebraic numbers (in the creation of which a great part was played by the Russian mathematician E. I. Zolotarev).

In this second period also, in connection with the same problems of solution of an algebraic equation, and with the theory of algebraic varieties of higher order (which were then being studied in analytic geometry) the algebraic apparatus was developed in different directions, e.g., the theory of determinants and matrices, the algebraic theory of quadratic forms and linear transformations, and, in particular, the theory of invariants. During almost the entire second half of the 19th century, the theory of invariants was a central theme in algebra. In turn, the development of group theory and the theory of invariants exerted in this period a great influence on the development of geometry.†)

A new, third point of view as to what algebra is came into existence chiefly in the following connection. In the second half of the last century, in mechanics, physics, and mathematics itself, scientists began more and more often to investigate objects for which it was natural to consider operations of addition and subtraction, and sometimes multiplication and division, but for which these operations were subjected to altogether different laws from those for rational numbers.

We have already spoken of vectors. Other sorts of mathematical objects with different laws of operation can only be mentioned here: e.g., matrices, tensors, spinors, hypercomplex numbers. All these quantities are denoted by letters, but their laws of operation differ from one another. If for some set of objects (denoted by letters) certain operations are defined together with the laws or rules that they must satisfy, then we say that an algebraic system is defined. The third point of view on what algebra is consists of regarding the whole of algebra as the study of various algebraic systems. This is the so-called axiomatic or abstract algebra. It is abstract because

---

* See Chapter XX.
† See Chapter XVII.

at a given step in the calculation we are not all concerned with what the letters in the algebraic system denote, the only important thing is the axioms or laws satisfied by the operations; and it is called axiomatic, because it is constructed exclusively from the axioms stated at the beginning. It is as though we have returned, but on a higher level, to the first or Viète point of view on algebra, that algebra is the theory of symbolic calculations. Although it makes no difference what the letters denote and only the rules of operation are important, it is still true, of course, that only those algebraic systems are interesting which have great significance either in mathematics itself or in its applications.

The great amount of algebraic material collected in the previous period served as the actual basis for the construction of contemporary abstract algebra.

The early 1930's saw the appearance of van der Waerden's well-known book "Modern Algebra," which has played a great role in the propagation of this third point of view as to what algebra is. The text of A. G. Kuroš on algebra is oriented in the same direction.

In the present century algebra has found deep applications to geometry (topology and the theory of Lie groups) and, as mentioned earlier, to contemporary physics, especially to functional analysis and quantum mechanics.

Particularly important at the present time are the problems of mechanization of algebraic calculations by means of various mathematical computing machines, especially high-speed electronic machines. The questions connected with this type of computational mathematics raise new distinctive problems in algebra.

In the present work, there are two chapters (not counting the present one) that are devoted to algebra: linear algebra (Chapter XVI) and the theory of groups and other algebraic systems (Chapter XX).

§2. Algebraic Solution of an Equation

An algebraic equation of the nth-degree with one unknown is an equation of the form

$$x^n + a_1 x^{n-1} + a_2 x^{n-2} + \cdots + a_{n-1} x + a_n = 0,$$

where a_1, a_2, $\cdots$, a_n are given coefficients.*

---

* We assume that all terms of the equation are transferred to the left-hand side and that the equation is divided by the coefficient of the highest power of the unknown.

Equations of the first- and second-degree. If the equation is of the
the first-degree, then it has the form

$$x + a = 0$$

and is solved at once

$$x = -a.$$

The second-degree equation

$$x^2 + px + q = 0$$

was solved in early antiquity. Its solution is very simple: If we transfer q
with the opposite sign to the right-hand side and then add $p^2/4$ to both
sides we have

$$x^2 + px + \frac{p^2}{4} = \frac{p^2}{4} - q.$$

But

$$x^2 + px + \frac{p^2}{4} = \left(x + \frac{p}{2}\right)^2,$$

hence

$$x + \frac{p}{2} = \pm \sqrt{\frac{p^2}{4} - q},$$

from which we obtain the well-known formula for the solutions of a
quadratic equation

$$x = -\frac{p}{2} \pm \sqrt{\frac{p^2}{4} - q}.$$

Third-degree equation. It was completely different with equations of
degree higher than 2. Already the general equation of the third-degree
required quite profound considerations and resisted all the efforts of the
mathematicians of antiquity. It was only solved at the beginning of the
1500's, in the era of the Renaissance in Italy, by the Italian mathematician
Scipio del Ferro. Del Ferro, following the custom of his time, did not
publish his own discoveries but communicated them to one of his pupils.
After the death of del Ferro this pupil challenged to competition one of
the great Italian mathematicians Tartaglia and proposed to him for
solution a series of third-degree equations. Tartaglia (1500–1557) accepted
the challenge and eight days before the end of the competition found a
method of solving any cubic equation of the form $x^3 + px + q = 0$.

In two hours he solved all problems of his opponent. A professor of
physics and mathematics in Milan, Cardan (1501–1576), learning of
Tartaglia's discoveries, began to entreat Tartaglia to inform him of his

secret. Tartaglia finally agreed, but with the condition that Cardan keep his method a deep secret. Cardan violated his promise and published Tartaglia's result in his work "The great art" (*Ars Magna*).

The formula for the solution of a cubic equation has since then been called Cardan's formula, although it would be correct to call it Tartaglia's formula.

Cardan's formula is derived as follows.

In the first place, the solution of the general cubic equation

$$y^3 + ay^2 + by + c = 0 \tag{1}$$

can easily be reduced to the solution of the cubic equation of the form

$$x^3 + px + q = 0, \tag{2}$$

not containing a term with the square of the unknown. To do this it is sufficient to set $y = x - a/3$. Indeed, substituting this expression into equation (1) and removing the parentheses, we obtain

$$\left(x - \frac{a}{3}\right)^3 + a\left(x - \frac{a}{3}\right)^2 + b\left(x - \frac{a}{3}\right) + c = x^3 - 3x^2\frac{a}{3} + \cdots + ax^2 + \cdots,$$

where the dots indicate those terms in which x is raised to first power or does not appear at all. We see that the terms containing x^2 cancel each other out.

Let us now consider the following equation

$$x^3 + px + q = 0.$$

We set $x = u + v$, i.e., in place of one unknown we put two, u and v, and thereby turn the whole problem into a problem with two unknowns. We have

$$(u + v)^3 + p(u + v) + q = 0,$$

or

$$u^3 + v^3 + q + (3uv + p)(u + v) = 0.$$

Whatever is the sum of the two numbers $u + v$, it is always possible to require that their product uv be equal to some quantity given beforehand. If $u + v = A$, and we require $uv = B$, then since $v = A - u$, we obtain

$$u(A - u) = B,$$

so that it is sufficient that u be a solution of the quadratic equation

$$u^2 - Au + B = 0,$$

and we know that every quadratic equation has real or complex roots, given by the well-known formula. In our case, $u + v$ is equal to the desired root x of our cubic equation and we require that

$$uv = -\frac{p}{3},$$

i.e., that $3uv + p = 0$. With this choice of u and v we obtain

$$
\begin{aligned}
u^3 + v^3 + q &= 0, \\
3uv + p &= 0.
\end{aligned}
\tag{3}
$$

Consequently, if we find the numbers u and v, satisfying this system of equations then the number $x = u + v$ will be the root of our equation.

From system (3) it is easy to form a quadratic equation whose roots will be u^3 and v^3. Indeed, it gives

$$
\begin{aligned}
u^3 + v^3 &= -q, \\
u^3 v^3 &= -\frac{p^3}{27},
\end{aligned}
$$

and, consequently by a theorem already used earlier u^3 and v^3 are the roots of the quadratic equation

$$z^2 + qz - \frac{p^3}{27} = 0.$$

Solving it by the usual formula, we obtain

$$u^3 = -\frac{q}{2} + \sqrt{\frac{q^2}{4} + \frac{p^3}{27}}, \quad v^3 = -\frac{q}{2} - \sqrt{\frac{q^2}{4} + \frac{p^3}{27}},$$

and, consequently,

$$x = \sqrt[3]{-\frac{q}{2} + \sqrt{\frac{q^2}{4} + \frac{p^3}{27}}} + \sqrt[3]{-\frac{q}{2} - \sqrt{\frac{q^2}{4} + \frac{p^3}{27}}},$$

this is the formula of Cardan.

Fourth-degree equation. Soon after the solution of the cubic equation the general fourth-degree equation was solved by Ferrari (1522-1565). For the solution of the third-degree equation we have seen that the preliminary solution of the auxiliary quadratic equation,

$$z^2 + qz - \frac{p^3}{27} = 0,$$

was necessary, where $z = u^3$ or v^3; analogously, the solution of a fourth-degree equation can be based on the preliminary solution of an auxiliary cubic equation.

Ferrari's method consists of the following. Let the general fourth-degree equation be given

$$x^4 + ax^3 + bx^2 + cx + d = 0.$$

Let us rewrite it as:

$$x^4 + ax^3 = -bx^2 - cx - d$$

and add to both sides $a^2x^2/4$; then on the left we obtain a perfect square

$$\left(x^2 + \frac{ax}{2}\right)^2 = \left(\frac{a^2}{4} - b\right)x^2 - cx - d.$$

Adding now to both sides of the equation the terms

$$\left(x^2 + \frac{ax}{2}\right)y + \frac{y^2}{4},$$

where y is a new variable, on which we later impose a necessary condition, on the left we obtain a perfect square

$$\left(x^2 + \frac{ax}{2} + \frac{y}{2}\right)^2 = \left(\frac{a^2}{4} - b + y\right)x^2 + \left(\frac{ay}{2} - c\right)x + \left(\frac{y^2}{4} - d\right). \qquad (4)$$

Thus we have reduced the problem to one with two unknowns.

On the right of equation (4) we have a quadratic trinomial in x, whose coefficients depend on y. We select y such that this trinomial will be the square of the first-degree binomial $\alpha x + \beta$.

In order that the quadratic trinomial $Ax^2 + Bx + C$ be the square of the binomial $\alpha x + \beta$ it is sufficient that

$$B^2 - 4AC = 0.$$

Indeed, if $B^2 - 4AC = 0$, then

$$Ax^2 + Bx + C = (\sqrt{A}x + \sqrt{C})^2,$$

i.e.,

$$Ax^2 + Bx + C = (\alpha x + \beta)^2,$$

where

$$\alpha = \sqrt{A}, \quad \beta = \sqrt{C}.$$

Consequently, if we select y such that

$$\left(\frac{ay}{2} - c\right)^2 - 4\left(\frac{a^2}{4} - b + y\right)\left(\frac{y^2}{4} - d\right) = 0,$$

then the first part of equation (4) will be the complete square $(\alpha x + \beta)^2$. Removing the parentheses, we obtain a cubic equation in y

$$y^3 - by^2 + (ac - 4d)y - [d(a^2 - 4b) + c^2] = 0.$$

Solving this auxiliary cubic equation (for example, by Cardan's formula) we find α and β in terms of its solution y_0, namely

$$\left(x^2 + \frac{ax}{2} + \frac{y_0}{2}\right)^2 = (\alpha x + \beta)^2,$$

from which

$$x^2 + \frac{ax}{2} + \frac{y_0}{2} = \alpha x + \beta \quad \text{or} \quad x^2 + \frac{ax}{2} + \frac{y_0}{2} = -\alpha x - \beta.$$

From these two quadratic equations we can find all four roots of the given fourth-degree equation.

This is how third- and fourth-degree algebraic equations were solved by Italian mathematicians in the 1500's.

The success of the Italian mathematicians produced a very great effect. It was the first instance when modern science had exceeded the achievements of the ancients. Until then, in the whole course of the Middle Ages, the aim had always been only to understand the work of the ancients, and now, finally, certain questions were solved which the ancients had not succeeded in conquering. And this happened in the 1500's, i.e., in the century before the invention of new branches of mathematics: analytic geometry, differential calculus, and integral calculus, which finally affirmed the superiority of the new science over the old. After this there was no important mathematician who did not attempt to extend the achievements of the Italians and to solve equations of fifth, sixth, and higher degree in an analogous way by means of radicals.

The prominent algebraist of the 17th century, Tschirnhausen (1651–1708) even believed that he had finally found a general method of solution. His method was based on the transformation of an equation to a simpler one, but this very transformation required the solution of some auxiliary equations. Subsequently, by a deeper analysis it was shown that Tschirnhausen's method of transformation indeed gives the solution of second- third-, and fourth-degree equations, but already for a fifth-degree equation it requires the preliminary solution of an auxiliary equation of the sixth-degree, whose solution in turn was not known.

Factorization of a polynomial and Viète's formulas. If we accept without proof the so-called fundamental theorem of algebra* that every equation

$$f(x) = 0,$$

where $\qquad\qquad f(x) = x^n + a_1 x^{n-1} + \cdots + a_n$

is a polynomial in x of given degree n and the coefficients $a_1, a_2, \cdots, a_n$ are given real or complex numbers, has at least one real or complex root, and take into consideration that all computations with complex numbers are carried out by the same rules as with rational numbers, then it is easy to show that the polynomial $f(x)$ can be represented (and in only one way) as a product of first-degree factors

$$f(x) = (x - a)(x - b) \cdots (x - l),$$

where $a, b, \cdots, l$ are real or complex numbers.

Indeed, let a be a root of $f(x)$; we divide $f(x)$ by $x - a$; since the divisor is of the first-degree, the remainder will be a constant number R, i.e., we will have the identity

$$f(x) = (x - a) f_1(x) + R,$$

where $f_1(x)$ is a polynomial of degree $n - 1$ and R is a constant. Substituting here in place of x the number a, we obtain

$$f(a) = (a - a) f_1(a) + R = R.$$

But since a is a root of $f(x)$, we have $f(a) = 0$, and hence $R = 0$, i.e., a polynomial can always be divided by $(x - a)$ without remainder, where a is a root of this polynomial. Thus

$$f(x) = (x - a) f_1(x).$$

But if the fundamental theorem of algebra is true, then in turn the polynomial $f_1(x)$ has a root b, and we obtain analogously

$$f_1(x) = (x - b) f_2(x),$$

where the polynomial $f_2(x)$ is already of degree $(n - 2)$, etc. This factorization, as can easily be shown, is unique.

Every nth-degree polynomial $f(x)$ has in this sense n and only n roots $a, b, \cdots, l$. These roots may be all distinct but it can happen that some among them are identical. Then we say that the corresponding root of

---

* The proof of the fundamental theorem of algebra is difficult and was given considerably later. We devote §3 to it. But its validity was assumed long before it was rigorously proved.

the polynomial $f(x)$ is a multiple root with such and such a multiplicity.
Multiplying out the expression

$$(x - a)(x - b)(x - c) \cdots (x - l)$$

and comparing the coefficients of the same powers of x, we see immediately
that

$$-a_1 = a + b + c + \cdots + l,$$
$$a_2 = ab + ac + \cdots + kl,$$
$$-a_3 = abc + abd + \cdots,$$
$$\cdots\cdots\cdots\cdots\cdots\cdots\cdots\cdots\cdots\cdots$$
$$\pm a_n = abc \cdots l$$

which are Viète's formulas.

A theorem on symmetric polynomials.   Viète's formulas are polynomials in the n letters a, b, $\cdots$, l which do not vary under any permutation
of these letters. Indeed, $a + b + \cdots + k + l = b + a + \cdots + k + l$,
etc. In general, any such polynomials in n letters, which do not change
under any permutations of these letters, are called symmetric polynomials
in n letters. For example, $5x^2 + 5y^2 - 7xy$ is a symmetric polynomial in x
and y. It is possible to prove the theorem that every integral symmetric
polynomial in n letters with arbitrary coefficients A, B, $\cdots$ can be expressed
integral rationally, i.e., with the operations of addition, subtraction, and
multiplication, in terms of the coefficients A, B, $\cdots$ and of Viète's polynomials in the letters. If a, b, $\cdots$, l are the roots of an nth-degree equation
$x^n + a_1 x^{n-1} + \cdots + a_n = 0$, then every symmetric polynomial in
a, b, $\cdots$, l with arbitrary coefficients A, B, $\cdots$ can thus be expressed integral
rationally in terms of these coefficients A, B, $\cdots$ and the coefficients
a_1, a_2, $\cdots$, a_n of the equation. This is the so-called fundamental theorem
of symmetric polynomials.

Lagrange's contributions.   The famous French mathematician Lagrange
in his great work "Reflections on the solution of algebraic equations"
published in 1770–1771 (with more than 200 pages), critically examined
all the solutions of second-, third- and fourth-degree equations that were
known up to his time and showed that their success was always based on
properties which did not hold for equations of degree 5 or higher. From
del Ferro's time until this work of Lagrange more than two and a half
centuries had passed by and nobody during this long interval had doubted
the possibility of solving equations of degree 5 and higher by radicals,
i.e., of finding formulas involving only the operations of addition, subtraction, multiplication, division, and radicals with integral positive

exponents, which would express the solution of an equation in terms of its coefficients, that is, formulas similar to those by which the quadratic equation had been solved in antiquity and the third- and fourth-degree equations in the 1500's by the Italians. They regarded this situation as being due only to their own inability to find a valid but apparently deeply hidden solution.

Lagrange says in his memoir: "The problem of solving (by radicals) equations whose degree is higher than four is one of those problems which have not been solved although nothing proves the impossibility of solving them" and two pages later he supplements this: "From our reasoning we see that it is very doubtful that the methods which we have considered could give a complete solution of equations of the fifth-degree."

In his investigations, Lagrange introduced the expression

$$a + \epsilon b + \epsilon^2 c + \cdots + \epsilon^{n-1} l$$

in the roots a, b, $\cdots$, l of an equation, where ϵ is an nth root of unity,* having established that such expressions are closely connected with the solution of equations by radicals. These expressions are now called "Lagrange resolvents."

In addition, Lagrange observed that the theory of permutations of roots of an equation is of great importance in the theory of solution of equations. He even expressed the thought that the theory of permutations is the "true philosophy of the whole question," in which he was completely right, as was shown in the later investigations of Galois.

Lagrange's method of solution of second-, third- and fourth-degree equations were not the same as those of the Italians, which in every case were based on special transformations of a complicated and so to speak accidental kind. Lagrange's methods were altogether orderly and developed from one general idea involving the theory of symmetric polynomials, the theory of permutations, and the theory of resolvents.

---

* I.e., a complex number which raised to the nth power is equal to one. For example, the cube roots of unity can have the values

$$1, \quad -\frac{1}{2} + \frac{\sqrt{3}}{2} i, \quad -\frac{1}{2} - \frac{\sqrt{3}}{2} i,$$

where $i = \sqrt{-1}$ (see §3). Indeed,

$$\left(-\frac{1}{2} + \frac{\sqrt{3}}{2} i\right)^3 = -\frac{1}{8} - \frac{3}{8}\sqrt{3} i + \frac{9}{8} + \frac{3\sqrt{3}}{8} i = 1,$$

and analogously

$$\left(-\frac{1}{2} - \frac{\sqrt{3}}{2} i\right)^3 = 1.$$

Let us consider, for example, the solution by Lagrange's method of the general fourth-degree equation

$$x^4 + mx^3 + nx^2 + px + q = 0.$$

Let the roots of this equation be a, b, c, d. Consider the resolvent

$$a + b - c - d,$$

i.e.,

$$a + \epsilon c + \epsilon^2 b + \epsilon^3 d,$$

where $\epsilon = -1$. If we permute a, b, c, d in all $1 \cdot 2 \cdot 3 \cdot 4 = 24$ different ways, we obtain altogether six different expressions

$$\begin{aligned}
a + b - c - d, \\
a + c - b - d, \\
a + d - c - b, \\
c + d - a - b, \\
b + d - a - c, \\
b + c - a - d.
\end{aligned} \tag{5}$$

An equation of the sixth-degree, whose roots are these six expressions, will thus have coefficients that do not vary with all 24 permutations of a, b, c, d, since any of the 24 permutations can only permute these expressions among themselves and the coefficients of the sixth-degree equation do not depend on the order in which we take its roots. Thus, these coefficients are symmetric polynomials in a, b, c, d. But then, by virtue of the fundamental theorem on symmetric polynomials, these coefficients are expressed integral rationally in terms of the coefficients m, n, p, q of the equation. In addition, since expressions (5) are pairwise of opposite signs, this sixth-degree equation will contain only terms of even powers. Indeed, if expressions (5) are denoted by $\alpha, \beta, \gamma, -\alpha, -\beta, -\gamma$ respectively, then the left-hand side of the sixth-degree equation will be equal to

$$(y - \alpha)(y + \alpha)(y - \beta)(y + \beta)(y - \gamma)(y + \gamma)$$
$$= (y^2 - \alpha^2)(y^2 - \beta)(y_2 - \gamma^2).$$

Direct computation gives the sixth-degree equation

$$y^6 - (3m^2 - 8n)y^4 + 3(m^4 - 16m^2n - 16n^2 + 16mp - 64q)y^2$$
$$- (m^2 - 4m + 8p)^2 = 0.$$

Letting $y^2 = t$, we obtain a cubic equation in t, and if t', t'', t''' are its roots, then

$$a + b - c - d = \sqrt{t'},$$

$$a + c - b - d = \sqrt{t''},$$

$$a + d - b - c = \sqrt{t'''}.$$

We also have

$$a + b + c + d = -m.$$

Adding these equations after multiplication by 1, 1, 1, 1 or 1, −1, −1, 1, or −1, 1, −1, 1, or −1, −1, 1, 1, we obtain

$$a = \frac{1}{4}(-m + \sqrt{t'} + \sqrt{t''} + \sqrt{t'''}),$$

$$b = \frac{1}{4}(-m + \sqrt{t'} - \sqrt{t''} - \sqrt{t'''}),$$

$$c = \frac{1}{4}(-m - \sqrt{t'} + \sqrt{t''} - \sqrt{t'''}),$$

$$d = \frac{1}{4}(-m - \sqrt{t'} - \sqrt{t''} + \sqrt{t'''}).$$

Thus, the solution of a fourth-degree equation is reduced to the solution of a cubic equation; and third- and second-degree equations are solved analogously.

Lagrange achieved a great deal in the theory of algebraic equations. However, even after his persistent efforts the problem of solution in radicals of algebraic equations with degree higher than 4 remained to be settled. This problem, on which mathematicians had worked in vain for almost three centuries, constituted, in the expression of Lagrange, "a challenge to the human mind."

Abel's discovery. Consequently it was a great surprise to all mathematicians when in 1824 the work of a young Norwegian genius Abel (1802–1829) came to light, in which a proof was given that if the coefficients of an equation a_1, a_2, $\cdots$, a_n are regarded simply as letters, then there does not exist any radical expression in these coefficients that is a root of the corresponding equation, if its degree $n \geq 5$. Thus, for three centuries the efforts of the greatest mathematicians of all countries to solve equations of degree 5 or higher in radicals did not lead to success for the simple reason that this problem simply does not have a solution.

Such a formula is known for second-degree equations, and as we saw analogous formulas exist for third- and fourth-degree equations, but for equations of degree 5 or greater there are no such formulas.

Abel's proof is difficult and we will not give it here.

Galois theory. But this was not yet all. A very remarkable result in the theory of algebraic equations still remained to come. The point is that there are arbitrarily many special forms of equations of any degree that are solvable in radicals, and many of them are exactly those equations that are important in the applications. Such, for instance, are the binomial equations $x^n = A$. Abel found another very broad class of such equations, the so-called cyclic equations and still more general "Abelian" equations. In connection with the problem of construction by ruler and compass of regular polygons, Gauss explicitly considered the so-called cyclotomic equations, i.e., equations of the form

$$x^{p-1} + x^{p-2} + \cdots + x + 1 = 0,$$

where p is a prime number, and showed that they can always be reduced to a chain of equations of lower degree; moreover, he found necessary and sufficient conditions that such an equation can be solved in square roots. The necessity of these conditions was rigorously proved only by Galois.

Thus, after Abel's work the situation was the following: Although, as was shown by Abel, the general equation of degree higher than 4 cannot be solved by radicals, there are arbitrarily many different special equations of arbitrary degree, all of which can be solved by radicals. The whole question of solving equations in radicals was placed by these discoveries on completely new ground. It became clear that the task now was to determine exactly which equations can be solved by radicals, or in other words, what are the necessary and sufficient conditions for the solvability of an equation in radicals. This problem, the answer to which gave in some sense the final elucidation of the whole problem, was solved by the ingenious French mathematician Evariste Galois.

Galois (1811–1832) perished at the age of 20 in a duel. In the last two years of his life he could not devote much time to mathematics, since he was carried away by the stormy whirl of political life at the time of the 1830 Revolution and languished in jail for his speech against the reactionary regime of Louis Philippe. Nevertheless, in his short life, Galois made discoveries far ahead of his time in various parts of mathematics and in particular produced some very remarkable results in the theory of algebraic equations. In a small publication "Memoir on the conditions of solvability of equations in radicals" which remained in manuscript form after his death and was first published in 1846 by Liouville, Galois started from some very simple but profound concepts and finally untangled the whole complex of difficulties surrounding the solution of equations in radicals, difficulties with which the most outstanding mathematicians had struggled unsuccessfully up to his time. The success of Galois was based on the fact that for the first time he introduced into the theory of equations a series of exceedingly important new general concepts, which subsequently played a great role in mathematics as a whole.

Let us consider the Galois theory for a special case, namely when the coefficients a_1, a_2, $\cdots$, a_n of the given nth-degree equation

$$x^n + a_1 x^{n-1} + \cdots + a_{n-1} x + a_n = 0 \tag{6}$$

are rational numbers. This case is particularly interesting and already involves essentially all the difficulties of the general Galois theory. We will also assume that the roots a, b, c, $\cdots$ of this equation are distinct.

Galois begins, like Lagrange, with considering a first-degree expression in a, b, c, $\cdots$

$$V = Aa + Bb + Cc + \cdots,$$

although he does not require that the coefficients A, B, C, $\cdots$ of this expression should be the roots of unity, but takes for A, B, C, $\cdots$ any integral rational numbers such as to give numerically distinct values for all the $n! = 1 \cdot 2 \cdot 3 \cdots n$ quantities V, V', V'', $\cdots$, $V^{(n!-1)}$ obtained from V by permuting the roots a, b, c, $\cdots$ in all $n!$ possible ways. This can always be done. Then Galois constructs the equation of degree $n!$ whose roots are V, V', V'', $\cdots$, $V^{(n!-1)}$. The theorem on symmetric polynomials shows that the coefficients of this equation $\Phi(x) = 0$ of degree $n!$ will be rational numbers.

Up to now everything is quite similar to what Lagrange did.

Next Galois introduced the first important new concept, the concept of irreducibility of a polynomial in a given field of numbers. If a polynomial in x is given, whose coefficients, for example, are rational numbers, then the polynomial is called reducible in the field of rational numbers if it can be represented in the form of a product of polynomials of lower degrees with rational coefficients. If not, then the polynomial is called irreducible in the field of rational numbers. The polynomial $x^3 - x^2 - 4x - 6$ is reducible in the rational number field, since it is equal to $(x^2 + 2x + 2)(x - 3)$, but for instance, the polynomial $x^3 + 3x^2 + 3x - 5$ is irreducible, as can be shown, in the field of rational numbers.

There exist methods, admittedly requiring long computations, of factoring any given polynomial with rational coefficients into irreducible polynomials in the field of rational numbers.

Galois then factors the above polynomial $\Phi(x)$ into irreducible factors in the field of rational numbers.

Let $F(x)$ be one of these irreducible polynomials (which one of them is immaterial for what follows) and let it be of degree m.

The polynomial $F(x)$ will then be the product of m of the $n!$ first-degree factors $x - V$, $x - V'$, $\cdots$, $x - V^{(n!-1)}$, into which the $n!$th-degree polynomial $\Phi(x)$ was decomposed. Let these m factors be $x - V$, $x - V'$, $\cdots$, $x - V^{(m-1)}$. We enumerate in any order the roots, a, b, c, $\cdots$, l

of the given nth-degree equation (6) by giving them the indices, $1, 2, \cdots, n$. Then the quantities $V, V', \cdots, V^{(n!-1)}$ correspond to all possible $n!$ permutations of the numbers $1, 2, \cdots, n$, corresponding to permutations of the roots, and the $V, V', \cdots, V^{(m-1)}$ correspond to only m of these permutations. The set G of these m permutations of the numbers $1, 2, \cdots, n$ is called the Galois group of the given equation (6).*

Then Galois introduces some new concepts and develops simple but truly remarkable arguments by which he proves that a necessary and sufficient condition for the solvability of equation (6) in radicals is that the group G of permutations of the numbers $1, 2, \cdots, n$ satisfies a certain definite condition.

Thus, Lagrange's prophecy that at the basis of the whole problem lay the theory of permutations proved to be true.

In particular, Abel's theorem on the nonsolvability of a general fifth-degree equation in radicals can now be proved as follows. It can be shown that there exist arbitrarily many fifth-degree equations, even with integral rational coefficients for which the corresponding 120th-degree polynomial $\Phi(x)$ is irreducible, i.e., whose Galois group is the group of all $5! = 120$ permutations of the indices $1, 2, 3, 4, 5$ of its roots. But this group, as can be shown, does not satisfy the Galois criterion, and therefore these fifth-degree equations cannot be solved in the radicals.

For instance, it can be shown that the equation $x^5 + x - a = 0$, where a is a positive whole number, in most cases cannot be solved by radicals. For example, it is not solvable in radicals for $a = 3, 4, 5, 7, 8, 9, 10, 11, \cdots$

The application of Galois theory to the problem of solvability of geometric problems by ruler and compass. One of the most remarkable special applications of Galois theory is the following. Many problems of plane geometry can be solved by constructions with ruler and compass alone. For example, we can construct with ruler and compass a regular triangle, square, pentagon, hexagon, octagon, decagon, etc., but it is impossible to construct a regular polygon of seven, nine, or eleven sides. Which problems can be solved by ruler and compass, and which not? Before Galois it was an unsolved problem. From the Galois theory we obtain the following answer.

The simultaneous solution of equations of two lines, a line and a circle, or two circles can be reduced to the solution of equations of first- or second- degree. For a line and a circle it is clear, and in the case of two circles $(x - a_1)^2 + (y - b_1)^2 = r_1^2$ and $(x - a_2)^2 + (y - b_2)^2 = r_2^2$ if we

---

* More will be said about Galois groups in §5, Chapter XX.

substract one equation from the other, the x^2 and y^2 cancel out, and we obtain a first-degree equation, which is to be solved simultaneously with the equation of one of the circles, so that again we have a quadratic equation. Therefore, every step of the problem to be solved by ruler and compass is reduced to an equation of first- or second-degree, and consequently, all problems solvable with ruler and compass are reduced to an algebraic equation with one unknown, whose solution involves the extraction of a chain of square roots. Conversely, if the solution of a geometric problem is reduced to such an algebraic equation, then it can be solved by ruler and compass, since square roots, as is well known, can be constructed by ruler and compass.

If a geometric problem is given, we must first set up an algebraic equation equivalent to the given problem. If it is impossible to set up such an equation, the problem is obviously not solvable by ruler and compass. If the equation has been set up, then we must select that one of its irreducible factors that is connected with the solution of the problem, and determine whether this irreducible equation can be solved in square roots. As the Galois theory shows, for this it is necessary and sufficient that the number m of permutations that constitute its Galois group be a power of 2.

With this test we can prove the theorem stated by Gauss that a regular polygon with a prime number p of sides can be constructed by ruler and compass if and only if the prime number p has the form $2^{2n} + 1$, i.e., for $p = 3, 5, 17, 257$ but not for $p = 7, 11, 13, 19, 23, 29, 31, \cdots$, etc. Gauss proved only the "if" part of this assertion.

By the same method we can prove that it is impossible to divide an arbitrary angle into three equal parts by ruler and compass, or to duplicate the cube, i.e., from the edge of a given cube to find the edge of a cube with twice as great a volume, and so forth.

The impossibility of squaring the circle, i.e., of constructing with ruler and compass the side of a square equal in area to a circle with given radius, is proved in a different way. Namely, it can be shown that the side of such a square is not connected with the radius by any algebraic equation, i.e., it is so to speak transcendental relative to the radius, and consequently it is *a fortiori* not expressible in terms of the radius by a chain of square roots. This proof is difficult and it does not follow from Galois theory.

Two fundamental unsolved problems connected with Galois theory. In Galois theory there remain two further basic problems which have not yet been solved in their general form although many excellent mathematicians have been working on them almost uninterruptedly.

The first of these is the problem of the so-called Hilbert–Čebotarev resolvents (not to be confused with the Lagrange resolvents) which is a direct generalization of the problem of solution of equations in radicals. The idea is this: Saying that an equation is solvable in radicals is exactly the same as saying that its solution is reduced to a chain of successive binomial equations, since the radical $\sqrt[n]{A}$ is a root of the binomial equation $x^n = A$. But it may happen that although the equation cannot be reduced to a chain of such simple equations as the binomial ones, it can nevertheless be reduced to a chain of certain other very simple equations. Back at the end of the 18th century, it had been shown that the general fifth-degree equation can be reduced to a chain of binomial equations together with one further equation of the form $x^5 + x + A = 0$, which, although not binomial, has like the binomial equations, only one parameter A.

Later on it was proved that even a sixth-degree equation cannot be reduced to a chain of one-parameter equations. For equations of any degree we require to solve the problem: what kind of simpler equations, i.e., with a minimum number of parameters, make up the chain to which our equation can be reduced.

If the given equation is reduced to a chain of one-parameter equations of a definite type, then for each of these one-parameter equations we can compute a table, giving its roots as a function of its parameter. Then the solution of the given equation is reduced to the use of a chain of such tables.

Second, a still deeper problem consists of the converse of Galois theory. Galois proved that the properties of the solutions of an equation depend on its group. But conversely, can any group of permutations be the Galois group of some equation and can we set up all the equations whose Galois group is a given group?

As to the first of these two questions only partial results are known, although such outstanding mathematicians as Klein and Hilbert worked on it persistently; the first general theorems were given by the remarkable Soviet algebraist H. G. Čebotarev.

The second question for the so-called solvable groups, i.e., groups satisfying Galois' criterion was solved in the affirmative in recent years by the Soviet mathematician I. R. Šafarevič.

§3. The Fundamental Theorem of Algebra

In the previous section we considered the attempts, lasting three centuries, to solve by radicals an nth-degree equation. The problem turned out to be very deep and difficult and led to the creation of new concepts, important not only for algebra but also for mathematics as a whole.

As for the practical solution of equations, the result of all this work was the following. It became clear that solution by radicals is far from being available for all algebraic equations, and even when it is available, it is of little practical value because of its complexity, except in the case of the quadratic equation.

In view of this, mathematicians long ago began to work on the theory of algebraic equations in three completely different directions, namely: (1) on the problem of the existence of a root; (2) on the problem of how can we learn from the coefficients of the equation something about its roots without solving it; for example, does it have real roots and how many; and finally, (3) on the approximate calculation of the roots of an equation.

First of all, it was necessary to prove that in general any nth-degree algebraic equation with real or complex coefficients always has at least one real or complex root.*

This theorem, which is one of the most important in the whole of mathematics, remained for a long time without rigorous proof. In view of its importance and difficulty, it is generally called the "fundamental theorem of algebra," although the majority of the methods by which it has been proved are as closely related to infinitesimal analysis as to algebra. The first proof was given by d'Alembert. One point in d'Alembert's proof, as was later made clear, turned out to be defective. Namely, d'Alembert assumed as trivial the general proposition of analysis that a continuous function, given on a bounded and closed set of points, has somewhere on the set a minimum. This is true but it had to be proved. A rigorous proof of this property was obtained only in the second half of the 19th century, i.e., a hundred years after d'Alembert's investigations.

It is generally considered that the first rigorous proof of the fundamental theorem of algebra were given by Gauss; however, some of his proofs require for full rigor no lesser additions than those required for d'Alembert's proof. Today a number of different completely rigorous proofs of this theorem are known.

In the present section we consider the proof of the fundamental theorem of algebra based on the so-called lemma of d'Alembert, and we also give a complete proof of the aforementioned proposition from analysis.

The theory of complex numbers. Before considering the proof of the fundamental theorem of algebra, we must first of all recall the theory of complex numbers as studied in high school. The difficulties which led to the creation of the theory of complex numbers are first encountered in

---

* The point is that there exist nonalgebraic equations, for example, $a^x = 0$, which definitely do not have roots, either real or complex.

solving quadratic equations. What should we do, if the number $p^2/4 - q$ under the square root in the formula for the solution of the quadratic equation is negative? There exists no real number, positive or negative, which is the square root of a negative number, since the square of any real number is either positive or zero.

After long doubts, lasting more than a century, mathematicians arrived at the conclusion that it is necessary to introduce a new form of numbers, the so-called complex numbers, with the following laws of operations on them.

Conventionally, a number of new character is introduced: $i = \sqrt{-1}$ such that $i^2 = -1$, and numbers of the form $a + bi$ are considered, where a and b are ordinary real numbers. The numbers $a + bi$ are called complex. Two such numbers $a + bi$ and $c + di$ are regarded as equal, if $a = c$, $b = d$. The sum of two such numbers is defined to be the number $(a + c) + (b + d)i$, and their difference is the number $(a - c) + (b - d)i$. In multiplication we agree to multiply these numbers as if they were binomials but to take into consideration that $i^2 = -1$, i.e.,

$$(a + bi)(c + di) = ac + bci + adi + bdi^2 = (ac - bd) + (bc + ad)i.$$

If a and b are regarded as rectangular coordinates of a point, and the point is associated with the complex number $a + bi$, then the addition and subtraction of complex numbers corresponds to the addition and subtraction of vectors, i.e., of directed segments from the origin to the points with coordinates (a, b) and (c, d), since in addition of vectors their corresponding coordinates are added. As to the geometrical meaning

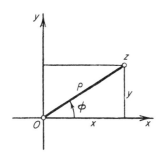

of a product in the so-called plane of complex numbers, we can see it more easily if we consider the length ρ of the vector from the origin of the coordinate system to the point (x, y) (this length is called the *modulus* of the complex number $z = x + iy$) and the angle ϕ which the vector makes with the Ox-axis (this angle is called the *argument* of the complex number $z = x + iy$); in other words, if we consider not the Cartesian coordinates x and y but the so-called polar coordinates

FIG. 1.

ρ and ϕ (figure 1). Then $x = \rho \cos \phi$, $y = \rho \sin \phi$ and consequently the complex number itself can be written as

$$x + iy = \rho(\cos \phi + i \sin \phi).$$

If

$$a + bi = \rho_1(\cos \phi_1 + i \sin \phi_1), \quad c + di = \rho_2(\cos \phi_2 + i \sin \phi_2),$$

then

$$ac - bd = \rho_1\rho_2(\cos \phi_1 \cos \phi_2 - \sin \phi_1 \sin \phi_2) = \rho_1\rho_2 \cos (\phi_1 + \phi_2),$$
$$bc + ad = \rho_1\rho_2(\sin \phi_1 \cos \phi_2 + \cos \phi_1 \sin \phi_2) = \rho_1\rho_2 \sin (\phi_1 + \phi_2),$$

from this we see that in multiplication of two complex numbers their moduli ρ_1 and ρ_2 are multiplied, and the arguments ϕ_1 and ϕ_2 are added. In division, since it is the inverse operation of multiplication, one modulus is divided by the other, and the arguments are subtracted

$$\rho_1(\cos \phi_1 + i \sin\phi_1) \, \rho_2(\cos \phi_2 + i \sin \phi_2)$$
$$= \rho_1\rho_2[\cos(\phi_1 + \phi_2) + i \sin (\phi_1 + \phi_2)]$$

and

$$\frac{\rho_1(\cos \phi_1 + i \sin \phi_1)}{\rho_2(\cos \phi_2 + i \sin \phi_2)} = \frac{\rho_1}{\rho_2} [\cos (\phi_1 - \phi_2) + i \sin (\phi_1 - \phi_2)].$$

In raising to a power with positive integral exponent n, consequently, the modulus is raised to the same nth power, and the argument is multiplied by n

$$[\rho(\cos \phi + i \sin \phi)]^n = \rho^n(\cos n\phi + i \sin n\phi).$$

Conversely, taking roots

$$\sqrt[n]{\rho(\cos \phi + i \sin \phi)} = \sqrt[n]{\rho} \left(\cos \frac{\phi}{n} + i \sin \frac{\phi}{n}\right).$$

However, in taking roots a special situation arises. Let n be a positive integral exponent. Then

$$\sqrt[n]{\rho(\cos \phi + i \sin \phi)}$$

is equal to the number

$$\sqrt[n]{\rho} \left(\cos \frac{\phi}{n} + i \sin \frac{\phi}{n}\right)$$

since raising this number to the nth power gives the radicand.

But this is only one value of the root. The point is that the complex number

$$\sqrt[n]{\rho} \left[\cos \left(\frac{\phi}{n} + \frac{2k\pi}{n}\right) + i \sin \left(\frac{\phi}{n} + \frac{2k\pi}{n}\right)\right],$$

where k is any of the numbers $1, 2, \cdots, n - 1$, will also be an nth root of the number

$$\rho(\cos\phi + i\sin\phi).$$

Indeed, according to the rule for raising to a power, if we raise this number to the nth power, we obtain the number

$$(\sqrt[n]{\rho})^n \left[\cos n\left(\frac{\phi}{n} + \frac{2k\pi}{n}\right) + i\sin n\left(\frac{\phi}{n} + \frac{2k\pi}{n}\right)\right]$$

$$= \rho[\cos(\phi + 2k\pi) + i\sin(\phi + 2k\pi)],$$

where the addend $2k\pi$, because of the properties of sines and cosines, can be neglected, since it changes neither sine nor cosine. Thus the nth power of this number is also

$$\rho(\cos\phi + i\sin\phi),$$

i.e., this number is

$$\sqrt[n]{\rho(\cos\phi + i\sin\phi)}.$$

It is easy to see that no other complex number, besides these n numbers for $k = 0, 1, 2, \cdots, n - 1$ is an nth root of

$$\rho(\cos\phi + i\sin\phi).$$

Geometrically, the extraction of nth roots can be described as follows. The points of the complex plane corresponding to the values of the $\sqrt[n]{}$ of the number $\rho(\cos\phi\, i\sin\phi)$ lie at the vertices of the regular n-sided polygon inscribed in a circle drawn about the origin with radius $\sqrt[n]{\rho}$ and so rotated that one of the vertices of this n-sided polygon has argument ϕ/n (figure 2).

We make the following observation. If

$$f(z) = z^n + c_1 z^{n-1} + \cdots + c_{n-1}z + c_n$$

is a polynomial in z with given real or complex coefficients $c_1, c_2, \cdots, c_n$ and we change z continuously, i.e., continuously shift the point $z = x + iy$ in the complex plane, then the complex point $Z = X + iY = f(z)$ will also move continuously in the complex plane. This is clear from the fact that if we substitute in $f(z)$ the value of $z = x + iy$, $c_1 = a_1 + b_1 i$, $c_2 = a_2 + b_2 i, \cdots, c_n = a_n + b_n i$, and perform all computations, we find that

$$f(z) = X + iY,$$

where

$$X = P(x, y), \quad Y = Q(x, y)$$

arc nth-degree polynomials in x and y with real coefficients expressed in terms of a_i and b_i. Under continuous change of x and y, these polynomials will also change continously.

We also note that, since the modulus $\rho = |f(z)|$ is equal to $\sqrt{X^2 + Y^2}$, during a continuous shift of the point z in the complex plane, the modulus $|f(z)|$ will also change continuously. In other words, if the point z is sufficiently close to the point α then the difference $|f(z)| - |f(\alpha)|$ of absolute values is smaller than any preassigned positive number.

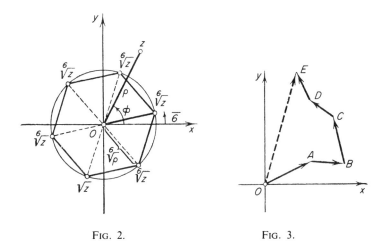

FIG. 2.   FIG. 3.

Let us also remark that the modulus of a sum of several complex numbers is always smaller than or equal to the sum of the moduli of these numbers, which is equivalent to saying that the rectilinear segment OE (figure 3) is shorter than or equal to the polygonal line $OABCDE$, being equal to it if and only if all of its segments lie on one line and in one direction.

We recall finally that to say "a complex number is equal to zero" is the same as to say that "its modulus is equal to zero," since the modulus ρ of a complex number is the distance from the origin to the corresponding point.

We now apply the theory of complex numbers to the proof of the fundamental theorem of algebra; though it must be remarked that the significance of the theory of complex numbers goes far beyond the limits of algebra. In many parts of mathematics other than algebra, we cannot get along without them. In many applications, for example in the theory of alternating currents, numerous problems are most simply solved by means of complex numbers. But what is most important is the application of complex numbers, or more precisely the theory of functions of a complex variable, to the theory of certain special functions of two real variables which are called *harmonic*. By means of these functions, important problems in the theory of airplane flight, of heat conduction in a plate, of plane electric fields, and of elasticity can be solved. A famous theorem on the lifting force on an airplane wing was obtained by the founder of contemporary aerodynamics, N. E. Žukovskiĭ, through investigations of functions of a complex variable.*

We now pass to the proof of the fundamental theorem of algebra.

Theorem. *Any polynomial*

$$f(z) = a_0 z^n + a_1 z^{n-1} + \cdots + a_{n-1} z + a_n ,$$

whose coefficients

$$a_0 , a_1 , \cdots, a_{n-1} , a_n$$

are any given real or complex numbers, has at least one real or complex root.

We will assume that the given polynomial is of degree n, i.e., that $a_0 \neq 0$.

The surface of the modulus of a polynomial.   We consider the whole problem geometrically. Above each point z of the complex plane, we erect a perpendicular altitude t, equal in length to the modulus $|f(z)|$ of the polynomial $f(z)$ at this point z. The ends of these altitudes define a surface M, which can be called the modulus surface of the polynomial $f(z)$. We see that this surface: (1) nowhere drops below the complex plane, since the modulus of any complex number (in this case, the number $f(z)$) is nonnegative; (2) for any given point z of the complex plane, the surface has one and only one point which either lies vertically above this point or else coincides with it, i.e., the surface M extends in one sheet above the whole complex plane and may at some points touch the plane itself; (3) the surface is continuous in the sense that a continuous change in the position of the point z on the complex plane produces a continuous change in the value of $t = |f(z)|$, i.e., in the altitude t of points of the surface. (This was shown in the last subsection.)

The fundamental theorem of algebra consists in proving that the surface M touches the complex plane in one point at least and does not remain everywhere at a positive distance above it.

On the growth of the modulus of a polynomial with increasing distance from the origin.   We show that no matter how large a positive number G

---

* See Chapter IX.

is given, we can find a radius R such that for all points z of the complex plane, lying outside of the circle of radius R with center at the origin, the altitude t of points of the surface M above the complex plane is greater than G.

For let us write the polynomial $f(z)$ as

$$a_0 z^n \left[1 + \left(\frac{a_1}{a_0 z} + \frac{a_2}{a_0 z^2} + \cdots + \frac{a_n}{a_0 z^n} \right) \right].$$

The modulus of the expression

$$\left(\frac{a_1}{a_0 z} + \frac{a_2}{a_0 z^2} + \cdots + \frac{a_n}{a_0 z^n} \right)$$

is not greater than the sum of the moduli of the summands

$$\left| \frac{a_1}{a_0 z} \right| + \left| \frac{a_2}{a_0 z^2} \right| + \cdots + \left| \frac{a_n}{a_0 z^n} \right|,$$

and, with an increase in the modulus of z, every one of these summands decreases, so that the sum also decreases. Therefore, for all z whose moduli are greather than some number R', the modulus of this expression in parentheses is smaller, for example, than $\frac{1}{2}$.

But then for all such z, the expression

$$\Omega = \left[1 + \left(\frac{a_1}{a_0 z} + \frac{a_2}{a_0 z^2} + \cdots + \frac{a_n}{a_0 z^n} \right) \right]$$

will have modulus greater than $\frac{1}{2}$. The modulus of the first factor $a_0 z^n$ is equal to $|a_0| \cdot |z|^n$, so that it increases with increasing modulus of z; moreover, it increases beyond all bounds. Therefore, no matter how large a positive number G is given, there exists a positive number R such that for all z, whose moduli are greater than R, $|f(z)| = |a_0| \cdot |z|^n \cdot |\Omega|$ is greater than G.

The existence of minima of the surface M. We will say that at a point α of the complex plane the surface M has a minimum if the value of the altitude t of the point of the surface M at this point α is smaller than or equal to its values at all points of some neighborhood of the point α, i.e., at all points of some circle, however small, with center at the point α.

Let the altitude t of the point of the surface M corresponding to the origin, i.e., to the point $z = 0$ of the complex plane, be equal to g, i.e., $|f(0)| = g$. We take $G > g$. All altitudes t of points of the surface M are nonnegative and continuously change during continuous movement of the point z in the complex plane. The surface M has altitude $t > G$ outside of a circle drawn about the origin with radius R and altitude $t = g < G$ at the center of the circle. D'Alembert regarded it as an obvious consequence that somewhere in the interior of the circle R there is a point where the altitude is a minimum; more precisely, where the value of t is smaller than or equal to its values at all remaining points of the circle R, i.e., the surface M has at least one minimum.

The rigorous proof of the existence of such a minimum is based on the following *axiom of continuity* of the set of real numbers.

If two sequences of real numbers are given: $a_1 \leqslant a_2 \leqslant \cdots \leqslant a_n \cdots$ and $b_1 \geqslant b_2 \geqslant \cdots \geqslant b_n \geqslant \cdots$, such that $b_n > a_n$ for all n and $b_n - a_n \to 0$ as $n \to \infty$, then there exists one and only one real number c, such that $a_n \leqslant c \leqslant b_n$ for all n.

Geometrically, this continuity property means that if on the line a sequence of interval $[a_n, b_n]$ (figure 4) is given, such that every successive interval is contained in the preceding interval, and the lengths of the intervals become arbitrarily small, then there exists a point c belonging to all intervals of the sequence. In other words, the intervals "shrink" to a point, and not to "an empty place."

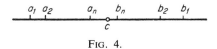

FIG. 4.

Since the length of the segment $[a_n, b_n]$ approaches zero with increasing n, there is only one such point c. From the property of continuity for the set of all points on the number axis immediately follows the property of continuity for complex numbers, i.e., for points of the plane. We give a geometrical formulation of this property.

If in the plane a sequence of rectangles $\Delta_1, \Delta_2, \cdots, \Delta_n, \cdots$ is given, with sides parallel to the coordinate axes, such that every rectangle is contained in the previous one, and such that the length of their diagonals decreases indefinitely, then there exists one and only one point which is contained in all the rectangles of the sequence. This property of continuity of the plane directly follows from the continuity property of the line. For the proof it is sufficient to project the rectangles on the coordinate axes.

Now it is easy to establish the so-called Bolzano-Weierstrass theorem.

If in a rectangle an infinite sequence of points $z_1, z_2, \cdots, z_n, \cdots$ is given, then in the interior or on the boundary of the rectangle there exists a point z_0 such that in any arbitrarily small neighborhood of z_0, i.e., in the interior of an arbitrarily small circle with center at z_0, there are infinitely many points of the sequence $z_1, z_2, \cdots, z_n, \cdots$.

For the proof we denote the given rectangle by Δ_1. We divide it into four equal parts by lines parallel to the coordinate axes. At least one of the parts necessarily contains infinitely many points of the given sequence. This part will be denoted by Δ_2. We again subdivide the rectangle Δ_2 into four equal parts and select among them a Δ_3 which contains infinitely many points of the given sequence, and so on.

We obtain a sequence of imbedded rectangles Δ_1, Δ_2, Δ_3, $\cdots$, whose diagonals decrease indefinitely. By the continuity property we can find a point z_0 contained in all these rectangles. Then this z_0 is the desired point. For, no matter how small a neighborhood of z_0 we take, the rectangles of the sequence Δ_1, Δ_2, Δ_3, $\cdots$, beginning with some one of them will be inside this neighborhood, as soon as their diagonals become smaller than the radius of the neighborhood, and any one of the rectangles contains infinitely many points of the sequence z_1, z_2, $\cdots$, z_n, $\cdots$. Thus the Bolzano-Weierstrass theorem is proved.

Now it is easy to prove the theorem on the minimum of the modulus $|f(z)|$ of a polynomial. As before, let $|f(0)| = g$, let G be a number greater than g, and let R be such that for $z > R$, we have $|f(z)| > G$.

If $g = 0$, i.e., $f(0) = 0$, then the modulus $|f(z)|$ of the polynomial has a minimum at the point 0, since at all points it is $\geqslant 0$.

If $g > 0$ and $|f(z)| \geqslant g$ for all points z then $|f(z)|$ still has a minimum at the point 0. Let $g > 0$ and let points z exist, in which $|f(z)| < g$; then in the sequence of numbers

$$0, \frac{g}{n}, \frac{2g}{n}, \cdots, \frac{ng}{n} = g \qquad (*)$$

we find the greatest $c_n = (i/n)\,g$, such that all values $|f(z)| \geqslant c_n$. For the next number $c_n' = [(i+1)/n]\,g$ the sequence $(*)$ contains at least one point z_n such that $|f(z_n)| < c_n'$.

Let n increase to infinity. For all n we have $|z_n| \leqslant R$, since if $|z_n| > R$, then $|f(z_n)|$ would be greater than G and consequently greater also than g.

Thus all points z_n lie inside a rectangle with sides $2R$, and with center at the origin. It is possible that some of these points coincide.

By the Bolzano-Weierstrass theorem there exists a point z_0 such that every neighborhood of z_0 contains infinitely many points of the sequence z_1, z_2, $\cdots$, z_n, $\cdots$.

We establish that the point z_0 furnishes the desired minimum of $|f(z)|$. For at any point z we have

$$|f(z)| > c_n = c_n' - \frac{g}{n} > |f(z_n)| - \frac{g}{n}$$
$$= |f(z_0)| + [|f(z_n)| - |f(z_0)|] - \frac{g}{n}.$$

This inequality is valid for any n. If we take for n a sequence of values for which z_n indefinitely approaches z_0, then on account of the continuity of $|f(z)|$, the difference $|f(z_n)| - |f(z_0)|$ becomes arbitrarily small in absolute value with g/n.

Consequently, $|f(z)| \geqslant |f(z_0)|$, i.e., $|f(z)|$ actually has a minimum at the point z_0.

D'Alembert's lemma. In view of the fact that all the altitudes t of points on the modulus surface M are nonnegative, it is clear that any root of the polynomial $f(z)$, i.e., any point z of the complex plane where the polynomial $f(z)$ itself (and consequently its modulus $|f(z)|$ also) is equal to zero, corresponds to a minimum of the modulus surface M. However, as d'Alembert showed, the converse is also true: At any minimum the surface M extends down to the complex plane itself, and consequently at that point there is a root of the polynomial $f(z)$. In other words, at any point at which the altitude t is positive and not zero, there is no minimum of the surface M. This follows from the so-called d'Alembert's lemma:

If α is a given complex number such that $f(\alpha) \neq 0$, then a complex number h can always be found with arbitrarily small modulus, such that $|f(\alpha + h)| < |f(\alpha)|$.

Proof. We consider the polynomial

$$f(\alpha + h) = a_0(\alpha + h)^n + a_1(\alpha + h)^{n-1} + \cdots + a_{n-1}(\alpha + h) + a_n$$

in two indeterminates α and h and arrange it in ascending powers of h. In this polynomial there will be a term not containing h at all, namely

$$a_0\alpha^n + a_1\alpha^{n-1} + \cdots + a_{n-1}\alpha + a_n = f(\alpha) \neq 0,$$

since it was assumed that $f(\alpha) \neq 0$. There will also be a term with h^n, namely $a_0 h^n$, since it was assumed that $a_0 \neq 0$. As to the terms with intermediate powers of h, some of them, and in some cases all of them may be missing. Let the lowest power of h which occurs in this polynomial be m, where $1 \leqslant m \leqslant n$, i.e., this expression will have the form

$$f(\alpha + h) = f(\alpha) + Ah^m + Bh^{m+1} + Ch^{m+2} + \cdots + a_0 h^n.$$

Let us write this as:

$$f(\alpha + h) = f(\alpha) + Ah^m + Ah^m \left(\frac{B}{A} h + \frac{C}{A} h^2 + \cdots + \frac{a_0}{A} h^{n-m} \right),$$

where $A \neq 0$, and B, C, etc., may or may not be equal to zero.

After this preparation the proof of d'Alembert's lemma runs as follows. For h it is sufficient to take a complex number with modulus so small that the length of the vector Ah^m is smaller than the length of the vector

$f(\alpha)$ and with argument such that the direction of the vector Ah^m is opposite to the direction of the vector $f(\alpha)$. Then the vector $f(\alpha) + Ah^m$ will be shorter than the vector $f(\alpha)$. But if the modulus of h is taken sufficiently small, the modulus of the expression

$$\left(\frac{B}{A} h + \frac{C}{A} h^2 + \cdots + \frac{a_0}{A} h^{n-m} \right)$$

can be made arbitrarily small, for example, smaller than one, and consequently, the vector

$$\Delta = Ah^m \left(\frac{B}{A} h + \frac{C}{A} h^2 + \cdots + \frac{a_0}{A} h^{n-m} \right)$$

is shorter than the vector Ah^m and therefore, the vector $f(\alpha + h) = f(\alpha) + Ah^m + \Delta$, as is seen in (figure 5), is also shorter than the vector $f(\alpha)$, even if the direction of the vector Δ is in the opposite direction of the vector Ah^m.

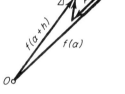

FIG. 5.

The details of this proof are as follows:

1. Since in multiplication, the arguments of the factors are added, we have to take the argument of h such that

$$\arg A + m \cdot \arg h = \arg f(\alpha) + 180°,$$

i.e., it is necessary to take

$$\arg h = \frac{\arg f(\alpha) - \arg A + 180°}{m}.$$

2. The modulus of

$$\left(\frac{B}{A} h + \frac{C}{A} h^2 + \cdots + \frac{a_0}{A} h^{n-m} \right)$$

is not greater than the sum of moduli of its summands

$$T = \left| \frac{B}{A} h \right| + \left| \frac{C}{A} h^2 \right| + \cdots + \left| \frac{a_0}{A} h^{n-m} \right|;$$

moreover, with decreasing modulus of h, each of the summands of this sum can be arbitrarily decreased and consequently so can the whole sum. Therefore, if h is a complex number with the above given argument, and h_0 is a modulus such that if h has a smaller modulus than h_0 and satisfies the two conditions $|Ah^m| < |f(\alpha)|$ and $T < 1$, then for such h we will have $|f(\alpha + h)| < |f(\alpha)|$, which proves d'Alembert's lemma.

From d'Alembert's lemma it immediately follows that every minimum of the modulus surface M of the polynomial $f(z)$ gives a root of this polynomial. Indeed, if at the point α, $f(\alpha) \neq 0$, then by virtue of d'Alembert's lemma at arbitrarily close points $\alpha + h$ we would have $|f(\alpha + h)| < |f(\alpha)|$, i.e., there would not exist a circle with center at α, at all of whose points the modulus of $f(z)$ is not smaller than the modulus

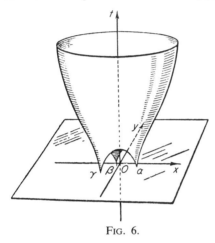

of $f(\alpha)$, and therefore at the point α we would not have a minimum of the modulus of $f(z)$. With this the fundamental theorem of algebra is proved.

The general form of the modulus surface M. The modulus surface M of the polynomial $f(z)$ lies above the complex plane z. It has the form shown in figure 6. It can be shown that at greater altitudes t, the surface M differs very little from the surface obtained by revolving the nth degree parabola $t = |a_0| x^n$ about the Ot-axis.

FIG. 6.

But for small t the surface M has minima, whose number is equal to the number of distinct roots of the equation $f(z) = 0$. At all these minima the surface M touches the complex plane z itself.

§4. Investigation of the Distribution of the Roots of a Polynomial on the Complex Plane

A number of problems important in practice are connected with this question: Without solving a given equation, obtain some information about the distribution of its roots on the complex plane. The first such problem, historically, was to determine the number of real roots of an equation. That is, if an equation with real coefficients is given, then by some test depending on its coefficients, to determine, without solving the equation, whether it has real roots and if it does, how many; or how many positive and how many negative roots it has; or how many real roots lying between given limits a and b.

Derivatives of a polynomial. In this section an essential role will be played by the derivative of a polynomial. The definition of the derivative of a function was given in Chapter II.

For the polynomial $a_0x^n + a_1x^{n-1} + \cdots + a_{n-1}x + a_n$ the derivative is given, as is well known, by the polynomial

$$na_0x^{n-1} + (n-1)a_1x^{n-2} + \cdots + a_{n-1}.$$

The concept of derivative in Chapter II was considered only for functions of a real variable. In algebra it is necessary to consider the variable as taking on arbitrary complex values and to introduce polynomials with complex coefficients.

However, the former definition of derivative can be retained, namely as the limit of the ratio of the increment of the function to the increment of the independent variable. The formula for computing the derivative of a polynomial with complex coefficients, and the basic laws for the derivative of sum, product, and power remain the same as before.*

Simple and multiple roots of a polynomial. In §2 of this chapter it was established that if the number a is a root of the polynomial $f(x)$, then $f(x)$ is divisible by $x - a$ without remainder. If $f(x)$ is not divisible by $(x - a)^2$, then the number a is called a simple root of the polynomial $f(x)$. Generally, if the polynomial $f(x)$ is divisible by $(x - a)^k$ but not by $(x - a)^{k+1}$, then the number a is called a root of multiplicity k.

A root a of multiplicity k is often regarded as k different roots. The basis for this is that the factor $(x - a)^k$, present in the factorization of $f(x)$ into linear factors, is the product of k factors, each equal to $(x - a)$.

By virtue of the fact that every polynomial of degree n can be factored into the product of n linear factors, the number of roots of the polynomial is equal to its degree, if we take into account the multiplicity of each root.

The following theorems are true:

1. A simple root of a polynomial is not a root of its derivative.

2. A multiple root of a polynomial is a root of its derivative of multiplicity one less.

For, let $f(x) = (x - a)^k f_1(x)$ and let $f_1(x)$ not be divisible by $(x - a)$, i.e., $f_1(a) \neq 0$. Then

$$\begin{aligned}
f'(x) &= k(x - a)^{k-1} f_1(x) + (x - a)^k f_1'(x) \\
&= (x - a)^{k-1}[kf_1(x) + (x - a)f_1'(x)] = (x - a)^{k-1} F(x).
\end{aligned}$$

The polynomial $F(x) = kf_1(x) + (x - a)f_1'(x)$ is not divisible by $(x - a)$, since $F(a) = kf_1(a) \neq 0$.

_____

* See Chapter IX.

Consequently, $f'(x)$ for $k = 1$ is not divisible by $x - a$, and for $k > 1 f'(x)$ is divisible by $(x - a)^{k-1}$ but not by $(x - a)^k$. With this both theorems are proved.

Rolle's theorem and some of its consequences. According to the well-known theorem of Rolle,* if the real numbers a and b are roots of a polynomial with real coefficients, then there exists a number c lying between a and b which is a root of the derivative.

From Rolle's theorem the following interesting theorems follow:

1. If all roots of the polynomial $f(x) = a_0 x^n + \cdots + a_n$ are real, then all roots of its derivative are also real. In addition, between two adjacent roots of $f(x)$ there exists one root of $f'(x)$ and this root is simple. Indeed, let $x_1 < x_2 \cdots < x_k$ be the roots of $f(x)$ with multiplicities $m_1, m_2, \cdots, m_k$, respectively. Clearly, $m_1 + m_2 + \cdots + m_k = n$.

Then the derivative $f'(x)$, by the above theorem on multiple roots, will have roots $x_1, x_2, \cdots, x_k$ with multiplicities $m_1 - 1, m_2 - 1, \cdots, m_k - 1$, and by Rolle's theorem there is at least one root $y_1, y_2, \cdots, y_{k-1}$ in the interior of each of the intervals $(x_1, x_2), (x_2, x_3), \cdots, (x_{k-1}, x_k)$ between two successive roots of $f(x)$. Thus, the number of real roots of $f'(x)$ is equal (with regard to multiplicities) to at least $(m_1 - 1) + (m_2 - 1) + \cdots + (m_k - 1) + k - 1 = n - 1$. But $f'(x)$ as an $(n-1)$th-degree polynomial has (with regard to multiplicities) $n - 1$ roots. Consequently, all roots of $f'(x)$ are real, $y_1, y_2, \cdots, y_{k-1}$ are simple roots, and roots other than $x_1, x_2, \cdots, x_k$ and $y_1, y_2, \cdots, y_{k-1}$ of the polynomial $f'(x)$ do not exist.

2. If all roots of a polynomial $f(x)$ are real and of these p are positive, then $f'(x)$ has p or $p - 1$ positive roots.

For, let $x_1 < x_2 < \cdots < x_k$ be all positive roots of the polynomial $f(x)$ with multiplicities $m_1, m_2, \cdots, m_k$, respectively. Then $m_1 + m_2 + \cdots + m_k = p$. The derivative $f'(x)$ will have the following positive roots: $x_1, x_2, \cdots, x_k$ with multiplicities $m_1 - 1, m_2 - 1, \cdots, m_k - 1$; simple roots $y_1, y_2, \cdots, y_{k-1}$ lying in the intervals $(x_1, x_2), \cdots, (x_{k-1}, x_k)$; and it can also have a simple root y_0 lying in the interval (x_0, x_1) where x_0 is the largest nonpositive root of $f(x)$. Consequently, the number of positive roots is equal to $(m_1 - 1) + \cdots + (m_k - 1) + k - 1 = p - 1$ or $(m_1 - 1) + \cdots + (m_k - 1) + (k - 1) + 1 = p$ which was required to be proved.

Descartes' law of signs. In his significant book of 1637 "Geometry," in which the first presentation of analytic geometry was given, Descartes,

---

* This theorem is the simplest form of the mean value theorem, which was mentioned in Chapter II.

among other things, gave the first significant algebraic theorem concerning the distribution of roots of a polynomial on the complex plane, the so-called "Descartes law of signs." It can be stated as follows:

If the coefficients of an equation are real and all its roots are also known to be real, then the number of its positive roots, with account taken of multiplicities, is equal to the number of changes of sign in the sequence of its coefficients. If it also has complex roots, then this number is equal to or an even number less than the number of these changes in sign.

We first explain what we mean by the number of changes of sign in the sequence of coefficients of the equation. To obtain this number we write down all coefficients of the equation, for example in the order of decreasing powers of the unknown, including the coefficient of x^n and the constant term, but omitting coefficients equal to zero, and consider all pairs of successive numbers of the sequence so obtained. If in such a pair the signs of the numbers are different, then we call this a change of sign. For example, if the given equation is

$$x^7 + 3x^5 - 5x^4 - 8x^2 + 7x + 2 = 0$$

then the sequence of its coefficients is

$$1, 3, -5, -8, 7, 2$$

and there are 2 changes of sign.

Now we pass to the proof of the first part of the theorem.*

Without loss of generality we can assume that the leading coefficient a_0 of the polynomial $f(x) = a_0 x^n + \cdots + a_n$ is positive.

First of all, we establish that if $f(x)$ has only real roots and of these p are positive (counting multiplicities) then $(-1)^p$ is the sign of the last coefficient of $f(x)$ different from zero.

Indeed, let

$$f(x) = a_0 x^n + \cdots + a_k x^{n-k}$$
$$= a_0 x^{n-k}(x - x_1) \cdots (x - x_p)(x - x_{p+1}) \cdots (x - x_{n-k}),$$

where $x_1, \cdots, x_p$ are the positive roots of $f(x)$, $x_{p+1}, \cdots, x_{n-k}$ are the negative roots of $f(x)$, account being taken of the multiplicity of each root. Then $a_k = a_0(-1)^p x_1 \cdots x_p (-x_{p+1}) \cdots (-x_{n+k})$ and, since all the numbers $a_0, x_1, \cdots, x_p, -x_{p+1}, \cdots, -x_{n-k}$ are positive, the sign of a_k is $(-1)^p$.†

The subsequent proof is based on the method of mathematical induction.

---

* We could give another, direct proof, not involving derivatives, but it would be somewhat longer.

†We note that this assertion is also correct for the case when some of the roots of $f(x)$ are complex.

For first-degree polynomials the theorem is trivial. Indeed, a first-degree polynomial $a_0 x + a_1$ has a unique root $-a_1/a_0$, which is positive if and only if a_0 and a_1 have opposite signs.

Let us assume now that the theorem is proved for all polynomials of $(n-1)$th degree with real roots, and with this assumption we will prove it for any polynomial $f(x) = a_0 x^n + \cdots + a_{n-1} x + a_n$ of degree n.

1. $a_n = 0$. We consider the polynomial $f_1(x) = a_0 x^{n-1} + \cdots + a_{n-1}$. The positive roots of the polynomials $f(x)$ and $f_1(x)$ are the same; the number of changes of sign in the sequence of their coefficients is also the same. For the polynomial $f_1(x)$ Descartes' law is valid; consequently it is valid for the polynomial $f(x)$.

2. $a_n \neq 0$. We consider the derivative

$$f'(x) = na_0 x^{n-1} + (n-1) a_1 x^{n-2} + \cdots + a_{n-1}.$$

It is clear that the number of changes of sign in the sequence of coefficients of the polynomial $f(x)$ is equal to the analogous number for the derivative $f'(x)$, if the signs of a_n and the last nonzero coefficient of the derivative coincide, or it is one more, if the signs are opposite.

By what was said above at the beginning of the proof, in the first case the number of positive roots of $f(x)$ and of $f'(x)$ have the same parity (are both even or both odd), and in the second case they have opposite parity. But as we deduced from Rolle's theorem, the number of positive roots of a polynomial, if all its roots are real, can be either equal to the number of positive roots of its derivative, or be one more. Taking this into consideration, we note that in the first case $f(x)$ has the same number of positive roots as $f'(x)$, and in the second case one more. For $f'(x)$ Descartes' law is valid by the induction assumption, i.e., the number of positive roots of $f'(x)$ is equal to the number of changes of sign in the sequence of its coefficients. Consequently, in both cases the number of positive roots of $f(x)$ is equal to the number of changes of sign in the sequence of coefficients, and this is the required proof.

The second part of Descartes' law is not more complicated to establish, and we will omit the proof here.·

Remark 1. The first assertion of Descartes' theorem is particularly important, since in many practical problems it is automatically known whether all roots of a given equation are positive. In this case it can be quickly determined, how many roots are positive and how many negative. Also it can be seen at once, how many zero roots the equation has.

Remark 2. If in the given polynomial we set $x = y + a$ where a is an arbitrary given real number, i.e., we form the polynomial $f(y + a)$, then the positive roots y of this polynomial will be those and only those that are obtained from the roots x of the given polynomial $f(x)$ that are greater than a. Therefore the number of roots of the given polynomial $f(x)$, all of whose roots are real, lying between given limits a and b ($b > a$), is equal to the number of changes of sign for the polynomial $f(y + a)$ minus the number of changes of sign for the polynomial $f(z + b)$. If, however, not all roots of $f(x)$ are real, then it can be shown that this number is equal to this difference or some even number less. This is the so-called Budan theorem.

Sturm's theorem. Descartes' law of signs, as well as Budan's theorem do not, however, give an answer to the problem: Does a given equation with real coefficients have at least one real root, how many real roots does it have altogether, and how many real roots does it have lying between given limits a and b? For more than two centuries mathematicians attempted to solve these problems but without result. A long series of efforts in this direction were made by Descartes, Newton, Sylvester, Fourier, and many others, but they did not succeed in solving even the first of these problems, until, finally in 1835 the French mathematician Sturm suggested a method that solved all three problems.

Sturm's method is really not very complicated, but it is of such a character that one might seek it for a long time and not find it. Sturm himself was very happy that he had succeeded in solving this remarkable and exceedingly important pratical problem of algebra. In his lectures, when he came to the presentation of his result, he usually said: "Here is the theorem whose name I bear." But it must be said that Sturm did not solve this problem by mere chance; he pondered for many years on questions related to it.

Let $f(z)$ be a polynomial with real coefficients and $f_1(z)$ be the derivative $f'(z)$. Let us divide the polynomial $f(z)$ by $f_1(z)$ and denote the remainder in this division by $f_2(z)$, taking it with the opposite sign. Then, divide $f_1(z)$ by $f_2(z)$ and denote the remainder, taken with opposite sign, by $f_3(z)$, etc.

It can be shown that the last nonzero polynomial $f_1(z)$ of the constructed sequence will be a constant number c.

Sturm's theorem is as follows: If $a < b$ are two real numbers, which are not roots of the polynomial $f(z)$. then substituting in the polynomials

$$f(z), f_1(z), \cdots, f_{s-1}(z), c$$

$z = a$ and $z = b$, we obtain two sequences of real numbers

$$f(a), f_1(a), f_2(a), \cdots, f_{s-1}(a), c, \qquad (I)$$

$$f(b), f_1(b), f_2(b), \cdots, f_{s-1}(b), c, \qquad (II)$$

such that the number of changes of sign in sequence (I) is greater than or equal to the number of changes of sign in sequence (II) and the difference between these numbers of changes of sign is exactly equal to the number of real roots of $f(z)$ lying between a and b, or in other words, the number of these roots is equal to the loss of changes of sign in sequence (I) in going from a to b.

The proof of Sturm's theorem is not more difficult than the proof of Descartes' theorem, but we will not give it here.

Sturm's theorem enables us to compute the number of roots of a polynomial with real coefficients on any segment of the real axis. Therefore the application of Sturm's theorem to any given polynomial gives a clear picture of distribution of roots of a polynomial on the real axis, in particular it enables us to *separate* the roots, i.e., to construct segments in each of which only one root of the polynomial is contained.

In many applications, the solution of the analogous problem for the complex roots of a polynomial is equally important. Since complex numbers are represented by points not on the line but in the plane, it is impossible to speak of "segments" in which complex roots are contained; instead of a segment, we have to consider a region, i.e., a part of the plane, chosen in one way or another.

Thus, with respect to complex roots the following problem arises:

Given a polynomial $f(z)$ and a region in the complex plane, it is required to find the number of roots of the polynomial inside this region.

We assume that the region is bounded by a closed contour (figure 7) and that on the contour the polynomial $f(z)$ does not have roots.

Imagine that the point z goes around the contour of the region once in the positive direction. Every value of the polynomial is also represented by points on the plane. With continuous change of z the polynomial $f(z)$ also changes continuously. Therefore, while z goes once around the

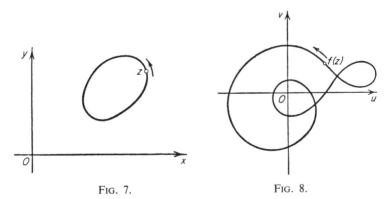

FIG. 7.   FIG. 8.

contour of the region, $f(z)$ describes some closed curve. This curve will not go through the origin of the coordinate system, since $f(z)$ by assumption does not reduce to zero at any of the points of the contour (figure 8).

The answer to the above mentioned problem is given by the following theorem:

Principle of the argument. The number of roots of the polynomial $f(z)$ inside the region bounded by a closed curve C is equal to the number of times the point $f(z)$ winds around the origin as z goes around the contour C once in the positive direction.

For the proof we decompose $f(z)$ into linear factors

$$f(z) = a_0 z^n + a_1 z^{n-1} + \cdots + a_n = a_0(z - z_1)(z - z_2) \cdots (z - z_n).$$

We know that the argument of the product of several complex numbers is equal to the sum of the arguments of the factors. Consequently,

$$\arg f(z) = \arg a_0 + \arg(z - z_1) + \arg(z - z_2) + \cdots + \arg(z - z_n).$$

Let us denote by $\varDelta \arg f(z)$ the increment of the argument of $f(z)$, computed under the assumption that z goes once around the contour C. It is clear that $\varDelta \arg f(z)$ is 2π multiplied by the number of times the point $f(z)$ winds around the origin.

Clearly,

$$\varDelta \arg f(z) = \varDelta \arg a_0 + \varDelta \arg(z - z_1)$$
$$+ \varDelta \arg(z - z_2) + \cdots + \varDelta \arg(z - z_n).$$

It is clear that $\varDelta \arg a_0 = 0$ since a_0 is a constant. Then, $z - z_1$ is represented by the vector going from the point z_1 to the point z. Let us assume that z_1 is in the interior of the region. Geometrically it is clear (figure 9) that as the point z goes around the contour C the vector $z - z_1$ makes a complete revolution about its initial point, so that $\varDelta \arg(z - z_1) = 2\pi$. We assume now that the point z_2 is in the exterior of the region. In this case the vector "oscillates" to one side and back, and returns to its original position without making a revolution about its initial point, so that $\varDelta \arg(z - z_2) = 0$. We can reason the same way about all the roots. Consequently $\varDelta \arg f(z)$ is equal to 2π multiplied by the number of roots of $f(z)$ lying in the interior of the region. Hence the number of roots of $f(z)$ inside the region is equal to the number of times the point $f(z)$ winds around the origin, and this is the required proof.

This theorem enables us to solve the problem in every particular case, and to draw the curve traced by the point $f(z)$ with any degree of accuracy. To do this it is necessary to take a sufficiently dense set of points z on the contour C, to compute the corresponding values $f(z)$ and to join them by a continuous curve. However, in some cases we can get by without these tedious computations. We indicate one of the methods with a numerical example.

Example. Let us find the number of roots of the polynomial $f(z) = z^{11} + 5z^2 - 2$ inside a circle of radius 1 with center at the origin.

On the indicated circle $|z| = 1$, one of the three terms which make up the polynomial $f(z)$, namely $5z^2$, dominates the others. Indeed, $|5z^2| = 5$, but $|z^{11} - 2| \leqslant |z|^{11} + 2 = 3$. This property allows us to reason thus. Let us denote $z^{11} + 5z^2 - 2$ by w, $5z^2$ by N_1, and $z^{11} - 2$ by N_2. While the point z goes once around the unit circle, $N_1 = 5z^2$ winds around a circle of radius 5 twice, since $|N_1| = 5$ and $\arg N_1 = 2 \arg z$. The point w is "tethered" to the point N_1 by a vector whose length is $|N_2| \leqslant 3$, i.e., the distance from the point w to the point N_1 is at all times smaller than the distance from N_1 to the origin of the coordinate system.

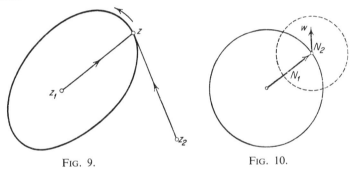

FIG. 9.   FIG. 10.

Consequently, the point w, however it may "wind" around N_1 (figure 10) cannot "independently" go around the origin, and therefore winds around the origin exactly as many times as the point N_1 does, i.e., twice. Consequently, the number of roots of $f(z)$ in the interior of the region in question is equal to two.

Hurwitz's problem. In mechanics, particularly in the theory of oscillations and control, an important role is played by the conditions that permit us to decide whether all the roots of a given polynomial $f(z) = a_0 z^n + a_1 z^{n-1} + \cdots + a_n$ (with real coefficients) have negative real parts, i.e., lie in the half plane left of the imaginary axis.

One of the criteria for solving this problem is easy to obtain from reasons similar to the principle of the argument.

We will assume that $a_0 > 0$.

Let the point z (figure 11) move on the imaginary axis downward from above, i.e., let $z = iy$ as y changes from $+\infty$ to $-\infty$, remaining real. Then $f(z)$ describes a curve with infinite branches. For our investigation the closely related curve described by the function

$$f_1(z) = (i)^{-n} f(z) = a_0 y^n - a_2 y^{n-2} + a_4 y^{n-4} + \cdots - i(a_1 y^{n-1} - a_3 y^{n-3} + \cdots)$$
$$= \phi(y) - i\psi(y),$$

where

$$\phi(y) = a_0 y^n - a_2 y^{n-2} + \cdots,$$
$$\psi(y) = a_1 y^{n-1} - a_3 y^{n-3} + \cdots$$

is more convenient.

Since $\arg i = \pi/2$, therefore $\arg f_1(z) = -n\pi/2 + \arg f(z)$, and consequently the increments of the arguments of $f(z)$ and $f_1(z)$ are the same.

Let us compute the increment of the argument of the point $f_1(z)$ as z moves on the imaginary axis downward.

Let $f(z) = a_0(z - z_1)(z - z_2) \cdots (z - z_n)$. Then

$$\arg f_1(z) = \arg(a_0 i^{-n}) + \arg(z - z_1) + \arg(z - z_2) + \cdots + \arg(z - z_n).$$

It is clear geometrically that the increment of $\arg(z - z_k)$ is equal to π, if z_k lies in the right half plane and to $-\pi$, if z lies in the left half plane (figure 11).

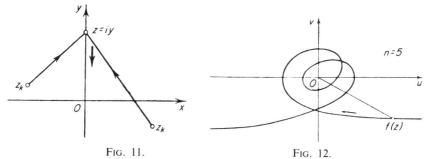

FIG. 11. FIG. 12.

Therefore the increment of the argument of $f_1(z)$ is equal to $\pi(N_1 - N_2)$, where N_1 is the number of roots of $f(z)$ in the right half plane and N_2 is the number of roots in the left half plane. For all the roots to lie in the left half plane it is necessary and sufficient that the increment of the argument of the point $f_1(z)$ be equal to $-\pi n$, i.e., that the point $f_1(z)$ make n half revolutions clockwise about the origin (figure 12).

We note that the point $f_1(z) = \phi(y) - i\psi(y)$ intersects the imaginary axis for those values of y that are roots of $\phi(y)$ and the real axis for the roots of $\psi(y)$. Since $\phi(y)$ has no more than n real roots and the number of real roots of $\psi(y)$ is not more than $n - 1$, it is easy to see geometrically that $f_1(z)$ can make n complete revolutions in the clockwise direction if and only if the curve comes from the fourth quadrant and intersects in turn the negative half of the imaginary axis, the negative half of the real axis, the positive half of the imaginary axis, the positive half of the real axis, etc., so that the total number of points of intersection with the imaginary axis is equal to n (one for every half revolution), and with the real axis it is equal to $n - 1$ (one less than the number of half revolutions). Therefore the coefficient a_1 must be positive, and the roots of the polynomials $\phi(y)$ and $\psi(y)$ must be all real and alternating. This last statement means that if $y_1 > y_2 > \cdots > y_n$ are the roots of $\phi(y)$ arranged in decreasing order, and $\eta_1 > \eta_2 > \cdots > \eta_{n-1}$ are the roots of $\psi(y)$, then $y_1 > \eta_1 > y_2 > \eta_2 > \cdots > y_{n-1} > \eta_{n-1} > y_n$.

Thus, in order that all roots of the polynomial $f(z) > z^n + a_1 z^{n-1} + \cdots + a_n$ with real coefficients and $a_0 > 0$ lie in the left half plane, it is necessary and sufficient that the coefficient a_1 be positive and the roots of the polynomials $\phi(y) = a_0 y^n - a_2 y^{n-2} + a_4 y^{n-4} - \cdots$ and $\psi(y) = a_1 y^{n-1} - a_3 y^{n-3} + \cdots$ be all real and alternating.

This condition is equivalent to the well-known condition of Hurwitz to the effect that all the following determinants are positive:

$$a_1 , \quad \begin{vmatrix} a_1 & a_0 \\ a_3 & a_2 \end{vmatrix} , \quad \begin{vmatrix} a_1 & a_0 & a_{-1} \\ a_3 & a_2 & a_1 \\ a_5 & a_4 & a_3 \end{vmatrix} , \quad \cdots, \quad \begin{vmatrix} a_1 & a_0 & a_{-1} & \cdots & a_{2-n} \\ a_3 & a_2 & a_1 & \cdots & a_{4-n} \\ \cdot & \cdot & \cdot & \cdots & \\ \cdot & \cdot & \cdot & \cdots & \cdot \\ \cdot & \cdot & \cdot & \cdots & \\ a_{2n-1} & a_{2n-2} & a_{2n-3} & \cdots & a_n \end{vmatrix} ,$$

where all a_i with indices less than 0 or greater than n are replaced by zero (on determinants, see Chapter XVI, §3).

§5. Approximate Calculation of Roots

Sturm's method in combination with the lower limit of the difference of two distinct real roots allows us to construct the "separation" of real roots of a polynomial with real coefficients, i.e., allows us to determine for each root limits a and b between which only this one root can be found. It remains to discover a suitable method for finding, in the segment $a < b$, numbers $\alpha_1 < \alpha_2 < \alpha_3 < \cdots$ and $\beta_1 > \beta_2 > \beta_3 > \cdots$, which converge as rapidly as possible to the desired root, the first sequence being an approximation by defect and the second by excess. Each of the two approximations α_k and β_k clearly differs from the desired root x by less than their difference $\beta_k - \alpha_k$, since the root lies between them. Thus we can find upper bounds for the error when we stop at any given approximation.

Graph of a polynomial. Let the given nth-degree polynomial with real coefficients be

$$f(x) = a_0 x^n + a_1 x^{n-1} + \cdots + a_{n-1} x + a_n.$$

Let us consider the curve that represents in rectangular coordinates the equation $y = f(x)$, i.e., the graph of this polynomial. This curve is sometimes called an nth-order parabola. First of all, it is clear that for any real x there is one and only one definite $y = f(x)$; consequently, the graph f ranges arbitrarily far to the right and to the left. In addition, for continuous change of x, $f(x)$ as well as $f'(x)$ change continuously, i.e., without jumps. Therefore, the graph f is a smooth curve. For x large in absolute value the first term $a_0 x^n$ exceeds in absolute value the sum of all remaining terms, since they are all of lower degree. From this it follows that if n is even and $a_0 > 0$, then the graph f on the right and on the left goes to infinity upward (and if $a_0 < 0$, downward); but if n is odd and $a_0 > 0$, then on the right it goes upward and on the left downward (if $a_0 < 0$, then conversely).

The points of intersection of the graph f with the Ox-axis, i.e., those points where $y = f(x) = 0$, correspond to the real roots of the equation $f(x) = 0$; there are no more than n of them. At the maxima and minima of the graph $y = f(x)$, the derivative $f'(x) = 0$; consequently, the number of maxima and minima is not greater than $n - 1$. If on some section $f''(x) > 0$, the first derivative increases there, i.e., the graph is concave upward; if $f''(x) < 0$, then the graph is concave downward. Because some roots of $f'(x) = 0$ may be complex, the number of maxima and minima of the graph f may be smaller than $n - 1$.

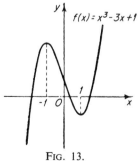

$f(x) = x^3 - 3x + 1$

FIG. 13.

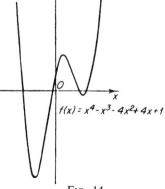

$f(x) = x^4 - x^3 - 4x^2 + 4x + 1$

FIG. 14.

Here are examples of the graphs of polynomials

$$f(x) = x^3 - 3x + 1 \quad \text{(figure 13)},$$
$$f(x) = x^4 - x^3 - 4x^2 + 4x + 1 \quad \text{(figure 14)}.$$

After constructing the graph of a polynomial it is easy to find approximations to its roots. Namely, the roots are the abscissas of the points of intersection of the graph with the Ox-axis.

The method of "undershot" and "overshot." Let us substitute in the polynomial $f(x)$ some integral rational number, for example 3, and then substitute 4, 5, $\cdots$. If in substituting 4, 5, 6 we still obtain the same sign as for 3, but for 7 the opposite sign, then it is clear that between 6 and 7 the polynomial $f(x)$ has at least one root. Now we substitute 6, 6.1, 6.2, $\cdots$ and find two neighbors of this sequence of numbers, for example 6.4 and 6.5 which when substituted give different signs. Accordingly, there will be at least one root between them. Then we substitute 6.4, 6.41, 6.42, 6.43, $\cdots$ and find even closer limits for the root, for example, 6.42 and 6.43, etc. This is the method of "undershooting and overshooting." The method can be considerably simplified by applying at each step of the calculations a supplementary transformation of the polynomial, and then at each step after the first, it will be necessary to substitute only whole numbers and not fractions, and moreover, only the whole numbers 1, 2, $\cdots$, 9. But we will not dwell on this simplification.

The method of tangents and the method of chords. The method of tangents, called Newton's method, and the method of chords, or of linear interpolation, called also the method of false position (*regula falsi*), are used either separately or together to obtain estimates of error. Suppose between a and b we have only one root of the polynomial $f(x)$, so that $f(a)$ and $f(b)$ are of opposite sign, and let us also suppose that the second derivative $f''(x)$ between a and b is of constant sign. In this case the part of the graph of $f(x)$ between a and b has one of four forms (figure 15).

In cases I and II in figure 15, the tangent to the graph at the point with abscissa a intersects the Ox-axis at a point with abscissa α_1 lying between the desired root and a. If we calculate the abscissa α_1 and consider now the tangent to the graph from the point with abscissa α_1, we analogously find a point α_2 lying between the point α_1 and the desired root, and then find a corresponding α_3 and so on. In this way we will obtain better and better approximations with defect. As can be seen from the diagram, these values approach the desired root with great rapidity.

In cases III and IV it is necessary, on the other hand, to start with the abscissa b, and then obtain points $\beta_1, \beta_2, \beta_3, \cdots$, i.e., better and better approximations with excess. Which of the four cases actually occurs, is easy to determine by the signs of $f(a)$, $f(b)$, and $f''(x)$ for $a < x < b$.

Since the equation of the tangent to the curve $y = f(x)$ at its points with abscissa a is

$$y - f(a) = f'(a)(x - a),$$

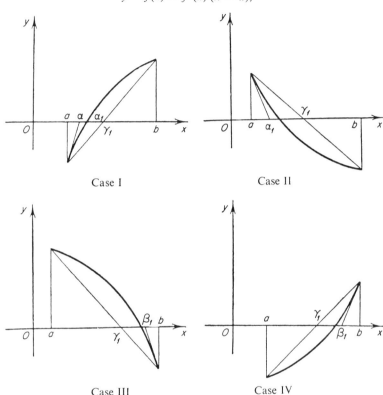

Case I

Case II

Case III

Case IV

Fig. 15.

the abscissa α_1 of the point of its intersection with the Ox-axis is obtained from the equality

$$0 - f(a) = f'(a)(\alpha_1 - a)$$

that is

$$\alpha_1 = a - \frac{f(a)}{f'(a)}.$$

Then

$$\alpha_2 = \alpha_1 - \frac{f(\alpha_1)}{f'(\alpha_1)}, \quad \alpha_3 = \alpha_2 - \frac{f(\alpha_2)}{f'(\alpha_2)}$$

and so on.

Analogously,

$$\beta_1 = b - \frac{f(b)}{f'(b)} , \beta_2 = \beta_1 - \frac{f(\beta_1)}{f'(\beta_1)} , \beta_3 = \beta_2 - \frac{f(\beta_2)}{f'(\beta_2)}$$

and so on.

This is Newton's method.*

The method of linear interpolation or false position, consists of the following. The equation of a chord, as the equation of a line passing through two given points, has the form

$$\frac{x - a}{b - a} = \frac{y - f(a)}{f(b) - f(a)} ,$$

and the abscissa γ_1 of the point of its intersection with the Ox-axis, as obtained from the equation

$$\frac{x - a}{b - a} = \frac{0 - f(a)}{f(b) - f(a)}$$

is equal to

$$\gamma_1 = -\frac{(b - a)f(a)}{f(b) - f(a)} + a = \frac{af(b) - bf(a)}{f(b) - f(a)} .$$

Taking this number for the new b in cases I and II and for the new a in cases III and IV, we find in cases I and II

$$\gamma_2 = \frac{af(\gamma_1) - \gamma_1 f(a)}{f(\gamma_1) - f(a)} , \quad \gamma_3 = \frac{af(\gamma_2) - \gamma_2 f(a)}{f(\gamma_2) - f(a)}$$

and so forth.

In cases III and IV, taking γ_1 for the new a we find

$$\gamma_2 = \frac{\gamma_1 f(b) - bf(\gamma_1)}{f(b) - f(\gamma_1)} , \quad \gamma_3 = \frac{\gamma_2 f(b) - bf(\gamma_2)}{f(b) - f(\gamma_2)}$$

and so forth.

The combination of these two methods is particularly important, since (as may be seen from the diagrams) it allows us, if the approximations from above and below are known, to estimate the error, which is clearly

---

* From these formulas we also obtain a rigorous proof of the two assertions made from a consideration of the diagrams. Namely, the values α_n (or β_n) with increasing n change monotonically, for example in case I they increase and are bounded, i.e., by virtue of the Weierstrass lemma, they approach some limit α. Replacing α_n in these formulas by its limit α, we obtain $\alpha = \alpha - [f(\alpha)/f'(\alpha)]$ from which $f(\alpha) = 0$, i.e., α is a root of f.

not greater than the difference between these approximations, since the desired root is between them.

Remark. It is important to note that the fact that $f(x)$ is a polynomial, and not some other function of x, does not play any role at all either in Newton's method, or in the method of linear interpolation, i.e., both of these methods and their combination can be adapted, under the aforementioned conditions, to transcendental equations.

Lobačevskiĭ's method. One of the most widely used methods of calculation of roots, especially of complex roots, is the method* proposed by N. I. Lobačevskiĭ in his book "Algebra," published in 1834. The basic idea of this method goes back to Bernoulli.

We note, first of all, that if we are given a polynomial whose roots are $x_1, x_2, \cdots, x_n$, then it is easy to write down the polynomial, also of the nth-degree, whose roots are $x_1^2, x_2^2, \cdots, x_n^2$, i.e., the squares of the roots of the given polynomial. Indeed, if $x_1, x_2, \cdots, x_n$ are the roots of the polynomial

$$x^n + a_1 x^{n-1} + a_2 x^{n-2} + \cdots + a_n,$$

then it may be written as

$$(x - x_1)(x - x_2) \cdots (x - x_n),$$

and the polynomial

$$x^n - a_1 x^{n-1} + a_2 x^{n-2} - \cdots \pm a_n,$$

whose roots are the roots of the given polynomial taken with opposite sign, may be written as

$$(x + x_1)(x + x_2) \cdots (x + x_n).$$

The product of these two polynomials is consequently

$$(x^2 - x_1^2)(x^2 - x_2^2) \cdots (x^2 - x_n^2)$$

and therefore contains only even powers of x. Setting $x^2 = y$, we obtain an nth-degree polynomial in y

$$y^n + b_1 y^{n-1} + b_2 y^{n-2} + \cdots + b_n,$$

which may be written as

$$(y - x_1^2)(y - x_2^2) \cdots (y - x_n^2),$$

---

* This method was discovered independently by Dandelin (1826), N. I. Lobačevskiĭ (1834), and Graeffe (1837).

since its roots are $x_1^2, x_2^2, \cdots, x_n^2$. Instead of directly multiplying the polynomial

$$x^n + a_1 x^{n-1} + a_2 x^{n-2} + \cdots + a_n$$

by the polynomial

$$x^n - a_1 x^{n-1} + a_2 x^{n-2} - \cdots \pm a_n,$$

we can obtain the coefficients b_k according to the following scheme. In the first row above a horizontal line, we write $1, a_1, a_2, \cdots, a_n$ and then below the line, under each of these coefficients a_k, we write first its square a_k^2, then minus twice the product of its neighbors

$$- 2a_{k-1} a_{k+1},$$

then plus twice the product of the coefficients

$$+ 2a_{k-2} a_{k+2},$$

symmetric with respect to a_k, etc., alternating in sign until all further coefficients on one side or the other are equal to zero. The coefficients b_k are then obtained as the sum of the corresponding columns of numbers written under the line.

After obtaining these coefficients $1, b_1, b_2, \cdots, b_n$ of the polynomial whose roots are $1, x_1^2, x_2^2, \cdots, x_n^2$, we next construct the coefficients $1, c_1, c_2, \cdots, c_n$ of the polynomial whose roots are the squares of the roots of the polynomial

$$y^n + b_1 y^{n-1} + b_2 y^{n-2} + \cdots + b_n,$$

i.e., $x_1^4, x_2^4, \cdots, x_n^4$. Then analogously we obtain the coefficients $1, d_1, d_2, \cdots, d_n$ of the polynomial whose roots are $x_1^8, x_2^8, \cdots, x_n^8$; and then the polynomial whose roots are $x_1^{16}, x_2^{16}, \cdots, x_n^{16}$, and so forth.

Let us consider only the fundamental idea of Lobačevskiĭ's method; moreover, we restrict ourselves for simplicity to the case when all roots of the equation are real and distinct in absolute value. Let

$$|x_1| > |x_2| > \cdots > |x_n|,$$

i.e., let x_1 be the root largest in absolute value, x_2 the next largest, and so on. Let N be a sufficiently large number and let the polynomial

$$X^n + A_1 X^{n-1} + A_2 X^{n-2} + \cdots + A_n$$

have roots equal to the Nth power of the roots x_1, x_2, $\cdots$, x_n of the given polynomial, i.e.,

$$- A_1 = x_1^N + x_2^N + \cdots + x_n^N,$$

$$A_2 = x_1^N x_2^N + x_1^N x_3^N + \cdots + x_{n-1}^N x_n^N,$$

$$\cdots\cdots\cdots\cdots\cdots\cdots\cdots\cdots\cdots\cdots\cdots\cdots\cdots\cdots\cdots$$

$$\pm A_n = x_1^N x_2^N \cdots x_n^N.$$

Then in the sequence of numbers $|x_1^N|$, $|x_2^N|$, $\cdots$, $|x_n^N|$ for large N each sucessive number is so much smaller than its predecessor that in these expressions for A_1, A_2, $\cdots$, A_n we may retain only the first summand, the sum of all remaining summands being neglected in comparison with the first. We thus obtain the approximate formulas

$$x_1^N \approx - A_1, \qquad x_1^N x_2^N \approx A_2,$$

$$x_1^N x_2^N x_3^N \approx A_3, \cdots, x_1^N x_2^N x_3^N \cdots x_n^N \approx \pm A_n,$$

or, dividing pairwise and taking the Nth roots, we have the following formulas for x_k:

$$x_1 = \sqrt[N]{-A_1}, \; x_2 = \sqrt[N]{- \frac{A_2}{A_1}}, \; x_3 = \sqrt[N]{- \frac{A_3}{A_2}}, \cdots, x_n = \sqrt[N]{- \frac{A_n}{A_{n-1}}}.$$

It can be shown that it is sufficient to extend the computation up to the polynomial whose coefficients taken with signs $+ - + - \cdots$ will be equal with the necessary degree of exactness, to the squares of the corresponding coefficients of the preceding polynomial.

A detailed exposition of Lobačevskiǐ's method can be found in the well-known book of Academician A. N. Krylov "Lectures on approximate calculations."

Suggested Reading

E. Artin, *Galois theory*, 2nd ed., University of Notre Dame, 1944.

——, *Geometric algebra*, Interscience, New York, 1957.

G. Birkhoff and S. MacLane, *A survey of modern algebra*, 2nd ed., Macmillan, New York, 1953.

F. Klein, *Famous problems of elementary geometry. The duplication of the cube, the trisection of an angle, the quadrature of the circle*, Dover, New York, 1956.

——, *Lectures on the icosahedron and the solution of equations of the fifth degree*, Dover, New York, 1956.

M. Marden, *The geometry of the zeros of a polynomial in a complex variable*, American Mathematical Society, Providence, R. I., 1949.

CHAPTER V

ORDINARY
DIFFERENTIAL EQUATIONS

§1. Introduction

Examples of differential equations. The equations that we have encountered up to now have been for the most part concerned with finding the numerical value of one magnitude or another. When, for example, in the search for maxima and minima of functions, we solved an equation and found those points for which the rate of change of a function vanishes, or when in Chapter IV we considered the problem of finding the roots of polynomials, we were in each case looking for isolated numbers. But in the applications of mathematics there often arise problems of a qualitatively different sort, in which the unknown is itself a function, a law expressing the dependence of certain variables on others. For example, in investigating the process of the cooling of a body, our task is to determine how its temperature will change in the course of time; to describe the motion of a planet or a star we must determine the dependence of their coordinates on time, and so forth.

We can quite often construct an equation for finding the required unknown functions, such equations being called functional equations. The nature of these may, generally speaking, be extremely varied; in fact, it may be said that we have already met the simplest and most primitive functional equations when we were considering implicit functions.

The problem of finding unknown functions will concern us in Chapters V, VI, and VII. In the present chapter, and in the following one, we will consider the most important class of equations serving to determine such functions, namely *differential equations*; that is, equations in which not only the unknown function occurs, but also its derivatives of various orders.

311

The following equations may serve as examples:

$$\frac{dx}{dt} + P(t)\,x = Q(t), \frac{d^2x}{dt^2} + m^2x = A \sin \omega t, \frac{d^2x}{dt^2} = tx,$$

$$\frac{\partial u}{\partial t} = \frac{\partial^2 u}{\partial x^2}, \frac{\partial^2 u}{\partial t^2} = \frac{\partial^2 u}{\partial x^2}, \frac{\partial^2 u}{\partial x^2} + \frac{\partial^2 u}{\partial y^2} = 0. \tag{1}$$

In the first three of these, the unknown function is denoted by the letter x and the independent variable by t; in the last three, the unknown function is denoted by the letter u and it depends on two arguments, x and t, or x and y.

The great importance of differential equations in mathematics, and especially in its applications, is due chiefly to the fact that the investigation of many problems in physics and technology may be reduced to the solution of such equations.

Calculations involved in the construction of electrical machinery or of radiotechnical devices, computation of the trajectory of projectiles, investigation of the stability of an aircraft in flight, or of the course of a chemical reaction, all depend on the solution of differential equations.

It often happens that the physical laws governing a phenomenon are written in the form of differential equations, so that the differential equations themselves provide an exact quantitative (numerical) expression of these laws. The reader will see in the following chapters how the laws of conservation of mass and of heat energy are written in the form of differential equations. The laws of mechanics discovered by Newton allow one to investigate the behavior of any mechanical system by means of differential equations.

Let us illustrate by a simple example. Consider a material particle of mass m moving along an axis Ox, and let x denote its coordinate at the instant of time t. The coordinate x will vary with the time, and knowledge of the entire motion of the particle is equivalent to knowledge of the functional dependence of x on the time t. Let us assume that the motion is caused by some force F, the value of which depends on the position of the particle (as defined by the coordinate x), on the velocity of motion $v = dx/dt$ and on the time t, i.e., $F = F(x, dx/dt, t)$. According to the laws of mechanics, the action of the force F on the particle necessarily produces an acceleration $w = d^2x/dt^2$ such that the product of w and the mass m of the particle is equal to the force, and so at every instant of the motion we have the equation

$$m\frac{d^2x}{dt^2} = F\left(x, \frac{dx}{dt}, t\right). \tag{2}$$

This is the differential equation that must be satisfied by the function $x(t)$ describing the behavior of the moving particle. It is simply a representation of laws of mechanics. Its significance lies in the fact that it enables us to reduce the mechanical problem of determining the motion of a particle to the mathematical problem of the solution of a differential equation.

Later in this chapter, the reader will find other examples showing how the study of various physical processes can be reduced to the investigation of differential equations.

The theory of differential equations began to develop at the end of the 17th century, almost simultaneously with the appearance of the differential and integral calculus. At the present time, differential equations have become a powerful tool in the investigation of natural phenomena. In mechanics, astronomy, physics, and technology they have been the means of immense progress. From his study of the differential equations of the motion of heavenly bodies, Newton deduced the laws of planetary motion discovered empirically by Kepler. In 1846 Leverrier predicted the existence of the planet Neptune and determined its position in the sky on the basis of a numerical analysis of the same equations.

To describe in general terms the problems in the theory of differential equations, we first remark that every differential equation has in general not one but infinitely many solutions; that is, there exists an infinite set of functions that satisfy it. For example, the equation of motion for a particle must be satisfied by any motion induced by the given force $F(x, dx/dt, t)$, independently of the starting point or the initial velocity. To each separate motion of the particle there will correspond a particular dependence of x on time t. Since under a given force F there may be infinitely many motions the differential equation (2) will have an infinite set of solutions.

Every differential equation defines, in general, a whole class of functions that satisfy it. The basic problem of the theory is to investigate the functions that satisfy the differential equation. The theory of these equations must enable us to form a sufficiently broad notion of the properties of all functions satisfying the equation, a requirement which is particularly important in applying these equations to the natural sciences. Moreover, our theory must guarantee the means of finding numerical values of the functions, if these are needed in the course of a computation. We will speak later about how these numerical values may be found.

If the unknown function depends on a single argument, the differential equation is called an *ordinary differential equation*. If the unknown function depends on several arguments and the equation contains derivatives with respect to some or all of these arguments, the differential equation is called a *partial differential equation*. The first three of the equations in (1) are ordinary and the last three are partial.

The theory of partial differential equations has many peculiar features which make them essentially different from ordinary differential equations. The basic ideas involved in such equations will be presented in the next chapter; here we will examine only ordinary differential equations.

Let us consider some examples.

Example 1.   The law of decay of radium says that the rate of decay is proportional to the initial amount of radium present. Suppose we know that a certain time $t = t_0$ we had R_0 grams of radium. We want to know the amount of radium present at any subsequent time t.

Let $R(t)$ be the amount of undecayed radium at time t. The rate of decay is given by the value of $-(dR/dt)$. Since this is proportional to R, we have

$$-\frac{dR}{dt} = kR, \tag{3}$$

where k is a constant

In order to solve our problem, it is necessary to determine a function from the differential equation (3). For this purpose we note that the function inverse to $R(t)$ satisfies the equation

$$-\frac{dt}{dR} = \frac{1}{kR}, \tag{4}$$

since $dt/dR = (1/dR)/dt$. From the integral calculus it is known that equation (4) is satisfied by any function of the form

$$t = -\frac{1}{k}\ln R + C,$$

where C is an arbitrary constant. From this relation we determine R as a function of t. We have

$$R = e^{-kt+kC} = C_1 e^{-kt}. \tag{5}$$

From the whole set of solutions (5) of equation (3) we must select one which for $t = t_0$ has the value R_0. This solution is obtained by setting $C_1 = R_0 e^{kt_0}$.

From the mathematical point of view, equation (3) is the statement of a very simple law for the change with time of the function R; it says that the rate of decrease $-(dR/dt)$ of the function is proportional to the value of the function R itself. Such a law for the rate of change of a function is satisfied not only by the phenomena of radioactive decay but also by many other physical phenomena.

We find exactly the same law for the rate of change of a function, for example, in the study of the cooling of a body, where the rate of decrease in the amount of heat in the body is proportional to the difference between the temperature of the body and the temperature of the surrounding medium, and the same law occurs in many other physical processes. Thus the range of application of equation (3) is vastly wider than the particular problem of the radioactive decay from which we obtained the equation.

Example 2. Let a material point of a mass m be moving along the horizontal axis Ox in a resisting medium, for example in a liquid or a gas, under the influence of the elastic force of two springs, acting under Hooke's law (figure 1), which states that the elastic force acts toward the

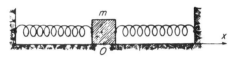

Fig. 1.

position of equilibrium and is proportional to the deviation from the equilibrium position. Let the equilibrium position occur at the point $x = 0$. Then the elastic force is equal to $-bx$ where $b > 0$.

We will assume that the resistance of the medium is proportional to the velocity of motion, i.e., equal to $-a(dx/dt)$, where $a > 0$ and the minus sign indicates that the resisting medium acts against the motion. Such an assumption about the resistance of the medium is confirmed by experiment.

From Newton's basic law that the product of the mass of a material point and its acceleration is equal to the sum of the forces acting on it, we have

$$m\frac{d^2x}{dt^2} = -bx - a\frac{dx}{dt}. \tag{6}$$

Thus the function $x(t)$, which describes the position of the moving point at any instant of time t, satisfies the differential equation (6). We will investigate the solutions of this equation in one of the later sections.

If, in addition to the forces mentioned, the material point is acted upon by still another force, F outside of the system, then the equation of motion (6) takes the form

$$m\frac{d^2x}{dt^2} = -bx - a\frac{dx}{dt} + F \tag{6'}$$

Example 3. A mathematical pendulum is a material point of mass m, suspended on a string whose length will be denoted by l. We will assume that at all stages the pendulum stays in one plane, the plane of the drawing (figure 2). The force tending to restore the pendulum to the vertical position OA is the force of gravity mg, acting on the material point. The position of the pendulum at any time t is given by the angle ϕ by which it differs from the vertical OA. We take the positive direction of ϕ to be counterclockwise. The arc $AA' = l\phi$ is the distance moved by the material point from the position of equilibrium A. The velocity of motion v will be directed along the tangent to the circle and will have the following numerical value:

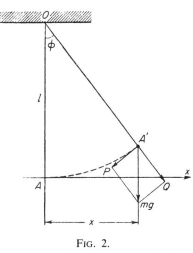

FIG. 2.

$$v = l\frac{d\phi}{dt}.$$

To establish the equation of motion, we decompose the force of gravity mg into two components Q and P, the first of which is directed along the radius OA' and the second along the tangent to the circle. The component Q cannot affect the numerical value of the rate v, since clearly it is balanced by the resistance of the suspension OA'. Only the component P can affect the value of the velocity v. This component always acts toward the equilibrium position A, i.e., toward a decrease in ϕ, if the angle ϕ is positive, and toward an increase in ϕ, if ϕ is negative. The numerical value of P is equal to $-mg \sin \phi$, so that the equation of motion of the pendulum is

$$m\frac{dv}{dt} = -mg \sin \phi$$

or

$$\frac{d^2\phi}{dt^2} = -\frac{g}{l}\sin \phi. \tag{7}$$

It is interesting to note that the solutions of this equation cannot be expressed by a finite combination of elementary functions. The set of elementary functions is too small to give an exact description of even such a simple physical process as the oscillation of a mathematical pendulum. Later we will see that the differential equations that are solvable by elementary functions are not very numerous, so that it very frequently happens that investigation of a differential equation encountered in physics or mechanics leads us to introduce new classes of functions, to subject them to investigation, and thus to widen our arsenal of functions that may be used for the solution of applied problems.

Let us now restrict ourselves to small oscillations of the pendulum for which, with small error, we may assume that the arc AA' is equal to its projection x on the horizontal axis Ox and $\sin \phi$ is equal to ϕ. Then $\phi \approx \sin \phi = x/l$ and the equation of motion of the pendulum will take on the simpler form

$$\frac{d^2x}{dt^2} = -\frac{g}{l} x. \tag{8}$$

Later we will see that this equation is solvable by trigonometric functions and that by using them we may describe with sufficient exactness the "small oscillations" of a pendulum.

Example 4. Helmholtz' acoustic resonator (figure 3) consists of an air-filled vessel V, the volume of which is equal to v, with a cylindrical neck F. Approximately, we may consider the air in the neck of the container as cork of mass

$$m = \rho s l, \tag{9}$$

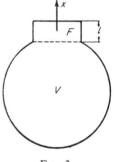

FIG. 3.

where ρ is the density of the air, s is the area of the cross section of the neck, and l is its length. If we assume that this mass of air is displaced from a position of equilibrium by an amount x, then the pressure of the air in the container with volume v is changed from the initial value p by some amount which we will call Δp.

We will assume that the pressure p and the volume v satisfy the adiabatic law $pv^k = C$. Then, neglecting magnitudes of higher order, we have

$$\Delta p \cdot v^k + pkv^{k-1} \cdot \Delta v = 0$$

and

$$\Delta p = -kp \frac{\Delta v}{v} = -\frac{kps}{v} x. \tag{10}$$

(In our case, $\Delta v = sx$.) The equation of motion of the mass of air in the neck may be written as:

$$m \frac{d^2x}{dt^2} = \Delta p \cdot s. \tag{11}$$

Here $\Delta p \cdot s$ is the force exerted by the gas within the container on the column of air in the neck. From (10) and (11) we get

$$\rho l \frac{d^2x}{dt^2} = -\frac{kps}{v} x, \tag{12}$$

where ρ, p, v, l, k, and s are constants.

Example 5.   An equation of the form (6) also arises in the study of electric oscillations in a simple oscillator circuit. The circuit diagram is given in (figure 4). Here on the left we have a condenser of capacity C, in series with a coil of inductance L, and a resistance R. At some instant let the condenser have a voltage across its terminals. In the absence of inductance from the circuit, the current would flow until such time as the terminals of the condenser were at the same potential. The presence of an inductance alters the situation, since the circuit will now generate electric oscillations. To find a law for these oscillations, we denote by $v(t)$, or simply by v, the voltage across the condenser at the instant t, by $I(t)$ the current at the instant t, and by R the resistance. From well-known laws of physics, $I(t)R$ remains constantly equal to the total electromotive force, which is the sum of the voltage across the condenser and the inductance $-L(dI/dt)$. Thus,

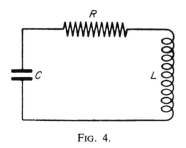

Fig. 4.

$$IR = -v - L \frac{dI}{dt}. \tag{13}$$

We denote by $Q(t)$ the charge on the condenser at time t. Then the current in the circuit will, at each instant, be equal to dQ/dt. The potential difference $v(t)$ across the condenser is equal to $Q(t)/C$. Thus $I = dQ/dt = C(dv/dt)$ and equation (13) may be transformed into

$$LC \frac{d^2v}{dt^2} + RC \frac{dv}{dt} + v = 0. \tag{14}$$

Example 6. The circuit diagram of an electron-tube generator of electromagnetic oscillations is shown in figure 5. The oscillator circuit consisting of a capacitance C, across a resistance R and an inductance L, represents the basic oscillator system. The coil L' and the tube shown in the center of figure 5 form a so-called "feedback." They connect a source of energy, namely the battery B, with the L-R-C circuit; K is the cathode of the tube, A the plate, and S the grid. In such an L-R-C circuit "self-oscillations" will arise. For any actual system in an oscillatory state the energy is transformed into heat or is dissipated in some other form to the surrounding bodies, so that to maintain a stationary state of oscillation it is necessary to have an outside source of energy. Self-oscillations differ from other oscillatory processes in that to maintain a stationary oscillatory state of the system the outside source does not have to be periodic. A self-oscillatory system is constructed in such a way that a constant source of energy, in our case the battery B, will maintain a stationary oscillatory state. Examples of self-oscillatory systems are a clock, an electric bell, a string and bow moved by the hand of the musician, the human voice, and so forth.

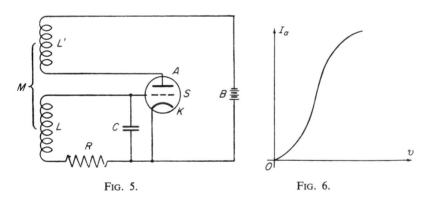

FIG. 5.   FIG. 6.

The current $I(t)$ in the oscillatory L-R-C circuit satisfies the equation

$$L \frac{dI}{dt} + RI + v = M \frac{dI_a}{dt}. \qquad (15)$$

Here $v = v(t)$ is the voltage across the condenser at the instant t, $I_a(t)$ is the plate current through the coil L'; M is the coupling coefficient between the coils L and L'. In comparison with equation (13), equation (15) contains the extra term $M(dI_a/dt)$.

We will assume that the plate current $I_a(t)$ depends only on the voltage between the grid S and the cathode of the tube (i.e., we will neglect the reactance of the anode), so that this voltage is equal to the voltage $v(t)$ across the condenser C. The character of the functional dependence of I_a on v is given in figure 6. The curve as sketched is usually taken to be a cubical parabola and we write an approximate equation for it by:

$$I_a = a_1 v + a_2 v^2 + a_3 v^3.$$

Substituting this into the right side of equation (15), and using the fact that

$$\frac{dv}{dt} = I,$$

we get for v the equation

$$L \frac{d^2 v}{dt^2} + [R - M(a_1 + 2a_2 v + 3a_3 v^2)] \frac{dv}{dt} + v = 0. \tag{16}$$

In the examples considered, the search for certain physical quantities characteristic of a given physical process is reduced to the search for solutions of ordinary differential equations.

Problems in the theory of differential equations.   We now give exact definitions. *An ordinary differential equation of order n in one unknown function y is a relation of the form*

$$F[x, y(x), y'(x), y''(x), \cdots, y^{(n)}(x)] = 0 \tag{17}$$

between the independent variable x and the quantities

$$y(x), y'(x) = \frac{dy}{dx}, y''(x) = \frac{d^2 y}{dx^2}, \cdots, y^{(n)}(x) = \frac{d^n y}{dx^n}.$$

The order of a differential equation is the order of the highest derivative of the unknown function appearing in the differential equation. Thus the equation in example 1 is of the first order, and those in examples 2, 3, 4, 5, and 6, are of the second order.

A function $\phi(x)$ is called a *solution of the differential equation* (17) if substitution of $\phi(x)$ for y, $\phi'(x)$ for y', $\cdots$, $\phi^{(n)}(x)$ for $y^{(n)}$ produces an identity.

Problems in physics and technology often lead to a system of ordinary differential equations with several unknown functions, all depending on the same argument and on their derivatives with respect to that argument.

For greater concreteness, the explanations that follow will deal chiefly with one ordinary differential equation of order not higher than the second and with one unknown function. With this example one may explain the essential properties of all ordinary differential equations and of systems of such equations in which the number of unknown functions is equal to the number of equations.

We have spoken earlier of the fact that, as a rule, every differential equation has not one but an infinite set of solutions. Let us illustrate this first of all by intuitive considerations based on the examples given in equations (2–6). In each of these, the corresponding differential equation is already fully defined by the physical arrangement of the system. But in each of these systems there can be many different motions. For example, it is perfectly clear that the pendulum described by equation (8) may oscillate with many different amplitudes. To each of these different oscillations of the pendulum there corresponds a different solution of equation (8), so that infinitely many such solutions must exist. It may be shown that equation (8) is satisfied by any function of the form

$$x = C_1 \cos \sqrt{\frac{g}{l}}\, t + C_2 \sin \sqrt{\frac{g}{l}}\, t, \tag{18}$$

where C_1 and C_2 are arbitrary constants.

It is also physically clear that the motion of the pendulum will be completely determined only in case we are given, at some instant t_0, the (initial) value x_0 of x (the initial displacement of the material point from the equilibrium position) and the initial rate of motion $x_0' = (dx/dt)\,|_{t=0}$. These intial conditions determine the constants C_1 and C_2 in formula (18).

In exactly the same way, the differential equations we have found in other examples will have infinitely many solutions.

In general, it can be proved, under very broad assumptions concerning the given differential equation (17) of order n in one unknown function that it has infinitely many solutions. More precisely: If for some "initial value" of the argument, we assign an "initial value" to the unknown function and to all of its derivatives through order $n - 1$, then one can find a solution of equation (17) which takes on these preassigned initial values. It may also be shown that such initial conditions completely determine the solution, so that there exists only one solution satisfying the initial conditions given earlier. We will discuss this question later in more detail. For our present aims, it is essential to note that the initial values of the function and the first $n - 1$ derivatives may be given arbitrarily. We have the right to make any choice of n values which define an "initial state" for the desired solution.

If we wish to construct a formula that will if possible include all solutions of a differential equation of order n, then such a formula must contain n

independent arbitrary constants, which will allow us to impose n initial conditions. Such solutions of a differential equation of order n, containing n independent arbitrary constants, are usually called *general solutions* of the equation. For example, a general solution of (8) is given by formula (18) containing two arbitrary constants; a general solution of equation (3) given by formula (5).

We will now try to formulate in very general outline the problems confronting the theory of differential equations. These are many and varied, and we will indicate only the most important ones.

If the differential equation is given together with its initial conditions, then its solution is completely determined. The construction of formulas giving the solution in explicit form is one of the first problems of the theory. Such formulas may be constructed only in simple cases, but if they are found, they are of great help in the computation and investigation of the solution.

The theory should provide a way to obtain some notion of the behavior of a solution: whether it is monotonic or oscillatory, whether it is periodic or approaches a periodic function, and so forth.

Suppose we change the initial values for the unknown function and its derivatives; that is, we change the initial state of the physical system. Then we will also change the solution, since the whole physical process will now run differently. The theory should provide the possibility of judging what this change will be. In particular, for small changes in the initial values will the solution also change by a small amount and will it therefore be stable in this respect, or may it be that small changes in the initial conditions will give rise to large changes in the solution so that the latter will be unstable?

We must also be able to set up a qualitative, and where possible, quantitative picture of the behavior not only of the separate solutions of an equation, but also of all of the solutions taken together.

In machine construction there often arises the question of making a choice of parameters characterizing an apparatus or machine that will guarantee satisfactory operation. The parameters of an apparatus appear in the form of certain magnitudes in the corresponding differential equation. The theory must help us make clear what will happen to the solutions of the equation (to the working of the apparatus) if we change the differential equation (change the parameters of the apparatus).

Finally, when it is necessary to carry out a computation, we will need to find the solution of an equation numerically, and here the theory will be obliged to provide the engineer and the physicist with the most rapid and economical methods for calculating the solutions.

§2. Linear Differential Equations with Constant Coefficients

For certain important classes of ordinary differential equations the general solution may be expressed in terms of simple well-known functions. One of these classes consists of those differential equations with constant coefficients that are linear with respect to the unknown function and its derivatives (in short, linear). The differential equations (3), (6), (8), and (14) are examples of such equations. A linear equation is called homogeneous if it has no term which does not contain the unknown variable, and nonhomogeneous if there is such a term.

Homogeneous linear equations of the second order with constant coefficients. Such equations have the form

$$m\frac{d^2x}{dt^2} + a\frac{dx}{dt} + bx = 0, \tag{6}$$

where m, a, and b are constants. We will assume that m is positive; this does not restrict the generality, since we can always ensure this situation if need be by changing the sign of all coefficients, provided that $m \neq 0$, which we will assume.

We will look for a solution of this equation in the form of an exponential function $e^{\lambda t}$ and ask how the constant λ should be chosen so that the function $x = e^{\lambda t}$ satisfies the equation. Putting $x = e^{\lambda t}$, $dx/dt = \lambda e^{\lambda t}$ and $d^2x/dt^2 = \lambda^2 e^{\lambda t}$ in the left side of equation (6), we get

$$e^{\lambda t}(m\lambda^2 + a\lambda + b).$$

Thus, in order that $x(t) = e^{\lambda t}$ be a solution of equation (6) it is necessary and sufficient that

$$m\lambda^2 + a\lambda + b = 0. \tag{19}$$

If λ_1 and λ_2 are two real roots of equation (19), then it is easy to prove that a solution of equation (6) is given by every function of the form

$$x = C_1 e^{\lambda_1 t} + C_2 e^{\lambda_2 t}, \tag{20}$$

where C_1 and C_2 are arbitrary constants.

Below we will show that formula (20) gives all solutions of equation (6) in the case that equation (19) has distinct real roots.

We note the following important properties of the solution of equation (6):

1. The sum of two solutions is also a solution.

2. A solution multiplied by a constant is also a solution.

In case λ_1 is a multiple root of equation (19), i.e., $m\lambda_1^2 + a\lambda_1 + b = 0$ and $2m\lambda_1 + a = 0$,* then a solution of equation (6) will also be given by the function $te^{\lambda_1 t}$, since if we substitute this function and its derivatives into the left side of equation (6) we get

$$te^{\lambda_1 t}(m\lambda_1^2 + a\lambda_1 + b) + e^{\lambda_1 t}(2m\lambda_1 + a),$$

which is seen from the previous equations to be identically zero.

The general solution of equation (6) in this case has the form

$$x = C_1 e^{\lambda_1 t} + C_2 t e^{\lambda_1 t}. \tag{21}$$

Now let equation (19) have complex roots. These roots will be complex conjugates of each other since m, a, and b are real numbers. Let $\lambda = \alpha \pm i\beta$ The equation

$$m(\alpha + i\beta)^2 + a(\alpha + i\beta) + b = 0$$

is equivalent to the two equations

$$m\alpha^2 - m\beta^2 + a\alpha + b = 0 \text{ and } 2m\alpha\beta + a\beta = 0. \tag{22}$$

It is easy to show that in this case the functions $x = e^{\alpha t} \cos \beta t$ and $x = e^{\alpha t} \sin \beta t$ are solutions of equation (6). Thus, for example, putting the function $x(t) = e^{\alpha t} \cos \beta t$ and its derivatives in the left side of equation (6), we get

$$e^{\alpha t} \cos \beta t(m\alpha^2 - m\beta^2 + a\alpha + b) - e^{\alpha t} \sin \beta t(2m\alpha\beta + a\beta).$$

By equation (22) this expression is identically equal to zero.

The general solution of equation (6), if equation (19) has complex roots, has the form

$$x = C_1 e^{\alpha t} \sin \beta t + C_2 e^{\alpha t} \cos \beta t, \tag{23}$$

where C_1 and C_2 are arbitrary constants.

In this way, if we know the roots of equation (19), called the *characteristic equation*, we can write down the general solution of equation (6).

We note that the general solution of a linear homogeneous equation of order n with constant coefficients

$$a_n \frac{d^n x}{dt^n} + a_{n-1} \frac{d^{n-1} x}{dt^{n-1}} + \cdots + a_1 \frac{dx}{dt} + a_0 x = 0$$

may be written in a similar manner as a polynomial in exponential and

---

* The sum of the roots λ_1 and λ_2 of the quadratic equation (19) is $\lambda_1 + \lambda_2 = -a/m$, and if the roots are the same, that is $\lambda_1 = \lambda_2$, then the second of the previous equations is true.

trigonometric functions, provided we know the roots of the algebraic equation

$$a_n\lambda^n + a_{n-1}\lambda^{n-1} + \cdots + a_0 = 0,$$

which again is called the *characteristic equation*. Thus, the problem of integrating a linear ordinary differential equation with constant coefficients is reduced to an algebraic problem.

We now show that formulas (20), (21), and (23) give all the solutions of equation (6). We note that C_1 and C_2 in these formulas may always be so chosen that the function $x(t)$ satisfies arbitrary initial conditions $x(t_0) = x_0$, $x'(t_0) = x_0'$. For this C_1 and C_2 need only to be determined from the system of equations

$$x_0 = C_1 e^{\lambda_1 t_0} + C_2 e^{\lambda_2 t_0},$$
$$x_0' = \lambda_1 C_1 e^{\lambda_1 t_0} + \lambda_2 C_2 e^{\lambda_2 t_0}.$$

in the case of formula (20), or by two similar equations in the case of formulas (21) and (23). Clearly, if there existed a solution of equation (6) not contained among the solutions we have constructed, then there would exist two distinct solutions of equation (6) satisfying the same initial conditions. Their difference $x_1(t)$ would not be identically zero and would satisfy the zero initial conditions $x_1(t_0) = 0$, $x_1'(t_0) = 0$. We will show that a solution of equation (6) which satisfies the zero initial conditions can only be $x_1(t) = 0$. Let us first show this under the assumption that $m > 0$, $a > 0$, and $b > 0$. We multiply the two sides of the equation

$$m\frac{d^2x_1}{dt^2} + a\frac{dx_1}{dt} + bx_1 = 0 \tag{24}$$

by $2(dx_1/dt)$. Since

$$2\frac{dx_1}{dt} \cdot \frac{d^2x_1}{dt^2} = \frac{d}{dt}\left(\frac{dx_1}{dt}\right)^2 \quad \text{and} \quad 2x_1(t)\frac{dx_1}{dt} = \frac{d}{dt}(x_1^2),$$

equation (24) may be put in the form

$$\frac{d}{dt}\left[m\left(\frac{dx_1}{dt}\right)^2\right] + 2a\left(\frac{dx_1}{dt}\right)^2 + b\frac{d}{dt}(x_1^2) = 0.$$

Integrating this identity between t_0 and t, we get

$$m\left(\frac{dx_1}{dt}\right)^2 + 2a\int_{t_0}^{t}\left(\frac{dx_1}{dt}\right)^2 dt + bx_1^2(t) = 0.$$

This equation is possible only if $x_1(t) \equiv 0$. Otherwise, for $t = t_0$, we would clearly have a positive quantity on the left and zero on the right, with a similar situation for $t < t_0$.

In order to establish our proposition for all constant coefficients m, a, and b, we consider the function $y_1(t) = x_1(t)e^{-\alpha t}$ which, as it is easy to show, also satisfies the zero boundary condtions. If the value of $\alpha > 0$ is chosen sufficiently large, then the function $y_1(t)$ will satisfy some equation of the form (6) for $a > 0$, $b > 0$, and $m > 0$. This equation is easily derived by substituting the function $x_1(t) = y_1(t)e^{\alpha t}$ and its derivatives into equation (6). Then, as was shown earlier, we have $y_1(t) \equiv 0$, which means that $x_1(t) = y_1(t)e^{\alpha t} \equiv 0$.

Thus we have shown that formulas (20), (21), and (23) give all the solutions of equation (6).

Let us see what information these formulas give about the character of the solutions of equations (6). To this end we note the formulas

$$\lambda_{1,2} = -\frac{a}{2m} \pm \sqrt{\frac{a^2}{4m^2} - \frac{b}{m}} \tag{25}$$

for the roots of equation (19). In accordance with the physical applications which led us to equation (6), we will assume $m > 0$, $a \geqslant 0$, and $b > 0$.

Case 1. $a^2 > 4bm$. The two roots of the characteristic equation (19) are real, negative, and distinct. In this case the function $x(t)$ given by by formula (20) is a general solution of equation (6). All the functions given by this formula together with their first derivatives tend to zero for $t \to +\infty$, and there is no more than one value of t for which they vanish. It follows that the function $x(t)$ has no more than one maximum or minimum. Physically, this means that the resistance of the medium is sufficiently large to prevent oscillations. The moving point cannot pass through the equilibrium position $x = 0$ more than once. From then on, after attaining a maximum distance from the point $x = 0$, it will begin a slow approach to the point but will never pass through it again.

Case 2. $a^2 = 4bm$. The two roots of equation (19) are equal to each other and the general solution of equation (6) given by formula (21). In this case again all solutions $x(t)$ and their first derivatives tend to zero for $t \to +\infty$. Here $x(t)$ and $x'(t)$ cannot vanish more than once. The character of the motion of the material point with abscissa $x(t)$ is the same as in the first case.

Case 3. $a^2 < 4bm$. The roots of the characteristic equation (19) have nonzero imaginary parts. The general solution of equation (6) is given by formula (23). The point x performs oscillations along the x-axis with a constant period $2\pi/\beta$, which is the same for all solutions of (6), and with amplitude $Ce^{\alpha t}$, where $\alpha = -(a/2m)$.

The oscillations of a physical system which take place without the action of an exterior force are called *characteristic oscillations* (eigenvibrations) of the system. From the previous discussion, it follows that the period of such oscillations for the systems discussed in examples 2, 3, 4 and 5, depends only on the structure of the system and will be the same for all oscillations which could possibly arise in it. In example 2 this period is equal to $2\pi\sqrt{b/m - a^2/4m^2}$; in example 4 to $2\pi\sqrt{kps/v\rho l}$; and example 5 to $2\pi\sqrt{1/LC - R^2/4L^2}$.

If $a = 0$, i.e., if the medium offers no resistance to the motion, then the amplitude of the oscillations is constant: the point oscillates harmonically. But if $a > 0$, i.e., if the medium offers resistance to the motion, although this resistance is small ($a^2 < 4bm$), then the amplitude of the oscillations tends to zero and the oscillations die out.

Finally, the solution $x(t) \equiv 0$ of equation (6) in all cases indicates a state of rest for the point x at the position $x = 0$, which is called the position of equilibrium.

If the real parts of both roots of equation (19) are negative, then it can be seen from formulas (20), (21), and (23), that all the solutions of equation (6), together with their derivatives, tend to zero for $t \to +\infty$; that is, the oscillations die out with the passage of time.

However, if the real part of even one of the roots of equation (19) is positive, then there are solutions of equation (6) not tending to zero for $t \to +\infty$, so that some of the solutions of (6) would not even be bounded for $t \to +\infty$. Such a case can occur only for negative b or negative a, if $m > 0$. Physically, this would correspond to the case in which the elastic force does not attract the point x to the equilibrium position but repels it or else that the resistance of the medium is negative. Such cases cannot be realized in the physical examples considered at the beginning of this chapter, but they are entirely realizable in other physical models.

If the real part of the roots λ_1 and λ_2 of equation (19) is equal to zero, which is possible only if the coefficient a in equation (19) is zero, then for $\alpha = 0$ the point $x(t)$, as can be seen from formula (23), carries out harmonic oscillations with bounded amplitude and bounded velocity.

Nonhomogeneous linear equations with constant coefficients. Let us consider in detail the equation

$$m\frac{d^2x}{dt^2} + a\frac{dx}{dt} + bx = A\cos\omega t. \qquad (26)$$

This is the equation of linear oscillations of a material point under the action of an elastic force, of the resistance of a medium and of an external periodic force $A \cos \omega t$ (see equation (6') in §1).

Equation (26) is a nonhomogeneous linear equation and (6) is the corresponding homogeneous equation.

We will now look for the general solution to equation (26).

We note that the sum of a solution of a nonhomogeneous equation and a solution of the corresponding homogeneous equation is also a solution of the nonhomogeneous linear equation. Thus, in order to find a general solution of equation (26), it is sufficient to find any one particular solution. The general solution of equation (26) will then be given in the form of the sum of this particular solution and a general solution of the corresponding homogeneous equation.

It is natural to expect that the motion will follow the rhythm of the external periodic force and to look for a particular solution of equation (26) in the form $x = B \cos (\omega t + \delta)$, where B and δ are as yet undetermined constants. We will attempt to determine B and δ in such a way that the function $x = B \cos (\omega t + \delta)$ will satisfy equation (26). Calculating the derivatives $dx/dt = -B\omega \sin (\omega t + \delta)$ and $d^2x/dt^2 = -B\omega^2 \cos (\omega t + \delta)$ and substituting them into equation (26), we get

$$m[-B\omega^2 \cos (\omega t + \delta)] + a[-B\omega \sin (\omega t + \delta)] + bB \cos (\omega t + \delta) = A \cos \omega t.$$

Applying well-known formulas, we have

$$B[(b - m\omega^2) \cos (\omega t + \delta) - a\omega \sin (\omega t + \delta)]$$
$$= B \sqrt{(b - m\omega^2)^2 + a^2\omega^2} \cos (\omega t + \delta') = A \cos \omega t,$$

where $\delta' = \delta + \gamma$ and $\gamma = \arctan a\omega/(b - m\omega^2)$. Obviously, if we set

$$\delta = -\arctan \frac{a\omega}{b - m\omega^2} \quad \text{and} \quad B = \frac{A}{\sqrt{(b - m\omega^2)^2 + a^2\omega^2}},$$

the function $x = B \cos (\omega t + \delta)$ will satisfy equation (26).

A solution of the form $B \cos (\omega t + \delta)$ will always exist if $(b - m\omega^2)^2 + a^2\omega^2 \neq 0$. In case $(b - m\omega^2)^2 + a^2\omega^2 = 0$, i.e., when $a = 0$ and $b = m\omega^2$, equation (26) has the form

$$m \frac{d^2x}{dt^2} + m\omega^2 x = A \cos \omega t.$$

A particular solution in this case, as is easily established, is $x = (At/2 \sqrt{mb}) \sin \omega t.$

Solutions of the nonhomogeneous equation (26) are called forced oscillations. The multiplier $\phi(\omega) = 1/\sqrt{(b - m\omega^2)^2 + a^2\omega^2}$ characterizes the relation of the amplitude B of the forced oscillation to the amplitude A of the disturbing force. The graph of the function $\phi(\omega)$ is called the resonance curve. The frequency ω for which $\phi(\omega)$ attains its maximum is called the resonant frequency. Let us calculate it. If $\phi(\omega)$ attains the maximum at $\omega_1 \neq 0$, then for this value of ω the derivative $\phi'(\omega)$ vanishes, i.e., $-4(b - m\omega_1^2)m\omega_1 + 2a^2\omega_1 = 0$, so that $\omega_1 = \sqrt{b/m - a^2/2m^2}$. For this value of ω_1

$$\phi(\omega_1) = \frac{1}{a\sqrt{b/m - a^2/4m^2}}.$$

Hence it can be seen that the amplitude of the forced oscillation for $\omega = \omega_1$ is greater for smaller values of a. For very small a, the frequency ω_1 is very close to the value $\sqrt{b/m}$, i.e., to the frequency the free oscillations. For $a = 0$ and $b = m\omega^2$, as we saw, the forced oscillation has the form

$$x = \frac{At}{2\sqrt{mb}}\sin \omega t,$$

i.e., the amplitude of this oscillation increases beyond all bounds as $t \to +\infty$, a situation which represents the mathematical meaning of resonance. Resonance will occur if the period of the external force is the same as the period of one of the characteristic oscillations of the system. In the practical world, in cases where the period of the external force and the period of the characteristic oscillations are close together, the displacements of the system may become extremly large.

The possibility of large oscillations is often made use of in the construction of various kinds of amplifiers, for example in radio technology. But large oscillations may also lead to the breaking up of structures such as bridges or the framework of machines. Thus it is very important to foresee the possibility of resonance or of oscillations close to it.

From the remarks made earlier, any solution of equation (26) can be written as a sum of the forced oscillation we have found and of one of the solutions of the homogeneous equation given in formulas (20), (21), and (23). For $a > 0$ and $b > 0$ the solution of the homogeneous equation tends to zero for $t \to +\infty$, i.e., any motion eventually approximates the forced oscillations. If $a = 0$ and $b > 0$, the forced oscillation is superposed on a nondecaying characteristic oscillation of the system. For $b = m\omega^2$ and $a = 0$, we have resonance.

If a periodic external force $f(t)$ is imposed on the sytem, the forced oscillations of the system may be found in the following manner. We may represent $f(t)$ with sufficient exactness as a segment of a trigonometric series*

$$\sum_{i=1}^{n} (a_i \cos \omega_i t + b_i \sin \omega_i t). \tag{27}$$

Let us find the forced oscillations corresponding to each term of this sum. Then the oscillation corresponding to the force $f(t)$ will be found by adding together the oscillations corresponding to the various terms of the sum (27). If any of these frequencies is identical with the frequency of a characteristic oscillation of the system, we will have resonance.

§3. Some General Remarks on the Formation and Solution of Differential Equations

There are not many differential equations with the property that all their solutions can be expressed explicitly in terms of simple functions, as is the case for linear equations with constant coefficients. It is possible to give simple examples of differential equations whose general solution cannot be expressed by a finite number of integral of known functions, or as one says, in quadratures.

As Liouville showed in 1841, the solution of the Riccati equation of the form $dy/dx + ay^2 = x^2$, for $a > 0$, cannot be expressed as a finite combination of integrals of elementary functions. So it becomes important to develop methods of approximation to the solutions of differential equations, which will be applicable to wide classes of equations.

The fact that in such cases we find not exact solutions but only approximations should not bother us. First of all, these approximate solutions may be calculated, at least in principle, to any desired degree of accuracy. Second, it must be emphasized that in most cases the differential equations describing a physical process are themselves not altogether exact, as can be seen in all the examples discussed in §1.

An especially good example is provided by the equation (12) for the acoustic resonator. In deriving this equation, we ignored the compressibility of the air in the neck of the container and the motion of the air in the container itself. As a matter of fact, the motion of the air in the neck sets into motion the mass of the air in the vessel, but these two motions have different velocities and displacements. In the neck the displacement of the particles of air is considerably greater than in the container. Thus we ignored the motion of the air in the container, and

---

* Cf. Chapter XII, §7.

took account only of its compression. For the air in the neck, however, we ignored the energy of its compression and took account only of the kinetic energy of its motion.

To derive the differential equation for a physical pendulum, we ignored the mass of the string on which it hangs. To derive equation (14) for electric oscillations in a circuit, we ignored the self-inductance of the wiring and the resistance of the coils. In general, to obtain a differential equation for any physical process, we must always ignore certain factors and idealize others. In view of this, A. A. Andronov drew especial attention to the fact that for physical investigations we are especially interested in those differential equations whose solutions do not change much for arbitrary small changes, in some sense or another, in the equations themselves. Such differential equations are called "intensive." These equations deserve particularly complete study.

It should be stated that in physical investigations not only are the differential equations that describe the laws of change of the physical quantities themselves inexactly defined but even the number of these quantities is defined only approximately. Strictly speaking, there are no such things as rigid bodies. So to study the oscillations of a pendulum, we ought to take into account the deformation of the string from which it hangs and the deformation of the rigid body itself, which we approximated by taking it as a material point. In exactly the same way, to study the oscillations of a load attached to springs, we ought to consider the masses of the separate coils of the springs. But in these examples it is easy to show that the character of the motion of the different particles, which make up the pendulum and its load together with the springs, has little influence on the character of the oscillation. If we wished to take this influence into account, the problem would become so complicated that we would be unable to solve it to any suitable approximation. Our solution would then bear no closer relation to physical reality than the solution given in §1 without consideration of these influences. Intelligent idealization of a problem is always unavoidable. To describe a process, it is necessary to take into account the essential features of the process but by no means to consider every feature without exception. This would not only complicate the problem a great deal but in most cases would result in the impossibility of calculating a solution. The fundamental problem of physics or mechanics, in the investigation of any phenomenon, is to find the smallest number of quantities, which with sufficient exactness describe the state of the phenomenon at any given moment, and then to set up the simplest differential equations that are good descriptions of the laws governing the changes in these quantities. This problem is often very difficult. Which features are the essential ones and which are non-essential is a question that in the final analysis can be decided only by long experience. Only by comparing the answers provided by an idealized argument with the results of experiment can we judge whether the idealization was a valid one.

The mathematical problem of the possibility of decreasing the number of quantities may be formulated in one of the simplest and most characteristic cases, as follows.

Suppose that to begin with we characterize the state of a physical system at time t by the two magnitudes $x_1(t)$ and $x_2(t)$. Let the differential equations expressing their rates of change have the form

$$\frac{dx_1}{dt} = f_1(t, x_1, x_2),$$

$$\epsilon \frac{dx_2}{dt} = f_2(t, x_1, x_2), \tag{28}$$

In the second equation the coefficient of the derivative is a small constant parameter ϵ. If we put $\epsilon = 0$, the second of equations (28) will cease to be a differential equation. It then takes the form

$$f_2(t, x_1, x_2) = 0.$$

From this equation, we define x_2 as a function of t and x_1 and we substitute it into the first of the equations (28). We then have the differential equation

$$\frac{dx_1}{dt} = F(t, x_1)$$

for the single variable x_1. In this way the number of parameters entering into the situation is reduced to one. We now ask, under what conditions will the error introduced by taking $\epsilon = 0$ be small. Of course, it may happen that as $\epsilon \to 0$ the value dx_2/dt grows beyond all bounds, so that the right side of the second of equations (28) does not tend to zero as $\epsilon \to 0$.

§4. Geometric Interpretation of the Problem of Integrating Differential Equations; Generalization of the Problem

For simplicity we will consider initially only one differential equation of the first order with one unknown function

$$\frac{dy}{dx} = f(x, y), \tag{29}$$

where the function $f(x, y)$ is defined on some domain G in the (x, y) plane.

This equation determines at each point of the domain the slope of the tangent to the graph of a solution of equation (29) at that point. If at each point (x, y) of the domain G we indicate by means of a line segment the the direction of the tangent (either of the two directions may be used) as determined by the value of $f(x, y)$ at this point, we obtain a field of directions. Then the problem of finding a solution of the differential equation (29) for the initial conditon $y(x_0) = y_0$ may be formulated thus: In the domain G we have to find a curve $y = \phi(x)$, passing through the point $M_0(x_0, y_0)$, which at each of its points has a tangent whose slope is given by equation (29), or briefly, which has at each of its points a preassigned direction.

From the geometric point of view this statement of the problem has two unnatural features:

1. By requiring that the slope of the tangent at any given point (x, y) of the domain G be equal to $f(x, y)$, we automatically exclude tangents parallel to Oy, since we generally consider only finite magnitudes; in particular, it is assumed that the function $f(x, y)$ on the right side of equation (29) assumes only finite values.

2. By considering only curves which are graphs of functions of x, we also exclude those curves which are intersected more than once by a line perpendicular to the axis Ox, since we consider only single-valued functions; in particular, every solution of a differential equation is assumed to be a single-valued function of x.

So let us generalize to some extent the preceding statement of the problem of finding a solution to the differential equation (29). Namely, we will now allow the tangent at some points to be parallel to the axis Oy. At these points, where the slope of the tangent with respect to the axis Ox has no meaning, we will take the slope with respect to the axis Oy. In other words, we consider, together with the differential equation (29), the equation

$$\frac{dx}{dy} = f_1(x, y), \tag{29$'$}$$

where $f_1(x, y) = 1/f(x, y)$, if $f(x, y) \neq 0$, using the second equation when the first is meaningless. The problem of integrating the differential equations (29) and (29$'$) then becomes: In the domain G to find all curves having at each point the tangent defined by these equations. These curves will be called integral curves (integral lines) of the equations (29) and (29$'$) or of the tangent field given by these equations. In place of the plural "equations (29), (29$'$)", we will often use the singular "equation (29), (29$'$)". It is clear that the graph of any solution of equation (29) will also be an integral curve of equation (29), (29$'$). But not every integral curve of equation (29), (29') will be the graph of a solution of equation (29). This case will occur, for example, if some perpendicular to the axis Ox intersects this curve at more than one point.

In what follows, if it can be clearly shown that

$$f(x, y) = \frac{M(x, y)}{N(x, y)},$$

then we will write only the equation

$$\frac{dy}{dx} = \frac{M(x, y)}{N(x, y)},$$

and omit writing

$$\frac{dx}{dy} = \frac{N(x, y)}{M(x, y)}.$$

Sometimes in place of these equations we introduce a parameter t, and write the system of equations

$$\frac{dx}{dt} = N(x, y), \frac{dy}{dt} = M(x, y),$$

where x and y are considered as functions of t.

Example 1. The equation

$$\frac{dy}{dx} = \frac{y}{x} \tag{30}$$

defines a tangent field everywhere except at the origin. This tangent field is sketched in figure 7. All the tangents given by equation (30) pass through the origin.

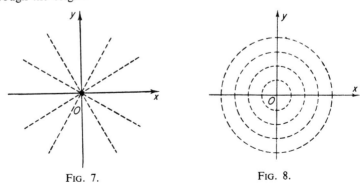

FIG. 7.   FIG. 8.

It is clear that for every k the function

$$y = kx \tag{31}$$

is a solution of equation (30). The collection of all integral curves of this equation is then defined by the relation

$$ax = by = 0, \tag{32}$$

where a and b are arbitrary constants, not both zero. The axis Oy is an integral curve of equation (30), but it is not the graph of a solution of it.

Since equation (30) does not define a tangent field at the origin, the curves (31) and (32) are, strictly speaking, integral curves everywhere except at the origin. Thus it is more correct to say that the integral curves of equation (30) are not straight lines passing through the origin but half lines issuing from it.

Example 2. The equation

$$\frac{dy}{dx} = -\frac{x}{y} \tag{33}$$

defines a field of tangents everywhere except at the origin, as sketched in figure 8. The tangents defined at a given point (x, y) by equations (30) and (33) are perpendicular to each other. It is clear that all circles centered at the origin will be integral curves of equation (33). However the solutions of this equation will be the functions

$$y = +\sqrt{R^2 - x^2},\, y = -\sqrt{R^2 - x^2},\, -R \leqslant x \leqslant R.$$

For brevity in what follows we will sometimes say "a solution passes through the point (x, y)" in place of the more exact statement "the graph of a solution passes through the point (x, y)."

§5. Existence and Uniqueness of the Solution of a Differential Equation; Approximate Solution of Equations

The question of existence and uniqueness of the solution of a differential equation. We return to the differential equation (17) of arbitrary order n. Generally, it has infinitely many solutions and in order that we may pick from all the possible solutions some one specific one, it is necessary to attach to the equation some supplementary conditions, the number of which should be equal to the order n of the equation. Such conditions may be of extremely varied character, depending on the physical, mechanical, or other significance of the original problem. For example, if we have to investigate the motion of a mechanical system beginning with some specific initial state, the supplementary conditions will refer to a specific (initial) value of the independent variable and will be called initial conditions of the problem. But if we want to define the curve of a cable in a suspension bridge, or of a loaded beam resting on supports at each end, we encounter conditions corresponding to different values of the independent variable, at the ends of the cable or at the points of support of the beam. We could give many other examples showing the variety of conditions to be fulfilled in connection with differential equations.

We will assume that the supplementary conditions have been defined and that we are required to find a solution of equation (17) that satisfies them. The first question we must consider is whether any such solution exists at all. It often happens that we cannot be sure of this in advance. Assume, say, that equation (17) is a description of the operation of some physical apparatus and suppose we want to determine whether periodic motion occurs in this apparatus. The supplementary conditions will then be conditions for the periodic repetition of the initial state in the apparatus, and we cannot say ahead of time whether or not there will exist a solution which satisfies them.

In any case the investigation of problems of existence and uniqueness of a solution makes clear just which conditions can be fulfilled for a given differential equation and which of these conditions will define the solution in a unique manner. But the determination of such conditions and the proof of existence and uniqueness of the solution for a differential equation corresponding to some physical problem also has great value for the physical theory itself. It shows that the assumptions adopted in setting up the mathematical description of the physical event are on the one hand mutually consistent and on the other constitute a complete description of the event.

The methods of investigating the existence problem are manifold, but among them an especially important role is played by what are called direct methods. The proof of the existence of the required solution is provided by the construction of approximate solutions, which are proved to converge to the exact solution of the problem. These methods not only establish the existence of an exact solution, but also provide a way, in fact the principal one, of approximating it to any desired degree of accuracy.

For the rest of this section we will consider, for the sake of definiteness, a problem with initial data, for which we will illustrate the ideas of Euler's method and the method of successive approximations.

Euler's method of broken lines. Consider in some domain G of the (x, y) plane the differential equation

$$\frac{dy}{dx} = f(x, y). \tag{34}$$

As we have already noted, equation (34) defines in G a field of tangents. We choose any point (x_0, y_0) of G. Through it there will pass a straight line L_0 with slope $f(x_0, y_0)$. On the straight line L_0 we choose a point (x_1, y_1), sufficiently close to (x_0, y_0); in figure 9 this point is indicated by the number 1. We draw the straight line L_1 through the point (x_1, y_1) with slope $f(x_1, y_1)$ and on it mark the point (x_2, y_2); in the figure this point is denoted by the number 2. Then on the straight line L_2

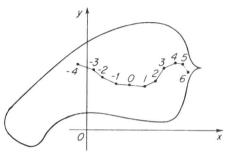

FIG. 9.

corresponding to the point (x_2, y_2) we mark the point (x_3, y_3), and continue in the same manner with $x_0 < x_1 < x_2 < x_3 < \cdots$. It is assumed, of course, that all the points (x_0, y_0), (x_1, y_1), (x_2, y_2), $\cdots$ are in the domain G. The broken line joining these points is called an Euler broken line. One may also construct an Euler broken line in the direction of decreasing x; the corresponding vertices on our figure are denoted by $-1, -2, -3$.

It is reasonable to expect that every Euler broken line through the point (x_0, y_0) with sufficiently short segments gives a representation of an integral curve l passing through the point (x_0, y_0), and that with decrease in the length of the links, i.e., when the length of the longest link tends to zero, the Euler broken line will approximate this integral curve.

Here, of course, it is assumed that the integral curve exists. In fact it is not hard to prove that if the function $f(x, y)$ is continuous in the domain G, one may find an infinite sequence of Euler broken lines, the length of the largest links tending to zero, which converges to an integral curve l. However, one usually cannot prove uniqueness: there may exist different sequences of Euler broken lines that converge to different integral curves passing through one and the same point (x_0, y_0). M. A. Lavrent'ev has constructed an example of a differential equation of the form (29) with a continuous function $f(x, y)$, such that in any neighborhood of any point P of the domain G there passes not one but at least two integral curves.

In order that through every point of the domain G there pass only one integral curve, it is necessary to impose on the function $f(x, y)$ certain conditions beyond that of continuity. It is sufficient, for example, to assume that the function $f(x, y)$ is continuous and has a bounded derivative with respect to y on the whole domain G. In this case it may be proved that through each point of G there passes one and only one integral curve and that every sequence of Euler broken lines passing through the point (x_0, y_0) converges uniformly to this unique integral curve, as the length of the longest link of the broken lines tends to zero. Thus for sufficiently small links the Euler broken line may be taken as an approximation to the integral curve of equation (34).

From the preceding it can be seen that the Euler broken lines are so constituted that small pieces of the integral curves are replaced by line segments tangent to these integral curves. In practice, many approximations to integral curves of the differential equation (34) consist not of straight-line segments tangent to the integral curves, but of parabolic segments that have a higher order of tangency with the integral curve. In this way it is possible to find an approximate solution with the same degree of accuracy in a smaller number of steps (with a smaller number of links in the approximating curve). The coefficients of the equation for the (higher order) parabola

$$ y = a_0 + a_1(x - x_k) + a_2(x - x_k)^2 + \cdots + a_n(x - x_k)^n, \qquad (35) $$

which at the point (x_k, y_k) has nth-order tangency with the integral curves of equation (34) through this point, are given by the following formulas:

$$ a_0 = y_k, \qquad (36) $$

$$ a_1 = \left(\frac{dy}{dx}\right)_{x=x_k} = f(x_k, y_k), \qquad (36') $$

$$ 2a_2 = \left(\frac{d^2y}{dx^2}\right)_{x=x_k} = \left[\frac{df(x, y)}{dx}\right]_{x=x_k} = f_x'(x_k, y_k) + f_y'(x_k, y_k)\left(\frac{dy}{dx}\right)_{x=x_k} $$

$$ = f_x'(x_k, y_k) + f_y'(x_k, y_k)f(x_k, y_k), \qquad (36'') $$

$$ 6a_3 = \left(\frac{d^3y}{dx^3}\right)_{x=x_k} = \left\{\frac{d}{dx}\left[f_x'(x, y(x)) + f_y'(x, y(x))f(x, y(x))\right]\right\}_{x=x_k} $$

$$ = f_{xx}''(x_k, y_k) + 2f_{xy}''(x_k, y_k)f(x_k, y_k) $$

$$ + f_{yy}''(x_k, y_k)f^2(x_k, y_k) + f_y'^2(x_k, y_k)f(x_k, y_k) $$

$$ + f_y'(x_k, y_k)f_x'(x_k, y_k). \qquad (36''') $$

The polynomial (35) is needed only in order to compute its value for $x = x_{k+1}$. The actual values of the coefficients a_0, a_1, a_2, $\cdots$, a_n themselves are not needed. There are many ways of computing the value for $x = x_{k+1}$ of the polynomial (35) whose coefficients are given by formula (36), without computing the coefficients a_0, a_1, $\cdots$, a_n themselves.

Other approximation methods exist for finding the solution of the differential equation (34), which are based on other ideas. One convenient method was developed by A. N. Krylov (1863–1945).

The method of successive approximations. We now describe another method of successive approximation, which is as widely used as the method of the Euler broken lines. We assume again that we are required to find a solution $y(x)$ of the differential equation (34) satisfying the initial condition

$$y(x_0) = y_0.$$

For the initial approximation to the function $y(x)$, we take an arbitrary function $y_0(x)$. For simplicity we will assume that it also satisfies the initial condition, although this is not necessary. We substitute it into the right side $f(x, y)$ of the equation for the unknown function y and construct a first approximation y_1 to the solution y from the following requirements:

$$\frac{dy_1}{dx} = f[x, y_0(x)], \; y_1(x_0) = y_0.$$

Since there is a known function on the right side of the first of these equations the function $y_1(x)$ may be found by integration:

$$y_1(x) = y_0 + \int_{x_0}^{x} f[t, y_0(t)] \, dt.$$

It may be expected that $y_1(x)$ will differ from the solution $y(x)$ by less than $y_0(x)$ does, since in the construction of $y_1(x)$ we made use of the differential equation itself, which should probably introduce a correction into the original approximation. One would also think that if we improve the first approximation $y_1(x)$ in the same way, then the second approximation

$$y_2(x) = y_0 + \int_{x_0}^{x} f[t, y_1(t)] \, dt$$

will be still closer to the desired solution.

Let us assume that this process of improvement has been continued indefinitely and that we have constructed the sequence of approximations

$$y_0(x), \; y_1(x), \; \cdots, \; y_n(x), \; \cdots.$$

Will this sequence converge to the solution $y(x)$?

More detailed investigations show that if $f(x, y)$ is continuous and f_y' is bounded in the domain G, the functions $y_n(x)$ will in fact converge to the exact solution $y(x)$ at least for all x sufficiently close to x_0 and that if we break off the computation after a sufficient number of steps, we will be able to find the solution $y(x)$ to any desired degree of accuracy.

Exactly in the same way as for the integral curves of equation (34), we may also find approximations to integral curves of a system of two or more differential equations of the first order. Essentially the necessary condition here is to be able to solve these equations for the derivatives of the unknown functions. For example, suppose we are given the system

$$\frac{dy}{dx} = f_1(x, y, z), \frac{dz}{dx} = f_2(x, y, z). \tag{37}$$

Asuming that the right sides of these equations are continuous and have bounded derivatives with respect to y and z in some domain G in space, it may be shown under these conditions that through each point (x_0, y_0, z_0) of the domain G, in which the right sides of the equations in (37) are defined, there passes one and only one integral curve

$$y = \phi(x), \quad z = \psi(x)$$

of the system (37). The functions $f_1(x, y, z)$ and $f_2(x, y, z)$ give the direction numbers at the point (x, y, z), of the tangent to the integral curve passing through this point. To find the functions $\phi(x)$ and $\psi(x)$ approximately, we may apply the Euler broken line method or other methods similar to the ones applied to the equation (34).

The process of approximate computation of the solution of ordinary differential equations with initial conditions may be carried out on computing machines. There are electronic machines that work so rapidly that if, for example, the machine is programmed to compute the trajectory of a projectile, this trajectory can be found in a shorter space time than it takes for the projectile to hit its target (cf. Chapter XIV).

The connection between differential equations of various orders and a system of a large number of equations of first order. A system of ordinary differential equations, when solved for the derivative of highest order of each of the unknown functions, may in general be reduced, by the introduction of new unknown functions, to a system of equations of the first order, which is solved for all the derivatives. For example, consider the differential equation

$$\frac{d^2y}{dx^2} = f\left(x, y, \frac{dy}{dx}\right). \tag{38}$$

We set

$$\frac{dy}{dx} = z. \tag{39}$$

Then equation (38) may be written in the form

$$\frac{dz}{dx} = f(x, y, z). \tag{40}$$

Hence, to every solution of equation (38) there corresponds a solution of the system consisting of equations (39) and (40). It is easy to show that to every solution of the system of equations (39) and (40) there corresponds a solution of equation (38).

Equations not explicitly containing the independent variable. The problems of the pendulum, of the Helmholtz acoustic resonator, of a simple electric circuit, or of an electron-tube generator considered in §1 lead to differential equations in which the independent variable (time) does not explicitly appear. We mention equations of this type here, because the corresponding differential equations of the second order may be reduced in each case to a single differential equation of the first order rather than to a system of first-order equations as in the paragraph above for the general equation of the second order. This reduction greatly simplifies their study.

Let us then consider a differential equation of the second order, not containing the argument t in explicit form

$$F\left(x, \frac{dx}{dt}, \frac{d^2x}{dt^2}\right) = 0. \tag{41}$$

We set

$$\frac{dx}{dt} = y \tag{42}$$

and consider y as a function of x, so that

$$\frac{d^2x}{dt^2} = \frac{d}{dt}\left(\frac{dx}{dt}\right) = \frac{dy}{dt} = \frac{dy}{dx} \cdot \frac{dx}{dt} = y\frac{dy}{dx}.$$

Then equation (41) may be rewritten in the form

$$F\left(x, y, y\frac{dy}{dx}\right) = 0. \tag{43}$$

In this manner, to every solution of equation (41) there corresponds a unique solution of equation (43). Also to each of the solutions $y = \phi(x)$

of equation (43) there correspond infinitely many solutions of equation (41). These solutions may be found by integrating the equation

$$\frac{dx}{dt} = \phi(x),$$ (44)

where x is considered as a function of t.

It is clear that if this equation is satisfied by a function $x = x(t)$, then it will also be satisfied by any function of the form $x(t + t_0)$, where t_0 is an arbitrary constant.

It may happen that not every integral curve of equation (43) is the graph of a single function of x. This will happen, for example, if the curve is closed. In this case the integral curve of equation (43) must be split up into a number of pieces, each of which is the graph of a function of x. For every one of these pieces, we have to find an integral of equation (44).

The values of x and dx/dt which at each instant characterize the state of the physical system corresponding to equation (41) are called the *phases* of the system, and the (x, y) plane is correspondingly called the *phase plane* for equation (41). To every solution $x = x(t)$ of this equation there corresponds the curve

$$x = x(t), \quad y = x'(t)$$

in the (x, y) plane; t here is considered as a parameter. Conversely, to every integral curve $y = \phi(x)$ of equation (43) in the (x, y) plane there corresponds an infinite set of solutions of the form $x = x(t + t_0)$ for equation (41); here t_0 is an arbitrary constant. Information about the behavior of the integral curves of equation (43) in the plane is easily transformed into information about the character of the possible solutions of equation (41). Every closed integral curve of equation (43) corresponds, for example, to a periodic solution of equation (41).

If we subject equation (6) to the transformation (42), we obtain

$$\frac{dy}{dx} = \frac{-ay - bx}{my}.$$ (45)

Setting $v = x$ and $dv/dt = y$ in equation (16), in like manner we get

$$L\frac{dy}{dx} = \frac{-[R - M(a_1 + 2a_2x + 3a_3x^2)]\, y - x}{y}.$$ (46)

Just as the state at every instant of the physical system corresponding to the second-order equation (41) is characterized by the two magnitudes*

---

* The values of d^2x/dt^2, d^3x/dt^3, $\cdots$ at the same instant of time are defined by the values of x and dx/dt from equation (41) and from the equations obtained from (45) by differentiation (cf. formula (36)).

(phases) x and $y = dx/dt$, the state of a physical system described by equations of higher order or by a system of differential equations is characterized by a larger number of magnitudes (phases). Instead of a phase plane, we then speak of a phase space.

§6. Singular Points

Let the point $P(x, y)$ be in the interior of the domain G in which we consider the differential equation

$$\frac{dy}{dx} = \frac{M(x, y)}{N(x, y)}. \tag{47}$$

If there exists a neighborhood R of the point P through each point of which passes one and only one integral curve (47), then the point P is called an *ordinary point* of equation (47). But if such a neighborhood does not exist, then the point P is called a *singular point* of this equation. The study of singular points is very important in the qualitative theory of differential equations, which we will consider in the next section.

Particularly important are the so-called *isolated singular points*, i.e., singular points in some neighborhood of each of which there are no other singular points. In applications one often encounters them in investigating equations of the form (47), where $M(x, y)$ and $N(x, y)$ are functions with continuous derivatives of high orders with respect to x and y. For such equations, all the interior points of the domain at which $M(x, y) \neq 0$ or $N(x, y) \neq 0$ are ordinary points. Let us now consider any interior point (x_0, y_0) where $M(x, y) = N(x, y) = 0$. To simplify the notation we will assume that $x_0 = 0$ and $y_0 = 0$. This can always be arranged by translating the original origin of coordinates to the point (x_0, y_0). Expanding $M(x, y)$ and $N(x, y)$ by Taylor's formula into powers of x and y and restricting ourselves to terms of the first order, we have, in a neighborhood of the point $(0, 0)$,

$$\frac{dy}{dx} = \frac{M'_x(0, 0)\, x + M'_y(0, 0)\, y + \phi_1(x, y)}{N'_x(0, 0)\, x + N'_y(0, 0)\, y + \phi_2(x, y)}, \tag{48}$$

where $\phi_1(x, y)$ and $\phi_2(x, y)$ are functions of x and y for which

$$\lim_{\substack{x \to 0 \\ y \to 0}} \frac{\phi_1(x, y)}{\sqrt{x^2 + y^2}} = 0 \quad \text{and} \quad \lim_{\substack{x \to 0 \\ y \to 0}} \frac{\phi_2(x, y)}{\sqrt{x^2 + y^2}} = 0.$$

Equations (45) and (46) are of this form. Equation (45) does not define either dy/dx or dx/dy for $x = 0$ and $y = 0$. If the determinant

$$\begin{vmatrix} M'_x(0,0) & M'_y(0,0) \\ N'_x(0,0) & N'_y(0,0) \end{vmatrix} \neq 0,$$

then, whatever value we assign to dy/dx at the origin, the origin will be a point of discontinuity for the values dy/dx and dx/dy, since they tend to different limits depending on the manner of approach to the origin. The origin is a singular point for our differential equation.

It has been shown that the character of the behavior of the integral curves near an isolated singular point (here the origin) is not influenced by the behavior of the terms $\phi_1(x, y)$ and $\phi_2(x, y)$ in the numerator and denominator, provided only that the real part of both roots of the equation

$$\begin{vmatrix} \lambda - M'_y(0,0) & -M'_x(0,0) \\ -N'_y(0,0) & \lambda - N'_x(0,0) \end{vmatrix} = 0 \qquad (49)$$

is different from zero. Thus, in order to form some idea of this behavior, we study the behavior near the origin of the integral curves of the equation

$$\frac{dy}{dx} = \frac{ax + by}{cx + dy} \qquad (50)$$

for which the determinant

$$\begin{vmatrix} a & b \\ c & d \end{vmatrix} \neq 0.$$

We note that the arrangement of the integral curves in the neighborhood of a singular point of a differential equation has great interest for many problems of mechanics, for example in the investigation of the trajectories of motions near the equilibrium position.

It has been shown that everywhere in the plane it is possible to choose coordinates ξ, η, connected with x, y by the equations

$$\begin{aligned} x &= k_{11}\xi + k_{12}\eta, \\ y &= k_{12}\xi + k_{22}\eta, \end{aligned} \qquad (51)$$

where the k_{ij} are real numbers such that equation (50) is transformed into one of the following three types:

$$1) \quad \frac{d\eta}{d\xi} = k\frac{\eta}{\xi}, \quad \text{where} \quad k = \frac{\lambda_2}{\lambda_1}. \tag{52}$$

$$2) \quad \frac{d\eta}{d\xi} = \frac{\xi + \eta}{\xi}. \tag{53}$$

$$3) \quad \frac{d\eta}{d\xi} = \frac{\beta\xi + \alpha\eta}{\alpha\xi - \beta\eta}. \tag{54}$$

Here λ_1 and λ_2 are the roots of the equation

$$\begin{vmatrix} c - \lambda & d \\ a & b - \lambda \end{vmatrix} = 0. \tag{55}$$

If these roots are real and different, then equation (50) is transformed into the form (52). If these roots are equal, then equation (50) is transformed either into the form (52) or into the form (53), depending on whether $a^2 + d^2 = 0$ or $a^2 + d^2 \neq 0$. If the roots of equation (55) are complex, $\lambda = \alpha \pm \beta i$, then equation (51) is transformed into the form (54).

We will consider each of the equations (52), (53), (54). To begin with, we note the following.

Even though the axes Ox and Oy were mutually perpendicular, the axes $O\xi$ and $O\eta$ need not, in general, be so. But to simplify the diagrams, we will assume they are perpendicular. Further, in the transformation (51) the scales on the $O\xi$ and $O\eta$ axes may be changed; they may not be the same as the ones originally chosen on the axes Ox and Oy. But again, for the sake of simplicity, we assume that the scales are not changed. Thus, for example, in place of the concentric circles, as in figure 8, there could in general occur a family of similar and similarly placed ellipses with common center at the origin.

All integral curves of equation (52) are given by a relation of the form

$$a\eta + b \mid \xi \mid^k = 0,$$

where a and b are arbitrary constants.

The integral curves of equation (52) are graphed in figure 10; here we we have assumed that $k > 1$. In this case all integral curves except one, the axis $O\eta$, are tangent at the origin to the axis $O\xi$. The case $0 < k < 1$ is the same as the case $k > 1$ with interchange of ξ and η, i.e., we have only to interchange the roles of the axes ξ and η. For $k = 1$, equation (52) becomes equation (30), whose integral curves were illustrated in figure 7.

An illustration of the integral curves of equation (52) for $k < 0$ is given in figure 11. In this case we have only two integral curves that pass through the point O: these are the axis $O\xi$ and the axis $O\eta$. All other integral

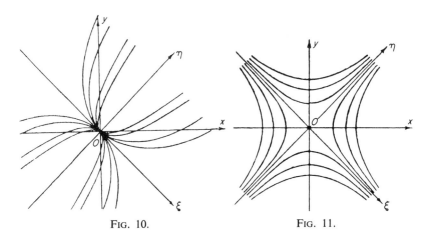

FIG. 10.   FIG. 11.

curves, after approaching the origin no closer than to some minimal distance, recede again from the origin. In this case we say that the point O is a *saddle point* because the integral curves are similar to the contours on a map representing the summit of a mountain pass (saddle).

All integral curves of equation (53) are given by the equation

$$b\eta = \xi(a + b \ln |\xi|),$$

where a and b are arbitrary constants. These are illustrated schematically in figure 12; all of them are tangent to the axis $O\eta$ at the origin.

If every integral curve entering some neighborhood of the singular point O passes through this point and has a definite direction there, i.e., has a definite tangent at the origin, as is illustrated in figures 10 and 12, then we say that the point O is a *node*.

Equation (54) is most easily integrated, if we change to polar coordinates ρ and ϕ, putting

$$\xi = \rho \cos \phi, \quad \eta = \rho \sin \phi.$$

Then this equation changes into the equation

$$\frac{d\rho}{d\phi} = k\rho, \quad \text{where} \quad k = \frac{\alpha}{\beta},$$

and hence,

$$\rho = Ce^{k\phi}. \tag{56}$$

If $k > 0$ then all the integral curves approach the point O, winding infinitely often around this point as $\phi \to -\infty$ (figure 13). If $k < 0$,

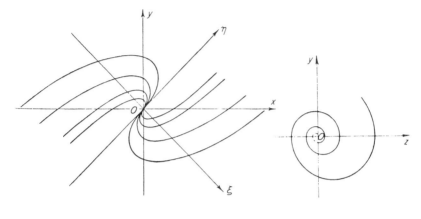

FIG. 12.   FIG. 13.

then this happens for $\phi \to +\infty$. In these cases, the point O is called a *focus*. If, however, $k = 0$, then the collection of integral curves of (56) consists of curves with center at the point O. Generally, if some neighborhood of the point O is completely filled by closed integral curves, surrounding the point O itself, then such a point is called a *center*.

A center may easily be transformed into a focus, if in the numerator and the denominator of the right side of equation (54) we add a term of arbitrarily high order; consequently, in this case the behavior of integral curves near a singular point is not given by terms of the first order.

Equation (55), corresponding to equation (45), is identical with the characteristic equation (19). Thus figures 10 and 12 schematically represent the behavior in the phase plane (x, y) of the curves

$$x = x(t), \quad y = x'(t),$$

corresponding to the solutions of equation (6) for real λ_1 and λ_2 of the same sign; Figure 11 corresponds to real λ_1 and λ_2 of opposite signs, and figures 13 and 8 (the case of a center) correspond to complex λ_1 and λ_2. If the real parts of λ_1 and λ_2 are negative, then the point $(x(t), y(t))$ approaches 0 for $t \to +\infty$; in this case the point $x = 0, y = 0$ corresponds to stable equilibrium. If, however, the real part of either of the numbers

λ_1 and λ_2 is positive, then at the point $x = 0$, $y = 0$, there is no stable equilibrium.

§7. Qualitative Theory of Ordinary Differential Equations

An important part of the general theory of ordinary differential equations is the qualitative theory of differential equations. It arose at the end of the last century from the requirements of mechanics and astronomy.

In many practical problems, it is necessary to establish the character of the solution of a differential equation describing some physical process and to describe the properties of its solutions as the independent variable ranges over a finite or infinite interval. For example, in celestial mechanics, which studies the motion of heavenly bodies, it is important to have information about the behavior of the solutions of differential equations describing the motion of the planets or other heavenly bodies for unbounded periods of time.

As we said earlier, for only a few particularly simple equations can a general solution be expressed in terms of integrals of known functions. So there arose the problem of investigating the properties of the solutions of a differential equation from the equation itself. Since the solution of a differential equation is given in the form of a curve in a plane or in space, the problem consisted of investigating the properties of integral curves, their distribution and their behavior in the neighborhood of singular points. For example, do they lie in a bounded part of the plane or do they have branches tending to infinity, are some of them closed curves, and so forth? The investigation of such questions constitutes the qualitative theory of differential equations.

The founders of the qualitative theory of differential equations are the Russian mathematician A M. Ljapunov and the French mathematician H. Poincaré.

In the preceding section, we considered in detail one of the important questions of the qualitative theory, namely the distribution of integral curves in a neighborhood of a singular point. We turn now to some other basic questions in qualitative theory.

Stability. In the examples considered at the beginning of the chapter, the question of stability or instability of the equilibrium of a system was easily answered from physical considerations, without investigating the differential equations. Thus in example 3 it is obvious that if the pendulum, in its equilibrium position OA, is moved by some external force to a nearby position OA', i.e., if a small change is made in the initial conditions, then the subsequent motion of the pendulum cannot carry it very far from the equilibrium position, and this deviation will be smaller for smaller original deviations OA', i.e., in this case the equilibrium position will be stable.

For other more complicated cases, the question of stability of the equilibrium position is considerably more complicated and can be dealt with only by investigating the corresponding differential equations. The problem of the stability of equilibrium is closely connected with the question of the stability of motion. Fundamental results in this field were established by A. M. Ljapunov.

Let some physical process be described by the system of equations

$$\frac{dx}{dt} = f_1(x, y, t),$$

$$\frac{dy}{dt} = f_2(x, y, t). \tag{57}$$

For simplicity, we consider only a system of two differential equations, although our conclusions remain valid for a system with a larger number of equations. Each particular solution of the system (57), consisting of two functions $x(t)$ and $y(t)$, will sometimes be called a motion, following the usage of Ljapunov. We will assume that $f_1(x, y, t)$ and $f_2(x, y, t)$ have continuous partial derivatives. It has been shown that, in this case, the solution of the system of differential equations (57) is uniquely defined if at any instant of time $t = t_0$ the initial values $x(t_0) = x_0$ and $y(t_0) = y_0$ are given.

We will denote by $x(t, x_0, y_0)$ and $y(t, x_0, y_0)$ the solution of the system of equations (57) satisfying the initial conditions

$$x = x_0 \text{ and } y = y_0 \text{ for } t = t_0.$$

A solution $x(t, x_0, y_0), y(t, x_0, y_0)$ is called *stable in the sense of Ljapunov* if for all $t > t_0$ the functions $x(t, x_0, y_0)$ and $y(t, x_0, y_0)$ have arbitrarily small changes for sufficiently small changes in the initial values x_0 and y_0.

More exactly, for a solution to be stable in the sense of Ljapunov, the differences

$$| x(t, x_0 + \delta_1, y_0 + \delta_2) - x(t, x_0, y_0) |,$$

$$| y(t, x_0 + \delta_1, y_0 + \delta_2) - y(t, x_0, y_0) | \tag{58}$$

may be made less than any previously given number ϵ for all $t > t_0$, if the numbers δ_1 and δ_2 are taken sufficiently small in absolute value.

Every motion that is not stable in the sense of Ljapunov is called *unstable*.

In his investigation, the motion $x(t, x_0, y_0)$ and $y(t, x_0, y_0)$ was called by Ljapunov unperturbed, and the motion $x(t, x_0 + \delta_1, y_0 + d_2)$, $y(t, x_0 + \delta_1, y_0 + \delta_2)$ with nearby initial conditions was called perturbed. In this way stability in the sense of Ljapunov for an unperturbed motion means that for all $t > t_0$ the perturbed motion must differ only a little from the unperturbed.

The stability of equilibrium is a special case of stability of motion, corresponding to the case in which the unperturbed motion is

$$x(t, x_0, y_0) \equiv 0 \text{ and } y(t, x_0, y_0) \equiv 0.$$

Conversely, the question of the stability of any motion $x = \phi_1(t)$ and $y = \phi_2(t)$ of the system (57) may be reduced to the question of the stability of equilibrium for some system of differential equations. To this end we replace the unknown functions $x(t)$ and $y(t)$ in the system (57) by the new unknown functions

$$\xi = x - \phi_1(t) \text{ and } \eta = y - \phi_2(t). \tag{59}$$

In the system (57) transformed in this way, the motion $x = \phi_1(t)$ and $y = \phi_2(t)$ will correspond to the motion $\xi \equiv 0$ and $\eta \equiv 0$, i.e., the position of equilibrium. In what follows we will everywhere assume that the transformation (59) has been made, so that we may consider stability in the sense of Ljapunov only for the solution $x = 0$, $y = 0$.

The condition of stability in the sense of Ljapunov now means that, for δ_1 and δ_2 sufficiently small and $t > t_0$, the trajectory in the (x, y) plane of a perturbed motion does not pass outside of the square with sides of length 2 parallel to the coordinate axes and with center at the point $x = 0, y = 0$.

We will be interested in those cases in which, without knowing an integral of the system (57), we can nevertheless arrive at conclusions about the stability or instability of a motion. Stability is a very important practical question in the motion of projectiles, or of aircraft; and the stability of orbits is important in celestial mechanics, where the motion of planets and other heavenly bodies leads to this kind of investigation.

We assume that the functions $f_1(x, y, t)$ and $f_2(x, y, t)$ may be represented in the form

$$f_1(x, y, t) = a_{11}x + a_{12}y + R_1(x, y, t),$$

$$f_2(x, y, t) = a_{21}x + a_{22}y + R_2(x, y, t), \tag{60}$$

where the a_{ij} are constants, and $R_1(x, y, t)$ and $R_2(x, y, t)$ are functions of x, y, and t such that

$$| R_1(x, y, t) | \leqslant M(x^2 + y^2) \text{ and } | R_2(x, y, t) | \leqslant M(x^2 + y^2), \quad (61)$$

where M is a positive constant.

If in the system (57) we substitute equations (60), neglecting $R_1(x, y, t)$ and $R_2(x, y, t)$, we get a system of differential equations with constant coefficients

$$\frac{dx}{dt} = a_{11}x + a_{12}y,$$

$$\frac{dy}{dt} = a_{21}x + a_{22}y, \quad (62)$$

which is called the *system of first approximation to the nonlinear system* (57).

Before the time of Ljapunov, researchers confined themselves to investigating stability of the first approximation, believing that the results obtained would carry over to the question of stability for the basic nonlinear system (57). Ljapunov was the first to show that in the general case this conclusion is false. On the other hand, he gave a series of very wide conditions under which the question of stability for the nonlinear system is completely solved by the first approximation. One of these conditions is the following. If the real parts of both the roots of the equation

$$\begin{vmatrix} a_{11} - \lambda & a_{12} \\ a_{21} & a_{22} - \lambda \end{vmatrix} = 0$$

are negative and the functions $R_1(x, y, t)$ and $R_2(x, y, t)$ fulfill condition (61), then the solution $x(t) \equiv 0$, $y(t) \equiv 0$ is stable in the sense of Ljapunov. If the real part of either of the roots is positive, then the solution $x(t) \equiv 0$, $y(t) \equiv 0$ of an equation satisfying the conditions (61) is unstable. Ljapunov also gave a series of other sufficient conditions for stability and instability of a motion.*

If the right sides of equations (57) do not depend on t, then dividing the first equation of the system (57) by the second we get

$$\frac{dy}{dx} = \frac{f_1(x, y)}{f_2(x, y)}. \quad (63)$$

The origin will be a singular point for this equation. In the case of stability of equilibrium, this point may be a focus, a node, or a center, but cannot be a saddle point.

---

* A. M. Ljapunov, *The general problem of stability of motion.*

Thus the character of a singular point may be determined from the stability or instability of the equilibrium position.

The behavior of integral curves in the large. It is sometimes important to construct a schematized representation of the behavior of the integral curves "in the large"; that is, in the entire domain of the given system of differential equations, without attempting to preserve the scale. We will consider a space in which this system defines a field of directions as the phase space of some physical process. Then the general scheme of the integral curves, corresponding to the system of differential equations, will give us an idea of the character of all processes (motions) which can possibly occur in this system. In figures 10–13 we have constructed approximate schematized representations of the behavior of the integral curves in the neighborhood of an isolated singular point.

One of the most fundamental problems in the theory of differential equations is the problem of finding as simple a method as possible for constructing such a scheme for the behavior of the family of integral curves of a given system of differential equations in the entire domain of definition, in order to study the behavior of the integral curves of this system of differential equations "in the large." This problem remains almost untouched for spaces of dimension higher than 2. It is still very far from being solved for the single equation of the form

$$\frac{dy}{dx} = \frac{M(x, y)}{N(x, y)} \tag{64}$$

even when $M(x, y)$ and $N(x, y)$ are polynomials.

In what follows, we will assume that the functions $M(x, y)$ and $N(x, y)$ have continuous partial derivatives of the first order.

If all the points of a simply connected domain G, in which the right side of the differential equation (64) is defined, are ordinary points, then the family of integral curves may be represented schematically as a family of segments of parallel straight lines; since in this case one integral curve will pass through each point, and no two integral curves can intersect. For an equation (64) of more general form, which may have singular points, the structure of the integral curves may be much more complicated. The case in which equation (64) has an infinite set of singular points (i.e., points where the numerator and the denominator both vanish) may be excluded, at least when $M(x, y)$ and $N(x, y)$ are polynomials. Thus we restrict our consideration to those cases in which equation (64) has a finite number of isolated singular points. The behavior of the integral curves that are near to one of these singular points forms the essential element in setting up a schematized representation of the behavior of all the integral curves of the equation.

A very typical element in such a scheme for the behavior of all the integral curves of equation (64) is formed by the so-called *limit cycles*. Let us consider the equation

$$\frac{d\rho}{d\phi} = \rho - 1, \tag{65}$$

where ρ and ϕ are polar coordinates in the (x, y) plane.

The collection of all integral curves of equation (65) is given by the formula

$$\rho = 1 + Ce^{\phi}, \tag{66}$$

where C is an arbitrary constant, different for different integral curves. In order that ρ be nonnegative, it is necessary that ϕ have values no larger than $- \ln | C |$, $C < 0$. The family of integral curves will consist of

1. the circle $\rho = 1 \ (C = 0)$;

2. the spirals issuing from the origin, which approach this circle from the inside as $\phi \to - \infty \ (C < 0)$;

3. the spirals, which approach the circle $\rho = 1$ from the outside as $\phi \to - \infty$ $(C > 0)$ (figure 14).

The circle $\rho = 1$ is called a limit. cycle for equation (65). In general a closed integral curve l is called a *limit cycle*, if it can be enclosed in a disc all points of which are ordinary for equation (64) and which is entirely filled by nonclosed integral curves.

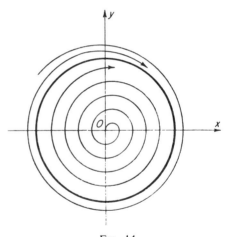

FIG. 14.

From equation (65) it can be seen that all points of the circle are ordinary. This means that a small piece of a limit cycle is not different from a small piece of any other integral curve.

Every closed integral curve in the (x, y) plane gives a periodic solution $[x(t), y(t)]$ of the system

$$\frac{dx}{dt} = N(x, y), \frac{dy}{dt} = M(x, y), \tag{67}$$

describing the law of change of some physical system. Those integral curves in the phase plane that as $t \to +\infty$ approximate a limit cycle are motions that as $t \to +\infty$ approximate periodic motions.

Let us suppose that for every point (x_0, y_0) sufficiently close to a limit cycle l, we have the following situation: If (x_0, y_0) is taken as initial point (i.e., for $t = t_0$) for the solution of the system (67), then the corresponding integral curve traced out by the point $[x(t), y(t)]$, as $t \to +\infty$ approximates the limit cycle l in the (x, y) plane. (This means that the motion in question is approximately periodic.) In this case the corresponding limit cycle is called *stable*. Oscillations that act in this way with respect to a limit cycle correspond physically to self-oscillations. In some self-oscillatory systems, there may exist several stable oscillatory processes with different amplitudes, one or another of which will be established by the initial conditions. In the phase plane for such "self-oscillatory systems," there will exist corresponding limit cycles if the processes occuring in these systems are described by an equation of the form (67).

The problem of finding, even if only approximately, the limit cycles of a given differential equation has not yet been satisfactorily solved. The most widely used method for solving this problem is the one suggested by Poincaré of constructing "cycles without contact." It is based on the following theorem. We assume that on the (x, y) plane we can find two closed curves L_1 and L_2 (cycles) which have the following properties:

1. The curve L_2 lies in the region enclosed by L_1.

2. In the annulus Ω, between L_1 and L_2, there are no singular points of equation (64).

3. L_1 and L_2 have tangents everywhere, and the directions of these tangents are nowhere identical with the direction of the field of directions for the given equation (64).

4. For all points of L_1 and L_2 the cosine of the angle between the interior normals to the boundary of the domain Ω and the vector with components $[N(x, y), M(x, y)]$ never changes sign.

Then between L_1 and L_2, there is at least one limit cycle of equation (64).

Poincaré called the curves L_1 and L_2 *cycles without contact*.

The proof of this theorem is based on the following rather obvious fact. We assume that for decreasing t (or for increasing t) all the integral curves

$$x = x(t), \quad y = y(t)$$

of equation (64) (or, what amounts to the same thing, of equations (67), where t is a parameter), which intersect L_1 or L_2, enter the annulus Ω

between L_1 and L_2. Then they must necessarily tend to some closed curve l lying between L_1 and L_2, since none of the integral curves lying in the annulus can leave it, and there are no singular points there.

But the problem of finding cycles without contact is also a complicated one and no general methods are known for solving it. For particular examples it has been possible to find cycles without contact, thereby proving the existence of limit cycles.

In radio technology it is important to find limit cycles (self-oscillatory processes) for equation (16) for the electron-tube generator. For equations of the type of (16), N. M. Krylov and N. N. Bogoljubov gave a method, about twenty years ago, for approximate computation of a certain limit cycle that exists for this equation. At about the same time the Soviet physicists L. I. Mandel'stam, N. D. Papaleksi, and A. A. Andronov gave a proof of the possibility of applying what is called the method of the small parameter, a method that to some extent had been used earlier in practice, though without any rigorous justification. Andronov was also the first to make systematic practical use, in the analysis of self-oscillatory systems, of the theoretical methods already developed by Ljapunov and Poincaré. In this manner he obtained a whole series of important results.

As was mentioned earlier, an important role is played in physics by "insensitive" systems (cf. §3). Andronov, together with L. S. Pontrjagin, set up a catalogue of the elements from which one could construct a complete chart of the behavior of the integral curves in the (x, y) plane for an insensitive differential equation of the form (64). It had been long known, for example, that a center near a singular point is easily destroyed by small changes in the equations (64). Thus in the construction of a chart of the behavior of the integral curves of equation (64), we cannot have a center, i.e., a family of closed integral curves surrounding a singular point, if the equation is "insensitive."

The question of the behavior of the integral curves in the large is still far from its final solution. We note that the analogous and probably simpler question of the form of real algebraic curves in the plane, i.e., curves defined by the equation

$$P(x, y) = 0,$$

where $P(x, y)$ is a polynomial of degree n, is also far from a complete solution. The form of these curves is completely known only for $n < 6$.

The solutions of the system (64) define motions in the plane. If we replace each point (x_0, y_0) in the plane by the corresponding point $[x(t, x_0, y_0), y(t, x_0, y_0)]$, where $x(t, x_0, y_0)$ and $y(t, x_0, y_0)$ are the solution of the system (64) with initial conditions $x = x_0$ and $y = y_0$ for $t = t_0$, we obtain a transformation of the points of the plane depending on the parameter t. Similar transformations depending on a parameter, together with the motions they generate, may be considered on a sphere, a torus, or other manifolds. The properties of these motions are studied in the theory of dynamical systems. In a neighborhood of every point these motions are the solutions of some system of differential equations. In the past decade the theory of dynamical systems has been developed on a broad basis in the works of V. V. Stepanov, A. Ja. Hinčin, N. N. Bogoljubov, N. M. Krylov, A. A. Markov, V. V. Nemyckiĭ and others, and also in the works of G. D. Birkhoff and other mathematicians.

In this chapter we have given a brief outline of the present state of the theory of ordinary differential equations and have attempted to describe the problems that are considered in this theory. Our study in no sense pretends to be complete. We have had to omit consideration of many branches of the theory that arise in the study of more special problems or that require broader mathematical knowledge than the reader of this book is assumed to possess. For example, we have nowhere touched upon the general and important area in which the theory of differential equations with complex arguments is considered. We have had no opportunity to examine the theory of boundary-value problems and in particular, of eigenfunctions, which is of great importance in the applications.

We have also been able to pay very little attention to approximative methods for the numerical or analytical solution of differential equations. For these questions, we recommend that the reader consult the specialized literature.

Suggested Reading

R. P. Agnew, *Differential equations*, 2nd ed., McGraw-Hill, New York, 1960.

E. A. Coddington, *An introduction to ordinary differential equations*, Prentice-Hall, Englewood Cliffs, N. J., 1961.

E. A. Coddington and N. Levinson, *Theory of ordinary differential equations*, McGraw-Hill, New York, 1955.

W. Hurewicz, *Lectures on ordinary differential equations*, Technology Press and Wiley, New York, 1958.

S. Lefschetz, *Differential equations: geometric theory*, Interscience, New York, 1957.

VOLUME TWO

PART 3

CHAPTER VI

PARTIAL
DIFFERENTIAL EQUATIONS

§1. Introduction

In the study of the phenomena of nature, partial differential equations are encountered just as often as ordinary ones. As a rule this happens in cases where an event is described by a function of several variables. From the study of nature there arose that class of partial differential equations that is at the present time the most thoroughly investigated and probably the most important in the general structure of human knowledge, namely the equations of mathematical physics.

Let us first consider oscillations in any kind of medium. In such oscillations every point of the medium, occupying in equilibrium the position (x, y, z), will at time t be displaced along a vector $u(x, y, z, t)$, depending on the initial position of the point (x, y, z) and on the time t. In this case the process in question will be described by a vector field. But it is easy to see that knowledge of this vector field, namely the field of displacements of points of the medium, is not sufficient in itself for a full description of the oscillation. It is also necessary to know, for example, the density $\rho(x, y, z, t)$ at each point of the medium, the temperature $T(x, y, z, t)$, and the internal stress, i.e., the forces exerted on an arbitrarily chosen volume of the body by the entire remaining part of it.

Physical events and processes occuring in space and time always consist of the changes, during the passage of time, of certain physical magnitudes related to the points of the space. As we saw in Chapter II these quantities can be described by functions with four independent variables, x, y, z, and t, where x, y, and z are the coordinates of a point of the space, and and t is the time.

Physical quantities may be of different kinds. Some are completely characterized by their numerical values, e.g., temperature, density, and the like, and are called scalars. Others have direction and are therefore vector quantities: velocity, acceleration, the strength of an electric field, etc. Vector quantities may be expressed not only by the length of the vector and its direction but also by its "components" if we decompose it into the sum of three mutually perpendicular vectors, for example parallel to the coordinate axes.

In mathematical physics a scalar quantity or a scalar field is presented by one function of four independent variables, whereas a vector quantity defined on the whole space or, as it is called, a vector field is described by three functions of these variables. We can write such a quantity either in the form

$$\boldsymbol{u}(x, y, z, t),$$

where the bold face type indicates the $\boldsymbol{u}$ is a vector, or in the form of three functions

$$u_x(x, y, z, t), \quad u_y(x, y, z, t), \quad u_z(x, y, z, t),$$

where u_x, u_y, and u_z denote the projections of the vector on the coordinate axes.

In addition to vector and scalar quantities, still more complicated entities occur in physics, for example the state of stress of a body at a given point. Such quantities are called tensors; after a fixed choice of coordinate axes, they may be characterized everywhere by a set of functions of the same four independent variables.

In this manner, the description of widely different kinds of physical phenomena is usually given by means of several functions of several variables. Of course, such a description cannot be absolutely exact.

For example, when we describe the density of a medium by means of one function of our independent variables, we ignore the fact that at a given point we cannot have any density whatsoever. The bodies we are investigating have a molecular structure, and the molecules are not contiguous but occur at finite distances from one another. The distances between molecules are for the most part considerably larger than the dimensions of the molecules themselves. Thus the density in question is the ratio of the mass contained in some small, but not extremely small, volume to this volume itself. The density at a point we usually think of as the limit of such ratios for decreasing volumes. A still greater simplification and idealization is introduced in the concept of the temperature of a medium. The heat in a body is due to the random motion of its molecules.

The energy of the molecules differs, but if we consider a volume containing a large collection of molecules, then the average energy of their random motions will define what is called temperature.

Similarly, when we speak of the pressure of a gas or a liquid on the wall of a container, we should not think of the pressure as though a particle of the liquid or gas were actually pressing against the wall of the container. In fact, these particles, in their random motion, hit the wall of the container and bounce off it. So what we describe as pressure against the wall is actually made up of a very large number of impulses received by a section of the wall that is small from an everyday point of view but extremely large in comparison with the distances between the molecules of the liquid or gas. It would be easy to give dozens of examples of a similar nature. The majority of the quantities studied in physics have exactly the same character. Mathematical physics deals with idealized quantities, abstracting them from the concrete properties of the corresponding physical entities and considering only the average values of these quantities.

Such an idealization may appear somewhat coarse but, as we will see, it is very useful, since it enables us to make an excellent analysis of many complicated matters, in which we consider only the essential elements and omit those features which are secondary from our point of view.

The object of mathematical physics is to study the relations existing among these idealized elements, these relations being described by sets of functions of several independent variables.

§2. The Simplest Equations of Mathematical Physics

The elementary connections and relations among physical quantities are expressed by the laws of mechanics and physics. Although these relations are extremely varied in character, they give rise to more complicated ones, which are derived from them by mathematical argument and are even more varied. The laws of mechanics and physics may be written in mathematical language in the form of partial differential equations, or perhaps integral equations, relating unknown functions to one another. To understand what is meant here, let us consider some examples of the equations of mathematical physics.

Equations of conservation of mass and of heat energy. Let us express in mathematical form the basic physical laws governing the motions of a medium.

1. First of all we express the law of conservation of the matter contained in any volume Ω which we mentally mark off in a space and keep fixed.

For this purpose we must calculate the mass of the matter contained in this volume. The mass $M_\Omega(t)$ is expressed by the integral

$$M_\Omega(t) = \iiint\limits_\Omega \rho(x, y, z, t)\, dx\, dy\, dz.$$

This mass will not, of course, be constant; in an oscillatory process the density at each point will be changing in view of the fact that the particles of matter in their oscillations will at one time enter this volume and at another leave it. The rate of change of the mass can be found by differentiation with respect to time and is given by the integral

$$\frac{dM_\Omega}{dt} = \iiint\limits_\Omega \frac{\partial \rho}{\partial t}\, dx\, dy\, dz.$$

This rate of change of the mass contained in the volume may also be calculated in another way. We may express the amount of matter which passes through the surface S, bounding our volume Ω, at each second of time, where the matter leaving Ω must be taken with a minus sign. To this end we consider an element ds of the surface S sufficiently small that it may be assumed to be plane and have the same displacement for all its points. We will follow the displacement of points on this segment of the surface during the interval of time from t to $t + dt$. First of all we compute the vector

$$\boldsymbol{v} = \frac{du}{dt},$$

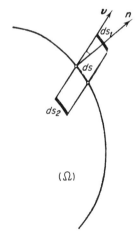

which represents the velocity of each particle. In the time dt the particles on ds move along the vector $\boldsymbol{v}\, dt$, and take up a position ds_1, while the position ds will now be occupied by the particles which were formerly at the position ds_2 (figure 1). So during this time the column of matter leaving the volume Ω will be that which was earlier contained between ds_2 and ds_1. The altitude of this small column is equal to $v\, dt \cos(\boldsymbol{n}, \boldsymbol{v})$, where $\boldsymbol{n}$ denotes the exterior normal to the surface; the volume of the small column will thus be equal to

FIG. 1. $v \cos(\boldsymbol{n}, \boldsymbol{v})\, ds\, dt,$

and the mass equal to

$$\rho v \cos (\boldsymbol{n}, \boldsymbol{v}) \, ds \, dt.$$

Adding together all these small pieces, we get for the amount of matter leaving the volume during the time dt the expression

$$\iint_S \rho v \cos (\boldsymbol{n}, \boldsymbol{v}) \, ds \, dt.$$

At those points where the velocity is directed toward the interior of Ω the sign of the cosine will be negative, which means that in this integral the matter entering Ω is taken with a minus sign. The product of the velocity of motion of the medium with its density is called its flux. The flux vector of the mass is $\boldsymbol{q} = \rho \boldsymbol{v}$.

In order to find the rate of flow of matter out of the volume Ω it is sufficient to divide this expression by dt, so that for the rate of flow we have

$$\iint_S \rho v_n \, ds = \iint_S q_n \, ds,$$

where

$$v_n = v \cos (\boldsymbol{n}, \boldsymbol{v}), \quad q_n = q \cos (\boldsymbol{n}, \boldsymbol{q}).$$

The normal component of the vector $\boldsymbol{v}$ may be replaced by its expression in terms of the components of the vectors $\boldsymbol{v}$ and $\boldsymbol{n}$ along the coordinate axes. From analytic geometry we know that

$$v_n = v \cos (\boldsymbol{n}, \boldsymbol{v}) = v_x \cos (\boldsymbol{n}, x) + v_y \cos (\boldsymbol{n}, y) + v_z \cos (\boldsymbol{n}, z),$$

hence we can rewrite the expression for the rate of flow in the form

$$\iint_S \rho (v_x \cos (\boldsymbol{n}, x) + v_y \cos (\boldsymbol{n}, y) + v_z \cos (\boldsymbol{n}, z)) \, ds.$$

From the law of conservation of matter, these two methods of computing the change in the amount of matter must give the same result, since all change in the mass included in Ω can occur only as a result of the entering or leaving of mass through the surface S.

Hence, equating the rate of change of the amount of matter contained in the volume with the rate of flow of matter into the volume, we get

$$\iiint_\Omega \frac{\partial \rho}{\partial t} \, dx \, dy \, dz$$

$$= - \iint_S [\rho v_x \cos (\boldsymbol{n}, x) + \rho v_y \cos (\boldsymbol{n}, y) + \rho v_z \cos (\boldsymbol{n}, z)] \, ds$$

$$= - \iint_S [q_x \cos (\boldsymbol{n}, x) + q_y \cos (\boldsymbol{n}, y) + q_z \cos (\boldsymbol{n}, z)] \, ds.$$

This integral relation, as we have said, is true for any volume Ω. It is called "the equation of continuity."

The integral occurring on the right side of the last equation may be transformed into a volume integral by using Ostrogradskiĭ's formula. This formula, derived in Chapter II gives

$$\iint\limits_{S} (\rho v_x \cos(\boldsymbol{n}, \boldsymbol{x}) + \rho v_y \cos(\boldsymbol{n}, \boldsymbol{y}) + \rho v_z \cos(\boldsymbol{n}, \boldsymbol{z}))\, ds$$
$$= \iiint\limits_{\Omega} \left[\frac{\partial(\rho v_x)}{\partial x} + \frac{\partial(\rho v_y)}{\partial y} + \frac{\partial(\rho v_z)}{\partial z} \right] d\Omega.$$

Hence it follows that

$$\iiint\limits_{\Omega} \left[\frac{\partial \rho}{dt} + \frac{\partial(\rho v_x)}{\partial x} + \frac{\partial(\rho v_y)}{\partial y} + \frac{\partial(\rho v_z)}{\partial z} \right] d\Omega = 0.$$

So we get the following result; the integral of the function

$$\frac{\partial \rho}{\partial t} + \frac{\partial(\rho v_x)}{\partial x} + \frac{\partial(\rho v_y)}{\partial y} + \frac{\partial(\rho v_z)}{\partial z} \quad \text{or} \quad \frac{\partial \rho}{\partial t} + \frac{\partial q_x}{\partial x} + \frac{\partial q_y}{\partial y} + \frac{\partial q_z}{\partial z}$$

over any volume Ω is equal to zero. But this is possible only if the function is identically zero. We thus obtain the equation of continuity in differential form

$$\frac{\partial \rho}{\partial t} + \frac{\partial(\rho v_x)}{\partial x} + \frac{\partial(\rho v_y)}{\partial y} + \frac{\partial(\rho v_z)}{\partial z} = 0. \tag{1}$$

Equation (1) is a typical example of the formulation of a physical law in the language of partial differential equations.

2. Let us consider another such problem, namely the problem of heat conduction.

In any medium whose particles are in motion on account of heat, the heat flows from some points to others. This flow of heat will occur through every element of surface ds lying in the given medium. It can be shown that the process may be described numerically by a single vector quantity, the heat-conduction vector, which we denote by τ. Then the amount of heat flowing per second through an element of area ds will be expressed by $\tau_n\, ds$, in the same way as $q_n\, ds$ earlier expressed the amount of material passing per second through an area ds. In place of the flux of liquid $\boldsymbol{q} = \rho \boldsymbol{v}$ we have the heat flow vector τ.

In the same way as we obtained the equation of continuity, which for the motion of a liquid expresses the law of conservation of mass, we may obtain a new partial differential equation expressing the law of conservation of energy, as follows.

The volume density of heat energy Q at a given point may be expressed by the formula

$$Q = CT,$$

where C is the heat capacity and T is the temperature.

Here it is easy to establish the equation

$$C \frac{\partial T}{\partial t} + \frac{\partial \tau_x}{\partial x} + \frac{\partial \tau_y}{\partial y} + \frac{\partial \tau_z}{\partial z} = 0. \qquad (2)$$

The derivation of this equation is identical with the derivation of the equation of continuity, if we replace "density" by "density of heat energy" and flow of mass by flow of heat. Here we have assumed that the heat energy in the medium never increases. But if there is a source of heat present in the medium, equation (2) for the balance of heat energy must be modified. If q is the productivity density of the source, that is the amount of heat energy produced per unit of volume in one second, then the equation of conservation of heat energy has the following more complicated form:

$$C \frac{\partial T}{\partial t} + \frac{\partial \tau_x}{\partial x} + \frac{\partial \tau_y}{\partial y} + \frac{\partial \tau_z}{\partial z} = q. \qquad (3)$$

3. Still another equation of the same type as the equation of continuity may be derived by differentiating equation (1) with respect to time. Let us do this for the equation of small oscillations of a gas near a position of equilibrium. We will assume that for such oscillations changes of the density are not great and the quantities $\partial \rho / \partial x$, $\partial \rho / \partial y$, $\partial \rho / \partial z$, and $\partial \rho / \partial t$ are sufficiently small that their products with v_x, v_y, and v_z may be ignored. Then

$$\frac{\partial \rho}{\partial t} + \rho \left(\frac{\partial v_x}{\partial x} + \frac{\partial v_y}{\partial y} + \frac{\partial v_z}{\partial z} \right) = 0.$$

Differentiating this equation with respect to time and ignoring the products of $\partial \rho / \partial t$ with $\partial v_x / \partial x$, $\partial v_y / \partial y$, and $\partial v_z / \partial z$, we obtain

$$\frac{\partial^2 \rho}{\partial t^2} + \rho \left[\frac{\partial \left(\dfrac{dv_x}{dt} \right)}{\partial x} + \frac{\partial \left(\dfrac{dv_y}{dt} \right)}{\partial y} + \frac{\partial \left(\dfrac{dv_z}{dt} \right)}{\partial z} \right] = 0. \qquad (4)$$

Equation of motion.

1. An important example of the expression of a physical law by a differential equation occurs in the equations of equilibrium or of motion of a medium. Let the medium consist of material particles, moving with

various velocities. As in the first example, we mentally mark off in space a volume Ω, bounded by the surface S and filled with particles of matter of the medium, and write Newton's second law for the particles in this volume. This law states that for every motion of the medium the rate of change of momentum, summed up for all particles, in the volume is equal to the sum of all the forces acting on the volume. The momentum, as is known from mechanics, is represented by the vector quantity

$$\boldsymbol{P} = \iiint\limits_{\Omega} \rho \boldsymbol{v} \, d\Omega.$$

The particles occupying a small volume $d\Omega$ with density ρ will, after time Δt, fill a new volume $d\Omega'$ with density ρ', although the mass will be unchanged

$$\rho' \, d\Omega' = \rho \, d\Omega.$$

If velocity $\boldsymbol{v}$ changes during this time to a new value $\boldsymbol{v}'$, i.e., by the amount $\Delta \boldsymbol{v} = \boldsymbol{v}' - \boldsymbol{v}$, the corresponding change of momentum will be

$$\rho' \boldsymbol{v}' \, d\Omega' - \rho \boldsymbol{v} \, d\Omega = \rho \boldsymbol{v}' \, d\Omega - \rho \boldsymbol{v} \, d\Omega = \rho \, \Delta \boldsymbol{v} \, d\Omega,$$

or in the unit of time:

$$\rho \, \frac{\Delta \boldsymbol{v}}{\Delta t} \, d\Omega \approx \rho \, \frac{d\boldsymbol{v}}{dt} \, d\Omega.$$

Adding over all particles in the volume Ω, we find that the rate of change of momentum is equal to

$$\iiint\limits_{\Omega} \rho \, \frac{d\boldsymbol{v}}{dt} \, d\Omega$$

or, in other words

$$\iiint\limits_{\Omega} \rho \, \frac{dv_x}{dt} \, d\Omega, \iiint\limits_{\Omega} \rho \, \frac{dv_y}{dt} \, d\Omega, \iiint\limits_{\Omega} \rho \, \frac{dv_z}{dt} \, d\Omega.$$

(Here the derivatives dv_x/dt, dv_y/dt, and dv_z/dt denote the rate of change of the components of $\boldsymbol{v}$ not at a given point of the space but for a given particle. This is what is meant by the notation d/dt instead of $\partial/\partial t$. As is well known, $d/dt = \partial/\partial t + v_x(\partial/\partial x) + v_y(\partial/\partial y) + v_z(\partial/\partial z)$.)

The forces acting on the volume may be of two kinds: volume forces acting on every particle of the body, and surface forces or stresses on the surface S bounding the volume. The former are long-range forces, while the latter are short-range.

To illustrate these remarks, let us assume that the medium under consideration is a fluid. The surface forces acting on an element of the surface ds will in this case have the value $p\,ds$, where p is the pressure on the fluid, and will be exerted in a direction opposite to that of the exterior normal.

If we denote the unit vector in the direction of the normal to the surface S by $\boldsymbol{n}$, then the forces acting on the section ds will be equal to

$$-p\boldsymbol{n}\,ds.$$

If we let $\boldsymbol{F}$ denote the vector of the external forces acting on a unit of volume, our equation takes the form

$$\iiint_{\Omega} \rho \frac{d\boldsymbol{v}}{dt}\,d\Omega = \iiint_{\Omega} \boldsymbol{F}\,d\Omega - \iint_{S} p\boldsymbol{n}\,ds.$$

This is the equation of motion in integral form. Like the equation of continuity, this equation also may be transformed into differential form. We obtain the system:

$$\rho \frac{dv_x}{dt} + \frac{\partial p}{\partial x} = F_x,\ \rho \frac{dv_y}{dt} + \frac{\partial p}{\partial y} = F_y,\ \rho \frac{\partial v_z}{dt} + \frac{\partial p}{\partial z} = F_z. \tag{5}$$

This system is the differential form of Newton's second law.

2. Another characteristic example of the application of the laws of mechanics in differential form is the equation of a vibrating string. A string is a long, very slender body of elastic material that is flexible because of its extreme thinness, and is usually tightly stretched. If we imagine the string divided at any point x into two parts, then on each of the parts there is exerted a force equal to the tension in the direction of the tangent to the curve of the string.

Let us examine a short segment of the string. We will denote by $u(x, t)$ the displacement of a point of the string from its position of equilibrium. We assume that the oscillation of the string occurs in one plane and consists of displacements perpendicular to the axis Ox, and we represent the displacement $u(x, t)$ graphically at some instant of time (figure 2). We will investigate the behavior of the segment of the string between the points x_1 and x_2. At these points there are two forces acting, which are equal to the tension T in the

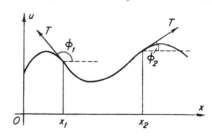

FIG. 2.

direction of the corresponding tangent to $u(x, t)$.

If the segment is curved, the resolvent of these two forces will not be equal to zero. This resolvent, from the laws of mechanics, must be equal to the rate of change of momentum of the segment.

Let the mass contained in each centimeter of length of the string be equal to ρ. Then the rate of change of momentum will be

$$\rho \int_{x_1}^{x_2} \frac{d^2u}{dt^2}\, dx.$$

If the angle between the tangent to the string and the axis Ox is denoted by ϕ, we will have

$$T \sin \phi_2 - T \sin \phi_1 = \int_{x_1}^{x_2} \rho\, \frac{\partial^2 u}{\partial t^2}\, dx.$$

This is the usual equation expressing the second law of mechanics in integral form. It is easy to transform it into differential form. We have obviously

$$\rho\, \frac{\partial^2 u}{\partial t^2} = \frac{\partial}{\partial x}\, (T \sin \phi).$$

From well-known theorems of differential calculus, it is easy to relate $T \sin \phi$ to the unknown function u. We get

$$\tan \phi = \frac{\partial u}{\partial x}, \; \sin \phi = \frac{\tan \phi}{\sqrt{1 + \tan^2 \phi}} = \frac{\partial u/\partial x}{\sqrt{1 + (\partial u/\partial x)^2}}$$

and under the assumption that $(\partial u/\partial x)^2$ is small, we have

$$\sin \phi \approx \frac{\partial u}{\partial x}.$$

Then

$$T\, \frac{\partial^2 u}{\partial x^2} = \rho\, \frac{\partial^2 u}{\partial t^2}. \tag{6}$$

This last equation is the *equation of the vibrating string* in differential form.

Basic forms of equations of mathematical physics. As mentioned previously, the various partial differential equations describing physical phenomena usually form a system of equations in several unknown variables. But in the great majority of cases it is possible to replace this system by one equation, as may easily be shown by very simple examples.

For instance, let us turn to the equations of motion considered in the preceding paragraph. It is required to solve these equations along with the equation of continuity. The actual methods of solution we will consider somewhat later.

1. We begin with the equation for steady flow of an idealized fluid.

All possible motions of a fluid can be divided into rotational and irrotational, the latter also being called *potential*. Although irrotational motions are only special cases of motion and, generally speaking, the motion of a liquid or a gas is always more or less rotational, nevertheless experience shows that in many cases the motion is irrotational to a high degree of exactness. Moreover, it may be shown from theoretical considerations that in a fluid with viscosity equal to zero a motion which is initially irrotational will remain so.

For a potential motion of a fluid, there exists a scalar function $U(x, y, z, t)$, called the *velocity potential*, such that the velocity vector $\boldsymbol{v}$ is expressed in terms of this function by the formulas

$$v_x = \frac{\partial U}{\partial x}, \quad v_y = \frac{\partial U}{\partial y}, \quad v_z = \frac{\partial U}{\partial z}.$$

In all the cases we have studied up to now, we have had to deal with systems of four equations in four unknown functions or, in other words, with one scalar and one vector equation, containing one unknown scalar function and one unknown vector field. Usually these equations may be combined into one equation with one unknown function, but this equation will be of the second order. Let us do this, beginning with the simplest case.

For potential motion of an incompressible fluid, for which $\partial \rho / \partial t = 0$, we have two systems of equations: the equation of continuity

$$\rho \left(\frac{\partial v_x}{\partial x} + \frac{\partial v_y}{\partial y} + \frac{\partial v_z}{\partial z} \right) = 0$$

and the equations of potential motion

$$v_x = \frac{\partial U}{\partial x}, \quad v_y = \frac{\partial U}{\partial y}, \quad v_z = \frac{\partial U}{\partial z}.$$

Substituting in the first equation the values of the velocity as given in the second we have

$$\frac{\partial^2 U}{\partial x^2} + \frac{\partial^2 U}{\partial y^2} + \frac{\partial^2 U}{\partial z^2} = 0. \tag{7}$$

2. The vector field of "heat flow" can also be expressed, by means of differential equations, in terms of one scalar quantity, the temperature.

It is well known that heat "flows" in the direction from a hot body to a cold one. Thus the vector of the flow of heat lies in the direction opposite to that of the so-called temperature-gradient vector. It is also natural to assume, as is justified by experience, that to a first approximation the length of this vector is directly proportional to the temperature gradient.

The components of the temperature gradient are

$$\frac{\partial T}{\partial x}, \frac{\partial T}{\partial y}, \frac{\partial T}{\partial z}.$$

Taking the coefficient of proportionality to be k, we get three equations

$$\tau_x = -k\,\frac{\partial T}{\partial x}, \quad \tau_y = -k\,\frac{\partial T}{\partial y}, \quad \tau_z = -k\,\frac{\partial T}{\partial z}.$$

These are to be solved, together with the equation for the conservation of heat energy

$$C\,\frac{\partial T}{\partial t} + \frac{\partial \tau_x}{\partial x} + \frac{\partial \tau_y}{\partial y} + \frac{\partial \tau_z}{\partial z} = q.$$

Replacing τ_x, τ_y, and τ_z by their values in terms of T, we get

$$C\,\frac{\partial T}{\partial t} = k\left(\frac{\partial^2 T}{\partial x^2} + \frac{\partial^2 T}{\partial y^2} + \frac{\partial^2 T}{\partial z^2}\right) + q. \tag{8}$$

3. Finally, for small vibrations in a gaseous medium, for example the vibrations of sound, the equation

$$\frac{\partial^2 \rho}{\partial t^2} + \rho\,\frac{\partial}{\partial x}\left(\frac{dv_x}{dt}\right) + \rho\,\frac{\partial}{\partial y}\left(\frac{dv_y}{dt}\right) + \rho\,\frac{\partial}{\partial z}\left(\frac{dv_z}{dt}\right) = 0$$

and the equations of dynamics (5), give

$$\rho\,\frac{dv_x}{dt} + \frac{\partial p}{\partial x} = F_x, \quad \rho\,\frac{dv_y}{dt} + \frac{\partial p}{\partial y} = F_y, \quad \rho\,\frac{dv_z}{dt} + \frac{\partial p}{\partial z} = F_z,$$

and, assuming the absence of external forces ($F_x = F_y = F_z = 0$) we get

$$\frac{\partial^2 p}{\partial t^2} = a^2\left(\frac{\partial^2 p}{\partial x^2} + \frac{\partial^2 p}{\partial y^2} + \frac{\partial^2 p}{\partial z^2}\right) \tag{9}$$

(to obtain this equation it is sufficient to substitute the expression for the accelerations in the equation of continuity and to eliminate the density ρ by using the Boyle-Mariotte law: $p = a^2\rho$).

Equations (7), (8), and (9) are typical for many problems of mathematical physics in addition to the ones considered here. The fact that they have been investigated in detail enables us to gain an understanding of many physical situations.

§3. Initial-Value and Boundary-Value Problems; Uniqueness of a Solution

With partial differential equations as with ordinary ones, it is the case, with rare exceptions, that every equation has infinitely many particular solutions. Thus to solve a concrete physical problem, i.e., to find an unknown function satisfying some equation, we must know how to choose the required solution from an infinite set of solutions. For this purpose it is usually necessary to know not only the equation itself but a certain number of supplementary conditions. As we saw previously, partial differential equations are the expression of elementary laws of mechanics or physics, referring to small particles situated in a medium. But it is not enough to know only the laws of mechanics, if we wish to predict the course of some process. For example, to predict the motion of the heavenly bodies, as is done in astronomy, we must know not only the general formulation of Newton's laws but also, assuming that the masses of these bodies are known, we must know the initial state of the system, i.e., the position of the bodies and their velocities at some initial instant of time. Supplementary conditions of this kind are always encountered in solving the problems of mathematical physics.

Thus, the problems of mathematical physics consist of finding solutions of partial differential equations that satisfy certain supplementary conditions.

The equations (7), (8), (9) differ in structure among themselves. Correspondingly different are the physical problems that may be solved by means of these equations.

The Laplace and Poisson equations; harmonic functions and uniqueness of solution of boundary-value problems for them. Let us analyze these problems a little more in detail. We begin with the Laplace and Poisson equations. The *Poisson equation* is *

$$\Delta u = -4\pi\rho,$$

where ρ is usually the density. In particular, ρ may vanish. For $\rho \equiv 0$ we get the *Laplace equation*

$$\Delta u = 0.$$

---

* The symbol Δu is an abbreviation for the expression $\partial^2 u/\partial x^2 + \partial^2 u/\partial y^2 + \partial^2 u/\partial z^2$ and is called the *Laplacian* of the function u.

It is not difficult to see that the difference between any two particular solutions u_1 and u_2 of the Poisson equation is a function satisfying the Laplace equation, or in other words is a *harmonic function*. The entire manifold of solutions of the Poisson equation is thus reduced to the manifold of harmonic functions.

If we have been able to construct even one particular solution u_0 of the Poisson equation, and if we define a new unknown function w by

$$u = u_0 + w,$$

we see that w must satisfy the Laplace equation; and in exactly the same way, we determine the corresponding boundary conditions for w. Thus it is particularly important to investigate boundary value problems for the Laplace equation.

As is most often the case with mathematical problems, the proper statement of the problem for an equation of mathematical physics is immediately suggested by the practical situation. The supplementary conditions arising in the solution of the Laplace equation come from the physical statement of the problem.

Let us consider, for example, the establishment of a steady temperature in a medium, i.e., the propagation of heat in a medium where the sources of heat are constant and are situated either inside or outside the medium. Under these conditions, with the passage of time the temperature attained at any point of the medium will be independent of the time. Thus to find the temperature T at each point, we must find that solution of the equation

$$\frac{\partial T}{\partial t} = \Delta T + q,$$

where q is the density of the sources of heat distribution, which is independent of t. We get

$$\Delta T + q = 0.$$

Thus the temperature in our medium satisfies the Poisson equation. If the density of heat sources q is zero, then the Poisson equation becomes the Laplace equation.

In order to find the temperature inside the medium, it is necessary, from simple physical considerations, to know also what happens on the boundary of the medium.

Obviously the physical laws previously considered for interior points of a body call for quite another formulation at boundary points.

In the problem of establishing the steady-state temperature, we can prescribe either the distribution of temperature on the boundary, or the rate of flow of heat through a unit area of the surface, or finally, a law connecting the temperature with the flow of heat.

Considering the temperature in a volume Ω, bounded by the surface S, we can write these three conditions as:

$$T\,|_S = \phi(Q), \tag{10}$$

or

$$\left.\frac{\partial T}{\partial n}\right|_S = \psi(Q), \tag{10'}$$

or finally, in the most general case

$$\alpha\left.\frac{\partial T}{\partial n}\right|_S + \beta T\,|_S = \chi(Q), \tag{10''}$$

where Q denotes an arbitrary point of the surface S. Conditions of the form (10) are called *boundary conditions*. Investigation of the Laplace or Poisson equation under boundary conditions of one of these types will show that as a rule the solution is uniquely determined.

Thus, in our search for a solution of the Laplace or Poisson equation it will usually be necessary and sufficient to be given one arbitrary function on the boundary of the domain.* Let us examine the Laplace equation a little more in detail. We will show that a harmonic function u, i.e., a function satisfying the Laplace equation, is completely determined if we know its values on the boundary of the domain.

First of all we establish the fact that a harmonic function cannot take on values inside the domain that are larger than the largest value on the boundary. More precisely, we show that the absolute maximum, as well as the absolute minimum of a harmonic function are attained on the boundary of the domain.

From this it will follow at once that if a harmonic function has a constant value on the boundary of a domain Ω, then in the interior of this domain it will also be equal to this constant. For if the maximum and minimum value of a function are both the same constant, then the function will be everywhere equal to this constant.

We now establish the fact that the absolute maximum and minimum of a harmonic function cannot occur inside the domain. First of all, we note that if the Laplacian Δu of the function $u(x, y, z)$ is positive for the whole domain, then this function cannot have a maximum inside the domain, and if it is negative, then the function cannot have a minimum inside the

---

* The words "arbitrary function" here and in what follows mean that no special conditions, other than certain requirements of regularity, are imposed on the functions.

domain. For at a point where the function u attains its maximum it must have a maximum as a function of each variable separately for fixed values of the other variables. Thus it follows that every partial derivative of second order with respect to each variable must be nonpositive. This means that their sum will be nonpositive, whereas the Laplacian is positive, which is impossible. Similarly it may be shown that if the function has a minimum at some interior point, then its Laplacian cannot be negative at this point. This means that if the Laplacian is negative everywhere in the domain, then the function cannot have a minimum in this domain.

If a function is harmonic, it may always be changed by an arbitrarily small amount in such a way that it will have a positive or negative Laplacian; to this end it is sufficient to add to it the quantity

$$\pm \eta r^2 = \pm \eta(x^2 + y^2 + z^2),$$

where η is an arbitrarily small constant:

The addition of a sufficiently small quantity cannot change the property that the function has an absolute maximum or absolute minimum within the domain. If a harmonic function were to have a maximum inside the domain, then by adding $+\eta r^2$ to it, we would get a function with a positive Laplacian which, as was shown above, could not have a maximum inside the domain. This means that a harmonic function cannot have an absolute maximum inside the domain. Similarly, it can be shown that a harmonic function cannot have an absolute minimum inside the domain.

This theorem has an important corollary. Two harmonic functions that agree on the boundary of a domain must agree everywhere inside the domain. For then the difference of these functions (which itself will be a harmonic function) vanishes on the boundary of the domain and thus is everywhere equal to zero in the interior of the domain.

So we see that the values of a harmonic function on the boundary completely determine the function. It may be shown (although we cannot give the details here) that for arbitrarily preassigned values on the boundary one can always find a harmonic function that assumes these values.

It is somewhat more complicated to prove that the steady-state temperature established in a body is completely determined, if we know the rate of flow of heat through each element of the surface of the body or a law connecting the flow of heat with the temperature. We will return to some aspects of this question when we discuss methods of solving the problems of mathematical physics.

The boundary-value problem for the heat equation.   A completely different situation occurs in the problem of the heat equation in the nonstationary case. It is physically clear that the values of the temperature on the boundary or of the rate of the flow of heat through the boundary are not sufficient in themselves to define a unique solution of the problem. But if in addition we know the temperature distribution at some initial instant of time, then the problem is uniquely determined. Thus to determine the solution of the equation of heat conduction (8) it is usually necessary and sufficient to assign one arbitrary function $T_0(x, y, z)$ describing the initial distribution of temperature and also one arbitrary function on the boundary of the domain. As before, this may be either the temperature on the surface of the body, or the rate of heat flow through each element of the surface, or a law connecting the flow of heat with the temperature.

In this manner, the problem may be stated as follows. We seek a solution of equation (8) under the condition

$$T\big|_{t=0} = T_0(x, y, z) \tag{11}$$

and one of three following conditions

$$T\big|_S = \phi(Q), \tag{12}$$

$$\frac{\partial T}{\partial n}\bigg|_S = \psi(Q), \tag{12'}$$

$$\alpha \frac{\partial T}{\partial n}\bigg|_S + \beta T\big|_S = \chi(Q), \tag{12''}$$

where Q is any point of the surface S.

Condition (11) is called an *initial condition*, while conditions (12) are *boundary conditions*.

We will not prove in detail that every such problem has a unique solution but will establish this fact only for the first of these problems; moreover, we will consider only the case where there are no heat sources in the interior of the medium. We show that the equation

$$\Delta T = \frac{1}{a^2} \frac{\partial T}{\partial t}$$

under the conditions

$$T\big|_{t=0} = T_0(x, y, z),$$
$$T\big|_S = \phi(Q)$$

can have only one solution.

The proof of this statement is very similar to the previous proof for the uniqueness of the solution of the Laplace equation. We show first of all that if

$$\Delta T - \frac{1}{a^2} \frac{\partial T}{\partial t} < 0,$$

then the function T, as a function of four variables, x, y, z, and $t(0 \leqslant t \leqslant t_0)$, assumes its minimum either on the boundary of the domain Ω or else inside Ω, but in the latter case necessarily at the initial instant of time, $t = 0$.

For if not, then the minimum would be attained at some interior point. At this point all the first derivatives, including $\partial T/\partial t$, will then be equal to zero, and if this minimum were to occur for $t = t_0$, then $\partial T/\partial t$ would be nonpositive. Also, at this point all second derivatives with respect to the variables x, y, and z will be nonnegative. Consequently $\Delta T - (1/a^2)$ $(\partial T/\partial t)$ will be nonnegative, which in our case is impossible.

In exactly the same way we can establish that if $\Delta T - (1/a^2) (\partial T/\partial t) > 0$, then inside Ω for $0 < t \leqslant t_0$ there cannot exist a maximum for the function T.

Finally, if $\Delta T - (1/a^2) (\partial T/\partial t) = 0$, then inside Ω for $0 < t \leqslant t_k$ the function T cannot attain its absolute maximum nor its absolute minimum, since if the function T were to have, for example, such an absolute minimum, then by adding to it the term $\eta(t - t_0)$ and considering the function $T_1 = T + \eta(t - t_0)$, we would not destroy the absolute minimum if η were sufficiently small, and then $\Delta T_1 - (1/a^2) (\partial T_1/\partial t)$ would be negative, which is impossible.

In the same way we can also show the absence of an absolute maximum for T in the domain under consideration.

However, an absolute maximum, as well as an absolute minimum of temperature may occur either at the initial instant $t = 0$ or on the boundary S of the medium. If $T = 0$ both at the initial instant and on the boundary, then we have the identity $T = 0$ throughout the interior of the domain for all $t \leqslant t_0$. If any two temperature distributions T_1 and T_2 have identical values for $t = 0$ and on the boundary then their difference $T_1 - T_2 = T$ will satisfy the heat equation and will vanish for $t = 0$ and on the boundary. This means that $T_1 - T_2$ will be everywhere equal to zero, so that the two temperature distributions T_1 and T_2 will be everywhere identical.

In the investigation given later of methods of solving the equations of mathematical physics we will see that the value of T for $t = 0$ and the right side of one of the equations (12) may be given arbitrarily, i.e., that the solution of such a problem will exist.

The energy of oscillations and the boundary-value problem for the equation of oscillation. We now consider the conditions under which the third of the basic differential equations has a unique solution, namely equation (9).

For simplicity we will consider the equation for the vibrating string $\partial^2 u/\partial x^2 = (1/a^2)\,(\partial^2 u/\partial t^2)$, which is very similar to equation (9), differing from it only in the number of space variables. On the right side of this equation there is the quantity $\partial^2 u/\partial t^2$ expressing the acceleration of an arbitrary point of the string. The motion of any mechanical system for which the forces, and consequently the accelerations, are expressed by the coordinates of the moving bodies, is completely determined if we are given the initial positions and velocities of all the points of the system. Thus for the equation of the vibrating string, it is natural to assign the positions and velocities of all points at the initial instant.

$$u\,|_{t=0} = u_0(x)$$

$$\frac{\partial u}{\partial t}\bigg|_{t=0} = u_1(x).$$

But as was pointed out earlier, at the ends of the string the formulas expressing the laws of mechanics for interior points cease to apply. Thus at both ends we must assign supplementary conditions. If, for example, the string is fixed in a position of equilibrium at both ends, then we will have

$$u\,|_{x=0} = u\,|_{x=l} = 0.$$

These conditions can sometimes be replaced by more general ones, but a change of this sort is not of basic importance.

The problem of finding the necessary solutions of equation (9) is analogous. In order that such a solution be well defined, it is customary to assign the conditions

$$p\,|_{t=0} = \phi_0(x, y, z),$$

$$\frac{\partial p}{\partial t}\bigg|_{t=0} = \phi_1(x, y, z), \tag{13}$$

and also one of the "boundary conditions"

$$p\,|_S = \phi(Q), \tag{14}$$

$$\frac{\partial p}{\partial n}\bigg|_S = \psi(Q), \tag{14'}$$

$$\alpha\frac{\partial p}{\partial n}\bigg|_S + \beta p|_S = \chi(Q).^* \tag{14''}$$

---

* If the right-hand sides in conditions (13) and (14) are equal to zero, such conditions are called "homogeneous."

The difference from the preceding case is simply that instead of the one initial condition in equation (11) we have the two conditions (13).

Equations (14) obviously express the physical laws for the particles on the boundary of the volume in question.

The proof that in the general case the conditions (13) together with an arbitrary one of the conditions (14) uniquely define a solution of the problem will be omitted. We will show only that the solution can be unique for one of the conditions in (14).

Let it be known that a function u satisfies the equation

$$\frac{\partial^2 u}{\partial x^2} = \frac{1}{a^2}\frac{\partial^2 u}{\partial t^2},$$

with initial conditions

$$u\,|_{t=0} = 0, \frac{\partial u}{\partial t}\bigg|_{t=0} = 0$$

and boundary condition

$$\frac{\partial u}{\partial n}\bigg|_{s} = 0.$$

(It would be just as easy to discuss the case in which $u\,|_{s} = 0$.)

We will show that under these conditions the function u must be identically zero.

To prove this property it will not be sufficient to use the arguments introduced earlier to establish the uniqueness of the solution of the first two problems. But here we may make use of the physical interpretation.

We will need just one physical law, the "law of conservation of energy." We restrict ourselves again for simplicity to the vibrating string, the displacement of whose points $u(x, t)$ satisfies the equation

$$T\frac{\partial^2 u}{\partial x^2} = \rho\,\frac{\partial^2 u}{\partial t^2}.$$

The kinetic energy of each particle of the string oscillating from x to $x + dx$ is expressed in the form

$$\frac{1}{2}\left(\frac{\partial u}{\partial t}\right)^2 \rho\, dx.$$

Along with its kinetic energy, the string in its displaced position also possesses potential energy created by its increase of length in comparison with the straight-line position. Let us compute this potential energy. We concern ourselves with an element of the string between the points x and

$x + dx$. This element has an inclined position with respect to the axis Ox, such that its length is approximately equal to

$$\sqrt{(dx)^2 + \left(\frac{\partial u}{\partial x} dx\right)^2} \; ;$$

so its elongation is

$$\sqrt{1 + \left(\frac{\partial u}{\partial x}\right)^2} \, dx - dx \approx \frac{1}{2} \left(\frac{\partial u}{\partial x}\right)^2 dx.$$

Multiplying this elongation by the tension T, we find the potential energy of the elongated element of the string

$$\frac{1}{2} T \left(\frac{\partial u}{\partial x}\right)^2 dx.$$

The total energy of the string of length l is obtained by summing the kinetic and potential energies over all of the points of the string. We get

$$E = \frac{1}{2} \int_0^l \left[T \left(\frac{\partial u}{\partial x}\right)^2 + \rho \left(\frac{\partial u}{\partial t}\right)^2 \right] dx.$$

If the forces acting on the end of the string do no work, in particular if the ends of the string are fixed, then the total energy of the string must be constant.

$$E = \text{const.}$$

Our expression for the law of conservation of energy is a mathematical corollary of the basic equations of mechanics and may be derived from them. Since we have already written the laws of motion in the form of the differential equation of the vibrating string with conditions on the ends, we can give the following mathematical proof of the law of conservation of energy in this case. If we differentiate E with respect to time, we have, from basic general rules,

$$\frac{dE}{dt} = \int_0^l \left(T \frac{\partial u}{\partial x} \frac{\partial^2 u}{\partial x \, \partial t} + \rho \frac{\partial u}{\partial t} \frac{\partial^2 u}{\partial t^2} \right) dx.$$

Using the wave equation (6) and replacing $\rho(\partial^2 u/\partial t^2)$ by $T(\partial^2 u/\partial x^2)$, we get dE/dt in the form

$$\frac{dE}{dt} = \int_0^l T \left[\left(\frac{\partial u}{\partial x} \frac{\partial^2 u}{\partial x \, \partial t}\right) + \frac{\partial u}{\partial t} \frac{\partial^2 u}{\partial x^2} \right] dx$$

$$= \int_0^l T \frac{\partial}{\partial x} \left(\frac{\partial u}{\partial x} \frac{\partial u}{\partial t}\right) dx = T \frac{\partial u}{\partial x} \frac{\partial u}{\partial t} \Big|_{x=l} - T \frac{\partial u}{\partial x} \frac{\partial u}{\partial t} \Big|_{x=0}.$$

If $(\partial u/\partial x)\,|_{x=0}$ or $u\,|_{x=0}$ vanishes, and also $(\partial u/\partial x)\,|_{x=l}$ or $u\,|_{x=l}$ vanishes, then

$$\frac{dE}{dt} = 0,$$

which shows that E is constant.

The wave equation (9) may be treated in exactly the same way to prove that the law of conservation of energy holds here also. If p satisfies equation (9) and the condition

$$p\,|_S = 0 \quad \text{or} \quad \left.\frac{\partial p}{\partial n}\right|_S = 0,$$

then the quantity

$$E = \iiint \left[\left(\frac{\partial p}{\partial x}\right)^2 + \left(\frac{\partial p}{\partial y}\right)^2 + \left(\frac{\partial p}{\partial z}\right)^2 + \frac{1}{a^2}\left(\frac{\partial p}{\partial t}\right)^2\right] dx\, dy\, dz$$

will not depend on t.

If, at the initial instant of time, the total energy of the oscillations is equal to zero, then it will always remain equal to zero, and this is possible only in the case that no motion occurs. If the problem of integrating the wave equation with initial and boundary conditions had two solutions p_1 and p_2, then $v = p_1 - p_2$ would be a solution of the wave equation satisfying the conditions with zero on the right-hand side, i.e., homogeneous conditions.

In this case, when we calculated the "energy" of such an oscillation, described by the function v, we would discover that the energy $E(v)$ is equal to zero at the initial instant of time. This means that it is always equal to zero and thus that the function v is identically equal to zero, so that the two solutions p_1 and p_2 are identical. Thus the solution of the problem is unique.

In this way we have convinced ourselves that all three problems are correctly posed.

Incidentally, we have been able to discover some very simple properties of the solutions of these equations. For example, solutions of the Laplace equation have the following maximum property: Functions satisfying this equation have their largest and smallest values on the boundaries of their domains of definition.

Functions describing the distribution of heat in a medium have a maximum property of a different form. Every maximum or minimum of temperature occuring at any point gradually disperses and decreases with time. The temperature at any point can rise or fall only if it is lower or higher than at nearby points. The temperature is smoothed out with the passage of time. All unevennesses in it are leveled out by the passage of heat from hot places to cold ones.

But no smoothing-out process of this kind occurs in the propagation of the oscillations considered here. These oscillations do not decrease or level out, since the sum of their kinetic and potential energies must remain constant for all time.

§4. The Propagation of Waves

The properties of oscillations can be very clearly demonstrated by the simplest examples. Let us consider two characteristic cases.

Our first example is the equation of the vibrating string

$$\frac{\partial^2 u}{\partial x^2} = \frac{1}{a^2} \frac{\partial^2 u}{\partial t^2}. \tag{15}$$

This equation, as may be proved, has two particular solutions of the form

$$u_1 = \phi_1(x - at), \quad u_2 = \phi_2(x + at),$$

where ϕ_1 and ϕ_2 are arbitrary twice-differentiable functions.

By direct differentiation it is easy to show that the functions u_1 and u_2 satisfy equation (15). It may be shown that

$$u = u_1 + u_2$$

is a general solution of this equation.

The general form of the oscillations described by the functions u_1 and u_2 is of considerable interest. To consider it in the most convenient fashion, we mentally carry out the following experiment. Let the observer of the vibrating string himself be not stationary but moving along the axis Ox with velocity a. For such an observer the position of a point on the string will be defined not by a stationary coordinate system but by a moving one. Let ξ denote the x-coordinate of this system. Then $\xi = 0$ will obviously correspond at each instant of time to the value $x = at$. Hence it is clear that

$$\xi = x - at.$$

We can represent an arbitrary function $u(x, t)$ in the form

$$u(x, t) = \phi(\xi, t).$$

For the solution u_1 we will have

$$u_1(x, t) = \phi_1(\xi),$$

so that in this coordinate system the solution $u_1(x, t)$ turns out to be independent of time. Consequently, for an observer moving with velocity a, the string looks like a stationary curve. For a stationary observer, however, the string appears to have a wave flowing along the axis Ox with velocity a.

In exactly the same way the solution $u_2(x, t)$ may be considered as a wave travelling in the opposite direction with velocity a. With an infinite string both waves will be propagated infinitely far. Moving in different directions they may, by their superposition, produce quite strange shapes in the string. The resultant displacement may be increasing at certain times and decreasing at others.

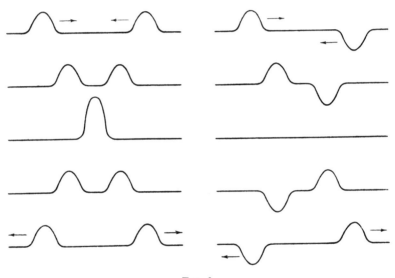

FIG. 3.

If u_1 and u_2, as they arrive at a given point from opposite sides, have the same sign, then they augment each other, but if they have opposite signs, they counteract each other. Figure 3 shows several successive positions of the string for two particular displacements. Initially the waves move independently toward each other, and then begin to interact. In the second case in figure 3 there will be an instant of complete annihilation of the oscillations, after which the waves again separate.

Another example that easily lends itself to qualitative investigation is the propagation of waves in space.

The equation

$$\Delta u = \frac{1}{a^2} \frac{\partial^2 u}{\partial t^2} \tag{16}$$

derived earlier, has two particular solutions of the form

$$u_1 = \frac{1}{r}\phi_1(r - at), \quad u_2 = \frac{1}{r}\phi_2(r + at), \tag{17}$$

where r denotes the distance of a given point from the origin of the coordinate system $r^2 = x^2 + y^2 + z^2$, and ϕ_1 and ϕ_2 are arbitrary, twice-differentiable functions.

The proof that u_1 and u_2 are solutions would take considerable time and is omitted here.

The form of the waves described by these solutions is in general the same as for the string. If we pay no attention to the factor $1/r$ occuring on the right, then the first solution represents a wave travelling in the direction of increasing r. This wave is spherically symmetric; it is identical at all points that have the same value of r.

The factor $1/r$ produces the result that the amplitude of the wave is inversely proportional to the distance from the origin. Such an oscillation is called a diverging spherical wave. A good picture of it is given by the circles that spread out over the surface of the water when a stone is thrown into it, except that in this case the waves are circular rather than spherical.

This second solution of (17) is also of great interest; it is called a converging wave, travelling in the direction of the origin. Its amplitude grows with time to infinity as it approaches the origin. We see that such a concentration of the disturbance at one point may lead, even though the initial oscillations are small, to an immense upheaval.

§5. Methods of Constructing Solutions

On the possibility of decomposing any solution into simpler solutions. Solutions of the problems of mathematical physics formulated previously may be derived by various devices, which are different specific problems. But at the basis of these methods there is one general idea. As we have seen, all the equations of mathematical physics are, for small values of the unknown functions, linear with respect to the functions and their derivatives. The boundary conditions and initial conditions are also linear.

If we form the difference between any two solutions of the same equation, this difference will also be a solution of the equation with the right-hand terms equal to zero. Such an equation is called the corresponding homogeneous equation. For example, for the Poisson equation $\Delta u = -4\pi\rho$, the corresponding homogeneous equation is the Laplace equation $\Delta u = 0$.

If two solutions of the same equation also satisfy the same boundary conditions, then their difference will satisfy the corresponding homogeneous condition: The values of the corresponding expression on the boundary will be equal to zero.

Hence the entire manifold of the solutions of such an equation, for given boundary conditions, may be found by taking any particular solution that satisfies the given nonhomogeneous condition together with all possible solutions of the homogeneous equation satisfying homogeneous boundary conditions (but not, in general, satisfying the initial conditions).

Solutions of homogeneous equations, satisfying homogeneous boundary conditions may be added, or multiplied by constants, without ceasing to be solutions.

If a solution of a homogeneous equation with homogeneous conditions is a function of some parameter, then integrating with respect to this parameter will also give us such a solution. These facts form the basis of the most important method of solving linear problems of all kinds for the equations of mathematical physics, the method of superposition.

The solution of the problem is sought in the form

$$u = u_0 + \sum_k u_k,$$

where u_o is a particular solution of the equation satisfying the boundary conditions but not satisfying the initial conditions, and the u_k are solutions of the corresponding homogeneous equation satisfying the corresponding homogeneous boundary conditions. If the equation and the boundary conditions were originally homogeneous, then the solution of the problem may be sought in the form

$$u = \sum u_k.$$

In order to be able to satisfy arbitrary initial conditions by the choice of particular solutions u_k of the homogeneous equation, we must have available a sufficiently large arsenal of such solutions.

The method of separation of variables.   For the construction of the necessary arsenal of solutions there exists a method called *separation of variables* or *Fourier's method*.

Let us examine this method, for example, for solving the problem

$$\Delta u = \frac{\partial^2 u}{\partial t^2}, \tag{18}$$

$$u\mid_S = 0, \quad u\mid_{t=0} = f_0(x, y, z), \quad u_t\mid_{t=0} = f_1(x, y, z).$$

In looking for any particular solution of the equation, we first of all assume that the desired function u satisfies the boundary condition $u\mid_S = 0$ and can be expressed as the product of two functions, one of which depends only on the time t and the other only on the space variables:

$$u(x, y, z, t) = U(x, y, z)\, T(t).$$

Substituting this assumed solution into our equation, we have

$$T(t)\, \Delta U = T''(t)\, U.$$

Dividing both sides by TU gives

$$\frac{T''}{T} = \frac{\Delta U}{U}.$$

The right side of this equation is a function of the space variables only and the left is independent of the space coordinates. Hence it follows that the given equation can be true only if the left and right sides have the same constant value. We are led to a system of two equations

$$\frac{T''}{T} = -\lambda_k^2, \quad \frac{\Delta U}{U} = -\lambda_k^2.$$

The constant quantity on the right is denoted here by $-\lambda_k^2$ in order to emphasize that it is negative (as may be rigorously proved). The subscript k is used here to note that there exist infinitely many possible values of $-\lambda_k^2$, where the solutions corresponding to them form a system of functions complete in a well-known sense.

Cross-multiplying in both equations, we get

$$T'' + \lambda_k^2 T = 0; \quad \Delta U + \lambda_k^2 U = 0.$$

The first of these equations has, as we know, the simple solution

$$T = A_k \cos \lambda_k t + B_k \sin \lambda_k t,$$

where A_k and B_k are arbitrary constants. This solution may be further simplified by introducing the auxiliary angle ϕ. We have

$$\frac{A_k}{\sqrt{A_k^2 + B_k^2}} = \sin \phi_k, \quad \frac{B_k}{\sqrt{A_k^2 + B_k^2}} = \cos \phi_k, \quad \sqrt{A_k^2 + B_k^2} = M_k.$$

Then

$$T = \sqrt{A_k^2 + B_k^2} \sin (\lambda_k t + \phi_k) = M_k \sin (\lambda_k t + \phi_k).$$

The function T represents a harmonic oscillation with frequency λ_k, shifted in phase by the angle ϕ_k.

More difficult and more interesting is the problem of finding a solution of the equation

$$\Delta U + \lambda_k^2 U = 0 \qquad (19)$$

for given homogeneous boundary conditions; for example, for the conditions

$$U \mid_S = 0$$

(where S is the boundary of the volume Ω under consideration), or for any other homogeneous condition. The solution of this problem is not always easy to construct as a finite combination of known functions, although it always exists and can be found to any desired degree of accuracy.

The equation $\Delta U + \lambda_k^2 U = 0$ for the condition $U \mid_S = 0$ has first of all the obvious solution $U \equiv 0$. This solution is trivial and completely useless for our purposes. If the λ_k are any randomly chosen numbers, then in general there will not be any other solution to our problem. However, there usually exist values of λ_k for which the equation does have a nontrivial solution.

All possible values of the constant λ_k^2 are determined by the requirement that equation (19) have a nontrivial solution, i.e., distinct from the identically vanishing function, which satisfies the condition $U \mid_S = 0$. From this it also follows that the numbers denoted by $-\lambda_k^2$ must be negative.

For each of the possible values of λ_k in equation (19), we can find at least one function U_k. This allows us to construct a particular solution of the wave equation (18) in the form

$$u_k = M_k \sin (\lambda_k t + \phi_k)\, U_k(x, y, z).$$

Such a solution is called a *characteristic oscillation* (or *eigenvibration*) of the volume under consideration. The constant λ_k is the frequency of the characteristic oscillation, and the function $U_k(x, y, z)$ gives us its form. This function is usually called an *eigenfunction* (*characteristic function*). For all instants of time, the function u_k, considered as a function of the variables x, y, and z, will differ from the function $U_k(x, y, z)$ only in scale.

We do not have space here for a detailed proof of the many remarkable properties of characteristic oscillations and of eigenfunctions; therefore we will restrict ourselves merely to listing some of them.

The first property of the characteristic oscillations consists of the fact that for any given volume there exists a countable set of characteristic frequencies. These frequencies tend to infinity with increasing k.

Another property of the characteristic oscillations is called *orthogonality*. It consists of the fact that the integral over the domain Ω of the product of eigenfunctions corresponding to different values of λ_k is equal to zero.*

$$\iiint\limits_{\Omega} U_k(x, y, z)\, U_j(x, y, z)\, dx\, dy\, dz = 0 \ (j \neq k).$$

For $j = k$ we will assume

$$\iiint\limits_{\Omega} U_k(x, y, z)^2\, dx\, dy\, dz = 1.$$

This can always be arranged by multiplying the functions $U_k(x, y, z)$ by an appropriate constant, the choice of which does not change the fact that the function satisfies equation (19) and the condition $U|_S = 0$.

Finally, a third property of the characteristic oscillations consists of the fact that, if we do not omit any value of λ_k, then by means of the eigenfunctions $U_k(x, y, z)$, we can represent with any desired degree of exactness a completely arbitrary function $f(x, y, z)$, provided only that it satisfies the boundary condition $f|_S = 0$ and has continuous first and second derivatives. Any such function $f(x, y, z)$ may be represented by the convergent series

$$f(x, y, z) = \sum_{k=1}^{\infty} C_k U_k(x, y, z). \tag{20}$$

The third property of the eigenfunctions provides us in principle with the possibility of representing any function $f(x, y, z)$ in a series of eigenfunctions of our problem, and from the second property we can find all

---

* If to one and the same value of λ there correspond several essentially different (linearly independent) functions U, then this value of λ is considered as occurring a corresponding number of times in the set of eigenvalues λ_k. The condition of orthogonality for functions corresponding to the same value of λ_k may be ensured by proper choice of these functions.

the coefficients of this series. In fact, if we multiply both sides of equation (20) by $U_j(x, y, z)$ and integrate over the domain Ω, we get

$$\iiint\limits_{\Omega} f(x, y, z)\, U_j(x, y, z)\, dx\, dy\, dz$$

$$= \sum_{k=1}^{\infty} C_k \iiint U_k(x, y, z)\, U_j(x, y, z)\, dx\, dy\, dz.$$

In the sum on the right, all the terms in which $k \neq j$ disappear because of the orthogonality, and the coefficient of C_j is equal to one. Consequently we have

$$C_j = \iiint\limits_{\Omega} f(x, y, z)\, U_j(x, y, z)\, dx\, dy\, dz.$$

These properties of the characteristic oscillations now allow us to solve the general problem of oscillation for any initial conditions.

For this we assume that we have a solution of the problem in the form

$$u = \sum U_k(x, y, z)\, (A_k \cos \lambda_k t + B_k \sin \lambda_k t) \tag{21}$$

and try to choose the constants A_k and B_k so that we have

$$u \big|_{t=0} = f_0(x, y, z),$$

$$\frac{\partial u}{\partial t}\bigg|_{t=0} = f_1(x, y, z).$$

Putting $t = 0$ in the right side of (21), we see that the sine terms disappear and $\cos \lambda_k t$ becomes equal to one, so that we will have

$$f_0(x, y, z) = \sum_{k=1}^{\infty} A_k U_k(x, y, z).$$

From the third property, the characteristic oscillations can be used for such a representation, and from the second property, we have

$$A_k = \iiint\limits_{\Omega} f_0(x, y, z)\, U_k(x, y, z)\, dx\, dy\, dz.$$

In the same way, differentiating formula (21) with respect to t and putting $t = 0$, we will have

$$\frac{\partial u}{\partial t}\bigg|_{t=0} = f_1(x, y, z) = \sum_{k=1}^{\infty} \lambda_k (B_k \cos \lambda_k t - A_k \sin \lambda_k t) \big|_{t=0} U_k(x, y, z)$$

$$= \sum_{k=1}^{\infty} \lambda_k B_k U_k(x, y, z).$$

Hence, as before, we obtain the values of B_k as

$$B_k = \frac{1}{\lambda_k} \iiint_\Omega f_1(x, y, z)\, U_k(x, y, z)\, dx\, dy\, dz.$$

Knowing A_k and B_k, we in fact know both the phases and the amplitudes of all the characteristic oscillations.

In this way we have shown that by addition of characteristic oscillations it is possible to obtain the most general solution of the problem with homogeneous boundary conditions.

Every solution thus consists of characteristic oscillations, whose amplitude and phase we can calculate if we know the initial conditions.

In exactly the same way, we may study oscillations with a smaller number of independent variables. As an example let us consider the vibrating string, fixed at both ends. The equation of the vibrating string has the form

$$\frac{\partial^2 u}{\partial t^2} = a^2 \frac{\partial^2 u}{\partial x^2}.$$

Let us suppose that we are looking for a solution of the problem for a string of length l, fixed at the ends

$$u\,|_{x=0} = u\,|_{x=l} = 0.$$

We will look for a collection of particular solutions

$$u_k = T_k(t)\, U_k(x).$$

We obviously obtain, just as before,

$$T_k'' U_k = a^2 U_k'' T_k\,,$$

or

$$\frac{T_k''}{T_k} = a^2 \frac{U_k''}{U_k} = -\lambda_k^2\,.$$

Hence

$$T_k = A_k \cos \lambda_k t + B_k \sin \lambda_k t,$$

$$U_k = M_k \cos \frac{\lambda_k}{a} x + N_k \sin \frac{\lambda_k}{a} x.$$

We use the boundary conditions in order to find the values of λ_k. For general λ_k it is not possible to satisfy both the boundary conditions. From the condition $U_k|_{x=0} = 0$ we get $M_k = 0$, and this means that $U_k = N_k$ sin $(\lambda_k/a)\, x$. Putting $x = l$, we get sin $(\lambda_k l/a) = 0$. This can only happen if $\lambda_k l/a = k\pi$, where k is an integer. This means that

$$\lambda_k = \frac{ak\pi}{l}.$$

The condition $\int_0^l U_k^2\, dx = 1$ shows that $N_k = \sqrt{\dfrac{2}{l}}$.   Finally

$$U_k(x) = \sqrt{\frac{2}{l}} \sin \frac{k\pi x}{l}, \quad T_k = A_k \cos \frac{ak\pi t}{l} \ B_k \sin \frac{ak\pi t}{l}.$$

In this manner the characteristic oscillations of the string, as we see, have sinusoidal form with an integral number of half waves on the entire string. Every oscillation has its own frequency, and the frequencies may be arranged in increasing order

$$\frac{a\pi}{l}, 2\frac{a\pi}{l}, 3\frac{a\pi}{l}, \cdots, k\frac{a\pi}{l}, \cdots.$$

It is well known that these frequencies are exactly those that we hear in the vibrations of a sounding string. The frequency is called the *fundamental frequency*, and the remaining frequencies are *overtones*. The eigenfunctions $\sqrt{2/l} \sin (k\pi x/l)$ on the interval $0 \leqslant x \leqslant l$ change sign $k - 1$ times, since $k\pi x/l$ runs through values from 0 to $k\pi$, which means that its sine changes sign $k - 1$ times. The points where the eigenfunctions U_k vanish are called *nodes* of the oscillations.

If we arrange in some way that the string does not move at a point corresponding to a node, for example of the first overtone, then the fundamental tone will be suppressed, and we will hear only the sound of the first overtone, which is an octave higher. Such a device, called stopping, is made use of on instruments played with a bow: the violin, viola, and violoncello.

We have analyzed the method of separating variables as applied to the problem of finding characteristic oscillations. But the method can be applied much more widely, to problems of heat flow and to a whole series of other problems.

For the equation of heat flow

$$\Delta T = \frac{\partial T}{\partial t}$$

with the condition

$$T\,|_S = 0$$

we will have, as before,

$$T = \Sigma \, F_k(t) \, U_k(x, y, z).$$

Here

$$\frac{F'_k(t)}{F_k(t)} = -\lambda_k^2, \quad \Delta U_k + \lambda_k^2 U_k = 0.$$

The solution is obtained in the form

$$T = \sum_{k=1}^{\infty} e^{-\lambda_k^2 t} \, U_k(x, y, z).$$

This method has also been used with great success to solve some other equations. Consider, for example, the Laplace equation

$$\Delta u = 0$$

in the circle

$$x^2 + y^2 \leqslant 1,$$

and assume that we have to construct a solution satisfying the condition

$$u \mid_{r=1} = f(\vartheta),$$

where r and ϑ denote the polar coordinates of a point in the plane.

The Laplace equation may be easily transformed into polar coordinates. It then has the form

$$\frac{\partial^2 u}{\partial r^2} + \frac{1}{r} \frac{\partial u}{\partial r} + \frac{1}{r^2} \frac{\partial^2 u}{\partial \vartheta^2} = 0.$$

We want to find a solution of this equation in the form

$$u = \sum_{k=1}^{\infty} R_k(r) \, \theta_k(\vartheta).$$

If we require that every term of the series individually satisfy the equation, we have

$$\left[R''_k(r) + \frac{1}{r} R'_k(r) \right] \theta_k(\vartheta) + \frac{1}{r^2} \theta''_k(\vartheta) \, R_k(r) = 0.$$

Dividing the equation by $R_k(r) \, \theta_k(\vartheta)/r^2$, we get

$$\frac{r^2 \left[R''_k(r) + \dfrac{1}{r} R'_k(r) \right]}{R_k(r)} = -\frac{\theta'_k(\vartheta)}{\theta_k(\vartheta)}.$$

Again setting

$$\frac{\theta_k''(\vartheta)}{\theta_k(\vartheta)} = -\lambda_k^2 ,$$

we have

$$r^2 \left[R_k'' + \frac{1}{r} R_k' \right] - \lambda_k^2 R_k = 0.$$

It is easy to see that the function $\theta_k(\vartheta)$ must be a periodic function of ϑ with period 2π. Integrating the equation $\theta_k''(\vartheta) + \lambda_k^2 \theta_k(\vartheta) = 0$, we get

$$\theta_k = a_k \cos \lambda_k \vartheta + b_k \sin \lambda_k \vartheta.$$

This function will be periodic with the required period only if λ_k is an integer. Putting $\lambda_k = k$, we have

$$\theta_k = a_k \cos k\vartheta + b_k \sin k\vartheta.$$

The equation for R_k has a general solution of the form

$$R_k = Ar^k + \frac{B}{r^k} .$$

Retaining only the term that is bounded for $r \to 0$, we get the general solution of the Laplace equation in the form

$$u = a_0 + \sum_{k=1}^{\infty} (a_k \cos k\vartheta + b_k \sin k\vartheta) \, r^k.$$

This method may often be used to find nontrivial solutions of the equation $\Delta U_k + \lambda_k^2 U_k = 0$ that satisfy homogeneous boundary conditions. In case the problem can be reduced to problems of solving ordinary differential equations, we say that it allows a complete separation of variables. This complete separation of variables by the Fourier method can be carried out, as was shown by the Soviet mathematician V. V. Stepanov, only in certain special cases. The method of separation of variables was known to mathematicians a long time ago. It was used essentially by Euler, Bernoulli, and d'Alembert. Fourier used it systematically for the solution of problems of mathematical physics, particularly in heat conduction. However, as we have mentioned, this method is often inapplicable; we must use other methods, which we will now discuss.

The method of potentials.   The essential feature of this method is, as before, the superposition of particular solutions for the construction of a solution in general form. But this time for the particular fundamental solutions, we use functions that become infinite at one point. Let us illustrate with the Laplace and Poisson equations.

Let M_0 be a point of our space. We denote by $r(M, M_0)$ the distance from the point M_0 to a variable point M. The function $1/r(M, M_0)$ for a fixed M_0 is a function of the variable point M. It is easy to establish the fact that this function is a harmonic function of the point M in the entire space,* except of course, at the point M_0, where the function becomes infinite, together with its derivatives.

The sum of several functions of this form

$$\sum_{i=1}^{N} A_i \frac{1}{r(M, M_i)},$$

where the points M_1, M_2, $\cdots$, M_N are any points in the space, is again a harmonic function of the point M. This function will have singularities at all the points M_i. If we choose the points M_1, M_2, $\cdots$, M_N as densely distributed as we please in some volume Ω, and at the same time multiply by coefficients A_i, we may pass to the limit in this expression and get a new function

$$U = \lim \sum_{i=1}^{N} \frac{A_i}{r(M, M_i)} = \iiint_{\Omega} \frac{A(M')}{r(M, M')} \, d\Omega,$$

where the points M' range over all of the volume Ω. The integral in this form is called a *Newtonian potential*. It may be shown, although we will not do it here, that the function U thus constructed satisfies the equation $\Delta U = -4\pi A$.

The Newtonian potential has a simple physical meaning. To understand it, we will begin with the function $A_i/r(M, M_i)$.

The partial derivatives of this function with respect to the coordinates are

$$A_i \frac{x_i - x}{r^3} = X, \quad A_i \frac{y_i - y}{r^3} = Y, \quad A_i \frac{z_i - z}{r^3} = Z.$$

At the point M_i we place a mass A_i, which will attract all bodies with a force directed toward the point M_i and inversely proportional to the square of the distance from M_i. We decompose this force into its components along the coordinate axes. If the magnitude of the force acting on a material point of unit mass is A_i/r^2, the cosines of the angles between the direction of this force and the coordinate axis will be $(x_i - x)/r$, $(y_i - y)/r$, $(z_i - z)/r$. Thus the components of the force exerted on a unit mass at the point M by an attracting center M_i will be equal to X, Y, and Z, the partial derivatives of the function A_i/r with respect to the coordinates. If

---

* That is, the function satisfies the Laplace equation.

we place attracting masses at points M_1, M_2, $\cdots$, M_N, then every material point with unit mass placed at a point M will be acted on by a force equal to the resultant of all the forces acting on it from the given points M_i. In other words

$$X = \frac{\partial}{\partial x} \sum \frac{A_i}{r(M, M_i)}, \quad Y = \frac{\partial}{\partial y} \sum \frac{A_i}{r(M, M_i)}, \quad Z = \frac{\partial}{\partial z} \sum \frac{A_i}{r(M, M_i)}.$$

Passing to the limit and replacing the sum by an integral, we get

$$\bar{X} = \frac{\partial U}{\partial x}, \quad \bar{Y} = \frac{\partial U}{\partial y}, \quad \bar{Z} = \frac{\partial U}{\partial z}, \quad \text{where} \quad U = \iiint_{\Omega} \frac{A}{r} d\Omega.$$

The function U, with partial derivatives equal to the components of the force acting on a point, is called the *potential* of the force. Thus the function $A_i/r(M, M_i)$ is the potential of the attraction exerted by the point M_i, the function $\sum [A_i/r(M, M_i)]$ is the potential of the attraction exerted by the group of points M_1, M_2, $\cdots$, M_N, and the function $U = \iiint_{\Omega} (A/r) d\Omega$ is the potential of the attraction exerted by the masses continuously distributed in the volume Ω.

Instead of distributing the masses in a volume, we may place the points M_1, M_2, $\cdots$, M_N on a surface S. Again increasing the number of these points, we get in the limit the integral

$$V = \iint_{S} \frac{A(Q)}{r} ds, \tag{22}$$

where Q is a point on the surface S.

It is not difficult to see that this function will be harmonic everywhere inside and outside the surface S. On the surface itself the function is continuous, as can be proved, although its partial derivatives of the first order have finite discontinuities.

The functions $\partial(1/r)/\partial x_i$, $\partial(1/r)/\partial y_i$, and $\partial(1/r)/\partial z_i$ also are harmonic functions of the point M for fixed M_i. From these functions in turn, we may form the sums

$$\sum A_i \frac{\partial \frac{1}{r}}{\partial x_i} + \sum B_i \frac{\partial \frac{1}{r}}{\partial y_i} + \sum C_i \frac{\partial \frac{1}{r}}{\partial z_i},$$

which will be harmonic functions everywhere except perhaps at the points M_1, M_2, $\cdots$, M_N.

Of particular importance is the integral

$$W = \iint\limits_S \mu(Q) \left[\frac{\partial \frac{1}{r}}{\partial x'} \cos(\boldsymbol{n}, \boldsymbol{x}) + \frac{\partial \frac{1}{r}}{\partial y'} \cos(\boldsymbol{n}, \boldsymbol{y}) + \frac{\partial \frac{1}{r}}{\partial z'} \cos(\boldsymbol{n}, \boldsymbol{z}) \right] ds$$

$$= \iint\limits_S \mu(Q) \, K(Q, M) \, ds, \tag{23}$$

in which x', y', and z' are the coordinates of a variable point Q on the surface S, $\boldsymbol{n}$ is the direction of the normal to the surface S at the point Q while x, y, and z are the directions of the coordinate axes, and r is the distance from Q to the point M at which the value of the function W is defined.

The integral (22) is called the *potential of a simple layer*, and the integral (23) the *potential of a double layer*.* The potential of a double layer and the potential of a simple layer represent a function harmonic inside and outside of the surface S.

Many problems in the theory of harmonic functions may be solved by using potentials. By using the potential of a double layer, we may solve the problem of constructing, in a given domain, a harmonic function u, having given values $2\pi\phi(Q)$ on the boundary S of the domain. In order to construct such a function, we only need to choose the function $\mu(Q)$ in a suitable way.

This problem is somewhat reminiscent of the similar problem of finding the coefficients in the series

$$\phi = \sum a_k U_k$$

so that it may represent the function on the left side.

A remarkable property of the integral W consists of the fact that its limiting value as the point M approaches Q_o from the inner side of the surface has the form

$$\lim_{M \to Q_0} W = 2\pi\mu(Q_0) + \iint\limits_S K(Q, Q_0)\,\mu(Q)\,ds.$$

---

* The names of these potentials are connected with the following physical fact. We assume that on the surface S, we have introduced electrical charges. They create in the space an electric field. The potential of this field will be represented by the integral (22), which is therefore called the potential of a simple layer.

We now assume that the surface S is a thin nonconducting film. On one side of it we distribute, according to some law, electric charges of one sign (for example, positive). On the other side of S we distribute, with the same law, electric charges of opposite sign. The action of these two electric layers also generates in the space an electric field. As can be calculated, the potential of this field will be represented by the integral (23).

Equating this expression to the given function $2\pi\phi(Q_0)$, we get the equation

$$\mu(Q_0) + \frac{1}{2\pi} \iint_S K(Q, Q_0)\, \mu(Q)\, ds = \phi(Q_0).$$

This equation is called an *integral equation of the second kind*. The theory of such equations has been developed by many mathematicians. If we can solve this equation by any method, we obtain a solution of our original problem.

In exactly the same way, we may find a solution of other problems in the theory of harmonic functions. After choice of a suitable potential, the density, i.e., the value of an arbitrary function appearing in it, is defined in such a way that all the prescribed conditions are fulfilled.

From a physical point of view, this means that every harmonic function may be represented as the potential of a double electric layer, if we distribute this layer over a surface S with appropriate density.

Approximate construction of solutions; Galerkin's method and the method of nets.   1. We have discussed two methods for solving equations of mathematical physics: the method of complete separation of variables and the method of potentials. These methods were developed by scientists of the 18th and 19th centuries, Fourier, Poisson, Ostrogradskiĭ, Ljapunov, and others. In the 20th century they were augmented by a series of other methods. We will examine two of them, Galerkin's method and the method of finite differences, or the method of nets.

The first method was proposed by the Academician B. G. Galerkin for the solution of equations of the form

$$\sum \sum \sum \sum A_{ijkl} \frac{\partial^4 U}{\partial x_i\, \partial x_j\, \partial x_k\, \partial x_l} + \sum \sum \sum B_{ijk} \frac{\partial^3 U}{\partial x_i\, \partial x_j\, \partial x_k}$$
$$+ \sum \sum C_{ij} \frac{\partial^2 U}{\partial x_i\, \partial x_j} + \sum D_i \frac{\partial U}{\partial x_i} + EU + \lambda U = 0,$$

containing an unknown parameter λ, where the indices i, j, k, and l independently take on the values 1, 2, and 3. These equations are derived from equations containing an independent variable t, by using the method of separation of variables in the same way as the wave equation

$$\Delta u = \frac{\partial^2 u}{\partial t^2}$$

leads to the equation $\Delta U + \lambda^2 U = 0$. The problem consists of finding those values of λ for which the homogeneous boundary-value problem has a nonzero solution and then constructing that solution.

The essence of Galerkin's method is as follows. The unknown function is sought in the approximate form

$$U \approx \sum_{m=1}^{N} a_m \omega_m(x_1, x_2, x_3),$$

where the $\omega_m(x_1, x_2, x_3)$ are arbitrary functions satisfying the boundary conditions.

The assumed solution is substituted in the left side of the equation, resulting in the approximate equation

$$\sum_{m=1}^{N} a_m \left[\sum \sum \sum \sum A_{ijkl} \frac{\partial^4 \omega_m}{\partial x_i \, \partial x_j \, \partial x_k \, \partial x_l} + \sum \sum \sum B_{ijk} \frac{\partial^3 \omega_m}{\partial x_i \, \partial x_j \, \partial x_k} \right.$$

$$\left. + \sum \sum C_{ij} \frac{\partial^2 \omega_m}{\partial x_i \, \partial x_j} + \sum D_i \frac{\partial \omega_m}{\partial x_i} + E \omega_m \right] + \lambda \sum_{m=1}^{N} a_m \omega_m \approx 0.$$

For brevity we denote the expression inside the brackets by $L\omega_m$, and write the equation in the form

$$\sum a_m L \omega_m + \lambda \sum a_m \omega_m \approx 0.$$

Now we multiply both sides of our approximate equation by ω_n and integrate over the domain Ω in which the solution is sought. We get

$$\iiint_{\Omega} \sum a_m \omega_n L \omega_m \, d\Omega + \lambda \iiint_{\Omega} \sum a_m \omega_m \omega_n \, d\Omega \approx 0,$$

which may be rewritten in the form

$$\sum_{m=1}^{N} a_m \iiint_{\Omega} \omega_n L \omega_m \, d\Omega + \lambda \sum_{m=1}^{N} a_m \iiint_{\Omega} \omega_m \omega_n \, d\Omega \approx 0.$$

If we set ourselves the aim of satisfying these equations exactly, we will have a system of algebraic equations of the first degree for the unknown coefficients a_m. The number of equations in the system will be equal to the number of unknowns, so that this system will have a nonvanishing solution only if its determinant is zero. If this determinant is expanded, we get an equation of the Nth degree for the unknown number λ.

After finding the value of λ and substituting it in the system, we solve this system to obtain approximate expressions of the function U.

Galerkin's method is not only suitable for equations of the fourth order, but may be applied to equations of different orders and different types.

2. The last of the methods that we will examine is called the method of finite differences or the method of nets.

The derivative of the function u with respect to the variable x is defined as the limit of the quotient

$$\frac{u(x + \Delta x) - u(x)}{\Delta x}.$$

This quotient in its turn may be represented in the form

$$\frac{1}{\Delta x} \int_x^{x+\Delta x} \frac{\partial u}{\partial x_1} \, dx_1 \, ,$$

and from the well-known theorem of the mean value (cf. Chapter II, §8):

$$\frac{u(x + \Delta x) - u(x)}{\Delta x} = \frac{\partial u}{\partial x}\bigg|_{x=\xi} \, ,$$

where ξ is a point in the interval

$$x < \xi < x + \Delta x.$$

All the second derivatives of u, both the mixed derivatives and the derivatives with respect to one variable, may also be approximately represented in the form of difference quotients. Thus the difference quotient

$$\frac{u(x + \Delta x) - 2u(x) + u(x - \Delta x)}{(\Delta x)^2}$$

is represented in the form

$$\frac{1}{\Delta x}\left[\frac{u(x + \Delta x) - u(x)}{\Delta x} - \frac{u(x) - u(x - \Delta x)}{\Delta x}\right]$$
$$= \frac{1}{\Delta x}\left\{\left[\frac{u(x_1 + \Delta x) - u(x_1)}{\Delta x}\right]\bigg|_{x_1=x-\Delta x}^{x_1=x}\right\}.$$

From the mean-value theorem the difference quotient of the function

$$\phi(x_1) = \frac{u(x_1 + \Delta x) - u(x_1)}{\Delta x}$$

may be replaced by the value of the derivative. Consequently

$$\frac{\phi(x_1) - \phi(x_1 - \Delta x)}{\Delta x} = \phi'(\xi),$$

where ξ is some intermediate value in the interval

$$x - \Delta x < \xi < x.$$

Thus

$$\left(\frac{1}{\Delta x}\right)^2 [u(x + \Delta x) - 2u(x) + u(x - \Delta x)]$$

$$= \frac{1}{\Delta x} [\phi(x) - \phi(x - \Delta x)] = \phi'(\xi).$$

On the other hand

$$\phi(\xi) = \frac{u(\xi + \Delta x) - u(\xi)}{\Delta x},$$

which means that

$$\phi'(\xi) = \frac{u'(\xi + \Delta x) - u'(\xi)}{\Delta x}.$$

Once more using the formula for finite increments, we see that

$$\phi'(\xi) = u''(\eta),$$

where

$$\xi < \eta < \xi + \Delta x.$$

Consequently,

$$\left(\frac{1}{\Delta x}\right)^2 [u(x + \Delta x) - 2u(x) + u(x - \Delta x)] = u''(\eta),$$

where $x - \Delta x < \eta < x + \Delta x$.

If the derivative $u''(x)$ is continuous and the value of Δx is sufficiently small, then $u''(\eta)$ will be only slightly different from $u''(x)$. Thus our second derivative is arbitrarily close to the difference quotient in question. In exactly the same way it may be shown, for example, that the mixed second derivative

$$\frac{\partial^2 u}{\partial x\, \partial y}$$

can be approximately represented by the formula

$$\frac{\partial^2 u}{\partial x\, \partial y} = \frac{1}{\Delta x\, \Delta y} [u(x + \Delta x, y + \Delta y) - u(x + \Delta x, y)$$

$$-u(x, y + \Delta y) + u(x, y)].$$

We return now to our partial differential equation.

For definiteness, let us assume that we are dealing with the Laplace equation in two independent variables

$$\frac{\partial^2 u}{\partial x^2} + \frac{\partial^2 u}{\partial y^2} = 0.$$

Further, let the unknown function u be given on the boundary S of the domain Ω. As an approximation we assume that

$$\frac{\partial^2 u}{\partial x^2} = \frac{u(x + \Delta x, y) - 2u(x, y) + u(x - \Delta x, y)}{(\Delta x)^2},$$

$$\frac{\partial^2 u}{\partial y^2} = \frac{u(x, y + \Delta y) - 2u(x, y) + u(x, y - \Delta y)}{(\Delta y)^2}$$

If we put $\Delta x = \Delta y = h$, then

$$\frac{\partial^2 u}{\partial x^2} + \frac{\partial^2 u}{\partial y^2} = \frac{1}{h^2} [u(x + h, y) + u(x, y + h) + u(x - h, y)$$
$$+ u(x, y - h) - 4u(x, y)].$$

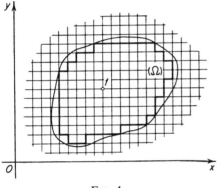

Now let us cover the domain Ω with a square net with vertices at the points $x = kh$, $y = bh$ (figure 4). We replace the domain by the polygon consisting of those squares of our net that fall inside Ω, so that the boundary of the domain is changed into a broken line. We take the values of the unknown function on this broken line to be those given on the boundary of S. The Laplace equation

FIG. 4.

is then approximated by the equation

$$u(x + h, y) + u(x, y + h) + u(x - h, y) + u(x, y - h) - 4u(x, y) = 0$$

for all interior points of the domain. This equation may be rewritten in the form

$$u(x, y) = \tfrac{1}{4} [u(x + h, y) + u(x, y + h) + u(x - h, y) + u(x, y - h)].$$

Then the value of u at any point of the net, for example the point 1 in figure 4, is equal to the arithmetic mean of its values at the four adjacent points.

We assume that inside the polygon there are N points of our net. At every such point we will have a corresponding equation. In this manner we get a system of N algebraic equations in N unknowns, the solution of which gives us the approximate values of the function u on the domain Ω.

It may be shown that for the Laplace equation the solution may be found to any desired degree of accuracy.

The method of finite differences reduces the problem to the solution of a system of N equations in N unknowns, where the unknowns are the values of the desired function at the knots of some net.

Further the method of finite differences can be shown to be applicable to other problems of mathematical physics: to other differential equations and to integral equations. However its application in many cases involves a number of difficulties.

It may turn out that the solution of the system of N algebraic equations in N unknowns, constructed by the method of nets, either does not exist in general or gives a result that is quite far from the true one. This happens when the solution of the system of equations leads to accumulation of errors; the smaller we take the length of the sides of the squares in the net the more equations we get, so that the accumulated error may become greater.

In the example given previously of the Laplace equation, this does not happen. The errors in solving this system do not accumulate but, on the contrary, steadily decrease if we solve the system, for example, by a method of successive approximations. For the equation of heat flow and for the wave equation it is essential to choose the nets properly. For these equations we may get both good and bad results.

If we are going to solve either of these equations by the method of nets, after choosing the net for the values of t, we must not choose too fine a net for the space variables. Otherwise we get a very unsatisfactory system of equations for the values of the unknown function; its solution gives a result that oscillates rapidly with large amplitudes and is thus very far from the true one.

The great variety of possible results may best be seen in a simple numerical example. Consider the equation

$$\frac{\partial u}{\partial t} = \frac{\partial^2 u}{\partial x^2}$$

for the equation of heat flow in the case in which the temperature does not depend on y or z. We take the mesh width of the net along the values of t equal to k and along the values of x equal to h

$$\frac{\partial u}{\partial t} \approx \frac{u(t+k, x) - u(t, x)}{k},$$

$$\frac{\partial^2 u}{\partial x^2} \approx \frac{u(t, x+h) - 2u(t, x) + u(t, x-h)}{h^2}.$$

Then our equation may be written approximately in the form

$$u(t + k, x) = \frac{k}{h^2} u(t, x + h) + \left(1 - 2\frac{k}{h^2}\right) u(t, x) + \frac{k}{h^2} u(t, x - h).$$

If, for a certain mesh-point value of t, we know the values of u at the points $x - h$, x, and $x + h$, it is easy to find the value of u at the point x and the next mesh point $t + k$. Assume that the constant k, i.e., the mesh width in the net with respect to t, is already chosen. Let us consider two cases for the choice of h. We put $h^2 = k$ in the first case and $h^2 = 2k$ in the second and solve the following problem by the method of nets.

At the initial instant, $u = 0$ for all negative values of x, and $u = 1$ for all nonnegative values of x. We will have, writing in one line the values of the unknown function u for the given instant, two tables:

Table 1

t \\ x	$-5h$	$-4h$	$-3h$	$-2h$	$-h$	0	h	$2h$	$3h$	$4h$	$5h$
0	0	0	0	0	0	1	1	1	1	1	1
k	0	0	0	0	1	0	1	1	1	1	1
$2k$	0	0	0	1	-1	2	0	1	1	1	1
$3k$	0	0	1	-2	4	-3	3	0	1	1	1
$4k$	0	1	-3	7	-9	10	-6	4	0	1	1
$5k$	1	-4	11	-19	26	-25	20	-10	5	0	1

Table 2

t \\ x	$-5h$	$-4h$	$-3h$	$-2h$	$-h$	0	h	$2h$	$3h$	$4h$	$5h$
0	0	0	0	0	0	1	1	1	1	1	1
k	0	0	0	0	$\frac{1}{2}$	$\frac{1}{2}$	1	1	1	1	1
$2k$	0	0	0	$\frac{1}{4}$	$\frac{1}{4}$	$\frac{3}{4}$	$\frac{3}{4}$	1	1	1	1
$3k$	0	0	$\frac{1}{8}$	$\frac{1}{8}$	$\frac{1}{2}$	$\frac{1}{2}$	$\frac{7}{8}$	$\frac{7}{8}$	1	1	1
$4k$	0	$\frac{1}{16}$	$\frac{1}{16}$	$\frac{5}{16}$	$\frac{5}{16}$	$\frac{11}{16}$	$\frac{11}{16}$	$\frac{15}{16}$	$\frac{15}{16}$	1	1
$5k$	$\frac{1}{32}$	$\frac{1}{32}$	$\frac{3}{16}$	$\frac{3}{16}$	$\frac{1}{2}$	$\frac{1}{2}$	$\frac{13}{16}$	$\frac{13}{16}$	$\frac{31}{32}$	$\frac{31}{32}$	1

In Table 2 we obtain values, for any given instant of time, which vary smoothly from point to point. This table gives a good approximation to the solution of the heat-flow equation. On the other hand, in Table 1, in which, as it would seem, the exactness should have been increased because of our finer division for the x-interval, the values of u oscillate very rapidly from positive values to negative ones and attain values that are much greater than the initially prescribed ones. It is clear that in this table the values are extraordinarily far from those that correspond to the true solution.

From these examples it is clear that if we wish to use the method of nets to get sufficiently accurate and reliable results, we must exercise great discretion in our choice of intervals in the net and must make preliminary investigations to justify the application of the method.

The solutions obtained by using the equations of mathematical physics for these or other problems of natural science give us a mathematical description of the expected course or the expected character of the physical events described by these equations.

Since the construction of a model is carried out by means of the equations of mathematical physics, we are forced to ignore, in our abstractions, many aspects of these events, to reject certain aspects as non-essential and to select others as basic, from which it follows that the results we obtain are not absolutely true. They are absolutely true only for that scheme or model that we have considered, but they must always be compared with experiment, if we are to be sure that our model of the event is close to the event itself and represents it with a sufficient degree of exactness.

The ultimate criterion of the truth of the results is thus practical experience only. In the final analysis, there is just one criterion, namely practical experience, although experience can only be properly understood in the light of a profound and well-developed theory.

If we consider the vibrating string of a musical instrument, we can understand how it produces its tones only if we are acquainted with the laws for superposition of characteristic oscillations. The relations that hold among the frequencies can be understood only if we investigate how these frequencies are determined by the material, by the tension in the string, and by the manner of fixing the ends. In this case the theory not only provides a method of calculating any desired numerical quantities but also indicates just which of these quantities are of fundamental importance, exactly how the physical process occurs, and what should be observed in it.

In this way a domain of science, namely mathematical physics, not only grew out of the requirements of practice but in turn exercised its own influence on that practice and pointed out paths for further progress.

Mathematical physics is very closely connected with other branches of mathematical analysis, but we cannot discuss these connections here, since they would lead us too far afield.

§6. Generalized Solutions

The range of problems in which a physical process is described by continuous, differentiable functions satisfying differential equations may be extended in an essential way by introducing into the discussion discontinuous solutions of these equations.

In a number of cases it is clear from the beginning that the problem under consideration cannot have solutions that are twice continuously differentiable; in other words, from the point of view of the classical statement of the problem given in the preceding section, such a problem has no solution. Nevertheless the corresponding physical process does occur, although we cannot find functions describing it in the preassigned class of twice-differentiable functions. Let us consider some simple examples.

1. If a string consists of two pieces of different density, then in the equation

$$\frac{\partial^2 u}{\partial t^2} = a^2 \frac{\partial^2 u}{\partial x^2} \tag{24}$$

the coefficient will be equal to a different constant on each of the corresponding pieces, and so equation (24) will not, in general, have classical (twice continuously differentiable) solutions.

2. Let the coefficient a be a constant, but in the initial position let the string have the form of a broken line given by the equation $u|_{i=0} = \phi(x)$. At the vertex of the broken line, the function $\phi(x)$ obviously cannot have a first derivative. It may be shown that there exists no classical solution of equation (24) satisfying the initial conditions

$$u|_{t=0} = \phi(x), \quad u_t|_{t=0} = 0$$

(here and in what follows u_t denotes $\partial u/\partial t$).

3. If a sharp blow is given to any small piece of the string, the resulting oscillations are described by the equation

$$\frac{\partial^2 u}{\partial t^2} = a^2 \frac{\partial^2 u}{\partial x^2} + f(x, t),$$

where $f(x, t)$ corresponds to the effect produced and is a discontinuous function, differing from zero only on the small piece of the string and during a short interval of time. Such an equation also, as can be easily established, cannot have classical solutions.

These examples show that requiring continuous derivatives for the desired solution strongly restricts the range of the problems we can solve. The search for a wider range of solvable problems proceeded first of all in the direction of allowing discontinuities of the first kind in the derivatives of highest order, for the functions serving as solutions to the problems, where these functions must satisfy the equations except at the points of discontinuity. It turns out that the solutions of an equation of the type $\Delta u = 0$ or $\partial u / \partial t - \Delta u = 0$ cannot have such (so-called weak) discontinuities inside the domain of definition. Solutions of the wave equation can have weak discontinuities in the space variables x, y, z, and in t only on surfaces of a special form, which are called characteristic surfaces. If a solution $u(x, y, z, t)$ of the wave equation is considered as a function defining, for $t = t_1$, a scalar field in the x, y, z space at the instant t_1, then the surfaces of discontinuity for the second derivatives of $u(x, y, z, t)$ will travel through the (x, y, z) space with a velocity equal to the square root of the coefficient of the Laplacian in the wave equation.

The second example for the string shows that it is also necessary to consider solutions in which there may be discontinuous first derivatives; and in the case of sound and light waves, we must even consider solutions that themselves have discontinuities.

The first question that comes up in investigating the introduction of discontinuous solutions consists in making clear exactly which discontinuous functions can be considered as physically admissible solutions of an equation or of the corresponding physical problem. We might, for example, assume that an arbitrary piecewise constant function is "a single solution" of the Laplace equation or the wave equation, since it satisfies the equation outside of the lines of discontinuity.

In order to clarify this question, the first thing that must be guaranteed is that in the wider class of functions, to which the admissible solutions must belong, we must have a uniqueness theorem. It is perfectly clear that if, for example, we allow arbitrary piecewise smooth functions, then this requirement will not be satisfied.

Historically, the first principle for selection of admissible functions was that they should be the limits (in some sense or other) of classical solutions of the same equation. Thus, in example 2, a solution of equation (24) corresponding to the function $\phi(x)$, which does not have a derivative at an angular point may be found as the uniform limit of classical solutions $u_n(x, t)$ of the same equation corresponding to the initial conditions $u_n|_{t=0} = \phi_n(x)$, $u_{n_t}|_{t=0} = 0$, where the $\phi_n(x)$ are twice continuously differentiable functions converging uniformly to $\phi(x)$ for $n \to \infty$.

In what follows, instead of this principle we will adopt the following: An admissible solution u must satisfy, instead of the equation $Lu = f$, an integral identity containing an arbitrary function Φ.

This identity is found as follows: We multiply both sides of the equation $Lu = f$ by an arbitrary function Φ, which has continuous derivatives with respect to all its arguments of orders up through the order of the equation and vanishes outside of the finite domain D in which the equation is defined. The equation thus found is integrated over D and then transformed by integration by parts so that it does not contain any derivatives of u. As a result we get the identity desired. For equation (24), for example, it has the form

$$\iint\limits_{D} u \left[\frac{\partial^2 \Phi}{\partial t^2} - \frac{\partial^2 (a^2 \Phi)}{\partial x^2} \right] dx \, dt = 0.$$

S. L. Sobolev has shown that for equations with constant coefficients these two principles for the selection of admissible (or as they are now usually called, generalized) solutions, are equivalent to each other. But for equations with variable coefficients, the first principle may turn out to be inapplicable, since these equations may in general have no classical solutions (cf. example 1). The second of these principles provides the possibility of selecting generalized solutions with very broad assumptions on the differentiability properties of the coefficients of the equations. It is true that this principle seems at first sight to be overly formal and to have a purely mathematical character, which does not directly indicate how the problems ought to be formulated in a manner similar to the classical problems.

We give here a modification that, it seems to us, is more appropriate physically, since it is directly connected with the well-known principle of Hamilton.

As is well known, analysis of the methods of deducing various equations of mathematical physics led in the first half of the 19th century to the discovery of a new law known as Hamilton's principle. Starting from this principle, it was possible to obtain in a uniform manner all the known equations of mathematical physics. We will illustrate this by the example of the problem considered in §3 for the oscillations of a string of finite length with fixed ends.

First of all we construct the so-called Lagrange function $L(t)$ for our string, namely the difference between the kinetic and potential energies. From what was said in §3 it follows that

$$L(t) = \int_0^l \left(\frac{1}{2} \rho u_t^2 - \frac{T}{2} u_x^2 \right) dx.$$

According to Hamilton's principle, the integral

$$S = \int_{t_1}^{t_2} L(t)\, dt$$

assumes its minimum value for the function $u(x, t)$, corresponding to the true motion of the string compared with all other functions $v(x, y)$ which are equal to zero for $x = 0$ and $x = l$ and coincide with $u(x, t_1)$ and $u(x, t_2)$ for $t = t_1$ and $t = t_2$. Here t_1 and t_2 are fixed arbitrarily, and the functions v must have finite integrals S. As a result of this principle the so-called first variation of S (cf. Chapter VIII) must be equal to zero, ie.,

$$\delta S = \int_{t_1}^{t_2} \int_0^l (\rho u_t \Phi_t - T u_x \Phi_x)\, dx\, dt = 0, \tag{25}$$

where $\Phi(x, t)$ is an arbitrary function differentiable with respect to x and t and equal to zero on the edges of the rectangle $0 \leqslant x \leqslant l$, $t_1 \leqslant t \leqslant t_2$.

Equation (25) is also the condition that must be met by the desired function $u(x, t)$. If we know that $u(x, t)$ has derivatives of the second order, then condition (25) may be put in a different form. Integrating (25) by parts and applying the fundamental lemma of the calculus of variations, we find that $u(x, t)$ must satisfy the equation

$$\frac{\partial}{\partial t}\left(\rho\, \frac{\partial u}{\partial t}\right) - \frac{\partial}{\partial x}\left(T\, \frac{\partial u}{\partial x}\right) = 0, \tag{26}$$

which is identical with (24), if ρ and T are constants and $T/\rho = a^2$.

It is not difficult to see that any solution $u(x, t)$ of equation (26) satisfies the identity (25) for all given Φ. The converse turns out to be false, since $u(x, t)$ may in general not have second derivatives. So we are extending the range of solvable problems, if we replace equation (26) by the identity (25).

To determine a specific oscillation of the string, we must add to the boundary conditions

$$u(0, t) = u(l, t) = 0, \tag{27}$$

the initial conditions

$$u(x, 0) = \phi_0(x),$$
$$u_t(x, 0) = \phi_1(x). \tag{28}$$

If a solution is sought in the class of continuously differentiable functions, then conditions (27) and (28) may be stated separately from (25) as requirements to be met. But if we allow the proposed solution to be "worse," then these conditions lose their meaning in the form given and they must be partly or wholly included in the integral identity (25).

For example, let $u(x, t)$ be continuous for $0 \leqslant x \leqslant l, 0 \leqslant t \leqslant T$, but let its first derivatives have discontinuities. The second equation in (28) then loses its meaning as a limiting condition. In this case the problem can be stated as follows: to find a continuous function u which fulfills condition (27) and the first of the conditions (28) for which the equation

$$\int_0^T \int_0^l (\rho u_t \Phi_t - T u_x \Phi_x) \, dx \, dt + \int_0^l \phi_1 \Phi(x, 0) \, dx = 0 \qquad (29)$$

is identically satisfied for all continuous $\Phi(x, t)$ equal to zero for $x = 0$, $x = l$ and $t = T$. Here the functions u and Φ must both have first derivatives whose squares are integrable in the sense of Lebesgue on the rectangle $0 \leqslant x \leqslant l$, $0 \leqslant t \leqslant T$. This last requirement for u means that the mean value with respect to time of the total energy of the string

$$\frac{1}{2T} \int_0^T \int_0^l (\rho u_t^2 + T u_x^2) \, dx \, dt$$

must be finite. Such a restriction on the function u, and thus also on its possible variations Φ, is a natural result of Hamilton's principle.

The identity (29) is precisely the condition that the first variation of the functional

$$\tilde{S} = \int_0^T \int_0^l \left(\frac{\rho}{2} u_t^2 - \frac{T}{2} u_x^2 \right) dx \, dt + \int_0^l \phi_1 u \mid_{t=0} dx$$

be equal to zero. Thus the problem of the vibration of a fixed string in the case considered may be stated as the problem of finding the minimum of the functional $\tilde{S}$ for all functions $v(x, t)$ which are continuous, satisfy condition (27), and are equal to $u(x, T)$ for $t = T$. Moreover, the desired function must satisfy the first of conditions (28).

This modification of Hamilton's principle allows us not only to widen the class of admissible solutions of equation (24) but also to state a well-defined boundary-value problem for them.

The fact that these generalized solutions or some of their derivatives are not defined at all points of the space does not lead to any contradiction with experiment, as was repeatedly pointed out by N. M. Gjunter, whose investigations were chiefly instrumental in establishing a new point of view for the concept of the solution of an equation of mathematical physics.

For example, if we wish to determine the flow of liquid in a channel, then in the classical presentation we must compute the velocity vector and the pressure at every point of the flow. But in practice we are never dealing with the pressure at a point but rather with the pressure on a certain arca and never with the velocity vector at a given point but rather with the amount of the liquid passing through some area in a unit of time. The definition of generalized solution thus proposes essentially the computation of just those quantities that have direct physical meaning.

In order that a larger number of problems may be solvable, we must seek the solutions among functions belonging to the widest possible class of functions for which uniqueness theorems still hold. Frequently such a class is dictated by the physical nature of the problem. Thus, in quantum mechanics it is not the state function $\psi(x)$, defined as a solution of the Schrödinger equation, that has physical meaning but rather the integral $a_\nu = \int_E \psi(x)\,\psi_\nu(x)\,dx$, where the ψ_ν are certain functions for which $\int_E \psi_\nu^2\,dx < \infty$. Thus the solution ψ is to be sought not among the twice continuously differentiable functions but among the ones with integrable square. In the problems of quantum electrodynamics, it is still an open question which classes of functions are the ones in which we ought to seek solutions for the equations considered in that theory.

Progress in mathematical physics during the last thirty years has been closely connected with this new formulation of the problems and with the creation of the mathematical apparatus necessary for their solution. One of the central features of this apparatus is the so-called embedding theorem of S. L. Sobolev.

Particularly convenient methods of finding generalized solutions in one or another of these classes of functions are: the method of finite differences, the direct methods in the calculus of variations (Ritz method and Trefftz method), Galerkin's method, and functional-operator methods. These latter methods basically depend on a study of transformations generated by these problems. We have already spoken in §5 of the method of finite differences and of Galerkin's method. Here we will explain the basic ideas of the direct methods of the calculus of variations.

Let us consider the problem of defining the position of a uniformly stretched membrane with fixed boundary. From the principle of minimum potential energy in a state of stable equilibrium the function $u(x, y)$ must give the least value of the integral

$$J(u) = \iint_D (u_x^2 + u_y^2)\,dx\,dy$$

in comparison with all other continuously differentiable functions $v(x, y)$ satisfying the same condition on the boundary, $v|_S = \phi$, as the function u does. With some restrictions on ϕ and on the boundary S it can be shown that such a minimum exists and is attained by a harmonic function, so that the desired function u is a solution of the Dirichlet problem

$\Delta u = 0$, $u|_S = \phi$. The converse is also true: The solution of the Dirichlet problem gives a minimum to the integral J with respect to all v satisfying the boundary condition.

The proof of the existence of the function u, for which J attains its minimum, and its computation to any desired degree of accuracy may be carried out, for example, in the following manner (Ritz method). We choose an infinite family of twice continuously differentiable functions $\{v_n(x, y)\}$, $n = 0, 1, 2, \cdots$, equal to zero on the boundary for $n > 0$ and equal to ϕ for $n = 0$. We consider J for functions of the form

$$v = \sum_{k=1}^{n} C_k v_k + v_0 \,,$$

where n is fixed and the C_k are arbitrary numbers. Then $J(v)$ will be a polynomial of second degree in the n independent variables C_1, C_2, $\cdots$, C_n. We determine the C_k from the condition that this polynomial should assume its minimum. This leads to a system of n linear algebraic equations in n unknowns, the determinant of which is different from zero. Thus the numbers C_k are uniquely defined. We denote the corresponding v by $v^n(x, y)$. It can be shown that if the system $\{v_n\}$ satisfies a certain condition of "completeness" the functions v^n will converge, as $n \to \infty$, to a function which will be the desired solution of the problem.

In conclusion, we note that in this chapter we have given a description of only the simplest linear problem of mechanics and have ignored many further questions, still far from completely worked out, which are connected with more general partial differential equations.

Suggested Reading

H. Bateman, *Partial differential equations of mathematical physics*, Dover, New York, 1944.

R. Courant and D. Hilbert, *Methods of mathematical physics*. II, *Partial differential equations*, Interscience, New York, 1962.

G. F. D. Duff, *Partial differential equations*, University of Toronto Press, 1956.

G. E. Forsythe and W. R. Wasow, *Finite-difference methods for partial differential equations*, Wiley, New York, 1960.

L. Hopf, *Introduction to the differential equations of physics*, Dover, New York, 1948.

I. G. Petrovskii, *Lectures on partial differential equations*, Interscience, New York, 1954.

H. Sagan, *Boundary and eigenvalue problems in mathematical physics*, Wiley, New York, 1961.

I. A. Sneddon, *Elements of partial differential equations*, McGraw-Hill, New York, 1957.

A. J. W. Sommerfeld, *Partial differential equations in physics*, Academic Press, New York, 1949.

A. G. Webster, *Partial differential equations of mathematical physics*, Dover, New York, 1955.

CHAPTER VII

CURVES
AND SURFACES

§1. Topics and Methods in the Theory of Curves and Surfaces

In a school course, geometry involves only the simplest curves: straight lines, broken lines, and circumferences and arcs of circles; and as for surfaces, merely planes, surfaces of polyhedra, spheres, cones, and cylinders. In more extended courses other curves are considered, chiefly the conic sections: ellipses, parabolas, and hyperbolas. But the study of an arbitrary curve or surface is completely alien to elementary geometry. At first sight it is even unclear how any general properties could be selected for investigation when we are speaking of arbitrary curves and surfaces. Yet such an investigation is completely natural and necessary.

In every kind of practical activity and experience of nature, we constantly encounter curves and surfaces of widely different forms. The path of a planet in space, of a ship at sea, or of a projectile in the air, the track of a chisel on metal, of a wheel on the road, of a pen on the tape of a recording device, the shape of a camshaft governing the valves of a motor, the contours of an artistic design, the form of a dangling rope, the shape of a spiral spring coiled for some specific purpose, such examples are endless. The surfaces of various objects, thin shells, cisterns, the framework of an airplane, casings, sheetlike materials, provide an endless diversity of surfaces. Methods for the processing of products, the optical properties of various objects, the streamlining of bodies, the rigidity or deformability of thin shells, these and many other features depend to a great extent on the geometric form of the surfaces of objects.

Of course, the gouge left by a chisel on metal is not a mathematical curve. A cistern, even with thin walls, is not a mathematical surface. But to a first approximation, which is sufficient for the study of many questions, actual objects may be represented mathematically by curves and surfaces.

In introducing the concept of a mathematical curve, we disregard all the reasons why we cannot decrease the thickness without limit. By means of this abstract concept, we succeed in representing those (completely concrete) properties of an object that are preserved when its thickness and breadth are decreased in comparison with its length.

Similarly, if we disregard the limitations on our ability to decrease the thickness of a shell or to determine precisely the actual boundaries of a given object, we are led to the concept of a mathematical surface. We will not give a rigorous description of these well-known concepts but will only remark that the exact mathematical definitions are not simple and belong to topology.

Finally, an important source of interest in various curves and surfaces has been the development of mathematical analysis. It is sufficient to remember, for example, that a curve is the geometric representation of a function, which is the most important concept of analysis. Moreover, every one is familiar with graphs quite apart from any study of analysis.

In elementary geometry as created by the ancient Greeks, there was nothing about arbitrary curves or surfaces, but even in elementary analytic geometry we are accustomed to say "every curve is represented by an equation" or "every equation in the two variables x and y represents a curve in the coordinate plane." Similarly the coordinates of surfaces are given by the equations $z = f(x, y)$ or $F(x, y, z) = 0$, and in general the coordinate method, by establishing a close connection between elementary geometry and analysis, enables us to define many different curves and surfaces.

But analytic geometry, being restricted to the methods of algebra and elementary geometry, goes no further than the investigation of certain specific types of figures. The study of arbitrary curves and surfaces represents a new branch of mathematics, known as *differential geometry*.

It must be admitted at once that differential geometry imposes on its curves and surfaces certain conditions arising from the methods of analysis. However, this is not an essential limitation on the diversity of the allowable curves and surfaces, since in the great majority of cases they are capable of representing actual objects with the necessary degree of precision. The name "differential geometry" itself gives an indication of the methods of the theory; its basic tool is the differential calculus and it primarily investigates the "differential" properties of the curves and surfaces, i.e., their properties "at a point."* Thus, the direction of a curve at a point is determined by its tangent at that point and the amount by which it twists is described by its curvature (the exact definition of this term will be given below). Differential geometry investigates the properties of small segments of curves and surfaces and only in its later developments does it proceed to the study of their properties "in the large," i.e., in their entire extent.

The development of differential geometry is inseparably connected with the development of analysis. The basic operations of analysis, namely differentiation and integration, have a direct geometric meaning. As was mentioned in Chapter II, differentiating a function $f(x)$ corresponds to drawing a tangent to the curve

$$y = f(x).$$

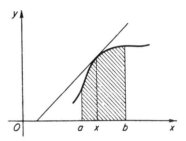

The slope of the tangent line (i.e., the trigonometric tangent of the angle it makes with the axis Ox) is precisely the derivative $f'(x)$ of the function $f(x)$ at the corresponding point (figure 1), and the area "under the curve"

$$y = f(x)$$

Fig. 1.

is precisely the integral $\int_a^b f(x)\ dx$ of this function, evaluated between the corresponding limits. Just as in analysis we investigate arbitrary functions, so in differential geometry we examine arbitrary curves and surfaces. In analysis, the first object of study is the general course of a curve on a plane, its rise and fall, its greater or smaller curvature, the direction of its convexity, its points of inflection, and so forth. The close connection between analysis and the curves is indicated by the name of the first textbook in analysis, by the French mathematician l'Hôpital in 1695: "Infinitesimal analysis applied to the study of curves."

By the middle of the 18th century, the differential and integral calculus had been sufficiently developed by the immediate successors of Newton and Leibnitz that the way was open for more profound applications to geometry. Indeed, it is only from this moment that one may properly

---

* The properties of curves and surfaces "at a point" are those properties that depend only on an arbitrarily small neighborhood of the point. Properties of this sort are defined in terms of the derivatives (at the given point) of the functions occurring in the equations of the curve or surface. It is for this reason that differential geometry imposes conditions guaranteeing that the differential calculus is applicable; it is required that the curve or surface be defined by functions with a sufficient number of derivatives.

speak of a theory of curves and surfaces. For surfaces, and for curves in space, the analogous problems are immeasurably richer in content than for plane curves, so that with the passage of time these problems outgrew the framework of a simple application of analysis to geometry and led to the formation of an independent theory. During the second half of the 18th century, many mathematicians shared in building up the elements of this theory: Clairaut, Euler, Monge, and others, among whom Euler must be considered as the founder of the general theory of surfaces. The first comprehensive work on curves and surfaces was the book of Monge "Application of analysis to geometry," published in 1795.* From the investigations of these mathematicians, and, in particular, from the book of Monge, we can easily understand the upsurge of interest in differential geometry. This upsurge was due to the demands of mechanics, physics, and astronomy, i.e., in the final analysis to the needs of technology and industry, for which the available results of elementary geometry were completely insufficient.

The classical work of Gauss (1777-1855) in the theory of surfaces is also related to practical questions. His "General investigations concerning curved surfaces," published in 1827, is basic for the differential geometry of surfaces as an independent branch of mathematics. His general methods and problems, discussed later in §4, originated to a great degree in the practical needs of map making. The problem of cartography consists of finding as exact a representation as possible of parts of the surface of the earth on a plane. A completely exact representation here is impossible, the mutual relations of various lengths being necessarily distorted because of the curvature of the earth. Thus one has the problem of finding the most nearly exact methods possible. The drawing of maps goes back to remote antiquity, but the creation of a general theory is an achievement of recent times and would not have been possible without the general theory of surfaces and the general methods of mathematical analysis. We note that one of the difficult mathematical problems of cartography was investigated by P. L. Čebyšev (1821-1894), who obtained important results relating to nets of curved lines on surfaces. His investigations also arose from purely practical problems.

The general questions of deforming one surface so that it can be mapped on another still constitute one of the main branches of geometry. Important results in this direction were obtained in 1838 by F. Minding (1806-1885), professor at the University of Dorpat (now Tartu).

---

* Gaspard Monge (1746-1828) was not only an outstanding scientist but also an active French revolutionary (minister of naval affairs, and then director of the manufacture of cannon and powder). He followed the path, characteristic of the French bourgeois of the time, from Jacobin to adherent of the emperor Napoleon.

By the second half of the last century, the theory of curves and surfaces was already well established in its basic features, provided we are speaking of "classical differential geometry" in contrast with the newer directions discussed later in §5. The basic equations in the theory of curves, namely the so-called Frenet formulas, had already been obtained, and in 1853 K. M. Peterson (1828-1881), a student of Minding's at Tartu University, discovered and investigated in his dissertation the basic equations of the theory of surfaces, rediscovered 15 years later and published by the Italian mathematician Codazzi, with whose name these equations are usually associated. Peterson, after graduating from the university at Tartu, lived and worked in Moscow, as a teacher in a gymnasium. Though he never held any academic position corresponding to his outstanding scientific achievements, he was nevertheless one of the founders of the Moscow Mathematical Society and of the journal "Matematičeskiĭ Sbornik," published in Moscow from 1866 up to the present day. The Moscow school of differential geometry begins with Peterson.

The results to date of the "classical" differential geometry were summarized by the French geometer Darboux in his four-volume "Lectures on the general theory of surfaces," issued from 1887 to 1896. In the present century classical differential geometry continues to be studied, but the center of interest in curves and surfaces has largely shifted to new directions in which the class of figures under study has been even more widely extended.

§2. The Theory of Curves

Various methods of defining curves in differential geometry. From analysis and analytic geometry we are accustomed to the idea of defining curves by means of equations. In a rectangular coordinate system on the plane, a curve may be given either by the equation

$$y = f(x),$$

or by the more general equation

$$F(x, y) = 0.$$

However, this method of definition is suitable only for a plane curve, i.e., a line in the plane. We also require a method of writing equations of space curves not lying in any plane. An example of such a curve may be seen in the helix (figure 2).

For the purposes of differential geometry, and for many other questions
as well, it is most convenient to repre-
sent a curve as the trace of a continuous
motion of a point. Of course, the given
curve may have originated in some
entirely different way, but we can always
think of it as the path of a point
moving along it.

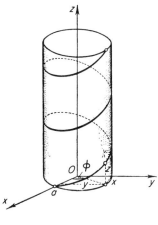

Let us assume that we have a fixed
Cartesian coordinate system in space. If
a moving point X traces out a curve
from time $t = a$ to $t = b$, then the
coordinates of this moving point are
given by the functions of the time
$x(t)$, $y(t)$, and $z(t)$; the flight of an
airplane or a projectile are examples.
Conversely, if we are initially given
the functions $x(t)$, $y(t)$, and $z(t)$, we

FIG. 2.

may let them define the coordinates of a moving point X, which traces
out some curve. Consequently, curves in space may be given by three
equations of the form

$$x = x(t), \quad y = y(t), \quad z = z(t).$$

In the same way a plane curve is defined by two equations

$$x = x(t), \quad y = y(t).$$

This is the most general manner of defining curves.

As an example we consider the helix. It is produced by the spiral
motion of a point that revolves uniformly around a straight line, the axis
of the helix, and at the same time moves uniformly in a direction parallel
to this axis. Let us take the axis of the helix as the axis Oz and suppose
that at time $t = 0$ the point lies on the axis Ox. We now wish to find
how its coordinates depend on the time. If the motion parallel to the
axis Oz has velocity c, then obviously the distance travelled in this direction
at time t will be

$$z = ct.$$

Also, if ϕ is the angle of rotation around the axis Oz and a is the distance
from the point to this axis, then, as can be seen in figure 2,

$$x = a \cos \phi, \quad y = a \sin \phi.$$

Since the rotation is uniform, the angle ϕ is proportional to time; that is, $\phi = \omega t$, where ω is the angular velocity of the rotation. In this manner we get

$$x = a \cos \omega t, \quad y = a \sin \omega t, \quad z = ct.$$

So these are the equations of the helix, which as t changes will be traced out by the moving point.

Of course the variable t or, as it is usually called, the parameter, need not be thought of as representing the time. Also, the given parameter t may be replaced by another; for example we may introduce a parameter u by the formula $t = u^3$, or, in general, by $t = f(u)$.* In geometry the most natural choice of parameter is the length s of the arc of the curve measured from some fixed point A on it. Every possible value of the length s represents a corresponding arc AX. Thus the position of X is fully determined by the value of s and the coordinates of the point X are given by the functions of arc length s

$$x = x(s), \quad y = y(s), \quad z = z(s).$$

All these ways of defining curves, as well as other possible ones,† open up the possibility of numerical computation. Only when curves have been defined by equations can their properties be investigated by mathematical analysis.

In the differential geometry of plane curves, there are three basic concepts: length, tangent, and curvature. For space curves, there are in addition the osculating plane and the torsion. We now proceed to explain the meaning and significance of these concepts.

Length. Everyone has in mind a natural idea of what is meant by length, but this idea must be converted into an exact definition of the length of a mathematical curve, a definition with a specific numerical character, which will enable us to compute the length of a curve with any desired degree of accuracy and consequently to argue about lengths in a rigorous way. The same remarks apply to all mathematical concepts. The transition from informal ideas to exact measurements and definitions represents the transition from a prescientific understanding of objects to

---

* Here, strictly speaking, it is necessary that the function f be monotone.

† A curve in space may also be given as the intersection of two surfaces, defined by the equations: $F(x, y, z) = 0$, $G(x, y, z) = 0$, i.e., the curve is given by this pair of equations. In theoretical discussions a curve is most frequently given by a variable vector, i.e., the position of the point X of the curve is defined by the vector $\mathbf{r} = \overrightarrow{OX}$, extending from the origin to this point. As the vector $\mathbf{r}$ changes, its end point X moves along the given curve (figure 3).

a scientific theory. The need for a precise definition of length arose in the final analysis from the requirements of technology and the natural sciences, whose development demanded investigation of the properties of lengths, areas, and other geometric entities.

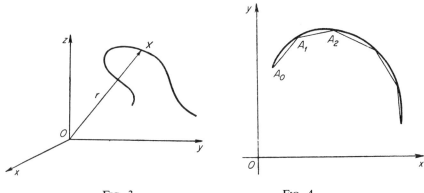

FIG. 3.   FIG. 4.

A simple and most useful definition of length is the following: The *length* of a curve is the limit of the length of broken lines inscribed in the curve under the condition that their vertices cluster closer and closer together on the curve.

This definition arises naturally from our everyday methods of measuring. On the curve we take a sequence of points A_0, A_1, A_2, $\cdots$ (figure 4) and measure the distances between them. The sum of these distances (which is the length of the broken line) expresses approximately the length of the curve. In order to define the length more exactly, it is natural to take the points A closer together, so that the broken line follows the twists of the curve more closely. Finally, the exact value of the length is defined as the limit of these approximations as the points A are chosen arbitrarily close together.* Thus the earlier definition of length is a generalization, based on taking finer and finer steps, of a completely practical manner of measuring length.

From this definition of length, it is easy to derive a formula for computing lengths when the curve is given analytically. We note, however, that mathematical formulas are useful for more than just computation.

---

* The existence of the indicated limit, i.e., the length of the curve, is not initially clear, even for curves lying in a bounded domain. If the curve is very twisted, its length may be very great, and it is possible mathematically to construct a plane curve which is so "twisted" that none of its arcs has a finite length since the lengths of broken lines inscribed in it increase beyond all bounds.

They are a brief statement of theorems that establish connections between different mathematical entities. The theoretical significance of such connections may far exceed the computational value of the formula. For example, the importance of the Pythagorean theorem, expressed by the formula

$$c^2 = a^2 + b^2,$$

is not confined to the computation of the square of the hypotenuse c but lies chiefly in the fact that it expresses a relation among the sides of a right triangle.

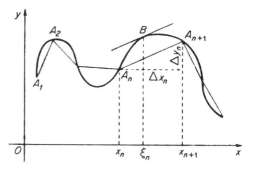

FIG. 5.

Let us now introduce a formula for the length of a plane curve, given in Cartesian coordinates by the equation $y = f(x)$, assuming that the function $f(x)$ has a first derivative.

We inscribe a broken line in the curve (figure 5). Let A_n, A_{n+1} be two of its adjacent vertices with coordinates x_n, y_n and x_{n+1}, y_{n+1}. The line segment A_nA_{n+1} is the hypotenuse of a right triangle the legs of which are equal to

$$\varDelta x_n = |x_{n+1} - x_n|, \quad \varDelta y_n = |y_{n+1} - y_n|.$$

Thus, by the Pythagorean theorem,

$$\overline{A_nA_{n+1}} = \sqrt{(\varDelta x_n)^2 + (\varDelta y_n)^2} = \sqrt{1 + \left(\frac{\varDelta y_n}{\varDelta x_n}\right)^2}\, \varDelta x_n .$$

It is easy to see that if the straight line drawn through the points A_n and A_{n+1} is translated parallel to itself, then at the instant when the line leaves the curve it will assume the position of a tangent to this curve at some point B, i.e., on the arc of the curve A_nA_{n+1}, there is at least one point at which the tangent has the same direction as the chord A_nA_{n+1}. (This obvious conclusion can easily be given a rigorous proof.)

Thus we may replace the ratio $\varDelta y_n/\varDelta x_n$ by the slope of the tangent at B, i.e., by the derivative $y'(\xi_n)$, where ξ_n is the abscissa of the point B. Now the length of one link of the broken line is expressed by

$$\overline{A_nA_{n+1}} = \sqrt{1 + y'^2(\xi_n)}\, \varDelta x_n .$$

The entire length of the broken line is the sum of the lengths of its pieces. Denoting the addition by the symbol Σ, we have

$$S_n = \sum \sqrt{1 + y'^2(\xi_n)}\, \Delta x_n .$$

To obtain the length of the curve, we must pass to the limit under the condition that the greatest of the values Δx_n tends to zero,

$$s = \lim_{\Delta x \to 0} \sum \sqrt{1 + y'^2(\xi_n)}\, \Delta x_n .$$

But this limit is exactly the integral defined in Chapter II, namely the integral of the function $\sqrt{1 + y'^2}$. Thus the length of a plane curve is expressed by the formula

$$s = \int_a^b \sqrt{1 + y'^2}\, dx, \tag{1}$$

where the limits of integration a and b are the values of x at the ends of the arc of the curve.

The corresponding, but somewhat different, formula for the length of a space curve is derived in basically the same way.

The actual computation of a length by means of these formulas is, of course, not always simple. Thus the calculation of the circumference of a circle from formula (1) is rather complicated. However, as we have said, the interest of formulas is not confined to computation; in particular, formula (1) is also important for investigating the general properties of length, its relations with other concepts, and so forth. We will have an opportunity to make use of formula (1) in Chapter VIII.

Tangent. The tangent to a plane curve was already considered in Chapter II. Its meaning for a space curve is completely analogous. In order to define the tangent at a point A, we choose a point X on the curve, distinct from A, and consider the secant AX. Then we allow X to approach A along the curve. If the secant AX converges to some limiting position, then the straight line in this limiting position is called the tangent at the point A.*

If we distinguish between the initial point and the end point of the curve and thereby establish an order in which the points of the curve

---

* The limiting position of the secant may not exist, as can be seen from the example in figure 13, Chapter II. The curve represented by $y = x \sin 1/x$ oscillates near zero in such a way that the secant OA, as A approaches O, constantly oscillates between the straight lines OM and OL.

are traversed, then we may say which of the points A and X comes first and which comes second. (For example, if a train travels from Moscow to Vladivostok, then Omsk obviously precedes Irkutsk.) So we may define a direction along the secant from the first point to the second. The limit of such "directed secants" gives us a "directed tangent." In figure 6, the arrow shows the direction in which the point A is passed through. For the motion of a point along the curve, the velocity at each instant is directed along the tangent to the curve.

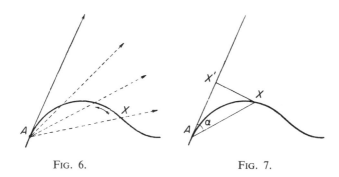

Fig. 6.   Fig. 7.

The tangent has an important geometric property: Near the point of tangency the curve departs less, in a well-defined sense, from this straight line than from any other. In other words, the distance from the points of the curve to the tangent is very small in comparison with their distance to the point of tangency. More precisely, the ratio XX'/AX (figure 7) tends to zero as X approaches A.* So a small segment of the curve may be replaced by a corresponding segment of the tangent with an error that is small in comparison with length of the segment. This procedure often allows us to simplify proofs, since in a passage to the limit it gives completely exact results.

It is interesting to observe that for a curve which is not a straight line, i.e., does not have a direction in the elementary sense, we have been able, by associating it with a straight line, to define its direction at each point. Thus the concept of direction has been extended; it has been given a meaning which it did not previously have. This new concept of direction reflects the actual nature of motion along a curve; at each instant the point is moving in some definite direction, which changes continuously.

---

* This result follows immediately from the definition of the tangent itself. Evidently, as is shown in figure 7, $XX'/AX = \sin \alpha$, where α is the angle between the tangent and the secant AX. Thus, as $\alpha \to 0$, XX'/AX also tends to zero.

Curvature. To be able to judge by eye whether a path, a thin rod, or a line in a drawing is more or less curved it is not necessary to be a mathematician. But for even the simplest problems of mechanics, a casual glance is not sufficient; we need an exact quantitative description of the curvature. This is obtained by giving precise expression to our intuitive impression of the curvature as the rapidity of change of direction of the curve.

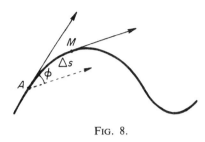

FIG. 8.

Let A be a point on the curve and M a point near A (figure 8). The angle between the tangents at these points expresses how much the curve has changed direction in the segment from A to M. Let us denote this angle by ϕ. The average rate of change of direction (more precisely, the average change per unit length of path along the segment AM of length Δs) will obviously be $\phi/\Delta s$. Then the curvature, namely the rate of change of direction of the curve at the point A itself, is naturally defined as the limit of the ratio $\phi/\Delta s$ as $M \to A$; in other words, as $\Delta s \to 0$. Thus the curvature is defined by the formula

$$k = \lim_{\Delta s \to 0} \frac{\phi}{\Delta s}.$$

As a particular example, let us consider the curvature of the circumference of a circle (figure 9). Obviously, the angle ϕ between the radii OA and OM is equal to the angle ϕ between the tangents at the points A and M, since the tangents are perpendicular to the radii. The arc AM, subtending the angle ϕ, has length $\Delta s = \phi r$, so that

$$\frac{\phi}{\Delta s} = \frac{1}{r}.$$

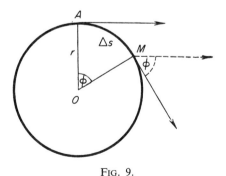

FIG. 9.

This means that the ratio $\phi/\Delta s$ is constant, so that the curvature of the circumference of a circle, as the limiting value of this ratio, is equal at all points to the reciprocal of the radius.*

Let us derive the formula for the curvature of a plane curve given by the equation $y = f(x)$. As the initial point for arc length we take a fixed point N (figure 10). The angle ϕ between the tangents at the points A

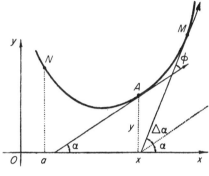

FIG. 10.

and M is obviously equal to the difference in the angle of inclination of the tangents at A to M.

$$\phi = |\varDelta\alpha|.$$

Since the angle α may decrease, we take the absolute value $|\varDelta\alpha|$. We are interested in the value

$$k = \lim_{\varDelta s \to 0} \frac{\phi}{\varDelta s} = \lim_{\varDelta s \to 0} \frac{|\varDelta\alpha|}{\varDelta s} = \lim_{\varDelta x \to 0} \frac{\dfrac{|\varDelta\alpha|}{\varDelta x}}{\dfrac{\varDelta s}{\varDelta x}} = \frac{|\alpha'|}{s'}.$$

The length of the arc of the curve NA is expressed by the integral

$$s = \int_{a}^{x} \sqrt{1 + y'^2}\, dx,$$

so that

$$s' = \sqrt{1 + y'^2}.$$

---

* We note that in general the concept of the curvature of a curve at a point may be defined by comparing the curve with the circumference of a certain circle, which plays the role of a model or standard for the curvature. For in fact, the curvature of the given curve proves to be equal to the reciprocal of the radius of the (unique) circle which fits the curve most closely in the neighborhood of the point.

It remains to find α'. We know that $\tan \alpha = y'$; thus $\alpha = \arctan y'$. Differentiating this last equation with respect to x, we get

$$\alpha' = \frac{1}{1 + y'^2} \, y''.$$

Thus, finally

$$k = \frac{|\alpha'|}{s'} = \frac{|y''|}{(1 + y'^2)^{3/2}}.$$

The corresponding formulas for other methods of representing plane and space curves are given in the usual courses in analysis or differential geometry.

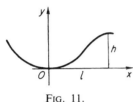

This formula allows us to give another geometric interpretation of curvature, which is useful in many questions. Namely, the curvature of a curve at a point is given by the formula

$$k = \lim_{l \to 0} \frac{2h}{l^2},$$

FIG. 11.

where h is the distance of a second point on the curve to the tangent at the given point and l is the length of the segment of the tangent between the point of tangency and the projection on the tangent of the other point on the curve (figure 11).

To prove this we choose a rectangular coordinate system such that the origin falls at the given point of the curve and the axis Ox is tangent to the curve at this point (figure 11). (For simplicity we assume that the curve is plane.) Then $y' = 0$ and $k = |y''|$. Expanding the function $y = f(x)$ by Taylor's formula, we get $y = \frac{1}{2} y'' x^2 + \epsilon x^2$ (where we have taken into account that $y' = 0$). Here $\epsilon \to 0$ as $x \to 0$. Hence it follows that $k = |y''| = \lim_{i \to 0} 2|y|/x^2$, and thus, since $|y| = h$, $x^2 = l^2$, we have

$$k = \lim_{l \to 0} \frac{2h}{l^2}.$$

This formula shows that the curvature describes the rate at which the curve leaves the tangent.

Let us now turn to some very important applications of curvature to problems of mechanics.

First we consider the following problem. Let a flexible string be stretched over a support (figure 12) in such a way that the string remains in one plane. We wish to find the pressure of the string on the support at every point, or to be more exact, to define the limit

$$p = \lim_{\Delta s \to 0} \frac{P}{\Delta s}, \tag{2}$$

where P is the magnitude of the force $\boldsymbol{P}$ acting on the support along a piece of length Δs containing the given point. We assume for simplicity that the magnitude T of the tension $\boldsymbol{T}$ is the same at all points of the string.

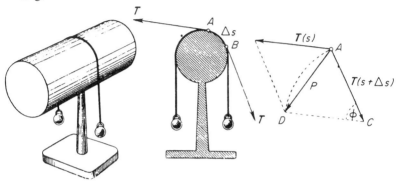

FIG. 12.

Now consider the point A and a segment of the string AB.* On this segment AB of length Δs, in addition to the reaction of the support, only two external forces are acting, namely the tensions at the ends, which are equal in magnitude and are directed along the tangents at the ends of the segment. Thus the force $\boldsymbol{P}$ exerted by the string on the support is equal to the geometric sum of the tensions at the ends. As can be seen from figure 12, the vector $\boldsymbol{P}$ is the base AD of the isosceles triangle CAD. The two equal sides of this triangle have length T and the angle at the vertex C is equal to the change of direction of the tangent in passing from A to B.

With decreasing Δs the angle ϕ decreases and the angle between $\boldsymbol{P}$ and the tangent at the point A approaches a right angle. Thus the pressure is perpendicular to the tangent.

To find the magnitude of the pressure, we make use of the fact that a small arc of the circumference has approximately the same length as

---

* It would be more natural to choose a segment with the point A in its interior; this would not change the result but would make the computation somewhat more complicated.

the chord subtending it. Thus we replace the length of the chord AD, i.e., the magnitude P, by the length $T\phi$ of the arc AD. Then by formula (2) we get

$$p = \lim_{\Delta s \to 0} \frac{P}{\Delta s} = \lim_{\Delta s \to 0} \frac{T\phi}{\Delta s} = T \lim_{\Delta s \to 0} \frac{\phi}{\Delta s} = Tk.$$

Hence the pressure at each point is equal to the product of the curvature and the tension on the string and is exerted perpendicularly to the tangent at this point.

Consider a second problem. Let a mathematical point (i.e., a very small body) move along a plane curve with a velocity of constant magnitude v. What is its acceleration at a given point A? By definition, the acceleration is equal to the limit of the ratio of the change in velocity (during the time Δt) to the increment Δt of the time. The velocity involves not only magnitude but also direction, i.e., we consider the change in the velocity vector. Therefore the mathematical problem of finding the magnitude of the acceleration consists of finding the limit

$$w = \lim_{\Delta t \to 0} \frac{|\, \boldsymbol{v}(t + \Delta t) - \boldsymbol{v}(t)\,|}{\Delta t},$$

where $\boldsymbol{v}(t)$ is the velocity at the point A itself, and $|\, \boldsymbol{v}(t + \Delta t) - \boldsymbol{v}(t)|$ is the length of the vector difference of the velocities. The limit which concerns us may also be represented as

$$\lim_{\Delta s \to 0} \frac{|\,-\boldsymbol{v}(t) + \boldsymbol{v}(t + \Delta t)\,|}{\Delta s} \lim_{\Delta t \to 0} \frac{\Delta s}{\Delta t},$$

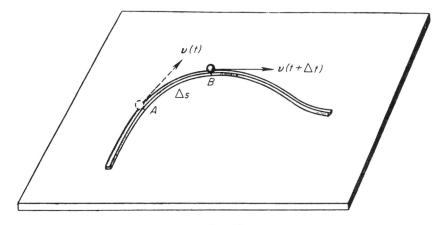

Fig. 13.

where $\varDelta s$ is the length of the arc AB traversed during time $\varDelta t$. Turning to figure 13 and noting that the velocity at each point is directed along the tangent while remaining constant in magnitude, we see geometrically that finding the sum $-v(t) + v(t + \varDelta t)$ is identical with finding the vector P in the preceding problem. So we may avail ourselves of the result there and, replacing tension by velocity, write

$$\lim_{\varDelta s \to 0} \frac{|-v(t) + v(t + \varDelta t)|}{\varDelta s} = vk.$$

Moreover, $\lim_{\varDelta t \to 0} \varDelta s / \varDelta t = v$. So we have the final result that the acceleration of a body in uniform motion along the curve is equal to the product of the curvature and the square of the velocity

$$w = kv^2 \tag{3}$$

and is directed along the normal to the curve, i.e., along a straight line perpendicular to the tangent.

Our recourse here to a geometric analogy, enabling us to use the solution of the problem of the pressure exerted by a string in order to solve a problem of the acceleration of a particle, shows once again how useful it is to make an abstraction from the particular concrete properties of a phenomenon to corresponding mathematical concepts and results; for we can then make use of these results in the most varied situations.

We also note that the curvature, which from a mechanical point of view reflects the change in the direction of motion, is seen to be closely connected with the forces causing this change. The equation which expresses this connection is easily derived if we multiply equation (3) by the mass m of the moving point. We have

$$F_n = mw = v^2 mk.$$

Here F_n is the magnitude of the normal component of the force acting on the point.

Osculating plane.   Although a space curve does not lie in one plane, still with each point A of the curve it is possible, as a rule, to associate a plane P which in the neighborhood of this point lies closer to the curve than any other plane. This plane is called the *osculating plane* of the curve at the point.

Naturally the osculating plane, as the plane closest to the given curve, passes through the point A and contains the tangent T to the curve. But there are many planes containing the point A and the straight line T.

In order to choose from among them the one plane that least deviates from the curve, we investigate the deviation of the curve from the tangent. For this purpose let us see how the curve runs along the tangent T; in other words, let us project our curve onto the *normal plane Q*, which

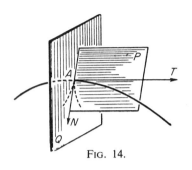

is perpendicular to T at the point A (figure 14). The projection on the plane Q of a segment of our curve containing A forms a new curve, indicated in figure 14 by a dotted line. Usually it has a cusp at the point A. If the curve so obtained has a tangent N at the point A, then the plane P determined by T and N will naturally be closest to the original curve in the neighborhood of the point A, i.e., it will be the

FIG. 14.

osculating plane at the point A. It may be shown that when the functions defining the original curve have second derivatives and the curvature of the curve at the point A is not zero, then the osculating plane necessarily exists, and its equation may be expressed very simply in terms of the first and second derivatives of the functions defining the curve.

We saw earlier that the properties of the tangent allow us to consider a small segment of a plane curve as though it were straight, thereby making an error which is small in comparison with the length of the segment; similarly the properties of the osculating plane allow us to consider a small segment of a space curve as though it were a plane curve, namely its projection on the osculating plane, and here the error will be small in comparison with the square of the length of the segment of the curve.

There are many straight lines in space that are perpendicular to the tangent; they form the normal plane at the given point of the curve. Among these straight lines there is one, the line N, which lies in the osculating plane. This line is called the *principal normal* to the curve. Usually we also fix a direction for it, namely the direction of the concavity of the projection of the curve on the osculating plane. The principal normal plays the same role for a space curve as the ordinary (unique) normal for a plane curve. In particular, if a thin string under tension T is stretched in the form of a space curve over a support, then the pressure of the string on the support has at each point the magnitude Tk and is directed along the principal normal. If a material point is moving along a space curve with a velocity of constant magnitude v, then its acceleration is equal to kv^2 and is directed along the principal normal.

Torsion. From point to point along a curve the position of the osculating plane will probably change. Just as the rate of change of direction of the tangent characterized the curvature, so the rate of change of direction of the osculating plane characterizes a new quantity, the *torsion* of the curve. Here, as in the case of curvature, the rate is taken with respect to arc length; that is, if ψ is the angle between the osculating planes at a fixed point A and at a nearby point X, and if Δs is the length of the arc AX, then the torsion τ at the point A is defined as the limit*

$$\tau = \lim_{\Delta s \to 0} \frac{\psi}{\Delta s}.$$

The sign of the torsion depends on the side of the curve toward which the osculating plane turns as it moves along the curve.

We may imagine the osculating curve as the blade of a fan with the two lines, the tangent and the principal normal, drawn on it. At each moment the tangent is turning in the direction of the normal at a rate determined by the curvature, while the osculating plane rotates around the tangent with a speed and direction determined by the torsion.

The simplest results of the theory of differential equations may be used to prove a fundamental theorem that states, roughly speaking, that two curves with the same curvature and the same torsion are identical with each other. Let us make this idea clearer. If we move along the curve to various distances A from our initial point, we will arrive at points where the curvature k and the torsion τ will have various values, depending on s. Thus $k(s)$ and $\tau(s)$ will be certain well-defined functions of the arc length s.

The theorem in question states that if two curves have identical curvature and torsion as functions of arc length, then the curves are identical (i.e., one of them may be rigidly moved so as to coincide with the other). In this manner curvature and torsion as functions of arc length define a curve completely except for its position in space; they describe all the properties of the curve by stating the relationship between its length, its curvature, and its torsion. In this way the three concepts constitute a sort of ultimate basis for questions concerning curves. With their help we can also express the simplest concepts in the theory of surfaces, to which we now turn.

---

* It may be shown that a helix has the same torsion at all its points and consequently that we may define the torsion of a curve by comparing the curve with the (unique) helix which best approximates the curve in the neighborhood of the given point. The torsion also characterizes the way in which a given space curve differs from a plane curve. With a certain analogy to curvature, it characterizes the rate at which the curve leaves its osculating plane.

Of course, the theory of curves has not been exhausted by our present remarks. There are many other concepts relating to curves: special types of curves, families of curves, the position of curves on surfaces, questions of the form of a curve as a whole, etc. These questions and the methods of answering them are connected with almost every branch of mathematics. The range of problems that may be solved by the theory of curves is extremely rich and varied.

§3. Basic Concepts in the Theory of Surfaces

The basic methods of defining a surface. If we wish to study surfaces by means of analysis we must, of course, define them analytically. The simplest way is by an equation

$$z = f(x, y),$$

in which x, y, and z are Cartesian coordinates of a point lying on the surface. Here the function $f(x, y)$ need not necessarily be defined for all x, y; its domain may have various shapes. Thus, the surface illustrated in figure 15 is given by the function $f(x, y)$ defined inside an annulus. Examples of surfaces given by equations of the form $z = f(x, y)$ are also familiar from analytic geometry. We know, for example, that the equation $z = Ax + By + C$ represents a plane, and $z = x^2 + y^2$ a paraboloid of revolution (figure 16). For the application of differential calculus it is necessary that the function $f(x, y)$ have first, second, and sometimes even higher derivatives. A surface given by such an equation is called *regular*. Geometrically this means (though not quite precisely) that

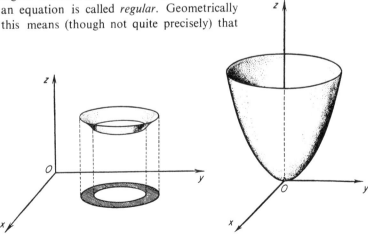

FIG. 15.   FIG. 16.

the surface curves continuously without breaks or other singularities. Surfaces that do not have this property, for example, those with cusps, breaks, or other singularities, require a new kind of investigation (cf. §5).

However, not every surface, even without singularities, can be entirely represented by an equation of the form $z = f(x, y)$. If every pair of values of x, y in the domain of $f(x, y)$ gives a completely determined z, then every straight line parallel to the axis Oz must intersect the surface at no more than one point (figure 17). Even such simple surfaces as

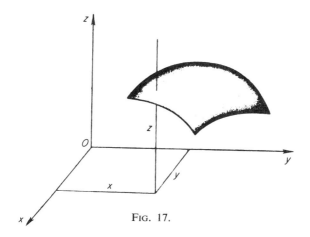

FIG. 17.

spheres or cylinders cannot be represented in the large by an equation of the form $z = f(x, y)$. In these cases the surface is defined in some other manner, for example by an equation of the form $F(x, y, z) = 0$. Thus a sphere of radius R with center at the origin has the equation

$$x^2 + y^2 + z^2 = R^2.$$

The equation $x^2 + y^2 = r^2$ gives a cylinder of radius r.

So when the investigation is concerned only with small segments of the surface, as is usually the case in classical differential geometry, the definition of a surface by an equation $z = f(x, y)$ is perfectly general, since every sufficiently small segment of a smooth surface can be represented in this form. We take this way as basic, and leave other methods of defining surfaces to be considered later in §§4 and 5.

Tangent plane. Just as at each point a smooth *curve* has a tangent line which is close to the curve in a neighborhood of the point, so also surfaces may have, at each of their points, a *tangent plane*.

The exact definition is as follows. A plane P, passing through a point M on a surface F, is said to be tangent to the surface F at this point if the angle α between the plane P and the secant MX, drawn from M to a point X of the surface, converges to zero as the point X approaches the point M (figure 18). All tangents to curves passing through the point M and lying on the surface obviously lie in the tangent plane.

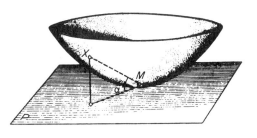

Fig. 18.

A surface F is called *smooth* if it has a tangent plane at each point and if, as we pass from point to point, the position of this plane varies continuously.

Near the point of tangency, the surface departs very little from its tangent plane: If the point X approaches the point M along the surface, then the distance of the point X from the tangent plane becomes smaller and smaller, even in comparison with its distance from the point M (the reader can easily verify this by considering how X approaches M in figure 18). In this way, the surface near the point M may be said to merge into the tangent plane. In the first approximation a small segment or, as it is called, an "element" of the surface may be replaced by a segment of the tangent plane. The perpendicular to the tangent plane which passes through the point of tangency acts as a perpendicular to the surface at this point and is called a *normal*.

This possibility of replacing an element of the surface by a segment of the tangent plane is useful in many situations. For example, the reflection of light on a curved surface takes place in the same way as the reflection on a plane, i.e., the direction of the reflected ray is defined by the usual law of reflection: The incident ray and the reflected ray lie in one plane together with the normal to the surface and they make equal angles with this normal (figure 19), just as if the reflection were occurring in the tangent plane. Similarly for the refraction of light in a curved surface, each ray is refracted by an element of the surface with the usual law of refraction, just as if the element were plane. These facts are the basis for all calculations of reflection and refraction of light in optical apparatus. Further, for example, solid bodies in contact with each other have a common tangent plane at their point of contact. The bodies are in contact over an element of their surface, and the pressure

of one body on the other, in the absence of friction, is directed along
the normal at the point of contact. This is also true when the bodies
are tangent at more than one point, in which case the pressure is directed
along the respective normals at each point of contact.

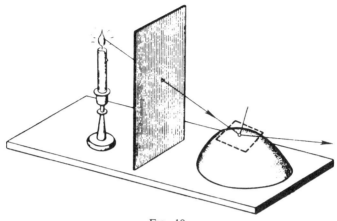

FIG. 19.

The replacement of elements of a surface by segments of the tangent
planes can also serve as the basis of a definition of the area of various
surfaces. The surface is decomposed into small pieces $F_1, F_2, \cdots, F_n$ and
each piece is projected onto a plane tangent to the surface at some point
of this piece (figure 20). We thus obtain a number of plane regions
$P_1, P_2, \cdots, P_n$, the sum of whose areas gives an approximation to the

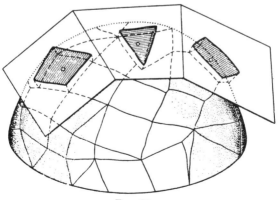

FIG. 20.

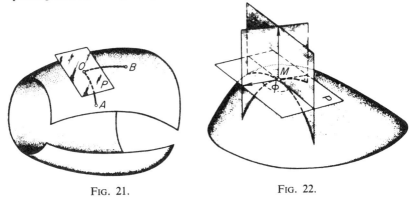

area of the surfaces. The area of the surface itself is defined as the limit of the sums of the areas of the segments $P_1, P_2, \cdots, P_n$ under the condition that the partitions of the surface become finer.* From this we can derive an exact expression for the area in the form of a double integral.

These remarks clearly demonstrate the significance of the concept of the tangent plane. However, in many questions the approximate representation of an element of a surface by means of a plane is inadequate and it is necessary to consider the curvature of the surface.

Curvature of curves on a surface. The curvature of a surface at a given point is characterized by the rate at which the surface leaves its tangent plane. But in different directions, the surface may leave its tangent plane at different rates. Thus the surface illustrated in figure 21 leaves the plane P in the direction OA at a faster rate than in the direction OB. So it is natural to define the curvature of a surface at a given point by means of the set of curvatures of all curves lying in the surface and passing through the given point in different directions.

Fig. 21.   Fig. 22.

This is done as follows. We construct the tangent plane P through the point M and choose a specific direction for the normal (figure 22). Then we consider curves which are sections of the surface cut by planes passing through the normal at the point M; these curves are called *normal sections*. The curvature of a normal section is given a sign, which is plus if the section is concave in the direction of the normal and minus if it is concave in the opposite direction. Thus, in a surface which is saddle-shaped, as illustrated in figure 23 with the arrow indicating the

---

* This is exactly the expression for the area which was used in §1, Chapter VIII.

direction of the normal to the surface, the curvature of the section MA is positive and that of the section MB is negative.

A normal section is defined by the angle ϕ by which its plane is rotated from some initial ray in the tangent plane (figure 22). If we know the curvature of the normal section $k(\phi)$ in terms of the angle ϕ, we will have a rather complete picture of the behavior of the surface in the vicinity of the point M.

A surface may be curved in many different ways and thus it would appear that the dependence of the curvature k on the angle ϕ may be arbitrary. In fact this is not so. For the surfaces studied in differential geometry, there exists a simple law, due to Euler, that establishes the connection between the curvatures of the normal sections passing through a given point in various directions.

It is shown that at each point of a surface there exist two particular directions such that

1. They are mutually perpendicular;

2. The curvatures k_1 and k_2 of the normal sections in these directions are the smallest and largest values of the curvatures of all normal sections;*

3. The curvature $k(\phi)$ of the normal section rotated from the section with curvature k_1 by the angle ϕ is expressed by the formula

$$k(\phi) = k_1 \cos^2 \phi + k_2 \sin^2 \phi. \tag{4}$$

Such directions are called the *principal directions* and the curvatures k_1 and k_2 are called the *principal curvatures of the surface* at the given point.

This theorem of Euler shows that in spite of the diversity of surfaces, their form in the neighborhood of each point must be one of a very few completely defined types, with an accuracy to within magnitudes of the second order of smallness in comparison with the distance from the given point. In fact, if k_1 and k_2 have the same sign, then the sign of $k(\phi)$ is constant and the surface near the point has the form illustrated in figure 22. If k_1 and k_2 have opposite signs, for example $k_1 > 0$ and $k_2 < 0$, then the curvature of the normal section obviously changes sign. This is seen from the fact that for $\phi = 0$ the curvature $k = k_1 > 0$ and for $\phi = \pi/2$ we have $k = k_2 < 0$.

From formula (4) for $k(\phi)$, it is not difficult to prove that as ϕ changes

---

* In the particular case $k_1 = k_2$ the curvature of all sections is the same; as, for example, on a sphere.

from 0 to π the sign of $k(\phi)$ changes twice,* so that near the point the surface has a saddle-shaped form (figure 23).

When one of the numbers k_1 and k_2 is equal to zero, the curvature always has the same sign, except for the one value of ϕ, for which it vanishes. This occurs, for example, for every point on a cylinder (figure 24).

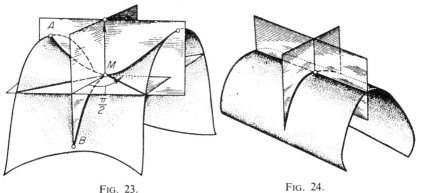

FIG. 23.   FIG. 24.

In the general case the surface near such point has a form close to that of a cylinder.

Finally, for $k_1 = k_2 = 0$ all normal sections have zero curvature. Near such a point the surface is especially "close" to its tangent plane. Such points are called *flat points*. One example of such a point is given in figure 25 (the point M). The properties of a surface near a flat point may be very complicated.

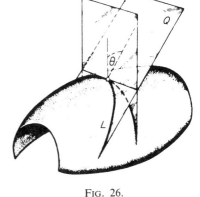

FIG. 25.   FIG. 26.

---

* It is a simple matter to show that $k(\phi) = k_1 \cos^2 \phi + k_2 \sin^2 \phi$ vanishes for $\phi = \arctan \sqrt{-k_1/k_2}$ and $\phi = \pi - \arctan \sqrt{-k_1/k_2}$, changing sign the first time from plus to minus and the second from minus to plus.

Let us now consider a section of the surface cut by an arbitrary plane Q (figure 26) not passing through the normal. The curvature k_L of such a curve L, as Meusnier showed,* is connected by a simple relation with the curvature k_N of the normal section in the same direction, i.e., the one that intersects the tangent plane in the same straight line. This connection is expressed by the formula

$$k_L = \frac{|k_N|}{\cos \theta},$$

where θ is the angle between the normal and the plane Q. The correctness of this formula may be visualized very conveniently on a sphere.

Finally, the curvature of *any* curve lying in the surface and having the plane Q as its osculating plane may be shown to be identical with the curvature of the intersection of Q with the surface.

Thus, if we know k_1 and k_2, the curvature of any curve in the surface is defined by the direction of its tangent and the angle between its osculating plane and the normal to the surface. Consequently, the character of the curvature of a surface at a given point is defined by the two numbers k_1 and k_2. Their absolute values are equal to the curvatures of two mutually perpendicular normal sections, and their signs show the direction of the concavity of the respective normal sections with respect to a chosen direction on the normal.

Let us now prove the theorems of Euler and Meusnier mentioned earlier.

1. For the proof of Euler's theorem we need the following lemma. If the function $f(x, y)$ has continuous second derivatives at a given point, then the coordinate axes may be rotated through an angle α such that in the new coordinate system the mixed derivative $f_{x'y'}$ will be equal to zero at this point.† We recall that after rotation of axes the new variables x', y' are connected with x and y by the formulas

$$x = x' \cos \alpha - y' \sin \alpha; \quad y = x' \sin \alpha + y' \cos \alpha$$

(cf. Chapter III, §7). For the proof of the lemma we note that

$$\frac{\partial x}{\partial x'} = \cos \alpha, \ \frac{\partial y}{\partial x'} = \sin \alpha, \ \frac{\partial x}{\partial y'} = -\sin \alpha, \ \frac{\partial y}{\partial y'} = \cos \alpha.$$

---

* Meusnier (1754–1793) was a French mathematician, a student of Monge; he was a general in the revolutionary army and died of wounds received in battle.

† We will denote partial derivatives by subscripts; for example, in place of $\partial f/\partial x$ we write f_x, in place of $\partial^2 f/\partial y^2$ we write f_{yy}, etc.

Computing the derivative $f_{x'y'}$ by the chain rule, we arrive after some calculation at the result

$$f_{x'y'} = f_{xy} \cos 2\alpha + \tfrac{1}{2}(f_{yy} - f_{xx}) \sin 2\alpha,$$

from which it readily follows that for

$$\cot 2\alpha = \frac{1}{2} \frac{f_{xx} - f_{yy}}{f_{xy}}$$

we will have

$$f_{x'y'} = 0.$$

We now consider the surface F, given by the equation $z = f(x, y)$, in which the origin is at the point M under consideration and the axes Ox and Oy are so chosen in the tangent plane that $f_{xy}(0, 0) = 0$. In the surface P we take an arbitrary straight line making an angle ϕ with the

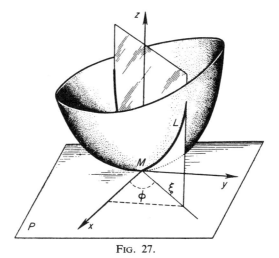

Fig. 27.

axis Ox and consider the normal section L in the direction of this straight line (figure 27). From the formula derived in §2, the curvature of L at the point M, taking its sign into account, is equal to

$$k_L = \lim_{\xi \to 0} \frac{2f(x, y)}{\xi^2}.$$

Here $f(x, y)$ is the distance (again taking its sign into account) of a point on L to the chosen straight line. Expanding $f(x, y)$ by Taylor's formula

(Chapter II, §9) and noting that $f_x(0, 0) = f_y(0, 0) = 0$ (since the axes Ox and Oy lie in the tangent plane) we get

$$f(x, y) = \tfrac{1}{2}(f_{xx}x^2 + f_{yy}y^2) + \epsilon(x^2 + y^2),$$

where $\epsilon \to 0$ as $x \to 0$, $y \to 0$. For a point on L, we have $x = \xi \cos\phi$, $y = \xi \sin\phi$, $\xi^2 = x^2 + y^2$ (figure 27), and thus

$$k_L = \lim_{\xi \to 0} \frac{f_{xx}\xi^2 \cos^2\phi + f_{yy}\xi^2 \sin^2\phi + 2\epsilon\xi^2}{\xi^2} = f_{xx}\cos^2\phi + f_{yy}\sin^2\phi.$$

Putting $\phi = 0, \phi = \pi/2$, we find that f_{xx} and f_{yy} are the curvatures k_1 and k_2 of the normal sections in the direction of the axes Ox and Oy. Thus the formula derived is actually Euler's formula: $k = k_1 \cos^2\phi + k_2 \sin^2\phi$. The fact that k_1 and k_2 are the maximal and minimal curvatures also follows from this formula.

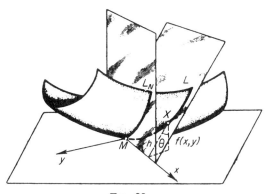

Fig. 28.

2. For the proof of Meusnier's theorem we consider a normal section L_N and a section L whose plane forms an angle θ with the plane of the section L_N, as in figure 28. The axes Ox and Oy lie in the tangent plane, and we also take the axis Ox to be tangent to the curves L_N and L at the origin. The distance $h(x, y)$ to the Ox axis of a point X on L with coordinates x, y, $f(x, y)$ is obviously equal to $h(x, y) = |f(x, y)|/\cos\theta$ (figure 28). Using Taylor's formula, we express the curvature k_L of the curve L in the following manner:

$$
\begin{aligned}
k_L &= \lim_{x \to 0} \frac{2h(x, y)}{x^2} = \lim_{x \to 0} 2\frac{|f(x, y)|}{x^2 \cos\theta} \\
&= \lim_{x \to 0} \frac{|f_{xx}x^2 + 2f_{xy}xy + f_{yy}y^2 + 2\epsilon(x^2 + y^2)|}{x^2 \cos\theta},
\end{aligned}
\tag{5}
$$

where $\epsilon \to 0$ as $x, y \to 0$. Since the axis Ox is tangent to the curve L, obviously $\lim_{x\to 0} y/x = 0$. Thus, taking the limit in formula (5), we get

$$k_L = \frac{|f_{xx}|}{\cos \theta}.$$

But for the chosen coordinate system the curve L_N has the equation $z = f(x, 0)$, for which $|k_N| = |f_{xx}|$. Thus $k_L = |k_N|/\cos \theta$ and Meusnier's theorem is proved.

Mean curvature. In many questions of the theory of surfaces, the most important role is played not by the principal curvatures themselves but by certain quantities dependent on them, namely the *mean curvature* and the *Gaussian* or *total curvature* of the surface at a given point. Let us examine them in detail.

The mean curvature of a surface at a given point is the average of the principal curvatures

$$K_{\mathrm{av}} = \tfrac{1}{2}(k_1 + k_2).$$

As an example of the usefulness of this concept, we consider the following mechanical problem. We assume that over the surface of some body F there is stretched a taut elastic rubber film. We ask about the pressure exerted by this film on each point of the surface of F.

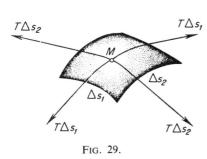

FIG. 29.

The pressure at a point M is measured by the force exerted by the film on a segment of the surface of unit area containing the point M; to be more exact, the pressure "at the point" M is defined as the limit of the ratio of this force to the area of the segment as the latter shrinks to the point M.

We surround the point M on the surface with a small curvilinear rectangle whose sides have lengths Δs_1 and Δs_2 and are perpendicular to the first and second principal directions at M (figure 29).* On each side of the rectangle there is exerted a force that is proportional (from the assumed uniformity of the tension) to the length of the side and the tension T acting on the film. Thus, on the sides perpendicular to

---

* Our reasoning here is not rigorous. However, by making estimates of the errors introduced, it is possible to give a rigorous proof of the result.

the first principal direction, there are exerted forces that are approximately equal to $T\Delta s_1$ and have the direction of the tangent to the surface. Similarly, forces equal to $T\Delta s_2$ act on the other pair of sides of the rectangle. In order to find the pressure at the point M, we must divide the resultant of these four forces by the area of the rectangle (approximately equal to $\Delta s_1 \Delta s_2$) and pass to the limit for Δs_1, $\Delta s_2 \to 0$. Let us begin by dividing the resultant of the first two forces by $\Delta s_1 \Delta s_2$.

If we examine the rectangle from its side (figure 30), we see that these forces are directed along tangents to the curve of the first normal section and that the distance between their points of application is exactly Δs_2. So we have the same problem here as in §2 for the pressure of a string on a support. Using the earlier result,

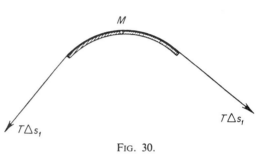

FIG. 30.

we find that the desired limit is equal to $k_1 T$, where k_1 is the curvature of the first normal section. With a similar expression for the other two forces, we obtain the formula:

$$P_M = T(k_1 + k_2) = 2TK_{\text{av}}.$$

This result has many important consequences. Let us consider an example.

It is known that the surface film of a liquid is under a tension that is the same in all directions on the surface. For a mass of liquid bounded by a curved surface, this tension, by the previous result, exerts a pressure on the surface which is proportional to its mean curvature at the given point.

So in drops of very small diameter the pressures are very large, a fact that hinders the formation of such drops. In a cooling vapor the drops begin to form, as a rule, around specks of dust and around charged particles. In a completely pure, slightly cooled vapor, the formation of drops is delayed. But if, for example, a particle passes through the vapor at high speed, causing ionization of the molecules, then around the ions formed in its path there will momentarily appear small drops of vapor, constituting a visible track of the particle. This is the basis for construction of the Wilson chamber, widely used in nuclear physics for observing the motions of various charged particles.

Since the pressure exerted by a liquid is the same in all directions, a drop of liquid in the absence of other sources of pressure must assume a form for which at all points of the surface the mean curvature is the same. In the experiment of Plateau, we take two liquids of the same specific weight, so that a clot of one of them will float in equilibrium in the other. It may be assumed that the floating liquid is acted on only by surface tension,* and it turns out that the "floating" liquid always takes the form of a sphere. This result suggests that every closed surface with constant mean curvature is a sphere, a theorem that is in fact true, although the strict mathematical proof of it is very difficult.

It is possible to approach the question from still another side. In view of the fact that the surface tension tends to decrease the area of the surface, while the volume of the liquid cannot change, it is natural to expect that the floating mass of liquid will have the smallest surface for a given volume. It can be proved that a body with this property is a sphere.

The relation between the lateral pressure of the film and its mean curvature can also be used to determine the form of a soap film suspended in a contour. Since the lateral pressure over the surface of the film, being directed along the normal to the surface, is not opposed by any reaction of the support (the support in this case is simply not there), it must be equal to zero, so that for the desired surface we have the condition

$$K_{\mathrm{av}} = 0. \tag{6}$$

From the analytic expression for mean curvature, we obtain a differential equation, and the problem consists of solving this equation under the condition that the desired surface passes through the given contour.† There have been many investigations of this difficult problem.

The same equation (6) arises from the problem of finding the surface of least area bounded by a given contour. From a physical point of view, the identity of these two problems is a natural one, since the film tends to decrease its area and reaches a position of stable equilibrium only when it attains the minimal area possible under the given conditions. Surfaces of zero mean curvature, by reason of their connection with this problem, are called *minimal*.

The mathematical investigation of minimal surfaces is of great interest, partly because of their wide variety of essentially different shapes, as

---

* The increase of pressure with depth may be ignored, since it is the same for both liquids because of their having the same specific weight. So on their common boundary the additional internal and external pressures caused by the depth are neutralized by each other.

† For a surface given by the equation $z = z(x, y)$, equation (6) assumes the form

$$(1 + z_y'^2)z_{xx}'' - 2z_x'z_y'z_{xy}'' + (1 + z_x'^2)z_{yy}'' = 0.$$

discovered by experiments with soap film. Figure 31 illustrates two soap films suspended from different contours.

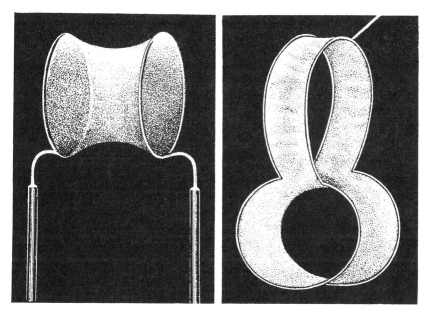

FIG. 31.

Gaussian curvature. The *Gaussian curvature* of a surface at a given point is the product of the principal curvatures

$$K = k_1 k_2 .$$

The sign of the Gaussian curvature defines the character of the surface near the point under consideration. For $K > 0$ the surface has the form of a bowl (k_1 and k_2 have the same sign) and for $K < 0$, when k_1 and k_2 have different signs, the surface is like a saddle. The remaining cases, discussed earlier, correspond to zero Gaussian curvature. The absolute value of the Gaussian curvature gives the degree of curvature of the surface in general, as a sort of abstraction from the various curvatures in different directions. This becomes particularly clear if we consider a different definition of Gaussian curvature, which does not depend on investigating curves on the surface.

Let us consider a small segment G of the surface F, containing the point M in its interior, and at each point of this segment let us erect a normal to the surface.

If we translate the initial points of all these normals to one point, then they fill out a solid angle (figure 32). The size of this solid angle will depend on the area of the segment G and on the extent to which the surface is curved on this segment. Thus the degree of curvature of

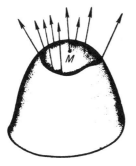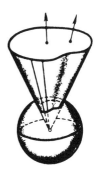

FIG. 32.

the segment G may be characterized by the ratio of the size of the solid angle to the area of G; so it is natural to define the curvature of the surface at a given point as the limit of this ratio when the segment G shrinks to the point M.* It turns out that this limit is equal to the absolute value of the Gaussian curvature at the point M.

The most remarkable property of the Gaussian curvature, which explains its great significance in the theory of surfaces, is the following. Let us suppose that the surface has been stamped out from a flexible but inextensible material, say a very thin sheet of tin, so that we can bend it into various shapes without stretching or tearing it. During this process the principal curvatures will change but, as Gauss showed, their product $k_1 k_2$ will remain unchanged at every point. This fundamental result shows that two surfaces with different Gaussian curvatures are inherently distinct from each other, the distinction consisting of the fact that if we deform them in every possible way, without stretching or tearing, we can never superpose them on each other. For example, a segment of the surface of a sphere can never be distorted so as to lie on a plane or on the surface of a sphere of different radius.

We have now considered certain basic concepts in the theory of surfaces. As for the methods used in this theory, they consist, as was stated previously, primarily in the application of analysis and above all of

---

* To measure the solid angle itself, we construct a sphere of unit radius with center at its vertex. The area of the region in which the sphere intersects the solid angle is then taken as the size of the solid angle (figure 32).

differential equations. Simple examples of the use of analysis are to be found in the proofs for the theorems of Euler and Meusnier. For more complicated questions, we require a special method of relating problems in the theory of surfaces to problems in analysis. This method is based on the introduction of so-called curvilinear coordinates and was first widely used in the work of Gauss on problems of the type discussed in the following section.

§4. Intrinsic Geometry and Deformation of Surfaces

Intrinsic geometry. As indicated previously, a deformation of a surface is defined as a change of shape that preserves the lengths of all curves lying in the surface. For example, rolling up a sheet of paper into a cylindrical tube represents, from the geometric point of view, a deformation of part of the plane, since in fact the paper undergoes practically no stretching, and the length of any curve drawn on it is not changed by its being rolled up. Certain other geometric quantities connected with the surface are also preserved; for example, the area of figures on it. All properties of a surface that are not changed by deformations make up what is called the *intrinsic geometry* of the surface.

But just which are these properties? It is clear that in a deformation only those properties can be preserved which in the final analysis depend entirely on lengths of curves, i.e., which may be determined by measurements carried out on the surface itself. A deformation is a change of shape preserving the length of curves, and any property which cannot change under *any* deformation must be definable in one way or another in terms of length. Thus intrinsic geometry is simply called *geometry on a surface*. The very meaning of the words "intrinsic geometry" is that it studies intrinsic properties of the surface itself, independent of the manner in which the surface is embedded in the surrounding space.* Thus, for example, if we join two points on a sheet of paper by a straight line and then bend the paper (figure 33), the segment becomes a curve but its property of being the shortest of all lines joining the given points on the surface is preserved; so this property belongs to intrinsic geometry. On the other hand, the curvature of this line will depend on how the paper was bent and thus is not a part of intrinsic geometry.

In general, since the proofs of plane geometry make no reference to the properties of the surrounding space, all its theorems belong to the

---

* We note that the ideas of intrinsic geometry have led to a wide generalization of the mathematical concept of space and have thereby played a very important role in contemporary physics; for details see Chapter XVII.

intrinsic geometry of any surface obtainable by deformation of a plane. One may say that plane geometry is the intrinsic geometry of the plane.

Another example of intrinsic geometry is familiar to everyone, namely geometry on the surface of a sphere, with which we usually have to deal

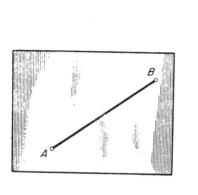 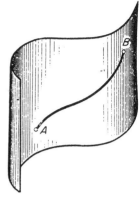

FIG. 33.

in making measurements on the surface of the earth. This example is a particularly good one to illustrate the essential nature of intrinsic geometry; because of the large radius of the earth, any immediately visible area of its surface appears to us as part of a plane, so that the deviations from plane geometry observable in the measurements of large distances impresses us as resulting not from the curvature of the earth's surface in space but from the inherent laws of "terrestrial geometry," expressing the geometric properties of the surface of the earth itself.

It remains to note that the idea of studying intrinsic geometry occurred to Gauss in connection with the problems of geodesy and cartography. Both these applied sciences are concerned in an essential way with the intrinsic geometry of the earth's surface. Cartography deals, in particular, with distortions in the ratios of distances when part of the surface of the earth is mapped on a plane and thus with distinguishing between plane geometry and the intrinsic geometry of the surface of the earth.

The intrinsic geometry of any surface may be pictured in the same way. Let us imagine that on a given surface there exist creatures so small that within the limits of their range of vision the surface appears to be plane (we know that a sufficiently small segment of any smooth surface differs very little from a tangent plane); then these creatures will not notice that the surface is curved in space, but in measuring large distances they will nevertheless convince themselves that in their geometry certain nonplanar laws prevail, corresponding to the intrinsic geometry of the surface on which they live. That these laws are actually different for different surfaces may easily be seen from the following simple discussion. Let us choose a point O on the surface and consider a curve L such that the distance of each of its points from the point O, measured on the surface (i.e., along the shortest curve connecting this point to the point O) is equal to a fixed number r (figure 34). The curve L, from the

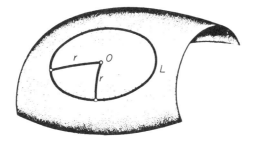

Fig. 34.

point of view of the intrinsic geometry of the surface, is simply the circumference of a circle of radius r. A formula expressing the length $s(r)$ in terms of r is part of the intrinsic geometry of the given surface. But such a formula may vary widely in character, depending on the nature of the surface: Thus on a plane, $s(r) = 2\pi r$; on a sphere of radius R, as can easily be shown, $s(r) = 2\pi R \sin r/R$; on the surface illustrated in figure 35, beginning with a certain value of r, the length of the cir-

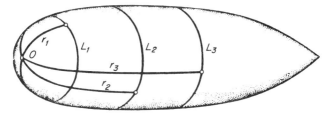

Fig. 35.

cumference with center O and radius r is at first independent of r but then begins to decrease. Consequently, all these surfaces have different intrinsic geometries.

The basic concepts of intrinsic geometry. To illustrate the wide range of concepts and theorems in intrinsic geometry, we may turn to plane geometry which, as we have seen, is the intrinsic geometry of the plane. Its subject matter consists of plane figures and their properties, which are usually expressed in the form of relations among basic geometric quantities such as length, angle, and area. For a rigorous proof that angle and area belong to the intrinsic geometry of the plane, it is necessary to show that they can be expressed in terms of length. But this is certainly so; in fact, an angle may be computed if we know the length of the sides of a triangle containing it, and the area of a triangle can also be computed in terms of its sides, while to compute the area of a polygon we need only divide it into triangles.

In considering plane geometry as the intrinsic geometry of the plane, there is no need to restrict ourselves to ideas learned in school. On the contrary, we may develop it as far as we like and study many new problems, provided only they can be stated, in the final analysis, in terms of length. Thus, in plane geometry we may successively introduce the length of a curve, the area of a surface bounded by curves, and so forth; they are all a part of the intrinsic geometry of the plane.

The same concepts are introduced in the intrinsic geometry of an arbitrary surface. The length of a curve is the initial concept; the definition of angles and areas is somewhat more complicated. If the intrinsic geometry of a given surface differs from plane geometry, we cannot use the customary formulas to define an angle or an area in terms of length. However, as we have seen, a surface near a given point differs little from its tangent plane. Speaking more precisely, the following is true: If a small segment of a surface containing a given point M is projected on the tangent plane at this point, then the distance between points, measured on the surface, differs from the distance between their projections by an infinitesimal of higher than the second order in comparison with distances from the point M. Thus in defining geometric quantities

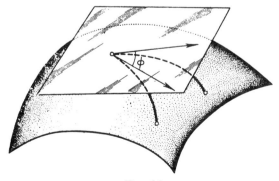

Fig. 36.

at a given point of a surface by taking a limit in which infinitesimals occur of order no higher than the second, we may replace a segment of the surface by its projection on the tangent plane. Thus the quantities determined by measurement in the tangent plane turn out to belong to the intrinsic geometry of the surface. This possibility of considering a small segment of the surface as a plane is the basis of the definitions of all the concepts of intrinsic geometry.

As an example let us consider the definitions of angle and area. Following the general principle, we define the angle between curves on a surface as the angle between their projections on the tangent plane (figure 36). Obviously the angle defined in this manner is identical with the angle between the tangents to the curves. The definition of area given in §3 is based on the same principle. Finally, in order that the tendency of a curve to twist in space may be defined "within" the surface itself, we introduce the concept of "geodesic curvature" the name being reminiscent of measurements on the surface of the earth. The *geodesic curvature* of a curve at a given point is defined as the curvature of its projection on the tangent plane (figure 37).

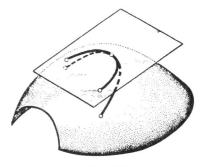

Fig. 37.

In this manner we see that the basic concepts of plane geometry may be introduced into the intrinsic geometry of an arbitrary surface.

In any arbitrary surface it is also easy to define figures analogous to the basic figures on the plane. For example, we have been dealing previously with circumferences of circles, which are defined precisely as in the case of the plane. Similarly, we may define the analogue of a line segment, namely a geodesic segment, as the shortest curve on the surface joining two given points. Further, it is natural to define a triangle as a figure bounded by three geodesic segments and similarly for a polygon, and so forth. Since the properties of all these figures and magnitudes depend on the surface, there exist in this sense infinitely many different intrinsic geometries. But intrinsic geometry, as a special branch of the theory of surfaces, pays particular attention to certain general laws holding for the intrinsic geometry of any surface and makes clear how these laws are expressed in terms of the quantities which characterize a given surface.

Thus, as we have noted earlier, one of the most important characteristics of a surface, its Gaussian curvature, is not changed by deformation, i.e., depends only on the intrinsic geometry of the surface. But it turns out that in general the Gaussian curvature already characterizes, to a remarkable degree, the extent to which the intrinsic geometry of the surface near a given point differs from plane geometry. As an example let us consider on a surface a circle L of very small radius r, with center at a given point O. On a plane the length $s(r)$ of its circumference is expressed by the formula $s(r) = 2\pi r$. On a surface differing from a plane, the dependence of the circumference on the radius is different; here the deviation of $s(r)$ from $2\pi r$, depends essentially, for small r, on the Gaussian curvature K at the center of the circle, namely;

$$s(r) = 2\pi r - \frac{\pi}{3} Kr^3 + \epsilon r^3,$$

where $\epsilon \to 0$ as $r \to 0$. In other words, for small r the circumference may be computed by the usual formula if we disregard terms of the third degree of smallness, and in this case the error (with accuracy to terms of higher than the third order) is proportional to the Gaussian curvature. In particular, if $K > 0$, then the circumference of a circle of small radius is smaller than the circumference of a circle with the same radius in a plane, and if $K < 0$, it is larger. These latter facts are easy to visualize: Near a point with positive curvature the surface has the shape of a bowl so that circumferences are reduced, whereas near a point with negative curvature the circumference, being situated on a "saddle," has a wavelike shape and is thus considerably lengthened (figure 38).

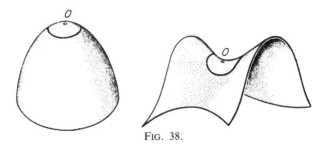

FIG. 38.

From the theorem just mentioned, it follows that a surface with varying Gaussian curvature is extremely inhomogeneous from a geometric point of view; the properties of its intrinsic geometry change from point to point. The general character of the problems of intrinsic geometry causes it to resemble plane geometry, but this inhomogeneity, on the other hand, makes it profoundly different from plane geometry. On the plane, for example, the sum of the angles of a triangle is equal to two right angles; but on an arbitrary surface the sum of the angles of a triangle, (with geodesics for sides) is undetermined even if we are told that it lies on a known surface and has sides of given length. However, if we know the Gaussian curvature K at every point of the triangle, then the sum of its angles, α, β, γ, can be computed by the formula

$$\alpha + \beta + \gamma = \pi + \iint K \, d\sigma,$$

where the integral is taken over the surface of the triangle. This formula contains as a special case the well-known theorems on the sum of the angles of a triangle in the plane and on the unit sphere. In the first case $K = 0$ and $\alpha + \beta + \gamma = \pi$, while in the second $K = 1$ and $\alpha + \beta + \gamma = \pi + S$, where S is the area of the spherical triangle.

It may be proved that every sufficiently small segment of a surface with zero Gaussian curvature may be deformed, or, as it is customary to say, developed into a plane, since it has the same intrinsic geometry as the plane. Such surfaces are called *developable*. And if the Gaussian curvature is near zero, then although the surface cannot be developed into a plane, still its intrinsic geometry differs little from plane geometry, which indicates once again that the Gaussian curvature acts as a measure of the extent to which the intrinsic geometry of a surface deviates from plane geometry.

Geodesic lines. In the intrinsic geometry of a surface the role of straight lines is played by geodesic lines, or, as they are usually called, "geodesics."

A straight line in a plane may be defined as a line made up of intervals overlapping one another. A geodesic is defined in exactly the same way, with geodesic segments taking the place of intervals. In other words, a *geodesic* is a curve on a surface such that every sufficiently small piece of it is a shortest path. Not every geodesic is a shortest path in the large, as may be noted on the surface of a sphere, where every arc of a great circle is a geodesic, although this arc will be the shortest path between its end points only if it is not greater than a semicircle. A geodesic, as we see, may even be a closed curve.

To illustrate certain important properties of geodesics, let us consider the following mechanical model.* On the surface F let there be stretched

---

* As noted previously, our reasoning here is not a strict proof of the properties of geodesic curves. It is given only to illustrate the most important of these properties.

a rubber string with fixed ends (figure 39).* The string will be in equilibrium

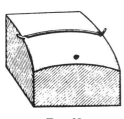

when it has the shortest possible length, since any change in its position will then involve an increase of length, which could be produced only by external forces. In other words, the string will be in equilibrium if it is lying along a geodesic. But for equilibrium, it is necessary that the elastic forces on each segment of the string be counterbalanced by the resistance of the surface, directed along the normal to it. (We assume that the surface is smooth and that there is no friction between it and the string.)

FIG. 39.

But it was proved in §2 that the pressure on the support caused by the tension of the string is directed along the principal normal to the curve along which the string lies. Thus we are led to the following result: The principal normal to a geodesic at each point coincides in direction with the normal to the surface. The converse of this theorem is also true: Every curve on a regular surface which has this property is a geodesic.

This property of a geodesic allows us to deduce the following important fact: If a material point is moving on a surface in such a way that there are no forces acting on it except for the reaction of the surface, then it follows a geodesic. For, as we know from §2, the normal acceleration of a point is directed along the principal normal to the trajectory and since the reaction of the surface is the only force acting on the point, the principal normal to the trajectory is identical with the normal to the surface, so that from the preceding theorem the trajectory is a geodesic. This last property of geodesics increases their resemblance to straight lines. Just as the motion of a free point, because of inertia, is along a straight line, so the motion of a point forced to stay on a surface, but not affected by external forces, will be along a geodesic.†

From the same property of geodesics comes the following theorem. If two surfaces are tangent along a curve that is a geodesic on one of them, then this curve will also be a geodesic on the other. For at each point of the curve, the surfaces have a common tangent plane and consequently a common normal, and since the curve is a geodesic on one of the surfaces, this normal coincides with the principal normal to the curve, so that on the second surface also the curve will be a geodesic.

---

* A stretched string will not remain on a surface unless the surface is convex; so in order not to make exceptions, it is better to imagine that the surface is in two layers, with the string running between them.

† Here by "external" forces we mean all forces except the reaction of the surface.

From these results follow two further intuitive properties of geodesic curves. In the first place, if an elastic rectangular plate (for example a steel ruler) lies with its median line completely on a surface, then it is tangent to this surface along a geodesic. (Evidently the line of contact is a geodesic on the ruler, so that it must be a geodesic also on the surface.) Second, if a surface rolls along a plane in such a way that the point of contact traces a straight line on the plane, then the trace of this straight line on the surface is a geodesic.* Both these properties are readily demonstrated on a cylinder, where it is easy to convince oneself by experiment that the median line of a straight plane strip lying on the cylinder (figure 40) coincides with either a generator or the circumference of a

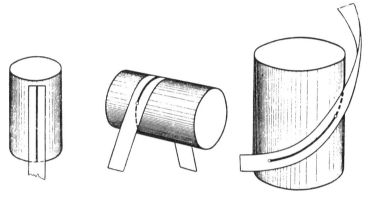

FIG. 40.

circle or a helix, and it is not difficult to prove that a geodesic curve on a cylinder can be only one of these three. The same curves will be traced out on a cylinder if we roll it on a plane on which we have drawn a straight line in chalk.

The analogy between geodesics and straight lines in a plane may be supplemented by still another important property, taken directly from the definition of a geodesic. Namely, straight lines in the plane may be defined as curves of zero curvature and geodesics on a surface as curves of zero geodesic curvature. (We recall that the geodesic curvature is the curvature of the projection of the curve on the tangent plane, cf. figure 37.) It is quite natural that our present definition of a geodesic should coincide with the earlier one; for if at every point of the curve the curvature of

---

* This proposition does not differ essentially from the preceding one, since the rolling of a surface on a plane is equivalent in a well-defined sense to the unwinding of a plane strip along the surface.

its projection on the tangent plane is equal to zero, then the curve departs from its tangent essentially in the direction of the normal to the surface, so that the principal normal to the curve is directed along the normal to the surface and the curve is a geodesic in the original sense. Conversely, if a curve is a geodesic, then its principal normal, and so also its deviation from the tangent line, are directed along the normal to the surface, so that in projecting on the tangent plane we get a curve in which the deviation from the tangent is essentially smaller than for the original curve, and the curvature of the projection so formed turns out to be equal to zero.

The course of a geodesic may vary widely for different surfaces. As an example, in figure 41 we trace some geodesics on a hyperboloid of revolution.

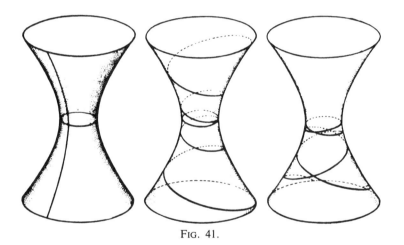

Fig. 41.

Deformation of surfaces. Since intrinsic geometry studies the properties of surfaces that are invariant under deformation, it naturally investigates these deformations themselves. The theory of deformation of surfaces is one of the most interesting and difficult branches of geometry and includes many problems which, although simple to state, have not yet been finally solved.

Certain questions about the deformation of surfaces were already considered by Euler and Minding, but general results for arbitrary surfaces were not derived until later.

In the general theory of deformation, we first of all raise the question whether deformation is possible for all surfaces and, if so, to what extent.

For analytic surfaces, i.e., surfaces defined by functions of the coordinates that can be expanded in a Taylor series, this question was solved at the end of the last century by the French mathematician Darboux. In particular, he showed the following: If on such a surface we consider any geodesic and assign in space an arbitrary (analytic) curve with the same length, and with curvature nowhere equal to zero, then a sufficiently narrow strip of the surface, containing the given geodesic, can be deformed so that the geodesic coincides with the given curve.* This theorem shows that a strip of the surface may be deformed rather arbitrarily. However, it has been proved that if a geodesic is to be transformed into a preassigned curve, then the surface may be deformed in no more than two ways. For example, if the curve is plane, then the two positions of the surface will be mirror images of each other in the plane. If the geodesic is a straight line, then this last proposition is not true, as can be shown by deforming a cylindrical surface.

We have defined a deformation as a transformation of the surface that preserves the lengths of all curves on the surface. Here we have considered only the final result of the transformation; the question of what happens to the curve during the process did not enter. However, in considering a surface as made from a flexible but unstretchable material, it is natural to consider a continuous transformation, at each instant of which the lengths remain unchanged (physically this corresponds to the unstretchability of the material). Such transformations are called *continuous* deformations.

At first glance it may seem that every deformation can be realized in a continuous manner, but this is not so. For example, it has been shown that a surface in the form of a circular trough (figure 42), does not admit

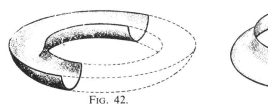 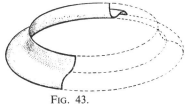

FIG. 42.   FIG. 43.

continuous deformations (this explains, among other things, the familiar fact that a pail with a curved rim is considerably stronger than one with a plain rim) although deformations of such a surface are possible: for

---

* The case of transforming a geodesic into a curve with zero curvature is excluded, since it is easy to show that for surfaces of positive Gaussian curvature this is impossible.

example, one may cut the trough along the circle on which it rests on a horizontal plane and replace one half of it by its mirror image in this plane (compare figure 43 with figure 42; to aid visualization we have drawn only the left half of the surface). It is intuitively clear that the impossibility of a continuous deformation is due to the circular shape of the trough; for a straight trough such a deformation can be performed continuously.

If we restrict ourselves to a sufficiently small segment of the surface, then there are no obvious hindrances to its continuous deformation, and we might expect that every deformation of a small segment of the surface can be realized by a continuous transformation, followed perhaps by a mirror reflection. This is in fact true, but only under the condition that on the given small segment of the surface the Gaussian curvature never vanishes (excepting the case that it vanishes everywhere). But if the Gaussian curvature vanishes at isolated points, then, as N. V. Efimov showed in 1940, even arbitrarily small segments of a regular surface may not admit any continuous deformation without loss of regularity. For example, the surface defined by the equation $z = x^9 + \lambda x^6 y^3 + y^9$, where λ is a transcendental number, has the property that no segment containing the origin, no matter how small it may be, admits sufficiently regular continuous deformations. Efimov's theorem is a new and somewhat unexpected result in classical differential geometry.

In addition to these general questions about deformation, a great deal of attention is being paid to special types of deformation of surfaces.

The connection of the intrinsic geometry of a surface with the form of the surface in space. We already know that certain properties of a surface, and of the figures on it, are defined by the intrinsic geometry of the surface even though these properties are very closely related to other properties that depend on how the surface is embedded in the surrounding space, properties that are, as they say, "extrinsic" to the surface. For example, the principal curvatures are extrinsic properties of a surface, but their product (the Gaussian curvature) is intrinsic. Another example, in order that the principal normal of a curve lying on a surface should coincide with the normal to the surface, it is necessary and sufficient that this curve have a property defined by its intrinsic geometry, namely that it be a geodesic.

Consequently, the intrinsic geometry of a surface will determine its space form only to a certain extent.

The dependence of the space form of a surface on its intrinsic geometry may be expressed analytically in the form of equations containing certain quantities that characterize the intrinsic geometry and certain other quantities that characterize the way in which the curved surface is embedded in space. One of these equations is the formula expressing the Gaussian curvature in intrinsic terms and is due to Gauss. Two other such equations are those of Peterson and Codazzi, mentioned in §1.

The equations of Gauss, Peterson, and Codazzi completely express the connection between the intrinsic geometry of a surface and the character of its curvature in space, since all possible interrelations between intrinsic and extrinsic properties of an arbitrary surface are included, at least in implicit form, in these equations.

Since the form of a surface in space is not completely defined by its intrinsic geometry, we naturally ask, What extrinsic properties must still be assigned in order to determine the surface completely? It turns out that if two surfaces have the same intrinsic geometry and if, at corresponding points and in corresponding directions, the curvatures of the normal sections of these surfaces have the same sign, then the surfaces are congruent; that is, they can be translated so as to coincide with each other. We note that Peterson discovered this theorem 15 years earlier than Bonnet, with whose name it is usually associated.

Analytic apparatus in the theory of surfaces. The systematic application of analysis to the theory of surfaces led to the building up of an analytic apparatus especially suitable for this purpose. The decisive step in this direction was taken by Gauss, who introduced the method of representing surfaces by so-called curvilinear coordinates. This method is a natural generalization of the idea of Cartesian coordinates on the plane and is closely connected with the intrinsic geometry of the surface, for which the presentation of the surface by an equation of the form $z = f(x, y)$ is not convenient. The inconvenience consists of the fact that the x, y coordinates of a point on the surface change when the surface is deformed. To eliminate this difficulty, the coordinates are chosen on the surface itself; they define each point by two numbers u and v, which are associated with the given point and remain associated with it even after deformation of the surface. The space coordinates x, y, z of the point will in each case be functions of u and v. The numbers u and v defining the point on a surface are called its *curvilinear coordinates*. The choice of name is to be understood as follows: If we fix the value of one of these coordinates, say v, and vary the other, then we get a coordinate curve on the surface. The coordinate curves form a curvilinear net on the surface, similar to the coordinate net on a plane. We note that the familiar method of describing the position of a point on the surface of the earth by means of longitude and latitude consists simply of introducing curvilinear coordinates on the surface of a sphere; the coordinate net in this case consists of circles, namely the meridians and parallels* (figure 44). To describe the spatial position of a surface by means of curvilinear coordinates, we need to define the position of each point in terms of u and v, for example by giving, as a function of u and v, the vector $r = r(u, v)$, issuing from some fixed origin to the points on the surface and called the radius vector of the surface. (This is equivalent to giving the x, y, and z components of the vector r as functions of u and v.)† To define a curve lying on a given surface, we need to give the coordinates u, v as functions of one parameter t; then the radius vector to a point moving along this curve is expressed as a composite function

$$r[u(t), v(t)].$$

For vector functions the concepts of derivative and differential may be generalized word for word; from the definition of the derivative as the limit of $\Delta r / \Delta t$ when $\Delta t \to 0$ (r is a function of the parameter t) it follows

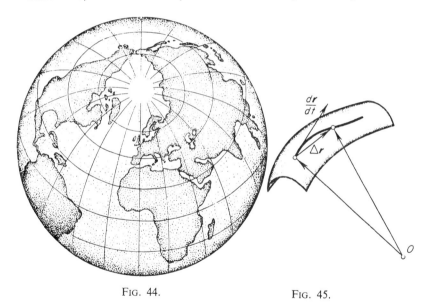

FIG. 44.   FIG. 45.

---

* It is characteristic that geographic coordinates and their practical applications were known long in advance of Descartes' introduction of the usual coordinates in the plane.

† Of course Gauss did not use vector notation, but defined the three coordinates x, y, z of the points of the surface separately as functions of u and v. Vectors, which were introduced as a result of the work of Hamilton and Grassmann, were at first used widely in physics and only later (in fact, in the 20th century) became the traditional apparatus for analytic and differential geometry.

at once that the derivative of the radius vector of a curve is a vector directed along the tangent to the curve (figure 45). For vector functions the basic properties of ordinary derivatives are still valid; for example, the chain rule

$$\frac{dr[u(t), v(t)]}{dt} = \frac{\partial r}{\partial u}\frac{du}{dt} + \frac{\partial r}{\partial v}\frac{dv}{dt} = r_u u_t' + r_v v_t' , \tag{7}$$

where r_u and r_v are the partial derivatives of the vector function $r(u, v)$.

The length of a curve, as can be shown, is expressed by the integral

$$s = \int \sqrt{x'^2(t) + y'^2(t) + z'^2(t)} \, dt.$$

Thus, the differential of the length of a curve is equal to

$$ds = \sqrt{x'^2(t) + y'^2(t) + z'^2(t)} \, dt.$$

But since $x'(t)$, $y'(t)$, and $z'(t)$ are components of the vector $dr/dt = r_t'$, we may write $ds = |r_t'| \, dt$, where $|r_t'|$ denotes the length of the vector r_t'. For curves lying on a surface, we get from (7)

$$ds = |r_u u_t' + r_v v_t'| \, dt.$$

Computing the square of the length of the vector on the right we obtain, by the rules of vector algebra,*

$$ds^2 = [r_u^2 u_t'^2 + 2r_u r_v u_t' v_t' + r_v^2 v_t'^2] \, dt^2 .$$

Passing to differentials and introducing the notation

$$r_u^2 = E(u, v), \quad r_u r_v = F(u, v), \quad r_v^2 = G(u, v),$$

we have

$$ds^2 = E \, du^2 + 2F \, du \, dv + G \, dv^2.$$

We see that the square of the differential of arc length on a surface is a quadratic form in the differentials du and dv with coefficients depending on the point of the surface. This form is called the *first fundamental quadratic form* of the surface. Given the coefficients E, F, and G of this

---

* The square of the length of a vector is the scalar product of the vector with itself, and for scalar multiplication (cf. Chapter III, §9) the usual rules hold for the removal of brackets.

form at each point on a surface we may compute the length of any curve on the surface by the formula

$$s = \int_{t_1}^{t_2} \sqrt{E u_t'^2 + 2F u_t' v_t' + G v_t'^2}\, dt,$$

so that its intrinsic geometry is thereby completely determined.

We show, as an example, how to express angle and area in terms of E, F, and G. Let two curves issue from a given point, one of them given by the equations $u = u_1(t)$, $v = v_1(t)$ and the other by the equations $u = u_2(t)$, $v = v_2(t)$. Then the tangents to these curves are given by the vectors

$$r_1 = r_u \frac{du_1}{dt} + r_v \frac{dv_1}{dt},$$

$$r_2 = r_u \frac{du_2}{dt} + r_v \frac{dv_2}{dt}.$$

The cosine of the angle between these vectors is equal to the scalar product $r_1 r_2$ divided by the product of the lengths $r_1 r_2$

$$\cos \alpha = \frac{r_1 r_2}{r_1 r_2}$$

$$= \frac{r_u^2 \dfrac{du_1}{dt} \dfrac{du_2}{dt} + r_u r_v \left(\dfrac{du_1}{dt} \dfrac{dv_2}{dt} + \dfrac{du_2}{dt} \dfrac{dv_1}{dt} \right) + r_v^2 \dfrac{dv_1}{dt} \cdot \dfrac{dv_2}{dt}}{r_1 r_2}.$$

Recalling that $r_u^2 = E$, $r_u r_v = F$, $r_v^2 = G$, we get

$$\cos \alpha =$$

$$\frac{E \dfrac{du_1}{dt} \dfrac{du_2}{dt} + F \left(\dfrac{du_1}{dt} \dfrac{dv_2}{dt} + \dfrac{du_2}{dt} \dfrac{dv_1}{dt} \right) + G \dfrac{dv_1}{dt} \dfrac{dv_2}{dt}}{\sqrt{E\left(\dfrac{du_1}{dt}\right)^2 + 2F \dfrac{du_1}{dt} \dfrac{dv_1}{dt} + G\left(\dfrac{dv_1}{dt}\right)^2} \sqrt{E\left(\dfrac{du_2}{dt}\right)^2 + 2F \dfrac{du_2}{dt} \dfrac{dv_2}{dt} + G\left(\dfrac{dv_2}{dt}\right)^2}}.$$

To obtain a formula for area, we consider a curvilinear rectangle bounded by the coordinate curves $u = u_0$, $v = v_0$, $u = u_0 + \Delta u$, $v = v_0 + \Delta v$, and we take as an approximation to it the parallelogram lying in the tangent plane and bounded by the vectors $r_u\, \Delta u$, $r_v\, \Delta v$, tangent to the coordinate curves (figure 46). The area of this parallelogram is

$\Delta s = |\, r_u \,|\, |\, r_v \,|\, \Delta u \, \Delta v \sin \phi$, where ϕ is the angle between r_u and r_v. Since $\sin \phi = \sqrt{1 - \cos^2 \phi}$, it follows that $\Delta s = |\, r_u \,|\, |\, r_v \,|\, \Delta u \, \Delta v \sqrt{1 - \cos^2 \phi}$ $= \sqrt{r_u^2 r_v^2 - |\, r_u \,|^2 \, |\, r_v \,|^2 \cos^2 \phi} \; \Delta u \, \Delta v$. Recalling that $r_u^2 = E$, $r_v^2 = G$, $|\, r_u \,| \cdot |\, r_v \,| \cos \phi = r_u r_v = F$, we get $\Delta s = \sqrt{EG - F^2} \, \Delta u \, \Delta v$. Summing up the areas of the parallelograms and taking the limit as $\Delta u, \Delta v \to 0$ we obtain the formula for area $S = \iint_D \sqrt{EG - F^2} \, du \, dv$, where the integration is taken over the domain D of the variables u and v which describe the given segment of the surface.

In this way, curvilinear coordinates are very convenient for studying the intrinsic geometry of a surface.

It also turns out that the manner in which a curved surface is embedded in the surrounding space can be characterized by a certain quadratic form in the differentials du, dv. Thus if n is a unit vector normal to the surface at the point M, and Δr is the increment in the radius vector to

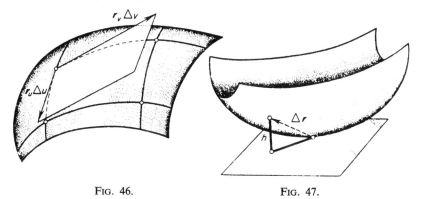

Fig. 46.   Fig. 47.

the surface as we move from this point, then the deviation h of the surface from the tangent plane (figure 47) is equal to $n \, \Delta r$. Expanding the increment Δr by Taylor's formula, we get

$$h = n \, dr + \tfrac{1}{2} n \, d^2 r + \epsilon (du^2 + dv^2),$$

where $\epsilon \to 0$ as $\sqrt{du^2 + dv^2} \to 0$. Since the vector dr lies in the tangent plane, we have $n \, dr = 0$. The last term, $\epsilon (du^2 + dv^2)$ is small in comparison with the squares of the differentials du and dv. There remains the principal term $\tfrac{1}{2} n \, d^2 r$. Thus twice the principal part of h, namely $n \, d^2 r$, is a quadratic form with respect to du and dv

$$n \, d^2 r = n r_{uu} \, du^2 + 2 n r_{uv} \, du \, dv + n r_{vv} \, dv^2.$$

This form describes the character of the deviation of the surface from the tangent plane. It is called the *second fundamental quadratic form* of the surface. Its coefficients, which depend on u and v, are usually written:

$$n\boldsymbol{r}_{uu} = L, \quad n\boldsymbol{r}_{uv} = M, \quad n\boldsymbol{r}_{vv} = N.$$

Knowing the second fundamental quadratic form, we can compute the curvature of any curve on a surface. Thus, applying the formula $k = \lim_{l\to 0} 2h/l^2$, we obtain the result that the curvature of the normal section in the direction corresponding to the ratio du/dv is equal to

$$k_n = \frac{n\,d^2\boldsymbol{r}}{ds^2} = \frac{L\,du^2 + 2M\,du\,dv + N\,dv^2}{E\,du^2 + 2F\,du\,dv + G\,dv^2}.$$

If the curve is not a normal section, then by Meusnier's theorem it is sufficient to divide the curvature of the normal section in the same direction by the cosine of the angle between the principal normal to the curve and the normal to the surface.

The introduction of the second fundamental quadratic form provides an analytic approach to the study of how the surface is curved in space. In particular, one may derive the theorems of Euler and Meusnier, the expressions for the Gaussian and mean curvature, and so forth, in a purely analytic way.

Peterson's theorem, mentioned earlier, shows that the two quadratic forms, taken together, define a surface up to its position in space, so that the analytic study of any properties of a surface consists of the study of these forms. In conclusion, we note that the coefficients of the two quadratic forms are not independent; the connection mentioned earlier between the intrinsic geometry of a curved surface and the way in which it is embedded in space is expressed analytically by three relations (the equations of Gauss-Codazzi) between the coefficients of the first and the second fundamental quadratic forms.

§5. New Developments in the Theory of Curves and Surfaces

Families of curves and surfaces. Even though the basic theory of curves and surfaces was to a large degree complete by the middle of the last century, it has continued to develop in several new directions, which greatly extend the range of figures and properties investigated in contemporary differential geometry. There is one of these developments whose origins go back to the beginning of differential geometry, namely the theory of "families" or of continuous collections of curves and surfaces, but this theory may be considered new in the sense that its more profound aspects were not investigated until after the basic theory of curves and surfaces was already completely developed.

In general a continuous collection of figures is called an *n-parameter family* if each figure of the collection is determined by the values of *n* parameters and all the quantities characterizing the figure (in respect to its position, form, and so forth) depend on these parameters in a manner which is at least continuous. From the point of view of this general definition, a curve may be considered as a one-parameter family of points and a surface as a two-parameter family of points. The collection of all circles in the plane is an example of a three-parameter family of curves, since a circle in the plane is determined by three parameters: the two coordinates of its center and its radius.

The simplest question in the theory of families of curves or surfaces consists of finding the so-called envelope of the family. A surface is called the *envelope* of a given family of surfaces if at each of its points it is tangent to one of the surfaces of the family and is in this way tangent to every one of them. For example, the envelope of a family of spheres of

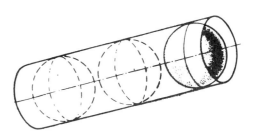

FIG. 48.

equal radius with centers on a given straight line will be a cylinder (figure 48), and the envelope of such spheres with centers on all points of a given plane will consist of two parallel planes. The envelope of a

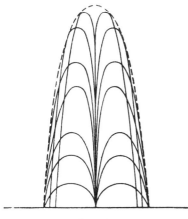

FIG. 49.

family of curves is defined similarly. Figure 49 diagrams jets of water issuing from a fountain at various angles; in any one plane they form a family of curves, which may be considered approximately as parabolas; their envelope stands out clearly as the general contour of the cascade of water. Of course, not every family of curves or surfaces has an envelope; for example, a family of parallel straight lines does not have one. There exists a simple general method of finding the envelope of any family; for a family of curves in the plane this method was given by Leibnitz.

Every curve is obviously the envelope of its tangents, and in exactly the same way every surface is the envelope of its tangent planes. Incidentally, this fact provides a new method of defining a curve or a surface by giving the family of its tangent lines or planes. For some problems this method turns out to be the most convenient.

Generally speaking, the tangent planes of a surface are different at different points, so that the family of tangents to the surface is obviously a two-parameter one. But in some cases, for example, a cylinder, it is one parameter. It can be shown that the following remarkable theorem holds. A one-parameter family of tangent planes occurs only for those surfaces that are developable into a plane, i.e., those in which any sufficiently small segment may be deformed into a plane segment; these are the developable surfaces noted in §4. Every analytic surface of this kind consists of segments of straight lines and is either cylindrical (parallel straight lines) or conical (straight lines passing through one point), or consists of the tangents to some space curve.

The theory of envelopes is particularly useful in engineering problems, for example in the theory of transmissions. We consider two gears A

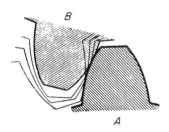

FIG. 50.

and B. To study their motion relative to each other, we may assume that gear A is stationary and gear B moves around it (figure 50). Then the contour of a cog on gear B, as it assumes various positions, traces out a family of curves in the plane of gear A, and the contour of gear A must at all times be tangent to them, i.e., must be the envelope of the family. Of course, this is not a complete statement of the situation, since in an actual transmission this engagement must be transferred from one pair of cogs to the next, but this condition is nevertheless the basic one which must be satisfied by every type of gear.

As we have said, the question of envelopes is a relatively simple one, solved long ago, in the theory of families of curves and surfaces. This theory is just as rich in interesting problems as, let us say, the theory of surfaces itself. Especially well developed is the theory of "congruences," i.e., two-parameter families of various curves (and in particular of straight lines: the so-called "straight-line" congruences). In this theory one applies essentially the same methods as in the theory of surfaces.

The theory of straight-line congruences originated in the paper of Monge, "On excavations and fills," the title of which already shows that

Monge undertook the investigation for practical purposes; the main idea was to find the most convenient way of transporting earth from an excavation to a fill.

The systematic development of the theory of congruences, beginning in the middle of the last century, is due in large measure to its connection with geometric optics; the set of rays of light in a homogeneous medium at any time constitutes a straight-line congruence.

Nonregular surfaces and geometry "in the large." The theory of curves and surfaces (and of families of them), as it had been constructed by the end of the last century, is usually called classical differential geometry; it has the following characteristic features.

First, it considers only "sufficiently smooth" (i.e., regular) curves and surfaces, namely those which are defined by functions with a sufficient number of derivatives. Thus, for example, surfaces with cusps or edges, such as polyhedral surfaces or the surface of a cone, are either excluded from the argument or are considered only on the parts where they remain smooth.

Second, classical differential geometry pays especial attention to properties of sufficiently small segments of curves and surfaces (geometry "in the small") and nowhere considers properties of an entire closed surface (geometry "in the large").

Typical examples, illustrating the distinction between geometry "in the small" and "in the large" are provided by the deformation of surfaces. For example, already in 1838 Minding showed that a sufficiently small segment of the surface of a sphere can be deformed, and this is a theorem "in the small." At the same time, he expressed the conjecture that the entire sphere cannot be deformed. This theorem was proved by other mathematicians as late as 1899. Incidentally, it is easy to confirm by experiment that a sphere of flexible but inextensible material cannot be deformed. For example, a ping-pong ball holds its shape perfectly well although the material it is made from is quite flexible. Another example, mentioned in §4, is the tin pail; it is rigid in the large, thanks to the presence of a curved flange, but separate pieces of it can easily be bent out of shape. As we see, there is an essential difference between properties of surfaces "in the small" and "in the large."

Other characteristic examples are provided by the theory of geodesics, discussed in §4. A geodesic "in the small," i.e., on a small segment of the surface, is a shortest path, but "in the large" it may not be so at all; for example, it may even be a closed curve, as was pointed out earlier for great circles of a sphere.

The reader will readily note that the theorems on geodesics formulated in §4 are basically theorems "in the small." Questions on the behavior of geodesic curves throughout their whole course will belong to geometry "in the large." It is known, for example, that on a regular surface two sufficiently adjacent points can be joined by a unique geodesic, remaining entirely in a certain small neighborhood of two points. But if we consider geodesics that during their course may depart as far as we like from the two points, then by a theorem of Morse any pair of points on a closed surface may be joined by an infinite number of geodesics. Thus, two points A and B on the lateral surface of a curved cylinder may be joined by very different geodesics: it is sufficient to consider helices which run from A to B but wind around the cylinder a different number of times. The theorem of Poincaré on closed geodesics, stated in §5 of Chapter XVIII, and proved by Ljusternik and Šnirelman, also belongs to geometry "in the large."

The proofs for these theorems, as for many theorems of geometry "in the large," were inaccessible with the usual tools of classical differential geometry and required the invention of new methods.

When these problems of geometry "in the large" were inevitably attracting the attention of mathematicians, the restriction to regular surfaces could no longer be maintained, if only because we are continually encountering surfaces that are not regular but have discontinuous curvature; for example, convex lenses with a sharp edge, and so forth. Moreover, there are many analytic surfaces that cannot be extended in any natural way without acquiring "singularities" in the form of edges or cusps and thus becoming nonregular.

Thus, a segment of the surface of a cone cannot be extended in a natural way without leading to the vertex, a cusp where the smoothness of the surface is destroyed.

This last result is only a particular case of the following remarkable theorem. Every developable surface other than a cylinder will lead, if naturally extended, to an edge (or a cusp in the case of a cone) beyond which it cannot be continued without losing its regularity.

Thus there is a profound connection between the behavior of a surface "in the large" and its singularities. This is the reason why the solution of problems "in the large" and the study of surfaces with "singularities" (edges, cusps, discontinuous curvature and the like) must be worked out together.

Similar new directions were taken in analysis. For example, the qualitative theory of differential equations mentioned in §7 of Chapter V, studies the properties of solutions of a differential equation in its entire domain of definition, i.e., "in the large," paying particular attention to "singularities," i.e., to violations of regularity, and to singular points of the equation. Moreover, contemporary analysis includes the study of nonregular functions which did not occur in classical analysis (cf. Chapter XV) and thereby provides geometry with a new means of studying more general surfaces. Finally, in the calculus of variations, where we are usually looking for curves or surfaces with some extremal property, it sometimes happens that the limit curve, for which the extreme is attained, is not regular. For such problems it is necessary that the class of curves or surfaces under consideration should be closed (that is, should include all its limit curves or surfaces), a fact which necessarily led to the study of at least the simplest nonregular curves and surfaces. In a word, the new directions taken by geometry did not originate in isolation but in close connection with the whole development of mathematics.

The turning of attention to problems "in the large" and nonregular surfaces began about 50 years ago and was shared by many mathematicians. The first essential step was taken by Hermann Minkowski (1864-1909), who laid the foundation for an extensive branch of geometry, the theory of convex bodies. Incidentally, one of the questions which started Minkowski on his investigations was the problem of regular lattices, which is closely connected with the theory of numbers and geometric crystallography.

A body is called *convex* if through each point of its surface we may pass a plane that does not intersect the body, i.e., at any point of its surface the body may rest on a plane (figure 51). A convex body is defined

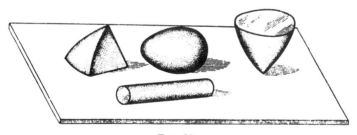

FIG. 51.

by its surface alone, so that for the most part it makes no difference whether we speak of the theory of convex bodies or of closed convex surfaces. The general theorems on convex bodies are proved, as a rule, without any additional assumptions about the smoothness or "regularity" of their surfaces. Thus these theorems are usually concerned with the whole convex body or surface, so that the restrictions of classical differential geometry are automatically removed. However, the two theories (of convex bodies and of nonregular surfaces) were at first very little connected with each other, the combination of the two taking place considerably later.

Beginning in 1940, A. D. Aleksandrov developed the theory of general curves and surfaces, including both the regular surfaces of classical differential geometry and also such nonsmooth surfaces as polyhedra, arbitrary convex sets, and others. In spite of the great generality of this theory, it is chiefly based on intuitive geometric concepts and methods, although it also makes essential use of contemporary analysis. One of the basic methods of the theory consists of approximating general surfaces by means of polyhedra (polyhedral surfaces). This device in its simplest form is known to every schoolboy, for example, in computing the area of the lateral surface of a cylinder as the limit of the areas of prisms. In a number of cases the method produces strong results that either cannot be derived in another way or else, if they are to be proved by an analytic method, require the introduction of complicated ideas. Its essential feature consists of the fact that the result is first obtained for polyhedra and is then extended to general surfaces by a limit process.

One of the beginnings of the theory of general convex surfaces was the theorem on the conditions under which a given evolute (cf. figure 52) may be pasted together to form a convex polyhedron. This theorem, completely elementary in its formulation, has a nonelementary proof and leads to far-reaching corollaries for general convex surfaces. The reader is, of course, familiar with the pasting together of a polyhedral surface from segments; for example, the assembling of a cube from the cross-shaped pattern in figure 52, or of a cylinder from a rectangle and two circles.

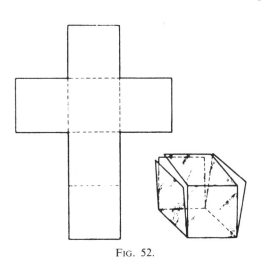

FIG. 52.

This simple example of assembling surfaces from segments of them is converted into a general method of "cutting apart and pasting together," which has produced profound results in various questions of the theory of surfaces and has found practical applications.

Deep-lying results in this theory were obtained by A. V. Pogorelov. In particular, he showed that every closed convex surface cannot be deformed as a whole with preservation of its convexity. This result, achieved in 1949, completes the efforts of many well-known mathematicians, who for the preceding 50 years had tried to prove it but had been successful only under various additional hypotheses. The results of Pogorelov, in conjunction with the "method of pasting together," not only provided a complete solution for the problem, but almost completely cleared up the whole question of the deformability or nondeformability of closed and nonclosed convex surfaces. They also established a close connection between the new theory and "classical" differential geometry.

In this way a theory of surfaces was constructed that included the classical theory as well as the theory of polyhedra, of arbitrary convex surfaces, and of very general nonconvex surfaces. Lack of space does not allow us to discuss in detail the results or the still unsolved problems of the theory, although this could readily be done, since they are for the most part quite easily visualized and, in spite of the difficulty of exact proofs, do not require any special knowledge.

In §4, in speaking of the deformation of surfaces, we had in mind deformations of a regular (continuously curved) surface that preserved its regularity. But in the theorem of Pogorelov, on the contrary, there is no requirement of regularity for either the initial or the deformed surface, although the requirement of convexity is imposed on both surfaces.

It is obvious that deformation of a sphere, for example, becomes possible if we allow breaks in the surface and violation of the convexity. It is sufficient to cut out a segment of the surface and then replace it after the deformation; that is, so to speak, to push a segment of the surface into the interior. Considerably more unexpected is the result obtained recently by the American mathematician Nash and the Dutch mathematician Kuiper. They showed that if we preserve only the smoothness of a surface and allow the appearance of any number of sharp jumps in the curvature of the surface (i.e., if we eliminate any requirement of continuity, boundedness, or even existence of the second derivatives of the functions defining the surface) then it turns out to be possible to deform the surface as a whole with a very great degree of arbitrariness. In particular a sphere may be deformed into an arbitrarily small ball, which has a smooth surface consisting of very shallow wavelike creases. Some idea of a deformation of this sort may be gained by the easily imagined possibility of rumpling up into almost any shape a spherical cover made of very soft cloth. On the other hand, a small celluloid ball behaves very differently. The elastic material of its surface resists not only extension but also sharp bending, so that such a ball is very rigid.

Differential geometry of various groups of transformations. At the beginning of this century, there arose from classical differential geometry a series of new developments based on one general idea, namely the study of properties of curves, surfaces, and families of curves and surfaces which remain invariant under various types of transformations. Classical differential geometry investigated properties invariant under translation; but of course there is nothing to prevent us from considering other geometric transformations. For example, a *projective transformation* is one in which straight lines remain straight, and projective geometry, which has been in existence for a long time, studies those properties of figures that remain invariant under projective transformations. Ordinary projective geometry remains similar, in the problems it investigates, to the usual elementary and analytic geometry, whereas "projective differential geometry" (the theory of curves, surfaces, and families developed at the beginning of the present century) is similar to classical differential geometry, except that it studies properties that are invariant under projective transformations. Fundamental in this last direction were the contributions of the American Wilczynski, the Italian Fubini, and the Czech mathematician, Čech.

In the same way arose "affine differential geometry," which studies the properties of curves, surfaces, and families invariant under affine transformations, i.e., under transformations that not only take straight lines into straight lines but also preserve parallelism. The work of the German mathematician Blaschke and his students developed this branch of geometry into a general theory. Let us also mention "conformal geometry," in which one studies the properties of figures invariant under transformations that do not change the angles between curves.

In general, the possible "geometries" are very diverse in character, since essentially any group of transformations may serve as the basis of a "geometry," which then studies just those properties of figures that are left unchanged by the transformations of the group. This principle for the definition of geometries will be discussed further in Chapter XVII.

Other new directions in differential geometry are being successfully developed by Soviet geometers, S. P. Finikov, G. F. Laptev, and others. But in our present outline it is not possible to give an account of all the various investigations that are taking place nowadays in the different branches of differential geometry.

Suggested Reading

H. Busemann, *Convex surfaces*, Interscience, New York, 1958.

H. S. M. Coxeter, *Introduction to geometry*, Wiley, New York, 1961.

H. G. Eggleston, *Convexity*, Cambridge University Press, New York, 1958.

D. Hilbert and S. Cohn-Vossen, *Geometry and the imagination*, Chelsea, New York, 1952.

I. M. Yaglom and V. G. Boltyanskiĭ, *Convex figures*, Holt, Rinehart and Winston, New York, 1961.

VIII

THE CALCULUS OF VARIATIONS

§1. Introduction

Examples of variational problems. We will be able to give a clearer description of the general range of problems studied in the calculus of variations,* if we first consider certain special problems.

1. The curve of fastest descent. The problem of the brachistochrone, or the curve of fastest descent, was historically the first problem in the development of the calculus of variations.

Among all curves connecting the points M_1 and M_2, it is required to find that one along which a mathematical point, moving under the force of gravity from M_1, with no initial velocity, arrives at the point M_2 in the least time.

To solve this problem we must consider all possible curves joining M_1 and M_2. If we choose a definite curve l, then to it will correspond some definite value T of the time taken for the descent of a material point along it. The time T will depend on the choice of l, and of all curves joining M_1 and M_2 we must choose the one which corresponds to the least value of T.

The problem of the brachistochrone may be expressed in the following way.

We draw a vertical plane through the

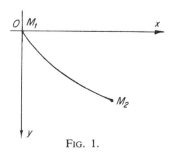

FIG. 1.

---

* The derivation of the name "calculus of variations" is explained later.

points M_1 and M_2. The curve of fastest descent must obviously lie in it, so that we may restrict ourselves to such curves. We take the point M_1 as the origin, the axis Ox horizontal, and the axis Oy vertical and directed downward (figure 1). The coordinates of the point M_1 will be $(0, 0)$; the coordinates of the point M_2 we will call (x_2, y_2). Let us consider an arbitrary curve described by the equation

$$y = f(x), \quad 0 \leqslant x \leqslant x_2, \tag{1}$$

where f is a continuously differentiable function. Since the curve passes through M_1 and M_2, the function f at the ends of the segment $[0, x_2]$ must satisfy the condition

$$f(0) = 0, \quad f(x_2) = y_2. \tag{2}$$

If we take an arbitrary point $M(x, y)$ on the curve, then the velocity v of a material point at this point of the curve will be connected with the y-coordinate of the point by the well-known physical relation

$$\tfrac{1}{2} v^2 = gy,$$

or

$$v = \sqrt{2gy}.$$

The time necessary for a material point to travel along an element ds of arc of the curve has the value

$$\frac{ds}{v} = \frac{\sqrt{1 + y'^2}}{\sqrt{2gy}} \, dx,$$

and thus the total time of the descent of the point along the curve from M_1 to M_2 is equal to

$$T = \frac{1}{\sqrt{2g}} \int_0^{x_2} \frac{\sqrt{1 + y'^2}}{\sqrt{y}} \, dx. \tag{3}$$

Finding the brachistochrone is equivalent to the solution of the following minimal problem: Among all possible functions (1) that satisfy conditions (2), find that one which corresponds to the least value of the integral (3).

2. The surface of revolution of the least area. Among the curves joining two points of a plane, it is required to find that one whose arc, by rotation around the axis Ox, generates the surface with the least area.

We denote the given points by $M_1(x_1, y_1)$ and $M_2(x_2, y_2)$ and consider an arbitrary curve given by the equation

$$y = f(x). \tag{4}$$

If the curve passes through M_1 and M_2, the function f will satisfy the condition

$$f(x_1) = y_1, \quad f(x_2) = y_2. \tag{5}$$

When rotated around the axis Ox this curve describes a surface with area numerically equal to the value of the integral

$$S = 2\pi \int_{x_1}^{x_2} y \sqrt{1 + y'^2} \, dx. \tag{6}$$

This value depends on the choice of the curve, or equivalently of the function $y = f(x)$. Among all functions (4) satisfying condition (5) we must find that function which gives the least value to the integral (6).

3. Uniform deformation of a membrane. By a membrane we usually mean an elastic surface that is plane in the state of rest, bends freely, and does work only against extension. We assume that the potential energy of a deformed membrane is proportional to the increase in the area of its surface.

In the state of rest let the membrane occupy a domain B of the Oxy plane (figure 2). We deform the boundary of the membrane in a direction perpendicular to Oxy and denote by $\phi(M)$ the displacement of the point M of the boundary. Then the interior of the membrane is also deformed, and we are required to find the position of equilibrium of the membrane for a given deformation of its boundary.

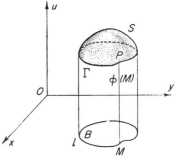

FIG. 2.

With a great degree of accuracy we may assume that all points of the membrane are displaced perpendicularly to the plane Oxy. We denote by $u(x, y)$ the displacement of the point (x, y). The area of the membrane in its displaced position will be*

$$\iint_B (1 + u_x^2 + u_y^2)^{1/2} \, dx \, dy.$$

---

* Here and everywhere in this chapter we use subscripts to denote the arguments with respect to which the partial derivatives are taken.

If the deformations of the elements of the membrane are so small that we can legitimately ignore higher powers of u_x and u_y, this expression for the area may be replaced by a simpler one:

$$\iint\limits_{B} \left[1 + \frac{1}{2}(u_x^2 + u_y^2)\right] dx\, dy.$$

The change in the area of the membrane is equal to

$$\frac{1}{2} \iint\limits_{B} (u_x^2 + u_y^2)\, dx\, dy;$$

so that the potential energy of the deformation will have the value

$$\frac{\mu}{2} \iint\limits_{B} (u_x^2 + u_y^2)\, dx\, dy, \tag{7}$$

where μ is a constant depending on the elastic properties of the membrane.

Since the displacement of the points on the edge of the membrane is assumed to be given, the function $u(x, y)$ will satisfy the condition

$$u\,|_l = \phi(M) \tag{8}$$

on the boundary of the domain B.

In the position of equilibrium the potential energy of the deformation must have the smallest possible value, so that the function $u(x, y)$, describing the displacement of the points of the membrane, is to be found by solving the following mathematical problem: Among all functions $u(x, y)$ that are continuously differentiable on the domain B and satisfy condition (8) on the boundary, find the one which gives the least value to the integral (7).

Extreme values of functionals and the calculus of variations. These examples allow us to form some impression of the kind of problems considered, but to define exactly the position of the calculus of variations in mathematics, we must become acquainted with certain new concepts. We recall that one of the basic concepts of mathematical analysis is that of a function. In the simplest case the concept of functional dependence may be described as follows. Let M be any set of real numbers. If to every number x of the set M there corresponds a number y, we say that there is defined on the set M a function $y = f(x)$. The set M is often called the domain of definition of the function.

The concept of a functional is a direct and natural generalization of the concept of a function and includes it as a special case.

Let M be a set of objects of any kind. The nature of these objects is immaterial at this time. They may be numbers, points of a space, curves, functions, surfaces, states or even motions of a mechanical system. For brevity we will call them elements of the set M and denote them by the letter x.

If to every element x of the set M there corresponds a number y, we say that there is defined on the set M a functional $y = F(x)$.

If the set M is a set of numbers x, the functional $y = F(x)$ will be a function of one argument. When M is a set of pairs of numbers (x_1, x_2) or a set of points of a plane, the functional will be a function $y = F(x_1, x_2)$ of two arguments, and so forth.

For the functional $y = F(x)$, we state the following problem:

Among all elements x of M find that element for which the functional $y = F(x)$ has the smallest value.

The problem of the maximum of the functional is formulated in the same way.

We note that if we change the sign in the functional $F(x)$ and consider the functional $-F(x)$, the maximum (minimum) of $F(x)$ becomes the minimum (maximum) of $-F(x)$. So there is no need to study both maxima and minima; in what follows we will deal chiefly with minima of functionals.

In the problem of the curve of fastest descent, the functional whose minimum we seek will be the integral (3), the time of descent of a material point along a curve. This functional will be defined on all possible functions (1), satisfying condition (2).

In the problem of the position of equilibrium of a membrane, the functional is the potential energy (7) of the deformed membrane, and we must find its minimum on the set of functions $u(x, y)$ satisfying the boundary condition (8).

Every functional is defined by two factors: the set M of elements x on which it is given and the law by which every element x corresponds to a number, the value of the functional. The methods of seeking the least and greatest values of a functional will certainly depend on the properties of the set M.

The calculus of variations is a particular chapter in the theory of functionals. In it we consider functionals given on a set of functions, and our problem consists of the construction of a theory of extreme values for such functionals.

This branch of mathematics became particularly important after the discovery of its connection with many situations in physics and mechanics. The reason for this connection may be seen as follows. As will be made clear later, it is necessary, in order that a function provide an extreme value for a functional, that it satisfy a certain differential equation. On the other hand, as was mentioned in the chapters describing differential equations, the quantitative laws of mechanics and physics are often written in the form of differential equations. As it turned out, many equations of this type also occurred among the differential equations of the calculus of variations. So it became possible to consider the equations of mechanics and physics as extremal conditions for suitable functionals and to state the laws of physics in the form of requiring an extreme value, in particular a minimum, for certain quantities. New points of view could thus be introduced into mechanics and physics, since certain laws could be replaced by equivalent statements in terms of "minimal principles." This in turn opened up a new method of solving physical problems, either exactly or approximately, by seeking the minima of corresponding functionals.

§2. The Differential Equations of the Calculus of Variations

The Euler differential equation. The reader will recall that a necessary condition for the existence of an extreme value of a differentiable function f at a point x is that the derivative f' be equal to zero at this point: $f'(x) = 0$; or what amounts to the same thing, that the differential of the function be equal to zero here: $df = f'(x)\,dx = 0$.

Our immediate goal will be to find an analogue of this condition in the calculus of variations, that is to say, to set up a necessary condition that a function must satisfy in order to provide an extreme value for a functional.

We will show that such a function must satisfy a certain differential equation. The form of the equation will depend on the kind of functional under consideration. We begin with the so-called simplest integral of the calculus of variations, by which we mean a functional with the following integral representation:

$$I(y) = \int_{x_1}^{x_2} F(x, y, y')\,dx. \tag{9}$$

The function F, occuring under the integral sign, depends on three arguments (x, y, y'). We will assume it is defined and is twice continuously differentiable with respect to the argument y' for all values of this argument, and with respect to the arguments x and y in some domain B of the Oxy plane. Below it is assumed that we always remain in the interior of this domain.

It is clear that y is a function of x

$$y = y(x), \qquad (10)$$

continuously differentiable on the seg-
ment $x_1 \leqslant x \leqslant x_2$, and that y' is its
derivative.

Geometrically the function $y(x)$ may
be represented on the Oxy plane by a
curve l over the interval $[x_1, x_2]$
(figure 3).

The integral (9) is a generalization of
the integrals (3) and (6), which we

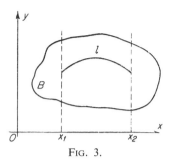

Fig. 3.

encountered in the problem of the curve of fastest descent and the
surface of revolution of least area. Its value depends on the choice of
the function $y(x)$ or in other words of the curve l, and the problem of
its minimum value is to be interpreted as follows:

Given some set M of functions (10) (curves l); among these we must
find that function (curve l) for which the integral $I(y)$ has the least value.

We must first of all define exactly the set of functions M for which
we will consider the value of the integral (9). In the calculus of variations
the functions of this set are usually called admissible for comparison.
We consider the problem with fixed boundary values. The set of admissible
functions is defined here by the following two requirements:

1. $y(x)$ is continuously differentiable on the segment $[x_1, x_2]$;

2. At the ends of the segment $y(x)$ has values given in advance

$$y(x_1) = y_1, \quad y(x_2) = y_2. \qquad (11)$$

Otherwise the function $y(x)$ may be completely arbitrary. In the language
of geometry, we are considering all possible smooth curves over the
interval $[x_1, x_2]$, which pass through the two points $A(x_1, y_1)$ and
$B(x_2, y_2)$ and can be represented by the equation (10). The function
giving the minimum of the integral will be assumed to exist and we will
call it $y(x)$.

The following simple and ingenious arguments, which can often be
applied in the calculus of variations, lead to a particularly simple form
of the necessary condition which $y(x)$ must satisfy. In essence they allow
us to reduce the problem of the minimum of the integral (9) to the problem
of the minimum of a function.

We consider the family of functions dependent on a numerical parameter α,

$$\bar{y}(x) = y(x) + \alpha\eta(x). \qquad (12)$$

In order that $\bar{y}(x)$ be an admissible function for arbitrary α, we must assume that $\eta(x)$ is continuously differentiable and vanishes at the ends of the interval $[x_1, x_2]$.

$$\eta(x_1) = \eta(x_2) = 0. \tag{13}$$

The integral (9) computed for $\bar{y}$ will be a function of the parameter α

$$I(\bar{y}) = \int_{x_1}^{x_2} F(x, y + \alpha\eta, y' + \alpha\eta')\, dx = \Phi(\alpha).^{*}$$

Since $y(x)$ gives a minimum to the value of the integral, the function $\Phi(\alpha)$ must have a minimum for $\alpha = 0$, so that its derivative at this point must vanish

$$\Phi'(0) = \int_{x_1}^{x_2} [F_y(x, y, y')\, \eta + F_{y'}(x, y, y')\, \eta']\, dx = 0. \tag{14}$$

This last equation must be satisfied for every continuously differentiable function $\eta(x)$ which vanishes at the ends of the segment $[x_1, x_2]$. In order to obtain the result which follows from this, it is convenient to transform the second term in condition (14) by integration by parts

$$\int_{x_1}^{x_2} F_{y'}\eta'\, dx = -\int_{x_1}^{x_2} \eta\, \frac{d}{dx} F_{y'}\, dx$$

so that condition (14) takes the new form

$$\Phi'(0) = \int_{x_1}^{x_2} \left(F_y - \frac{d}{dx} F_{y'}\right) \eta\, dx = 0. \tag{15}$$

It may be shown that the following simple lemma holds.

Let the following two conditions be fulfilled:

1. The function $f(x)$ is continuous on the interval $[a, b]$;

2. The function $\eta(x)$ is continuously differentiable on the interval $[a, b]$ and vanishes at the ends of this interval.

If for an arbitrary function $\eta(x)$ the integral $\int_a^b f(x)\, \eta(x)\, dx$ is equal to zero, then it follows that $f(x) \equiv 0$.

---

* The difference $\bar{y} - y = \alpha\eta$ is called the *variation* (change) *of the function y* and is denoted by δy, and the difference $I(\bar{y}) - I(y)$ is called the *total variation of the integral* (9). Hence we get the name calculus of variations.

For let us assume that at some point c the function f is different from zero and show that then a function $\eta(x)$ necessarily exists for which $\int_a^b f(x)\,\eta(x)\,dx \neq 0$, in contradiction to the condition of the lemma.

Since $f(c) \neq 0$ and f is continuous, there must exist a neighborhood $[\alpha, \beta]$ of c in which f will be everywhere different from zero and thus will have a constant sign throughout.

We can always construct a function $\eta(x)$ which is continuously differentiable on $[a, b]$, positive on $[\alpha, \beta]$, and equal to zero outside of $[\alpha, \beta]$ (figure 4).

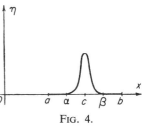

FIG. 4.

Such a function $\eta(x)$, for example, is defined by the equations

$$\eta(x) = \begin{cases} 0 & \text{on } [a, \alpha], \\ (x - \alpha)^2(\beta - x)^2 & \text{on } [\alpha, \beta], \\ 0 & \text{on } [\beta, b]. \end{cases}$$

But for such a function $\eta(x)$

$$\int_a^b f\eta\,dx = \int_a^\beta f\eta\,dx.$$

The latter of these integrals cannot be equal to zero since, in the interior of the interval of integration, the product $f\eta$ is different from zero and never changes its sign.

Since equation (15) must be satisfied for every $\eta(x)$ that is continuously differentiable and vanishes at the ends of the segment $[x_1, x_2]$, we may assert, on the basis of the lemma, that this can occur only in the case

$$F_y - \frac{d}{dx} F_{y'} = 0, \tag{16}$$

or, by computing the derivative with respect to x

$$F_y(x, y, y') - F_{xy'}(x, y, y') - F_{yy'}(x, y, y')y' - F_{y'y'}(x, y, y')y'' = 0. \tag{17}$$

This equation is a differential equation of the second order with respect to the function y. It is called *Euler's equation*.

We may state the following conclusion.

If a function $y(x)$ minimizes the integral $I(y)$, then it must satisfy Euler's differential equation (17). In the calculus of variations, this last statement has a meaning completely analogous to the necessary condition

$df = 0$ in the theory of extreme values of functions. It allows us immediately to exclude all admissible functions that do not satisfy this condition, since for them the integral cannot have a minimum, so that the set of admissible functions we need to study is very sharply reduced.

Solutions of equation (17) have the property that for them the derivative $[(d/d\alpha)I(y + \alpha\eta)]_{\alpha=0}$ vanishes for arbitrary $\eta(x)$, so that they are analogous in meaning to the stationary points of a function. Thus it is often said that for solutions of (17) the integral $I(y)$ has a stationary value.

In our problem with fixed boundary values, we do not need to find all solutions of the Euler equation but only those which take on the values y_1, y_2 at the points x_1, x_2.

We turn our attention to the fact that the Euler equation (17) is of the second order. Its general solution will contain two arbitrary constants

$$y = \phi(x, C_1, C_2).$$

These must be defined so that the integral curve passes through the points A and B, so we have the two equations for finding the constants C_1 and C_2

$$\phi(x_1, C_1, C_2) = y_1, \quad \phi(x_2, C_1, C_2) = y_2.$$

In many cases this system has only one solution and then there will exist only one integral curve passing through A and B.

The search for functions giving a minimum for this integral is thus reduced to the solution of the following boundary-value problem for differential equations: On the interval $[x_1, x_2]$ find those solutions of equation (17) that have the given values y_1, y_2 at the ends of the interval.

Frequently this last problem can be solved by using known methods in the theory of differential equations.

We emphasize again that every solution of such a boundary-value problem can provide only a suspected minimum and that it is necessary to verify whether or not it actually does give a minimum value to the integral. But in particular cases, especially in those occurring in the applications, Euler's equation completely solves the problem of finding the minimum of the integral. Suppose we know initially that a function giving a minimum for the integral exists, and assume, moreover, that the Euler equation (17) has only one solution satisfying the boundary conditions (11). Then only one of the admissible curves can be a suspected minimum, and we may be sure, under these circumstances, that the solution found for the equation (17) indeed gives a minimum for the integral.

Example. It was previously established that the problem of the curve of fastest descent may be reduced to finding the minimum of the integral

$$I(y) = \int_0^{x_2} \frac{\sqrt{1+y'^2}}{\sqrt{y}}\,dx$$

among the set of functions satisfying the boundary conditions

$$y(0) = 0, \quad y(x_2) = y_2\,.$$

In this problem

$$F = \frac{\sqrt{1+y'^2}}{\sqrt{y}}\,.$$

Euler's equation has the form

$$-\frac{1}{2}\,y^{-3/2}\sqrt{1+y'^2} - \frac{d}{dx}\left[y^{-1/2}\,\frac{y'}{\sqrt{1+y'^2}}\right] = 0.$$

After some manipulation it takes the form

$$\frac{2y''}{1+y'^2} = -\frac{1}{y}\,.$$

Multiplying both sides of the equation by y' and integrating, we get

$$\ln(1+y'^2) = -\ln y + \ln k,$$

or

$$y'^2 = \frac{k}{y} - 1,$$

$$\sqrt{\frac{y}{k-y}}\,dy = \pm\,dx.$$

Now letting

$$y = \frac{k}{2}(1 - \cos u), \quad dy = \frac{k}{2}\sin u\,du,$$

we find after substituting and simplifying

$$\frac{k}{2}(1 - \cos u)\,du = \pm\,dx,$$

from which, by integrating, we get: $x = \pm\,k/2\,(u - \sin u) + C$. Since the curve must pass through the origin, it follows that we must put $C = 0$.

In this way we see that the brachistochrone is the cycloid

$$x = \frac{k}{2}(u - \sin u), \quad y = \frac{k}{2}(1 - \cos u).$$

The constant k must be found from the condition that this curve passes through the point $M_2(x_2, y_2)$.

Functionals depending on several functions. The simplest functional in the calculus of variations (17) depended on only one function. In the applications such functionals will occur in those cases where the objects (or their behavior) are defined by only one functional dependence. For example, a curve in the plane is defined by the dependence of the ordinate of a point on its abscissa, the motion of a material point along an axis is defined by the dependence of its coordinate on time, etc.

But we must often deal with objects that cannot be defined so simply. In order to define a curve in space, we must know the functional dependence of two of its coordinates on the third. The motion of a point in space is defined by the dependence of its three coordinates on time, etc. Study of these more complicated objects leads to variational problems with several varying functions.

We will restrict ourselves to cases in which the functional depends on two functions $y(x)$ and $z(x)$, since the case of a larger number of functions does not differ in principle from this one.

We consider the following problem. Admissible pairs of functions $y(x)$ and $z(x)$ are defined by the conditions:

1. The functions
$$y = y(x), \quad z = z(x) \tag{18}$$
are continuously differentiable on the segment $[x_1, x_2]$;

2. At the ends of the segment these functions have given values
$$y(x_1) = y_1, \quad y(x_2) = y_2,$$
$$z(x_1) = z_1, \quad z(x_2) = z_2. \tag{19}$$

Among all possible pairs of functions $y(x)$ and $z(x)$, we must find the pair that gives the least value to the integral
$$I(y, z) = \int_{x_1}^{x_2} F(x, y, z, y', z') \, dx. \tag{20}$$

In the three-dimensional space x, y, z, each pair of admissible functions will correspond to a curve l, defined by equations (18) and passing through the points
$$M_1(x_1, y_1, z_1), \quad M_2(x_2, y_2, z_2).$$

We must find the minimum of the integral (20) on the set of all such curves.

We assume that the pair of functions giving the minimum of the integral (20) exists, and we will call these functions $y(x)$ and $z(x)$. Together with them we consider a second pair of functions

$$\bar{y} = y + \alpha\eta(x), \quad \bar{z} = z + \alpha\zeta(x),$$

where $\eta(x)$ and $\zeta(x)$ are any continuously differentiable functions vanishing at the ends x_1, x_2 of the segment; $\bar{y}$, $\bar{z}$ will also be admissible, and for $\alpha = 0$ they will coincide with the functions y, z. We substitute them in (20)

$$I(\bar{y}, \bar{z}) = \int_{x_1}^{x_2} F(x, y + \alpha\eta, z + \alpha\zeta, y' + \alpha\eta', z' + \alpha\zeta') \, dx = \Phi(\alpha).$$

The integral so derived will be a function of α. Since $\bar{y}$ and $\bar{z}$ coincide with y and z when $\alpha = 0$, the function $\Phi(\alpha)$ must have a minimum for $\alpha = 0$. But at a minimum point the derivative of Φ must vanish

$$\Phi'(0) = 0.$$

Computing the derivative gives

$$\int_{x_1}^{x_2} (F_y \cdot \eta + F_z \cdot \zeta + F_{y'} \cdot \eta' + F_{z'} \cdot \zeta') \, dx = 0,$$

or, if the terms in η' and ζ' are integrated by parts

$$\int_{x_1}^{x_2} \left[\left(F_y - \frac{d}{dx} F_{y'} \right) \eta(x) + \left(F_z - \frac{d}{dx} F_{z'} \right) \zeta(x) \right] dx = 0.$$

This last equation must be satisfied for any two continuously differentiable functions $\eta(x)$ and $\zeta(x)$ vanishing at the ends of the interval. Hence, from the basic lemma proved earlier, the following two conditions must be fulfilled:

$$F_y - \frac{d}{dx} F_{y'} = 0,$$

$$F_z - \frac{d}{dx} F_{z'} = 0. \tag{21}$$

Hence, if the functions y, z give a minimum for the integral (20), they must satisfy the system of Euler differential equations (21).

This result again allows us to replace a variational problem for the minimum of the integral (20) by a boundary-value problem in the theory of differential equations: On the interval $[x_1, x_2]$, we must find those solutions y, z of the system of differential equations (21) that satisfy the boundary conditions (19).

As in the preceding case, this opens up a possible path for the solution of the minimal problem.

As an example of an application of the Euler system (21), let us consider the variational principle of Ostrogradskiĭ-Hamilton in Newtonian mechanics. We restrict ourselves to the simplest form of this principle.

We consider a material body of mass m and assume that the dimensions and form of the body may be ignored, so that we may consider it as a material point.

We assume that the point moves from its position $M_1(x_1, y_1, z_1)$ at time t_1 to the position $M_2(x_2, y_2, z_2)$ at time t_2. We also assume that the motion occurs under the laws of Newtonian mechanics and is caused by application of a force $F(x, y, z, t)$ which depends on the position of the point and on the time t and possesses a potential function $U(x, y, z, t)$. This last condition means the following: the components F_x, F_y, F_z of the force F along the coordinate axes are the partial derivatives of a function U with respect to the corresponding coordinates

$$F_x = \frac{\partial U}{\partial x}, \quad F_y = \frac{\partial U}{\partial y}, \quad F_z = \frac{\partial U}{\partial z}.$$

We assume the motion to be free, that is, not subject to any kind of constraints.*

The equations of motion of Newton are

$$m\frac{d^2x}{dt^2} = \frac{\partial U}{\partial x}, \quad m\frac{d^2y}{dt^2} = \frac{\partial U}{\partial y}, \quad m\frac{d^2z}{dt^2} = \frac{\partial U}{\partial z}.$$

If the point obeys the laws of Newtonian mechanics, it moves in a completely determined manner. But together with these "Newtonian motions" of the point, let us consider other (non-Newtonian) motions, which for brevity we will call "admissible," and which will be defined by two requirements only, that at time t_1 the point is in the position M_1 and at time t_2 is in the position M_2.

How can we distinguish the "Newtonian motion" of the point from these other "admissible" motions? Such a possibility is given by the Ostrogradskiĭ-Hamilton principle.

We introduce the kinetic energy of the point

$$T = \tfrac{1}{2}m(x'^2 + y'^2 + z'^2)$$

---

* This is not essential for the Ostrogradskiĭ-Hamilton principle: We may impose any restraints we like on the mechanical system, even nonstationary ones, provided only that they are holonomic, i.e., that they may be described in the form of equations not containing derivatives of the coordinates with respect to time.

and form the so called action integral

$$I = \int_{t_1}^{t_2} (T + U) \, dt.$$

The principle states: The "Newtonian motion" of the point is distinguished among all its "admissible" motions by the fact that it gives the action integral a stationary value.

The action integral I depends on three functions: $x(t)$, $y(t)$, $z(t)$.

Since for all the motions under comparison the initial and final positions of the point are identical, the boundary values of these functions are fixed. We are dealing here with a variational problem for three varying functions with fixed values at the ends of the interval $[t_1, t_2]$.

Previously we agreed to say that the integral (17) has a stationary value for any curve which is an integral curve of the Euler equation. In our problem we are integrating a function

$$F = T + U = \tfrac{1}{2}m(x'^2 + y'^2 + z'^2) + U(x, y, z, t)$$

which depends on three functions, so that for a stationary value of the integral we must satisfy the system of three differential equations

$$F_x - \frac{d}{dt} F_{x'} = 0,$$

$$F_y - \frac{d}{dt} F_{y'} = 0,$$

$$F_z - \frac{d}{dt} F_{z'} = 0.$$

Since $F_x = \partial U / \partial x$, $F_{x'} = mx'$, $\cdots$, the system of Euler equations is identical with the equations of motion of Newtonian mechanics, which provides a verification of the Ostrogradskiĭ-Hamilton principle.

The minimum problem for a multiple integral. The last problem in the calculus of variations to which we wish to draw the attention of the reader is the problem of minimizing a multiple integral. Since the facts connected with the solution of such problems are similar for integrals of any multiplicity, we will confine ourselves to the simplest case, that of double integrals.

Let B be a domain in the Oxy plane, bounded by the contour l. The set of admissible functions is defined by the conditions:

1. $u(x, y)$ is continuously differentiable on the domain B,

2. On l the function u takes given values

$$u \mid_l = f(M). \tag{22}$$

Among all functions we must find the one which gives a minimum value for the integral

$$I(u) = \iint\limits_{B} F(x, y, u, u_x, u_y) \, dx \, dy. \tag{23}$$

The given boundary values (22) for the function u in the space (x, y, u) determine a given space curve Γ, lying above l (cf. figure 2, Chapter VII).

We consider all possible surfaces S passing through Γ and lying above B. Among these we want to find the one for which the integral (23) is minimal.

As before, we assume the existence of the minimizing function and denote it by u. At the same time we consider another function

$$\bar{u} = u + \alpha\eta(x, y),$$

where $\eta(x, y)$ is any continuously differentiable function vanishing on l. Then the function

$$I(\bar{u}) = \iint\limits_{B} F(x, y, u + \alpha\eta, u_x + \alpha\eta_x, u_y + \alpha\eta_y) \, dx \, dy = \Phi(\alpha)$$

must have a minimum for $\alpha = 0$. In this case its first derivative must be equal to zero for $\alpha = 0$

$$\Phi'(0) = 0,$$

or

$$\iint\limits_{B} (F_u \eta + F_{u_x} \eta_x + F_{u_y} \eta_y) \, dx \, dy = 0. \tag{24}$$

We transform the last two terms by Ostrogradskiĭ's formula

$$\iint\limits_{B} (F_{u_x} \eta_x + F_{u_y} \eta_y) \, dx \, dy$$

$$= \iint\limits_{B} \left[\frac{\partial}{\partial x} (F_{u_x} \eta) + \frac{\partial}{\partial y} (F_{u_y} \eta) \right] dx \, dy - \iint\limits_{B} \left(\frac{\partial}{\partial x} F_{u_x} + \frac{\partial}{\partial y} F_{u_y} \right) \eta \, dx \, dy$$

$$= \int\limits_{l} [F_{u_x} \cos(n, x) + F_{u_y} \cos(n, y)] \eta \, ds$$

$$- \iint\limits_{B} \left(\frac{\partial}{\partial x} F_{u_x} + \frac{\partial}{\partial y} F_{u_y} \right) \eta \, dx \, dy.$$

The contour integral along l must vanish, since on the contour l the function η is equal to zero, so that condition (24) may be put in the form

$$\iint\limits_{B} \left(F_u - \frac{\partial}{\partial x} F_{u_x} - \frac{\partial}{\partial y} F_{u_y} \right) \eta \, dx \, dy = 0.$$

This equation must be satisfied for every function η which is continuously differentiable and vanishes on the boundary l.

We may conclude, as before, that all points of the domain B the equation

$$F_u - \frac{\partial}{\partial x} F_{u_x} - \frac{\partial}{\partial y} F_{u_y} = 0 \tag{25}$$

must be satisfied.

So if the function u gives a minimum for the integral (23), it must satisfy the partial differential equation (25).

As in all the preceding problems, we have here established a connection between a variational problem of minimizing an integral and a boundary-value problem for a differential equation (in this case partial).

Example.   The displacement $u(x, y)$ of points of a membrane with a deformed boundary is to be found from the condition of the minimum of the potential energy

$$\frac{\mu}{2} \iint\limits_{B} (u_x^2 + u_y^2) \, dx \, dy$$

for the given boundary values $u \mid_l, = \phi$.

Omitting, for simplicity, the constant factor μ, we may set

$$F = \frac{1}{2} (u_x^2 + u_y^2),$$

so that equation (25) has the form

$$-\frac{\partial}{\partial x} u_x - \frac{\partial}{\partial y} u_y = 0,$$

or

$$\Delta u = \frac{\partial^2 u}{\partial x^2} + \frac{\partial^2 u}{\partial y^2} = 0.$$

Thus the problem of determining the displacement of the points of a membrane has been reduced to that of finding a harmonic function u with given values on the boundary of the domain (cf. Chapter VI, §3).

§3. Methods of Approximate Solution of Problems in the Calculus of Variations

We conclude the present chapter with an indication of the ideas involved in some of the approximation methods in the calculus of variations.

For definiteness we discuss the simplest functional

$$I(y) = \int_{x_1}^{x_2} F(x, y, y')\, dx$$

for fixed boundary values of the admissible functions.

Let $y(x)$ be an exact solution of the problem of minimizing I, with $m = I(y)$ the corresponding minimal value of the integral. It would appear that if we determine an admissible function $\bar{y}$ for which the value of the integral $I(\bar{y})$ is very near to m, we may assume that $\bar{y}$ will also differ little from the exact solution y. Moreover, if we are able to construct a sequence of admissible functions $\bar{y}_1, \bar{y}_2, \cdots$ for which $I(\bar{y}_n) \to m$, we may expect that such a sequence will converge in some sense or other to the solution y, so that computation of $\bar{y}_n$ with sufficiently large index will allow us to find the solution to any desired degree of accuracy.

Depending on how we go about choosing the "minimizing sequence" $\bar{y}_n (n = 1, 2, \cdots)$, we will have one or another of the various approximation methods in the calculus of variations.

Historically, the first of these was the method of broken lines, or Euler's method. We decompose the interval $[x_1, x_2]$ into a number of segments. For example, if we choose these segments of equal length, the points of division will be

$$x_1, x_1 + h, x_1 + 2h, \cdots, x_1 + nh = x_2, h = \frac{x_2 - x_1}{n}.$$

We now construct the broken line p_{n-1} with vertices lying above the points of division. The ordinates of the vertices we denote by

$$b_0, b_1, b_2, \cdots, b_{n-1}, b_n$$

and require that this broken line begin and end at the same points as the admissible curves, so that $b_0 = y_1$ and $b_n = y_2$. Then the broken line will be defined by the ordinates

$$b_1, b_2, \cdots, b_{n-1}.$$

The question now is to find out how to choose the broken line p_{n-1} (i.e., the ordinates b_i of its vertices) so as to approximate as closely as possible the exact solution of the problem.

To achieve this object it is natural to proceed as follows. We compute the integral I for the broken line. Its value will depend on the b_i

$$I(p_{n-1}) = \Phi(b_1, b_2, \cdots, b_{n-1})$$

and will therefore be a function of these ordinates. We now choose the b_i so that they give $I(p_{n-1})$ a minimum value. To define these b_i we will have the system of equations

$$\frac{\partial}{\partial b_i} I(p_{n-1}) = 0 \quad (i = 1, 2, \cdots, n - 1).$$

Since any admissible curve, and in particular the exact solution of the problem, may be approximated by broken lines with any desired accuracy, both in its position on the plane and in the directions of its tangents, it is clear that the sequence of broken lines p_{n-1} thus constructed will, in fact, be a minimizing sequence. By taking n sufficiently large, we may expect to approximate the solution with any desired degree of accuracy over the whole interval $[x_1, x_2]$. Of course, the fact of convergence must be investigated in each case.

The following method, which is very convenient for calculation, is widely used in physics and technology.

We choose any function $\phi_0(x)$ satisfying the boundary conditions $\phi_0(x_1) = y_1$ and $\phi_0(x_2) = y_2$, and a sequence of functions $\phi_1(x), \phi_2(x), \cdots$, vanishing at the ends of the interval $[x_1, x_2]$.

We then form the linear combination

$$s_n(x) = \phi_0(x) + a_1\phi_1(x) + \cdots + a_n\phi_n(x).$$

For arbitrary values of the numerical coefficients $a_1, a_2, \cdots, a_n$, the function $s_n(x)$ will be admissible.

Replacing y by $s_n(x)$ in the integral I and making the necessary computations, we obtain a certain function of the coefficients a_i.

We now choose the a_i so that this function has the least possible value. The coefficients must be found from the system

$$\frac{\partial}{\partial a_i} I(s_n) = 0 \quad (i = 1, 2, \cdots, n).$$

Solving this system, we obtain, in general, the values of the coefficients $a_1, \cdots, a_n$ producing a minimum value for $I(s_n)$ and with them we construct an approximation to the solution

$$\bar{s}_n(x) = \phi_0(x) + \bar{a}_1\phi_1(x) + \cdots + \bar{a}_n\phi_n(x).$$

The sequence of approximations $\bar{s}_n$ $(n = 1, 2, \cdots)$ constructed in this way will not be a minimizing sequence for arbitrary choice of the functions ϕ_i. The necessary condition for it to be so is that the sequence of functions ϕ_i satisfy a certain condition of "completeness" which we will not define here.

Suggested Reading

G. A. Bliss, *Calculus of variations*, Open Court, La Salle, Ill., 1925.

C. Lanczos, *The variational principles of mechanics*, University of Toronto Press, 1949.

G. Pólya and G. Szegö, *Isoperimetric inequalities in mathematical physics*, Princeton University Press, 1951.

R. Weinstock, *Calculus of variations, with applications to physics and engineering*, McGraw-Hill, New York, 1952.

CHAPTER IX

FUNCTIONS
OF A COMPLEX VARIABLE

§1. Complex Numbers and Functions of a Complex Variable

Complex numbers and their significance in algebra. Complex numbers were introduced into mathematics in connection with the solution of algebraic equations. The impossibility of solving the algebraic equation

$$x^2 + 1 = 0 \tag{1}$$

in the domain of real numbers led to the introduction of a conventional number, the imaginary unit i, defined by the equation

$$i^2 = -1. \tag{2}$$

Numbers of the form $a + bi$, where a and b are real numbers, were called *complex numbers*. These numbers were manipulated like real numbers, being added and multiplied as binomials. If we also make use of equation (2), the basic operations of arithmetic when carried out on complex numbers produce other complex numbers.* The division of complex numbers being defined as the inverse of multiplication, it turns out that this operation also is uniquely defined, provided only that the denominator is not equal to zero. In this manner, the introduction of complex numbers first brought to light the interesting, though for the time being purely formal, fact that in addition to the real numbers there exist other numbers, the complex ones, on which all the arithmetic operations can be performed.

---

* Complex numbers are known to the reader from secondary school. See also Chapter IV, §3.

The next step consists of the geometric representation of complex numbers. Every complex number $a + bi$ may be represented by a point in the Oxy plane with coordinates (a, b), or by a vector issuing from the origin to the point (a, b). This led to a new point of view concerning complex numbers. Complex numbers are pairs (a, b) of real numbers for which there are established definitions of the operations of addition and multiplication, obeying the same laws as for real numbers. Here we discover a remarkable situation: The sum of two complex numbers

$$(a + bi) + (c + di) = (a + c) + (b + d)i$$

is represented geometrically by the diagonal of the parallelogram constructed from the vectors representing the summands (figure 1). In this

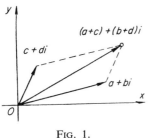

FIG. 1.

way, complex numbers are added by the same law as the vector quantities found in mechanics and physics: forces, velocities, and accelerations. This was a further reason for considering that complex numbers are not merely formal generalizations but may be used to represent actual physical quantities.

We will see later how this point of view is very successful in various problems of mathematical physics.

However, the introduction of complex numbers had its first successes in the discovery of the laws of algebra and analysis. The domain of real numbers, closed with respect to arithmetic operations, was seen to be not sufficiently extensive for algebra. Even such a simple equation as (1) does not have a root in the domain of real numbers, but for complex numbers we have the following remarkable fact, the so-called fundamental theorem of algebra: Every algebraic equation

$$z^n + a_1 z^{n-1} + \cdots + a_{n-1} z + a_n = 0$$

with complex coefficients has n complex roots.*

This theorem shows that the complex numbers form a system of numbers which, in a well-known sense, is complete with respect to the operations of algebra. It is not at all trivial that adjoining to the domain of real numbers a root of the single equation (1) leads to the numbers $a + bi$ in whose domain any algebraic equation is solvable. The fundamental theorem of algebra showed that the theory of polynomials, even

---

* Cf. Chapter IV, §3.

with real coefficients, may be given a finished form only when we consider the values of the polynomial in the whole complex plane. The further development of the theory of algebraic polynomials supported this point of view more and more. The properties of polynomials are discovered only by considering them as functions of a complex variable.

Power series and functions of a complex variable. The development of analysis brought to light a series of facts showing that the introduction of complex numbers was significant not only in the theory of polynomials but also for another very important class of functions, namely those which are expandable in a power series

$$f(x) = a_0 + a_1(x - a) + a_2(x - a)^2 + \cdots. \tag{3}$$

As was already mentioned in Chapter II, the development of the infinitesimal analysis required the establishment of a more precise point of view for the concept of a function and for the various possibilities of defining functions in mathematics. Without pausing here to discuss these interesting questions, we recall only that at the very beginning of the development of analysis it turned out that the most frequently encountered functions could be expanded in a power series in the neighborhood of every point in their domain of definition. For example, this property holds for all the so-called *elementary functions*.

The majority of the concrete problems of analysis led to functions that are expandable in power series. On the other hand, there was a desire to connect the definition of a "mathematical" function with a "mathematical" formula, and the power series represented a very inclusive kind of "mathematical" formula. This situation even led to serious attempts to restrict analysis to the study of functions that are expandable in power series and thus are called *analytic functions*. The development of science showed that such a restriction is inexpedient. The problems of mathematical physics began to extend beyond the class of analytic functions, which does not even include, for example, functions represented by curves with a sharp corner. However, the class of analytic functions, in view of its remarkable properties and numerous applications, proved to be the most important of all the classes of functions studied by mathematicians.

Since the computation of each term of a power series requires only arithmetic operations, the values of a function represented by a power series may be computed also for complex values of the argument, at least for those values for which the series is convergent. When we thus extend the definition of a function of a real variable to complex arguments, we speak of the "continuation" of the function into the complex domain.

Thus an analytic function, in the same way as a polynomial, may be considered not only for real values of the argument but also for complex. Further, we may also consider power series with complex coefficients. The properties of analytic functions, as also of polynomials, are fully revealed only when they are considered in the complex domain. To illustrate we turn now to an example.

Consider the two functions of a real variable

$$e^x \quad \text{and} \quad \frac{1}{1+x^2}.$$

Both these functions are finite, continuous, and differentiable an arbitrary number of times on the whole axis Ox. They may be expanded in a Taylor series, for example, around the origin $x = 0$

$$e^x = 1 + \frac{x}{1!} + \frac{x^2}{2!} + \cdots, \tag{4}$$

$$\frac{1}{1+x^2} = 1 - x^2 + x^4 - x^6 + \cdots. \tag{5}$$

The first of the series so obtained converges for all values of x, while the second series converges only for $-1 < x < +1$. Consideration of the function (5) for real values of the argument does not show why its Taylor series diverges for $|x| \geqslant 1$. Passing to the complex domain allows us to clear up the situation. We consider the series (5) for complex values of the argument

$$1 - z^2 + z^4 - z^6 + \cdots. \tag{6}$$

The sum of n terms of this series

$$s_n = 1 - z^2 + z^4 - z^6 + \cdots + (-1)^{n-1}z^{2n-2}$$

is computed in the same way as for real values of z:

$$s_n + z^2 s_n = 1 + (-1)^n z^{2n},$$

hence

$$s_n = \frac{1 + (-1)^n z^{2n}}{1 + z^2}.$$

This expression shows that for $|z| < 1$

$$\lim_{n \to \infty} s_n = \frac{1}{1+z^2},$$

since $|z|^{2n} \to 0$. Thus for complex z satisfying the inequality $|z| < 1$ the series (6) converges and has the sum $1/(1 + z^2)$. For $|z| \geqslant 1$ the series (6) diverges, since in this case the difference $s_n - s_{n-1} = (-1)^{n-1}z^{2n-2}$ does not converge to zero.

The inequality $|z| < 1$ shows that the point z is located at a distance from the origin which is less than one. Thus the points at which the series (6) converges form a circle in the complex plane with center at the origin. On the circumference of this circle there lie two points i and $-i$ for which the function $1/(1 + z^2)$ becomes infinite; the presence of these points determines the restrictions on the domain of convergence of the series (6).

The domain of convergence of a power series. The domain of convergence of the power series

$$a_0 + a_1(z - a) + a_2(z - a)^2 + \cdots + a_n(z - a)^n + \cdots \qquad (7)$$

in the complex plane is always a circle with center at the point a.

Let us prove this proposition, which is called *Abel's theorem*.

First of all we note that a series whose terms are the complex numbers w_n

$$w_1 + w_2 + \cdots + w_n + \cdots, \qquad (8)$$

may be considered as two series, consisting of the real parts and the imaginary parts of the number $w_n = u_n + iv_n$

$$u_1 + u_2 + \cdots + u_n + \cdots, \qquad (9)$$

$$v_1 + v_2 + \cdots + v_n + \cdots. \qquad (10)$$

A partial sum s_n of the series (8) is expressed by the partial sums σ_n and τ_n of the series (9) and (10)

$$s_n = \sigma_n + i\tau_n,$$

so that convergence of the series (8) is equivalent to convergence of both the series (9) and (10), and the sum s of the series (8) is expressed by the sums σ and τ of the series (9) and (10)

$$s = \sigma + i\tau.$$

After these remarks the following lemma is obvious:

If the terms of the series (8) are less in absolute value than the terms of a convergent geometric progression

$$A + Aq + \cdots + Aq^n + \cdots$$

with positive A and q, where $q < 1$, then the series (8) converges.

For if $|w_n| < Aq^n$, then

$$|u_n| \leqslant |w_n| < Aq^n,$$

$$|v_n| \leqslant |w_n| < Aq^n,$$

so that (cf. Chapter II, §14) the series (9) and (10) converge and thus the series (8) also converges.

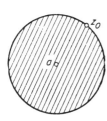

We now show that if the power series (7) converges at some point z_0, then it converges at all points lying inside the circle with center at a and having z_0 on its boundary (figure 2). From this proposition it follows readily that the domain of convergence of the series (7)

$$a_0 + a_1(z - a) + \cdots + a_n(z - a)^n + \cdots$$

Fig. 2.

is either the entire plane, or the single point $z = a$, or some circle of finite radius.

For let the series (7) converge at the point z_0; then the general term of the series (7) for $z = z_0$ converges to zero for $n \to \infty$, and this means that all the terms in the series (7) lie inside some circle; let A be the radius of such a circle, so that for any n

$$|a_n(z_0 - a)^n| < A. \qquad (11)$$

We now take any point z closer than z_0 to a and show that at the point z the series converges.

Obviously

$$|z - a| < |z_0 - a|,$$

so that

$$q = \frac{|z - a|}{|z_0 - a|} < 1. \qquad (12)$$

Let us estimate the general term of the series (7) at the point z

$$|a_n(z - a)^n| = \left| a_n(z_0 - a)^n \left(\frac{z - a}{z_0 - a} \right)^n \right| = |a_n(z_0 - a)^n| \left(\frac{|z - a|}{|z_0 - a|} \right)^n;$$

from inequalities (11) and (12) it follows that

$$|a_n(z - a)^n| < Aq^n;$$

i.e., the general term of the series (7) at the point z is less than the general term of a convergent geometric progression. From the basic lemma above, the series (7) converges at the point z.

The circle in which a power series converges, and outside of which it diverges, will be called the *circle of convergence*; the radius of this circle is called the *radius of convergence* of the power series. The boundary of the circle of convergence, as may be shown, always passes through the point of the complex plane nearest to a at which the regular behavior of the function ceases to hold.

The power series (4) converges on the whole complex plane; the power series (5), as was shown above, has a radius of convergence equal to one.

Exponential and trigonometric functions of a complex variable. A power series may serve to "continue" a function of a real variable into the complex domain. For example, for a complex value of z we define the function e^z by the power series

$$e^z = 1 + \frac{z}{1!} + \frac{z^2}{2!} + \cdots. \tag{13}$$

In like manner the trigonometric functions of a complex variable are introduced by

$$\sin z = \frac{z}{1!} - \frac{z^3}{3!} + \frac{z^5}{5!} - \cdots, \tag{14}$$

$$\cos z = 1 - \frac{z^2}{2!} + \frac{z^4}{4!} - \cdots. \tag{15}$$

These series converge on the whole plane.

It is interesting to note the connection which occurs between the exponential and trigonometric functions when we turn to the complex domain.

If in (13) we replace z by iz, we get

$$e^{iz} = 1 + i\frac{z}{1!} - \frac{z^2}{2!} - i\frac{z^3}{3!} + \frac{z^4}{4!} + \cdots.$$

Grouping everywhere the terms without the multiplier i and the terms with multiplier i, we have

$$e^{iz} = \cos z + i \sin z. \tag{16}$$

Similarly we can derive

$$e^{-iz} = \cos z - i \sin z. \tag{16'}$$

Formulas (16) and (16′) are called *Euler's formulas.* Solving (16) and (16′) for cos z and sin z, we get

$$\cos z = \frac{e^{iz} + e^{-iz}}{2},$$

$$\sin z = \frac{e^{iz} - e^{-iz}}{2i}. \tag{17}$$

It is very important that for complex values the simple rule of addition of exponents continue to hold

$$e^{z_1} \cdot e^{z_2} = e^{z_1 + z_2}. \tag{18}$$

Since for complex values of the argument we define the function e^z by the series (13), formula (18) must be proved on the basis of this definition. We give the proof:

$$e^{z_1} \cdot e^{z_2} = \left(1 + \frac{z_1}{1!} + \frac{z_1^2}{2!} + \cdots\right) \cdot \left(1 + \frac{z_2}{1!} + \frac{z_2^2}{2!} + \cdots\right).$$

We will carry out the multiplication of series termwise. The terms obtained in this multiplication of series may be written in the form of a square table

$$1 \cdot 1 + 1 \quad \cdot \frac{z_2}{1!} + 1 \quad \cdot \frac{z_2^2}{2!} + 1 \quad \cdot \frac{z_2^3}{3!} +$$

$$\cdots + \frac{z_1}{1!} \cdot 1 + \frac{z_1}{1!} \cdot \frac{z_2}{1!} + \frac{z_1}{1!} \cdot \frac{z_2^2}{2!} + \frac{z_1}{1!} \cdot \frac{z_2^3}{3!} +$$

$$\cdots + \frac{z_1^2}{2!} \cdot 1 + \frac{z_1^2}{2!} \cdot \frac{z_2}{1!} + \frac{z_1^2}{2!} \cdot \frac{z_2^2}{2!} + \frac{z_1^2}{2!} \cdot \frac{z_2^3}{3!} +$$

$$\cdots + \frac{z_1^3}{3!} \cdot 1 + \frac{z_1^3}{3!} \cdot \frac{z_2}{1!} + \frac{z_1^3}{3!} \cdot \frac{z_2^2}{3!} + \frac{z_1^3}{3!} \cdot \frac{z_2^3}{3!} + \cdots.$$

We now collect the terms which have the same sum of powers of z_1 and z_2. It is easy to see that such terms lie on the diagonals of our table. We get

$$e^{z_1} \cdot e^{z_2} = 1 + \left(\frac{z_2}{1!} + \frac{z_1}{1!}\right) + \left(\frac{z_2^2}{2!} + \frac{z_1}{1!} \frac{z_2}{1!} + \frac{z_1^2}{2!}\right) + \cdots. \tag{19}$$

The general term of this series will be

$$\frac{z_2^n}{n!} + \frac{z_2^{n-1}}{(n-1)!}\frac{z_1}{1!} + \frac{z_2^{n-2}}{(n-2)!}\frac{z_1^2}{2!} + \cdots + \frac{z_1^n}{n!}$$

$$= \frac{1}{n!}\left(z_2^n + \frac{n!}{1!(n-1)!}z_2^{n-1}z_1 + \frac{n!}{2!(n-2)!}z_2^{n-2}z_1^2 + \cdots + z_1^n\right).$$

Applying the binomial formula of Newton, we get the general term in the form

$$\frac{(z_1 + z_2)^n}{n!}.$$

So the general term of the series (19) is identical with the general term of the series for $e^{z_1+z_2}$, which proves the theorem on the rule for multiplication (18).

The multiplication theorem and Euler's formula allow us to derive an expression for the function e^z in terms of functions of real variables in finite form (without series). Thus, putting

$$z = x + iy,$$

we get

$$e^z = e^{x+iy} = e^x \cdot e^{iy},$$

and since

$$e^{iy} = \cos y + i \sin y,$$

we find that

$$e^z = e^x(\cos y + i \sin y). \tag{20}$$

The formula so derived is very convenient for investigating the properties of the function e^z. We note two of its properties: (1) the function e^z vanishes nowhere; for in fact, $e^x \neq 0$ and the functions $\cos y$ and $\sin y$ in formula (20) never vanish simultaneously; (2) the function e^z has period $2\pi i$, i.e.,

$$e^{z+2\pi i} = e^z.$$

This last statement follows from the multiplication theorem and the equality

$$e^{2\pi i} = \cos 2\pi + i \sin 2\pi = 1.$$

The formulas (17) allow us to investigate the functions $\cos z$ and $\sin z$ in the complex domain. We leave it as an exercise for the reader to prove that in the complex domain $\cos z$ and $\sin z$ have period 2π and that the theorems about the sine and cosine of a sum continue to hold for them.

The general concept of a function of a complex variable and the differentiability of functions. Power series allow us to define analytic functions of a complex variable. However, it is of interest to study the basic operations of analysis for an arbitrary function of a complex variable and in particular the operation of differentiation. Here we uncover very deep-lying facts connected with the differentiation of functions of a complex variable. As we will see on the one hand, a function, having a first derivative at all points in a neighborhood of some point z_0, necessarily has derivatives of all orders at z_0, and further, it can be expanded in a power series centered at this point; i.e., it is analytic. Thus, if we consider differentiable functions of a complex variable, we return immediately to the class of analytic functions. On the other hand, a study of the derivative uncovers the geometric behavior of functions of a complex variable and the connections of the theory of these functions with problems in mathematical physics.

In view of what has been said, we will, in what follows, call a function *analytic at the point* z_0 if it has a derivative at all points of some neighborhood of z_0.

We will say, following the general definition of a function, that a complex variable w is a function of the complex variable z if some law exists which allows us to find the value of w, given the value of z.

Every complex number $z = x + iy$ is represented by a point (x, y) on the Oxy plane, and the numbers $w = u + iv$ will also be represented by points on an Ouv plane, the plane of the function. Then from the geometric point of view a function of a complex variable $w = f(z)$ defines a law of correspondence between the points of the Oxy plane of the argument z and points of the Ouv plane of the value w of the function. In other words, a function of a complex variable determines a transformation of the plane of the argument to the plane of the function. To define a function of a complex variable means to give the correspondence between the pairs of numbers (x, y) and (u, v); defining a function of a complex variable is thus equivalent to defining two functions

$$u = \phi(x, y), \qquad v = \psi(x, y),$$

for which, obviously

$$w = u + iv = \phi(x, y) + i\psi(x, y).$$

For example, if

$$w = z^2 = (x + iy)^2 = x^2 - y^2 + 2ixy,$$

then

$$u = \phi(x, y) = x^2 - y^2, \qquad v = \psi(x, y) = 2xy.$$

The derivative of a function of a complex variable is defined formally in the same way as the derivative of a function of a real variable. The derivative is the limit of the difference quotient of the function

$$f'(z) = \lim_{\Delta z \to 0} \frac{f(z + \Delta z) - f(z)}{\Delta z}, \qquad (21)$$

if this limit exists.

If we assume that the two real functions u and v, making up $w = f(z)$, have partial derivatives with respect to x and y, this is still not a sufficient condition that the derivative of the function $f(z)$ exists. The limit of the difference quotient, as a rule, depends on the direction in which the points $z' = z + \Delta z$ approximate the point z (figure 3). For the existence of the derivative $f'(z)$, it is necessary that the limit does not depend on the manner of approach of z' to z. Consider, for example, the case when z' approaches z parallel to the axis Ox or parallel to the axis Oy.

In the first case

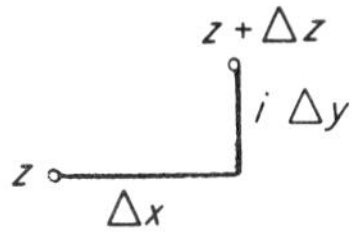

$$\Delta z = \Delta x,$$

$$f(z + \Delta z) - f(z) = u(x + \Delta x, y) - u(x, y)$$
$$+ i[v(x + \Delta x, y) - v(x, y)],$$

FIG. 3.

and the difference quotient

$$\frac{f(z + \Delta z) - f(z)}{\Delta z} = \frac{u(x + \Delta x, y) - u(x, y)}{\Delta x} + i\frac{v(x + \Delta x, y) - v(x, y)}{\Delta x}$$

for $\Delta x \to 0$ converges to

$$\frac{\partial u}{\partial x} + i\frac{\partial v}{\partial x}. \qquad (22)$$

In the second case

$$\Delta z = i\,\Delta y,$$

and the difference quotient

$$\frac{f(z + \Delta z) - f(z)}{\Delta z} = -i\frac{u(x, y + \Delta y) - u(x, y)}{\Delta y} + \frac{v(x, y + \Delta y) - v(x, y)}{\Delta y}$$

leads in the limit to

$$\frac{\partial v}{\partial y} - i\frac{\partial u}{\partial y}. \qquad (23)$$

If the function $w = f(x)$ has a derivative, these two expressions must be equal, and thus

$$\frac{\partial u}{\partial x} = \frac{\partial v}{\partial y},$$

$$\frac{\partial u}{\partial y} = -\frac{\partial v}{\partial x}. \tag{24}$$

Satisfying these equations is a necessary condition for the existence of the derivative of the function $w = u + iv$. It can be shown that condition (24) is not only necessary but also sufficient (if the functions u and v have a total differential). We will not give a proof of the sufficiency of conditions (24), which are called the *Cauchy-Riemann equations*.

It is easy to establish the fact that the usual rules for differentiating functions of a real variable carry over without alteration to functions of a complex variable. Certainly this is true for the derivative of the function z^n and for the derivative of a sum, a product, or a quotient. The method of proof remains exactly the same as for functions of a real variable, excepting only that in place of real quantities, complex ones are to be understood. This shows that every polynomial in z

$$w = a_0 + a_1 z + \cdots + a_n z^n$$

is an everywhere differentiable function. Any rational function, equal to the quotient of two polynomials

$$w = \frac{a_0 + a_1 z + \cdots + a_n z^n}{b_0 + b_1 z + \cdots + b_m z^m}$$

is differentiable at all points where the denominator is not zero.

In order to establish the differentiability of the function $w = e^z$, we may use the Cauchy-Riemann conditions. In this case, on the basis of formula (20)

$$u = e^x \cos y, \qquad v = e^x \sin y;$$

we substitute these functions in (24) and show that the Cauchy-Riemann equations are satisfied. The derivative may be computed, for example by formula (22). This gives

$$\frac{dw}{dz} = e^z.$$

On the basis of formula (17) it is easy to establish the differentiability of the trigonometric functions and the validity of the formulas known from analysis for the values of their derivatives.

The function Ln z. We will not give here an investigation of all the elementary functions of a complex variable. However, it is important for our purposes to become acquainted with some of the properties of the function Ln z. As in the case of the real domain, we set

$$w = \text{Ln } z,$$

if

$$z = e^w.$$

In order to analyze the function Ln z, we write the number z in trigonometric form

$$z = r(\cos \phi + i \sin \phi).$$

Applying the multiplication theorem to e^w, we get

$$z = e^w = e^{u+iv} = e^u e^{iv} = e^u(\cos v + i \sin v).$$

Equating the two expressions derived for z, we have

$$e^u = r, \tag{α}$$

$$\cos v + i \sin v = \cos \phi + i \sin \phi. \tag{β}$$

Since u and r are real numbers, from formula (α) we derive

$$u = \ln r,$$

where $\ln r$ is the usual value of the natural logarithm of a real number. Equation (β) can be satisfied only if

$$\cos v = \cos \phi, \quad \sin v = \sin \phi,$$

and in this case v and ϕ must differ by a number which is a multiple of 2π

$$v = \phi + 2\pi n,$$

where for any integer n equation (β) will be satisfied. On the basis of the expressions derived for u and v

$$\text{Ln } z = \ln r + i(\phi + 2\pi n). \tag{25}$$

Formula (25) defines the function Ln z for all values of the complex number z that are different from zero. It gives the definition of the logarithm not only for positive numbers but also for negative and complex numbers.

The expression derived for the function Ln z contains an arbitrary integer n. This means that Ln z is a multiple-valued function. For any value of n we get one of the possible values of the function Ln z. If we fix the value of n, we get one of the possible values of this function.

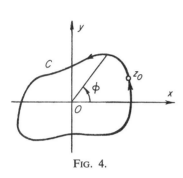

FIG. 4.

However, the different values of Ln z, as can be shown, are organically related to one another. In fact, let us fix, for example, the value $n = 0$ at the point z_0 and then let z move continuously around a closed curve C, which surrounds the origin and returns to the point z_0 (figure 4). During the motion of z, the angle ϕ will increase continuously and when z moves around the entire closed contour, ϕ will increase by 2π. In this manner, fixing the value of the logarithm at z_0

$$(\text{Ln } z)_0 = \ln r_0 + i\phi_0$$

and changing this value continuously while moving z along the closed curve surrounding the origin, we return to the point z_0 with another value of the function

$$(\text{Ln } z)_0 = \ln r_0 + i(\phi_0 + 2\pi).$$

This situation shows us that we may pass continuously from one value of Ln z to another. For this the point need only travel around the origin continuously a sufficient number of times. The point $z = 0$ is called a *branch point* of the function Ln z.

If we wish to restrict consideration to only one value of the function Ln z, we must prevent the point z from describing a closed curve surrounding the point $z = 0$. This may be done by drawing a continuous curve from the origin to infinity and preventing the point z from crossing this curve, which is called a *cut*. If z varies over the cut plane, then it never changes continuously from one value of Ln z to another and thus, starting from a specific value of logarithm at any point z_0, we get at each point only one value of the logarithm. The values of the function Ln z selected in this way constitute a single-valued branch of the function.

For example, if the cut lies along the negative part of the axis Ox, we get a single-valued branch of Ln z by restricting the argument to the limits

$$(2k - 1)\pi < \phi \leqslant (2k + 1)\pi,$$

where k is an arbitrary integer.

Considering a single-valued branch of the logarithm, we can study its differentiability. Putting

$$r = \sqrt{x^2 + y^2}, \qquad \phi = \arctan \frac{y}{x},$$

it is easy to show that Ln z satisfies the Cauchy-Riemann conditions and its derivative, calculated for example by formula (22), will be equal to

$$\frac{d \operatorname{Ln} z}{dz} = \frac{1}{z}.$$

We emphasize that the derivative of Ln z is also a single-valued function.

§2. The Connection Between Functions of a Complex Variable and the Problems of Mathematical Physics

Connection with problems of hydrodynamics. The Cauchy-Riemann conditions relate the problems of mathematical physics to the theory of functions of a complex variable. Let us illustrate this from the problems of hydrodynamics.

Among all possible motions of a fluid an important role is played by the *steady motions*. This name is given to motions of the fluid for which there is no change with time in the distribution of velocities in space. For example, an observer standing on a bridge and watching the flow of the river around a supporting pillar sees a steady flow. Sometimes a flow is steady for an observer in motion on some conveyance. In the case of a steamship travelling through rough water, the flow will appear nonsteady to an observer on the shore but steady to one on the ship. To a passenger seated in an airplane that is flying with constant velocity, the flow of the air as disturbed by the plane will still appear to be a steady one.

For steady motion the velocity vector V of the particle of the fluid passing through a given point of space does not change with time. If the motion is steady for a moving observer, then the velocity vector does not change with time at points having constant coordinates in a coordinate system which moves with the observer.

Among the motions of a fluid great importance has been attached to the class of *plane-parallel motions*. These are flows for which the velocity of the particles is everywhere parallel to some plane and the distribution of the velocities is identical on all planes parallel to the given plane.

If we imagine an infinitely extended mass of fluid, flowing around a cylindrical body in a direction perpendicular to a generator, the distribution of velocities will be the same on all planes perpendicular to the generator, so that the flow will be plane-parallel. In many cases the motion of a fluid is approximately plane-parallel. For example, if we consider the flow of air in a plane perpendicular to the wing of an airplane, the motion of the air may be considered as approximately plane-parallel, provided the plane in question is not very close either to the fuselage or to the tip of the wing.

We will show how the theory of functions of a complex variable may be applied to the study of steady plane-parallel flow.

Here we will assume that the liquid is incompressible, i.e., that its density does not change with change in pressure. This assumption holds, for example, for water, but it can be shown that even air may be considered incompressible in the study of its flow, if the velocity of the motion is not very large. The hypothesis of incompressibility of air will not produce a noticeable distortion if the velocities of motion do not exceed the range of 0.6 to 0.8 of the velocity of sound (330 m/sec).

The flow of a liquid is characterized by the distribution of the velocities of its particles. If the flow is plane-parallel, then it is sufficient to determine the velocities of the particles in one of the planes parallel to which the motion occurs.

We will denote by $V(x, y, t)$ the velocity vector of the particle passing through the point with coordinates x, y at the instant of time t. In the case of steady motion, V does not depend on time. The vector V will be given by its projections u and v on the coordinate axes. We consider the trajectories of particles of the fluid. In the case of steady motion, there is no change with time in the velocities of the successive particles issuing from a given point in space. If we know the field of the velocities, i.e., if we know the components of the velocity as functions of x, y, then the trajectories of the particles may be determined by using the fact that the velocity of a particle is everywhere tangent to the trajectory. This gives

$$\frac{dy}{dx} = \frac{v(x, y)}{u(x, y)} .$$

The equation so obtained is the differential equation for the trajectories. The trajectory of a particle in a steady motion is called a *streamline*. Through each point of the plane passes exactly one streamline.

An important role is played here by the so-called stream function. For a fixed streamline C_0 let us consider the imaginary channel bounded by the following four walls: One wall is the cylindrical surface (with generators perpendicular to the plane of the flow) passing through the streamline C_0; the second wall is the same cylindrical surface for a neighboring streamline C_1; the third is the plane of the flow; and the fourth is a parallel plane at unit distance (figure 5). If we consider two arbitrary cross sections of our channel, denoted by γ_1 and γ_2, then the quantity of fluid passing through the sections γ_1 and γ_2 in unit time will be the same, as follows from the fact that the quantity of fluid inside the part of the channel marked off by C_1, C_0 and γ_1, γ_2 cannot change, because of the constant density, since the side walls of the channel C_0 and C_1 are formed

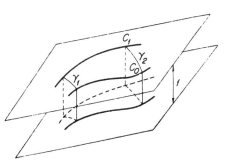

FIG. 5.

by streamlines, so that no fluid passes through them. Consequently the same amount of fluid must leave in unit time through γ_1 as enters through γ_2.

Now by the *stream function* we mean the function $\psi(x, y)$ that has a constant value on the streamline C_1 equal to the quantity of liquid passing in unit time through the cross section of the channel constructed on the curves C_0 and C_1.

The stream function is defined only up to an arbitrary constant, depending on the choice of the initial streamline C_0. If we know the stream function, then the equations for the streamlines are obviously

$$\psi(x, y) = \text{const.}$$

We now wish to express the components of the velocity of the flow at a given point $M(x, y)$ in terms of the derivatives of the stream function. To this end we consider the channel formed by the streamline C through the point $M(x, y)$ and a neighboring streamline C' through a nearby point $M'(x, y + \Delta y)$, together with the two planes parallel to the plane of flow and a unit distance apart. Let us compute the quantity of the liquid q passing through the section MM' of the channel during time dt.

On the one hand, from the definition of the stream function

$$q = (\psi' - \psi)\, dt.$$

On the other hand, q is equal (figure 6) to the volume of the solid formed by drawing the vector $V\,dt$ from each point of the section MM'. If MM' is small, we may assume that V is constant over the whole of

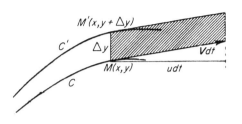

MM' and is equal to the value of V at the point M. The area of the base of the parallelepiped so constructed is $\varDelta y \times 1$ (in figure 6 the unit thickness is not shown), and the altitude is the projection of the vector $V\,dt$ on the Ox axis, i.e., $u\,dt$ so

Fig. 6.   that

$$q \approx u\,\varDelta y\,dt$$

and thus

$$u\,\varDelta y \approx \varDelta\psi.$$

Dividing this equation by $\varDelta y$, and passing to the limit, we get

$$u = \frac{\partial\psi}{\partial y}. \tag{26}$$

A similar argument gives for the second component

$$v = -\frac{\partial\psi}{\partial x}. \tag{26'}$$

To define the field of the velocity vectors, we introduce, in addition to the stream function, another function, which arises from considering the rotation of small particles of the liquid. If we imagine that a particular particle of the fluid were to become solidified, it would in general have a rotatory motion. However, if the motion of the fluid starts from rest and if there is no internal friction between particles, then it can be shown that rotation of the particles of the fluid cannot begin. Motions of a fluid in which there is no rotation of this sort are called irrotational; they play a fundamental role in the study of the motion of bodies in a fluid. In the theory of hydromechanics it is shown that for irrotational flow there exists a second function $\phi(x, y)$ such that the components of the velocity are expressed by the formulas

$$u = \frac{\partial\phi}{\partial x}, \qquad v = \frac{\partial\phi}{\partial y}; \tag{27}$$

the function ϕ is called the *velocity potential* of the flow. Later, we will consider motions with velocity potential.

Comparison of the formulas for the components of the velocity from the stream function and from the velocity potential gives the following remarkable result.

The velocity potential $\phi(x, y)$ and the stream function $\psi(x, y)$ for the flow of an incompressible fluid satisfy the Cauchy-Riemann equations

$$\frac{\partial \phi}{\partial x} = \frac{\partial \psi}{\partial y},$$
$$\frac{\partial \phi}{\partial y} = -\frac{\partial \psi}{\partial x}. \tag{28}$$

In other words, the function of a complex variable

$$w = \phi(x, y) + i\psi(x, y)$$

is a differentiable function of a complex variable. Conversely, if we choose an arbitrary differentiable function of a complex variable, its real and imaginary parts satisfy the Cauchy-Riemann conditions and may be considered as the velocity potential and the stream function of the flow of an incompressible fluid. The function w is called the *characteristic function of the flow.*

Let us now consider the significance of the derivative of w. Using, for example, formula (22), we have

$$\frac{dw}{dz} = \frac{\partial \phi}{\partial x} + i\frac{\partial \psi}{\partial x}.$$

From (27) and (26′) we find

$$\frac{dw}{dz} = u - iv$$

or, taking complex conjugates,

$$u + iv = \overline{\left(\frac{dw}{dz}\right)}, \tag{29}$$

where the bar over dw/dz denotes the complex conjugate.

Consequently, the velocity vector of the flow is equal to the conjugate of the value of the derivative of the characteristic function of the flow.

Examples of plane-parallel flow of a fluid.   We consider several examples. Let

$$w = Az, \tag{30}$$

where A is a complex quantity. From (29) it follows that

$$u + iv = \bar{A}.$$

Thus the linear function (30) defines the flow of a fluid with constant vector velocity. If we set

$$A = u_0 - iv_0 ,$$

then, decomposing into the real and imaginary parts of w, we have

$$\phi(x, y) = u_0 x + v_0 y,$$
$$\psi(x, y) = u_0 y - v_0 x,$$

so that the streamlines will be straight lines parallel to the velocity vector (figure 7).

As a second example we consider the function

$$w = Az^2,$$

where the constant A is real. In order to graph the flow, we first determine the streamlines. In this case

$$\psi(x, y) = 2Axy,$$

and the equations of the streamlines are

$$xy = \text{const.}$$

These are hyperbolas with the coordinate axes as asymptotes (figure 8). The arrows show the direction of motion of the particles along the streamlines for $A > 0$. The axes Ox and Oy are also streamlines.

If the friction in the liquid is very small, we will not disturb the rest of the flow if we replace any streamline by a rigid wall, since the fluid

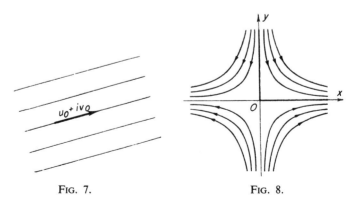

FIG. 7.   FIG. 8.

will glide along the wall. Using this principle to construct walls along the positive coordinate axes (in figure 8 they are represented by heavy lines), we have a diagram of how the fluid flows irrotationally, in this case around a corner.

An important example of a flow is given by the function

$$w = a \left(z + \frac{R^2}{z} \right), \tag{31}$$

where a and R are positive real quantities.

The stream function will be

$$\psi = a \left(y - \frac{R^2 y}{x^2 + y^2} \right),$$

and thus the equation for the streamlines is

$$y - \frac{R^2 y}{x^2 + y^2} = \text{const.}$$

In particular, taking the constant equal to zero, we have either $y = 0$ or $x^2 + y^2 = R^2$; thus, a circle of radius R is a streamline. If we replace the interior of this streamline by a solid body, we obtain the flow around a circular cylinder. A diagram of the streamlines of this flow is shown in figure 9. The velocity of the flow may be defined from formula (29) by

$$u + iv = a \left(1 - \frac{R^2}{\bar{z}^2} \right).$$

At a great distance from the cylinder we find

$$\lim_{z \to \infty} (u + iv) = a;$$

i.e., far from the cylinder the velocity tends to a constant value and thus the flow tends to be uniform. Consequently, formula (29) defines the flow which arises from the passage around a circular cylinder of a fluid which is in uniform motion at a distance from the cylinder.

The basic ideas of the theory of an airplane wing; theorem of Žukovskiĭ.

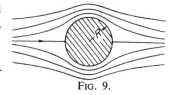

FIG. 9.

The application of the theory of functions of a complex variable to the study of plane-parallel flows of a fluid was the source of several remarkable discoveries in aerodynamics by Žukovskiĭ and Čaplygin. The study of streamlines of bodies led them to discover the law for the formation of lifting force on the wing on an airplane. In order to present the ideas which led to this discovery, we need to consider one more concrete example of fluid flow. Let us consider the characteristic function

$$w = \frac{\Gamma}{2\pi i} \ln z,$$

where Γ is a real constant. Although w is a multiple-valued function, its derivative

$$\frac{dw}{dz} = \frac{\Gamma}{2\pi i} \frac{1}{z} \tag{32}$$

is single valued, so that our function uniquely defines the velocity field of some fluid flow. If we set $z = re^{i\theta}$, the velocity potential and the stream function may be computed from (25) as

$$\phi = \frac{\Gamma}{2\pi} \theta, \qquad \psi = -\frac{\Gamma}{2\pi} \ln r.$$

The second of these formulas shows that the streamlines are the circles $r = $ const (figure 10).

The velocity of the flow is defined by formula (29) as

$$u + iv = -\frac{\Gamma}{2\pi i} \frac{1}{\bar{z}}.$$

In particular, it follows that the value of the velocity vector will be

$$V = |u + iv| = \frac{|\Gamma|}{2\pi} \frac{1}{r},$$

i.e., the velocity is constant on every streamline. A more detailed investigation shows that the flow goes counterclockwise for $\Gamma > 0$ and clockwise for $\Gamma < 0$.

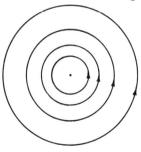

FIG. 10.

If we replace one of the streamlines by a rigid boundary, we obtain the circular motion of a fluid around a cylinder. Such a motion is called *circulatory.*

However, the potential of our motion is not a single-valued function. In one passage over a closed contour around the cylinder the potential is changed by an amount Γ. This change in potential is called the *circulation of the flow.*

If to the characteristic function of a flow past a cylinder (31), we add the characteristic function of a circulatory flow (with clockwise circuit), we get a new characteristic function

$$w = a \left(z + \frac{R^2}{z} \right) - \frac{\Gamma}{2\pi i} \operatorname{Ln} z. \tag{33}$$

This characteristic function also represents the flow around a cylinder of radius R. In fact, the stream function will be constant on a circumference of radius R, since there the coefficients of the imaginary parts of both terms are constant. The velocity of the flow, defined by the function (33), will again converge to a as $z \to \infty$. This shows that the characteristic function (33) defines, for any value of Γ, the streamlines of a translational flow past a cylinder. Figure 11 illustrates the character of the flow for

$\Gamma > 0$. This flow will not be symmetric, since the stagnation points a and b where the streams meet and leave the cylinder are displaced downward. The potential of the flow under consideration will be a multiple-valued function. As the result of one circuit around the cylinder it will change by an amount equal to Γ.

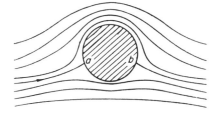

Fig. 11.

Because of symmetry, the flow around a cylinder will usually be of the form defined by the functions (32), but for nonsymmetric bodies the flow which arises usually has a multiple-valued potential. Later we will discuss the physical significance of this fact. The methods of the theory of functions of a complex variable allow us to define the possible flows around bodies of arbitrary shape. These methods will be discussed in the following section. With their help we can make use of the flow around a cylinder to construct the flow, with single-valued or multiple-valued potential, around any body.

In studying the streamlines of the wing of an airplane, we are dealing with a body with a sharp edge at the rear. The profile of the wing of an airplane always narrows toward the rear. If for such a profile we construct a flow with a single-valued potential, then the stagnation point where the stream leaves the wing proves not to be at the edge (figure 12a). But it

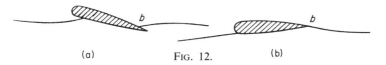

(a)   Fig. 12.   (b)

turns out that such a flow is physically impossible. (Infinite velocity, with consequent infinite rarefaction of the fluid would occur at the sharp edge.) The flow for which the point b falls on the edge of the wing (figure 12b) is the uniquely possible flow, and this flow, as a rule, will have a multiple-valued potential, i.e., will be a circulatory flow. The circulation Γ of such a flow again is defined as the change in the potential for a circuit of a closed contour around the wing.

The physical realizability of a flow around the profile of a wing with a stream leaving the rear edge is called *Čaplygin's postulate*.

The remarkable discovery of Žukovskiĭ consists of the fact that the existence of circulation in the flow causes a lifting force on the wing, in a direction perpendicular to the velocity a of the oncoming flow and equal in magnitude to the quantity

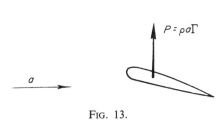

FIG. 13.

$$\rho a \Gamma,$$

where ρ is the density of the medium and Γ is the circulation (figure 13).

This theorem of Žukovskiĭ about the lifting force on a wing is basic for all contemporary aerodynamics. We will not give the proof here, merely noting that the usual proofs are based on the theory of integrals of functions of a complex variable.

The basic results in aerodynamics as established by Žukovskiĭ and Čaplygin have been extensively developed by the work of Soviet scientists.

Applications to other problems of mathematical physics.   The theory of functions of a complex variable has found wide application not only in wing theory but in many other problems of hydrodynamics.

However, the domain of application of the theory of functions is not restricted to hydrodynamics; it is much wider than that, including many other problems of mathematical physics. To illustrate, we return to the Cauchy-Riemann conditions

$$\frac{\partial u}{\partial x} = \frac{\partial v}{\partial y},$$

$$\frac{\partial u}{\partial y} = -\frac{\partial v}{\partial x}$$

and deduce from them an equation which is satisfied by the real part of an analytic function of a complex variable. If the first of these equations is differentiated with respect to x, and the second with respect to y, we obtain by addition

$$\frac{\partial^2 u}{\partial x^2} + \frac{\partial^2 u}{\partial y^2} = 0.$$

This equation (which we have already met in Chapter VI) is known as the *Laplace equation*. A large number of problems of physics and mechanics involve the Laplace equation. For example, if the heat in a body is in equilibrium, the temperature satisfies the Laplace equation. The study of magnetic or electrostatic fields is connected with this equation. In the investigation of the filtration of a liquid through a porous medium, we also arrive at the Laplace equation. In all these problems involving the solution of the Laplace equation the methods of the theory of functions have found wide application.

Not only the Laplace equation but also the more general equations of mathematical physics can be brought into connection with the theory of functions of a complex variable. One of the most remarkable examples is provided by planar problems in the theory of elasticity. The foundations of the application of functions of a complex variable to this domain were laid by the Soviet scientists G. B. Kolosov and N. I. Mushelišvili.

§3. The Connection of Functions of a Complex Variable with Geometry

Geometric properties of differentiable functions. As in the case of functions of a real variable, a great role is played in the theory of analytic functions of a complex variable by the geometric interpretation of these functions. Broadly speaking, the geometric properties of functions of a complex variable have not only provided a natural means of visualizing the analytic properties of the functions but have also given rise to a special set of problems. The range of problems connected with the geometric properties of functions has been called the *geometric theory of functions.* As we said earlier, from the geometric point of view a function of a complex variable $w = f(z)$ is a transformation from the z-plane to the w-plane. This transformation may also be defined by two functions of two real variables

$$u = u(x, y),$$

$$v = v(x, y).$$

If we wish to study the character of the transformation in a very small neighborhood of some point, we may expand these functions into

Taylor series and restrict ourselves to the leading terms of the expansion

$$u - u_0 = \left(\frac{\partial u}{\partial x}\right)_0 (x - x_0) + \left(\frac{\partial u}{\partial y}\right)_0 (y - y_0) + \cdots,$$

$$v - v_0 = \left(\frac{\partial v}{\partial x}\right)_0 (x - x_0) + \left(\frac{\partial v}{\partial y}\right)_0 (y - y_0) + \cdots,$$

where the derivatives are taken at the point (x_0, y_0). Thus, in the neighborhood of a point, any transformation may be considered approximately as an affine transformation*

$$u - u_0 = a(x - x_0) + b(y - y_0),$$
$$v - v_0 = c(x - x_0) + d(y - y_0),$$

where

$$a = \left(\frac{\partial u}{\partial x}\right)_0, \qquad b = \left(\frac{\partial u}{\partial y}\right)_0,$$

$$c = \left(\frac{\partial v}{\partial x}\right)_0, \qquad d = \left(\frac{\partial v}{\partial y}\right)_0.$$

Let us consider the properties of the transformation effected by the analytic function near the point $z = x + iy$. Let C be a curve issuing from the point z; on the w-plane the corresponding points trace out the curve Γ, issuing from the point w. If z' is a neighboring point and w' is the point corresponding to it, then for $z' \to z$ we will have $w' \to w$ and

$$\frac{w' - w}{z' - z} \to f'(z). \tag{34}$$

In particular, it follows that

$$\frac{|w' - w|}{|z' - z|} \to |f'(z)|. \tag{35}$$

This fact may be formulated in the following manner.

The limit of the ratio of the lengths of corresponding chords in the w-plane and in the z-plane at the point z is the same for all curves issuing from the given point z, or as it is also expressed, the ratio of linear elements on the w-plane and on the z-plane at a given point does not depend on the curve issuing from z.

The quantity $|f'(z)|$, which characterizes the magnification of linear elements at the point z, is called the *coefficient of dilation* at the point z.

---

* Cf. Chapter III, §11.

We now suppose that at some point z the derivative $f'(z) \neq 0$, so that $f'(z)$ has a uniquely determined argument.* Let us compute this argument, using (34)

$$\arg \frac{w' - w}{z' - z} = \arg (w' - w) - \arg (z' - z),$$

but $\arg (w' - w)$ is the angle β' between the chord ww' and the real axis, and $\arg (z' - z)$ is the angle α' between the chord zz' and the real axis.

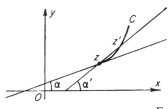

FIG. 14.

If we denote by α and β the corresponding angles for the tangents to the curves C and Γ at the points z and w (figure 14), then for $z' \to z$

$$\alpha' \to \alpha, \quad \beta' \to \beta,$$

so that in the limit we get

$$\arg f'(z) = \beta - \alpha. \tag{36}$$

This equation shows that $\arg f'(z)$ is equal to the angle ϕ through which the direction of the tangent to the curve C at the point z must be turned to assume the direction of the tangent to the curve Γ at the point w. From this property $\arg f'(z)$ is called the *rotation of the transformation* at the point z.

From equation (36) the reader can easily derive the following propositions.

As we pass from the z-plane to the w-plane, the tangents to all curves issuing from a given point are rotated through the same angle.

If C_1 and C_2 are two curves issuing from the point z, and Γ_1 and Γ_2 are the corresponding curves from the point w, then the angle between Γ_1 and Γ_2 at the point w is equal to the angle between C_1 and C_2 at the point z.

---

* Cf. Chapter IV, §3.

In this manner, for the transformation effected by an analytic function, at each point where $f'(z) \neq 0$, all linear elements are changed by the same ratio, and the angles between corresponding directions are not changed.

Transformations with these properties are called *conformal* transformations.

From the geometric properties just proved for transformations near a point at which $f'(z_0) \neq 0$, it is natural to expect that in a small neighborhood of z_0 the transformation will be one-to-one; i.e., not only will each point z correspond to only one point w, but also conversely each point w will correspond to only one point z. This proposition can be rigorously proved.

To show more clearly how conformal transformations are distinguished from various other types of transformations, it is useful to consider an arbitrary transformation in a small neighborhood of a point. If we consider the leading terms of the Taylor expansions of the functions u and v effecting the transformation, we get

$$u - u_0 = \left(\frac{\partial u}{\partial x}\right)_0 (x - x_0) + \left(\frac{\partial u}{\partial y}\right)_0 (y - y_0) + \cdots,$$

$$v - v_0 = \left(\frac{\partial v}{\partial x}\right)_0 (x - x_0) + \left(\frac{\partial v}{\partial y}\right)_0 (y - y_0) + \cdots.$$

If in a small neighborhood of the point (x_0, y_0) we ignore the terms of higher order, then our transformation will act like an affine transformation. This transformation has an inverse if its determinant does not vanish

$$\Delta = \left(\frac{\partial u}{\partial x}\right)_0 \left(\frac{\partial v}{\partial y}\right)_0 - \left(\frac{\partial u}{\partial y}\right)_0 \left(\frac{\partial v}{\partial x}\right)_0 \neq 0.$$

If $\Delta = 0$, then to describe the behavior of the transformation near the point (x_0, y_0) we must consider terms of higher order.*

In case $u + iv$ is an analytic function, we can express the derivatives with respect to y in terms of the derivatives with respect to x by using the Cauchy-Riemann conditions, from which we get

$$\Delta = \left(\frac{\partial u}{\partial x}\right)_0^2 + \left(\frac{\partial v}{\partial x}\right)_0^2 = \left| \left(\frac{\partial u}{\partial x}\right)_0 + i \left(\frac{\partial v}{\partial x}\right)_0 \right|^2 = |f'(z_0)|^2,$$

---

* In this last case, i.e., for $\Delta = 0$, the transformation is not called affine. For affine transformations see also Chapter III, §11.

i.e., the transformation has an inverse when $f'(z_0) \not= 0$. If we set $f'(z_0) = r(\cos \phi + i \sin \phi)$, then

$$\left(\frac{\partial u}{\partial x} \right)_0 = \left(\frac{\partial v}{\partial y} \right)_0 = r \cos \phi,$$

$$\left(\frac{\partial u}{\partial y} \right)_0 = - \left(\frac{\partial v}{\partial x} \right)_0 = -r \sin \phi,$$

and the transformation near the point (x_0, y_0) will have the form

$$u - u_0 = r[(x - x_0) \cos \phi - (y - y_0) \sin \phi] + \cdots,$$

$$v - v_0 = r[(x - x_0) \sin \phi + (y - y_0) \cos \phi] + \cdots.$$

These formulas show that in the case of an analytic function $w = u + iv$, the transformation near the point (x_0, y_0) consists of rotation through the angle ϕ and dilation with coefficient r. In fact, the expressions inside the brackets are the well-known formulas from analytic geometry for rotation in the plane through an angle ϕ, and multiplication by r gives the dilation.

To form an idea of the possibilities when $f'(z) = 0$ it is useful to consider the function

$$w = z^n. \tag{37}$$

The derivative of this function $w' = nz^{n-1}$ vanishes for $z = 0$. The transformation (37) is most conveniently considered by using polar coordinates or the trigonometric form of a complex number. Let

$$z = r(\cos \phi + i \sin \phi),$$

$$w = \rho(\cos \theta + i \sin \theta).$$

Using the fact that in multiplying complex numbers the moduli are multiplied and the arguments added, we get

$$z^n = r^n(\cos n\phi + i \sin n\phi),$$

and thus

$$\rho = r^n,$$

$$\theta = n\phi.$$

From the last formula we see that the ray $\phi = \text{const.}$ of the z-plane transforms into the ray $\theta = n\phi = \text{const.}$ in the w-plane. Thus an angle α between two rays in the z-plane will transform into an angle of magnitude $\beta = n\alpha$. The transformation of the z-plane into the w-plane ceases to be one-to-one. In fact, a given point w with modulus ρ and argument θ

may be obtained as the image of each of the n points with moduli $r = \sqrt[n]{\rho}$ and arguments

$$\phi = \frac{\theta}{n}, \frac{\theta}{n} + \frac{2\pi}{n}, \cdots, \frac{\theta}{n} + \frac{2\pi}{n}(n-1).$$

When raised to the power n, the moduli of the corresponding points will all be equal to ρ and their arguments will be equal to

$$\theta, \theta + 2\pi, \cdots, \theta + 2\pi(n-1),$$

and since changing the value of the argument by a multiple of 2π does not change the geometric position of the point, all the images on the w-plane are identical.

Conformal transformations.   If an analytic function $w = f(z)$ takes a domain D of the z-plane into a domain Δ of the w-plane in a one-to-one manner, then we say that it effects a conformal transformation of the domain D into the domain Δ.

The great role of conformal transformations in the theory of functions and its applications is due to the following almost trivial theorem.

If $\zeta = F(w)$ is an analytic function on the domain Δ, then the composite function $F[f(z)]$ is an analytic function on the domain D. This theorem results from the equation

$$\frac{\Delta\zeta}{\Delta z} = \frac{\Delta\zeta}{\Delta w} \cdot \frac{\Delta w}{\Delta z}.$$

In view of the fact that the functions $\zeta = F(w)$ and $w = f(z)$ are analytic, we conclude that both factors on the right side have a limit, and thus at each point of the domain D the quotient $\Delta\zeta/\Delta z$ has a unique limit $d\zeta/dz$. This shows that the function $\zeta = F[f(z)]$ is analytic.

The theorem just proved shows that the study of analytic functions on the domain Δ may be reduced to the study of analytic functions on the domain D. If the geometric structure of the domain D is simpler, this fact simplifies the study of the functions.

The most important class of domains in which it is necessary to study analytic functions is the class of *simply connected* domains. This is the name given to domains whose boundary consists of one piece (figure 15a) as opposed to domains whose

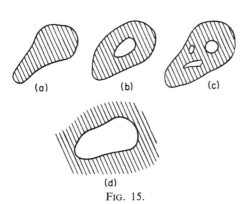

(a)   (b)   (c)

(d)

FIG. 15.

boundary falls into several pieces (for example, the domains illustrated in figures 15b and 15c).

We note that sometimes we are interested in investigating functions on a domain lying outside a curve rather than inside it. If the boundary of such a domain consists of only one piece, then the domain is also called simply connected (figure 15d).

At the foundations of the theory of conformal transformations lies the following remarkable theorem of Riemann.

For an arbitrary simply connected domain Δ, it is possible to construct an analytic function which effects a conformal transformation of the circle with unit radius and center at the origin into the given domain in such a way that the center of the circle is transformed into a given point w_0 of the domain Δ, and a curve in an arbitrary direction at the center of the circle transforms into a curve with an arbitrary direction at the point w_0. This theorem shows that the study of functions of a complex variable on arbitrary simply connected domains may be reduced to the study of functions defined, for example, on the unit circle.

We will now explain in general outline how these facts may be applied to problems in the theory of the wing of an airplane. Let us suppose that we wish to study the flow around a curved profile of arbitrary shape.

If we can construct a conformal transformation of the domain outside the profile to the domain outside the unit circle, then we can make use of the characteristic function for the flow around the circle to construct the characteristic function for the flow around the profile.

Let ζ be the plane of the circle, z the plane of the profile, and $\zeta = f(z)$ a function effecting the transformation of the domain outside the profile to the domain outside the circle, where

$$\lim_{z \to \infty} \zeta = \infty.$$

We denote by a the point of the circle corresponding to the edge of the profile A and construct the circulatory flow past the circle with one of the streamlines leaving the circle at a (figure 16). This function will be denoted by $W(\zeta)$:

$$W(\zeta) = \Phi + i\Psi.$$

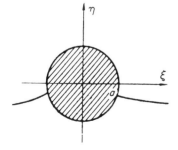

Fig. 16.

The streamlines of this flow are defined by the equation

$$\Psi = \text{const.}$$

We now consider the function

$$w(z) = W[f(z)],$$

and set

$$w = \phi + i\psi.$$

We show that $w(z)$ is the characteristic function of the flow past the profile with a streamline leaving the profile at the point A. First of all the flow defined by the function $w(z)$ is actually a flow past the profile. To prove this, we must show that the contour of the profile is a streamline curve, i.e., that on the contour of the profile

$$\psi(x, y) = \text{const.}$$

But this follows from the fact that

$$\psi(x, y) = \Psi(\xi, \eta),$$

and the points (x, y) lying on the profile correspond to the points (ξ, η) lying on the circle, where $\Psi(\xi, \eta) = \text{const.}$

It is also simple to show that A is a stagnation point for the flow, and it may be proved that by suitable choice of velocity for the flow past the circle, we may obtain a flow past the profile with any desired velocity.

The important role played by conformal transformations in the theory of functions and their applications gave rise to many problems of finding the conformal transformation of one domain into another of a given geometric form. In a series of simple but useful cases this problem may be solved by means of elementary functions. But in the general case the elementary functions are not enough. As we saw earlier, the general theorem in the theory of conformal transformations was stated by Riemann, although he did not give a rigorous proof. In fact, a complete proof required the efforts of many great mathematicians over a period of several decades.

In close connection with the different approaches to the proof of Riemann's theorem came approximation methods for the general construction of conformal transformations of domains. The actual construction of the conformal transformation of one domain onto another is sometimes a very difficult problem. For investigation of many of the general properties of functions, it is often not necessary to know the actual transformation of one domain onto another, but it is sufficient to exploit some of its geometric properties. This fact has led to a wide study of the geometric properties of conformal transformations. To illustrate the nature of theorems of this sort we will formulate one of them.

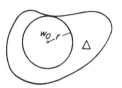

FIG. 17.

Let the circle of unit radius on the z-plane with center at the origin be transformed into some domain (figure 17). If we consider a completely arbitrary transformation of the circle into the domain Δ, we cannot make any statements about its behavior at the point $z = 0$. But for conformal transformations we have the following remarkable theorem.

The dilation at the origin does not exceed four times the radius of the circle with center at w_0, inscribed in the domain

$$|f'(0)| \leqslant 4r.$$

Various questions in the theory of conformal transformations were considered in a large number of studies by Soviet mathematicians. In these works exact formulas were derived for many interesting classes of conformal transformations, methods for approximate calculation of conformal transformations were developed, and many general geometric theorems on conformal transformations were established.

Quasi-conformal transformations. Conformal transformations are closely connected with the investigation of analytic functions, i.e., with the study of a pair of functions satisfying the Cauchy-Riemann conditions

$$\frac{\partial u}{\partial x} = \frac{\partial v}{\partial y},$$

$$\frac{\partial u}{\partial y} = -\frac{\partial v}{\partial x}.$$

But many problems in mathematical physics involve more general systems of differential equations, which may also be connected with transformations from one plane to another, and these transformations will have specific geometric properties in the neighborhood of points in the Oxy

plane. To illustrate, we consider the following example of differential equations

$$\frac{\partial u}{\partial x} = p(x, y) \frac{\partial v}{\partial y},$$

$$\frac{\partial v}{\partial x} = -p(x, y) \frac{\partial u}{\partial y}. \tag{38}$$

If $p(x, y) = 1$, this is the system of Cauchy-Riemann equations. In the general case of an arbitrary function $p(x, y)$, we can also consider every solution of the system (38) as a transformation of the Oxy plane to the Ouv plane. Let us examine the geometric properties of this transformation in the neighborhood of a point (x_0, y_0). Taking a small neighborhood of (x_0, y_0), we retain only the first terms in the expansion of the functions u and v in powers of $x - x_0$ and $y - y_0$, and thereby consider the following affine transformation

$$u - u_0 = \left(\frac{\partial u}{\partial x}\right)_0 (x - x_0) + \left(\frac{\partial u}{\partial y}\right)_0 (y - y_0),$$

$$v - v_0 = \left(\frac{\partial v}{\partial x}\right)_0 (x - x_0) + \left(\frac{\partial v}{\partial y}\right)_0 (y - y_0). \tag{39}$$

If the functions u and v satisfy the system of equations (38), then for this affine transformation we have the following property.

Ellipses with center at the point (x_0, y_0) with principal axes parallel to the coordinate axes, and with ratio of semiaxes

$$\frac{b}{a} = p(x_0, y_0)$$

are transformed in the Ouv plane to circles with center at the point (u_0, v_0).

Let us prove this proposition. The equation of the circle with center (u_0, v_0) in the Ouv plane will be

$$(u - u_0)^2 + (v - v_0)^2 = \rho^2.$$

Replacing $u - u_0$ and $v - v_0$ by their expressions in terms of x and y, we get the equation for the corresponding curve in the Oxy plane:

$$\left[\left(\frac{\partial u}{\partial x}\right)_0^2 + \left(\frac{\partial v}{\partial x}\right)_0^2\right] (x - x_0)^2$$

$$+ 2\left[\left(\frac{\partial u}{\partial x}\right)_0 \left(\frac{\partial u}{\partial y}\right)_0 + \left(\frac{\partial v}{\partial x}\right)_0 \left(\frac{\partial v}{\partial y}\right)_0\right] (x - x_0)(y - y_0)$$

$$+ \left[\left(\frac{\partial u}{\partial y}\right)_0^2 + \left(\frac{\partial v}{\partial y}\right)_0^2\right] (y - y_0)^2 = \rho^2.$$

Using the equations in (38) to express the derivatives of v in terms of the derivatives of u, we get

$$\left[\left(\frac{\partial u}{\partial x}\right)_0^2 + p^2 \left(\frac{\partial u}{\partial y}\right)_0^2\right] (x - x_0)^2 + \frac{1}{p^2} \left[\left(\frac{\partial u}{\partial x}\right)_0^2 + p^2 \left(\frac{\partial u}{\partial y}\right)_0^2\right] (y - y_0)^2 = \rho^2.$$

If we set

$$a = \frac{\rho}{\sqrt{\left(\frac{\partial u}{\partial x}\right)_0^2 + p^2 \left(\frac{\partial u}{\partial y}\right)_0^2}},$$

$$b = \frac{p\rho}{\sqrt{\left(\frac{\partial u}{\partial x}\right)_0^2 + p^2 \left(\frac{\partial u}{\partial y}\right)_0^2}},$$

this equation takes the form

$$\frac{(x - x_0)^2}{a^2} + \frac{(y - y_0)^2}{b^2} = 1.$$

Thus the curve that is transformed into a circle is in fact an ellipse with the indicated properties.

If we do not consider the affine transformation given by the first terms of the expansion but rather the exact transformation itself, then the above property of the transformation will hold more and more exactly for smaller and smaller ellipses, so that we may say that the property holds for infinitely small ellipses.

In this manner, from equations (38) it follows that at every point the infinitesimal ellipse that is transformed into a circle has its semiaxes completely determined by the transformation, both with respect to their direction and to the ratio of their lengths. It can be shown that this geometric property completely characterizes the system of differential equations (38); i.e., if the functions u and v effect a transformation with the given geometric property, then they satisfy this system of equations. In this way, the problem of investigating the solutions of equations (38) is equivalent to investigating transformations with the given properties.

We note, in particular, that for the Cauchy-Riemann equations this property is formulated in the following manner.

An infinitesimal circle with center at the point (x_0, y_0) is transformed into an infinitesimal circle with center at the point (u_0, v_0).

A very wide class of equations of mathematical physics may be reduced to the study of transformations with the following geometric properties.

For each point (x, y) of the argument plane, we are given the direction of the semiaxes of two ellipses and also the ratio of the lengths of these semiaxes. We wish to construct a transformation of the Oxy plane to the Ouv plane such that an infinitesimal ellipse of the first family transforms into an infinitesimal ellipse of the second with center at the point (u, v).

Transformations connected with such general systems of equations were introduced by the Soviet mathematician M. A. Lavrent'ev and have received the name *quasi-conformal*. The idea of studying transformations defined by systems of differential equations made it possible to extend the methods of the theory of analytic functions to a very wide class of problems. Lavrent'ev and his students developed the study of quasi-conformal transformations and found a large number of applications to various problems of mathematical physics, mechanics, and geometry. It is interesting to note that the study of quasi-conformal transformations has proved very fruitful in the theory of analytic functions itself. Of course, we cannot dwell here on all the various applications of the geometric method in the theory of functions of a complex variable.

§4. The Line Integral; Cauchy's Formula and Its Corollaries

Integrals of functions of a complex variable. In the study of the properties of analytic functions the concept of a complex variable plays a very important role. Corresponding to the definite integral of a function of a real variable, we here deal with the integral of a function of a complex variable along a curve. We consider in the plane a curve C beginning at the point z_0 and ending at the point z, and a function $f(z)$ defined on a domain containing the curve C. We divide the curve C into small segments (figure 18) at the points

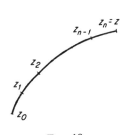

FIG. 18.

$$z_0, z_1, \cdots, z_n = z$$

and consider the sum

$$S = \sum_{k=1}^{n} f(z_k)\,(z_k - z_{k-1}).$$

If the function $f(z)$ is continuous and the curve C has finite length, we can prove, just as for real functions, that as the number of points of division is increased and the distance between neighboring points decreases to zero, the sum S approaches a completely determined limit.

This limit is called the *integral along the curve* C and is denoted by

$$\int_C f(z)\, dz.$$

We note that in this definition of the integral we have distinguished between the beginning and the end of the curve C; in other words, we have chosen a specific direction of motion on the curve C.

It is easy to prove a number of simple properties of the integral.

1. The integral of the sum of two functions is equal to the sum of the integrals of the individual functions:

$$\int_C [f(z) + g(z)]\, dz = \int_C f(z)\, dz + \int_C g(z)\, dz.$$

2. A constant multiple may be taken outside the integral sign:

$$\int_C A f(z)\, dz = A \int_C f(z)\, dz.$$

3. If the curve C is the sum of the curves C_1 and C_2, then

$$\int_C f(z)\, dz = \int_{C_1} f(z)\, dz + \int_{C_2} f(z)\, dz.$$

4. If $\bar{C}$ is the curve C with opposite orientation, then

$$\int_{\bar{C}} f(z)\, dz = - \int_C f(z)\, dz.$$

All these properties are obvious for the approximating sums and carry over to the integral in passing to the limit.

5. If the length of the curve C is equal to L and if everywhere on C the inequality

$$|f(z)| \leqslant M$$

is satisfied, then

$$\left| \int_C f(z)\, dz \right| \leqslant ML.$$

Let us prove this property. It is sufficient to prove the inequality for the sum S, since then it will carry over in the limit for the integral also.

For the sum

$$|S| = \left| \sum f(z_k)(z_k - z_{k-1}) \right| \leqslant \sum |f(z_k)| \, |z_k - z_{k-1}| \leqslant M \sum |z_k - z_{k-1}|.$$

But the sum in the second factor is equal to the sum of the lengths of the segments of the broken line inscribed in the curve C with vertices at the points z_k. The length of the broken line, as is well known, is not greater than the length of the curve, so that

$$|S| \leqslant ML.$$

We consider the integral of the simplest function $f(z) = 1$. Obviously in this case

$$S = (z_1 - z_0) + (z_2 - z_1) + \cdots + (z_n - z_{n-1}) = z_n - z_0 = z - z_0 \, .$$

This proves that

$$\int_C 1 \cdot dz = z - z_0 \, .$$

This result shows that for the function $f(z) = 1$ the value of the integral for all curves joining the points z_0 and z is the same. In other words, the value of the integral depends only on the beginning and end points of the path of integration. But it is easy to show that this property does not hold for arbitrary functions of a complex variable. For example, if $f(z) = x$, then a simple computation shows that

$$\int_{C_1} x \, dz = \frac{x^2}{2} + iyx, \quad \int_{C_2} x \, dz = \frac{x^2}{2}, \quad z = x + iy,$$

where C_1 and C_2 are the paths of integration shown in figure 19.

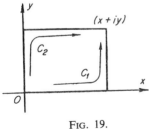

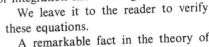

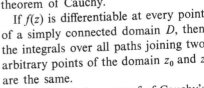

Fig. 19.

We leave it to the reader to verify these equations.

A remarkable fact in the theory of analytic functions is the following theorem of Cauchy.

If $f(z)$ is differentiable at every point of a simply connected domain D, then the integrals over all paths joining two arbitrary points of the domain z_0 and z are the same.

We will not give a proof of Cauchy's

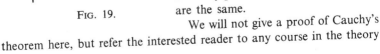

theorem here, but refer the interested reader to any course in the theory of functions of a complex variable. Let us mention here some important consequences of this theorem.

First of all, Cauchy's theorem allows us to introduce the indefinite integral of an analytic function. For let us fix the point z_0 and consider the integral along curves connecting z_0 and z:

$$F(z) = \int_{z_0}^{z} f(\zeta)\, d\zeta.$$

Here we may take the integral over any curve joining z_0 and z, since changing the curve does not change the value of the integral, which thus depends only on z. The function $F(z)$ is called an *indefinite integral* of $f(z)$.

An indefinite integral of $f(z)$ has a derivative equal to $f(z)$.

In many applications it is convenient to have a slightly different formulation of Cauchy's theorem, as follows.

If $f(z)$ is everywhere differentiable in a simply connected domain, then the integral over any closed contour lying in this domain is equal to zero:

$$\int_{\Gamma} f(z)\, dz = 0.$$

This is obvious since a closed contour has the same beginning and end, so that z_0 and z may be joined by a null path.

By a closed contour we will understand a contour traversed in the counterclockwise direction. If the contour is traversed in the clockwise direction we will denote it by $\bar{\Gamma}$.

The Cauchy integral. On the basis of the last theorem we can prove the following fundamental formula of Cauchy that expresses the value of a differentiable function at interior points of a closed contour in terms of the values of the function on the contour itself

$$f(z) = \frac{1}{2\pi i} \int_C \frac{f(\zeta)\, d\zeta}{\zeta - z}.$$

We give a proof of this formula. Let z be fixed and ζ be an independent variable. The function

$$\phi(\zeta) = \frac{f(\zeta)}{\zeta - z}$$

will be continuous and differentiable at every point ζ inside the domain D, with the exception of the point $\zeta = z$, where the denominator vanishes, a circumstance that prevents the application of Cauchy's theorem to the function $\phi(\zeta)$ on the contour C.

We consider a circle K_ρ with center at the point z and radius ρ and show that

$$\int_C \phi(\zeta)\, d\zeta = \int_{K_\rho} \phi(\zeta)\, d\zeta. \tag{40}$$

To this end we construct the auxiliary closed contour Γ_ρ, consisting of

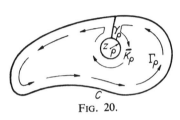

FIG. 20.

the contour C, the path γ_ρ connecting C with the circle, and the circle $\bar{K}_\rho$, taken with the opposite orientation (figure 20). The contour Γ_ρ is indicated by arrows. Since the point $\zeta = z$ is excluded, the function $\phi(\zeta)$ is differentiable everywhere inside Γ_ρ and thus

$$\int_{\Gamma_\rho} \phi(\zeta)\, d\zeta = 0. \tag{41}$$

But the contour Γ_ρ is divided into four parts: C, γ_ρ, $\bar{K}_\rho$ and $\bar{\gamma}_\rho$, so that from property 3 in the last subsection, we have

$$\int_{\Gamma_\rho} \phi(\zeta)\, d\zeta = \int_C \phi(\zeta)\, d\zeta + \int_{\gamma_\rho} \phi(\zeta)\, d\zeta + \int_{\bar{K}_\rho} \phi(\zeta)\, d\zeta + \int_{\bar{\gamma}_\rho} \phi(\zeta)\, d\zeta = 0.$$

Replacing the integrals along $\bar{K}_\rho$ and $\bar{\gamma}_\rho$ by integrals along K_ρ and γ_ρ, and using property 4, we get

$$\int_{\Gamma_\rho} \phi(\zeta)\, d\zeta = \int_C \phi(\zeta)\, d\zeta - \int_{K_\rho} \phi(\zeta)\, d\zeta = 0,$$

which proves formula (40).

To compute the right side of (40), we set

$$\int_{K_\rho} \phi(\zeta)\, d\zeta = \int_{K_\rho} \frac{f(\zeta)}{\zeta - z}\, d\zeta = \int_{K_\rho} \frac{f(\zeta) - f(z)}{\zeta - z}\, d\zeta + \int_{K_\rho} \frac{f(z)\, d\zeta}{\zeta - z}$$

$$= \int_{K_\rho} \frac{f(\zeta) - f(z)}{\zeta - z}\, d\zeta + f(z) \int_{K_\rho} \frac{d\zeta}{\zeta - z}. \tag{42}$$

We compute the second term first. On the circle K_ρ,

$$\zeta = z + \rho(\cos\theta + i\sin\theta).$$

Using the fact that z and ρ are constant, we get

$$d\zeta = \rho(-\sin\theta + i\cos\theta)\,d\theta = i\rho(\cos\theta + i\sin\theta)\,d\theta,$$

and thus

$$\zeta - z = \rho(\cos\theta + i\sin\theta),$$

so that

$$\int_{K_\rho}\frac{d\zeta}{\zeta - z} = \int_{K_\rho} i\,d\theta = 2\pi i,$$

since for a circuit of the circumference the total change in θ is equal to 2π. From (40) and (42) we have

$$\int_C \frac{f(\zeta)\,d\zeta}{\zeta - z} = 2\pi i f(z) + \int_{K_\rho}\frac{f(\zeta) - f(z)}{\zeta - z}\,d\zeta.$$

In this equation let us take limits as $\rho \to 0$. The left side and the first term of the right side will remain unchanged. We will show that the limit of the second term is equal to zero. Then for $\rho \to 0$ our equation gives us Cauchy's formula. In order to prove that the second term tends to zero as $\rho \to 0$ we note that

$$\lim_{\zeta \to z}\frac{f(\zeta) - f(z)}{\zeta - z} = f'(\zeta),$$

i.e., the expression under the integral sign has a finite limit, and thus is bounded

$$\left|\frac{f(\zeta) - f(z)}{\zeta - z}\right| < M.$$

Applying property 5 of the integral, we have

$$\left|\int_{K_\rho}\frac{f(\zeta) - f(z)}{\zeta - z}\,d\zeta\right| \leqslant M2\pi\rho \to 0.$$

This completes the proof of Cauchy's formula. Cauchy's formula is one of the basic tools of investigation in the theory of functions of a complex variable.

Expansion of differentiable functions in a power series. We apply Cauchy's theorem to establish two basic properties of differentiable functions of a complex variable.

Every function of a complex variable that has a first derivative in a domain D has derivatives of all orders.

In fact, inside a closed contour our function may be expressed by the Cauchy integral formula

$$f(z) = \frac{1}{2\pi i} \int_C \frac{f(\zeta)}{\zeta - z}\, d\zeta.$$

The function of z under the sign of integration is a differentiable function; thus, differentiating under the integral sign, we get

$$f'(z) = \frac{1}{2\pi i} \int_C \frac{f(\zeta)}{(\zeta - z)^2}\, d\zeta.$$

Under the integral sign there is again a differentiable function; thus we can again differentiate, obtaining

$$f''(z) = \frac{1 \cdot 2}{2\pi i} \int_C \frac{f(\zeta)\, d\zeta}{(\zeta - z)^3}.$$

Continuing the differentiation, we get the general formula

$$f^{(n)}(z) = \frac{n!}{2\pi i} \int_C \frac{f(\zeta)\, d\zeta}{(\zeta - z)^{n+1}}.$$

In this manner we may compute the derivative of any order. To make this proof completely rigorous, we need also to show that the differentiation under the integral sign is valid. We will not give this part of the proof.

The second property is the following:

If $f(z)$ is everywhere differentiable on a circle K with center at the point a, then $f(z)$ can be expanded in a Taylor series

$$f(z) = f(a) + \frac{f'(a)}{1!}(z - a) + \cdots + \frac{f^{(n)}(a)}{n!}(z - a)^{n+1} + \cdots,$$

which converges inside K.

In §1 we defined analytic functions of a complex variable as functions that can be expanded in power series. This last theorem says that every differentiable function of a complex variable is an analytic function. This is a special property of functions of a complex variable that has no analogue in the real domain. A function of a real variable that has a first derivative may fail to have a second derivative at every point.

We prove the theorem formulated in the previous paragraphs.

Let $f(z)$ have a derivative inside and on the boundary of the circle K with center at the point a. Then inside K the function $f(z)$ can be expressed by the Cauchy integral

$$f(z) = \frac{1}{2\pi i} \int_C \frac{f(\zeta)\, d\zeta}{\zeta - z}. \tag{43}$$

We write

$$\zeta - z = (\zeta - a) - (z - a),$$

then

$$\frac{1}{\zeta - z} = \frac{1}{(\zeta - a) - (z - a)} = \frac{1}{\zeta - a}\, \frac{1}{1 - \dfrac{z - a}{\zeta - a}}. \tag{44}$$

Using the fact that the point z lies inside the circle, and ζ is on the circumference we get

$$\left| \frac{z - a}{\zeta - a} \right| < 1,$$

so that from the basic formula for a geometric progression

$$\frac{1}{1 - \dfrac{z - a}{\zeta - a}} = 1 + \left(\frac{z - a}{\zeta - a}\right) + \cdots + \left(\frac{z - a}{\zeta - a}\right)^n + \cdots, \tag{45}$$

and the series on the right converges. Using (44) and (45), we can represent formula (43) in the form

$$f(z) = \frac{1}{2\pi i} \int_C \left[\frac{f(\zeta)}{\zeta - a} + (z - a)\frac{f(\zeta)}{(\zeta - a)^2} + \cdots \right.$$
$$\left. + (z - a)^n \frac{f(\zeta)}{(\zeta - a)^{n+1}} + \cdots \right] d\zeta.$$

We now apply term-by-term integration to the series inside the brackets. (The validity of this operation can be established rigorously.) Removing the factor $(z - a)^n$, which does not depend on ζ, from the integral sign in each term, we get

$$f(z) = \frac{1}{2\pi i} \int_C \frac{f(\zeta)\, d\zeta}{\zeta - a} + \frac{z - a}{2\pi i} \int_C \frac{f(\zeta)\, d\zeta}{(\zeta - a)^2} + \cdots$$
$$+ \frac{(z - a)^n}{2\pi i} \int_C \frac{f(\zeta)\, d\zeta}{(\zeta - a)^{n+1}} + \cdots.$$

Now using the integral formulas for the sequence of derivatives, we may write

$$\frac{1}{2\pi i} \int_C \frac{f(\zeta)\, d\zeta}{(\zeta - a)^{n+1}} = \frac{f^{(n)}(a)}{n!},$$

so that we get

$$f(z) = f(a) + \frac{f'(a)}{1!}(z - a) + \cdots + \frac{f^{(n)}(a)}{n!}(z - a)^n + \cdots.$$

We have shown that differentiable functions of a complex variable can be expanded in power series. Conversely, functions represented by power series are differentiable. Their derivatives may be found by term-by-term differentiation of the series. (The validity of this operation can be established rigorously.)

Entire functions. A power series gives an analytic representation of a function only in some circle. This circle has a radius equal to the distance to the nearest point at which the function ceases to be analytic, i.e., to the nearest singular point of the function.

Among analytic functions it is natural to single out the class of functions that are analytic for all finite values of their argument. Such functions are represented by power series, converging for all values of the argument z, and are called *entire functions* of z. If we consider expansions about the origin, then an entire function will be expressed by a series of the form

$$G(z) = c_0 + c_1 z + c_2 z^2 + \cdots + c_n z^n + \cdots.$$

If in this series all the coefficients, from a certain one on, are equal to zero, the function is simply a polynomial, or an entire rational function

$$P(z) = c_0 + c_1 z + \cdots + c_n z^n.$$

If in the expansion there are infinitely many terms that are different from zero, then the entire function is called *transcendental*.

Examples of such functions are:

$$e^z = 1 + \frac{z}{1!} + \frac{z^2}{2!} + \cdots,$$

$$\sin z = \frac{z}{1!} - \frac{z^3}{3!} + \frac{z^5}{5!} - \cdots,$$

$$\cos z = 1 - \frac{z^2}{2!} + \frac{z^4}{4!} - \cdots.$$

In the study of properties of polynomials, an important role is played by the distribution of the roots of the equation

$$P(z) = 0,$$

or, more generally speaking, we may raise the question of the distribution of the points for which the polynomial has a given value A

$$P(z) = A.$$

The fundamental theorem of algebra says that every polynomial takes a given value A in at least one point. This property cannot be extended to an arbitrary entire function. For example, the function $w = e^z$ does not take the value zero at any point of the z-plane. However, we do have the following theorem of Picard: Every entire function assumes every arbitrarily preassigned value an infinite number of times, with the possible exception of one value.

The distribution of the points of the plane at which an entire function takes on a given value A is one of the central questions in the theory of entire functions.

The number of roots of a polynomial is equal to its degree. The degree of a polynomial is closely related to the rapidity of growth of $|P(z)|$ as $|z| \to \infty$. In fact, we can write

$$|P(z)| = |z|^n \cdot \left| a_n + \frac{a_{n-1}}{z} + \cdots + \frac{a_0}{z^n} \right|,$$

and since for $|z| \to \infty$, the second factor tends to $|a_n|$, a polynomial of degree n, for large values of $|z|$, grows like $|a_n| \cdot |z|^n$. So it is clear that for larger values of n, the growth of $|P_n(z)|$ for $|z| \to \infty$ will be faster and also the polynomial will have more roots. It turns out that this principle is also valid for entire functions. However, for an entire function $f(z)$, generally speaking, there are infinitely many roots, and thus the question of the number of roots has no meaning. Nevertheless, we can consider the number of roots $n(r, a)$ of the equation

$$f(z) = a$$

in a circle of radius r, and investigate how this number changes with increasing r. The rate of growth of $n(r, a)$ proves to be connected with the rate of growth of the maximum $M(r)$ of the modulus of the entire function on the circle of radius r. As stated earlier, for an entire function there may exist one exceptional value of a for which the equation may not have even one root. For all other values of a, the rate of growth of the number $n(r, a)$ is comparable to the rate of growth of the quantity $\ln M(r)$. We cannot give more exact formulations here for these laws.

The properties of the distribution of the roots of entire functions are connected with problems in the theory of numbers and have enabled mathematicians to establish many important properties of the Riemann zeta functions,* on the basis of which it is possible to prove many theorems about prime numbers.

Fractional or meromorphic functions. The class of entire functions may be considered as an extension of the class of algebraic polynomials. From the polynomials we may derive the wider class of rational functions

$$R(z) = \frac{P(z)}{Q(z)},$$

which are the quotients of two polynomials.

Similarly it is natural to form a new class of functions by means of entire functions. A function $f(z)$ which is the quotient of two entire functions $G_1(z)$ and $G_2(z)$

$$f(z) = \frac{G_1(z)}{G_2(z)}$$

is called a *fractional* or *meromorphic* function. The class of functions arising in this way plays a large role in mathematical analysis. Among the elementary functions contained in the class of meromorphic functions are, for example:

$$\tan z = \frac{\sin z}{\cos z}, \qquad \cot z = \frac{\cos z}{\sin z}.$$

A meromorphic function will not be analytic on the whole complex plane. At those points where the denominator $G_2(z)$ vanishes, the function $f(z)$ becomes infinite. The roots of $G_2(z)$ form a set of isolated points in the plane. In neighborhoods of these points, the function $f(z)$ naturally cannot be expanded in a Taylor series; in a neighborhood of such a point a, however, a meromorphic function may be represented by a power series that also contains a certain number of negative powers of $(z - a)$:

$$f(z) = \frac{C_{-m}}{(z - a)^m} + \cdots + \frac{C_{-1}}{z - a} + C_0 + C_1(z - a) + \cdots + C_n(z - a)^n + \cdots.$$
$$\tag{46}$$

---

* Cf. Chapter X on the theory of numbers.

As z approaches the point a, the value of $f(z)$ tends to infinity. An isolated singular point at which an analytic function goes to infinity is called a *pole*. The loss of analyticity of the function at the point a comes from the terms with negative powers of $z - a$ in the expansion (46). The expression

$$\frac{C_{-m}}{(z - a)^m} + \cdots + \frac{C_{-1}}{(z - a)}$$

characterizes the behavior of a meromorphic function near a singular point and is called the *principal part* of the expansion (46). The behavior of a meromorphic function is determined by its principal part in a neighborhood of a pole. In many cases, if we know the principal part of the expansion of a meromorphic function in the neighborhood of all its poles, we may construct the function. Thus, for example, if $f(z)$ is rational and vanishes at infinity, then it is equal to the sum of the principal parts of its expansions about all of its poles, the number of which, for a rational function, is finite:

$$f(z) = \sum_{(k)} \left[\frac{C_{-m_k}^{(k)}}{(z - a_k)^{m_k}} + \cdots + \frac{C_{-1}^{(k)}}{z - a_k} \right].$$

In the general case a rational function may be represented as the sum of all of its principal parts and a polynomial

$$f(z) = \sum_{(k)} \left[\frac{C_{-m_k}^{(k)}}{(z - a_k)^{m_k}} + \cdots + \frac{C_{-1}^{(k)}}{z - a_k} \right] + C_0 + C_1 z + \cdots + C_m z^m.$$

(47)

Formula (47) gives an expression for a rational function in which the role played by its singular points is clear. Expression (47) for a rational function is very convenient for various applications of rational functions and also has great theoretical interest as showing how the singular points of the function define its structure everywhere. It turns out that, just as in the case of a rational function, every meromorphic function may be constructed from the principal parts of its poles. We introduce without proof the appropriate expression, for example, for the function $\cot z$. The poles of the function $\cot z$ are obtained as the roots of the equation

$$\sin z = 0$$

and are situated at the points: $\cdots, -k\pi, \cdots, -\pi, 0, \pi, \cdots, k\pi, \cdots$. It may be shown that the principal part of the expansion of the function $\cot z$ in a power series at the pole $z = k\pi$ will be

$$\frac{1}{z - k\pi},$$

and the function cot z is equal to the sum of the principal parts with respect to all poles

$$\cot z = \frac{1}{z} + \sum_{k=1}^{\infty} \left(\frac{1}{z - k\pi} + \frac{1}{z + k\pi} \right). \qquad (48)$$

The expansion of a meromorphic function in a series of the principal parts is noteworthy in that it clearly shows the position of all the singular points and also allows us to compute the function on the whole of its domain of definition.

The theory of meromorphic functions has become fundamental for the study of many classes of functions that are of great importance in analysis. In particular, we must emphasize its significance for the equations of mathematical physics. The creation of the theory of integral equations, providing answers to many important questions in the theory of the equations of mathematical physics, was based to a great extent on the fundamental theorems for meromorphic functions.

Since that time the development of that part of functional analysis which is most closely connected with mathematical physics, namely the theory of operators, has very often depended on facts from the theory of analytic functions.

On analytic representation of functions. We saw previously that in a neighborhood of every point where a function is differentiable it may be defined by a power series. For an entire function the power series converges on the whole plane and gives an analytic expression for the function wherever it is defined. In case the function is not entire, the Taylor series, as we know, converges only in a circle whose circumference passes through the nearest singular point of the function. Consequently the power series does not allow us to compute the function everywhere, and so it may happen that an analytic function cannot be given by a power series on its whole domain of definition. For a meromorphic function an analytic expression giving the function on its whole domain of definition is the expansion in principal parts.

If a function is not entire but is defined in some circle or if we have a function defined in some domain but we want to study it only in a circle, then the Taylor series may serve to represent it. But when we study the function in domains that are different from circles, there arises the question of finding an analytic expression for the function suitable for representing it on the whole domain. A power series giving an expression for an analytic function in a circle has as its terms the simplest polynomials $a_n z^n$. It is natural to ask whether we can expand an analytic function in an arbitrary domain in a more general series of polynomials. Then every term of the series can again be computed by arithmetic operations, and we obtain a method for representing functions that is once more based on the simplest operations of arithmetic. The general answer to this question is given by the following theorem.

An analytic function, given on an arbitrary domain, the boundary of which consists of one curve, may be expanded in a series of polynomials

$$f(z) = P_0(z) + P_1(z) + \cdots + P_n(z) + \cdots.$$

The theorem formulated gives only a general answer to the question of expanding a function in a series of polynomials in an arbitrary domain but does not yet allow us to construct the series for a given function, as was done earlier in the case of the Taylor series. This theorem raises rather then solves the question of expanding functions in a series of polynomials. Questions of the construction of the series of polynomials, given the function or some of its properties, questions of the construction of more rapidly converging series or of series closely related to the behavior of the function itself, questions of the structure of a function defined by a given series of polynomials, all these questions represent an extensive development of the theory of approximation of functions by series of polynomials. In the creation of this theory a large role has been played by Soviet mathematicians, who have derived a series of fundamental results.

§5. Uniqueness Properties and Analytic Continuation

Uniqueness properties of analytic functions. One of the most remarkable properties of analytic functions is their uniqueness, as expressed in the following theorem.

If in the domain D two analytic functions are given that agree on some curve C lying inside the domain, then they agree on the entire domain.

The proof of this theorem is very simple. Let $f_1(z)$ and $f_2(z)$ be the two functions analytic in the domain D and agreeing on the curve C. The difference

$$\phi(z) = f_1(z) - f_2(z)$$

will be an analytic function on the domain D and will vanish on the curve C. We now show that $\phi(z) = 0$ at every point of the domain D. In fact, if in the domain D there exists a point z_0 (figure 21) at which $\phi(z_0) \neq 0$, we extend the curve C to the point z_0 and proceed along the extended curve toward z_0 as long as the function remains equal to zero on Γ. Let ζ be the last point of Γ that is accessible in this way. If $\phi(z_0) \neq 0$, then $\zeta \neq z_0$, and on a segment of the curve Γ beyond ζ the function $\phi(z)$, by the definition of the point ζ, will not be equal to zero. We show that this is impossible. In fact, on the part Γ_ζ of the curve Γ up to the point ζ, we have $\phi(z) = 0$. We may compute all derivatives of the function $\phi(z)$ on Γ_ζ using only the values of $\phi(z)$ on Γ_ζ, so that on Γ_ζ all derivatives of $\phi(z)$ are equal to zero. In particular, at the point ζ

$$\phi(\zeta) = \phi'(\zeta) = \cdots = \phi^{(n)}(\zeta) = \cdots = 0.$$

Let us expand the function $\phi(\zeta)$ in a Taylor series at the point ζ. All the coefficients of the expansion vanish, so that we get

$$\phi(z) = 0$$

in some circle with center at the point ζ, lying in the domain D. In

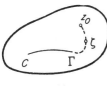

FIG. 21.

particular, it follows that the equation $\phi(z) = 0$ must be satisfied on some segment of the curve Γ lying beyond ζ. The assumption $\phi(z_0) \neq 0$ gives us a contradiction.

This theorem shows that if we know the values of an analytic function on some segment of a curve or on some part of a domain, then the values of the function are uniquely determined everywhere in the given domain. Consequently, the values of an analytic function in various parts of the argument plane are closely connected with one another.

To realize the significance of this uniqueness property of an analytic function, it is only necessary to recall that the general definition of a function of a complex variable allows any law of correspondence between values of the argument and values of the function. With such a definition there can, of course, be no question of determining the values of a function at any point by its values in another part of the plane. We see that the single requirement of differentiability of a function of a complex variable is so strong that it determines the connection between values of the function at different places.

We also emphasize that in the theory of functions of a real variable the differentiability of a function does not in itself lead to any similar consequences. In fact, we may construct examples of functions that are infinitely often differentiable and agree on some part of the Ox axis but differ elsewhere. For example, a function equal to zero for all negative values of x may be defined in such a manner that for positive x it differs from zero and has continuous derivatives of every order. For this it is sufficient, for example, to set, for $x > 0$

$$f(x) = e^{-1/x}.$$

Analytic continuation and complete analytic functions. The domain of definition of a given function of a complex variable is often restricted by the very manner of defining the function. Consider a very elementary example. Let the function be given by the series

$$f(z) = 1 + z + z^2 + \cdots + z^n + \cdots. \qquad (49)$$

This series, as is well known, converges in the unit circle and diverges outside this circle. Thus the analytic function given by formula (49) is defined only in this circle. On the other hand, we know that the sum of the series (49) in the circle $|z| < 1$ is expressed by the formula

$$f(z) = \frac{1}{1-z}. \qquad (50)$$

Formula (50) has meaning for all values of $z \neq 1$. From the uniqueness theorem it follows that expression (50) represents the unique analytic function, agreeing with the sum of the series (49) in the circle $|z| < 1$. So this function, given at first only in the unit circle, has been extended to the whole plane.

If we have a function $f(z)$ defined inside some domain D, and there exists another function $F(z)$ defined in a domain Δ, containing D, and agreeing with $f(z)$ in D, then from the uniqueness theorem the value of $F(z)$ in Δ is defined in a unique manner.

The function $F(z)$ is called the *analytic continuation* of $f(z)$. An analytic function is called *complete* if it cannot be continued analytically beyond the domain on which it is already defined. For example, an entire function, defined for the whole plane, is a complete function. A meromorphic function is also a complete function; it is defined everywhere except at its poles. However there exist analytic functions whose entire domain of definition is a bounded domain. We will not give these more complicated examples.

The concept of a complete analytic function leads to the necessity of considering multiple-valued functions of a complex variable. We show this by the example of the function

$$\text{Ln } z = \ln r + i\phi,$$

where $r = |z|$ and $\phi = \arg z$. If at some point $z_0 = r_0(\cos \phi_0 + i \sin \phi_0)$ of the z-plane we consider some initial value of the function

$$(\text{Ln } z)_0 = \ln r_0 + i\phi_0 ,$$

then our analytic function may be extended continuously along a curve C.

As was mentioned earlier, it is easy to see that if the point z describes

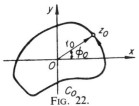

FIG. 22.

a closed path C_0, issuing from the point z_0 and circling around the origin (figure 22), and then returning to the point z_0, we find at the point z_0 the original value of $\ln r_0$ but the angle ϕ is increased by 2π. This shows that if we extend the function $\operatorname{Ln} z$ in a continuous manner along the path C, we increase its value by $2\pi i$ in one circuit of the

contour C. If the point z moves along this closed contour n times, then in place of the original value

$$(\operatorname{Ln} z)_0 = \ln r_0 + i\phi_0$$

we obtain the new value

$$(\operatorname{Ln} z)_n = \ln r_0 + (2\pi n + \phi_0)i.$$

If the point z describes the contour m times in the opposite direction, we get

$$(\operatorname{Ln} z)_{-m} = \ln r_0 + (-2\pi m + \phi_0)i.$$

These remarks show that on the complex plane we are unavoidably compelled to consider the connection between the various values of $\operatorname{Ln} z$. The function $\operatorname{Ln} z$ has infinitely many values. With respect to its multiple-valued character, a special role is played by the point $z = 0$, around which we pass from one value of the function to another. It is easy to establish that if z describes a closed contour not surrounding the origin, the value of $\operatorname{Ln} z$ is not changed. The point $z = 0$ is called a branch point of the function $\operatorname{Ln} z$.

In general, if for a function $f(z)$, in a circuit around the point a, we pass from one of its values to another, then the point a is called a *branch point* of the function $f(z)$.

Let us consider a second example. Let

$$w = \sqrt[n]{z}.$$

As noted previously, this function is also multiple-valued and takes on n values

$$\sqrt[n]{r}\left(\cos\frac{\phi}{n} + i\sin\frac{\phi}{n}\right), \qquad \sqrt[n]{r}\left(\cos\frac{\phi + 2\pi}{n} + i\sin\frac{\phi + 2\pi}{n}\right),$$

$$\cdots, \sqrt[n]{r}\left(\cos\frac{\phi + 2\pi(n-1)}{n} + i\sin\frac{\phi + 2\pi(n-1)}{n}\right).$$

All the various values of our function may be derived from the single one

$$w_0 = \sqrt[n]{r_0}\left(\cos\frac{\phi_0}{n} + i\sin\frac{\phi_0}{n}\right)$$

by describing a closed curve around the origin, since for each circuit around the origin the angle ϕ will be increased by 2π.

In describing the closed curve $(n-1)$ times, we obtain from the first value of $\sqrt[n]{z}$, all the remaining $(n-1)$ values. Going around the contour the nth time leads back to the value

$$\sqrt[n]{z_0} = \sqrt[n]{r_0}\left(\cos\frac{\phi_0 + 2\pi n}{n} + i\sin\frac{\phi_0 + 2\pi n}{n}\right) = \sqrt[n]{r_0}\left(\cos\frac{\phi_0}{n} + i\sin\frac{\phi_0}{n}\right),$$

i.e., we return to the original value of the root.

Riemann surfaces for multiple-valued functions. There exists an easily visualized geometric manner of representing the character of a multiple-valued function.

We consider again the function Ln z, and on the z-plane we make a cut along the positive part of the axis Ox. If the point z is prevented from crossing the cut, then we cannot pass continuously from one value of Ln z to another. If we continue Ln z from the point z_0, we can arrive only at the same value of Ln z.

The single-valued function found in this manner in the cut z-plane is called a *single-valued branch* of the function Ln z. All the values of Ln z are distributed on an infinite set of single-valued branches

$$\ln r + i\phi, \quad 2\pi n < \phi \leqslant 2\pi(n+1).$$

It is easy to show that the nth branch takes on the same value on the lower side of the cut as the $(n+1)$th branch has on the upper side.

To distinguish the different branches of Ln z, we imagine infinitely many examples of the z-plane, each of them cut along the positive part of the axis Ox, and map onto the nth sheet the values of the argument z corresponding to the nth branch. The points lying on different examples of the plane but having the same coordinates will here correspond to one and the same number $x + iy$; but the fact that this number is mapped on the nth sheet shows that we are considering the nth branch of the logarithm.

In order to represent geometrically the fact that the nth branch of the logarithm, on the lower part of the cut of the nth plane, agrees with the

$(n + 1)$th branch of the logarithm on the upper part of the cut in the $(n + 1)$th plane, we paste together the nth plane and the $(n + 1)$th, connecting the lower part of the cut in the nth plane with the upper part of the cut in the $(n + 1)$th plane. This construction leads us to a many-sheeted surface, having the form of a spiral staircase (figure 23). The role of the central column of the staircase is played by the point $z = 0$.

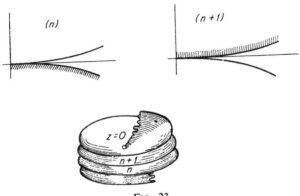

FIG. 23.

If a point passes from one sheet to another, then the complex number returns to its original value, but the function Ln z passes from one branch to another.

The surface so constructed is called the *Riemann surface* of the function Ln z. Riemann first introduced the idea of constructing surfaces representing the character of multiple-valued analytic functions and showed the fruitfulness of this idea.

Let us also discuss the construction of the Riemann surface for the function $w = \sqrt{z}$. This function is double-valued and has a branch point at the origin.

We imagine two examples of the z-plane, placed one on top of the other and both cut along the positive part of the axis Ox. If z starts from z_0 and describes a closed contour C containing the origin, then $\sqrt{z}$ passes from one branch to the other, and thus the point on the Riemann surface passes from one sheet to the other. To arrange this, we paste the lower border of the cut in the first sheet to the upper border of the cut in the second sheet. If z describes the closed contour C a second time, then $\sqrt{z}$ must return to its original value, so that the point in the Riemann surface must return to its original position on the first sheet. To arrange this, we must now attach the lower border of the second sheet to the upper border of the first sheet. As a result we get a two-sheeted surface, intersecting itself along the positive part of the axis Ox. Some idea of this surface may be obtained from figure 24, showing the neighborhood of the point $z = 0$.

In the same way we can construct a many-sheeted surface to represent the character of any given multiple-valued function. The different sheets of such a surface are connected with one another around branch points

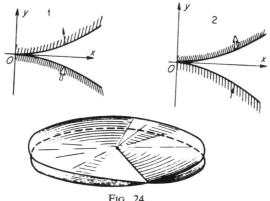

FIG. 24.

of the function. It turns out that the properties of analytic functions are closely connected with the geometric properties of Riemann surfaces. These surfaces are not only an auxiliary means of illustrating the character of a multiple-valued function but also play a fundamental role in the study of the properties of analytic functions and the development of methods of investigating them. Riemann surfaces formed a kind of bridge between analysis and geometry in the region of complex variables, enabling us not only to relate to geometry the most profound analytic properties of the functions but also to develop a whole new region of geometry, namely topology, which investigates those geometric properties of figures which remain unchanged under continuous deformation.

One of the clearest examples of the significance of the geometric properties of Riemann surfaces is the theory of algebraic functions, i.e., functions obtained as the solution of an equation

$$f(z, w) = 0$$

the left side of which is a polynomial in z and w. The Riemann surface of such a function may always be deformed continuously into a sphere or else into a sphere with handles (figure 25). The characteristic property of these surfaces is the number of handles. This number is called the genus of the surface and of the algebraic function from which the surface was obtained. It turns out that the genus of an algebraic function determines its most important properties.

  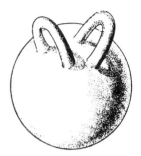

FIG. 25.

§6. Conclusion

The theory of analytic functions arose in connection with the problem of solving algebraic equations. But as it developed it came into constant contact with newer and newer branches of mathematics. It shed light on the fundamental classes of functions occurring in analysis, mechanics, and mathematical physics. Many of the central facts of analysis could at last be made clear only by passing to the complex domain. Functions of a complex variable received an immediate physical interpretation in the important vector fields of hydrodynamics and electrodynamics and provided a remarkable apparatus for the solution of problems arising in these branches of science. Relations were discovered between the theory of functions and problems in the theory of heat conduction, elasticity, and so forth.

General questions in the theory of differential equations and special methods for their solution have always been based to a great extent on the theory of functions of a complex variable. Analytic functions entered naturally into the theory of integral equations and the general theory of linear operators. Close connections were discovered between the theory of analytic functions and geometry. All these constantly widening connections of the theory of functions with new areas of mathematics and science show the vitality of the theory and the continuous enrichment of its range of problems.

In our survey we have not been able to present a complete picture of all the manifold ramifications of the theory of functions. We have tried only to give some idea of the widely varied nature of its problems by indicating the basic elementary facts for some of the various fundamental directions in which the theory has moved. Some of its most important aspects, its connection with the theory of differential equations and special functions, with elliptic and automorphic functions, with the theory of trigonometric series, and with many other branches of mathematics, have been completely ignored in our discussion. In other cases we have had to restrict ourselves to the briefest indications. But we hope that this survey will give the reader a general idea of the character and significance of the theory of functions of a complex variable.

Suggested Reading

R. V. Churchill, *Introduction to complex variables and applications*, McGraw-Hill, New York, 1948.

P. Franklin, *Functions of complex variables*, Prentice-Hall, Englewood Cliffs, N. J., 1958.

G. N. Watson, *Complex integration and Cauchy's theorem*, Hafner, New York, 1960.

PART 4

CHAPTER X

PRIME NUMBERS

§1. The Study of the Theory of Numbers

Whole numbers. As the reader knows from the introduction to Chapter I, mankind had to deal even in the most ancient times with whole numbers, but the passage of many centuries was necessary to produce the concept of the infinite sequence of natural numbers

$$1, 2, 3, 4, 5, \cdots. \tag{1}$$

Nowadays, in the most various questions of practical activity, we are constantly faced with problems involving whole numbers. Whole numbers reflect many quantitative relations in nature; in all questions connected with discrete objects, they form the necessary mathematical apparatus.

Moreover, whole numbers play an important role in the study of the continuous. Thus, for example, in mathematical analysis one considers the expansion of an analytic function in a power series with integral powers of x

$$f(x) = a_0 + a_1x + a_2x^2 + \cdots + a_nx^n + \cdots.$$

All computations are essentially carried out with *whole* numbers, as is immediately obvious from even a superficial examination of automatic computing machines or desk calculators, or of mathematical tables, such as tables of logarithms. After these operations on whole numbers have been carried out, decimal points are inserted in well-defined positions, corresponding to the formation of decimal fractions; such fractions, like all rational fractions, represent quotients of two *whole* numbers. In dealing with any *real* number in practical work (for example, π), we replace it in fact by a rational fraction (for example, we assume that $\pi = 22/7$, or that $\pi = 3.14$).

While the establishment of rules for operating on numbers is the concern of arithmetic, the deeper properties of the sequence of natural numbers (1), extended to include zero and the negative integers, are studied in the *theory of numbers*, which is the science of the system of integers and, in an extended sense, also of systems of numbers constructed in some definite manner from the integers (see, in particular, §5 of this chapter). It is understood that the theory of numbers considers integers not as isolated one from another but as interdependent; the theory of numbers studies properties of integers that are defined by certain relations among them.

One of the basic questions in the theory of numbers concerns *divisibility* of one number by another; if the result of dividing the integer a by the integer b (not equal to zero) is an integer, i.e., if

$$a = b \cdot c$$

(a, b, c are integers) then we say that a is *divisible* by b or that b *divides a*. If the result of dividing the integer a by the integer b is a fraction, then we say that a is not divisible by b. Questions of divisibility of numbers are encountered constantly in practice and also play an important role in some questions of mathematical analysis. For example, if the expansion of a function in integer powers of x

$$f(x) = a_0 + a_1 x + a_2 x^2 + \cdots + a_n x^n + \cdots \tag{2}$$

is such that all odd coefficients (with indices not divisible by 2) are equal to zero, i.e., if

$$f(x) = a_0 + a_2 x^2 + \cdots + a_{2k} x^{2k} + \cdots,$$

then the function satisfies the condition

$$f(-x) = f(x);$$

such a function is called an even function, and its graph is symmetric with respect to the axis of ordinates. But if in the expansion (2) all the even coefficients (with indices divisible by 2) are equal to zero, in other words, if

$$f(x) = a_1 x + a_3 x^3 + \cdots + a_{2k+1} x^{2k+1} + \cdots,$$

then

$$f(-x) = -f(x);$$

in this case the function is called odd, and its graph is symmetric with respect to the origin.

Thus, for example

$$\sin x = x - \frac{x^3}{3!} + \frac{x^5}{5!} - \cdots \quad \text{(odd function)};$$

$$\cos x = 1 - \frac{x^2}{2!} + \frac{x^4}{4!} - \cdots \quad \text{(even function)}.$$

The geometric question of the possibility of construction of a regular n-polygon with ruler and compass turns out to depend on the arithmetic nature of the number n.*

A *prime number* is any integer (greater than one) that has only the two positive integer divisors, one and itself. One is not considered as a prime number since it does not have two different positive divisors.

Thus the prime numbers are

$$2, 3, 5, 7, 11, 13, 17, 19, 23, 29, \cdots. \tag{3}$$

Prime numbers play a fundamental role in the theory of numbers because of the basic theorem: Every integer $n > 1$ may be represented as the product of prime numbers (with possible repetition of factors), i.e., in the form

$$n = p_1^{a_1} p_2^{a_2} \cdots p_k^{a_k}, \tag{4}$$

where $p_1 < p_2 < \cdots < p_k$ are primes and $a_1, a_2, \cdots, a_k$ are integers not less than one; furthermore, the representation of n in the form (4) is unique.

The properties of numbers connected with the representation of numbers as a sum of terms are called *additive*; the properties of numbers relating to their representation in the form of a product are called *multiplicative*. The connection between additive and multiplicative properties of numbers is extraordinarily complicated; it has given rise to a series of basic problems in the theory of numbers.

The existence of these difficult problems in the theory of numbers together with the fact that the whole number is not only the simplest and clearest of all mathematical concepts but is closely related to objective reality have led to the creation, for use in the theory of numbers, of profound new ideas and powerful methods, many of which have become important in other branches of mathematics as well. For example, a vast influence on all developments of mathematics has been exerted by the idea of the infinite sequence of natural numbers, reflecting the infiniteness of the material world in space and time. Of great significance also is the fact the terms in the sequence of natural numbers are ordered. Study of the

---

* See Chapter IV.

operations on integers has led to the concept of an algebraic operation, which plays a basic role in several different branches of mathematics.

Of immense importance in mathematics has been the concept, particularly applicable to arithmetical questions, of an algorithm, a process of solving problems based on the repeated carrying out of a strictly defined procedure; in particular, the role of the algorithm is fundamental to the use of mathematical machines. The essential nature of the algorithmic method for solving a problem is clearly illustrated by the Euclidean algorithm for finding the greatest common divisor of two natural numbers a and b.

Suppose $a > b$. We divide a by b and find the quotient q_1 and, if b does not divide a, the remainder r_2

$$a = bq_1 + r_2, \quad 0 < r_2 < b. \tag{5_1}$$

Further, if $r_2 \neq 0$, we divide b by r_2

$$b = r_2 q_2 + r_3, \quad 0 < r_3 < r_2. \tag{5_2}$$

Then we divide r_2 by r_3 and continue until we get to a zero remainder, which must necessarily happen for a decreasing set of nonnegative integers $r_2, r_3, \cdots$. Let

$$r_{n-2} = r_{n-1} q_{n-1} + r_n, \tag{5_{n-1}}$$

$$r_{n-1} = r_n q_n, \tag{5_n}$$

then r_n is at once seen to be the greatest common divisor of a and b. For if two integers l and m have a common divisor d, then for any integers h and k the number $hl + km$ will also be divisible by d. Let us denote the greatest common divisor of a and b by δ. From equation (5_1) we see that δ is a divisor of r_2; from (5_2) it follows that δ is a divisor of r_3, $\cdots$; from (5_{n-1}) that δ is a divisor of r_n. But r_n itself is a common divisor of a and b, since in (5_n) we see that r_n divides r_{n-1}; from (5_{n-1}) that r_n divides r_{n-2}, etc. Thus δ is identical with r_n and the problem of finding the greatest common divisor of a and b is solved. We have here a well-defined procedure, of the same type for all a and b, which leads us automatically to the desired result and is thus a characteristic example of an algorithm.

The theory of numbers has exerted an influence on the development of many mathematical disciplines: mathematical analysis, geometry, classical and contemporary algebra, the theory of summability of series, the theory of probability, and so forth.

Methods of the theory of numbers. In its methods, the theory of numbers is divided into four parts: elementary, analytic, algebraic, and geometric.

The elementary theory of numbers studies the properties of integers without calling on other mathematical disciplines. Thus, starting from Euler's identity

$$(x_1^2 + x_2^2 + x_3^2 + x_4^2)(y_1^2 + y_2^2 + y_3^2 + y_4^2) = (x_1y_1 + x_2y_2 + x_3y_3 + x_4y_4)^2$$
$$+ (x_1y_2 - x_2y_1 + x_3y_4 - x_4y_3)^2 + (x_1y_3 - x_3y_1 + x_4y_2 - x_2y_4)^2$$
$$+ (x_1y_4 - x_4y_1 + x_2y_3 - x_3y_2)^2, \tag{6}$$

we may very simply prove that every integer $N > 0$ may be expressed as the sum of the squares of four integers; i.e., every integer is representable in the form

$$N = x^2 + y^2 + z^2 + u^2,$$

where x, y, z, and u are integers.*

The analytic theory of numbers makes use of mathematical analysis for problems of the theory of numbers. Its foundations were laid by Euler and it was developed by P. L. Čebyšev, Dirichlet, Riemann, Ramanujan, Hardy, Littlewood, and other mathematicians, its most powerful methods being due to Vinogradov. This part of the theory of numbers is closely connected with the theory of functions of a complex variable (a theory that is very rich in practical applications), and also with the theory of series, the theory of probability, and other branches of mathematics.

The basic concept of the algebraic theory of numbers is the concept of an algebraic number, i.e., a root of the equation

$$a_0x^n + a_1x^{n-1} + a_2x^{n-2} + \cdots + a_{n-1}x + a_n = 0,$$

where a_0, a_1, a_2, $\cdots$, a_n are integers.†

The greatest contributions to this branch of the theory of numbers were made by Lagrange, Gauss, Kummer, E. I. Zolotarev, Dedekind, A. O. Gel'fond, and others.

The basic objects of study in the geometric theory of numbers are "space lattices"; that is, systems consisting entirely of "integral" points, all of whose coordinates in a given rectilinear coordinate system, rectangular or oblique, are integers. Space lattices have great significance in geometry and in crystallography, and are intimately connected with important questions in the theory of numbers; in particular, with the

---

* We have here an example of an indeterminate equation, to be investigated from the point of view of its solvability in integers.

† If $a_0 = 1$, the algebraic number is called an algebraic integer. A number which is not algebraic is called transcendental.

arithmetic theory of quadratic forms, i.e., the theory of quadratic forms with integer coefficients and integer variables. Basic work in the geometric theory of numbers is due to H. Minkowski and G. F. Voronoĭ.

It is·to be noted that the methods of the analytic theory of numbers have important applications in the other two branches, the algebraic and the geometric. Particularly noteworthy is the problem of counting the number of integral points in a given domain, a problem which is important in certain branches of physics. Various means of approach to this problem were indicated by G. F. Voronoĭ and methods for its solution were developed by I. M. Vinogradov.

The deep-lying reason for the power of analytic methods in the theory of numbers is that they enrich our study of the interrelations among discrete integers by summoning to our aid new relations among continuous magnitudes.

We must emphasize that in this chapter we are considering only certain selected questions in the theory of numbers.

§2. The Investigation of Problems Concerning Prime Numbers

The number of primes is infinite. In considering the sequence (3) of prime numbers

$$2, 3, 5, 7, 11, 13, 17, 19, \cdots$$

it is natural to ask the question: Is this sequence infinite? The fact that any integer can be represented in the form (4) does not yet solve the problem, since the exponents $a_1, \cdots, a_k$ may take on an infinite set of values. An affirmative answer to the question was given by Euclid, who proved that the number of primes cannot be equal to any finite integer k.

Let $p_1, p_2, \cdots, p_k$ be primes; then the number

$$m = p_1 p_1 \cdots p_k + 1,$$

since it is an integer greater than one, is either itself a prime or has a prime factor. But m is not divisible by any one of the primes $p_1, p_2, \cdots, p_k$ since, if it were, the difference $m - p_1 p_2 \cdots p_k$ would also be divisible by this number; which is impossible, since this difference is equal to one. Thus, either m itself is a prime or it is divisible by some prime p_{k+1}, different from $p_1, \cdots, p_k$. So the set of primes cannot be finite.

The sieve of Eratosthenes. The Greek mathematician Eratosthenes in the 3rd century B.C. described the following "sieve" method for finding all the primes not exceeding a given natural number N. We write all the integers from 1 through N

$$1, 2, 3, 4, \cdots, N,$$

and then cross out, from the left, first the number 1, then all numbers except 2 that are multiples of 2, then all except 3 that are multiples of 3, and then all except 5 that are multiples of 5 (the multiples of four have already been crossed out), and so forth; the remaining numbers will then be primes. It is worthy of note that the process of crossing out needs to be continued only to the point where we have found all primes less than $\sqrt{N}$, since every composite number (i.e., not prime) that is not greater than N will necessarily have a prime divisor not exceeding $\sqrt{N}$.

Examination of the sequence of prime numbers in the sequence of all positive integers would lead us to believe that the law of distribution of prime numbers must be very complicated; for example, we encounter primes such as 8,004,119 and 8,004,121 (the so-called twin primes) whose difference is two, and also primes that are far from each other, such as 86,629 and 86,677, between which there is no other prime. But the tables show that "on the average" prime numbers occur more and more rarely as we traverse the sequence of integers.

Euler's identity; his proof that the number of primes is infinite. The great 18th century mathematician L. Euler, a member of the Russian Academy of Sciences, introduced the following function, with argument $s > 1$, which at the present time is denoted by $\zeta(s)$:

$$\zeta(s) = 1 + \frac{1}{2^s} + \frac{1}{3^s} + \cdots + \frac{1}{n^s} + \cdots. \tag{7}$$

As we know from Chapter II, this series converges for $s > 1$ (and diverges for $s \leqslant 1$). Euler derived a remarkable identity that plays a very important role in the theory of prime numbers:

$$\sum_{n=1}^{\infty} \frac{1}{n^s} = \prod_p \frac{1}{1 - \dfrac{1}{p^s}}, \tag{8}$$

where the symbol $\prod_p$ means that we must multiply together the expressions $1/[1 - (1/p^s)]$ for all primes p. To see how the proof of this identity goes, we note that $1/(1 - q) = 1 + q + q^2 + \cdots$ for $|q| < 1$, so that

$$\frac{1}{1 - \dfrac{1}{p^s}} = 1 + \frac{1}{p^s} + \frac{1}{p^{2s}} + \cdots.$$

Multiplying these series for the various primes p and recalling that every n is uniquely representable as the product of primes, we find that

$$\prod_p \left(1 + \frac{1}{p^s} + \frac{1}{p^{2s}} + \cdots\right) = 1 + \frac{1}{2^s} + \frac{1}{3^s} + \cdots + \frac{1}{n^s} + \cdots.$$

For a rigorous proof, of course, we must establish the validity of our limit process, but this presents no particular difficulty.

From identity (8) we may derive as a corollary the fact that the series $\Sigma_p\, 1/p$, consisting of the reciprocals of all the primes, diverges (this provides a new proof of the fact already known to us that the prime numbers cannot be finite in number), and also that the quotient of the number of prime numbers not exceeding x, divided by x itself, converges to zero for unboundedly increasing x.

The investigations of P. L. Čebyšev on the distribution of the prime numbers in the sequence of natural numbers. We denote by $\pi(x)$, as is now customary, the number of prime numbers not exceeding x; for example, $\pi(10) = 4$, since 2, 3, 5, and 7 are all the primes not exceeding 10, $\pi(\pi) = 2$, since 2 and 3 are all the primes not exceeding π. As noted earlier

$$\lim_{x \to \infty} \frac{\pi(x)}{x} = 0.$$

But just how does the ratio $\pi(x)/x$ decrease; in other words what is the law of growth for $\pi(x)$? May we look for a fairly simple, well-known function that differs only a little from $\pi(x)$? The famous French mathematician Legendre, in considering tables of prime numbers, stated that such a function will be

$$\frac{x}{\ln x - A}, \tag{9}$$

where $A = 1.08\cdots$, but he did not give a proof of this proposition. Gauss, who also considered the question of the distribution of the prime numbers, conjectured that $\pi(x)$ differs comparatively little from $\int_2^x dt/\ln t$ (we note that the following relation holds:

$$\lim_{x \to \infty} \frac{\displaystyle\int_2^x \frac{dt}{\ln t}}{\displaystyle\frac{x}{\ln x}} = 1, \tag{10}$$

which is established by integrating by parts and finding estimates for the new integral).

The first mathematician since the time of Euclid to make real progress in the very difficult question of the distribution of the prime numbers was P. L. Čebyšev. In 1848, basing his work on a study of Euler's function $\zeta(s)$ for real s, Čebyšev showed that for arbitrarily large positive n and arbitrarily small positive α there exist arbitrarily large values of x for which

$$\pi(x) > \int_2^x \frac{dt}{\ln t} - \frac{\alpha x}{\ln^n x},$$

and also arbitrarily large x for which

$$\pi(x) < \int_2^x \frac{dt}{\ln t} + \frac{\alpha x}{\ln^n x},$$

which is in good agreement with Gauss's assumption. In particular, taking $n = 1$ and applying (10), Čebyšev established the fact that

$$\lim_{x \to \infty} \frac{\pi(x)}{\dfrac{x}{\ln x}} = 1, \tag{11}$$

provided that the limit in (11) exists.

Čebyšev also refuted Legendre's assumption concerning the value of the constant A which occurs in expression (9) as giving the best approximation to $\pi(x)$; he showed that this value can only be $A = 1$.

The well-known French mathematician Bertrand was led by his investigations in the theory of groups to the following conjecture, which he verified empirically from the tables up to quite large values of n: If $n > 3$, then between n and $2n - 2$ there is at least one prime. All the attempts of Bertrand, and of other mathematicians, to prove this conjecture proved fruitless until 1850, when Čebyšev published his second article on prime numbers, in which he not only proved the conjecture ("Bertrand's postulate") but also showed that for sufficiently large x

$$A_1 < \frac{\pi(x)}{\dfrac{x}{\ln x}} < A_2, \tag{12}$$

where

$$0.92 < A_1 < 1 \text{ and } 1 < A_2 < 1.1.$$

In §3 we give a simplified presentation of Čebyšev's method, which leads, however, to considerably less precise results than those of Čebyšev himself.

Čebyšev's works had a great influence on many mathematicians, in particular Sylvester and Poincaré. In the course of more than forty years a number of scientists busied themselves with the improvement of Čebyšev's inequality (12) (increasing the constant on the left side of the inequality.and decreasing the constant on the right side), but they were unable to establish the existence of the limit

$$\lim_{x \to \infty} \frac{\pi(x)}{\dfrac{x}{\ln x}}$$

(as was pointed out previously, we know from the work of Čebyšev that if this limit exists it is equal to one).

Only in 1896 did Hadamard, using arguments from the theory of functions of a complex variable, prove that the function $\Theta(x)$, introduced by Čebyšev and defined by the equation

$$\Theta(x) = \sum_{p \leqslant x} \ln p,$$

satisfies the condition

$$\lim_{x \to \infty} \frac{\Theta(x)}{x} = 1, \tag{13}$$

from which it is relatively easy to obtain the relation (11) without any further assumptions; this is the so-called asymptotic law for the distribution of primes.

The result (13) was found by Hadamard on the basis of the investigations by the famous German 19th century mathematician Riemann, who studied the $\zeta(s)$ function of Euler (7) for complex values of the variable $s = \sigma + it$ (Čebyšev himself had considered this function only for real values of the argument).*

Riemann showed that the function $\zeta(s)$, defined in the half plane $\sigma > 1$ by the series (7)

$$\zeta(s) = \sum_{n=1}^{\infty} \frac{1}{n^s}$$

has the property that

$$\zeta(s) - \frac{1}{s - 1}$$

---

* In 1949 A. Selberg gave an elementary proof (i.e., not using complex variables) of the asymptotic law of distribution of primes.

is an entire transcendental function (for $\sigma \leqslant 1$ the series (7) ceases to converge, but the values of $\zeta(s)$ in the half plane $\sigma \leqslant 1$ are defined by analytic continuation) (see Chapter IX). Riemann made the conjecture ("the Riemann hypothesis") that all roots of $\zeta(s)$ in the strip $0 \leqslant \sigma \leqslant 1$ have real part equal to $\frac{1}{2}$, i.e., lie on the straight line $\sigma = \frac{1}{2}$; the question of the correctness of this assumption remains open to this day.

An important step in the proof of (13) was the establishment of the fact that on the straight line $\sigma = 1$ there are no roots of $\zeta(s)$.

The investigation of the behavior of $\zeta(s)$ led to the development of an elegant theory of entire and meromorphic functions, with important practical applications.

The work of Vinogradov and his students in the theory of prime numbers. From equation (13), which by (10) may be written in the form

$$\lim_{x \to \infty} \frac{\pi(x)}{\int_2^x \dfrac{dt}{\ln t}} = 1, \tag{14}$$

there arose the question of the degree of exactness with which the function $\int_2^x dt/\ln t$ represents $\pi(x)$. The best results in this direction were found by N. G. Čudakov and were based on Vinogradov's *method of trigonometric sums* (this method will be described in §4), which also allowed Čudakov to decrease considerably the bounds between which we can find at least one prime. Namely, it had been established previously that if we consider the sequence

$$1^{250},\ 2^{250},\ 3^{250},\ \cdots,\ n^{250},\ (n+1)^{250},\ \cdots, \tag{15}$$

then, starting with some $n = n_0$, there must exist, between any two adjacent terms, i.e., between n^{250} and $(n + 1)^{250}$, at least one prime. We note that, as follows from the binomial formula

$$(n + 1)^{250} - n^{250} > 250n^{249},$$

this difference is very large. N. G. Čudakov succeeded in replacing the sequence (15) by

$$1^4,\ 2^4,\ 3^4,\ \cdots,\ n^4,\ (n+1)^4,\ \cdots, \tag{16}$$

whose terms lie considerably closer together than those of the sequence (15) but which also contains at least one prime between every two successive terms, i.e., between n^4 and $(n + 1)^4$, beginning at some $n = n_0$. Subsequently, this result has been improved by replacing the fourth powers by cubes.

If k and l are relatively prime, i.e., have no common divisor larger than one, then an arithmetic progression with general term $kt + l$ contains infinitely many prime numbers. This fact, a generalization of the result of Euclid, was established in the 19th century by Dirichlet. But can we find a bound that will certainly not be exceeded by the smallest prime in the progression $kt + l$? The Leningrad mathematician Ju. V. Linnik proved the existence of an absolute constant C with the property that in progression $kt + l$ (k and l relatively prime) there necessarily exists at least one prime less than k^C. Thus Linnik provided an essentially complete solution of the problem, raised many years before, of the least prime in an arithmetic progression; further investigators can only decrease the value of the constant C. Linnik also carried out very important investigations concerning the zeros of the function $\zeta(s)$ and more general functions.

As mentioned previously, the best results with regard to the distribution of primes were found by the method of Vinogradov for estimating trigonometric sums.

A *trigonometric sum* is a sum of the form

$$\sum_{A < x < B} e^{2\pi i f(x)},$$

where $f(x)$ is a real function of x, and x takes on all integral values between A and B, or some specific subset of these values, for example the primes between A and B. Since the modulus of $e^{2\pi i z}$ for real z is equal to one, and the modulus of a sum does not exceed the sum of the moduli of its terms, we have

$$\left| \sum_{x=1}^{p} e^{2\pi i f(x)} \right| \leqslant P. \tag{17}$$

This "trivial" estimate can be improved considerably in a number of cases; the decisive steps in this direction were taken by Vinogradov. For definiteness, let $f(x)$ be a polynomial

$$f(x) = \alpha_n x^n + \alpha_{n-1} x^{n-1} + \cdots + \alpha_1 x + \alpha_0.$$

If all the α are integers, then $e^{2\pi i f(x)} = 1$ for integral x, and in this case the estimate (17) obviously cannot be improved. But if $\alpha_1, \cdots, \alpha_n$ are not all integers then, as Vinogradov showed, the estimate (17) may be sharpened by approximating any of these coefficients by rational fractions with denominators not exceeding some bound (it may be shown that any α lying between 0 and 1 is representable in the form $\alpha = a/q + z$, where a and q are relatively prime integers, $q \leqslant \tau$, $|z| \leqslant 1/q\tau$ and τ is a preassigned integer greater than 1).

The creation of the method of trigonometric sums by Vinogradov allowed him to solve a series of very difficult problems in the theory of numbers. In particular, in 1937 he solved a famous problem stated by Goldbach, by proving that every sufficiently large odd N is representable as the sum of three primes

$$N = p_1 + p_2 + p_3 . \tag{18}$$

This problem arose in 1742 in correspondence between Euler and another member of the Russian Academy of Sciences, C. Goldbach, and remained unsolved for almost two centuries, despite the efforts of a number of eminent mathematicians.

As we have seen, the equation (4) shows that prime numbers play a fundamental role in the *multiplicative* representation of an odd number by means of primes. It is easy to show from (18) that one can represent a sufficiently large even number as the sum of no more than four primes.* In this manner, the Vinogradov-Goldbach theorem established a profound connection between additive and multiplicative properties of numbers.

The significance of the method of trigonometric sums created by Vinogradov is not restricted to the theory of numbers. In particular, it plays an important role in the theory of functions and in the theory of probability. Some idea of Vinogradov's method may be obtained from §4 of this chapter.

Readers who are interested in a more detailed treatment may consult Vinogradov's book "The method of trigonometric sums in the theory of numbers," after a preliminary reading of his book "Foundations of the theory of numbers."

§3. Čebyšev's Method

Čebyšev's Θ function and its estimates. We now give a simplified presentation of Čebyšev's method for computing the number of primes lying with given limits. For brevity we agree to use the following notation: if B is a positive variable quantity that may grow unboundedly, and A is another quantity such that $|A|$ grows "no more rapidly" than CB, where C is a positive constant (more precisely, if there exists a constant $C > 0$ such that starting from some instant we always have $|A|/B \leqslant C$), then we will write

$$A = O(B).$$

---

* The correctness of the conjecture that every sufficiently large *even* number N can be represented as the sum of two primes remains an open question to this day.

This is usually read as: "A is a quantity of the order of B." Thus, for example

$$\sin x = O(1),$$

since everywhere

$$\frac{|\sin x|}{1} \leqslant 1;$$

in exactly the same way

$$5x^3 \cos 2x = O(x^3).$$

We will also denote by $[x]$ the integral part of x, i.e., the largest integer not exceeding x; thus, for example

$$[\pi] = 3, \ [5] = 5, \ [-1.5] = -2, \ [0.999] = 0.$$

We now pose the following question: Let p be a prime, and n a natural number, and let $n!$, as usual, denote the product $1 \cdot 2 \cdot 3 \cdot \cdots \cdot n$; we note incidentally that as n increases the value of $n!$ grows very rapidly. What is the largest power a of the prime p that divides $n!$ with no remainder?

Among the numbers $1, 2, \cdots, n$, there will be precisely $[n/p]$ numbers divisible by p; the number of these which will also be divisible by p^2 is $[n/p^2]$; further, of these there will be $[n/p^3]$ divisible by p^3, etc. Hence it is easy to show that

$$a = \left[\frac{n}{p}\right] + \left[\frac{n}{p^2}\right] + \left[\frac{n}{p^3}\right] + \cdots$$

(where the series terminates, since $[n/p^s] > 0$ only for $n \geqslant p^s$). Thus, in the last sum every factor of the product $1 \cdot 2 \cdot 3 \cdot \cdots \cdot n$ such that the highest power of the number p by which it is divisible is equal to p^m will occur precisely m times, once as a multiple of p, once as a multiple of p^2, once as a multiple of p^3, $\cdots$, and finally once as a multiple of p^m.

From this result and from the representability of any natural number in the form (4) it follows that $n!$ will be the product of powers of the form

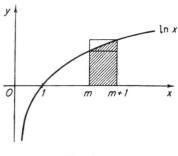

FIG. 1.

$$p^{\left[\frac{n}{p}\right]+\left[\frac{n}{p^2}\right]+\left[\frac{n}{p^3}\right]+\cdots},$$

taken for all primes $p \leqslant n$. Thus $\ln (n!)$ will be the sum of the logarithms of these powers, which can be concisely written in the form

$$\ln n! = \sum_{p \leqslant n} \left(\left[\frac{n}{p}\right] + \left[\frac{n}{p^2}\right] + \left[\frac{n}{p^3}\right] + \cdots \right) \ln p. \quad (19)$$

We simplify equation (19). Since $y = \ln x$ is an increasing function, we have

$$\ln m = \ln m \int_m^{m+1} dx < \int_m^{m+1} \ln x \, dx < \ln(m+1) \int_m^{m+1} dx = \ln(m+1)$$

as is clear from figure 1. Thus

$$\ln n! = \ln 1 + \ln 2 + \cdots + \ln n$$

$$< \int_1^2 \ln x \, dx + \int_2^3 \ln x \, dx + \cdots$$

$$+ \int_{n-1}^n \ln x \, dx + \ln n = \int_1^n \ln x \, dx + \ln n,$$

on the other hand

$$\ln n! > \ln 1 + \int_1^2 \ln x \, dx + \cdots$$

$$+ \int_{n-2}^{n-1} \ln x \, dx + \int_{n-1}^n \ln x \, dx = \int_1^n \ln x \, dx.$$

Using the formula for integration by parts, we find

$$\int_1^n \ln x \, dx = [x \cdot \ln x]_1^n - \int_1^n x \cdot \frac{1}{x} \cdot dx = n \ln n - (n-1).$$

Thus

$$n \ln n - n + 1 < \ln n! < n \ln n - n + 1 + \ln n,$$

and hence it follows that

$$\ln n! = n \ln n + O(n). \tag{20}$$

We note that $\ln n = O(n)$; further, for $n \to \infty$, the function $\ln n$ increases more slowly than any positive power of n, i.e., for any constant $\alpha > 0$

$$\lim_{n \to \infty} \frac{\ln n}{n^\alpha} = 0, \tag{21}$$

since by the rule for indeterminate forms (cf. Chapter II)

$$\lim_{n \to \infty} \frac{\ln n}{n^\alpha} = \lim_{n \to \infty} \frac{\dfrac{1}{n}}{\alpha n^{\alpha-1}} = \frac{1}{\alpha} \lim_{n \to \infty} \frac{1}{n^\alpha} = 0.$$

Further, we find

$$\sum_{p \leqslant n} \left(\left[\frac{n}{p^2} \right] + \left[\frac{n}{p^3} \right] + \cdots \right) \ln p \leqslant \sum_{p \leqslant n} \left(\frac{n}{p^2} + \frac{n}{p^3} + \cdots \right) \ln p$$

$$= \sum_{p \leqslant n} \frac{n \ln p}{p^2 \left(1 - \frac{1}{p} \right)} < 2n \sum_{p \leqslant n} \frac{\ln p}{p^2} < 2n \sum_{m=1}^{\infty} \frac{\ln m}{m^2} = 2nC_0 = O(n),$$
(22)

where C_0 is the sum of the convergent series $\displaystyle\sum_{m=1}^{\infty} \frac{\ln m}{m^2}$. The absolute convergence of this series is established by using (21), for example, for $\alpha = \frac{1}{2}$, by the comparison test and the so-called integral test for convergence (cf. Chapter II, §14). In view of (20) and (22), equation (19) may be put in the form

$$\sum_{p \leqslant n} \left[\frac{n}{p} \right] \ln p = n \ln n + O(n).$$
(23)

We now consider the function introduced by Čebyšev

$$\Theta(n) = \sum_{p \leqslant n} \ln p$$
(24)

(the logarithm of the product of all prime numbers not exceeding n).

Equation (23) can be rewritten as:

$$\Theta \left(\frac{n}{1} \right) + \Theta \left(\frac{n}{2} \right) + \Theta \left(\frac{n}{3} \right) + \Theta \left(\frac{n}{4} \right) + \cdots = n \ln n + O(n).$$
(25)

In fact, every given $\ln p$ enters into all the sums of the form $\Theta(n/s)$, where $p \leqslant n/s$, i.e., where $s \leqslant n/p$, and the number of such sums $\Theta(n/s)$ is equal to $[n/p]$.

Equation (25) is also valid for any noninteger n. To see this, it is obviously sufficient to prove that it is true for all x under the condition $n < x < n + 1$; and for this it is enough to prove that replacing n by x in the left side of (25) does not change that side, and that the first term in the right side may increase by an amount which is $O(n)$. But the first follows from the fact that such a replacement will not increase the value of any one of the terms of the left-hand side (such an increase would be possible only if n were increased by more than unity) and, of course, the left side is not decreased. The second follows from the fact that by the formula for the increment of a function (cf. Chapter II)

$$f(x) - f(a) = (x - a)f'(\xi), \quad a < \xi < x,$$

we have

$$x \ln x - n \ln n = (x - n) \cdot (\ln \xi + 1), \quad n < \xi < x,$$

and the right side of this last equation is less than $\ln (n + 1) + 1 = O(n)$, since $0 < x - n < 1$. From equation (25) let us subtract twice the equation derived from (25) by replacing n by $n/2$;

$$\Theta \left(\frac{n}{1} \right) + \Theta \left(\frac{n}{2} \right) + \Theta \left(\frac{n}{3} \right) + \Theta \left(\frac{n}{4} \right) + \cdots = n \ln n + O(n),$$

$$2\Theta \left(\frac{n}{2} \right) + 2\Theta \left(\frac{n}{4} \right) + \cdots = 2 \cdot \frac{n}{2} \cdot \ln \frac{n}{2} + O(n),$$

we obtain

$$\Theta \left(\frac{n}{1} \right) - \Theta \left(\frac{n}{2} \right) + \Theta \left(\frac{n}{3} \right) - \Theta \left(\frac{n}{4} \right) + \cdots = n \ln 2 + O(n) < C_1 n,$$

where C_1 is some positive constant. But $\Theta(n/1) - \Theta(n/2)$ is not larger than the whole left side, since the differences, $\Theta(n/3) - \Theta(n/4)$, $\Theta(n/5) - \Theta(n/6)$, $\cdots$ cannot be negative. Thus it follows from this last inequality that

$$\Theta \left(\frac{n}{1} \right) - \Theta \left(\frac{n}{2} \right) < C_1 n.$$

Inserting here the numbers $n/2$, $n/4$, $\cdots$ in place of n, we also get

$$\Theta \left(\frac{n}{2} \right) - \Theta \left(\frac{n}{4} \right) < C_1 \cdot \frac{n}{2},$$

$$\Theta \left(\frac{n}{4} \right) - \Theta \left(\frac{n}{8} \right) < C_1 \cdot \frac{n}{4},$$

$$\cdots \cdots \cdots \cdots \cdots \cdots \cdots,$$

hence, using the fact that $\Theta(n/2^k) = 0$ for sufficiently large k (when $n/2^k < 2$), addition of terms gives

$$\Theta(n) < C_1 \left(n + \frac{n}{2} + \frac{n}{4} + \cdots \right) = 2C_1 n. \tag{26}$$

Returning to equation (23), we find

$$0 \leqslant \sum_{p \leqslant n} \frac{n}{p} \ln p - \sum_{p \leqslant n} \left[\frac{n}{p} \right] \ln p \leqslant \sum_{p \leqslant n} \ln p = \Theta(n) \leqslant 2C_1 n = O(n),$$

so that equation (23) gives

$$\sum_{p \leqslant n} \frac{n}{p} \ln p = n \ln n + O(n),$$

$$\sum_{p \leqslant n} \frac{\ln p}{p} = \ln n + \theta C, \tag{27}$$

where C is a constant greater than zero and θ depends on the number n in such a manner that $|\theta| \leqslant 1$.

An estimate for the number of primes in a given interval. We now show that one may choose a positive constant M in such a manner that between n and Mn there will lie as many primes p as desired, if n is sufficiently large. Namely, we establish simple inequalities for the number T of primes in the interval $n < p \leqslant Mn$. Obviously,

$$\sum_{n < p \leqslant Mn} \frac{\ln p}{p} = \sum_{p \leqslant Mn} \frac{\ln p}{p} - \sum_{p \leqslant n} \frac{\ln p}{p}. \tag{28}$$

From equation (27), replacing n by Mn, we get

$$\sum_{p \leqslant Mn} \frac{\ln p}{p} = \ln(Mn) + \theta' C = \ln M + \ln n + \theta' C, \tag{29}$$

where $|\theta'| \leqslant 1$; thus, in view of equations (28), (29), and (27), we have

$$\sum_{n < p \leqslant Mn} \frac{\ln p}{p} = \ln M + \theta' C - \theta C = \ln M + 2\theta_0 C,$$

where $|\theta_0| \leqslant 1$, i.e.,

$$\ln M - 2C \leqslant \sum_{n < p \leqslant Mn} \frac{\ln p}{p} \leqslant \ln M + 2C. \tag{30}$$

On the other hand, since $y = \ln x / x$ for $x > e$ is a decreasing function (since $y' = (1 - \ln x)/x^2 < 0$ for $\ln x > 1$, i.e., $x > e$), it follows that for $n \geqslant 3$

$$T \frac{\ln Mn}{Mn} \leqslant \sum_{n < p \leqslant Mn} \frac{\ln p}{p} \leqslant T \frac{\ln n}{n},$$

hence, from (30), we have

$$T \frac{\ln n}{n} > \ln M - 2C \tag{31}$$

and

$$T \frac{\ln (Mn)}{Mn} < \ln M + 2C. \tag{32}$$

We now choose the constant M such that the right side of (31) is equal to one

$$\ln M - 2C = 1,$$

i.e.,

$$M = e^{2C+1},$$

and we set

$$L = M(\ln M + 2C).$$

Then for the number T of primes lying between n and Mn, we get from (31) and (32) the inequalities

$$\frac{n}{\ln n} < T < L \frac{n}{\ln n}, \tag{33}$$

which it was our purpose to establish. Since $n/\ln n \to \infty$ for unbounded increase in n, it follows that $T \to \infty$ also.

§4. Vinogradov's Method

Vinogradov's method in its application to the solution of Goldbach's problem. We attempt in this section to give some account of Vinogradov's method for the particular case of Goldbach's problem of representing an odd number as the sum of three prime numbers.

An expression in the form of an integral for the number of representations of N as the sum of three primes. Let N be a sufficiently large odd number. We denote by $I(N)$ the number of representations of N as the sum of three primes; in other words, the number of solutions of the equation

$$N = p_1 + p_2 + p_3 \tag{34}$$

in prime numbers p_1, p_2, and p_3.

Goldbach's problem will be solved if it can be established that $I(N) > 0$. Vinogradov's method allows us not only to establish this fact (for sufficiently large N), but also to find an approximating expression for $I(N)$.

$I(N)$ may be written in the following form

$$I(N) = \sum_{p_1 \leqslant N} \sum_{p_2 \leqslant N} \sum_{p_3 \leqslant N} \int_0^1 e^{2\pi i (p_1 + p_2 + p_3 - N)\alpha} \, d\alpha, \tag{35}$$

where the summations are taken over the prime numbers not exceeding N. In fact, for integer $n \neq 0$

$$\int_0^1 e^{2\pi i n\alpha}\, d\alpha = \frac{1}{2\pi ni}\left[e^{2\pi i n\alpha}\right]_0^1 = \frac{1}{2\pi ni}\left(e^{2\pi ni} - e^0\right) = 0,$$

since

$$e^{2\pi in} = \cos 2\pi n + i \sin 2\pi n = 1;$$

but if $n = 0$, then

$$\int_0^1 e^{2\pi i n\alpha}\, d\alpha = \int_0^1 d\alpha = 1.$$

Thus, every time the primes p_1, p_2, and p_3 have the sum N the integral inside the summation sign in (35) has the value one, and when the sum $p_1 + p_2 + p_3 \neq N$, this integral is equal to zero, which proves the validity of equation (35).

Since $e^{2\pi i a} \cdot e^{2\pi i b} = e^{2\pi i(a+b)}$ and the integral of a sum of terms is equal to the sum of the integrals of these terms, it follows from equation (35) that

$$I(N) = \int_0^1 \left(\sum_{p\leqslant N} e^{2\pi i\alpha p}\right)^3 e^{-2\pi i\alpha N}\, d\alpha.$$

Introducing the notation

$$T_\alpha = \sum_{p\leqslant N} e^{2\pi i\alpha p} \tag{36}$$

we then have

$$I(N) = \int_0^1 T_\alpha^3 e^{-2\pi i\alpha N}\, d\alpha. \tag{37}$$

Decomposition of the interval of integration into basic and complementary intervals. Let h be a quantity, chosen in an appropriate manner depending on N, which increases unboundedly with N but is small in comparison with N and even with $\sqrt[3]{N/2}$, and set $\tau = N/h$. Since the function integrated in (37) has a period equal to one, the interval of integration in (37) may be replaced by the segment from $-(1/\tau)$ to $1 - (1/\tau)$. Thus

$$I(N) = \int_{-1/\tau}^{1-1/\tau} T_\alpha^3 e^{-2\pi i\alpha N}\, d\alpha. \tag{38}$$

We now consider all proper irreducible fractions a/q with denominators not exceeding h, and distinguish in the segment $-(1/\tau) \leqslant \alpha \leqslant 1 - (1/\tau)$ the "basic" intervals corresponding to these fractions

$$\frac{a}{q} - \frac{1}{\tau} \leqslant \alpha \leqslant \frac{a}{q} + \frac{1}{\tau}; \tag{39}$$

for sufficiently large N these intervals, as may be proved,* will have no points in common. In this manner, the segment $-(1/\tau) \leqslant \alpha \leqslant 1 - (1/\tau)$ can be decomposed into basic intervals and "complementary" intervals.

We represent $I(N)$ as the sum of two terms

$$I(N) = I_1(N) + I_2(N), \tag{40}$$

where $I_1(N)$ denotes the sum of the integrals on the basic intervals and $I_2(N)$ is the sum of the integrals on the complementary intervals. As will be seen below, for unbounded growth of odd N we also have unbounded growth of $I_1(N)$, with

$$\lim_{N\to\infty} \frac{I_2(N)}{I_1(N)} = 0. \tag{41}$$

So we see from (40) that the number of representations of an odd N as the sum of three primes grows unboundedly with N, so that, in particular, we have proved Goldbach's conjecture for all sufficiently large odd N.

An expression for the integral on the basic intervals. Let α belong to one of the basic intervals; from (39), $\alpha = a/q + z$, where $1 \leqslant q \leqslant h$ and $|z| \leqslant 1/\tau$. We break up the sum (36)

$$T_\alpha = \sum_{p \leqslant N} e^{2\pi i p \alpha} = \sum_{p \leqslant N} e^{2\pi i (a/q+z)p},$$

extended over all primes not exceeding N into partial sums $T_{\alpha,M}$ of the form

$$T_{\alpha,M} = \sum_{M \leqslant p < M'} e^{2\pi i (a/q+z)p},$$

where M' is so chosen that $e^{2\pi i z p}$ differs "little" from $e^{2\pi i z M}$; since we intend to give only the *idea* of Vinogradov's method, and not a proof of the Goldbach-Vinogradov theorem, we will not state precisely what we mean

---

* If two such intervals surrounding the points a_1/q_1 and a_2/q_2 intersect, then at a common point we will have the equation

$$\frac{a_1}{q_1} + \frac{\theta_1}{\tau} = \frac{a_2}{q_2} + \frac{\theta_2}{\tau}, \quad \text{where} \quad |\theta_1| \leqslant 1, \quad |\theta_2| \leqslant 1,$$

or

$$\frac{a_1 q_2 - a_2 q_1}{q_1 q_2} = \frac{\theta_1 - \theta_2}{\tau}.$$

But the absolute value of the left side of this last equation is not less than $1/q_1 q_2$, i.e., is greater than $1/h^2$, and the right side is not greater than $2/\tau$, i.e., is less than $2h/N$. So if this last equation were true, it would imply the inequality $1/h^2 < 2h/N$ which contradicts the choice of h.

by the expression "differs little"; in his proof Vinogradov deals with
rigorously defined inequalities, involving a great deal of calculation. Thus

$$T_{\alpha,M} \approx e^{2\pi i Mz} \sum_{M \leqslant p < M'} e^{2\pi i (a/q) p} = e^{2\pi i Mz}\, T_{a/q,M}, \tag{42}$$

where the symbol $\approx$ means that the first of the three expressions on the
last relation differs "little" from the second.

We further break up each of the sums

$$T_{a/q,M} = \sum_{M \leqslant p < M'} e^{2\pi i (a/q) p} \tag{43}$$

into sums $T_{a/q,\, M'l}$, taken over all primes p_l satisfying the relation
$M \leqslant p_l < M'$ and belonging to arithmetic progressions $qx + l$, where l
takes on all values from 0 to $q - 1$ which are relatively prime to q. But

$$e^{2\pi i (a/q) pl} = e^{2\pi i x + 2\pi i (a/q) l} = e^{2\pi i (a/q) l},$$

and thus

$$T_{a/q,M'l} = e^{2\pi i (a/q) l} \cdot \pi(M, M', l), \tag{44}$$

where $\pi(M, M', l)$ is the number of primes satisfying the conditions
$M \leqslant p < M'$ and belonging to the arithmetic progression $qx + l$. In the
development of formula (14) for the number $\pi(x)$ of primes not exceeding
x, it was established that $\pi(M, M', l)$, for values of q which are "small"
in comparison with the difference $M' - M$, differs little from
$1/\phi(q) \int_M^{M'} dx/\ln x$, where $\phi(q)$ is *Euler's function*. This is a number-theoretic
function (i.e., a function defined for natural numbers q) representing the
number of positive integers not exceeding q and relatively prime to q.
From (44) we may thus derive

$$T_{a/q,M'l} \approx e^{2\pi i (a/q) l} \cdot \frac{1}{\phi(q)} \int_M^{M'} \frac{dx}{\ln x}. \tag{45}$$

In the expression on the right side of (45), only the first factor depends on
l, i.e., on the choice of the arithmetic progression $qx + l$ (we now consider
q as fixed). After summing on l, we obtain

$$T_{a/q,M} \approx \frac{1}{\phi(q)} \int_M^{M'} \frac{dx}{\ln x} \sum_l e^{2\pi i (a/q) l},$$

and further, from (42),

$$T_{\alpha,M} \approx e^{2\pi i Mz} \cdot \frac{1}{\phi(q)} \int_M^{M'} \frac{dx}{\ln x} \cdot \sum_l e^{2\pi i (a/q) l}, \tag{46}$$

where

$$e^{2\pi iMz} \int_M^{M'} \frac{dx}{\ln x} \approx \int_M^{M'} \frac{e^{2\pi izx}}{\ln x}\, dx,$$

which allows us to replace (46) by the relation

$$T_{\alpha,M} \approx \int_M^{M'} \frac{e^{2\pi izx}}{\ln x}\, dx \cdot \frac{1}{\phi(q)} \sum_l e^{2\pi i(a/q)l}. \tag{47}$$

After summing on M it is established that

$$T_\alpha \approx \int_2^N \frac{e^{2\pi izx}}{\ln x}\, dx \cdot \frac{1}{\phi(q)} \sum_l e^{2\pi i(a/q)l}. \tag{48}$$

The sum

$$\sum_l e^{2\pi i(a/q)l},$$

occuring on the right side of (48), with the summation taken over natural numbers l not exceeding q and relatively prime to q may be expressed as a number-theoretic function $\mu(q)$ defined in the following manner: $\mu(q) = 0$ if q is divisible by the square of an integer greater than one; $\mu(1) = 1$ and $\mu(q) = (-1)^n$ if $q = p_1 p_2 \cdots p_n$ where $p_1, p_2, \cdots, p_n$ are distinct primes. Thus, for relatively prime a and q

$$\sum_l e^{2\pi i(a/q)l} = \mu(q). \tag{49}$$

Thus equation (48) may be written in the form

$$T_\alpha \approx \frac{\mu(q)}{\phi(q)} \int_2^N \frac{e^{2\pi izx}}{\ln x}\, dx.$$

From the fact $\mu^3(q) = \mu(q)$ we have

$$T_\alpha^3 \approx \frac{\mu(q)}{(\phi(q))^3} \left(\int_2^N \frac{e^{2\pi izx}}{\ln x}\, dx \right)^3, \tag{50}$$

and from the definition of $I_1(N)$

$$I_1(N) = \sum_{1 \leqslant q < h} \sum_a \int_{a/q - 1/\tau}^{a/q + 1/\tau} T_\alpha^3\, e^{-2\pi i\alpha N}\, d\alpha, \tag{51}$$

where for a given q the summation is taken over all nonnegative a less than q. Since $\alpha = a/q = z$, we then have, as a result of (50),

$$I_1(N) \approx \sum_{1 \leqslant q < h} \frac{\mu(q)}{(\phi(q))^3} \sum_a e^{-2\pi i(a/q)N} \int_{-1/\tau}^{1/\tau} \left(\int_2^N \frac{e^{2\pi izx}}{\ln x} \, dx \right)^3 e^{-2\pi izN} \, dz. \tag{52}$$

We introduce the notation

$$R(N) = \int_{-1/\tau}^{1/\tau} \left(\int_2^N \frac{e^{2\pi izx}}{\ln x} \, dx \right)^3 e^{-2\pi izN} \, dz. \tag{53}$$

From relation (52) it follows that

$$I_1(N) \approx R(N) \sum_{1 \leqslant q < h} \frac{\mu(q)}{[\phi(q)]^3} \sum_a e^{-2\pi i(a/q)N}. \tag{54}$$

Here we must draw attention to the fact that $R(N)$ is an analytic expression, which can therefore be calculated approximately; in fact, it runs out that

$$R(N) \approx \frac{N^2}{2(\ln N)^3}. \tag{55}$$

The expression occuring as a factor of $R(N)$ on the right side of (54) differs "little" from the sum of the infinite series

$$S(N) = \sum_{q=1}^{\infty} \frac{\mu(q)}{[\phi(q)]^3} \sum_a e^{-2\pi i(a/q)N}, \tag{56}$$

so that, from (54) and (55), it can be established that

$$I_1(N) \approx \frac{N^2}{2(\ln N)^3} S(N), \tag{57}$$

or, more precisely,

$$I_1(N) = \frac{N^2}{2(\ln N)^3} [S(N) + \gamma_1(N)], \tag{58}$$

where

$$\lim_{N \to \infty} \gamma_1(N) = 0. \tag{59}$$

We note that number-theoretic expression $S(N)$ may be written in the form

$$S(N) = C \prod_p \left(1 - \frac{1}{p^2 - 3p + 3} \right), \tag{60}$$

where C is a constant, the multiplication is extended over all prime divisors of the number N, and, as the computations show,

$$S(N) > 0.6. \tag{61}$$

Estimate of the integral on the complementary intervals. We turn now to an estimate of the sum I_2 of the integrals on the complementary intervals. Since the modulus of the integral does not exceed the integral of the modulus of the function being integrated, and since $\mid e^{-2\pi i \alpha N} \mid = 1$ for real αN, we have

$$\mid I_2 \mid < \max \mid T_\alpha \mid \cdot \int_{-1/\tau}^{1-1/\tau} \mid T_\alpha \mid^2 d\alpha, \tag{62}$$

where $\max \mid T_\alpha \mid$ represents the largest value of $\mid T_\alpha \mid$ for α belonging to the complementary intervals (we have strengthened the inequality by taking as the factor of $\max \mid T_\alpha \mid$ the integral extended over the whole interval $-(1/\tau) \leqslant \alpha \leqslant 1 - 1/\tau$).

But the square of the modulus of a complex number is equal to the product of the number with its complex conjugate, so that

$$\mid T_\alpha \mid^2 = T_\alpha \cdot \bar{T}_\alpha,$$

where from (36) we have

$$\bar{T}_\alpha = \sum_{p \leqslant N} e^{-2\pi i \alpha p},$$

since $e^{-2\pi i \alpha p} = \cos 2\pi \alpha p - i \sin 2\pi \alpha p$. Thus, inequality (62) may be rewritten in the form

$$\mid I_2 \mid < \max \mid T_\alpha \mid \cdot \int_{-1/\tau}^{1-1/\tau} \sum_{p \leqslant N} e^{2\pi i \alpha p} \sum_{p_1 \leqslant N} e^{-2\pi i \alpha p_1} d\alpha$$

or in the form

$$\mid I_2 \mid < \max \mid T_\alpha \mid \cdot \int_{-1/\tau}^{1-1/\tau} \sum_{p \leqslant N} \sum_{p_1 \leqslant N} e^{2\pi i \alpha (p - p_1)} d\alpha. \tag{63}$$

But the integral in the inequality (63), from what was said at the beginning of the present section, represents the number of U of solutions in primes p, p_1, not exceeding N, of the equation $p - p_1 = 0$, or simply the number of primes not exceeding N, i.e., $\pi(N)$. From the result (12) of Čebyšev we have

$$\pi(N) < B \cdot \frac{N}{\ln N},$$

where B is a constant. In this manner

$$| I_2 | < B \cdot \frac{N}{\ln N} \cdot \max | T_\alpha |, \tag{64}$$

where, to repeat, $\max | T_\alpha |$ represents the largest value of $| T_\alpha |$ on the complementary intervals. From (58) and (59) it follows that in order to complete the proof of the Goldbach-Vinogradov theorem, we must now show that $\max | T_\alpha |$ has order less than $N/(\ln N)^2$; however, the establishment of this fact presents the greatest difficulty and constitutes the essential part of the whole proof of the theorem.

Every α belonging to a complementary interval can be represented in the form $\alpha = a/q + z$, where $h < q \leqslant \tau$ and $| z | \leqslant 1/q\tau$. The problem thus consists of estimating the modulus of the trigonometric sum

$$T_\alpha = \sum_{p \leqslant N} e^{2\pi i(a/q+z)p}$$

under the given conditions. Vinogradov established, in particular, that

$$\lim_{N \to \infty} \frac{\max T_\alpha}{\dfrac{N}{(\ln N)^3}} = 0 \; ; \tag{65}$$

here he made use of a very important identity which he discovered for the function $\mu(n)$ discussed previously.

Unfortunately, it is not possible here to give a proof of equation (65); the interested reader is referred to Chapter X in Vinogradov's book "Methods of trigonometric sums in the theory of numbers."

From (65) and (64), as we noted, it follows that

$$\lim_{N \to \infty} \frac{I_2(N)}{I_1(N)} = 0.$$

In this manner, from (40), (58), and (59) we have

$$I(N) = \frac{N^2}{2(\ln N)^3} [S(N) + \gamma(N)], \tag{66}$$

where

$$\lim_{N \to \infty} \gamma(N) = 0,$$

and $S(N)$ has the value (60), so that, from (61), $S(N) > 0.6$. This completes the proof of the theorem.

§5. Decomposition of Integers into the Sum of Two Squares; Complex Integers

The importance of the study of prime numbers is chiefly because of the central role they play in most of the laws of number theory: It frequently happens that questions which at first sight seem far removed from divisibility are nevertheless shown by more careful consideration to be intimately connected with the theory of prime numbers. We illustrate this statement by the following example.

One of the problems of number theory consists of finding those natural numbers that can be decomposed into the sum of the squares of two integers (not necessarily different from zero).

The rule for the sequence of numbers that are the sum of two squares is not immediately clear. From 1 to 50, for example, it consists of the numbers 1, 2, 4, 5, 8, 9, 10, 13, 16, 18, 20, 25, 26, 29, 32, 34, 36, 37, 40, 41, 45, 49, 50 a sequence which seems quite erratic. The 17th century French mathematician Fermat noticed that here everything depends on how the number can be represented as the product of primes, i.e., the question is inherently related to the theory of prime numbers.

Prime numbers, other than $p = 2$, are odd, so that division by 4 gives a remainder equal to 1 (for a prime number of the form $4n + 1$) or to 3 (for a prime number of the form $4n + 3$).

We will consider the question of expressing a given number as the sum of two squares under the following three headings.

1. A prime number p is the sum of two squares if and only if $p = 4n + 1$.

The proof of the fact that a number of the form $4n + 3$ cannot be expressed as the sum of two squares is almost obvious: The sum of the squares of two even numbers is divisible by 4, the sum of the square of two odd numbers gives a remainder of 2 when divided by 4, and the sum of the squares of an even and an odd number, when divided by 4, gives a remainder of 1.

Let us now prove a preliminary theorem, namely that if p is a prime, then $(p - 1)! + 1$ is divisible by p. The numbers not divisible by p, when divided by p give the remainder 1, 2, 3, $\cdots$, $p - 1$. We choose an integer r, $1 \leqslant r \leqslant p - 1$ and multiply r by 1, 2, $\cdots$, $p - 1$; when we divide the products so constructed by p we obtain, as is not difficult to prove, all these same remainders, but in general in a different order. In particular, among these remainders will be the number 1, that is to say, for every r one can find an r_1 such that $r \cdot r_1 = 1 + kp$. We note that $r = r_1$ only if $r = 1$ or $r = p - 1$. For if $r^2 = 1 + kp$, then $(r + 1)(r - 1)$ is divisible by p; but for numbers $1 \leqslant r \leqslant p - 1$ this is possible only for $r = 1$ and

$r = p - 1$. Let us find the remainders on dividing $(p - 1)! = 1 \cdot 2$ $\cdots (p - 1)$ by p. In this product, for every factor r, except 1 and $p - 1$, there occurs a corresponding r_1, distinct from r, such that $r \cdot r_1$ gives the remainder 1. Thus $(p - 1)!$ will give a remainder dividing by p which is the same as if only the two factors 1 and $p - 1$ were present, i.e., it gives the remainder $p - 1$. Thus, $(p - 1)! + 1$ is divisible by p.

Now let $p = 4n + 1$. Further, we write

$$(p - 1)! + 1 = \left\{ 1 \cdot 2 \cdots \frac{p - 1}{2} \right\} \cdot \left\{ \left(p - \frac{p - 1}{2} \right) \cdots (p - 2)(p - 1) \right\} + 1.$$

The second expression in braces, when divided by p, will leave the remainder $(-1)^{p-1/2} [(p - 1)/2]!$. But $(p - 1)/2 = 2n$ is an even number, so that in this case $[(p - 1)/2]!^2 + 1$ is also divisible by p. We denote by A the remainder on dividing $[(p - 1)/2]!$ by p. It is obvious that $A^2 + 1$ is also divisible by p.

We consider the expression $x - Ay$, in which x and y range independently over the numbers $0, 1, \cdots, [\sqrt{p}]$; (here $[x]$ denotes the largest integer not exceeding x). We thus obtain $([\sqrt{p}] + 1)^2 \geqslant p + 1$ numerical values for $x - Ay$, which may be distinct or may in some cases coincide. Since the various remainders on dividing by p can only be $p(0, 1, 2, \cdots, p - 1)$, while we here have at least $p + 1$ values for $x - Ay$, there must exist two distinct pairs (x_1, y_1) and (x_2, y_2) such that $x_1 - Ay_1$ and $x_2 - Ay_2$ leave the same remainder on dividing by p; i.e., $(x_1 - x_2) - A(y_1 - y_2)$ is divisible by p. We set $x_0 = x_1 - x_2$, $y_0 = y_1 - y_2$. Obviously, $|x_0| < \sqrt{p}, |y_0| < \sqrt{p}$. Since $A^2 + 1$ is divisible by p, it follows that $y_0^2(A^2 + 1) = (Ay_0)^2 + y_0^2$ is divisible by p; but since $x_0 - Ay_0$ is divisible by p, the number $x_0^2 - (Ay_0)^2 = (x_0 - Ay_0)$ $(x_0 + Ay_0)$ is divisible by p. Thus the quantity $x_0^2 + y_0^2$, which is equal to $(x_0^2 - (Ay_0)^2 + (Ay_0)^2 + y_0^2)$, is divisible by p. But $|x_0| < \sqrt{p}$, $|y_0| < \sqrt{p}$. Hence $x_0^2 + y_0^2 = 0$ or $x_0^2 + y_0^2 = p$. The first is impossible, since the pairs (x_1, y_1) and (x_2, y_2) were distinct. Thus a prime number of the form $4n + 1$ is representable as the sum of two squares.

2. We turn to the decomposition of an arbitrary integer into the sum of two squares. It is easy to establish the identity

$$(a^2 + b^2)(c^2 + d^2) = (ac - bd)^2 + (ad + bc)^2.$$

This identity shows that the product of two integers that are the sum of two squares is again the sum of two squares. Hence the product of any powers of prime numbers of the form $4n + 1$ (or which are equal to 2) is the sum of two squares. Since multipliying the sum of two squares by a square gives the sum of two squares, any number in which the prime factors of the $4n + 3$ occur in even powers is the sum of two squares.

3. We now show that if a prime number of the form $4n + 3$ enters into a number in an odd power, the number cannot be expressed as the sum of two squares. The original question will then be completely settled.

We will consider complex numbers of the form $a + bi$, where a and b are ordinary integers. Such a complex number will be called a *complex integer*. If an integer N is the sum of two squares $N = a^2 + b^2$, then $N = (a + bi)(a - bi) = \alpha \cdot \bar{\alpha}$ (where $\bar{\alpha}$ denotes the complex conjugate of the number α), i.e., N is factored in the domain of complex integers into complex conjugate factors.

In this domain of complex integers, we may construct a theory of divisibility completely analogous to the theory of divisibility in the domain of ordinary integers. We will say that the complex integer α is divisible by the complex integer β, if α/β is again a complex integer. There exist only four complex integers α which divide 1, namely 1, -1, i, and $-i$. We will say that a complex integer α is a prime, if it does not have any divisors other than 1, -1, i, $-i$, α, $-\alpha$, αi, $-\alpha i$. But now the problem solved under the first heading above will have a different meaning; it will now turn out that numbers of the form $4n + 1$ (or equal to 2) which in the previous case were prime will cease to be prime in the domain of complex integers, while it is easy to prove that primes of the form $4n + 3$ remain prime.

For, if $p = \alpha\beta$, then $p = \bar{\alpha}\bar{\beta}$ and $p^2 = \alpha\bar{\alpha}\beta\bar{\beta}$. But $\alpha\bar{\alpha}$ and $\beta\bar{\beta}$ are ordinary positive integers; and $p \neq \alpha\bar{\alpha}$, since prime numbers of the form $4n + 3$ are not the sum of two squares. This means that $\alpha\bar{\alpha} = 1$; thus α can be only ± 1 or $\pm i$, so that p has no divisors other than the obvious ones.

For complex integers the theorem on the unique decomposition into prime factors still holds. Uniqueness here means, of course, that the order of multiplication is ignored and also all factors of the form 1, -1, i, $-i$.

Let N be the sum of two squares, $N = \alpha\bar{\alpha}$. Let p be a prime number of the form $4n + 3$. Let us calculate what power of p appears in the number N. From the fact that p remains a prime in the complex domain, it is sufficient to calculate what power of p appears in α and in $\bar{\alpha}$. But these powers are equal, so that p necessarily appears in N to an even power, which proves the proposition.

The discovery that a rich theory of divisibility is possible elsewhere than in the domain of whole rational numbers greatly extended the field of vision of 19th century mathematicians. The development of these ideas called for the creation of new general concepts in mathematics, such as, for example, rings and ideals. The significance of these concepts at the present time has far outgrown the frame of number theory.

Suggested Reading

H. Davenport, *The higher arithmetic: an introduction to the theory of numbers*, Harper Brothers, New York, 1960.

T. Estermann, *Introduction to modern prime number theory*, Cambridge University Press, New York, 1952.

A. A. Fraenkel, *Integers and theory of numbers*, Scripta Mathematica, New York, 1955.

G. H. Hardy and F. M. Wright, *An introduction to the theory of numbers*, 2nd ed., Clarendon Press, Oxford, 1945.

A. Ia. Khinchin, *Three pearls of number theory*, Graylock Press, Rochester, N. Y., 1952.

O. Ore, *Number theory and its history*, McGraw-Hill, New York, 1948.

I. M. Vinogradov, *An introduction to the theory of numbers*, Pergamon Press, New York, 1955.

CHAPTER XI

THE THEORY
OF PROBABILITY

§1. The Laws of Probability

The simplest laws of natural science are those that state the conditions under which some event of interest to us will either certainly occur or certainly not occur; i.e., these conditions may be expressed in one of the following two forms:

1. If a complex (i.e., a set or collection) of conditions S is realized, then event A certainly occurs;

2. If a complex of conditions S is realized, then event A cannot occur.

In the first case the event A, with respect to the complex of conditions S, is called a "certain" or "necessary" event, and in the second an "impossible" event. For example, under atmospheric pressure and at temperature t between $0°$ and $100°$ (the complex of conditions S) water necessarily occurs in the liquid state (the event A_1 is certain) and cannot occur in a gaseous or solid state (events A_2 and A_3 are impossible).

An event A, which under a complex of conditions S sometimes occurs and sometimes does not occur, is called random with respect to the complex of conditions. This raises the question: Does the randomness of the event A demonstrate the absence of any law connecting the complex of conditions S and the event A? For example, let it be established that lamps of a specific type, manufactured in a certain factory (condition S) sometimes continue to burn more than 2,000 hours (event A), but sometimes burn out and become useless before the expiration of that time. May it not still be possible that the results of experiments to see whether a given lamp will or will not burn for 2,000 hours will serve to evaluate the production of the factory? Or should we restrict ourselves to indicating only the period (say 500 hours) for which in practice all lamps work without fail, and the period (say 10,000 hours) after which in practice all lamps do not work? It is clear that to describe the working life of a lamp by an inequality of the form $500 \leqslant T \leqslant 10,000$ is of little help to the consumer. He will receive much more valuable information if we tell him that in approximately 80% of the cases the lamps work for no less than 2,000 hours. A still more complete evaluation of the quality of the lamps

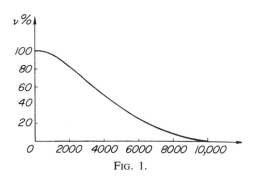

FIG. 1.

will consist of showing for any T the percent $\nu(T)$ of the lamps which work for no less than T hours, say in the form of the graph in figure 1.

The curve $\nu(T)$ is found in practice by testing with a sufficiently large sample (100-200) of the lamps. Of course, the curve found in such a manner is of real value only in those

where it truly represents an actual law governing not only the given sample but all the lamps manufactured with a given quality of material and under given technological conditions; that is, only if the same experiments conducted with another sample will give approximately the same results (i.e., the new curve $\nu(T)$ will differ little from the curve derived from the first sample). In other words, the statistical law expressed by the curves $\nu(T)$ for the various samples is only a reflection of the law of probability connecting the useful life of a lamp with the materials and the technological conditions of its manufacture.

This law of probability is given by a function $\mathbf{P}(T)$, where $\mathbf{P}(T)$ is the probability that a single lamp (made under the given conditions) will burn no less than T hours.

The assertion that the event A occurs under conditions S with a definite probability

$$\mathbf{P}(A/S) = p$$

amounts to saying that in a sufficiently long series of tests (i.e., realizations of the complex of conditions S) the frequencies

$$\nu_r = \frac{\mu_r}{n_r}$$

of the occurrence of the event A (where n_r is the number of tests in the rth series, and μ_r is the number of tests of this series for which event A occurs) will be approximately identical with one another and will be close to p.

The assumption of the existence of a constant $p = \mathbf{P}(A/S)$ (objectively determined by the connection between the complex of conditions S and the event A) such that the frequencies ν get closer "generally speaking" to p as the number of tests increases, is well borne out in practice for a wide class of events. Events of this kind are usuaully called *random* or *stochastic*.

This example belongs to the laws of probability for mass production. The reality of such laws cannot be doubted, and they form the basis of important practical applications in statistical quality control. Of a similar kind are the laws of probability for the scattering of missiles, which are basic in the theory of gunfire. Since this is historically one of the earliest applications of the theory of probability to technical problems, we will return below to some simple problems in the theory of gunfire.

What was said about the "closeness" of the frequency ν to the probability p for a large number n of tests is somewhat vague; we said nothing about how small the difference $\nu - p$ may be for any n. The degree of closeness of ν to p is estimated in §3. It is interesting to note that a certain indefiniteness in this question is quite unavoidable. The very statement itself that ν and p are close to each other has only a probabilistic character, as becomes clear if we try to make the whole situation precise.

§2. The Axioms and Basic Formulas of the Elementary Theory of Probability

Since it cannot be doubted that statistical laws are of great importance, we turn to the question of methods of studying them. First of all one thinks of the possibility of proceeding in a purely empirical way. Since a law of probability exhibits itself only in mass processes, it is natural to imagine that in order to discover the law we must conduct a mass experiment.

Such an idea, however, is only partly right. As soon as we have established certain laws of probability by experiment, we may proceed to deduce from them new laws of probability by logical means or by computation, under certain general assumptions. Before showing how this is done, we must enumerate certain basic definitions and formulas of the theory of probability.

From the representation of probability as the standard value of the frequency $\nu = m/n$, where $0 \leqslant m \leqslant n$, and thus $0 \leqslant \nu \leqslant 1$, it follows that the probability $\mathbf{P}(A)$ of any event A must be assumed to lie between zero and one*

$$0 \leqslant \mathbf{P}(A) \leqslant 1. \tag{1}$$

Two events are said to be mutually exclusive if they cannot both occur (under the complex of conditions S). For example, in throwing a die, the the occurrence of an even number of spots and of a three are mutually exclusive. An event A is called the *union* of events A_1 and A_2 if it consists of the occurrence of at least one of the events A_1, A_2. For example, in throwing a die, the event A, consisting of rolling 1, 2, or 3, is the union of the events A_1 and A_2, where A_1 consists of rolling 1 or 2 and A_2 consists of rolling 2 or 3. It is easy to see that for the number of occurrences m_1, m_2, and m of two mutually exclusive events A_1 and A_2 and their union $A = A_1 \cup A_2$, we have the equation $m = m_1 + m_2$, or for the corresponding frequencies $\nu = \nu_1 + \nu_2$.

This leads naturally to the following axiom for the addition of probabilities:

$$\mathbf{P}(A_1 \cup A_2) = \mathbf{P}(A_1) + \mathbf{P}(A_2), \tag{2}$$

if the events A_1 and A_2 are mutually exclusive and $A_1 \cup A_2$ denotes their union.

Further, for an event U which is certain, we naturally take

$$\mathbf{P}(U) = 1. \tag{3}$$

The whole mathematical theory of probability is constructed on the basis of simple axioms of the type (1), (2), and (3). From the point of view of pure mathematics, *probability* is a numerical function of "events," with a number of properties determined by axioms. The properties of probability, expressed by formulas (1), (2), and (3), serve as a sufficient basis for the construction of what is called the elementary theory of probability, if we do not insist on including in the axiomatization the concepts of an event itself, the union of events, and their intersection, as defined later. For the beginner it is more useful to confine himself to an intuitive understanding of the terms "event" and "probability," but to realize that although the meaning of these terms in practical life cannot be completely formalized, still this fact does not affect the complete formal precision of an axiomatized, purely mathematical presentation of the theory of probability.

The union of any given number of events A_1, A_2, $\cdots$, A_s is defined as the event A consisting of the occurrence of at least one of these events.

---

* For brevity we now change $\mathbf{P}(A/S)$ to $\mathbf{P}(A)$.

From the axiom of addition, we easily obtain for any number of pairwise mutually exclusive events A_1, A_2, $\cdots$, A_s and their union A,

$$\mathbf{P}(A) = \mathbf{P}(A_1) + \mathbf{P}(A_2) + \cdots + \mathbf{P}(A_s)$$

(the so-called *theorem of the addition of probabilities*).

If the union of these events is an event that is certain (i.e., under the complex of conditions S one of the events A_k must occur), then

$$\mathbf{P}(A_1) + \mathbf{P}(A_2) + \cdots + \mathbf{P}(A_s) = 1.$$

In this case the system of events A_1, $\cdots$, A_s is called a *complete system* of events.

We now consider two events A, and B, which, generally speaking, are not mutually exclusive. The event C is the intersection of the events A and B, written $C = AB$, if the event C consists of the occurrence of both A and B.*

For example, if the event A consists of obtaining an even number in the throw of a die and B consists of obtaining a multiple of three, then the event C consists of obtaining a six.

In a large number n of repeated trials, let the event A occur m times and the event B occur l times, in k of which B occurs together with the event A. The quotient k/m is called the conditional frequency of the event B under the condition A. The frequencies k/m, m/n, and k/n are connected by the formula

$$\frac{k}{m} = \frac{k}{n} : \frac{m}{n}$$

which naturally gives rise to the following definition:

The conditional probability $\mathbf{P}(B/A)$ of the event B under the condition A is the quotient

$$\mathbf{P}(B/A) = \frac{\mathbf{P}(AB)}{\mathbf{P}(A)} .$$

Here it is assumed, of course, that $\mathbf{P}(A) \neq 0$.

If the events A and B are in no way essentially connected with each other, then it is natural to assume that event B will not appear more often, or less often, when A has occurred than when A has not occurred, i.e., that approximately $k/m \sim l/n$ or

$$\frac{k}{n} = \frac{k}{m}\frac{m}{n} \sim \frac{l}{n}\frac{m}{n} .$$

---

* Similarly, the intersection C of any number of events A_1, A_2, $\cdots$, A_s consists of the occurrence of all the given events.

In this last approximate equation $m/n = \nu_A$ is the frequency of the event A, and $l/n = \nu_B$ is the frequency of the event B and finally $k/n = \nu_{AB}$ is the frequency of the intersection of the events A and B.

We see that these frequencies are connected by the relation

$$\nu_{AB} \sim \nu_A \nu_B .$$

For the probabilities of the events A, B and AB, it is therefore natural to accept the corresponding exact equation

$$\mathbf{P}(AB) = \mathbf{P}(A) \cdot \mathbf{P}(B). \tag{4}$$

Equation (4) serves to define the *independence* of two events A and B.

Similarly, we may define the independence of any number of events. Also, we may give a definition of the independence of any number of experiments, which means, roughly speaking, that the outcome of any part of the experiments do not depend on the outcome of the rest.*

We now compute the probability P_k of precisely k occurrences of a certain event A in n independent tests, in each one of which the probability p of the occurrence of this event is the same. We denote by $\bar{A}$ the event that event A does not occur. It is obvious that

$$\mathbf{P}(\bar{A}) = 1 - \mathbf{P}(A) = 1 - p.$$

From the definition of the independence of experiments it is easy to see that the probability of any specific sequence consisting of k occurrences of A and $n - k$ nonoccurrences of A is equal to

$$p^k(1 - p)^{n-k}. \tag{5}$$

Thus, for example, for $n = 5$ and $k = 2$ the probability of getting the sequence $A\bar{A}A\bar{A}\bar{A}$ will be $p(1 - p) p(1 - p) (1 - p) = p^2(1 - p)^3$,

By the theorem on the addition of probabilities, P_k will be equal to the sum of the probabilities of all sequences with k occurrences and $n - k$ nonoccurrences of the event A, i.e., P_k will be equal from (5) to the product of the number of such sequences by $p^k(1 - p)^{n-k}$. The number of such

---

* A more exact meaning of *independent experiments* is the following. We divide the n experiments in any way into two groups and let the event A consist of the result that all the experiments of the first group have certain preassigned outcomes, and the event B that the experiments of the second group have preassigned outcomes. The experiments are called independent (as a collection) if for arbitrary decomposition into two groups and arbitrarily preassigned outcomes the events A and B are independent in the sense of (4).

We will return in §4 to a consideration of the objective meaning in the actual world of the independence of events.

sequences is obviously equal to the number of combinations of n things taken k at a time, since the k positive outcomes may occupy any k places in the sequence of n trials.

Finally we get

$$P_k = C_n^k p^k (1 - p)^{n-k} \quad (k = 0, 1, 2, \cdots, n) \tag{6}$$

(which is called a binomial distribution).

In order to see how the definitions and formulas are applied, we consider an example that arises in the theory of gunfire.

Let five hits be sufficient for the destruction of the target. What interests us is the question whether we have the right to assume that 40 shots will insure the necessary five hits. A purely empirical solution of the problem would proceed as follows. For given dimensions of the target and for a given range, we carry out a large number (say 200) of firings, each consisting of 40 shots, and we determine how many of these firings produce at least five hits. If this result is achieved, for example, by 195 firings out of the 200, then the probability P is approximately equal to

$$P = \frac{195}{200} = 0.975.$$

If we proceed in this purely empirical way, we will use up 8,000 shells to solve a simple special problem. In practice, of course, no one proceeds in such a way. Instead, we begin the investigation by assuming that the scattering of the shells for a given range is independent of the size of the target. It turns out that the longitudinal and lateral deviations, from the mean point of landing of the shells, follow a law with respect to the frequency of deviations of various sizes that is illustrated in figure 2.

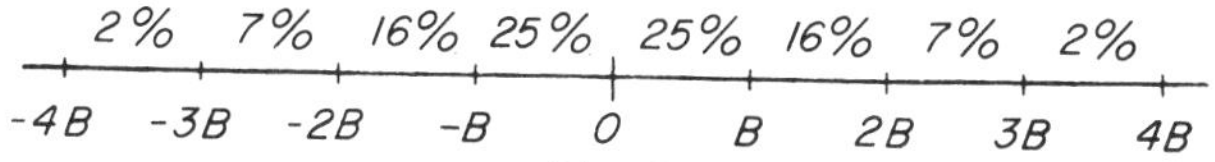

FIG. 2.

The letter B here denotes what is called the probable deviation. The probable deviation, generally speaking, is different for longitudinal and for lateral deviations and increases with increasing range. The probable deviations for different ranges for each type of gun and of shell are found empirically in firing practice on an artillery range. But the subsequent solution of all possible special problems of the kind described is carried out by calculations only.

For simplicity, we assume that the target has the form of a rectangle, one side of which is directed along the line of fire and has a length of two probable longitudinal deviations, while the other side is perpendicular to the line of fire and is equal in length to two probable lateral deviations. We assume further that the range has already been well established, so that the mean trajectory of the shells passes through its center (figure 3).

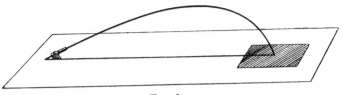

Fig. 3.

We also assume that the lateral and longitudinal deviations are independent.* Then for a given shell to fall on the target, it is necessary and sufficient that its longitudinal and lateral deviations do not exceed the corresponding probable deviations. From figure 2 each of these events will be observed for about 50% of the shells fired, i.e., with probability $\frac{1}{2}$. The intersection of the two events will occur for about 25% of the shells fired; i.e., the probability that a specific shell will hit the target will be equal to

$$p = \frac{1}{2} \cdot \frac{1}{2} = \frac{1}{4},$$

and the probability of a miss for a single shell will be

$$q = 1 - p = 1 - \frac{1}{4} = \frac{3}{4}.$$

Assuming that hits by the individual shells represent independent events, and applying the binomial formula (6), we find that the probability for getting exactly k hits in 40 shots will be

$$P_k = C_{40}^k p^k q^{40-k} = \frac{40 \cdot 39 \cdots (39-k)}{1 \cdot 2 \cdots k} \left(\frac{1}{4}\right)^k \left(\frac{3}{4}\right)^{40-k}.$$

What concerns us is the probability of getting no less than five hits, and this is now expressed by the formula

$$P = \sum_{k=5}^{40} P_k.$$

---

* This assumption of independence is borne out by experience.

But it is simpler to compute this probability from the formula $P = 1 - Q$, where

$$Q = \sum_{k=0}^{4} P_k$$

is the probability of getting less than five hits.

We may calculate that

$$P_0 = \left(\frac{3}{4}\right)^{40} \sim 0.00001,$$

$$P_1 = 40 \left(\frac{3}{4}\right)^{39} \frac{1}{4} \sim 0.00013,$$

$$P_2 = \frac{40 \cdot 39}{2} \left(\frac{3}{4}\right)^{38} \left(\frac{1}{4}\right)^2 \sim 0.00087,$$

$$P_3 = \frac{40 \cdot 39 \cdot 38}{2 \cdot 3} \left(\frac{3}{4}\right)^{37} \left(\frac{1}{4}\right)^3 \sim 0.0037,$$

$$P_4 = \frac{40 \cdot 39 \cdot 38 \cdot 37}{2 \cdot 3 \cdot 4} \left(\frac{3}{4}\right)^{36} \left(\frac{1}{4}\right)^4 \sim 0.0113,$$

so that

$$Q = 0.016, \quad P = 0.984.$$

The probability P so obtained is somewhat closer to certainty than is usually taken to be sufficient in the theory of gunfire. Most often it is considered permissible to determine the number of shells needed to guarantee the result with probability 0.95.

The previous example is somewhat schematized, but it shows in sufficient detail the practical importance of probability calculations. After establishing by experiment the dependence of the probable deviations on the range (for which we did not need to fire a large number of shells), we were then able to obtain, by simple calculation, the answers to questions of the most diverse kind. The situation is the same in all other domains where the collective influence of a large number of random factors leads to a statistical law. Direct examination of the mass of observations makes clear only the the very simplest statistical laws; it uncovers only a few of the basic probabilities involved. But then, by means of the laws of the theory of probability, we use these simplest probabilities to compute the probabilities of more complicated occurrences and deduce the statistical laws that govern them.

Sometimes we succeed in completely avoiding massive statistical material, since the probabilities may be defined by sufficiently convincing considerations of symmetry. For example, the traditional conclusion that a die, i.e., a cube made of a homogeneous material will fall, when thrown to a sufficient height, with equal probability on each of its faces was reached long before there was any systematic accumulation of data to verify it by observation. Systematic experiments of this kind have been carried out in the last three centuries, chiefly by authors of textbooks in the theory of probability, at a time when the theory of probability was already a well-developed science. The results of these experiments were satisfactory, but the question of extending them to analogous cases scarcely arouses interest. For example, as far as we know, no one has carried out sufficiently extensive experiments in tossing homogeneous dice with twelve sides. But there is no doubt that if we were to make 12,000 such tosses, the twelve-sided die would show each of its faces approximately a thousand times.

The basic probabilities derived from arguments of symmetry or homogeneity also play a large role in many serious scientific problems, for example in all problems of collision or near approach of molecules in random motion in a gas; another case where the successes have been equally great is the motion of stars in a galaxy. Of course, in these more delicate cases we prefer to check our theoretical assumptions by comparison with observation or experiment.

§3. The Law of Large Numbers and Limit Theorems

It is completely natural to wish for greater quantitative precision in the proposition that in a "long" series of tests the frequency of an occurrence comes "close" to its probability. But here we must form a clear notion of the delicate nature of the problem. In the most typical cases in the theory of probability, the situation is such that in an arbitrarily long series of tests it remains theoretically possible that we may obtain either of the two extremes for the value of the frequency

$$\frac{\mu}{n} = \frac{n}{n} = 1 \quad \text{and} \quad \frac{\mu}{n} = \frac{0}{n} = 0.$$

Thus, whatever may be the number of tests n, it is impossible to assert with complete certainty that we will have, say, the inequality

$$\left| \frac{\mu}{n} - p \right| < \frac{1}{10}.$$

For example, if the event A is the rolling of a six with a die, then in n trials, the probability that we will turn up a six on all n trials is $(\frac{1}{6})^n > 0$, in other words, with probability $(\frac{1}{6})^n$ we will obtain a frequency of rolling a six which is equal to *one*; and with probability $(1 - \frac{1}{6})^n > 0$ a six will not come up at all, i.e., the frequency of rolling a six will be equal to *zero*.

In all similar problems any nontrivial estimate of the closeness of the frequency to the probability cannot be made with complete certainty, but only with some probability less than one. For example, it may be shown that in independent tests,* with constant probability p of the occurrence of an event in each test the inequality

$$\left| \frac{\mu}{n} - p \right| < 0.02 \tag{7}$$

for the frequency μ/n will be satisfied, for $n = 10,000$ (and any p), with probability

$$P > 0.9999. \tag{8}$$

Here we wish first of all to emphasize that in this formulation the quantitative estimate of the closeness of the frequency μ/n to the probability p involves the introduction of a new probability P.

The practical meaning of the estimate (8) is this: If we carry out N sets of n tests each, and count the M sets in which inequality (7) is satisfied, then for sufficiently large N we will have approximately

$$\frac{M}{N} \approx P > 0.9999. \tag{9}$$

But if we wish to define the relation (9) more precisely, either with respect to the degree of closeness of M/N to P, or with respect to the confidence with which we may assert that (9) will be verified, then we must have recourse to general considerations of the kind introduced previously in discussing what is meant by the closeness of μ/n and p. Such considerations may be repeated as often as we like, but it is clear that this procedure will never allow us to be free of the necessity, at the last stage, of referring to probabilities in the primitive imprecise sense of this term.

It would be quite wrong to think that difficulties of this kind are peculiar in some way to the theory of probability. In a mathematical investigation of actual events, we always make a model of them. The discrepancies between the actual course of events and the theoretical model can, in its turn, be made the subject of mathematical investigation. But for these discrepancies we must construct a model that we will use without formal mathematical analysis of the discrepancies which again would arise in it in actual experiment.

---

* The proof of the estimate (8) is discussed later in this section.

We note, moreover, that in an actual application of the estimate*

$$\mathbf{P}\left\{\left|\frac{\mu}{n}-p\right|<0.02\right\}>0.9999 \tag{10}$$

to one series of n tests we are already depending on certain considerations of symmetry: inequality (10) shows that for a very large number N of series of tests, relation (7) will be satisfied in no less than 99.99% of the cases; now it is natural to expect with great confidence that inequality (7) will apply in particular to that one of the sequence of n tests which is of interest to us, but we may expect this only if we have some reason for assuming that the position of this sequence among the others is a regular one, that is, that it has no special features.

The probabilities that we may decide to neglect are different in different practical situations. We noted earlier that our preliminary calculations for the expenditure of shells necessary to produce a given result meet the standard that the problem is to be solved with probability 0.95, i.e., that the neglected probabilities do not exceed 0.05. This standard is explained by the fact that if we were to make calculations neglecting a probability of only 0.01, let us say, we would necessarily require a much greater expenditure of shells, so that in practice we would conclude that the task could not be carried out in the time at our disposal, or with the given supply of shells.

In scientific investigations also, we are sometimes restricted to statistical methods calculated on the basis of neglecting probabilities of 0.05, although this practice should be adopted only in cases where the accumulation of more extensive data is very difficult. As an example of such a method let us consider the following problem. We assume that under specific conditions the customary medicine for treating a certain illness gives positive results 50% of the time, i.e., with probability 0.5. A new preparation is proposed, and to test its advantages we plan to use it in ten cases, chosen without bias from among the patients suffering from the illness. Here we agree that the advantage of the new preparation will be considered as proved if it gives a positive result in no less than eight cases out of the ten. It is easy to calculate that such a procedure involves the neglect of probabilities of the order of 0.05 of getting a wrong result, i.e., of indicating an advantage for the new preparation when in fact it is only equally effective or even worse than the old. For if in each of the ten experiments, the probability of a positive outcome is equal to p, then the

---

* This is the accepted notation for estimate (8) of the probability of inequality (7).

probability of obtaining in ten experiments 10, 9, or 8 positive outcomes, is equal respectively to

$$P_{10} = p^{10}, \quad P_9 = 10p^9(1 - p), \quad P_8 = 45p^8(1 - p)^2.$$

For the case $p = \frac{1}{2}$ the sum of these is

$$P = P_{10} + P_9 + P_8 = \frac{56}{1024} \sim 0.05.$$

In this way, under the assumption that in fact the new preparation is exactly as effective as the old, we risk with probability of order 0.05 the error of finding that the new preparation is better than the old. To reduce this probability to about 0 01, without increasing the number of experiments $n = 10$, we will need to agree that the advantage of the new preparation is proved if it gives a positive result in no less than nine cases out of the ten. If this requirement seems too severe to the advocates of the new preparation, it will be necessary to make the number of experiments considerably larger than 10. For example, for $n = 100$, if we agree that the advantage of the new preparation is proved for $\mu > 65$, then the probability of error will only be $P \approx 0.0015$.

For serious scientific investigations a standard of 0.05 is clearly insufficient; but even in such academic and circumstantial matters as the treatment of astronomical observations, it is customary to neglect probabilities of error of 0.001 or 0.003. On the other hand, some of the scientific results based on the laws of probability are considerably more reliable even than that; i.e., they involve the neglect of smaller probabilities. We will return to this question later.

In the previous examples, we have made use of particular cases of the binomial formula (6)

$$P_m = C_n^m p^m (1 - p)^{n-m}$$

for the probability of getting exactly m positive results in n independent trials, in each one of which a positive outcome has probability p. Let us consider, by means of this formula, the question raised at the beginning of this section concerning the probability

$$P = \mathbf{P} \left\{ \left| \frac{\mu}{n} - p \right| < \epsilon \right\}, \tag{11}$$

where μ is the actual number of positive results.* Obviously, this prob-

---

* Here μ takes the values $m = 0, 1, \cdots, n$, with probability P_m; i.e.,
$$\mathbf{P}(\mu = m) = P_m.$$

ability may be written as the sum of those P_m for which m satisfies the inequality

$$\left| \frac{m}{n} - p \right| < \epsilon, \tag{12}$$

i.e., in the form

$$P = \sum_{m=m_1}^{m_2} P_m, \tag{13}$$

where m_1 is the smallest of the values of m satisfying inequality (12), and m_2 is the largest.

Formula (13) for fairly large n is hardly convenient for immediate calculation, a fact which explains the great importance of the asymptotic formula discovered by de Moivre for $p = \frac{1}{2}$ and by Laplace for general p. This formula allows us to find P_m very simply and to study its behavior for large n. The formula in question is

$$P_m \sim \frac{1}{\sqrt{2\pi np(1-p)}} e^{-(m-np)^2/2np(1-p)}. \tag{14}$$

If p is not too close to zero or one, it is sufficiently exact even for n of the order of 100. If we set

$$t = \frac{m - np}{\sqrt{np(1-p)}}, \tag{15}$$

then formula (14) becomes

$$P_m \sim \frac{1}{\sqrt{2\pi np(1-p)}} e^{-t^2/2}. \tag{16}$$

From (13) and (16) one may derive an approximate representation of the probability (11)

$$P \sim \frac{1}{\sqrt{2\pi}} \int_{-T}^{T} e^{-t^2/2}\, dt = F(T), \tag{17}$$

where

$$T = \epsilon \sqrt{\frac{n}{p(1-p)}}. \tag{18}$$

The difference between the left and right sides of (17) for fixed p, different from zero or one, approaches zero uniformly with respect to ϵ, as $n \to \infty$. For the function $F(T)$ detailed tables have been constructed. Here is a small excerpt from them

T	1	2	3	4
F	0.68269	0.95450	0.99730	0.99993

For $T \to \infty$ the values of the function $F(T)$ converge to one.

From formula (17) we derive an estimate of the probability

$$\mathbf{P} \left\{ \left| \frac{\mu}{n} - p \right| < 0.02 \right\}$$

for $n = 10,000$. Since

$$T = \frac{2}{\sqrt{p(1-p)}},$$

we have

$$P \approx F\left(\frac{2}{\sqrt{p(1-p)}} \right).$$

Since the function $F(T)$ is monotonic increasing with increasing T, it follows for an estimate of P from the following which is independent of p, we must take the smallest possible (for the various p) value of T. Such a smallest value occurs for $p = \frac{1}{2}$ and is equal to 4. Thus, approximately

$$P \geqslant F(4) = 0.99993. \tag{19}$$

In equality (19) no account is taken of the error arising from the approximate character of formula (17). By estimating the error involved here, we may show that in any case $P > 0.9999$.

In connection with this example of the application of formula (17), one should note that the estimates of the remainder term in formula (17) given in theoretical works on the theory of probability were for a long time unsatisfactory. Thus the applications of (17) and similar formulas to calculations based on small values of n, or with probabilities p very close to 0 or 1 (such probabilities are frequently of particular importance) were often based on experimental verification only of results of this kind for a restricted number of examples, and not on any valid estimates of the possible error. Also, it was shown by more detailed investigation that in many important practical cases the asymptotic formulas introduced previously require not only an estimate of the remainder term but also certain further refinements (without which the remainder term would be too large). In both directions the most complete results are due to S. N. Bernšteĭn.

Relations (11), (17), and (18) may be rewritten in the form

$$\mathbf{P} \left\{ \left| \frac{\mu}{n} - p \right| < t\sqrt{\frac{p(1-p)}{n}} \right\} \sim F(t). \tag{20}$$

For sufficiently large t the right side of formula (20), which does not contain n, is arbitrarily close to one, i.e., to the value of the probability which gives complete certainty. We see, in this way, that, *as a rule, the deviation of the frequency μ/n from the probability p is of order $1/\sqrt{n}$.* Such a proportionality between the exactness of a law of probability and the square root of the number of observations is typical for many other questions. Sometimes it is even said in popular simplifications that "the law of the square root of n" is the basic law of the theory of probability. Complete precision concerning this idea was attained through the introduction and systematic use by the great Russian mathematician P. L. Čebyšev of the concepts of "mathematical expectation" and "variance" for sums and arithmetic means of "random variables."

A *random variable* is the name given to a quantity which under given conditions S may take various values with specific probabilities. For us it is sufficient to consider random variables that may take on only a finite number of different values. To give the *probability distribution*, as it is called, of such a random variable ξ, it is sufficient to state its possible values $x_1, x_2, \cdots, x_n$ and the probabilities

$$P\{ = \mathbf{P}\{\xi = x_r\}.$$

The sum of these probabilities for all possible values of the variable ξ is always equal to one:

$$\sum_{r=1}^{s} P_r = 1.$$

The number investigated above of positive outcomes in n experiments may serve as an example of a random variable.

The *mathematical expectation* of the variable ξ is the expression

$$\mathbf{M}(\xi) = \sum_{r=1}^{s} P_r x_r ,$$

and the *variance* of ξ is the mathematical expectation of the square of the deviation $\xi - \mathbf{M}(\xi)$, i.e., the expression

$$\mathbf{D}(\xi) = \sum_{r=1}^{s} P_r [x_r - \mathbf{M}(\xi)]^2 .$$

The square root of the variance

$$\sigma\xi = \sqrt{\mathbf{D}(\xi)}$$

is called the *standard deviation* (of the variable from its mathematical expectation $\mathbf{M}(\xi)$).

At the basis of the simplest applications of variance and standard deviation lies the famous inequality of Čebyšev

$$\mathbf{P}\{|\ \xi - \mathbf{M}(\xi)| \leqslant t\ \sigma_\xi\} \geqslant 1 - \frac{1}{t^2}. \tag{21}$$

It shows that deviations of ξ from $\mathbf{M}(\xi)$ significantly greater than σ_ξ are rare.

As for the sum of random variables

$$\xi = \xi^{(1)} + \xi^{(2)} + \cdots + \xi^{(n)},$$

their mathematical expectations always satisfy the equation

$$\mathbf{M}(\xi) = \mathbf{M}(\xi^{(1)}) + \mathbf{M}(\xi^{(2)}) + \cdots + \mathbf{M}(\xi^{(n)}). \tag{22}$$

But the analogous equation for the variance

$$\mathbf{D}(\xi) = \mathbf{D}(\xi^{(1)}) + \mathbf{D}(\xi^{(2)}) + \cdots + \mathbf{D}(\xi^{(n)}) \tag{23}$$

is true only under certain restrictions. For the validity of equation (23) it is sufficient, for example, that the variables $\xi^{(i)}$ and $\xi^{(j)}$ with different indices not be "correlated" with one another, i.e., that for $i \neq j$ the equation*

$$\mathbf{M}\{[\xi^{(i)} - \mathbf{M}(\xi^{(i)})]\ [\xi^{(j)} - \mathbf{M}(\xi^{(j)})]\} = 0 \tag{24}$$

be satisfied.

In particular, equation (24) holds if the variables $\xi^{(i)}$ and $\xi^{(j)}$ are independent of each other.[†] Consequently, for mutually independent terms equation (23) always holds. For the arithmetic mean

$$\zeta = \frac{1}{n}(\xi^{(1)} + \xi^{(2)} + \cdots + \xi^{(n)})$$

it follows from (23) that

$$\mathbf{D}(\zeta) = \frac{1}{n^2}[\mathbf{D}(\xi^{(1)}) + \mathbf{D}(\xi^{(2)}) + \cdots + \mathbf{D}(\xi^{(n)})]. \tag{25}$$

---

* The *correlation coefficient* between the variables $\xi^{(i)}$ and $\xi^{(j)}$ is the expression

$$R = \frac{M\{[\xi^{(i)} - M(\xi^{(i)})][\xi^{(j)} - M(\xi^{(j)})]\}}{\sigma_{\xi^{(i)}}\sigma_{\xi^{(j)}}}.$$

If $\sigma_{\xi^{(i)}} > 0$ and $\sigma_{\xi^{(j)}} > 0$, then condition (24) is equivalent to saying that $R = 0$.

The correlation coefficient R characterizes the degree of dependence between random variables. $|R| \leqslant 1$ always, and $R = \pm 1$ only for a linear relationship

$$\eta = a\xi + b \qquad (a \neq 0).$$

For independent variables $R = 0$.

† The independence of two random variables ξ and η, which may assume, respectively, the values $x_1, x_2, \cdots, x_m$ and $y_1, y_2, \cdots, y_n$, is defined to mean that for any i and j the events $A_i = \{\xi = x_i\}$ and $B_j = \{\eta = y_j\}$ are independent in the sense of the definition given in §2.

We now assume that for each of these terms the variance does not exceed a certain constant

$$\mathbf{D}(\xi^{(i)}) \leqslant C^2.$$

Then from (25)

$$\mathbf{D}(\zeta) \leqslant \frac{C^2}{n},$$

and from Čebyšev's inequality for any t

$$\mathbf{P}\left\{| \zeta - \mathbf{M}(\zeta)| \leqslant \frac{tC}{\sqrt{n}}\right\} \geqslant 1 - \frac{1}{t^2}. \qquad (26)$$

Inequality (26) expresses what is called the law of large numbers, in the form established by Čebyšev: If the variables $\xi^{(i)}$ are mutually independent and have bounded variance, then for increasing n the arithmetic mean ζ will deviate more and more rarely from the mathematical expectation $\mathbf{M}(\zeta)$.

More precisely, the *sequence of variables*

$$\xi^{(1)}, \ \xi^{(2)}, \ \cdots, \ \xi^{(n)}, \ \cdots$$

is said to obey the law of large numbers if for the corresponding arithmetic means ζ and for any constant $\epsilon > 0$

$$\mathbf{P}\{| \zeta - \mathbf{M}(\zeta)| \leqslant \epsilon\} \rightarrow 1 \qquad (27)$$

for $n \rightarrow \infty$.

In order to pass from inequality (26) to the limiting relation (27) it is sufficient to put

$$t = \epsilon \frac{\sqrt{n}}{C}.$$

A large number of investigations of A. A. Markov, S. N. Bernšteĭn, A. Ja. Hinčin, and others were devoted to the question of widening as far as possible the conditions under which the limit relation (27) is valid, i.e., the conditions for the validity of the law of large numbers. These investigations are of basic theoretical significance, but still more important is an exact study of the probability distribution for the variable $\zeta - \mathbf{M}(\zeta)$.

One of the greatest services rendered by the classical Russian school of mathematicians to the theory of probability is the establishment of the fact that under very wide conditions the equation

$$\mathbf{P}\{t_1\sigma_\zeta < \zeta - \mathbf{M}(\zeta) < t_2\sigma_\zeta\} \sim \frac{1}{\sqrt{2\pi}} \int_{t_1}^{t_2} e^{-t^2/2} \, dt \qquad (28)$$

is asymptotically valid (i.e., with greater and greater exactness as n increases beyond all bounds).

Čebyšev gave an almost complete proof of this formula for the case of independent and bounded terms. Markov closed a gap in Čebyšev's argument and widened the conditions of applicability of formula (28). Still more general conditions were given by Ljapunov. The applicability of formula (28) to the sum of mutually dependent terms was studied with particular completeness by S. N. Bernšteĭn.

Formula (28) embraces such a large number of particular cases that it has long been called the central limit theorem in the theory of probability. Even though it has been shown lately to be included in a series of more general laws its value can scarcely be overrated even at the present time.

If the terms are independent and their variances are all the same, and are equal to

$$\mathbf{D}(\xi^{(i)}) = \sigma^2,$$

then it is convenient, using relation (25), to put formula (28) into the form

$$\mathbf{P}\left\{\frac{t_1\sigma}{\sqrt{n}} < \zeta - \mathbf{M}(\zeta) < \frac{t_2\sigma}{\sqrt{n}}\right\} \sim \frac{1}{\sqrt{2\pi}} \int_{t_1}^{t_2} e^{-t^2/2}\,dt. \tag{29}$$

Let us show that relation (29) contains the solution of the problem, considered earlier, of evaluating the deviation of the frequency μ/n from the probability p. For this we introduce the random variables $\xi^{(i)}$, defined as follows:

$$\xi^{(i)} = \begin{cases} 0, & \text{if the } i\text{th test has a negative outcome,} \\ 1, & \text{if the } i\text{th test has a positive outcome.} \end{cases}$$

It is easy to verify that then

$$\mu = \xi^{(1)} + \xi^{(2)} + \cdots + \xi^{(n)}, \qquad \frac{\mu}{n} = \zeta,$$

$$\mathbf{M}(\xi^{(i)}) = p, \quad \mathbf{D}(\xi^{(1)}) = p(1-p), \quad \mathbf{M}(\zeta) = p,$$

and formula (29) gives

$$\mathbf{P}\left\{t_1\sqrt{\frac{p(1-p)}{n}} < \frac{\mu}{n} - p < t_2\sqrt{\frac{p(1-p)}{n}}\right\} \sim \frac{1}{\sqrt{2\pi}} \int_{t_1}^{t_2} e^{-t^2/2}\,dt,$$

which for $t_1 = -t$, $t_2 = t$ leads again to formula (20).

§4. Further Remarks on the Basic Concepts of the Theory of Probability

In speaking of random events, which have the property that their frequencies tend to become stable, i.e., in a long sequence of experiments repeated under fixed conditions, their frequencies are grouped around some *standard level*, called their probability $\mathbf{P}(A/S)$, we were guilty, in §1, of a certain vagueness in our formulations, in two respects. In the first place, we did not indicate how long the sequence of experiments n_r must be in order to exhibit beyond all doubt the existence of the supposed stability; in other words, we did not say what deviations of the frequencies μ_r/n_r from one another or from their standard level p were allowable for sequences of trials $n_1, n_2, \cdots, n_s$ of given length. This inexactness in the first stage of formulating the concepts of a new science is unavoidable. It is no greater than the well-known vagueness surrounding the simplest geometric concepts of point and straight line and their *physical* meaning. This aspect of the matter was made clear in §3.

More fundamental, however, is the second lack of clearness concealed in our formulations; it concerns the manner of forming the sequences of trials in which we are to examine the stability of the frequency of occurrence of the event A.

As stated earlier, we are led to statistical and probabilistic methods of investigation in those cases in which an exact specific prediction of the course of events is impossible. But if we wish to create in some artificial way a sequence of events that will be, as far as possible, purely random, then we must take special care that there shall be no methods available for determining in advance those cases in which A is likely to occur with more than normal frequency.

Such precautions are taken, for example, in the organization of government lotteries. If in a given lottery there are to be M winning tickets in a drawing of N tickets, then the probability of winning for an individual ticket is equal to $p = M/N$. This means that in whatever manner we select, in advance of the drawing, a sufficiently large set of n tickets, we can be practically certain that the ratio μ/n of the number μ of winning tickets in the chosen set to the whole number n of tickets in this set will be close to p. For example, people who prefer tickets labeled with an even number will not have any systematic advantage over those who prefer tickets labeled with odd numbers, and in exactly the same way there will be no advantage in proceeding on the principle, say, that it is always better to buy tickets with numbers having exactly three prime factors, or tickets whose numbers are close to those that were winners in the preceding lottery, etc.

Similarly, when we are firing a well-constructed gun of a given type, with a well-trained crew and with shells that have been subjected to a standard quality control, the deviation from the mean position of the points of impact of the shells will be less than the previously determined probable deviation B in approximately *half* the cases. This fraction remains the same in a series of successive trials, and also in case we count separately the number of deviations that are less than B for even-numbered shots (in the order of firing) or for odd-numbered. But it is completely possible that if we were to make a selection of particularly homogeneous shells (with respect to weight, etc.), the scattering would be considerably decreased, i.e., we would have a sequence of firings for which the fraction of the deviations which are greater than the standard B would be considerably less than a half.

Thus, to say that an event A is "random" or "stochastic" and to assign it a definite probability

$$p = \mathbf{P}(A/S)$$

is possible only when we have already determined the class of allowable ways of setting up the series of experiments. The nature of this class will be assumed to be included in the conditions S.

For *given* conditions S the properties of the event A of being random and of having the probability $p = \mathbf{P}(A/S)$ express the objective character of the connection between the condition S and the event A. In other words, there exists no event which is absolutely random; an event is random or is predetermined depending on the connection in which it is considered, but under specific conditions an event may be random in a completely non-subjective sense, i.e., independently of the state of knowledge of any observer. If we imagine an observer who can master all the detailed distinctive properties and particular circumstances of the flight of shells, and can thus predict for each one of them the deviation from the mean trajectory, his presence would still not prevent the shells from scattering in accordance with the laws of the theory of probability, provided, of course, that the shooting was done in the usual manner, and not according to instructions from our imaginary observer.

In this connection we note that the formation of a series of the kind discussed earlier, in which there is a tendency for the frequencies to become constant in the sense of being grouped around a normal value, namely the probability, proceeds in the actual world in a manner completely independent of our intervention. For example, it is precisely by virtue of the random character of the motion of the molecules in a gas that the number of molecules which, even in a very small interval of time, strike an arbitrarily preassigned small section of the wall of the container (or of the surface of bodies situated in the gas) proves to be proportional with very great exactness to the area of this small piece of the wall and to the length of the interval of time. Deviations from this proportionality in cases where the number of hits is not large also follow the laws of the theory of probability and produce phenomena of the type of Brownian motion, of which more will be said later.

We turn now to the objective meaning of the concept of independence. We recall that the conditional probability of an event A under the condition B is defined by the formula

$$\mathbf{P}(A/B) = \frac{\mathbf{P}(AB)}{\mathbf{P}(B)} . \tag{30}$$

We also recall that events A and B are called independent if, as in (4),

$$\mathbf{P}(AB) = \mathbf{P}(A)\,\mathbf{P}(B).$$

From the independence of the events A and B and the fact that $\mathrm{P}(B) > 0$ it follows that

$$\mathbf{P}(A/B) = \mathbf{P}(A).$$

All the theorems of the mathematical theory of probability that deal with independent events apply to any events satisfying the condition (4), or to its generalization to the case of the mutual independence of several events. These theorems will be of little interest, however, if this definition bears no relation to the properties of objective events which are independent in the causal sense.

It is known, for example, that the probability of giving birth to a boy is, with sufficient stability, $\mathbf{P}(A) = 22/43$. If B denotes the condition that the birth occur on a day of the conjunction of Jupiter with Mars, then under the assumption that the position of the planets does not influence the fate of individuals, the conditional probability $\mathbf{P}(A/B)$ has the same value: $\mathbf{P}(A/B) = 22/43$; i.e., the actual calculation of the frequency of births of boys under such special astrological conditions would give just the same frequency $22/43$. Although such a calculation has probably never been carried out on a sufficiently large scale, still there is no reason to doubt what the result would be.

We give this example, from a somewhat outmoded subject, in order to show that the development of human knowledge consists not only in establishing valid relations among phenomena, but also in refuting imagined relations, i.e., in establishing in relevant cases the thesis of the independence of any two sets of events. This unmasking of the meaningless attempts of the astrologers to connect two sets of events that are not in fact connected is one of the classic examples.

Naturally, in dealing with the concept of independence, we must not proceed in too absolute a fashion. For example, from the law of universal graviation, it is an undoubted fact that the motions of the moons of Jupiter have a certain effect, say, on the flight of an artillery shell. But it is also obvious that in practice this influence may be ignored. From the philosophical point of view, we may perhaps, in a given concrete situation, speak more properly not of the independence but of the insignificance of the dependence of certain events. However that may be, the independence of events in the cited concrete and relative sense of this term in no way contradicts the principle of the universal interconnection of all phenomena; it serves only as a necessary supplement to this principle.

The computation of probabilities from formulas derived by assuming the independence of certain events is still of practical interest in cases where the events were originally independent but became interdependent as a result of the events themselves. For example, one may compute probabilities for the collision of particles of cosmic radiation with particles of the medium penetrated by the radiation, on the assumption that the motion of the particles of the medium, up to the time of the appearance near them of a rapidly moving particle of cosmic radiation, proceeds independently of the motion of the cosmic particle. One may compute the probability that a hostile bullet will strike the blade of a rotating propeller, on the assumption that the position of the blade with respect to the axis of rotation does not depend on the trajectory of the bullet, a supposition that will of course be wrong with respect to the bullets of the aviator himself, since they are fired between the blades of the rotating propeller. The number of such examples may be extended without limit.

It may even be said that wherever probabilistic laws turn up in any clear-cut way we are dealing with the influence of a large number of factors that, if not entirely independent of one another, are interconnected only in some weak sense.

This does not at all mean that we should uncritically introduce assumptions of independence. On the contrary, it leads us, in the first place, to be particularly careful in the choice of criteria for testing hypotheses of independence, and second, to be very careful in investigating the borderline cases where dependence between the facts must be assumed but is of such a kind as to introduce complications into the relevant laws of probability. We noted earlier that the classical Russian school of the theory of probability has carried out far-reaching investigations in this direction.

To bring to an end our discussion of the concept of independence, we note that, just as with the definition of independence of two events given in formula (4), the formal definition of the independence of several random variables is considerably broader than the concept of independence in the practical world, i.e., the absence of causal connection.

Let us assume, for example, that the point ξ falls in the interval [0, 1] in such a manner for

$$0 \leqslant a \leqslant b \leqslant 1$$

the probability that it belongs to the segment $[a, b]$ is equal to the length of this segment $b - a$. It is easy to prove that in the expansion

$$\xi = \frac{\alpha_1}{10} + \frac{\alpha_2}{100} + \frac{\alpha_3}{1000} + \cdots$$

of the abscissa of the point ξ in a decimal fraction, the digits α_k will be mutually independent, although they are interconnected by the way they are produced.* (From this fact follow many theoretical results, some of which are of practical interest.)

Such flexibility in the formal definition of independence should not be considered as a blemish. On the contrary it merely extends the domain of applicability of theorems established for one or another assumption of independence. These theorems are equally applicable in cases where the independence is postulated on the basis of practical considerations and in cases where the independence is proved by computation proceeding from previous assumptions concerning the probability distributions of the events and the random variables under study.

In general, investigation of the formal structure of the mathematical apparatus of the theory of probability has led to interesting results. It turns out that this apparatus occupies a very definite and clear-cut place in the classification, which nowadays is gradually becoming clear in outline, of the basic objects of study in contemporary mathematics.

We have already spoken of the concepts of intersection AB and union $A \cup B$ of the events A and B. We recall that events are called mutually exclusive if their intersection is empty, i.e., if $AB = N$, where N is the symbol for an impossible event.

The basic axiom of the elementary theory of probability consists of the requirement (cf. §2) that under the condition $AB = N$ we have the equation

$$\mathbf{P}(A \cup B) = \mathbf{P}(A) + \mathbf{P}(B).$$

The basic concepts of the theory of probability, namely random events and their probabilities, are completely analogous in their properties to plane figures and their areas. It is sufficient to understand by AB the intersection (common part) of two figures, by $A \cup B$ their union, by N the conventional "empty" figure, and by $\mathbf{P}(A)$ the area of the figure A, whereupon the analogy is complete.

---

* This is also valid, for any n, for the digits α_k in the expansion of the number ξ in the fraction

$$\xi = \frac{\alpha_1}{n} + \frac{\alpha_2}{n^2} + \frac{\alpha^3}{n^3} + \cdots.$$

The same remarks apply to the volumes of three-dimensional figures.

The most general theory of entities of such a type, which contains as special cases the theory of volume and area, is now usually called *measure theory*, discussed in Chapter XV in connection with the theory of functions of a real variable.

It remains only to note that in the theory of probability, in comparison with the general theory of measure or in particular with the theory of area and volume, there is a certain special feature: A probability is never greater than one. This maximal probability holds for a necessary event U.

$$\mathbf{P}(U) = 1.$$

The analogy is by no means superficial. It turns out that the whole mathematical theory of probability from the formal point of view may be constructed as a theory of measure, making the special assumption that the measure of "the entire space" U is equal to one.*

Such an approach to the matter has produced complete clarity in the formal construction of the mathematical theory of probability and has also led to concrete progress not only in this theory itself but in other theories closely related to it in their formal structure. In the theory of probability success has been achieved by refined methods developed in the metric theory of functions of a real variable and at the same time probabilistic methods have proved to be applicable to questions in neighboring domains of mathematics not "by analogy," but by a formal and strict transfer of them to the new domain. Wherever we can show that the axioms of the theory of probability are satisfied, the results of these axioms are applicable, even though the given domain has nothing to do with randomness in the actual world.

The existence of an axiomatized theory of probability preserves us from the temptation "to define" probability by methods that claim to construct a strict, purely formal mathematical theory on the basis of features of probability that are immediately suggested by the natural sciences. Such definitions roughly correspond to the "definition" in geometry of a point as the result of trimming down a physical body an infinite number of times, each time decreasing its diameter by a factor of 2.

With definitions of this sort, probability is taken to be the limit of the frequency as the number of experiments increases beyond all bounds. The very assumption that the experiments are probabilistic, i.e., that the frequencies tend to cluster around a constant value, will remain valid (and

---

* Nevertheless, because of the nature of its problems, the theory of probability remains an independent mathematical discipline; its basic results (presented in detail in §3) appear artificial and unnecessary from the point of view of pure measure theory.

the same is true for the "randomness" of any particular event) only if certain conditions are kept fixed for an unlimited time and with absolute exactness. Thus the exact passage to the limit

$$\frac{\mu}{n} \to p$$

cannot have any objective meaning. Formulation of the principle of stability of the frequencies in such a limit process demands that we define the allowable methods of setting up an infinite sequence of experiments, and this can only be done by a mathematical fiction. This whole conglomeration of concepts might deserve serious consideration if the final result were a theory of such distinctive nature that no other means existed of putting it on a rigorous basis. But, as was stated earlier, the mathematical theory of probability may be based on the theory of measure, in its present-day form, by simply adding the condition

$$\mathbf{P}(U) = 1.$$

In general, for any practical analysis of the concept of probability, there is no need to refer to its formal definition. It is obvious that concerning the purely formal side of probability, we can only say the following: The probability $\mathbf{P}(A/S)$ is a number around which, under conditions S determining the allowable manner of setting up the experiments, the frequencies have a tendency to be grouped, and that this tendency will occur with greater and greater exactness as the experiments, always conducted in such a way as to preserve the original conditions, become more numerous, and finally that the tendency will reach a satisfactory degree of reliability and exactness during the course of a practicable number of experiments.

In fact, the problem of importance, in practice, is not to give a formally precise definition of randomness but to clarify as widely as possible the conditions under which randomness of the cited type will occur. One must clearly understand that, in reality, hypotheses concerning the probabilistic character of any phenomenon are very rarely based on immediate statistical verification. Only in the first stage of the penetration of probabilistic methods into a new domain of science has the work consisted of purely empirical observation of the constancy of frequencies. From §3, we see that statistical verification of the constancy of frequencies with an exactness of ϵ requires a series of experiments, each consisting of $n = 1/\epsilon^2$ trials. For example, in order to establish that in a given concrete problem the probability is defined with an exactness of 0.0001, it is necessary to carry out a series of experiments containing approximately 100,000,000 trials in each.

The hypothesis of probabilistic randomness is much more often introduced from considerations of symmetry or of successive series of events, with subsequent verification of the hypothesis in some indirect way. For example, since the number of molecules in a finite volume of gas is of the order of 10^{20} or more, the number $\sqrt{n}$, corresponding to the probabilistic deductions made in the kinetic theory of gases, is very large, so that many of these deductions are verified with great exactness. Thus, the pressures on the opposite sides of a plate suspended in still air, even if the plate is of microscopic dimensions, turn out exactly the same, although an excess of pressure on one side of the order of a thousandth of one per cent can be detected in a properly arranged experiment.

§5. Deterministic and Random Processes

The principle of causal relation among phenomena finds its simplest mathematical expression in the study of physical processes by means of differential equations as demonstrated in a series of examples in §1 of Chapter V.

Let the state of the system under study be defined at the instant of time t by n parameters

$$x_1, x_2, \cdots, x_n.$$

The rates of change of these parameters are expressed by their derivatives with respect to time

$$\dot{x}_k = \frac{dx_k}{dt}.$$

If it is assumed that these rates are functions of the values of the parameters, then we get a system of differential equations

$$\dot{x}_1 = f_1(x_1, x_2, \cdots, x_n),$$
$$\dot{x}_2 = f_2(x_1, x_2, \cdots, x_n),$$
$$\cdots\cdots\cdots\cdots\cdots\cdots\cdots\cdots\cdots$$
$$\dot{x}_n = f_n(x_1, x_2, \cdots, x_n).$$

The greater part of the laws of nature discovered at the time of the birth of mathematical physics, beginning with Galileo's law for falling bodies, are expressed in just such a manner. Galileo could not express his discovery in this standard form, since in his time the corresponding mathematical concepts had not yet been developed, and this was first done by Newton.

In mechanics and in any other fields of physics, it is customary to express these laws by differential equations of the second order. But no new principles are involved here; for if we denote the rates $\dot{x}_k$ by the new symbols

$$v_k = \dot{x}_k,$$

we get for the second derivative of the quantities x_k the expressions

$$\frac{d^2 x_k}{dt^2} = \dot{v}_k,$$

and the equations of the second order for the n quantities $x_1, x_2, \cdots, x_n$ become equations of the first order for the $2n$ quantities $x_1, \cdots, x_n$, $v_1, v_2, \cdots, v_n$.

As an example, let us consider the fall of a heavy body in the atmosphere of the earth. If we consider only short distances above the surface, we may assume that the resistance of the medium depends only on the velocity and not on the height. The state of the system under study is characterized by two parameters: the distance z of the body from the surface of the earth, and its velocity v. The change of these two quantities with time is defined by the two differential equations

$$\begin{aligned} \dot{z} &= -v, \\ \dot{v} &= g - f(v), \end{aligned} \tag{31}$$

where g is the acceleration of gravity and $f(v)$ is some "law of resistance" for the given body.

If the velocity is not great and the body is sufficiently massive, say a stone of moderate size falling from a height of several meters, the resistance of the air may be neglected and equations (31) are transformed into the equations

$$\begin{aligned} \dot{z} &= -v, \\ \dot{v} &= g. \end{aligned} \tag{32}$$

If it is assumed that at the initial instant of time t_0 the quantities z and v have values z_0 and v_0, then it is easy to solve equations (32) to obtain the formula

$$z = z_0 - v(t - t_0) - g\left(\frac{t - t_0}{2}\right)^2,$$

which describes the whole process of falling. For example, if $t_0 = 0$, $v_0 = 0$ we get

$$z = z_0 - \frac{gt^2}{2},$$

found by Galileo.

In the general case, the integration of equations (31) is more difficult, although the basic result, with very general restrictions on the function $f(v)$, remains the same: Given the values z_0 and v_0 at the initial instant t_0, the values of z and v for all further instants t are computed uniquely, up to the time that the falling body hits the surface of the earth. Theoretically, this last restriction may also be removed, if we assume that the fall is extended to negative values of z. For problems set up in this manner, the following may be established: If the function $f(v)$ is monotone for increasing v and tends to infinity for $v \to \infty$, then if the fall continues unchecked, i.e., for unbounded growth of the variable t, the velocity v tends to a constant limiting value c, which is the solution of the equation

$$g = f(c).$$

From the intuitive point of view, this result of the mathematical analysis of the problem is quite understandable: The velocity of fall increases up to the time that the accelerative force of gravity is balanced by the resistance of the air. For a jump with an open parachute, the stationary velocity v of about five meters per second is attained rather quickly.* For a long jump with unopened parachute the resistance of the air is less, so that the stationary velocity is greater and is attained only after the parachutist has fallen a very long way.

For the falling of light bodies like a feather tossed into the air or a bit of fluff, the initial period of acceleration is very short, often quite unobservable. The stationary rate of falling is established very quickly, and to a standard approximation we may consider that throughout the fall $v = c$. In this case we have only one differential equation

$$\dot{z} = -c,$$

which is integrated very simply:

$$z = z_0 - c(t - t_0).$$

This is how a bit of fluff will fall in perfectly still air.

This deterministic conception is treated in a completely general way in the contemporary theory of dynamical systems, to which is dedicated a series of important works by Soviet mathematicians, N. N. Bogoljubov, V. V. Stepanov, and many others. This general theory also includes as special cases the mathematical formulation of physical phenomena in which the state of a system is not defined by a finite number of parameters

---

* This statement is to be taken in the sense that in practice v soon gets quite close to c.

as in the earlier case, but by one or more functions, for example, in the mechanics of continuous media. In such cases the elementary laws for change of state in "infinitely small" intervals of time are given not by ordinary but by partial differential equations or by some other means. But the features common to all deterministic mathematical formulations of actual processes are: first, that the state of the system under study is considered to be completely defined by some mathematical entity ω (a set of n real numbers, one or more functions, and so forth); and second, that the later values for instants of time $t > t_0$ are uniquely determined by the value ω_0 at the initial instant t_0

$$\omega = F(t_0, \omega_0, t).$$

For phenomena described by differential equations the process of finding the function ϕ consists, as we have seen, in integrating these differential equations with the initial conditions $\omega = \omega_0$ for $t = t_0$.

The proponents of mechanistic materialism assumed that such a formulation is an exact and direct expression of the deterministic character of the actual phenomena, of the physical principle of causation. According to Laplace, the state of the world at a given instant is defined by an infinite number of parameters, subject to an infinite number of differential equations. If some "universal mind" could write down all these equations and integrate them, it could then predict with complete exactness, according to Laplace, the entire evolution of the world in the infinite future.

But in fact this quantitative mathematical infinity is extremely coarse in comparison with the qualitatively inexhaustible character of the real world. Neither the introduction of an infinite number of parameters nor the description of the state of continuous media by functions of a point in space is adequate to represent the infinite complexity of actual events.

As was emphasized in §3 of Chapter V, the study of actual events does not always proceed in the direction of increasing the number of parameters introduced into the problem; in general, it is far from expedient to complicate the ω which describes the separate "states of the system" in our mathematical scheme. The art of the investigation consists rather in finding a very simple space Ω (i.e., a set of values of ω or in other words, of different possible states of the system),* such that if we replace the actual process by varying the point ω in a determinate way over this space, we can include all the *essential* aspects of the actual process.

---

* In the example given earlier of a falling body, the phase space is the system of pairs of numbers (z, v), i.e., a plane. For phase spaces in general, see Chapters XVII and XVIII.

But if from an actual process we abstract its essential aspects, we are left with a certain residue which we must consider to be random. The neglected random factors always exercise a certain influence on the course of the process. Very few of the phenomena that admit mathematical investigation fail, when theory is compared with observation, to show the influence of ignored random factors. This is more or less the state of affairs in the theory of planetary motion under the force of gravity: The distance between planets is so large in comparison with their size that the idealized representation of them as material points is almost perfectly satisfactory; the space in which they are moving is filled with such dispersed material that its resistance to their motion is vanishingly small; the masses of the planets are so large that the pressure of light plays almost no role in their motions. These exceptional circumstances explain the fact that the mathematical solution for the motion of a system of n material points, whose "states" are described by $6n$ parameters* which take into account only the force of gravity, agrees so astonishingly well with observation of the motion of the planets.

Somewhat similar to the case of planetary motion is the flight of an artillery shell under gravity and resistance of the air. This is also one of the classical regions in which mathematical methods of investigation were comparatively easy and quickly produced great success. But here the role of the perturbing random factors is significantly larger and the scattering of the shells, i.e., their deviation from the theoretical trajectory reaches tens of meters, or for long ranges even hundreds of meters. These deviations are caused partly by random deviations in the initial direction and velocity, partly by random deviations in the mass and the coefficient of resistance of the shell, and partly by gusts and other irregularities in the wind and the other random factors governing the extraordinarily complicated and changing conditions in the actual atmosphere of the earth.

The scattering of shells is studied in detail by the methods of the theory of probability, and the results of this study are essential for the practice of gunnery.

But what does it mean, properly speaking, to study random events? It would seem that, when the random "residue" for a given formulation of a phenomenon proves to be so large that it can not be neglected, then the only possible way to proceed is to describe the phenomenon more accurately by introducing new parameters and to make a more detailed study by the same method as before.

But in many cases such a procedure is not realizable in practice. For example, in studying the fall of a material body in the atmosphere, with

---

* The three coordinates and the three components of the velocity of each point.

account taken of an irregular and gusty (or, as one usually says, turbulent) wind flow, we would be required to introduce, in place of the two parameters z and v, an altogether unwieldy mathematical apparatus to describe this structure completely.

But in fact this complicated procedure is necessary only in those cases where for some reason we must determine the influence of these residual "random" factors in all detail and separately for each individual factor. Fortunately, our practical requirements are usually quite different; we need only estimate the total effect exerted by the random factors for a long interval of time or for a large number of repetitions of the process under study.

As an example, let us consider the shifting of sand in the bed of a river, or in a hydroelectric construction. Usually this shifting occurs in such a way that the greater part of the sand remains undisturbed, while only now and then a particularly strong turbulence near the bottom picks up individual grains and carries them to a considerable distance, where they are suddenly deposited in a new position. The purely theoretical motion of each grain may be computed individually by the laws of hydrodynamics, but for this it would be necessary to determine the initial state of the bottom and of the flow in every detail and to compute the flow step by step, noting those instants when the pressure on any particular grain of sand becomes sufficient to set it in motion, and tracing this motion until it suddenly comes to an end. The absurdity of setting up such a problem for actual scientific study is obvious. Nevertheless the average laws or, as they are usually called, the statistical laws of shifting of sand over river bottoms are completely amenable to investigation.

Examples of this sort, where the effect of a large number of random factors leads to a completely clear-cut statistical law, could easily be multiplied. One of the best known and at the same time most fascinating of these, in view of the breadth of its applications, is the kinetic theory of gases, which shows how the joint influence of random collisions of molecules gives rise to exact laws governing the pressure of a gas on the wall, the diffusion of one gas through another, and so forth.

§6. Random Processes of Markov Type

To A. A. Markov is due the construction of a probabilistic scheme which is an immediate generalization of the deterministic scheme of §5 described by the equation

$$\omega = F(t_0, \omega_0, t).$$

It is true that Markov considered only the case where the phase space of the system consists of a finite number of states $\Omega = (\omega_1, \omega_2, \cdots, \omega_n)$ and studied the change of state of the system only for changes of time t in discrete steps. But in this extremely schematic model he succeeded in establishing a series of fundamental laws.

Instead of a function F, uniquely defining the state ω at time $t > t_0$ corresponding to the state ω_0 at time t_0, Markov introduced the probabilities

$$(t_0, \omega_i; t, \omega_j)$$

of obtaining the state ω_j at time t under the condition that at time t_0 we had the state ω_i. These probabilities are connected for any three instants of time

$$t_0 < t_1 < t_2$$

by a relation, introduced by Markov, which may be called the basic equation for a Markov process

$$\mathbf{P}(t_0, \omega_i; t_2, \omega_j) = \sum_{k=1}^{n} \mathbf{P}(t_0, \omega_i; t_1, \omega_k)\mathbf{P}(t_1, \omega_k; t_2, \omega_j). \quad (33)$$

When the phase space is a continuous manifold, the most typical case is that a *probability density* $p(t_0, \omega_0; t, \omega)$ exists for passing from the state ω_0 to the state ω in the interval of time (t_0, t). In this case the probability of passing from the state ω_0 to any of the states ω belonging to a domain G in the phase space Ω is written in the form

$$\mathbf{P}(t_0, \omega_0; t, G) = \int_G p(t_0, \omega_0; t, \omega) \, d\omega, \quad (34)$$

where $d\omega$ is an element of volume in the phase space.* For the probability density $p(t_0, \omega_0; t, \omega)$, the basic equation (33) takes the form

$$p(t_0, \omega_0; t_2, \omega_2) = \int_\Omega p(t_0, \omega_0; t_1, \omega) \, p(t_1, \omega; t_2, \omega_2) \, d\omega. \quad (35)$$

Equation (35) is usually difficult to solve, but under known restrictions we may deduce from it certain partial differential equations that are easy to investigate. Some of these equations were derived from nonrigorous physical considerations by the physicists Fokker and Planck. In its complete form this theory of so-called stochastic differential equations

---

* Properly speaking, equation (34) serves to define the probability density. The quantity $p \, d\omega$ is equal (up to an infinitesimal of higher order) to the probability of passing in the time from t_0 to t from the state ω_0 to the element of volume $d\omega$.

was constructed by Soviet authors, S. N. Bernšteĭn, A. N. Kolmogorov, I. G. Petrovskiĭ, A. Ja. Hinčin, and others.

We will not give these equations here.

The method of stochastic differential equations allows us, for example, to solve without difficulty the problem of the motion in still air of a very small body, for which the mean velocity c of its fall is significantly less than the velocity of the "Brownian motion" arising from the fact, because of the smallness of the particle, its collisions with the molecules of the air are not in perfect balance on its various sides.

Let c be the mean velocity of fall, and D be the so-called coefficient of diffusion. If we assume that a particle does not remain on the surface of the earth ($z = 0$) but is "reflected", i.e., under the influence of the Brownian forces it is again sent up into the atmosphere, and if we also assume that at the instant t_0 the particle is at height z_0, then the probability density $p(t_0, z_0; t, z)$ of its being at height z at the instant t is expressed by the formula

$$
p(t_0, z_0 ; t, z) = \frac{1}{2\sqrt{\pi D(t - t_0)}}
$$

$$
\times \left[e^{-\frac{(z-z_0)^2}{4D(t-t_0)}} + e^{-\frac{(z+z_0)^2}{4D(t-t_0)}} \right] e^{-\frac{(cz-z_0)}{2D} - \frac{c^2(t-t_0)}{4D}}
$$

$$
+ \frac{c}{D\sqrt{5}} e^{-cz/D} \int_{\frac{z+z_0-c(t-t_0)}{2\sqrt{D(t-t_0)}}}^{\infty} e^{-z^2} dz.
$$

In figure 4 we illustrate how the curves $p(t_0, z_0; t, z)$ may change for a sequence of instants t.

We see that in the mean the height of the particle increases, and its position is more and more indefinite, more "random." The most interesting aspect of the situation is that for any t_0 and z_0 and for $t \to \infty$

$$
p(t_0, z_0 ; t, z) \to \frac{c}{D} e^{-cz/D} ; \tag{36}
$$

i.e., there exists a limit distribution for the height of the particle, and the mathematical expectation for this height with increasing t tends to a positive limit

$$
z^* = \frac{c}{D} \int_0^\infty z e^{-cz/D} dz = \frac{D}{c}. \tag{37}
$$

So in spite of the fact that as long as our particle is above the surface of the earth, it will always tend to fall because of the force of gravity, nevertheless, as this process (wandering in the atmosphere) continues, the particle will be found on the average at a definite positive height. If we take the initial z_0 smaller than z^*, it will turn out that in a sufficiently great interval of time the mean position of the particle will be higher than its initial position, as is shown in figure 5, where $z_0 = 0$.

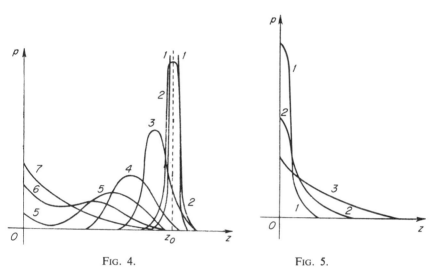

FIG. 4.   FIG. 5.

For individual particles the mean values z^* under discussion here are only mathematical expectations, but from the law of large numbers it follows that for a large number of particles they will actually be realized: The density of the distribution in height of such particles will follow from the indicated laws, and, in particular, after a sufficient interval of time this density will become stable in accordance with formula (36).

What has been said so far is immediately applicable only to gases, to smoke and the like, which occur in the air in small concentrations, since the quantities c and D were assumed to be defined by a preassigned state of the atmosphere. However with certain complications, the theory is applicable to the mutual diffusion of the gases that compose the atmosphere, and to the distribution in height of their densities arising from this mutual diffusion.

The quotient c/D increases with the size of the particles, so that the character of the motion changes from diffusion to regular fall in accordance with the laws considered in §5. The theory allows us to trace all transitions between purely diffusive motion and such laws of fall.

The problem of motion of particles suspended in a turbulent atmosphere is more difficult, but in principle it may be handled by similar probabilistic methods.

Suggested Reading

H. Cramer, *The elements of probability theory and some of its applications*, Wiley, New York, 1955.

W. K. Feller, *An introduction to probability theory and its applications*, 2nd ed., Wiley, New York, 1957.

B. V. Gnedenko and A. Ya. Khinchin, *An elementary introduction to the theory of probability*, W. H. Freeman and Co., San Francisco, Calif., 1961.

M. Kac, *Statistical independence in probability, analysis and number theory*, Wiley, New York, 1959.

J. G. Kemeny and J. L. Snell, *Finite Markov chains*, Van Nostrand, New York, 1960.

A. N. Kolmogorov, *Foundations of the theory of probability*, Chelsea, New York, 1950.

M. Loève, *Probability theory: foundations, random sequences*, Van Nostrand, New York, 1955.

E. Parzen, *Modern probability theory and its applications*, Wiley, New York, 1960.

W. A. Whitworth, *Choice and chance*, Stechert, New York, 1942.

CHAPTER XII

APPROXIMATIONS OF FUNCTIONS

§1. Introduction

In practical life we are constantly faced with the problem of approximating certain numbers by means of others. For example, our measurements of various concrete magnitudes, length, area, temperature, and so forth, lead us to numbers that are only approximations. In practice we use only rational numbers, i.e., numbers of the form p/q, where p and $q(q \neq 0)$ are integers. But, in addition to the rational numbers, the irrational numbers also exist, and although we do not use them in measuring, still our theoretical arguments often lead to them. We know, for example, that the length of the circumference of a circle of radius $r = \frac{1}{2}$ is equal to the irrational number π, and the length of the hypotenuse of a right triangle with unit sides is equal to $\sqrt{2}$. In actual computations with irrational numbers, one first of all approximates them by rational numbers with a required degree of exactness, usually by means of a terminating decimal fraction.

The same situation also occurs for functions. The quantitative laws of nature are expressed in mathematics by means of functions, not with absolute exactness, but approximately, with various degrees of precision. Further, in a vast number of cases we find it necessary, even for functions defined by completely mathematical rules, to approximate them by other functions with specified exactness so as to be able to compute them in practice.

However, these remarks do not refer to computations only. The problem of defining a function by means of other functions has great theoretical importance. Let us illustrate in a few words. The development of mathematical analysis has led to the discovery and study of very important classes of approximating functions that under known conditions have proved to be the natural means of approximating other more or less arbitrary functions. These classes turned out to be, above all, the algebraic and trigonometric polynomials, and also their various generalizations. It was shown that from the properties of the function to be approximated we may estimate, under certain conditions, the character of its deviation from a specific sequence of functions approximating it. Conversely, if we know how it deviates from its approximation by a sequence of functions, we can establish certain properties of the function. In this direction a theory of functions has been constructed that is based on their approximate representation by various classes of approximating functions. There is a similar theory in the theory of numbers. In it the properties of irrational numbers are studied on the basis of their approximations by rational numbers.

In Chapter II the reader has already met one very important method of approximation, namely Taylor's formula. With its help a function satisfying certain conditions is approximated by another function of the form $P(x) = a_0 + a_1 x + \cdots + a_n x^n$, which is called an algebraic polynomial. Here the a_k are constants, independent of x.

An algebraic polynomial has a very simple structure; in order to compute it for given coefficients a_k and given values of x we need to apply only the three arithmetic operations, addition, subtraction, and multiplication. The simplicity of this computation is extremely important in practice and is one of the reasons why algebraic polynomials are the most widespread means of approximating functions (another important reason is discussed later). It is sufficient to point out that especially at the present time technical computations must be carried out on computing machines on a massive scale. In their present state of perfection computing machines work very rapidly and tirelessly. However, machines can perform only relatively simple operations. They may be set to perform arithmetic operations on very large numbers, but never, for example, the infinite process of passage to the limit. A machine cannot compute log x exactly, but we can approximate log x by a polynomial $P(x)$ with any required degree of accuracy, and then compute the polynomial by a machine.

In addition to Taylor's formula, there are others of great practical importance in the approximation of functions by algebraic polynomials. Among them are the various interpolation formulas, which are widely used, in particular, in approximate computation of integrals, and also in approximate integration of differential equations. Well known also is the method of approximation in the sense of the mean square, which is very widely used with other functions as well as algebraic polynomials. For certain practical questions great importance is attached to the method of best uniform (or Čebyšev) approximation, originated by the great Russian mathematician Čebyšev, a method which arose, as we will see, from the solution of a problem connected with the construction of mechanisms.

Our present purpose is to give the reader some idea of these methods and, as far as possible, to state the conditions under which one method is preferable to another. No one of them is absolutely the best. Every method can be seen to be better than the others under certain conditions. For example, if we have a physical problem to solve, then some one method of approximating the functions that occur in the problem is particularly indicated by the character of the problem itself or, as one says, by physical considerations. Also we will see later that under well-known conditions one method of approximation may be applicable, and another not.

Each of the methods of computation arose in its own time and has its own characteristic theory and history. Newton was already familiar with a formula for interpolation and gave it a very convenient form for practical computation with what are called difference quotients. The method of approximation in the sense of the mean square is at least 150 years old. But, for a long time these methods did not give rise to a connected theory. They were only various practical methods of approximating functions, and furthermore, the restrictions on their applicability were not clear.

The present theory of approximations to functions arose from the work of Čebyšev, who introduced the important concept of best approximation, in particular best uniform approximation, made systematic use of it in practical applications and developed its theoretical basis. Best approximation is the fundamental concept in the contemporary theory of approximation. After Čebyšev, his ideas were developed further by his students E. I. Zolotarev, A. N. Korkin, and the brothers A. A. and V. A. Markov. In the Čebyšev period of the theory of approximation of functions, not only were the fundamental concepts introduced, but basic methods were found for obtaining the best approximations to arbitrary individual functions, methods which are in wide use at the present time; also, there were basic investigations of the properties of the approximating classes, particularly of algebraic and trigonometric polynomials, from the point of view of the requirements arising from practical problems.

The further development of the theory of approximation of functions was influenced by an important mathematical discovery, made at the end of the last century by the German mathematician Weierstrass. With complete rigor he proved the theoretical possibility of approximating an arbitrary continuous function by an algebraic polynomial with any given degree of accuracy. This is the second reason why algebraic polynomials are a universal means of approximating functions. The mere simplicity of construction of algebraic polynomials is not sufficient; we also require the possibility of approximating any continuous function by a polynomial with arbitrary prescribed error. This possibility was proved by Weierstrass.

The profound ideas of Čebyšev on best approximation and the theorem of Weierstrass served as a basis, at the beginning of the present century, of the present-day development in the theory of approximation. In this connection let us mention the names of S. N. Bernšteĭn, Borel, Jackson, Lebesgue, and de la Vallée-Poussin. Briefly, this development may be described as follows. Up to the time of Čebyšev (the beginning of the present century), the problems usually consisted of approximation of individual functions, but the characteristic problem of the present-day period is the approximation, by polynomials or otherwise, of entire classes of functions, analytic, differentiable, and the like.

The Russian school, and now the Soviet school, of the theory of approximation has played a leading role in this theory. Important contributions have been made by S. N. Bernšteĭn, A. N. Kolmogorov, M. A. Lavrent'ev, and their students. At the present time the theory has developed into an essentially distinct branch of the theory of functions.

In addition to algebraic polynomials, another very important means of approximation consists of the trigonometric polynomials. A *trigonometric polynomial* of order n is a function of the form

$$u_n(x) = \alpha_0 + \alpha_1 \cos x + \beta_1 \sin x + \alpha_2 \cos 2x + \beta_2 \sin 2x +$$
$$\cdots + \alpha_n \cos nx + \beta_n \sin nx,$$

or more concisely

$$u_n(x) = \alpha_0 + \sum_{k=1}^{n} (\alpha_k \cos kx + \beta_k \sin kx),$$

where α_k and β_k are constants.

There are various particular methods of approximation by trigonometric polynomials, which are usually connected in a rather simple way with the corresponding methods of approximation by algebraic polynomials. Among these methods an especially important role is played by the expansion of functions in a Fourier series (see §7). These series are known by the name of the French mathematician Fourier, who at the beginning of the last century made several theoretical discoveries concerning them, in his study of the conduction of heat. However, it should be noted that trigonometric series were investigated as early as the middle of the 18th century by the great mathematicians Leonhard Euler and Daniel Bernoulli. In Euler's work they were related to his researches in astronomy, and in Bernoulli's to his study of the oscillating string. We may remark that Euler and Bernouilli raised the fundamentally significant question of the possibility of representing a more or less arbitrary function by a trigonometric series, a question which was finally answered only in the middle of the last century. Its affirmative answer, discussed later, was anticipated by Bernoulli.

Fourier series are of great importance in physics, but we will give little attention to this aspect of them, since it has been considered in Chapter VI. In that chapter also the reader will find examples of physical problems that naturally lead to the expansion of a given function in series other than the trigonometric series but with great similarly to them. We refer to the so-called series of orthogonal functions.

Fourier series have had a history of two hundred years. So it is not surprising that by now their theory is extraordinarily broad, subtle, and profound and constitutes an independent discipline in mathematics. An especially remarkable role in this theory has been played by the Moscow school of the theory of functions of a real variable, N. N. Luzin, A. N. Kolmogorov, D. E. Men'šov, and others.

We note also that the significance of trigonometric polynomials in contemporary mathematics is hardly exhausted by their role as methods of approximation. For example, in Chapter X the reader became acquainted with the fundamental results of I. M. Vinogradov in the theory of numbers, which were derived on the basis of a suitably devised apparatus of trigonometric sums (polynomials).

§2. Interpolation Polynomials

A special case of the construction of interpolating polynomials. In practical computations the interpolation method of approximating a function is widely used. To introduce the reader to a range of questions of this type, we consider the following elementary problem.

Let the function $y = f(x)$ be given on the interval $[x_0, x_2]$, with graph as illustrated in figure 1. The appearance of this graph is reminiscent of an arc of a parabola. So if we wish to approximate our function by a simple function, it is natural to choose a polynomial of the second degree

$$P(x) = a_0 + a_1 x + a_2 x^2, \tag{1}$$

the graph of which is a parabola.

The interpolation method consists of the following. In the interval $[x_0, x_2]$ we choose an interior point x_1. The points x_0, x_1, x_2 give corresponding values of our function

$$y_0 = f(x_0), y_1 = f(x_1), y_2 = f(x_2).$$

We construct a polynomial (1) such that at the points x_0, x_1, x_2 it agrees with the function in question (its graph is shown by the dashed curve in figure 1). In other words, we must choose the coefficients a_0, a_1, a_2 in the polynomial (1) so that they satisfy the equations

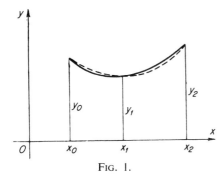

FIG. 1.

$$P(x_0) = y_0, P(x_1) = y_1,$$

$$P(x_2) = y_2. \qquad (2)$$

We note that our function $f(x)$ may be defined otherwise than by a formula; for example, its values may be given empirically as shown by the graph in figure 1. To solve the interpolation problem, we choose an approximating function in the form of an analytic expression, namely the polynomial $P(x)$. If the exactness of the approximation is satisfactory, the polynomial so chosen has the advantage over the original function that we can compute its intermediate values.

This interpolation problem could be solved as follows: We could set up the three equations

$$y_0 = a_0 + a_1 x_0 + a_2 x_0^2,$$

$$y_1 = a_0 + a_1 x_1 + a_2 x_1^2,$$

$$y_2 = a_0 + a_1 x_2 + a_2 x_2^2,$$

solve them for a_0, a_1, a_2 and substitute the values of these coefficients in equation (1). But let us solve it in a somewhat different way. We begin by constructing the polynomial $Q_0(x)$ of the second degree such that it satisfies the three conditions: $Q_0(x_0) = 1$, $Q_0(x_1) = 0$, $Q_0(x_2) = 0$. From the last two conditions it follows that this polynomial must have the form $A(x - x_1)(x - x_2)$, and from the first condition that

$$A = \frac{1}{(x_0 - x_1)(x_0 - x_2)}.$$

So, the desired polynomial has the form

$$Q_0(x) = \frac{(x - x_1)(x - x_2)}{(x_0 - x_1)(x_0 - x_2)}.$$

Similarly the polynomials

$$Q_1(x) = \frac{(x - x_0)(x - x_2)}{(x_1 - x_0)(x_1 - x_2)}, \qquad Q_2(x) = \frac{(x - x_0)(x - x_1)}{(x_2 - x_0)(x_2 - x_1)}$$

satisfy the conditions

$$Q_1(x_0) = Q_1(x_2) = 0, \quad Q_1(x_1) = 1,$$
$$Q_2(x_0) = Q_2(x_1) = 0, \quad Q_2(x_2) = 1.$$

Further, it is obvious that the polynomial $y_0 Q_0(x)$ has the value y_0 for $x = x_0$ and vanishes for $x = x_1$ and $x = x_2$, and corresponding properties hold for the polynomials $y_1 Q_1(x)$ and $y_2 Q_2(x)$.

Hence it readily follows that the desired interpolating polynomial is given by the formula

$$\begin{aligned}
P(x) &= y_0 Q_0(x) + y_1 Q_1(x) + y_2 Q_2(x) \\
&= y_0 \frac{(x - x_1)(x - x_2)}{(x_0 - x_1)(x_0 - x_2)} + y_1 \frac{(x - x_0)(x - x_2)}{(x_1 - x_0)(x_1 - x_2)} \\
&\qquad + y_2 \frac{(x - x_0)(x - x_1)}{(x_2 + x_0)(x_2 - x_1)}.
\end{aligned} \tag{3}$$

We note that the polynomial so obtained is the unique polynomial of the second degree which solves our interpolation problem. For if we assume that some other polynomial $P_1(x)$ of the second degree is also a solution of the problem, then the difference $P_1(x) - P(x)$, which is also a polynomial of the second degree, vanishes at the three points $x = x_0$, x_1, x_2. But we know from algebra that if a polynomial of the second degree vanishes for three values of x, then it is identically zero. So the polynomials $P(x)$ and $P_1(x)$ agree identically.

It is clear that in general the polynomial so obtained agrees with the given function only at the points x_0, x_1, x_2 and differs from it for other values of x.

If we take x_1 at the center of the interval $[x_0, x_2]$ and put $x_2 - x_1 = x_1 - x_0 = h$, then formula (3) is somewhat simplified:

$$P(x) = \frac{1}{2h^2} [y_0(x - x_1)(x - x_2) - 2y_1(x - x_0)(x - x_2) + y_2(x - x_0)(x - x_1)].$$

As an example let us interpolate the sine curve $y = \sin x$ (figure 2) by a polynomial of degree two, agreeing with it at the points $x = 0, \pi/2, \pi$. Obviously, the desired polynomial has the form

$$P(x) = \frac{4}{\pi^2} x(\pi - x) \approx \sin x.$$

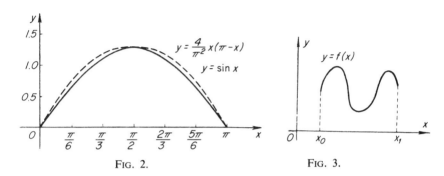

FIG. 2.   FIG. 3.

Let us compare $\sin x$ and $P(x)$ at two intermediate points:

$$P\left(\frac{\pi}{4}\right) = 0.75, \quad \text{while} \quad \sin\frac{\pi}{4} = \frac{\sqrt{2}}{2} \approx 0.71,$$

$$P\left(\frac{\pi}{6}\right) = \frac{10}{18}, \quad \text{while} \quad \sin\frac{\pi}{6} = \frac{9}{18}.$$

In this way we have approximated $\sin x$ on the interval $[0, \pi]$ with an accuracy* of about 0.05. On the other hand, the expansion of $\sin x$ in a Taylor series around the point $\pi/2$ gives

$$\sin x = \cos\left(\frac{\pi}{2} - x\right) = 1 - \frac{\left(\frac{\pi}{2} - x\right)^2}{2!} + \frac{\left(\frac{\pi}{2} - x\right)^4}{4!} - \cdots.$$

If we stop at the second term of the expansion, we have at the point $x = 0$, the approximation $\sin 0 = 1 - \pi^2/8 \approx 0.234$, i.e., an error greater than 0.2.

We see that our interpolation method has produced an approximation to $\sin x$ on the whole interval $[0, \pi]$ by a polynomial of degree two that

---

* However, for a complete justification of this statement, we need to prove that the difference $(4x/\pi^2)(\pi - x) - \sin x$ does not exceed in value 0.05, not only for $x = \pi/4$ and $x = \pi/6$, but also for all x on the interval $[0, \pi]$; we will not do this.

is more satisfactory than the Taylor expansion of second degree. However, we must not forget that Taylor's formula gives a very exact approximation close to the point $x = \pi/2$ around which it is taken, more exact in this neighborhood than the approximation obtained by interpolating.

The general solution of the problem. It is clear that a more complicated function $y = f(x)$, as illustrated in figure 3, is hardly suitable for approximation by a polynomial of degree two, since no parabola of degree two could follow all the bends of the curve $y = f(x)$. In this case it is natural to try an interpolation of the function with a polynomial of higher degree (not less than the fourth).

The general problem of interpolation consists of constructing a polynomial $P(x) = a_0 + a_1 x + a_2 x^2 + \cdots + a_n x^n$ of degree n which agrees with a given function at $n + 1$ equations:

$$P(x_0) = f(x_0), \, P(x_1) = f(x_1), \ldots, P(x_n) = f(x_n).$$

The points at which it is required that the function agree with its approximating polynomial are called the *points of interpolation*.

Reasoning in the same way as for a second-degree polynomial, we can easily prove that the desired polynomial may be written in the form

$$P_n(x) = \sum_{k=0}^{n} \frac{(x - x_0)(x - x_1) \cdots (x - x_{k-1})(x - x_{k+1}) \cdots (x - x_n)}{(x_k - x_0)(x_k - x_1) \cdots (x_k - x_{k-1})(x_k - x_{k+1}) \cdots (x_k - x_n)} f(x_k),$$

$$(4)$$

and further that this polynomial (of degree n) is unique. The formula so written is known as *Lagrange's formula*. It may also be put in various other forms; for example, it is widely used in practice in the form involving Newton's difference quotients.

The deviation of the interpolation polynomial from the generating function. The method of interpolation is a universal means of approximating functions. In principle, the function is not required to have any particular properties for interpolation to be possible; for example, it is not required to have derivatives over the whole interval of approximation. In this respect the method of interpolation has an advantage over Taylor's formula. It is interesting to note that there are cases when the function is even analytic at every point on an interval but cannot be approximated by its Taylor's formula over the interval. Suppose, for example, that we require a good approximation of the function $1/(1 + x^2)$ on the interval

[—2, 2] by means of an algebraic polynomial. At first glance it is natural to try its expansion in a Taylor series about the point $x = 0$

$$\frac{1}{1 + x^2} = 1 - x^2 + x^4 - x^6 + \cdots.$$

But it is easy to see that this series is convergent only in the interval $-1 < x < 1$. Outside the interval $[-1, 1]$, it diverges and consequently cannot approximate $1/(1 + x^2)$ on the whole interval $[-2, 2]$. Nevertheless, the interpolation method is completely applicable here.

Of course, the question arises in each case of choosing the number and distribution of the points of interpolation in such a way that the error will satisfy certain requirements. For functions with derivatives of sufficiently high order, the answer to this question of the possible magnitude of error is given by the following classical result, which we introduce without proof.

If on the interval $[x_0, x_n]$ the function $f(x)$ has a continuous derivative of order $n + 1$, then for any intermediate value of x the deviation of $f(x)$ from the Lagrange interpolation polynomial $P(x)$ with points of interpolation $x_0 < x_1 < \cdots < x_n$ is given by the formula

$$f(x) - P(x) = \frac{(x - x_0)(x - x_1) \cdots (x - x_n)}{n!} f^{(n+1)}(c),$$

where c is an intermediate point between x_0 and x_n. This formula is reminiscent of the corresponding formula for the remainder term in the Taylor expansion and is essentially a generalization of it. So, if it is known that the derivative $f^{(n+1)}(x)$ of order $n + 1$ on the interval $[x_0, x_n]$ nowhere exceeds the number M in absolute value, then the error of the approximation for any value of x on this interval is bounded by the following estimate:

$$|f(x) - P_n(x)| \leqslant \frac{|x - x_0| \cdots |x - x_n|}{n!} M.$$

The contemporary theory of approximation provides many other methods of estimating the error in interpolation. This question has been carefully studied and some interesting, completely unexpected facts have been discovered.

Consider, for example, a smooth function $y = f(x)$, defined on the interval $[-1, 1]$, i.e., one whose graph is a continuous curve with a continuously varying tangent. Our choice of the interval with specific end points —1 and 1 is unimportant; the facts described here remain valid for an arbitrary interval $[a, b]$ with inconsequential changes.

We assume now that on the interval $[-1, 1]$ we have chosen a system of $n + 1$ points

$$-1 \leqslant x_0 < x_1 < \cdots < x_n \leqslant 1 \qquad (5)$$

and have then constructed the polynomial $P(x) = a_0 + a_1 x + \cdots + a_n x^n$ of degree n that agrees with $f(x)$ at these points. We will assume temporarily that the points of the system (5) are equally spaced along the interval. If n increases indefinitely, then the corresponding interpolating polynomial $P_n(x)$ will agree with $f(x)$ at a greater and greater number of points, and we might think that at an intermediate point x, not belonging to the system (5), the difference $f(x) — P_n(x)$ would converge to zero as $n \to \infty$. This opinion was held even at the end of the last century, but it was afterwards discovered that the facts are far otherwise. It has been shown that for many smooth (even analytic) functions $f(x)$, in the case of evenly spaced points of division x_k, the interpolating polynomials $P_n(x)$ do not at all converge to $f(x)$ as $n \to \infty$. The graph of the interpolating polynomial certainly agrees with $f(x)$ at the given points of interpolation, but in spite of this it deviates strongly for large n from the graph of $f(x)$ at intermediate values of x and the deviation increases with increasing n. As further investigation showed, this situation may be avoided, at least for smooth functions, if the points of interpolation are distributed more sparsely near the center of the interval and more densely near the ends. Indeed, it has been shown that in a well-known sense the best distribution of the points of interpolation is the one in which the points x_k occur at the zeros* of the Čebyšev polynomials $\cos [(n + 1) \text{ arc cos } x]$ defined by the formula

$$x_k = \cos \frac{2k + 1}{2(n + 1)} \pi \qquad (k = 0, 1, \cdots, n).$$

The polynomials (called Čebyšev polynomials) which correspond to these points of interpolation have the property that they are uniformly convergent to the function which generated them, provided the latter is smooth, i.e., is itself continuous and has a continuous first derivative. The graph of such a function is a continuous curve with a continuously varying tangent. Figure 4 shows the distribution of the zeros of the Čebyšev polynomial for the case $n = 5$.

---

* A zero of the function $f(x)$ is a value x_k for which $f(x_k) = 0$. For details on Čebyšev polynomials see §5.

As for arbitrary nonsmooth continuous functions, the situation is worse; it can be shown that in general there is no sequence of points of interpolation such that the interpolating process will converge for any continuous function (Faber's theorem). In other words, however we may divide the interval [—1, 1] into parts, with the number of points of interpolation approaching infinity, we can always find a function $f(x)$, continuous in the interval, such that the successive polynomials with these points of interpolation will not converge to the function. Even for the mathematicians of the middle of the last century, this fact, had it been known, would have sounded paradoxical. Of course, the explanation is that among the continuous nonsmooth functions there are some extraordinarily "bad" ones, for example those which do not have a derivative at any point of the interval on which they are defined, and these supply examples for which a given interpolation process will not converge. Effective methods of approximation to these functions by polynomials can be suggested by making some changes in the previous interpolation process, but we will not take the time to do this here.

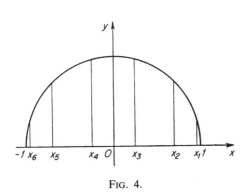

FIG. 4.

In conclusion we note that algebraic polynomials are not the only means available for interpolation. There are methods for interpolation by trigonometric polynomials, for example, which are well developed from the practical and also from the theoretical point of view.

§3. Approximation of Definite Integrals

Interpolation of functions has wide application in questions related to the approximate computation of integrals. As an example, we introduce an approximate formula for a definite integral, namely Simpson's rule, which is widely used in applied analysis.

Let it be required to compute an approximation to the definite integral on the interval $[a, b]$ of the function $f(x)$, whose graph is illustrated in figure 5. The exact value is given by the area of the curvilinear trapezoid $aABb$. Let C be the point of the graph with abscissa $c = (a + b)/2$.

Through the points A, B, and C, we pass a parabola of degree two. As we know from the preceding section, this parabola is the graph of a polynomial of the second degree, defined by the equation

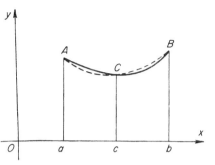

FIG. 5.

$$P(x) = \frac{1}{2h^2} [(x-c)(x-b)y_0$$

$$- 2(x-a)(x-b)y_1$$

$$+ (x-a)(x-c)y_2],$$

where

$$h = \frac{b-a}{2} \quad y_0 = f(a), \quad y_1 = f(c), \quad y_2 = f(b).$$

In the terminology of the preceding section, we may say that the second-degree polynomial $P(x)$ interpolates $f(x)$ at the points with abscissas a, c, b. If the graph of the function $f(x)$ on the interval $[a, b]$ does not change too violently and the interval is not large, then the polynomial $P(x)$ will everywhere differ little from $f(x)$; this, in turn, implies that their integrals taken over $[a, b]$ will also differ little from each other. On this basis we may assume these integrals are approximately equal,

$$\int_a^b f(x)\, dx \approx \int_a^b P(x)\, dx,$$

or, as it is customarily stated, the second integral is an approximation to the first. Simple computations, which we leave to the reader, show that

$$\int_a^b (x-c)(x-b)\, dx = \frac{2}{3} h^3, \quad -\int_a^b (x-a)(x-b)\, dx = \frac{4}{3} h^3,$$

$$\int_a^b (x-a)(x-c)\, dx = \frac{2}{3} h^3.$$

Hence

$$\int_a^b P(x)\, dx = \frac{h}{3} [f(a) + 4(fc) + f(b)].$$

Thus the definite integral may be computed by the following approximation formula:

$$\int_a^b f(x)\, dx \approx \frac{h}{3} [f(a) + 4f(c) + f(b)].$$

This is Simpson's formula.

As an example, let us use this formula to compute the integral of $\sin x$ on the interval $[0, \pi]$. In this case

$$h = \frac{\pi}{2}, \quad f(a) = \sin 0 = 0, \quad f(c) = \sin \frac{\pi}{2} = 1, \quad f(b) = \sin \pi = 0,$$

and consequently $(h/3)[f(a) + 4f(c) + f(b)] = \frac{2}{3}\pi = 2.09 \cdots$. On the other hand the integral can be found exactly

$$\int_0^\pi \sin x \, dx = -\cos x \Big|_0^\pi = 2.$$

The error does not exceed 0.1.

If the interval $[0, \pi]$ is decomposed into two equal parts and on each of these our formula is applied separately, then we get

$$\int_0^{\pi/2} \sin x \, dx \approx \frac{\pi}{12} \left[\sin 0 + 4 \sin \frac{\pi}{4} + \sin \frac{\pi}{2}\right] = \frac{\pi}{12}\left(4\frac{\sqrt{2}}{2} + 1\right) \approx 1.001,$$

$$\int_{\pi/2}^\pi \sin x \, dx \approx 1.001.$$

In this manner

$$\int_0^\pi \sin x \, dx \approx 2.002 \, ;$$

and now the error is considerably less than 0.002.

In practice, in order to compute approximately the definite integral of a function $f(x)$ on $[a, b]$ we divide the interval into an even number n of parts by the points $a = x_0 < x_1 < \cdots < x_n = b$ and successively apply Simpson's rule to the segment $[x_0, x_2]$, and then to the segment $[x_2, x_4]$ and so forth. As a result we have the following general formula of Simpson:

$$\int_a^b f(x) \, dx \approx \frac{b-a}{3n} [f(x_0) + 4f(x_1) + 2f(x_2) + 4f(x_3) + \cdots + f(x_n)]. \tag{6}$$

Let us now give without proof the classical estimate for the error. If on the interval $[a, b]$ the function $f(x)$ has a fourth derivative which satisfies the inequality $|f^{IV}(x)| \leqslant M$, then the following estimate holds

$$\left| \int_a^b f(x) \, dx - L(f) \right| \leqslant \frac{M(b-a)^5}{180n^4}. \tag{7}$$

Here by $L(f)$ we denote the right side of formula (6). In this case the error will be of order n^{-4}.*

---

* If a certain quantity α_n, depending on $n = 1, 2, \cdots$, satisfies the inequality $|\alpha_n| < C/n^k$, where C is constant independent of n, then we say that it is of order n^{-k}.

We could have decomposed the interval $[a, b]$ into n equal parts and taken as our approximation to the integral the sum of the areas of the rectangles drawn in figure 6. Then we would get an approximation formula from the rectangles*

$$\int_a^b f(x)\, dx \approx \frac{b-a}{n} [f(x_0) + f(x_1) + \cdots + f(x_{n-1})]. \qquad (8)$$

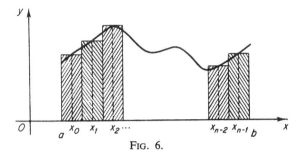

FIG. 6.

It may be shown that the order of error here is n^{-2}, provided the function has a second derivative that is bounded on the interval $[a, b]$. We may also take as an approximation the sum of the areas of the trapezoids drawn in figure 7 and get the trapezoidal formula

$$\int_a^b f(x)\, dx \approx \frac{b-a}{2n} [f(x_0) + 2f(x_1) + \cdots + 2f(x_{n-1}) + f(x_n)] \qquad (9)$$

with order of error n^{-2}, provided the function has a bounded second derivative.

It is usually said that Simpson's formula is more exact than the trapezoidal and rectangular formulas. This statement requires amplifi-

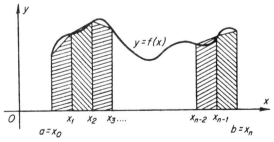

FIG. 7.

---

* In this case x_0, x_1, $\cdots$, x_{n-1} are the centers of the equal parts of the interval $[a, b]$, and not points of division as in formulas (6) and (9).

cation, without which it will not be true. If we know only that a function
has a first derivative, then the guaranteed order of approximation for
each of the three methods is alike equal to n^{-1}; in this case Simpson's
formula has no essential advantage over the rectangular and trapezoidal
formulas. For functions that have a second derivative, it is guaranteed
that the approximations by the trapezoidal formula and by Simpson
formulas are each of order n^{-2}. But if the function has a third and fourth
derivative, then the order of the error is still equal to n^{-2} for the rectangular and trapezoidal formulas, but for Simpson's formula it is equal to
n^{-3} and n^{-4} respectively. But the order n^{-4} for Simpson's formula proves
in its turn to be the best possible result; in other words, for functions
that have derivatives of higher order than the fourth, the order of error
remains equal to n^{-4}. Thus, if we are given a function that has a derivative
of fifth order and wish to make use of this fact to obtain an approximation
of order n^{-5}, we need a new method of approximation to the definite
integral, different from Simpson's formula. To explain how it must be
constructed, we note the following.

The trapezoidal and rectangular formulas, as is easily shown, are
exact for polynomials of the first degree; this means that substitution in (9)
of the function $A + Bx$, where A and B are constants, leads to exact
equality. In the same way Simpson's formula proves to be exact for
polynomials of the third degree $A + Bx + Cx^2 + Dx^3$. The gist of the
matter lies in this fact. Let us suppose that we have divided the interval
$[a, b]$ into n equal parts and on each part have used a method of approximation, the same on each part, which is exact for polynomials
$A + Bx + \cdots + Fx^{m-1}$ of degree $m - 1$. Then the error of the approximation for every function which has a bounded mth derivative will be
of order n^{-m}, and if this function is not a polynomial of degree $m - 1$,
then this order cannot be increased even for functions which have
derivatives of much higher order.

Our present remarks emphasize the importance of finding the simplest
possible approximate methods of integration that are exact for polynomials of a given degree. This question, on which the present-day
literature is quite large, has interested mathematicians for a long time.
Here we can only refer to certain classical results.

Let the function $p(x)$ be given. We are asked how to distribute on the
interval $[-1, 1]$ the points of division $x_1, \cdots, x_m$ and how to choose the
number K, so as to satisfy the equation

$$\int_{-1}^{1} f(x)\, p(x)\, dx = K \sum_{1}^{m} f(x_i),$$

for every polynomial $f(x)$ of degree m.

It can be shown that for $p(x) = (1 - x^2)^{\frac{1}{2}}$ the problem is solved if $K = \pi/m$, and x_i are the zeros of the Čebyšev polynomial $\cos m$ arc $\cos x$ (cf. §5).

For $p(x) = 1$, Čebyšev gave a solution of the problem for $m = 1, 2, \cdots, 7$. For $m = 8$ the problem has no solution: the points of division may be found but they are complex. For $m = 9$ it again has a solution. However, as S. N. Bernšteĭn showed, for any $m > 9$ the problem has no solution (the points of division lie outside the interval $[-1, +1]$).

A quadrature formula that is exact for polynomials of degree n can be constructed very simply by means of Lagrange's formula (4). If we integrate its left and right sides on the interval $[a, b]$, we obtain

$$\int_a^b P_n(x) \, dx = \sum_{k=0}^n p_k f(x_k), \qquad (10)$$

where

$$p_k = \int_a^b \frac{(x - x_0) \cdots (x - x_{k-1})(x - x_{k+1}) \cdots (x - x_n)}{(x_k - x_0) \cdots (x_k - x_{k-1})(x_k - x_{k+1}) \cdots (x_k - x_n)} \, dx$$
$$(k = 0, 1, \cdots, n).$$

Consequently, equation (10) is valid for all polynomials of degree n, and thus the quadrature formula

$$\int_a^b f(x) \, dx \approx \sum_0^n p_k f(x_k)$$

is exact for all polynomials of degree n.

When

$$x_0 = a, \qquad x_1 = \frac{a + b}{2}, \qquad x_2 = b,$$

this formula reduces, as we have seen earlier, to Simpson's formula.

The distribution of the points of interpolation x_k ($k = 0, 1, \cdots, n$) in the interval $[a, b]$ may be changed. For every distribution of the points there will be a corresponding quadrature formula.

Gauss, the famous German mathematician of the last century, showed that the interpolation points x_k may be distributed in such a manner that the formula will be exact for all polynomials not only of degree n, but also of degree $2n + 1$.

The polynomial

$$A_{n+1}(x) = (x - x_0)(x - x_1) \cdots (x - x_n)$$

of degree $n + 1$, arising from Gauss's points of division x_k, has a remarkable property: For any polynomial $P(x)$ of degree less than $n + 1$, we have the equation

$$\int_a^b A_{n+1}(x)\, P(x)\, dx = 0.$$

In other words, the polynomial $A_{n+1}(x)$ is orthogonal on the interval $[a, b]$ to all polynomials of degree not greater than n. The polynomials $A_{n+1}(x)$ we call the *Legendre polynomials* (corresponding to the interval $[a, b]$).

§4. The Čebyšev Concept of Best Uniform Approximation

Statement of the question. Čebyšev came to the idea of best uniform approximation from a purely practical problem, since he was not only one of the greatest mathematicians of the last century, creating the basis for a number of mathematical disciplines that are widely developed at present, but was also a leading engineer of his time. In particular, Čebyšev was very much interested in questions of the construction of mechanisms producing a given trajectory of motion. We will now explain this idea.

Let the curve $y = f(x)$ be given on the interval $a \leqslant x \leqslant b$. We wish to construct, subject to specific technical requirements, a mechanism such that a certain one of its points will describe this curve as exactly as possible when the mechanism is in operation. Čebyšev solved the problem as follows. First of all, looking for the solution as an engineer, he constructed the required mechanism in such a manner as to get a rough approximation to the required trajectory. Thus, a certain point A of the mechanism, admittedly not yet in its final form, would describe the curve

$$y = \phi(x), \tag{11}$$

resembling the required curve $y = f(x)$ only in its general features. The mechanism so constructed consists of separate parts, gears, levers of various kinds, and the like. All of these have specific measurements

$$\alpha_0,\, \alpha_1,\, \alpha_2,\, \cdots,\, \alpha_m, \tag{12}$$

which completely describe the mechanism, and consequently the curve (11). They are the parameters of the mechanism and of the curve (11).* Thus

---

* Details of the calculations for mechanisms of this sort may be found in the publication "The Scientific Heritage of P. L. Čebyšev," Volume II, Academy of Sciences of the USSR, 1945.

the curve (11) depends not only on the argument x, but also on the parameters (12). To any assigned system of values of the parameters will correspond a specific curve, whose equation may be conveniently written in the form

$$y = \phi(x; \alpha_0, \alpha_1, \cdots, \alpha_m). \tag{13}$$

It is customary to say in such cases that we have obtained a family of functions (13), defined on the interval $a \leqslant x \leqslant b$ and depending on the $m + 1$ parameters (12).

For the further solution of his problem Čebyšev worked as a pure mathematician. He proposed, in a perfectly natural way, to take as the measure of the deviation of the function $f(x)$ from the approximating function $\phi(x; \alpha_0, \alpha_1, \cdots, \alpha_m)$ the magnitude

$$\|f - \phi\| = \max_{a \leqslant x \leqslant b} |f(x) - \phi(x; \alpha_0, \alpha_1, \cdots, \alpha_m)|, \tag{14}$$

equal to the maximum of the absolute value of the difference $f(x) - \phi(x; \alpha_0, \alpha_1, \cdots, \alpha_m)$ on the interval $a \leqslant x \leqslant b$ (figure 8). This quantity is obviously a certain function

$$\|f - \phi\| = F(\alpha_0, \alpha_1, \cdots, \alpha_m) \tag{15}$$

of the parameters $\alpha_0, \alpha_1, \cdots, \alpha_m$. The problem is now to find those values of the parameters for which the function (15) is a minimum. These values define a function ϕ, which it is customary to describe as the best uniform approximation of the given function $y = f(x)$ among all possible functions of the given family (11). The magnitude $F(\alpha_0, \alpha_1, \cdots, \alpha_m)$ for these values of the parameters is called the *best uniform approximation of the function $f(x)$* on the interval $[a, b]$ by means of the functions of the family (13). It is usually denoted by the sym-

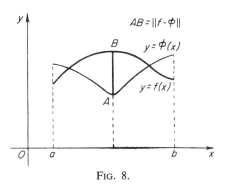

Fig. 8.

bol $E_m(f)$. The term "uniform" is often replaced, especially in non-Soviet literature, by the term "Čebyšev." They both emphasize the specific character of the approximation, since other types of approximation are of course possible; for example, one may speak of the best approximation to $f(x)$ by functions from a given family in the sense of the mean square. This subject will be discussed in §8.

Čebyšev first discovered the various laws which hold for the type of approximation we are discussing here and found that in many cases the function ϕ which is the best uniform approximation to $f(x)$ on the interval $[a, b]$ has the remarkable property that for it the maximum (15) of the absolute value of the difference

$$f(x) - \phi(x; \alpha_0, \alpha_1, \cdots, \alpha_m)$$

is attained for at least $m + 2$ points of the interval $[a, b]$ with successively alternating signs (figure 9).

We have no space here for an exact formulation of the conditions under which this proposition is valid and refer our better prepared readers to the article of V. L. Gončarov "The theory of the best approximation of functions" ("The Scientific Heritage of Čebyšev," Volume I).

The case of approximation of functions by polynomials. The cited investigations of Čebyšev are especially important for the general theory of approximation when applied to the question of approximating an arbitrary function $f(x)$ on a given interval $[a, b]$ by polynomials $P_n(x) = a_0 + a_1 x + a_2 x^2 + \cdots + a_n x^n$ of given degree n. The polynomials $P_n(x)$ of degree n constitute a family of functions depending on the $n + 1$ coefficients as parameters. As may be shown, the theory of Čebyšev is fully applicable to polynomials, so that if we wish to make the best uniform approximation to the function $f(x)$ on the segment $[a, b]$ by a polynomial $P_n(x)$ chosen from all possible polynomials of the given degree n, then we need only find all those values of x on this interval for which the function $|f(x) - P_n(x)|$ assumes its maximum L on $[a, b]$. If among them we can find $n + 2$ values $x_1, x_2, \cdots, x_{n+2}$, such that the difference $f(x) - P_n(x)$ successively changes sign

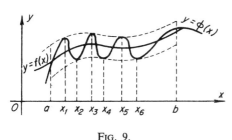

FIG. 9.

$$f(x_1) - P_n(x_1) = \pm L,$$

$$f(x_2) - P_n(x_2) = \pm L,$$

$$\cdots\cdots\cdots\cdots\cdots\cdots\cdots\cdots\cdots\cdots\cdots\cdots\cdots\cdots$$

$$f(x_{n+2}) - P_n(x_{n+2}) = \pm (-1)^{n+1} L,$$

then $P_n(x)$ is the best polynomial, and otherwise not. For example, the solution of the problem of best uniform approximation by polynomials $P_1(x) = p + qx$ of the first degree to the function $f(x)$ illustrated in figure 10 consists of the polynomial $p_0 + q_0x$ whose graph is a straight line parallel to the chord AB and dividing into equal parts the parallelogram enclosed between the chord and the tangent CD to the curve

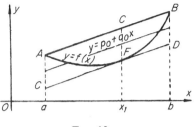

FIG. 10.

$y = f(x)$ which is parallel to that chord, since the absolute value of the difference $f(x) - (p_0 + q_0x)$ obviously assumes its maximum for the values $x_0 = a, x_1$, and $x_2 = b$, where x_1 is the abscissa of the point of tangency F, and for these values the difference itself successively changes sign. To avoid misunderstanding, we note that we are speaking of a curve that is convex downward and has a tangent at every point. In this example $E_1(f)$ is equal to half the length of any one of the (equal) segments AC, BD, or GF.

§5. The Čebyšev Polynomials Deviating Least from Zero

Let us consider the following problem. It is required to find a polynomial $P_{n-1}(x)$ of degree $n - 1$ which is the best uniform approximation on the interval $[-1, 1]$ to the function x^n.

It turns out that the desired polynomial satisfies the equation

$$x^n - P_{n-1}(x) = \frac{1}{2^{n-1}} \cos n \operatorname{arc} \cos x. \qquad (16)$$

This fact follows directly from Čebyšev's theorem, if we prove, first that the right side of (16) is an algebraic polynomial of degree n with the coefficient of x^n equal to one; second, that its absolute value on the interval $[-1, +1]$ assumes its maximum, equal to $L = 1/2^{n-1}$, at the $n + 1$ points $x_k = \cos k\pi/n$ $(k = 0, 1, \cdots, n)$; and third, that it changes sign successively at these points.

The fact that the right side of (16) is a polynomial of degree n with coefficient of x^n equal to one may be proved as follows.

Let us assume that for a given natural number n we have already proved that

$$\cos n \operatorname{arc} \cos x = 2^{n-1}[x^n - Q_{n-1}(x)];$$

$$-\sqrt{1 - x^2} \sin n \operatorname{arc} \cos x = 2^{n-1}[x^{n+1} - Q_n(x)],$$

where Q_{n-1} and Q_n are algebraic polynomials of degree $n-1$ and n, respectively. Then similar equations will also be valid for $n+1$, as is easily established by consideration of the following formulas:

$$\cos(n+1) \text{ arc cos } x = x \cos n \text{ arc cos } x - \sqrt{1-x^2} \sin n \text{ arc cos } x;$$

$$-\sqrt{1-x^2} \sin(n+1) \text{ arc cos } x$$
$$= -x\sqrt{1-x^2} \sin n \text{ arc cos } x + (x^2-1) \cos n \text{ arc cos } x.$$

But our equations for $n=1$ are true, since

$$\cos \text{ arc cos } x = x,$$

$$-\sqrt{1-x^2} \sin \text{ arc cos } x = x^2 - 1.$$

Consequently, they are true for any n.

The right side of (16) is called the *Čebyšev polynomial of degree n deviating least from zero*, since Čebyšev was the first to state and solve this problem. The first few of these polynomials are

$$T_0(x) = 1,$$
$$T_1(x) = x,$$
$$T_2(x) = \tfrac{1}{2}(2x^2 - 1),$$
$$T_3(x) = \tfrac{1}{4}(4x^3 - 3x),$$
$$T_4(x) = \tfrac{1}{8}(8x^4 - 8x^2 + 1),$$
$$T_5(x) = \tfrac{1}{16}(16x^5 - 20x^3 + 5x).$$

We have already seen the important role of the Čebyšev polynomials in questions of interpolation and of approximate methods of integration. Let us make some further remarks on interpolation.

From the fact that the difference $f(x) - P_n(x)$ between an arbitrary function $f(x)$ and its best approximating polynomial $P_n(x)$ changes sign at $n+2$ points, it follows from the properties of continuous functions that $P_n(x)$ agrees with $f(x)$ at $n+1$ specific points of the interval $[a, b]$; i.e., $P_n(x)$ is an interpolating polynomial of degree n for $f(x)$ with a certain choice of points of interpolation.

In this way the problem of the best uniform approximation of a continuous function $f(x)$ becomes one of choosing, on the interval $[-1, 1]$, a system $x_0, x_1, \cdots, x_n$ of points of interpolation such that the corresponding interpolating polynomial of degree n will have a deviation $\|f - Q\| = \max_x f(x) - Q(x)$ of least possible value. Unfortunately, the required points of division are often difficult to find in practice. Usually it is necessary to solve the problem in some approximate way, and here the Čebyšev polynomials play a special role. It turns out that if, in particular, the points of interpolation are taken to be zeros of the polynomial cos $(n + 1)$ arc cos x (i.e., the points where this polynomial is equal to zero), then the corresponding interpolating polynomial, at least for large n, will give a uniform deviation from the function (if it is sufficiently smooth) which differs little from the corresponding deviation of the best uniform approximation to the function by a polynomial. The somewhat vague expression "differs little" can be replaced, in a number of important characteristic cases, by very exact quantitative estimates, which we will not establish here.

Returning to the Čebyšev polynomial, let us consider it in the form $T_n(x) = M \cos n$ arc cos x $(-1 \leqslant x \leqslant 1)$, where M is some positive number. Obviously, on the interval $[-1, 1]$ its absolute value does not exceed the number M. Its derivative is

$$T_n'(x) = -\frac{nM \sin n \text{ arc cos } x}{\sqrt{1 - x^2}},$$

which on the interval $[-1, 1]$ satisfies the inequality

$$| T_n'(x)| \leqslant \frac{nM}{\sqrt{1 - x^2}}.$$

It turns out that this inequality is true for all polynomials $P_n(x)$ of degree n which do not exceed the number M in absolute value on the interval $[-1, 1]$; i.e., for the derivative of any such polynomial on the interval $[-1, 1]$ we have the inequality

$$| P_n'(x)| \leqslant \frac{nM}{\sqrt{1 - x^2}}.$$

This inequality is to be credited to A. A. Markov, since it follows directly from results of his which even go somewhat further. Markov himself obtained it in connection with a question suggested to him by D. I. Mendeleev.

In 1912, S. N. Bernšteĭn obtained a similar inequality, which bears his name, for trigonometric polynomials and by using these inequalities first showed how to establish the differentiability properties of a function if one knows how fast it is approached by its sequence of best approximations. Results of this kind concerning differentiable functions are given in §§6 and 7.

§6. The Theorem of Weierstrass; the Best Approximation to a
Function as Related to Its Properties of Differentiability

The Weierstrass theorem.   If we apply the general definition, given in
§4, of best approximation to a function to the case of approximating
polynomials, we are led to the following definition. The best uniform
approximation to the function $f(x)$ on the interval $[a, b]$ by polynomials
of degree n occurs when the (nonnegative) number $E_n(f)$, is equal to the
minimum of the expression

$$\max_{a \leqslant x \leqslant b} |f(x) - P_n(x)| = \|f - P_n\|,$$

taken over all possible polynomials $P_n(x)$ of degree n.

Independently of whether or not we are able to find the exact polynomial that best approximates the given function $f(x)$, it is of great
practical and theoretical interest to estimate the quantity $E_n(f)$ as closely
as possible. In fact, if we wish to approximate the function f by a polynomial with accuracy δ, in other words, in such a way that

$$|f(x) - P_n(x)| \leqslant \delta \tag{17}$$

for all x in the given interval, then there is no sense in choosing it from
the polynomials of degree n for which $E_n(f) > \delta$, since for this n there
will certainly not be any polynomial P_n for which (17) holds. On the
other hand, if it is known that $E_n(f) < \delta$, then it makes sense for such
n to look for a polynomial $P_n(x)$ which will approximate $f(x)$ with
accuracy δ, since such polynomials evidently exist.

The properties of the best approximating functions of various classes
have been the subject of deep and careful study. First of all we note the
following important fact.

*If a function $f(x)$ is continuous on the interval $[a, b]$, then its best
approximation $E_n(f)$ tends to zero as n increases to infinity.*

This is the theorem proved by Weierstrass at the end of the last century.
It has great significance, since it guarantees the possibility of approximating
an arbitrary continuous function by a polynomial with any desired
accuracy. As a result, the set of all polynomials of any degree bears to
the set of all continuous functions defined on the interval exactly the
same relation as the collection R of rational numbers bears to the collection
H of all real (rational and irrational) numbers. In fact, for every irrational
number α and arbitrarily small positive number ϵ, one can always find
a rational number r satisfying the inequality $|\alpha - r| < \epsilon$. On the other
hand, if $f(x)$ is a function continuous on $[a, b]$ and ϵ is an arbitrarily

small positive number, then by Weierstrass's theorem there will exist an algebraic polynomial $P_n(x)$ such that for all x from the interval $[a, b]$ we have $|f(x) - P_n(x)| < \epsilon$. Consequently, the best approximation $E_n(f)$ to a continuous function tends to zero for $n \to \infty$.

Let us illustrate the theorem of Weierstrass in the following way. Given the graph of an arbitrary continuous function (figure 9) defined on the interval $[a, b]$, and an arbitrarily small positive number ϵ, let us surround our graph with a strip of height 2ϵ in such a way that the graph passes through the center of the strip. Then it is always possible to choose an algebraic polynomial

$$P_n(x) = a_0 + a_1 x + \cdots + a_n x^n,$$

of sufficiently high degree such that its graph lies entirely inside the strip.

We make the following remark. As before, let $f(x)$ be an arbitrary function continuous on $[a, b]$, and let $P_n(x)$ $(n = 1, 2, \cdots)$ be the polynomials which are the best uniform approximation to it. It is easy to see that the function $f(x)$ may be represented in the form of a series $f(x) = P_1(x) + [P_2(x) - P_1(x)] + [P_3(x) - P_2(x)] + \cdots$, which is uniformly convergent to $f(x)$ on $[a, b]$. This follows from the fact that the sum of the first n terms of the series is equal to $P_n(x)$, and

$$\max_{a \leqslant x \leqslant b} |f(x) - P_n(x)| = E_n(f),$$

while $E_n(f) \to 0$ as $n \to \infty$.

As a result we have a new formulation of Weierstrass's theorem:

Every function continuous on the interval $[a, b]$ *may be represented by a series of algebraic polynomials converging uniformly to the function.*

This result has great theoretical significance. It guarantees the possibility of representing an arbitrary continuous function, however originally given (for example, by means of a graph), in the form of an analytic expression. (By an analytic expression we mean an elementary function or else a function derived from a sequence of elementary functions by means of a limit process.) Historically this result finally destroyed the notion of analytic expression that had existed in mathematics almost up to the middle of the last century. We say "finally," since Weierstrass's theorem had been preceded by a series of general results of similar type, relating chiefly to Fourier series. Until these results were obtained, it had been assumed that analytic expressions were the means of representing the especially desirable properties that were characteristic of analytic functions. For example, it was usually taken for granted that analytic expressions were infinitely differentiable and could even be expanded in power series. But these ideas all proved to be without foundation. A function may have no derivative anywhere in its interval of definition and yet be representable by an analytic expression.

Fom a methodological point of view, the value of this discovery lies in the fact that it enables us to realize with complete clarity that at least in principle the methods of mathematics are applicable to an immeasurably wider class of laws than had been realized before.

At the present time many different proofs of Weierstrass's theorem are known. For the most part they reduce to the construction of a sequence of polynomials for a given continuous function f, which approximate f uniformly as their degree increases. The simply constructed polynomial

$$B_n(x) = \sum_{k=1}^{n} C_n^k x^k (1-x)^{n-k} f\left(\frac{k}{n}\right),$$

will approximate a continuous function $f(x)$ on the interval $[0, 1]$. It is called the *Bernšteĭn polynomial*. With increasing n this polynomial converges uniformly on the interval $[0, 1]$ to the function which generated it.* Here C_n^k is the number of combinations of n elements taken k at a time.

We note that a theorem similar to Weierstrass's holds in the complex domain. Exhaustive results in this direction are due to M. A. Lavrent'ev, M. V. Keldyš, and S. N. Mergeljan.

The connection between the order of the best uniform approximation of a function and its differentiability properties. We note further the following results. If a function $f(x)$ on the interval $[a, b]$ has a derivative $f^{(r)}(x)$ of order r which does not exceed the number K in absolute value, then its best approximation $E_n(f)$ satisfies the inequality

$$E_n(f) \leqslant \frac{c_r K}{n^r}, \tag{18}$$

where c_r is a constant, depending only on r (Jackson's theorem). From inequality (18) it can be seen that with increasing n the quantity $E_n(f)$ converges to zero more rapidly for functions with derivatives of higher order. In other words, the better (smoother) the function, the faster the convergence to zero of its best approximation. Bernšteĭn proved that in a certain sense the converse to this proposition is also true.

Still better in this respect than the differentiable functions are the

---

* It must be remarked that, in spite of their simplicity, the Bernšteĭn polynomials are little used in practice. The explanation is that they converge very slowly, even for functions with good differentiability properties.

analytic functions. Bernšteĭn proved that for such functions, $E_n(f)$ satisfies the inequality

$$E_n(f) \leqslant cq^n, \qquad (19)$$

where c and q are constants depending on the function f, and $0 < q < 1$; i.e., $E_n(f)$ converges to zero more rapidly than a certain decreasing progression. He also proved that conversely the inequality (19) implies that the function f is analytic on $[a, b]$.

We have given certain very important results that were discovered at the beginning of this century and have been characteristic of the direction taken by contemporary research in the theory of approximation of functions. The practical value of these results may be seen from the following example.

If $Q_n(x)$ is a polynomial of degree n, which interpolates the function $f(x)$ on the interval $[-1, 1]$ at the $n + 1$ points of interpolation which are the zeros of the Čebyšev polynomial $\cos(n + 1)$ arc cos x, then on this interval one has the inequality $|f(x) - Q_n(x)| < c \ln n \, E_n(f)$, where c is a constant independent of n, and $E_n(f)$ is the best approximation to the function f on $[-1, 1]$. In this inequality we may replace $E_n(f)$ by the larger expressions, occurring in (18) or (19), provided f is sufficiently smooth, and obtain a good estimate of the approximation of our interpolating polynomial. Since $\ln n$ increases very slowly with increasing n, the order of the estimate in the given case differs little from the order of convergence to zero of $E_n(f)$. The advantage of interpolation by the Čebyšev points consists of the fact that for other points of interpolation the factor $c \ln n$ in the corresponding inequality is replaced by a more rapidly increasing factor; this is particularly true in the case of equally spaced points of interpolation.

§7. Fourier Series

The origin of Fourier series. Fourier series arose in connection with the study of certain physical phenomena, in particular, small oscillations of elastic media. A characteristic example is the oscillation of a musical string. Indeed, the investigation of oscillating strings was the origin historically of Fourier series and determined the direction in which their theory developed.

Let us consider (figure 11) a tautly stretched string, the ends of which are fixed at the points $x = 0$ and $x = l$ of the axis Ox. If we displace the string from its position of equilibrium, it will oscillate.

We will follow the motion of a specific point of the string, with abscissa x_0. Its deviation vertically from the position of equilibrium is a function

$\phi(t)$ of time. It can be shown that one can always give the string an initial position and velocity at $t = 0$ such that as a result the point which we have agreed to follow will perform harmonic oscillations in the vertical direction, defined by the function

$$\phi = \phi(t) = A \cos \alpha kt + B \sin \alpha kt. \qquad (20)$$

FIG. 11.

Here α is a constant depending only on the physical properties of the string (on the density, tension, and length), k is an arbitrary number, and A and B are constants.

We note that our discussion relates only to small oscillations of the string. This gives us the right to assume approximately that every point x_0 is oscillating only in the vertical direction, displacements in the horizontal direction being ignored.* We also assume that the friction arising from the oscillation of the string is so small that we may ignore it. As a result of these approximate assumptions, the oscillations will not die out.

The possibilities of oscillation for the point x_0 are of course, not exhausted by the periodic motions defined by the harmonic functions (20), but these functions do have the following remarkable property. Experiments and their accompanying theory show that every possible oscillation of the point x_0 is the result of combining certain harmonic oscillations of the form (20). Relatively simple oscillations are obtained by combining a finite number of such oscillations; i.e., they are described by functions of the form

$$\phi(t) = A_0 + \sum_{k=1}^{n} (A_k \cos \alpha kt + B_k \sin \alpha kt),$$

---

* This question is directly connected with the differential equation of the oscillating string

$$\frac{\partial^2 u}{\partial t^2} = a^2 \frac{\partial^2 u}{\partial x^2} \left(a = \frac{l}{\pi} \alpha \right),$$

which was discussed in Chapter VI.

where A_k and B_k are corresponding constants. These functions are called trigonometric polynomials. In more complicated cases, the oscillation will be the result of combining an infinite number of oscillations of the form (20), corresponding to $k = 1, 2, 3, \cdots$ and with suitably chosen constants A_k and B_k, depending on the number k. Consequently, we arrive at the necessity of representing a given function $\phi(t)$ of period $2\pi/\alpha$, which describes an arbitrary oscillation of the point x_0 in the form of a series

$$\phi(t) = A_0 + \sum_{k=1}^{\infty} (A_k \cos \alpha kt + B_k \sin \alpha kt). \qquad (21)$$

There are many other situations in physics where it is natural to consider a given function, even though it does not necessarily describe an oscillation, as the sum of an infinite trigonometric series of the form (21). Such a case arises, for example, in connection with the vibrating string itself. The exact law for the subsequent oscillation of a string, to which at the beginning of the experiment we have given a specific initial displacement (for example, as illustrated in figure 12) is easy to calculate, provided we know the expansion in a trigonometric series $f(x) = \sum_1^{\infty} a_k \sin (k\pi/l)x$, (a particular case of the series (21)), of the function $f(x)$ describing the initial position.

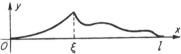

Expansion of functions in a trigonometric series. On the basis of what has been said there arises the

FIG. 12.

fundamental question: Which functions of period $2\pi/\alpha$ can be represented as the sum of a trigonometric series of the form (21)? This question was raised in the 18th century by Euler and Bernoulli in connection with Bernoulli's study of the vibrating string. Here Bernoulli took the point of view suggested by physical considerations that a very wide class of continuous functions, including in particular all graphs drawn by hand, can be expanded in a trigonometric series. This opinion received harsh treatment from many of Bernoulli's contemporaries. They held tenaciously to the idea prevalent at the time that if a function is represented as an analytic expression (such as a trigonometric series) then it must have good differentiability properties. But the function illustrated in figure 12 does not even have a derivative at the point ξ; in such a case, how can it be defined by one and the same analytic expression on the whole interval $[0, l]$?

We know now that the physical point of view of Bernoulli was quite right. But to put an end to the controversy it was necessary to wait an entire century, since a full answer to these questions required first of all that the concepts of a limit and of the sum of a series be put on an exact basis.

The fundamental mathematical investigations confirming the physical point of view but based on the older ideas concerning the foundations of analysis were completed in 1807-1822 by the French mathematician Fourier.

Finally, in 1829, the German mathematician Dirichlet showed, with all the rigor with which it would be done in present-day mathematics, that every continuous function of period $2\pi/\alpha$,* which for any one period has a finite number of maxima and minima, can be expanded in a unique trigonometric Fourier series, uniformly convergent† to the function.

Figure 13 illustrates a function satisfying Dirichlet's conditions. Its graph is continuous and periodic, with period 2π, and has one maximum and one minimum in the period $0 \leqslant x \leqslant 2\pi$.

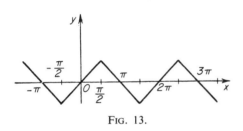

FIG. 13.

Fourier coefficients. In what follows we will consider functions of period 2π, which will simplify the formulas. We consider any continuous function $f(x)$ of period 2π satisfying Dirichlet's condition. By Dirichlet's theorem it may be expanded into a trigonometric series

$$f(x) = \frac{a_0}{2} + \sum_{k=1}^{\infty} (a_k \cos kx + b_k \sin kx), \qquad (22)$$

which is uniformly convergent to it. The fact that the first term is written as $a_0/2$ rather than a_0 has no real significance but is purely a matter of convenience, as we shall see later.

We pose the problem: to compute the coefficients a_k and b_k of the series for a given function $f(x)$.

---

* The function $f(x)$ has period ω if it satisfies the equation $f(x + \omega) = f(x)$.

† In fact, Dirichlet's theorem also applies to a certain class of discontinuous functions, the so-called functions of bounded variation. For discontinuous functions, of course, the corresponding series is nonuniformly convergent.

To this end we note the following equation:

$$\int_{-\pi}^{\pi} \cos kx \cos lx \, dx = 0 \quad (k \neq l; \; k, l = 0, 1, \cdots),$$

$$\int_{-\pi}^{\pi} \sin kx \sin lx \, dx = 0 \quad (k \neq l; \; k, l = 0, 1, \cdots),$$

$$\int_{-\pi}^{\pi} \sin kx \cos lx \, dx = 0 \quad (k, l = 0, 1, 2, \cdots), \tag{23}$$

$$\int_{-\pi}^{\pi} \cos^2 kx \, dx = \pi \quad (k = 1, 2, \cdots),$$

$$\int_{-\pi}^{\pi} \sin^2 kx \, dx = \pi \quad (k = 1, 2, \cdots),$$

which the reader may verify. These integrals are easy to compute by reducing the products of the various trigonometric functions to their sums and differences and their squares to expressions containing the corresponding trigonometric functions of double the angle. The first equation states that the integral, over a period of the function, of the product of two different functions from the sequence 1, $\cos x$, $\sin x$, $\cos 2x$, $\sin 2x$, $\cdots$ is equal to zero (the so-called orthogonality property of the trigonometric functions). On the other hand, the integral of the square of each of the functions of this sequence is equal to π. The first function, identically equal to one, forms an exception, since the integral of its square over the period is equal to 2π. It is this fact which makes it convenient to write the first term of the series (22) in the form $a_0/2$.

Now we can easily solve our problem. To compute the coefficient a_m, we multiply the left side and each term on the right side of the series (22) by $\cos mx$ and integrate term by term over a period 2π, as is permissible since the series obtained after multiplication by $\cos mx$ is uniformly convergent. By (23) all integrals on the right side, with the exception of the integral corresponding to $\cos mx$, will be zero, so that obviously

$$\int_{-\pi}^{\pi} f(x) \cos mx \, dx = a_m \pi,$$

hence

$$a_m = \frac{1}{\pi} \int_{-\pi}^{\pi} f(x) \cos mx \, dx \quad (m = 0, 1, 2, \cdots). \tag{24}$$

Similarly, multiplying the left and right sides of (22) by $\sin mx$ and integrating over the period, we get an expression for the coefficients

$$b_m = \frac{1}{\pi} \int_{-\pi}^{\pi} f(x) \sin mx \, dx \quad (m = 1, 2, \cdots), \tag{25}$$

and we have solved our problem. The numbers a_m and b_m computed by formulas (24) and (25) are called the *Fourier coefficients* of the function $f(x)$.

Let us take an example the function $f(x)$ of period 2π illustrated in figure 13. Obviously this function is continuous and satisfies Dirichlet's condition, so that its Fourier series converges uniformly to it.

It is easy to see that this function also satisfies the condition $f(-x) = -f(x)$. The same condition also clearly holds for the function $F_1(x) = f(x) \cos mx$, which means that the graph of $F_1(x)$ is symmetric with respect to the origin. From geometric arguments it is clear that $\int_{-\pi}^{\pi} F_1(x)\, dx = 0$, so that $a_m = 0$ ($m = 0, 1, 2, \cdots$). Further, it is not difficult to see that the functions $F_2(x) = f(x) \sin mx$ has a graph which is symmetric with respect to the axis Oy so that

$$b_m = \frac{1}{\pi} \int_{-\pi}^{\pi} F_2(x)\, dx = \frac{2}{\pi} \int_0^{\pi} F_2(x)\, dx.$$

But for even m this graph is symmetric with respect to the center $\pi/2$ of the segment $[0, \pi]$, so that $b_m = 0$ for even m. For odd $m = 2l + 1$ ($l = 0, 1, 2, \cdots$) the graph of $F_2(x)$ is symmetric with respect to the straight line $x = \pi/2$, so that

$$b_{2l+1} = \frac{4}{\pi} \int_0^{\pi/2} F_2(x)\, dx.$$

But, as can be seen from the sketch, on the segment $[0, \pi/2]$ we have simply $f(x) = x$, so that by integration by parts, we get

$$b_{2l+1} = \frac{4}{\pi} \int_0^{\pi/2} x \sin (2l + 1)x\, dx = \frac{4(-1)^l}{\pi(2l + 1)^2},$$

and consequently

$$f(x) = \frac{4}{\pi} \sum_{l=1}^{\infty} \frac{(-1)^l \sin (2l + 1)x}{(2l + 1)^2}.$$

Thus we have found the expansion of our function in a Fourier series.

Convergence of the Fourier partial sums to the generating function. In applications it is customary to take as an approximation to the function $f(x)$ of period 2π the sum

$$S_n = \frac{a_0}{2} + \sum_{1}^{n} (a_k \cos kx + b_k \sin kx)$$

of the first n terms of its Fourier series, and then there arises the question of the error of the approximation. If the function $f(x)$ of period 2π has a derivative $f^{(r)}(x)$ of order r which for all x satisfies the inequality $|f^{(r)}(x)| \leqslant K$, then the error of the approximation may be estimated as follows:

$$|f(x) - S_n(x)| \leqslant \frac{c_r K \ln n}{n^r},$$

where c_r is a constant depending only on r. We see that the error converges to zero with increasing n, the convergence being the more rapid the more derivatives the function has.

For a function which is analytic on the whole real axis there is an even better estimate, as follows:

$$|f(x) - S_n(x)| < cq^n, \tag{26}$$

where c and q are positive constants depending on f and $q < 1$. It is remarkable that the converse is also true, namely that if the inequality (26) holds for a given function, then the function is necessarily analytic. This fact, which was discovered at the beginning of the present century, in a certain sense reconciles the controversy between D. Bernoulli and his contemporaries. We can now state: If a function is expandable in a Fourier series which converges to it, this fact in itself is far from implying that the function is analytic; however, it will be analytic, if its deviation from the sum of the first n terms of the Fourier series decreases more rapidly than the terms of some decreasing geometric progression.

A comparison of the estimates of the approximations provided by the Fourier sums with the corresponding estimates for the best approximations of the same functions by trigonometric polynomials shows that for smooth functions the Fourier sums give very good approximations, which are in fact, close to the best approximations. But for nonsmooth continuous functions the situation is worse: Among these, for example, occur some functions whose Fourier series diverges on the set of all rational points.

It remains to note that in the theory of Fourier series there is a question which was raised long ago and has not yet been answered: Does there exist a continuous periodic function $f(x)$ whose Fourier series fails for all x to converge to the function as $n = \infty$? The best result in this direction is due to A. N. Kolmogorov, who proved in 1926 that there exists a periodic Lebesgue-integrable function whose Fourier series does not converge to it at any point. But a Lebesgue-integrable function may be discontinuous, as is the case with the function constructed by Kolmogorov. The problem still awaits its final solution.

To provide approximations by trigonometric polynomials to arbitrary continuous periodic functions, the methods of the so-called summation of Fourier series are in use at the present time. In place of the Fourier sums as an approximation to a given function we consider certain modifications of them. A very simple method of this sort was proposed by the Hungarian mathematician Fejér. For a continuous periodic function we first, in a purely formal way, construct its Fourier series, which may be divergent, and then form the arithmetic means of the first n partial sums

$$\sigma_n(x) = \frac{S_0(x) + S_1(x) + \cdots + S_n(x)}{n + 1}. \tag{27}$$

This is called the *Fejér sum* of order n corresponding to the given function $f(x)$. Fejér proved that as $n = \infty$ this sum converges uniformly to $f(x)$.

§8. Approximation in the Sense of the Mean Square

Let us return to the problem of the oscillating string. We assume that at a certain moment t_0 the string has the form $y = f(x)$. We can prove that its potential energy W, i.e., the work made available as it moves from the given position to its position of equilibrium, is equal (for small deviations of the string) to the integral $W = \int_0^l f'^2(x)\, dx$, at least up to a constant factor. Suppose now that we wish to approximate the function $f(x)$ by another function $\phi(x)$. Together with the given string, we will consider a string whose shape is defined by $\phi(x)$, and still a third string, defined by the function $f(x) - \phi(x)$. It may be proved that if the energy

$$\int_0^l [f'(x) - \phi'(x)]^2\, dx \tag{28}$$

of the third string is small, then the difference between the energy of the first two strings will also be small.* Thus, if it is important that the second string have an energy which differs little from the first, we must

---

* In fact, if

$$\int_0^l f'^2\, dx \leqslant M^2 \quad \text{and} \quad \int_0^l \phi'^2\, dx \leqslant M^2,$$

then

$$\left| \int_0^l f'^2\, dx - \int_0^l \phi'^2\, dx \right|$$

$$\leqslant \left(\sqrt{\int_0^l f'^2\, dx} + \sqrt{\int_0^l \phi'^2\, dx} \right) \left| \sqrt{\int_0^l f'^2\, dx} - \sqrt{\int_0^l \phi'^2\, dx} \right| \leqslant 2M \sqrt{\int_0^l (f' - \phi')^2\, dx}.$$

try to find a function $\phi'(x)$ for which the integral (28) will be as small as possible. We are thus led to the problem of approximation to a function (in this case $f'(x)$) in the sense of the mean square.

Here is how this problem is to be stated in the general case. On the interval $[a, b]$ we are given the function $F(x)$, and also the function

$$\Phi(x; \alpha_0, \alpha_1, \cdots, \alpha_n), \tag{29}$$

depending not only on x but also on the parameters $\alpha_0, \alpha_1, \cdots, \alpha_n$. It is required to choose these parameters in such a way as to minimize the integral

$$\int_a^b [F(x) - \Phi(x; \alpha_0, \alpha_1, \cdots, \alpha_n)]^2 \, dx. \tag{30}$$

This problem is very similar in idea to Čebyšev's problem. Here also the idea is to find the best approximation of the function $F(x)$ by functions of the family (29), but only in the sense of the mean square. It is now unimportant for us whether or not the difference $F - \Phi$ is small for all values of x on the interval $[a, b]$; on a small part of the interval the difference $F - \Phi$ may even be large provided only that the integral (30) is small, as is the case, for example, for the two graphs illustrated in figure 14. The smallness of the quantity (30) shows that the functions F and Φ are close to each other on by far the greater part on the interval.* As to the choice in practice of one method of approxi-

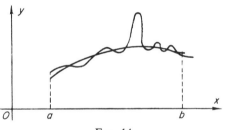

Fig. 14.

mation or another, everything depends on the purpose in view. In the earlier example of the string, it is natural to approximate the function $f'(x)$ in the sense of the mean square. On the other hand, the method of mean squares was unsatisfactory for Čebyšev in solving his problems in the construction of mechanisms, since a machine component projecting beyond the limits of tolerance, even if only over a very small part of the machine, would be quite intolerable: One such projection would spoil the whole machine. Thus Čebyšev had to develop a new mathematical method corresponding to the problem which confronted him.

---

* In Chapter XIX we will see that there is a profound analogy between the closeness of the functions in the sense of the mean square and the distance between points in ordinary space.

We should state that from the computational point of view the method of the mean square is more convenient, since it can be reduced to the application of well-developed methods of general analysis.

As an example let us consider the following characteristic problem.

We wish to make the best approximation in the sense of the mean square to a given continuous function $f(x)$ on the interval $[a, b]$ by sums of the form

$$\sum_{1}^{n} \alpha_k \phi_k(x),$$

where the α_k are constants and the functions $\phi_k(x)$ are continuous and form an orthogonal and normal system.

This last means that we have the following equations:

$$\int_a^b \phi_k \phi_l \, dx = 0 \quad k - l \quad (k, l = 1, 2, ..., n),$$

$$\int_a^b \phi_k^2 \, dx = 1 \quad (k = 1, 2, ..., n).$$

Let us introduce the numbers

$$a_k = \int_a^b f(x) \, \phi_k(x) \, dx \quad (k = 1, ..., n).$$

These numbers a_k are called the Fourier coefficients of f with respect to the ϕ_k.

For arbitrary coefficients α_k, on the basis of the properties of orthogonality and normality of ϕ_k, we have the equation

$$\int_a^b \left(f - \sum_{1}^{n} \alpha_k \phi_k \right)^2 dx = \int_a^b f^2 dx + \sum_{1}^{n} \alpha_k^2 - 2 \sum_{1}^{n} \alpha_k a_k$$

$$= \left(\int_a^b f^2 \, dx - \sum_{1}^{n} a_k^2 \right) + \sum_{1}^{n} (\alpha_k - a_k)^2.$$

The first term on the right side of the derived equation does not depend on the numbers α_k. Thus the right side will be smallest for those α_k which make the second term itself small, and obviously this can happen only if the numbers α_k are equal to the corresponding Fourier coefficients a_k.

Thus we have reached the following important result. If the functions ϕ_k form an orthogonal and normal system on the interval $[a, b]$, then the sum $\sum_1^n \alpha_k \phi_k(x)$ will be the best approximation, in the sense of the mean square, to the function $f(x)$ on this interval if and only if the numbers α_k are the Fourier coefficients of the function f with respect to $\phi_k(x)$.

On the basis of equation (23) it is easily established that the functions

$$\frac{1}{\sqrt{2\pi}}, \frac{\cos x}{\sqrt{\pi}}, \frac{\sin x}{\sqrt{\pi}}, \frac{\cos 2x}{\sqrt{\pi}}, \cdots$$

form an orthogonal and normal system on the interval $[0, 2\pi]$. Thus the stated proposition, as applied to the trigonometric functions, will have the following form.

The Fourier sum $S_n(x)$, computed for a given continuous function $f(x)$ of period 2π, is the best approximation, in the sense of the mean square, to the function $f(x)$ on the interval $[0, 2\pi]$, among all trigonometric polynomials

$$t_n(x) = \alpha_0 + \sum_{1}^{n} (\alpha_k \cos kx + \beta_k \sin kx)$$

of order n.

From this result and from Fejér's theorem, formulated in §7, we are led to another remarkable fact.

Let $f(x)$ be a continuous function of period 2π and $\sigma_n(x)$ be its Fejér sum of order n, defined in §7 by equation (27).

We introduce the notation

$$\max |f(x) - \sigma_n(x)| = \eta_n .$$

Since the Fourier sums $S_k(x)$ ($k = 0, 1, \ldots, n$) are trigonometric polynomials of order $k \leqslant n$, it is obvious that $\sigma_n(x)$ is a trigonometric polynomial of order n. Thus from the minimal property of the sum $S_n(x)$ shown previously, we have the inequality

$$\int_{-\pi}^{\pi} [f(x) - S_n(x)]^2 \, dx \leqslant \int_{-\pi}^{\pi} [f(x) - \sigma_n(x)]^2 \, dx \leqslant \int_{-\pi}^{\pi} \eta_n^2 \, dx = 2\pi\eta_n^2 .$$

Since, by Fejér's theorem, the quantity η_n converges to zero for $n \to \infty$ we obtain the following important result.

For any continuous function of period 2π we have the equation

$$\lim_{n\to\infty} \int_{-\pi}^{\pi} [f(x) - S_n(x)]^2 \, dx = 0.$$

In this case we say that the Fourier sum of order n of a continuous function $f(x)$ converges to $f(x)$ in the sense of the mean square, as n increases beyond all bounds.

In fact, this statement is true for a wider class of functions, namely those which are integrable, together with their square, in the sense of Lebesgue.

We will stop here and will not present other interesting facts from the theory of Fourier series and orthogonal functions, based on approximation in the sense of the mean square. Important physical applications of orthogonal systems of functions have already been introduced in Chapter VI. Finally, we note that these questions are also discussed from a somewhat different point of view in Chapter XIX.

Suggested Reading

N. I. Ahiezer, *Theory of approximation*, Frederick Ungar, New York, 1956.

D. Jackson, *The theory of approximation*, American Mathematical Society, Providence, R. I., 1930.

J. L. Walsh, *Interpolation and approximation by rational functions in the complex domain*, 2nd ed., American Mathematical Society, Providence, R. I., 1956.

CHAPTER XIII

APPROXIMATION METHODS
AND COMPUTING TECHNIQUES

§1. Approximation and Numerical Methods

Characteristic peculiarities of approximation methods. In many cases the application of mathematics to the study of events in the outside world is based on the fact that the laws governing these events have a quantitative character and can be described by certain formulas, equations, or inequaltities. This allows us to investigate the events numerically and to make the calculations which are so necessary in practical life.

As soon as a quantitative law has been found, purely mathematical methods may be used to investigate it. For definiteness, let us take some law which is described by an equation. This may be the law of motion of a body in Newtonian mechanics, the law of heat conduction or the propagation of electromagnetic oscillations, and so forth. Such equations are discussed in detail in Chapters V and VI. Usually the equation has adjoined to it certain conditions which its solution must satisfy (in Chapters V and VI these are the boundary and initial conditions) and which define a unique solution.

The first and most important mathematical tasks here will be the following:

1. To establish the existence of a solution. Even if it seems obvious from the physical point of view that the problem has a solution, a mathematical proof of the solvability of a rigorously formulated problem is usually considered as the necessary evidence that the mathematical formulation of the problem is a satisfactory one. In a wide class of problems it is possible to establish mathematically the existence of a solution.

2. To attempt to find an explicit expression or *formula* for the quantity

which characterizes the event under consideration. Usually such an expression can be found only in the simplest cases. It often happens that the explicit expression obtained is so complicated that to make use of it for the desired numerical results is very difficult or even impossible.

3. To find a procedure for constructing an *approximation formula*, which gives a solution with any desired degree of accuracy. This can be done in many cases.

4. But very often it will be possible to find one or more methods for direct *numerical calculation* of the solution.

The development of such numerical methods (many of which are approximate) of solving problems of science and technology has produced a particular branch of mathematics that at the present time is usually called mathematics of computation.

The methods of computational mathematics are naturally approximative, since every quantity is computed only to a certain number of significant figures; for example, to five, six, etc., decimal places.

For applications this is sufficient, since knowing the exact value of any quantity is often unnecessary. In technical questions, for example, the desired quantity usually serves to define the dimensions or other parameters of a manufactured article. Every manufacturing process is only approximate, so that technical computations with an exactness which goes beyond the allowed "tolerances" are obviously valueless.

So for computational purposes there is no need of exact formulas or of exact solutions of equations. Exact formulas and equations may be replaced by others that are admittedly inexact, provided they are close enough to the original ones that the error produced by such a change does not exceed given bounds.

Later we shall return to this question of replacing one problem by another. At the moment, however, we merely wish to emphasize the first characteristic feature of computational methods, namely that by their very nature they can, as a rule, produce only approximate results; but then only such results are needed in practice.

We now turn our attention to a second aspect of computational methods in mathematics. In any computation we can operate with only a finite number of digits and obtain all the results after a finite number of arithmetic operations. If we perform the computations according to some formula, then the latter must first have been transformed in such a way that it involves only a finite number of terms with a finite number of parameters. It is known, for example, that many functions may be represented as the sum of a power series

$$f(x) = c_0 + c_1 x + c_2 x^2 + \cdots. \tag{1}$$

Thus, the function $\sin x$, where x is the radian measure of an angle, may be expanded in the power series

$$\sin x = \frac{x}{1!} - \frac{x^3}{3!} + \frac{x^5}{5!} - \cdots.$$

To find the exact value of $f(x)$, we would need to sum up "all" the terms of the series (1), but generally speaking, this is impossible. To find $f(x)$ approximately, it is sufficient to take only a certain finite number of terms of the series. For example, it may be proved that to compute $\sin x$ with an accuracy of 10^{-5} for an angle from zero to half a right angle it is sufficient to take the terms through x^5, so that $\sin x$ is replaced by the polynomial

$$\frac{x}{1!} - \frac{x^3}{3!} + \frac{x^5}{5!}.$$

For the numerical solution of a problem of mathematical analysis that consists of determining some function, we must by one means or another replace this problem by the problem of finding certain numerical parameters, the knowledge of which enables us to make an approximate computation of the unknown function. We will illustrate this by an example.

Let it be required to solve, on the interval $a \leqslant x \leqslant b$, the boundary-value problem for the differential equation

$$L(y) - f(x) = y'' + p(x)\,y' + q(x)\,y - f(x) = 0 \qquad (2)$$

with boundary conditions $y(a) = 0$, $y(b) = 0$. In one of the possible methods of solution, namely Galerkin's method, we start with a system of linearly independent functions $\omega_1(x)$, $\omega_2(x)$, $\cdots$, which satisfy the boundary conditions (Chapter VI, §5). This system is so chosen as to be "complete" in the sense that a function which is integrable on $[a, b]$ and is orthogonal to all the ω_k ($k = 1, 2, \cdots$) will be equal to zero at all (more exactly, at "almost all") points of the interval. The condition that $y(x)$ satisfies the differential equation (2) may be described in the form of an orthogonality requirement

$$\int_a^b [L(y) - f]\omega_k \, dx = 0 \qquad (k = 1, 2, \cdots). \qquad (3)$$

Let us assume that the solution of the problem may be expanded in a series in the ω_k

$$y(x) = a_1\omega_1(x) + a_2\omega_2(x) + \cdots. \qquad (4)$$

We now seek to determine the conditions that must be satisfied by the coefficients a_k. For arbitrary a_k the sum of the series (4) will satisfy the boundary conditions. It remains to choose the a_k in such a way that equations (3) are satisfied. The coefficients a_k form an infinite set, and to compute all of them is generally speaking impossible. For simplification we retain only a finite number of terms on the right side of (4) and so obtain the expression

$$y(x) \approx a_1\omega_1(x) + \cdots + a_n\omega_n(x). \tag{5}$$

We cannot hope to satisfy equation (3) for all ω_k ($k = 1, 2, \cdots$) since we have only n arbitrary parameters a_k ($k = 1, 2, \cdots, n$). Thus we are forced to give up an exact solution of the differential equation (2). But it is natural to expect that the sum (5) will satisfy this differential equation with a small error if n is taken sufficiently large and condition (3) is satisfied for the first n of the functions ω_k. This leads to the equations of Galerkin's method

$$\int_a^b \left[L\left(\sum_{k=1}^n a_k\omega_k \right) - f \right] \omega_i \, dx = 0 \qquad (i = 1, 2, \cdots, n).$$

After finding the a_k from these equations, we construct an approximate expression for the function (5).

A similar simplified formula holds for the solution of variational problems by the Ritz method, in approximate harmonic analysis of functions and in many other questions.

We give another example of simplification of an equation. Let it be required to find a function y of one or several arguments by solving some functional equation, for example, a differential or an integral equation. As parameters defining the function y let us choose its values $y_1, y_2, \cdots, y_n$ at some system of points (on a net).

The functional equation must then be changed to a system of numerical equations containing n unknown quantities y_k ($k = 1, \cdots, n$). Such a replacement may, as a rule, be made in many ways. Here it is always necessary to take pains that the solution of the numerical system differs sufficiently little from the solution of the functional equation.

We give several examples of this sort of replacement. When we solve a differential equation of the first order $y' = f(x, y)$ by Euler's method, we replace this equation by a recursive numerical scheme which enables us to make an approximate calculation of each succeeding value of the unknown function from the previous value (Chapter V, §5):

$$y_{n+1} = y_n + (x_{n+1} - x_n)f(x_n, y_n).$$

For an approximate solution of the Laplace equation

$$\Delta u = \frac{\partial^2 u}{\partial x^2} + \frac{\partial^2 u}{\partial y^2} = 0$$

by the net method, we replace this equation by a linear algebraic system (Chapter VI, §5)

$$u(x + h, y) + u(x, y + h) + u(x - h, y) + u(x, y - h) - 4u(x, y) = 0.$$

Let us consider one more example of such a kind. Let it be required to solve numerically the integral equation

$$y(x) = f(x) + \int_a^b K(x, s)\, y(s)\, ds. \tag{6}$$

The points at which we wish to find the values of the unknown function $y(x)$ will be denoted by $x_1, x_2, \cdots, x_n$. In order to set up the system of numerical equations replacing (6), we require that equation (6) be satisfied not for all the x on the interval $a \leqslant x \leqslant b$ but only at the points x_i $(i = 1, 2, \cdots, n)$

$$y(x_i) = f(x_i) + \int_a^b K(x_i, s)\, y(s)\, ds.$$

Then we replace the integral by any approximate quadrature (by the trapezoidal rule, Simpson's rule, or some other)* with the points of division $x_1, \cdots, x_n$

$$\int_a^b K(x_i, s)\, y(s)\, ds \approx \sum_{j=1}^n A_{ij} K(x_i, x_j)\, y(x_j).$$

To determine the desired values of $y(x_i)$, we have the system of linear algebraic equations

$$y(x_i) = f(x_i) + \sum_{j=1}^n A_{ij} K(x_i, y_j)\, y(x_j) \qquad (i = 1, 2, \cdots, n). \tag{7}$$

We note that all the methods considered of seeking an unknown function have involved determining certain parameters which define it

---

* Cf. Chapter XII, §3.

approximately. Thus the exactness of these methods depends on how well the function is defined by this system of parameters; for example, how well it may be approximated by an expression of the form (7) or represented by its values at a certain system of points. Questions of this kind constitute a particular branch of mathematics, called the theory of approximation of functions (Chapter XII). From this it can be seen that the theory of approximation has very great value for applied mathematics.

Convergence of approximate methods and an estimate of error. Let us examine in more detail the requirements for a computational method. The simplest and most basic of these requirements is the possibility of finding the desired quantity with any chosen degree of accuracy.

The required exactness of a computation may change greatly from one problem to another. For certain rough technical computations, two or three decimal places will be sufficiently exact. Most engineering computations are carried out to three or four decimal places. But considerably greater exactness is often required in scientific calculations. Generally speaking, the need for greater accuracy has increased with the passage of time.

Particularly important, therefore, are the approximation methods and processes that allow one to get results with as great a degree of accuracy as desired. Such methods are called *convergent*. Since they are encountered most often in practice and since the requirements they must satisfy are typical, we will keep them in mind in what follows.

Let x be the exact value of a desired quantity. For every such method we may construct a sequence of approximations, $x_1, x_2, \cdots, x_n, \cdots$ to the solution x.

After showing how the approximations are constructed, the first problem in the theory of the method is to establish the convergence of the approximations to the solution $x_n \rightarrow x$, and if the method is not always convergent, to set out the conditions under which it will converge.

After the convergence is established there arises the more difficult and subtle problem of an estimate of the *rapidity of convergence*, i.e., an estimate of how rapidly x_n converges to the solution x for $n \rightarrow \infty$. Every convergent method theoretically guarantees the possibility of finding the solution with any desired degree of accuracy, if we take an approximation x_n with sufficiently large index n. But, as a rule, the larger the n, the greater the labor required to calculate x_n. Thus, if x_n converges slowly to x, then to get the needed accuracy it may be necessary to make enormous computations.

In mathematics itself, and especially in its applications, many cases are known of a convergent process for finding the solution x, which would require more computational work than can be carried out even on present-day high-speed computers.*

Insufficiently fast convergence is one of the criteria by which the disadvantages of a given method are judged. But this criterion is, of course, not the only one and in comparing methods one must consider many other sides of the question, in particular the convenience of making the computations on machines. Of two methods we sometimes prefer to use the one with somewhat slower convergence, if the computations by this method are easier to carry out on a computing machine.

The error produced by replacing x with its approximate value x_n is equal to the difference $x - x_n$. Its exact value is unknown, and in order to estimate the rapidity of convergence, we must find an upper bound for the absolute value of this difference, i.e., a quantity A_n, such that

$$| x - x_n | \leqslant A_n,$$

which we call an *error estimate*. Later we give examples of estimates A_n. Consequently, the usual method of judging the rapidity of convergence of a method is to examine how fast the estimate A_n decreases with increasing n. In order that the estimate reflects the actual degree of nearness of x_n to x, it is necessary that A_n differ little from $| x - x_n |$. Also the estimate A_n must be effective, i.e., be such that it can itself be found, otherwise it cannot be used.

Let x be a numerical variable whose value we wish to determine from some equation. We assume that our equation reduces to the form

$$x = \phi(x). \tag{8}$$

---

* Let us mention some simple examples of slowly converging computational processes. It is known that the series

$$\frac{1}{1} - \frac{1}{2} + \frac{1}{3} - \frac{1}{4} + \cdots$$

converges to the natural logarithm of the number 2. We can find ln 2 approximately by means of this series, by computing the sum

$$s_n = \frac{1}{1} - \frac{1}{2} + \cdots \pm \frac{1}{n}$$

of the first n terms for sufficiently large n. But it may be shown that to compute ln 2 with an error less than half of the fifth significant figure, we must take more than 100,000 terms of the series. To find the sum of such a number of terms, if we are using, for example, only a desk computer, would be very laborious. Another familiar example is the series

$$\frac{1}{\sqrt{2}} = 1 - \frac{1}{2 \cdot 1!} + \frac{1 \cdot 3}{2^2 \cdot 2!} - \frac{1 \cdot 3 \cdot 5}{2^4 \cdot 3!} + \frac{1 \cdot 3 \cdot 5 \cdot 7}{2^4 \cdot 4!} - \cdots.$$

Its convergence is so slow that to compute $1/\sqrt{2}$ with accuracy of 10^{-5}, we would need to take about 10^{10} terms, which is difficult even with high-speed machines.

To this equation we apply the *method of iteration*, which is also often called the *method of successive approximations*. To explain the method itself and the estimates connected with it, we will examine the case of one numerical equation, although the method also applies to systems of numerical equations, to differential equations, integral equations, and many other cases. The application of the method to ordinary differential equations has already been illustrated in Chapter V, §5.

We will assume that we have somehow found an approximate value x_0 for a root of the equation. If x_0 were an exact solution of equation (8), then after substituting it in the right side $\phi(x)$ of the equation we would get a result equal to x_0. But since x_0, generally speaking, is not an exact solution, the result of the substitution will differ from x_0. Let us denote it by $x_1 = \phi(x_0)$.

In order to establish in which cases x_1 will be nearer to the exact solution than x_0, we turn to a geometric interpretation of our problem. Let us consider the function

$$y = \phi(x). \tag{9}$$

We choose a numerical axis and represent the numbers x and y by points of this axis. Equation (9) assigns to every point x a corresponding point y on the same axis. It may be regarded as a rule that produces a point transformation of the numerical axis into itself.

Consider the segment $[x_1, x_2]$ on the numerical axis. By the transformation (9) the points x_1 and x_2 will be transformed into the points

$$y_1 = \phi(x_1) \text{ and } y_2 = \phi(x_2).$$

The segment $[x_1, x_2]$ is transformed into the segment $[y_1, y_2]$. The ratio

$$k = \frac{|y_2 - y_1|}{|x_2 - x_1|}$$

is called the "coefficient of dilation" of the segment under the transformation. If $k < 1$, we will have a contraction of the segment.

We return to equation (8). It says that the desired point x must be transformed into itself under the transformation (9). Thus solving equation (8) is equivalent to finding a point on the numerical axis which is transformed into itself under the transformation (9), i.e., remains fixed.

We now consider the segment $[x, x_0]$, one end of which lies at the fixed point x and the other at the point x_0. Under the given transformation x_0 goes into x_1 and the segment $[x, x_0]$ into the segment $[x, x_1]$. If the function ϕ has the property that under transformation (9) every segment is contracted, then x_1 will certainly be closer than x_0 to the root of equation (8).

Since we wish to obtain approximations which converge to the exact solution of (8), we make the same transformation many times in succession on the right side of (8) and construct the sequence of numbers

$$x_1 = \phi(x_0), \quad x_2 = \phi(x_1), \cdots, x_{n+1} = \phi(x_n), \cdots. \tag{10}$$

Here we will prove that the sequence of approximations (10) converges.*

Let us assume that the function $\phi(x)$ is defined on a certain segment $[a, b]$ and that equation (9) gives a transformation of $[a, b]$ into itself, i.e., for every x belonging to $[a, b]$, $y = \phi(x)$ will also belong to $[a, b]$. We will also assume that the initial approximation x_0 is in $[a, b]$; all the successive approximations (10) will then also lie in $[a, b]$. Under these conditions the following theorem is true. If $\phi(x)$ has a derivative ϕ' satisfying the condition

$$|\phi'| \leqslant q < 1$$

on $[a, b]$, then the following proposition holds. Equation (8) has a root x^* in the segment $[a, b]$. The sequence (10) converges to this root, and the rapidity of convergence is characterized by the estimate

$$|x^* - x_n| \leqslant \frac{m}{1 - q} q^n,$$

where $m = |x_0 - \phi(x_0)| = |x_0 - x_1|$. Equation (8) has a unique root in $[a, b]$.

To prove these statements, we estimate the difference $x_2 - x_1$. If Taylor's formula is applicable (Chapter II, §9, (26)), we obtain, for $n = 0$

$$x_2 - x_1 = \phi(x_1) - \phi(x_0) = \phi'(\xi_0)(x_1 - x_0).$$

Then ξ_0 lies between x_1 and x_0 and so belongs to the segment $[a, b]$. Thus $|\phi'(\xi_0)| \leqslant q$ and

$$|x_2 - x_1| \leqslant q |x_1 - x_0| = mq.$$

Similarly

$$|x_3 - x_2| = |\phi(x_2) - \phi(x_1)| = |\phi'(\xi_1)(x_2 - x_1)| \leqslant q |x_2 - x_1| \leqslant mq^2.$$

Continuing these estimates, we have, for every value of n, the inequality

$$|x_{n+1} - x_n| \leqslant mq^n. \tag{11}$$

---

* Because of the geometric interpretation, this theorem and others like it are often called contraction theorems.

We now establish the convergence of the sequence x_n. To this end we consider the auxiliary series

$$x_0 + (x_1 - x_0) + (x_2 - x_1) + \cdots + (x_n - x_{n-1}) + \cdots . \qquad (12)$$

The partial sum of the first $n + 1$ of its terms is equal to

$$s_{n+1} = x_0 + (x_1 - x_0) + \cdots + (x_n - x_{n-1}) = x_n .$$

Thus $\lim_{n\to\infty} s_{n+1} = \lim_{n\to\infty} x_n$ and the existence of a finite limit for x_n is equivalent to the convergence of the series (12). We compare the series (12) with the series

$$|x_0| + m + mq + \cdots + mq^{n-1} + \cdots .$$

From the estimate (11) the terms of the series (12) are not greater in absolute value than the corresponding terms in the latter series. But this series, except for its first term $|x_0|$, is a geometric progression with common ratio q, and since $q < 1$, the series converges. Series (12) is thus also convergent, and the sequence (10) is convergent to some finite limit x^*

$$\lim_{n\to\infty} x_n = x^*.$$

Obviously x^* belongs to the segment $[a, b]$, since all the x_n belong to it.

If in the equation $x_{n+1} = \phi(x_n)$ we pass to the limit as $n \to \infty$, then in the limit we get the equation $x^* = \phi(x^*)$, which shows that x^* actually satisfies equation (8). We now estimate how close x_n is to x^*. We choose x_n and any following approximation x_{n+p}

$$|x_{n+p} - x_n| = |(x_{n+p} - x_{n+p-1}) + (x_{n+p-1} - x_{n+p-2}) + \cdots + (x_{n+1} - x_n)|$$

$$\leqslant mq^{n+p-1} + mq^{n+p-2} + \cdots + mq^n$$

$$= \frac{mq^n - mq^{p+n}}{1 - q} .$$

Hence, for $p \to \infty$, from $x_{n+p} \to x^*$ and $q^{n+p} \to 0$ it follows that

$$|x^* - x_n| \leqslant \frac{m}{1 - q} q^n.$$

It remains to prove the statement on uniqueness. Let x' be any solution of the equation on $[a, b]$. We estimate the difference $x' - x^*$

$$|x' - x^*| = |\phi(x') - \phi(x^*)| = |\phi'(\xi)(x' - x^*)| \leqslant q |x' - x^*|,$$

from which
$$(1 - q)|\, x' - x^*\, | \leqslant 0.$$

Since $1 - q > 0$, this inequality is possible only for $|\, x' - x^*\, | = 0$, which means that x' is identical with x^*.

The theorem not only exhibits sufficient conditions for the convergence of the method of iteration but also allows us to estimate the necessary number of steps in the computation, i.e., how large n must be taken to obtain the required accuracy when the exact solution x^* is replaced by x_n. Such an estimate is effective, since the quantities m and q appearing in the inequality $|\, x^* - x_n\, | \leqslant (m/1 - q)q^n$ may in fact be found by investigating the function ϕ.

As an example let us consider the equation $x = k \tan x$, which has many practical applications. For definiteness, we consider the case $k = 0.5$. Let it be required to find the smallest positive root of the equation $x = \frac{1}{2} \tan x$. It must lie near the point 1 and be somewhat larger than 1, as can be easily established from any table or graph of the function $\tan x$.

To secure the condition $|\, \phi'\, | \leqslant q < 1$, which enters into the theorem on the convergence of the method of iteration, we invert the function $\tan x$ and consider the equation $x = \arctan 2x$, which is equivalent to the given one.

We give here the results of the computation. For the original approximation we have taken the value $x_0 = 1$. The following approximations are computed from a table of the function $\arctan x$, from which one finds the following numerical values

$$
\begin{aligned}
x_1 &= \arctan 2 &&= 1.10715, \\
x_2 &= \arctan 2.21430 &&= 1.14660, \\
x_3 &= \arctan 2.29320 &&= 1.15959, \\
x_4 &= \arctan 2.31918 &&= 1.16370, \\
x_5 &= \arctan 2.32740 &&= 1.16498, \\
x_6 &= \arctan 2.32996 &&= 1.16538, \\
x_7 &= \arctan 2.33076 &&= 1.16550, \\
x_8 &= \arctan 2.33100 &&= 1.16554, \\
x_9 &= \arctan 2.33108 &&= 1.16555, \\
x_{10} &= \arctan 2.33110 &&= 1.16556, \\
x_{11} &= \arctan 2.33112 &&= 1.16556.
\end{aligned}
$$

The computation may be stopped here, since further iterations will repeat the value of the root

$$x^* = 1.16556.$$

A geometric illustration of the approximations to the root is given in figure 1. Here x_n tends to x^* so rapidly that x_4 is already indistinguishable from x^* in the diagram.

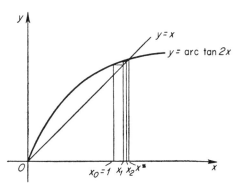

Let us give one more example of the method of iteration. We solve numerically the integral equation

$$y(x) = \frac{1}{6} \int_0^1 e^{xt} y(t)\, dt + e^x$$
$$- \frac{1}{6} \frac{1}{x+1} (e^{x+1} - 1). \quad (13)$$

Fig. 1.

Its exact solution is $y = e^x$.

First we replace the integral equation by a system of linear algebraic equations. To this end the interval of integration [0, 1] is divided into four equal parts at the points $t = 0, \frac{1}{4}, \frac{1}{2}, \frac{3}{4}, 1$. The values of the unknown function y at these points will be denoted by y_0, y_1, y_2, y_3, y_4, respectively. If we require that the equation be satisfied for $x_0 = 0, \frac{1}{4}, \frac{1}{2}, \frac{3}{4}, 1$, when the integral is replaced by Simpson's sum for four partial intervals (Chapter XII, §3, (6)), we have the following system of equations for y_k:

$$y_0 = \tfrac{1}{6}(0.083333 y_0 + 0.333333 y_1 + 0.166667 y_2$$
$$+ 0.333333 y_3 + 0.083333 y_4) + 0.713619,$$

$$y_1 = \tfrac{1}{6}(0.083333 y_0 + 0.354831 y_1 + 0.188858 y_2$$
$$+ 0.402077 y_3 + 0.107002 y_4) + 0.951980,$$

$$y_2 = \tfrac{1}{6}(0.083333 y_0 + 0.377716 y_1 + 0.214004 y_2$$
$$+ 0.484997 y_3 + 0.137393 y_4) + 1.261867,$$

$$y_3 = \tfrac{1}{6}(0.083333 y_0 + 0.402077 y_1 + 0.242499 y_2$$
$$+ 0.585018 y_3 + 0.176417 y_4) + 1.664181,$$

$$y_4 = \tfrac{1}{6}(0.083333 y_0 + 0.428008 y_1 + 0.274787 y_2$$
$$+ 0.705667 y_3 + 0.226523 y_4) + 2.185861.$$

This system is solved by the method of iteration. As our initial approximation to y_k ($k = 0, 1, 2, 3, 4$) we will take the constant terms of the corresponding equations: $y_0^{(0)} = 0.713619$, $y_1^{(0)} = 0.951980$, $\cdots$. The values found for the successive approximations are given in Table 1:

Table 1.

Number of Approximation	y_0	y_1	y_2	y_3	y_4
1	0.93428	1.20841	1.56129	2.01542	2.59972
2	0.98517	1.26699	1.62905	2.09419	2.69173
3	0.99667	1.28021	1.64433	2.11194	2.71245
4	0.99926	1.28319	1.64778	2.11595	2.71713
5	0.99985	1.28386	1.64856	2.11685	2.71818
6	0.99998	1.28402	1.64873	2.11705	2.71842
7	1.00001	1.28405	1.64877	2.11710	2.71847
Value of the exact solution	1.00000	1.28403	1.64872	2.11700	2.71828

At the end of Table 1 the value of the exact solution is given for comparison. Further approximations would not improve the values of y_k. The divergence in the last digits in the y_k comes from the error introduced by replacing the integral by Simpson's sum.

Stability of approximate methods. The needs of practical computation impose on approximative methods another general requirement that must be kept in mind because of its great importance. This is the requirement of the *stability* of the computational process. The essence of the matter is as follows: Every approximative method leads to some computational scheme, and it often turns out that to produce all the required numbers, we must carry out a long series of computational steps in accordance with the scheme. At each step the computation is not carried out exactly but only to some specific number of significant figures, and thus at each step we introduce a small error. All such errors will have their influence on the final results.

The computational scheme adopted may sometimes turn out to be so unsatisfactory that small errors made at the beginning may have a greater and greater influence as the calculations are carried further and may produce in the final stages a wide deviation from the exact values.

Let us consider the numerical solution of a differential equation

$$y' = f(x, y)$$

with the initial condition $y(x_0) = y_0$, where we are required to find the values of $y(x)$ for equally spaced values $x_k = x_0 + kh$ $(k = 0, 1, \cdots)$.

We assume that the computation has begun and has been carried out to step n with the results shown in Table 2.

Table 2.

x	y	$y' = f$
x_0	y_0	y'_0
x_1	y_1	y'_1
$\cdots$	$\cdots$	$\cdots$
x_{n-1}	y_{n-1}	y'_{n-1}
x_n	y_n	y'_n

We must now find y_{n+1}. By the Euler method of broken lines we make the approximation

$$y_{n+1} = y_n + hy'_n. \qquad (14)$$

Here y_{n+1} is calculated only from the numbers y_n and y'_n which occur in the last line of Table 2. Suppose we wish to increase the accuracy and for this purpose make use of all the quantities appearing in the last two lines. Then we may construct the computational formula

$$y_{n+1} = -4y_n + 5y_{n-1} + h(4y'_n + 2y'_{n-1}). \qquad (15)$$

We note that if the computation is absolutely exact, i.e., with an infinite number of significant digits, then formula (14) will give the exact result whenever y is a linear polynomial, and formula (15) will be exact for every polynomial of degree through the third. It would seem at first glance that the results produced by applying formula (15) must be more exact than those found by the method of broken lines. However, it can easily be seen that formula (15) is inappropriate for computation, since its application may produce a rapid increase in the error.

The values of the derivative y'_n and y'_{n-1} contain a small multiplier h, so that the errors in these values have less influence than the errors in y_n and y_{n-1}. For simplicity we will assume that the values of y' are found exactly so that we do not need to take them into account in the following attempt to estimate the error in general in the above two cases. Let us suppose that in finding y_{n-1}, we make an error of $+ \epsilon$, and in finding y_n an error of $- \epsilon$. Then, as equation (15) shows, in y_{n+1} we will make an error of the magnitude of $+ 9\epsilon$. In y_{n+2} the error will be $- 41\epsilon$ and will grow rapidly as we continue. Formula (15) leads to a computational process that is unstable with respect to errors and must be discarded.

The example given shows how badly the results may be distorted by an unstable computational scheme. Here we have solved the differential equation $y' = y$ with the initial condition $y_0 = 1$. The exact solution is

$y + e^x$. For the numerical solution we took equally spaced values of the independent variable x with steps $h = 0.01$, i.e., $x_k = 0.01\ k$. An approximate solution was computed in two ways: by the method of broken lines (14) and by formula (15). For comparison, Table 3 gives the value of the exact solution to seven decimal places.

The approximate values of the solution found by formula (15) are more exact for the first few steps than the results given by the method of broken lines. But after a small number of steps the instability of formula (15) begins to distort the approximate values of y_k quite strongly and leads to numbers which are very different from the true values of y_k.

Table 3.

x	Values of the Exact Solution	Values of the Approximation Solutions Computed	
		by Formula (14)	by Formula (15)
0.00	1.0000000	1.0000000	1.0000000
0.01	1.0100502	1.0100000	1.0100502
0.02	1.0202013	1.0201000	1.0202012
0.03	1.0304545	1.0303010	1.0304553
0.04	1.0408108	1.0406040	1.0408070
0.05	1.0512711	1.0510100	1.0512899
0.06	1.0618365	1.0615201	1.0617431
0.07	1.0725082	1.0721353	1.0729726
0.08	1.0832871	1.0828567	1.0809789
0.09	1.0941743	1.0936853	1.1056460
0.10	1.1051709	1.1046222	1.0481559
0.11	1.1162781	1.1156684	1.3996456
0.12	1.1274969	1.1268250	-0.2808540

Choice of computational methods. Every computation may in the final analysis be reduced to the four arithmetic operations of addition, subtraction, multiplication, and division. Describing a method of computation consists of stating the initial data with which one begins and then prescribing which arithmetical operations, and in which order, are to be performed in order to get the desired results. Let us show by a very simple example how much depends in the organization of the calculations on the experience and knowledge of the mathematician responsible for setting up the computational scheme and what excellent results can be obtained by a suitable choice of methods especially adapted to the situation.

Let it be required to solve the system of n equations in n unknowns $x_1, x_2, \cdots, x_n$

$$a_{11}x_1 + a_{12}x_2 + \cdots + a_{1n}x_n = b_1,$$
$$a_{21}x_1 + a_{22}x_2 + \cdots + a_{2n}x_n = b_2,$$
$$\dots\dots\dots\dots\dots\dots\dots\dots\dots\dots\dots\dots$$
$$a_{n1}x_1 + a_{n2}x_2 + \cdots + a_{nn}x_n = b_n.$$

From the theory of algebraic systems (Chapter XVI, §3) we have an explicit expression for the values of the unknowns by means of determinants

$$x_j = \frac{\Delta_j}{\Delta} \qquad (j = 1, 2, \cdots, n). \tag{16}$$

Here Δ is the determinant of the system

$$\Delta = \begin{vmatrix} a_{11} & a_{12} & \cdots & a_{1n} \\ a_{21} & a_{22} & \cdots & a_{2n} \\ \cdots\cdots\cdots\cdots\cdots \\ a_{n1} & a_{n2} & \cdots & a_{nn} \end{vmatrix},$$

and Δ_j is the determinant obtained from Δ by replacing its jth column by the column of constant terms in the system.

Let us assume that we wish to make use of formula (16) to solve the system and that we have begun to compute the determinants on the basis of their usual definition, without recourse to any simplifications. How many multiplications and divisions will be necessary? (Addition and subtraction will not be taken into account, since they are relatively simple operations.) We face the prospect of computing $n + 1$ determinants of order n. Each of them consists of $n!$ terms, each term being the product of n factors and consequently requiring $n - 1$ multiplications. For the computation of all the determinants, we must carry out $(n + 1)\, n! \times (n - 1)$ multiplications. The total number of multiplications and divisions will be equal to $(n^2 - 1)\, n! + n$.

We now choose another method of solving the system, namely successive elimination of the unknowns. The scheme of computation corresponding to this method is associated with the name of Gauss. We find x_1 from the first equation of the system

$$x_1 = \frac{b_1}{a_{11}} - \frac{a_{12}}{a_{11}} x_2 - \cdots - \frac{a_{1n}}{a_{11}} x_n.$$

For this we need n divisions. Substituting x_1 in each of the following $n - 1$ equations requires n multiplications. The elimination of x_1 and the setting up of $n - 1$ equations in the unknowns $x_2, \cdots, x_n$ will then require n^2

multiplications and divisions. Continuing in this way, we find that to compute all the values of x_j ($j = 1, \cdots, n$) the elimination method requires $n/6\,(2n^2 + 9n - 5)$ multiplications and divisions. Let us compare these two results. For the solution of a system of five equations in the first case we would need 2,885 multiplications and divisions, and in the second case 75.

For a system of ten equations the number of operations will be $(10^2 - 1)\,10! + 10 \approx 360,000,000$ and $10/6\,(2 \cdot 10^2 + 9 \cdot 10 - 5) = 475$, respectively. So we see that the amount of computational labor depends very strongly on the choice of the method of computing. In organizing the scheme of computation, it is often possible by a rational choice of the method to reduce the necessary amount of work very greatly.

§2. The Simplest Auxiliary Means of Computation*

Tables. The oldest auxiliary means of computation consists of tables. The simplest tables, e.g. the multiplication table and tables of logarithms or of the trigonometric functions, are certainly well known to the reader. The range of problems that are solvable in practical affairs is being continuously extended. New problems are often solved by the application of new formulas or may lead to new functions, so that the number of tables required is constantly increasing.

Every table, regardless of how it is constructed, contains the results of earlier computations and therefore represents a sort of mathematical memory. Printed or written tables are intended to be read by human beings. But we might also consider tables formed in some special manner, for example by holes punched in some special manner in cards, which are intended to be read by computing machines. But such tables are considerably rare and we will not discuss them here.

The tables in widest use are those of the values of functions. If a function y depends on only one argument x, then the simplest table corresponding to it has the form

x	y
x_1	y_1
x_2	y_2
$\cdots$	$\cdots$
x_n	y_n

(17)

---

* In this section we give a description only of the simplest auxiliary equipment and machines. The description of contemporary rapid computing machines is given in Chapter XIV. For lack of space we have also omitted graphical methods.

This is called a single-entry table.* From it we may take without further effort only the values corresponding to tabulated values of x. Values corresponding to x not in the table must be found by interpolation of various kinds, as described in Chapter XII.† Consequently the tables often contain, in addition to the values of the functions, certain auxiliary quantities which make the interpolation easier. Usually these are values of the first or second differences. More specialized tables require specially devised interpolation formulas for which they include the corresponding data.

In a table of a function of two arguments $u = f(x, y)$ the values of the function are distributed in a double-entry table of the following form

x \\ y	y_1	y_2	$\ldots$	y_m	
x_1	u_{11}	u_{12}	$\ldots$	u_{1m}	
x_2	u_{21}	u_{22}	$\ldots$	u_{2m}	(17')
$\ldots$	$\ldots$	$\ldots$	$\ldots$	$\ldots$	
x_n	u_{n1}	u_{n2}	$\ldots$	u_{nm}	

Each column of such a table is itself a single-entry table, so that (17') is a collection of many tables of the form (17). The size of a table for a function of two arguments is, as a rule, much greater than for a function of one argument with the same interval for the independent variables. In view of this, functions of two arguments are much less often tabulated than functions of one argument.

How quickly the size of a table can grow with an increase in the number of arguments is shown by the following simple example. Let it be required to tabulate a function of four arguments $f(x, y, z, t)$ for 100 values of each of the arguments. Let us assume that the function does not need to be computed very exactly, only to three significant figures. If under such conditions we tabulate a function of one argument, the whole table of values will consist of a hundred three-digit numbers and may easily be put on one page.

---

* Such a column may be very long and may therefore be broken up into many smaller columns for convenience of printing. But of course it is still called a single-entry table.

† Interpolation, as a rule, is more complicated if the tabulated values x_i are farther apart and simpler if they are closer together. Moreover, the requirement concerning rapidity of interpolation may vary widely. In tables designed for artillery use, interpolation must be done almost instantly, "at sight." But in tables of higher accuracy, designed for use in the sciences, we may allow interpolations which require a whole series of operations.

But in a four-entry table for the function $f(x, y, z, t)$, we will have 100^4 combinations of the values of x, y, z, t and as many values of f, from which it is easy to calculate that the table would fill more than 300 volumes.

Because such tables are so unwieldy, functions of many arguments are seldom tabulated and then only in particularly simple cases. In the last few years there has begun a systematic study of classes of functions of many variables for which tables may be formed with a number of entries less than the number of arguments. At the same time studies have been begun on the simplest possible construction of such tables.

We give a simple example of such a function.

Let it be required to tabulate the function u of three arguments x, y, z with the following structure

$$u = f[\phi(x, y), z].$$

It is perfectly clear that here one may restrict oneself to two double-entry tables if we introduce the auxiliary variable $t = \phi(x, y)$ and consider u as the composite function

$$u = f(t, z),$$
$$t = \phi(x, y).$$

For convenience in the use of these tables, we may combine them in the following manner. We consider the function $t = \phi(x, y)$ and solve this equation with respect to y

$$y = \Phi(x, t).$$

In theory it makes no difference which of the functions $t = \phi(x, y)$ or $y = \Phi(x, t)$ is tabulated, but it will be more convenient for us to tabulate the second of them. We construct two double-entry tables for the functions $y = \Phi(x, t)$ and $u = f(t, z)$ and combine them in the manner shown in Table 4.

Table 4.

x_1	x_2	$\cdots$	x_i	$\cdots$	t	z_1	z_2	$\cdots$	z_k	$\cdots$
					t_1					
					t_2					
					$\vdots$					
$\cdots$	$\cdots$	$\cdots$	y_j	$\cdots$	t_j	$\cdots$	$\cdots$	$\cdots$	u_{jk}	$\cdots$

The value of u which corresponds to given values x_i, y_j, z_k is found as follows: We find the column headed by x_i and running down it, pick out the value y_j (or one near it). In the horizontal row through it will be the corresponding value of t. Moving further along this horizontal row we find in column z_k the required value $u = f(x_i, y_j, z_k)$.

In this example we see that, rather than make a triple-entry table, we may restrict ourselves to two double-entry tables with a simple rule for operating with them.

The use of various possible methods of shortening tables allows us in certain cases to decrease the size of the tables by a factor of ten, a hundred, or even a thousand in comparison with tables in which the number of entries is equal to the number of independent arguments.

Desk computers. Almost as old as tables as an aid to computation are various computing devices. Some of them were used even in ancient Greece.

The first models of calculating machines were constructed in the 17th century by Pascal, Moreland, and Leibnitz. From that time on the machines were repeatedly changed and improved and were in wide use by the end of the last century and especially at the beginning of the present one.

We will only look at certain forms of machines and will consider the possibility of speeding up the computations which they perform. We begin with the small, so-called universal desk computers. Each of these, independently of its construction, is designed to perform the four arithmetic operations, with multiplication and division being done by repeated series of additions and subtractions.

A typical early model of such a machine is the wheeled arithmometer of Odner. Entering a number into the adjustable mechanism is accomplished by moving a lever the necessary number of notches corresponding to each digit of the number. In the process of addition each summand is entered into the adjustable mechanism and then, by one rotation of the handle, is transferred to the accumulator, where it is automatically added to the number already there. Subtraction corresponds to a rotation of the handle in the opposite direction. Multiplication is carried out by entering the multiplicand into the adjustable mechanism and then repeatedly adding it to itself for each digit of the multiplier. For example, to multiply by 45 corresponds to five repeated additions of the multiplicand and then four repeated additions of the same number moved over one place.

For division the dividend is placed in the accumulator and the quotient is found by repeated subtraction of the divisor, digit by digit. The result is determined by the number of rotations of the handle needed in each digit place to remove the number from the accumulator.

We have given this brief description of the computations here only in order to make clear the direction of further improvements in desk calculators. Some of these improvements have merely made the machines more convenient without changing the basic scheme of their construction. An improvement of this kind is the introduction of electricity, which accelerates the action of the machine and frees the operator from having to turn the handle.

To accelerate and simplify the entering of numbers into the adjusting mechanism, keys for receiving instructions were introduced. The entering of given digits is carried out, not by rotating a lever for the specific number of notches, but simply by punching the corresponding key. Calculators were invented on which it is sufficient for the operator to enter the number on which it is desired to perform a given operation and then to punch the key which tells which of the four operations is to be performed. The machine will carry on from there without further human intervention. The improvement of desk computers also brought about a remarkable increase in their rapidity, so that in the latest models the result of a multiplication is obtained within one second after punching the keys. Further acceleration in the action of such machines is obviously superfluous, since it takes considerably longer than that for the operator merely to punch the keys and record the results.

Digital (punched card) machines and relay machines. Digital machines were invented for statistical computations and for financial and industrial use. They are designed to carry out a large number of uncomplicated computations of the same kind. They are less convenient for technical and scientific calculations because of their very small operating "memory" and the restricted possibility of establishing computational programs for them. In spite of these deficiencies, digital machines, up to the appearance of fast-acting electronic machines, were quite widely used in complicated and large-scale calculations when the whole process could be reduced to a fairly short sequence of operations to be carried out on a massive scale (for example, in preparing tables).

The numbers with which the digital machine operates are entered on punched cards (figure 2). The digits and symbols are entered on the card by means of a punch in specific places. The card is introduced into the machine through a system of brushes. A brush under which a hole is passing closes an electrical circuit and sets in operation a given phase of the machine.

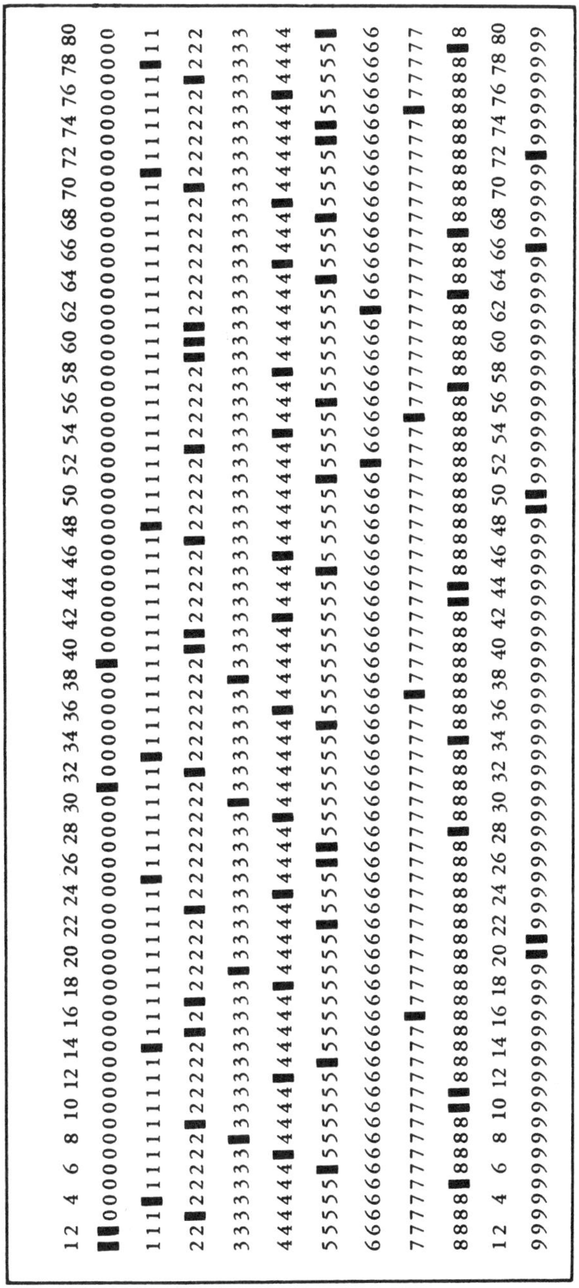

FIG. 2.

The different types of digital machines are designed to work in sets, each set containing at least the following machines:

A card-punch serves to punch the holes in the cards. The machine has a keyboard operated by hand and works at the speed of a typewriter.

A sorter is designed to arrange the cards in the order in which they are to be introduced into the calculating machines. The speed of the work is 450-650 cards per minute.

A reproducing punch or *reproducer* transfers punches from one card to another, compares two sets of cards, and selects from them cards with specific perforations. The speed of working is around 100 cards per minute.

A tabulator performs the operations of addition and subtraction and also prints out the results. It may handle 6,000 to 9,000 cards an hour.

A multiplying punch (*multiplier*) adds, subtracts, and multiplies numbers. The results are given in the form of punches on the cards. In working with numbers of 6 or 7 digits it may perform 700-1,000 multiplications an hour.

Digital machines work rather slowly. As a rough estimate of the amount of work they can perform, we may say that the above set of machines can replace 12 to 18 desk computers. The first attempts to create faster machines led to the construction of relay machines based on the application of electromechanical relays. The rate of work of such machines turned out to be about ten times as great as the speed of the simple digital machines. But the gains in other respects were remarkable: Relay machines carry out complicated computational programs and have a flexible control system that greatly extended the range of technical and scientific problems solvable on machines. However, the appearance of these machines almost coincided in time with the creation of the first models of electronic machines with programmed control, and these led to a further sharp increase in the working speed. As an indication of the great increases in speed which have been made possible by the invention of electronic machines, we may point out that the time required for a change of state in an electronic tube is measured in millionths of a second.

Mathematical machines with continuous action (analogue machines). Mathematical machines with continuous action are made up of physical systems (mechanical apparatus, electrical circuits, and so forth), constructed in such a manner that the same numerical interrelations occur among the continuously changing parameters of the system (displacements, angles of rotation, currents, voltages, and so forth) as among the corresponding magnitudes in the mathematical problem to be solved. Such machines are often called *simulating* (or *analogue*) *machines.*

Every machine with continuous action is especially designed for the solution of some narrow class of problems.

The accuracy with which the machine gives the solution depends on the quality of manufacture of the component parts, the assembling and calibration of the machine, the inertial errors in its operation, and so forth. On the basis of lengthy experience in using the machines, it has been established that as a rule they are capable of an accuracy of two or three significant digits. In this respect simulating machines are notably inferior to digital machines, whose accuracy is theoretically unlimited.

An important characteristic of machines with continuous action is that they are suitable for the solution of a large number of problems of one type. In addition, they often produce the solution with considerably greater rapidity than a digital machine. Their principal advantage consists of the fact that in many cases it is more convenient to introduce the initial data of the problem into them, and also the results are often obtained in a more convenient form.

There are many different types of simulating machines. It is possible to create machines, or parts of machines, that are models of various mathematical operations: addition, multiplication, integration, differentiation, and so forth. We may also simulate various formulas used in computation; for example, we can construct machines to compute the values of polynomials or the Fourier coefficients in harmonic analysis of functions. We may also simulate numerical or functional equations. The many analogies that exist between problems from completely different branches of science lead to the same differential equations. Identity of the equations involved allows us for example, to simulate heat phenomena by electrical means and to solve problems in heat engineering by means of electrical measurements, a procedure that is certainly convenient, since electrical measurements are more exact than measurements of heat and are much easier to make.

In view of the large number of simulating machines, it is impossible to describe in a few words the machines themselves or even the principles of their construction. To give the reader at least some idea of how mathematical problems may be simulated, let us give a short description of two simple mathematical machines, one of which is designed for integration of functions and the other for approximate solution of the Laplace equation.

The *friction integrator* (figure 3) is designed, as the name indicates, to integrate functions. It works by friction. The basic idea of its construction is shown in figure 4, where the component 1 is the base of the integrator, 2 is a horizontal friction disc with a vertical shaft, 3 is a friction roller, i.e., a roller with a smooth rim which can not only roll along the disc but also move in the plane perpendicular to the plane of rolling. Components 4 and 5 constitute a screw mechanism in which the screw 4 is connected with the carriage bearing the roller. If the pitch of the screw is denoted by h, then rotation of the screw through angle γ will transfer the roller over a distance $\rho = h\gamma$ in the plane of the drawing.

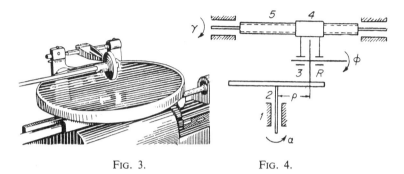

FIG. 3.   FIG. 4.

Let the shaft of the disc be rotated through angle $d\alpha$. The point of contact of the roller will then move through an arc of length $\rho \, d\alpha$. If the roller moves over the disc without slipping, the angle of rotation of the roller will be equal to

$$d\phi = \frac{\rho}{R} \, d\alpha = \frac{h}{R} \, \gamma \, d\alpha.$$

We assume that the rotation of the shaft of the disc began with angle α_0 and the initial angle of rotation of the roller was ϕ_0. From this equation we obtain by integration

$$\phi - \phi_0 = \frac{h}{R} \int_{\alpha_0}^{\alpha} \gamma \, d\alpha.$$

By suitable choice of the relation between the angles γ and α, we can use the friction integrator to compute a desired integral in a wide variety of cases. By means of integrating mechanisms it is possible to obtain a mechanical solution of many differential equations.

We turn to the second example. Let a domain Ω be given in the plane, bounded by a curve l. It is required to find a function u which inside the domain satisfies the Laplace equation

$$\Delta u = \frac{\partial^2 u}{\partial x^2} + \frac{\partial^2 u}{\partial y^2} = 0$$

and on the contour l takes given values

$$u \mid_l = f.$$

We introduce a square net of points

$$x_k = x_0 + kh, \ y_k = y_0 + kh, \ k = 0, \pm 1, \pm 2, \cdots,$$

and replace the domain Ω itself by a polygon composed of squares. Corresponding to the contour l we have a broken line. We transfer the boundary values of f on l to this broken line. The value of the unknown function u at a node (x_j, y_k) is denoted by u_{jk}. To secure an approximate solution of the Laplace equation in Ω, we replace it by an algebraic system, which must be satisfied for all interior points of the domain:

$$u_{jk} = \tfrac{1}{4}(u_{j+1,k} + u_{j,k+1} + u_{j-1,k} + u_{j,k-1}).$$

For a solution of this algebraic system, we may construct the following

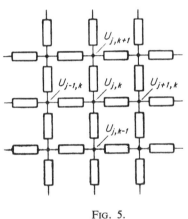

FIG. 5.

electrical model. We introduce in the plane a two-dimensional conduction net, the scheme of which is illustrated in figure 5. The resistance between two nodes is assumed to be everywhere the same. At the boundary nodes of the net, we now apply voltages equal to the boundary values of u at these nodes. These voltages will determine the voltage at all interior points of the net. We denote by $U_{j,k}$ the voltage at the node (x_j, y_k). If we apply Kirchhoff's law to the node (x_j, y_k), it is clear that at this

node the following equation will be satisfied

$$\frac{1}{R}\,[(U_{j+1,k} - U_{j,k}) + (U_{j,k+1} - U_{j,k})$$
$$+ (U_{j-1,k} - U_{j,k}) + (U_{j,k-1} - U_{j,k})] = 0,$$

which differs only in notation from the previous equation for our algebraic system. At the nodes of the net the values u_{jk} of the solution of the algebraic system must agree with the voltages U_{jk}, which can be obtained from the model by the usual electrical measurements.

Suggested Reading

V. N. Faddeeva, *Computational methods of linear algebra*, Dover, New York, 1959.

D. R. Hartree, *Numerical analysis*, 2nd ed., Oxford University Press, New York, 1958.

F. B. Hildebrand, *Introduction to numerical analysis*, McGraw-Hill, New York, 1956.

A. S. Householder, *Principles of numerical analysis*, McGraw-Hill, New York, 1953.

W. J. Karplus and W. W. Soroka, *Analog methods: computation and simulation*, 2nd ed., McGraw-Hill, New York, 1959.

W. E. Milne, *Numerical calculus: approximations, interpolation, finite differences, numerical integration and curve fitting*, Princeton University Press, Princeton, N. J., 1949.

——, *Numerical solution of differential equations*, Wiley, New York, 1953.

A. M. Ostrowski, *Solution of equations and systems of equations*, Academic Press, New York, 1960.

J. B. Scarborough, *Numerical mathematical analysis*, 5th ed., Johns Hopkins Press, Baltimore, Md., 1962.

R. G. Stanton, *Numerical methods for science and engineering*, Prentice-Hall, Englewood Cliffs, N. J., 1961.

CHAPTER XIV

ELECTRONIC
COMPUTING MACHINES

§1. Purposes and Basic Principles of the Operation of Electronic Computers

Mathematical methods are widely used in science and technology, but the solution of many important problems involves such a large amount of computation that with an ordinary desk calculator they are practically unsolvable. The advent of electronic computing machines, which perform computations with a rapidity previously unknown has completely revolutionized the application of mathematics to the most important problems of physics, mechanics, astronomy, chemistry, and so forth.

A contemporary universal electronic computing machine performs thousands and even tens of thousands of arithmetic and logical operations in one second and takes the place of several hundred thousand human computers. Such rapidity of computation allows us, for example, to compute the trajectory of a flying missile more rapidly than the missile itself flies.

In addition to their great rapidity in performing arithmetic and logical operations, universal electronic computing machines enable us to solve the most diverse problems on one and the same machine. These machines represent a qualitatively new method which, besides an enormously increased production of standard results, makes it possible to solve problems previously considered quite inaccessible.

In many cases the computations must be carried out with great rapidity if the results are to have any value. This is particularly obvious in the example of predicting the weather for the following day. With hand calculators the computations involved in a reliable weather forecast for the next day may themselves require several days, in which case they naturally lose all practical value. The use of electronic computing machines for this purpose makes it possible to secure the complete results in plenty of time.

The high-speed electronic computing machine. The high-speed electronic computing machine (BESM) which was constructed in the Institute for Exact Mechanics and Computing Technology of the Academy of Sciences of the USSR is an example of such a machine. In one second the machine performs between 8,000 and 10,000 arithmetic operations. We scarcely need to remind the reader that on a desk calculator an experienced operator can carry out only about 2,000 such operations in one working day. Consequently, the electronic computer can perform in a few hours computations that the experienced operator could not perform in his whole lifetime. One such machine would replace a colossal army of tens of thousands of such operators. Merely to give them a place to stand would take up several hundred thousand square yards.

These electronic machines have been used to solve a large number of problems from various domains of science and technology. As a result economies have been achieved amounting to hundreds of millions of dollars. We give several examples.

For the international astronomical calendar the orbits of approximately seven hundred asteriods were computed in the course of a few days, account being taken of the influence on them of Jupiter and Saturn. Their coordinates were determined for ten years ahead and their exact positions were given for every forty days. Up till now such computations would have required many months of labor by a large computing office.

In making maps from the data provided by a geodetic survey of a given locality, it is necessary to solve a system of algebraic equations with a large number of unknowns. Problems with 800 equations, requiring up to 250 million arithmetic operations, were solved on the electronic machine in less than twenty hours.

On the same machine tables were calculated to determine the steepest possible slope for which the banks of a canal would not crumble, and in this way large savings of time and material were effected in the construction of hydroelectric power stations. In previous attempts fifteen human computers had worked without success for several months in an effort to solve this problem for only one special case. On the electronic machine the computations for ten cases took less than three hours.

On the machine one may rapidly test many different solutions for given problems and choose the most appropriate. Thus one may determine, for example, the most appropriate mechanical construction of a bridge, the best shape for the wing of an airplane, or for the nozzle of a jet motor, the blade of a turbine, and so forth.

The practically infinite accuracy of the computations makes it possible to construct very rapidly all kinds of tables for the needs of science and technology. On the BESM the construction of a table containing 50,000 values of the Fresnel integral required only one hour.

Applications of electronic computing machines to problems of logic. In addition to handling mathematical problems, we may also solve logical problems on an electronic computing machine; for example, we may translate given texts from one language into another. In this case, instead of storing numbers in the machine, we store the words and numbers that take the place of a dictionary.

Comparing the words in the text with the words in the "dictionary," the machine finds the necessary words in the desired language. Then by means of grammatical and syntactical rules, which are described in the form of a program, the machine "processes" these words, changing them in case, number or tense, and setting them in the right order in a sentence. The translated text is printed on paper. For a successful translation a very large amount of painstaking work on the part of philologists and mathematicians is needed to set up the programs.

Experimental dictionaries and programs for the translation of a scientific-technical text from English into Russian were set up at the Academy of Sciences of the USSR, and at the end of 1955 the first experimental translation was produced on the BESM machine, even though this machine is not especially adapted for translation.

By way of experiment complicated logical problems were successfully solved on the BESM; for example, chess problems. A complete analysis of chess is not possible on present-day electronic machines in view of the enormous number of possible combinations. As an approximate method the relative values of the various pieces are estimated; for example, ten thousand points for the king, one hundred for the queen, fifty for a rook. Various positional advantages are also estimated to be worth a certain number of points; i.e., open files, passed pawns, and so forth. By a series of trials the machine chooses the course of action that after a specified number of moves produces the greatest number of points for all possible answers on the part of the opponent. However, in view of the enormous number of possible combinations the solution is necessarily restricted to trying a comparatively small number of moves, which excludes the study of strategic plans of play.

Basic principles of the operation of electronic computing machines.
A present-day electronic computing machine consists of a complicated complex of elements of electronic automation: electron tubes, germanium crystal elements, magnetic elements, photoelements, resistors, condensers, and other elements of radio technology.

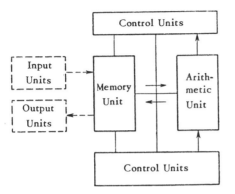

Arithmetic operations are performed with colossal rapidity by electronic computing devices, which are assembled in the arithmetic unit (figure 1).

But to guarantee high speed for computations it is not enough just to perform rapid arithmetic operations

FIG. 1.   Diagram of the basic units of an electronic digital computer.

on numbers. In the machine the whole computational process must be completely automatic. Access to the required numbers and establishment of a specific sequence of operations on them are set up automatically.

The numbers on which the operations are to be performed and also the results of intermediate calculations must be stored in the machine. An entire mechanism, the so-called "memory unit" is designed for this purpose; it allows access to any required number and also stores the result of the computation. The capacity of the memory unit, i.e., the number of numbers that may be stored in it, to a great extent determines the flexibility of the machine for the solution of various problems.

In present-day electronic machines the capacity of the memory unit is from 1,000 to 4,000 numbers.

The extraction of the required numbers from the memory unit, the operation that must be performed on these numbers, the storing of the result in the memory unit and the passage to the next operation are all guided in the electronic computing machine by a control unit. After the computing program and the initial data are introduced into the machine the control unit guarantees the fully automatic character of the computational process.

To introduce the initial data and the computational program into the machine, and also to print the results on paper, is the purpose of special input and output units.

When we are using the machine for making computations, we must have confidence in the correctness of the results produced; i.e., we must have some means of checking them. Verification of the correctness of the computations is effected either by means of special verification mechanisms or by the usual methods of logical or mathematical verification embodied in a special program. The simplest example of such a verification is the "duplication check" (the so-called "calculation on both hands"), which consists of computing twice and collating the results.

Before proceeding to the solution of a particular problem, we must first of all, on the basis of the physical process under investigation, state the problem in terms of algebraic formulas, or of differential or integral equations, or other mathematical relations. Then by applying well-developed methods of numerical analysis, we can almost always reduce the solution of such a problem to a specific sequence of arithmetic operations. In this way the most complicated problems are solved by means of the four operations of arithmetic.

To perform any arithmetic operation by hand computation it is necessary to take two numbers, perform the given arithmetic operation on them, and write down the result produced. This result may be necessary for further computations or may itself be the desired answer.

The same operations are also carried out in electronic computing machines. The memory unit of the machine consists of a series of locations or cells. The locations are all enumerated in order, and to select a number for calculation, we must give the location in which it "is stored."

To perform any one arithmetic operation on two numbers, we must give the locations in the memory unit from which the two numbers are to be taken, the operation to be performed on them, and the location in which the result is to be placed in the memory. Such information, presented in a specific code, is called an "instruction."

The solution of a problem consists of performing a sequence of instructions. These instructions constitute the program for the computation and usually they are also stored in the memory unit.

A computing program, i.e., a set of instructions effecting the sequence of arithmetic operations necessary for the solution of the problem, is prepared by mathematicians in advance.

Many problems require for their solution several hundred million arithmetic operations. So in electronic machines we use methods which allow a comparatively small number of initial instructions to govern a large number of arithmetic operations.

Together with the instructions governing arithmetic operations, electronic computers also provide for instructions governing logical operations; such a logical operation may consist, for example, of the comparison of two numbers with the purpose of choosing one of two possible further courses for the computation, depending on which of the two numbers is the larger.

The instructions of a program and also the initial data are written in terms of a prearranged code. Usually the description of the instruction is recorded on perforated cards or tape in the form of punched holes or else on magnetic tape in the form of pulses. Then these codes are introduced into the machine and placed in the memory unit, after which the machine automatically carries out the given program.

The results of the computation are again recorded, for example in the form of pulses on a magnetic tape. Special decoding and printing units translate the magnetic tape code into ordinary digits and print them in the form of a table.

The speed with which computers perform the most complicated calculations has produced a saving of mental labor which can only be compared with the saving in physical labor made possible by modern machinery. Of course, an electronic machine only carries out a program set up by its operator; it does not itself have any creative possibilities and cannot be expected to replace a human being.

The wide use of electronic computing machines in institutes of science and technology, in construction offices, and in planning organizations has opened up limitless possibilities in the solution of problems in the national economy. Engineers and mathematicians have before them rewarding prospects for further development in the operation and construction of computing machines and also in their application and exploitation.

Electronic computing machines are powerful tools in human hands. The significance of these machines for the national economy can hardly be overestimated.

§2. Programming and Coding for High-Speed Electronic Machines

The basic principles of programming; 1. Euler's method for differential equations. For computations on electronic machines the mathematical method selected for approximating the solution of a problem necessarily consists of a sequence of arithmetic operations. The execution of these operations by the machine is guaranteed by the program, which as we have said, consists of a sequence of instructions. Of course, if we were required to give a separate instruction for each one of the arithmetical operations, the program would be very lengthy and even to describe it would take about as much time as performing the operations themselves by hand. Thus in programming we must try to make a small number of instructions suffice for a large number of arithmetic operations.

To clarify the structure of a sequence of instructions and the methods of setting up a program, let us first examine the operations that must be performed when a very simple problem is solved by hand.

We will take as an example the solution by Euler's method of the following differential equation of the first order with the given initial conditions

$$\frac{dy}{dx} = ay, \qquad y\,|_{x_0} = y_0 \,. \tag{1}$$

In this method the range of values of x is divided up into a sequence of intervals of equal length $\Delta x = h$, and within each interval the derivative dy/dx is regarded as a constant, equal to its value at the beginning of the interval.* With these assumptions the computation for the kth interval is given by the formulas

$$\left(\frac{dy}{dx}\right)_k = ay_k \,,$$

$$\Delta y_k = \left(\frac{dy}{dx}\right)_k h = (ah)y_k \,,$$

$$y_{k+1} = y_k + \Delta y_k \,,$$

$$x_{k+1} = x_k + h.$$

After carrying out the calculation for the kth interval, we go on to the $(k+1)$th interval. The computation begins with the given initial values x_0 and y_0. The sequence of operations is shown in Table 1.

In hand computations only the first three operations are performed, the others being understood but not written down; this is true, for example, of the instruction to begin over again for the following interval, to end the computation, and so forth. In machine computation all these operations must be exactly formulated (operations 4-7). Consequently, in the machine, in addition to the arithmetic operations, we must also arrange in advance for the control operations (operations 4-7). The control operations have either a completely definite character (for example, operations 4 and 5) or a conditional character, which depends on the result just produced (for example, operations 6 and 7). Since the last two operations are mutually exclusive (we must perform either one or the other of them), these two operations are combined in the machine into one (a comparison operation), which is formulated in the following way: "If x is less than x_n, repeat the operations beginning with number 1; but if x is equal to or greater than x_n, stop the computation." In this

---

* In practice the solution of an ordinary differential equation is usually calculated by a more complicated and exact formula.

Table 1. Operations Necessary for the Solution of Equation (1) by Euler's Method

Number of the Operation	Quantity Defined	Formula	Computations*
1	$\varDelta y_k$	$(ah)y_k$	$(ah)(2)_{k-1}$
2	y_{k-1}	$y_k + \varDelta y_k$	$(2)_{k-1} + (1)_k$
3	x_{k+1}	$x_k + h$	$(3)_{k-1} + h$
4			Print the value found for x_{k+1}.
5			Print the value found for y_{k+1}.
6			Repeat the computation, beginning with operation no. 1 for the new values of x and y.
7			When x reaches the value x_n, stop the computation.

way, the sequence of further computations depends on the magnitude of the x already produced in the process of computing.

2. The three-address system. A glance at Table 1 shows that to perform any arithmetic operation it is necessary to indicate: First, which operation (addition, multiplication, etc.) is to be performed; second, which numbers is it to be performed on; and third, where to put the result, since it is to be used in further computation.

The code expressions for the numbers are stored in the memory unit of the machine; consequently the indexes of the corresponding locations in the memory must be given: namely, where the numbers are to be taken from and where the result is to be placed. This leads to the most natural "three-address system of instructions."

In the three-address system, a specific set of locations in the code is assigned to defining the operations; i.e., to stating which operation is to be performed on the given two numbers (the code of operations). The remaining locations in the instruction code are divided into three equal groups, called "instruction addresses" (figure 2). The code in the

---

* The digits (with subscripts) in parentheses in the column "Computations" indicate the operation whose result is to be used in the computation. For example, in the first operation (the first row) we have to multiply the quantity (ah) by the quantity found as a result of performing the second operation (the second row for the preceding interval $(2)_{k-1}$; in the second operation we have to add the quantity resulting from the operation for the preceding interval $(2)_{k-1}$ to the quantity resulting from the first operation for the present interval $(1)_k$.

At the beginning of the computation the initial data x_0 and y_0 are placed in the column "Quantity Defined" for operations 2 and 3.

first address shows the index of the location in the memory unit from which the first number is to be taken, the second address code is the index of the location from which the second number is to be taken, and

Code of the Operation	1st Address	2nd Address	3rd Address

FIG. 2. The structure of a three-address system of instructions.

the third address code is the index of the location of the memory unit in which the result is to be placed.

Code expressions for instructions referring to the control unit may also be put into the three-address system. Thus, the instruction "transfer a number to the print-out unit" must be represented in the code of operations by the number assigned to this operation; in the first address will appear the index of the location in the memory unit where the number to be printed is stored and in the third address the index of the printing unit (in the second address the code is blank). An instruction that either one course or another is to be followed is called a "comparison instruction." The code of operations of such an instruction states that it is necessary to compare two numbers, namely the ones indicated in the first and second addresses of the instruction. If the first number is smaller than the second, we must pass to the instruction indicated in the third address of the comparison command. But if the first number is greater than or equal to the second, then the given instruction consists simply of the command to pass to the next instruction.

Instruction codes, as well as number codes, are stored in the memory unit and follow one after the other in the order in which they are numbered provided there is no change indicated in the course of the computations (for example, by a comparison operation).

Let us consider how the program will look in the previous example. We set up the following distribution of number codes in the locations of the memory unit:

> The quantity ah is in the 11th location
> The quantity h is in the 12th location
> The quantity x_n is in the 13th location
> The quantity x is in the 14th location
> The quantity y is in the 15th location
> The operative location* is the 16th.

---

* A location in which intermediate values found in the course of the computation are placed is called operative.

Corresponding to the preceding table we get the following program (Table 2).

Table 2. Program for the Solution of Equation (1) by Euler's Method

Number of the Instruction	Instruction Code				Remarks
	Code of the Operation	1st Address	2nd Address	3rd Address	
1	Multiplication	11	15	16	$\Delta y_k = (ah)y_k$
2	Addition	15	16	15	$y_{k+1} = y_k + \Delta y_k$
3	Addition	14	12	14	$x_{k+1} = x_k + h$
4	Print	14	—	1	Print x_{k+1} in the first printing unit
5	Print	15	—	2	Print y_{k+1} in the second printing unit
6	Compare	14	13	1	If $x < x_k$, return to instruction no. 1; if $x \geqslant x_k$, pass to the following instruction, i.e., to instruction no. 7.
7	Stop	—	—	—	End of the computation.

The instruction code is placed in the memory unit (in Table 2, in the 1st through 7th locations). In the control unit we then place the instruction found in the first location of the memory unit. In obedience to this instruction the number in the 11th location is multiplied by the number in the 15th; i.e., the quantity $\Delta y_k = (ah)y_k$ is computed. The result is placed in the operative 16th location. With the completion of this operation the instruction from the next location of the memory unit, i.e., from the second location, enters the control unit. By this instruction the quantity $y_{k+1} = y_k + \Delta y_k$ is found, and is placed in the 15th memory location; i.e., it replaces the previous value of y. Similarly, by the third instruction the new value of x is found; the 4th and 5th instructions cause the printing of the newly found values of x and y; the 6th instruction defines the further course of the computational process. This instruction produces a comparison of the number found in the 14th memory location with the number in the 13th location, i.e., a comparison of the value x_{k+1} which has been produced with the final value x_n. If $x_{k+1} < x_n$, the computation must be repeated for the next interval; i.e., in the given example we must return to the first instruction. The index of this instruction, to which we must pass if the first number is less than the second, is shown in the third address of the comparison instruction. But if the computation has produced a value $x_{k+1} \geq x_n$, the comparison instruction causes passage to the next instruction, i.e., to the 7th, which stops the computing process.

Before beginning the computation, we must introduce in the memory unit the instruction codes (in locations 1-7), the code expressions for the constants (locations 11-13) and also the initial data, i.e., the values x_0 and y_0 (in locations 14 and 15).

After completion of the computation for the first interval, the 14th and 15th memory locations will contain, in place of x_0 and y_0, the quantities x_1 and y_1, i.e., the values of the variables for the beginning of the next interval. In this manner, the computations for the next interval will be produced by repetition of the same instruction program.

The example considered shows that, by carrying out a cyclical repetition of a series of instructions, we may carry out a large amount of computation with a comparatively small program. The method of cyclical repetition of separate parts of a program is widely used in programming the solution of problems.

3. Change of address of instructions. A second widely used method that allows one to make essential reductions in the size of a program consists of automatically changing the addresses of certain instructions. To explain the essence of this method, we take the example of computation of the values of a polynomial.

Let it be required to compute the value of the polynomial

$$y = a_0 x^6 + a_1 x^5 + a_2 x^4 + a_3 x^3 + a_4 x^2 + a_5 x + a_6 .$$

For machine computation this polynomial is more conveniently represented in the form

$$y = (((((a_0 x + a_1)x + a_2)x + a_3)x + a_4)x + a_5)x + a_6 .$$

Let the values of the coefficients $a_0, \cdots, a_6$ be placed in memory locations 20-26, and the value of x in the 31st location of the memory unit. The program is very easy to construct and is given in Table 3.

As can be seen, in this program the operations of multiplication and addition occur alternately. All the multiplication instructions, with the exception of the 1st, are completely alike: we have to multiply the number found in the 27th location by the number found in the 31st and put the result in the 27th. All the addition instructions have the same 1st and

Table 3. Program for Computing a Polynomial

Number of the Instruction	Code of the Operation	Instruction Code			Remarks
		1st Address	2nd Address	3rd Address	
1	Multiplication	20	31	27	a_0x
2	Addition	27	21	27	$a_0x + a_1$
3	Multiplication	27	31	27	$(a_0x + a_1)x$
4	Addition	27	22	27	$(a_0x + a_1)x + a_2$
5	Multiplication	27	31	27	$((a_0x + a_1)x + a_2)x$
6	Addition	27	23	27	$((a_0x + a_1)x + a_2)x + a_3$
7	Multiplication	27	31	27	$(((a_0x + a_1)x + a_2)x + a_3)x$
8	Addition	27	24	27	$(((a_0x + a_1)x + a_2)x + a_3)x + a_4$
9	Multiplication	27	31	27	$((((a_0x + a_1)x + a_2)x + a_3)x + a_4)x$
10	Addition	27	25	27	$((((a_0x + a_1)x + a_2)x + a_3)x + a_4)x + a_5$
11	Multiplication	27	31	27	$(((((a_0x + a_1)x + a_2)x + a_3)x + a_4)x + a_5)x$
12	Addition	27	26	27	$y = (((((a_0x + a_1)x + a_2)x + a_3)x + a_4)x + a_5)x + a_6$

3rd address. But the index of the location in the second address, in changing from one instruction of addition to the next, is increased each time by one: in the second instruction the number is found in the 21st location, in the fourth instruction in the 22nd, and so forth.

The computing program may be essentially shortened, if we arrange for an automatic change in the indexes (giving the memory location) in the second address of the addition instruction. The instruction codes are stored in the corresponding locations and they may themselves be considered as certain numbers. By the addition of suitable numbers to them, we can make an automatic change in the instruction addresses. In such a method the program for computing the values of a polynomial will have the form given in Table 4.

Table 4. Program for Computing a Polynomial

Number of the Instruction	Instruction Code			
	Code of the Operation	1st Address	2nd Address	3rd Address
1	Addition	20	—	27
2	Multiplication	27	31	27
3	Addition	27	21	27
4	Addition	3	28	3
5	Comparison	3	29	2
6	Stop			

The first instruction serves to transfer the number from the 20th location to the 27th in order to have the multiplication instruction in standard form. In performing the 2nd and 3rd instructions, we get the values of $a_0 x + a_1$. For further computation it is necessary as a preliminary to change by 1 the second address in the addition instruction (the 3rd instruction), and this change is made by the 4th instruction. According to this instruction we take the number found in the 3rd location, i.e., the addition instruction in question (the 3rd instruction) and add to it the quantity found in the 28th location. In order to change by 1 the 2nd address of the 3rd instruction, the 28th memory location must contain the following:

Code of the Operation	1st Address	2nd Address	3rd Address
—	—	1	—

After performing the instruction in this way, we have put the 3rd instruction into the following form:

Code of the Operation	1st Address	2nd Address	3rd Address
Addition	27	22	27

This new form is stored in the 3rd memory location in place of the previous form of the addition instruction.

Having obtained this new form by the addition instruction, we may repeat the computations, beginning with the multiplication instruction, i.e., with the 2nd instruction. The 5th comparison serves for this purpose. This instruction compares the newly found instruction in the 3rd location with the quantity stored in the 29th location. In the 29th location is stored the following:

Code of the Operation	1st Address	2nd Address	3rd Address
Addition	27	27	27

This comparison initially tells us that the first quantity (in the third location) is less than the second (in the 29th location), and so the process of computation passes to the 2nd instruction, shown in the 3rd address of the comparison instruction. Thus the multiplication instruction (the 2nd instruction) and the addition instruction (the 3rd instruction) will be automatically repeated, and each time the number of the location in the 2nd address of the addition instruction will be changed by one (as arranged for by the 4th instruction).

Repetition of the cycle will continue until the 2nd address of the addition instruction (the 3rd instruction) reaches the magnitude 27, which happens after six repetitions of the cycle. Here the 3rd instruction will have the form:

Code of the Operation	1st Address	2nd Address	3rd Address
Addition	27	27	27

i.e., the instruction code will be the same as in the 29th location. The comparison instruction (the 5th instruction) takes note at this stage of the equality of the quantities found in the 3rd and 29th location, so that the process of computation passes to the next instruction, i.e., the 6th, and herewith the computation of the polynomial is finished.

The method of automatically changing, as part of the program itself, the number of the location in the addresses of certain instructions is widely applied for the solution of many different problems. Together with the method of cyclic repetitions, it enables us to perform a very large volume of computations with a small number of instructions.

4. The one-address system.   In addition to the three-address system of instructions that we have considered, in many machines a one-address system is used. In a one-address system each instruction contains, in addition to the code of operation, only one address. Performing an arithmetic operation with two numbers and placing the result in the memory unit calls for three instructions: The first instruction puts one of the numbers of the memory unit into the arithmetic unit, the second puts in the second number and performs the given operation with the numbers, the third places the result in the memory unit. In the course of any computation, the result produced is often used only to perform the next following arithmetic operation. In these cases one does not need to put the result obtained into the memory unit, and for the performance of the following operation one does not need to recall the first number. Thus the number of instructions in a program with a one-address system is found to be roughly only twice as large as for a three-address system. Since a one-address instruction needs a smaller number of locations than a three-address system, the amount of space taken up in the memory unit by the program will be about the same for both systems of instructions (usually in a one-address system of instructions each location of the memory unit will contain two instructions). The differences in the two different systems of instructions must be taken into account in making a comparison of the rapidity of working of the machines. For the same rapidity of performing an operation, a one-address machine will perform computations about twice as slowly as a three-address machine.

In addition to these systems, certain machines have a two-address or a four-address system of instructions.

5. Subroutines.   Usually the solution of a problem is carried out in several stages. Many of these stages are common to a series of problems. Examples of such stages are: computing the value of an elementary function for a given argument, or determining the definite integral of a function already computed.

Naturally it is desirable for such typical stages to have standard subroutines worked out once and for all. If in the course of the solution of a problem we are required to carry out standard computations, we should transfer the computation at the appropriate moment to one of the standard subroutines. Then at the end of the computations involved in the subroutine, it is necessary to return to the basic program at the place where it was interrupted.

The existence of standard programs makes the task of the programmer considerably easier. With a library of such subroutines, recorded either on punched cards or on magnetic tape, the programming of many problems consists simply of setting up some short parts of the basic program linking together a sequence of standard subroutines.

6. Verification of results. On electronic computing machines problems are solved that require several million arithmetic operations. An error in even one of the operations may lead to incorrect results. Of course, it is practically impossible to set up a check system by hand over such a large number of computations. Thus the checks and verifications must be carried out by the machine itself. Apparatus exists that will verify the correctness of the machine's operations and bring it to an automatic stop if an error is discovered. However, this apparatus involves a considerable increase in the size and complexity of the machine and usually does not act on all its parts. More promising are the methods of verification that are included in advance in the program itself.

One such method of verification consists simply of repetition of the computation, as is so common in hand computation under the name of "duplication check." If an independent repetition of the computation produces the same results, we may be sure that there are no random errors but this method will naturally fail to reveal the presence of systematic errors. To exclude the latter we must carry out in advance some control computations with previously known answers, and these computations must involve all parts of the machine. Correctness of the results produced in the control computations serves to guarantee the absence of systematic errors.

In addition to this "duplication check," we may apply more complicated methods of verification, depending on the type of problem. For example, in calculating the trajectory of a projectile, we may first solve the system of differential equations for the two components of the velocity and then subsequently solve the single differential equation for the total velocity and at each step of the integration verify the formula:

$$v^2 = v_x^2 + v_y^2 .$$

For the solution of ordinary differential equations, in addition to the computation with steps of integration h, we may carry out a second computation with steps $h/2$. This will not only guarantee the absence of random errors in the computation but also will give an estimate of the validity of the choice of step size. In computing a table by a recurrence formula, we may sometimes compute certain key values by other methods. A correct result for the key values is a sufficient guarantee of the correctness of all intermediate values. In some cases verification may consist of noting the differences between the results produced.

In constructing a program it is necessary to provide in advance for some form of logical verification of the results obtained.

Coding of numbers and instructions. Numbers and instructions are placed in machines in the form of codes. In most cases the binary system of notation is used instead of the ordinary decimal system.

In the decimal system the number 10 is taken as the base. The digits in each position may take one of the ten values from 0 through 9. The unit in each successive position is ten times as large as the unit in the preceding position. Consequently, an integer in the decimal system may be written

$$N_{10} = k_0 10^0 + k_1 10^1 + k_2 10^2 + \cdots + k_n 10^n,$$

where $k_0, k_1, \cdots, k_n$ may take the values from 0 through 9.

In the binary system the number 2 is taken as the base. The digits in each position may take only the two values 0 and 1. A unit in each successive position is twice as large as a unit in the preceding position. Consequently, an integer in the binary system may be written

$$N_2 = k_0 2^0 + k_1 2^1 + \cdots + k_p 2^p,$$

where $k_0, k_1, \cdots, k_p$ may take the values 0 or 1.

The first few natural numbers in the binary and the decimal system are written,

Binary system	0	1	10	11	100	101	110	111	1000	1001	1010	1011 etc.
Decimal system	0	1	2	3	4	5	6	7	8	9	10	11 etc.

A noninteger is written analogously in terms of negative powers of the base. For example, $3\frac{1}{8}$ is written in the binary system as

$$11.001.$$

The transfer of numbers from one system of notation to another involves specific arithmetic operations that are usually carried out in the electronic computing machine itself by special programs.

Arithmetic operations on numbers in the binary system are carried out in exactly the same way as in the decimal system. Here the addition of two units in any position produces zero in the given position and carries one to the following position. For example,

$$1010 + 111 = 10001.$$

Multiplication and division in the binary system are simpler than in the decimal system, since the multiplication table is replaced by the rules for multiplying by 0 and 1. For example,

$$
\begin{array}{r}
1010\ (10) \\
\times\ \ 101\ \ (5) \\
\hline
1010 \\
0000 \\
1010 \\
\hline
110010\ (50)
\end{array}
$$

The choice of the binary system of notation in the majority of electronic computing machines is because the arithmetic unit is thereby greatly simplified (generally at the expense of brevity in the operations of multiplication and division) and also the digits in each position are conveniently represented, for example, by open or closed relays, the presence or absence of a signal in a circuit, and so forth (in the binary system the digits in each position can only have the two values: 0 or 1).

Every digit of a binary number may be represented in the form of the presence or absence of a signal in its circuit, or in the state of a relay. In this case it is necessary that every digit have its own circuit or relay (figure 3) and the number of such circuits will be equal to the number of digits (parallel system). A binary number may also be represented in the form of a time-pulse code. In this case each digit of a number is represented at specific intervals of time on one circuit (series system). The time intervals for each digit are created by synchronizing pulses, common to the entire machine.

Corresponding to these two principles, the methods of coding a number for an electronic computing machine fall into two categories: one for a machine with parallel operation and the other for a machine with series operation. In a machine with parallel operation all the digits of a number are transmitted at the same time and each digit requires its own circuit.

In a machine with series operation the number is transmitted by one circuit, but the time of the transmission is proportional to the number of digits. Thus machines with parallel operation are faster than machines with series operation, but they also require more apparatus.

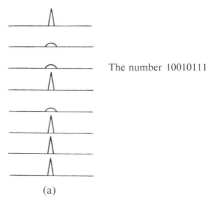

The number 10010111

(a)

Every electronic computing machine has a specific number of places for digits. All numbers to be dealt with in a computation must be included in that number of places, and the position of the decimal point, separating the integer part from the fractional, must naturally be included.

In certain machines the position of the decimal point is rigidly fixed; these are the so-called "fixed-point" machines. Usually the decimal point is put before the first place; i.e., all the numbers for the com-

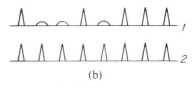

(b)

FIG. 3. Code systems:
(a) parallel; (b) series;
(1 is the code; 2 is the synchronizing pulse)

putation must be less than one, which is guaranteed by the choice of a suitable scale. For complicated computations it is difficult to determine in advance the range of the results to be expected, and thus we have to choose the scale so as to have something in reserve, a procedure which lowers the accuracy, or else we must arrange in the program itself for an automatic change of scale, which complicates the programming.

In certain machines the position of the decimal point is indicated for each number; these are machines which keep track of the exponents and they are usually called "floating-point" machines. Indicating the position of the decimal point is equivalent to representing the number in the form of its sequence of digits and its exponent, i.e.,

$$N_{10} = 10^k N'_{10} \text{ in the decimal system,}$$

$$N_2 = 2^p N'_2 \text{ in the binary system.}$$

Thus the number 97.35 may be represented as $10^2 \cdot 0.9735$. To represent the number in a machine we must indicate both its exponent (p or k)

and its sequence of digits. Thus all the digits in the number are made use of independently of its size; i.e., every number is represented by its entire set of significant digits with the same relative error. This increases the accuracy of the computation, especially for multiplication, so that in most cases one can dispense with a special choice of scale.

Increased accuracy and simplified programming in the floating-point machines are attained at the expense of some complication in the arithmetic unit, particularly in the operations of addition and subtraction. Since numbers may initially have different exponents, it is necessary to provide them with the same exponents before adding or subtracting them, in which process the final digits of the smaller number are discarded, thus:

$$10^2 \cdot 0.7587 + 10^0 \cdot 0.3743 = 10^2 \cdot 0.7587 + 10^2 \cdot 0.0037 = 10^2 \cdot 0.7624.$$

The code for a number in the binary system for a fixed-point machine consists simply of its sequences of digits (the number is assumed to be less than one); for example:

$$.00110110000000 = \frac{27}{128}.$$

In floating-point machines a specific part of the code describes the exponent, which is also coded in the binary system. An example of the way in which a number is expressed in such a code is

$$6\frac{3}{4} = 2^3 \cdot \frac{27}{32} = 0011.11011000000.$$

In addition, it is customary to reserve two places for the algebraic sign (for example, "+" in the form 0 or "—" in the form 1), one for the sign of the exponent and one for the sign of the number itself.

Instructions are coded the same way as numbers are, a specific part of the code being allotted to expressing the index (in the binary system) of the operation and another to the indexes of the memory location of each address.

§3. Technical Principles of the Various Units of a High-Speed Computing Machine

The order of performing the operations in electronic computing machines. The performance of each arithmetic operation in a machine in accordance with a given list of instructions may be reduced to the following successive steps (it is understood that we are talking about a three-address system of instructions).

1. Transfer of the first number from the memory unit to the arithmetic unit (the location of this number in the memory unit is given in the first address of the instruction code).

2. Transfer of the second number from the memory unit to the arithmetic unit (its location is given in the second address of the instruction code).

3. Performance by the arithmetic unit of the given operation on these numbers in accordance with the operation code.

4. Transfer of the result from the arithmetic unit to the corresponding location in the memory unit (the index of this location is given in the third address of the instruction code).

5. Selection from the memory unit of the next instruction, whereupon the machine begins to carry out the next operation.

In the machine the instruction code is accepted in the "instruction memory block" (IMB, figure 4). An electronic commutator (EC) trans-

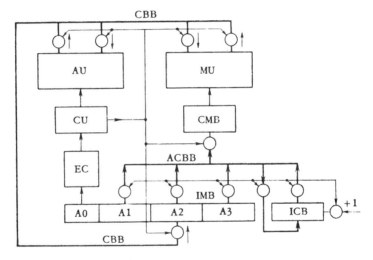

Fig. 4.   Structural diagram of an electronic digital computer.

forms the binary number of the operation code into an activating voltage in one of its output circuits corresponding to the given arithmetic operation. This voltage through the control unit (CU) prepares the circuits of the machine to perform the required operation.

In order to select the first number, the first address code of the instruction (A1), is transferred via the address code bus bars (ACBB) from the instruction memory block (IMB) to the control memory block (CMB). The signal for the transfer of this code is given by the control unit (CU) of the machine. From the location in the memory unit (MU) which corresponds to the code number transmitted the first number is selected and via the code bus bars (CBB) is placed in the arithmetic unit (AU). The opening of the input circuits in the arithmetic unit is effected by a corresponding signal from the control unit (CU) of the machine.

The second number is selected in a similar manner. A signal from the control unit (CU) of the machine transfers the code of the second instruction address (A2) from the instruction memory block (IMB) to the control memory block (CMB). The second number, taken in this way from the memory unit (MU), is transferred via the code bus bars (CBB) into the arithmetic unit (AU).

The arithmetic unit (AU) performs the given operation with the numbers in accordance with the operation code inserted in it previously.

In order to effect the transfer of the result thus obtained into the memory unit the third address code of the instruction (A3) is transferred via the address code bus bars (ACBB) from the instruction memory block (IMB) to the control memory block (CMB). The signal for the transfer of this code is given by the control unit (CU) of the machine. The memory location corresponding to the number thus obtained is then selected and its input circuits are opened. The rules for the selection or insertion of numbers are given by signals from the control unit (CU) of the machine. The signal from the control unit (CU) of the machine transfers the result obtained from the arithmetic unit (AU) to the code bus bars (CBB), via which the number is placed in the chosen location of the memory unit.

The instruction control block (ICB) is provided for the selection of the instructions. In this block is given the number of the chosen instruction. Usually the instructions go in numerical order so that, to give the number of the following instruction, it is necessary that the number found in the instruction control block (ICB) be increased by one. This is done by the control unit of the machine (circuit + 1). The instructions are stored in the memory unit. For selection of the next instruction the newly obtained number is transferred via the address code bus bars (ACBB) from the instruction control block (ICB) to the control memory block (CMB). The signal for this transfer comes from the control unit of the machine (CU). The new instruction taken from the memory unit (MU) is transferred via the code bus bars (CBB) into the instruction memory block (IMB), the output circuits of which are opened by a signal from the control block of the machine. This concludes one cycle of the operation of the machine. In the next cycle the machine performs the newly received instructions. The normal succession of instructions in numerical order may be altered by performing a control operation; for example, a comparison instruction. This instruction does not call for any arithmetic operation but specifies the course of the computational process. If the first number is less than the second, then it is necessary to go over to the instruction whose number is shown in the third address. But if the first number is greater than or equal to the second, then we pass on to the next instruction.

In transferring the comparison instruction code to the instruction memory block (IMB) an electronic commutator (EC) transforms the binary number of the operation code to an activating voltage in that one of its output circuits which corresponds to this operation. This voltage prepares the circuits of the machine for performing the operation of comparison.

The selection from the memory unit of the two numbers whose locations are given in the first and second addresses of the comparison instruction is carried out in exactly the same way as an arithmetic operation. The comparison of the numbers in the arithmetic unit (AU) may be carried out by subtracting the second number from the first. Depending on the sign of the result the control unit (CU) either transfers the code number of the next command from the third address (A3) via the address code bus bar (ACBB) to the instruction control block (ICB), or adds one to the number which is found in this block (circuit $+$ 1), exactly as in performing an arithmetic operation. After the number of the next command has been placed in the instruction control block (ICB), its selection from the memory unit is effected in the same way as in an arithmetic operation.

The arithmetic unit and the control unit. Electronic computing machines make use of present-day devices for electronic automatization. Basically the units of the machine work on the crude principle of "yes" or "no"; i.e., essentially there either is a signal or the signal is absent. Consequently, we may vary the parameters of an electronic circuit rather widely without affecting the operation of the machine.

One of the most widely used elements applied in electronic machines is the flip-flop or trigger cell. The simplest flip-flop (figure 5) consists of two amplifiers with plate resistors R_a, connected by the divider resistors R_1 and R_2. The bias established (O_B) is chosen so that one of the tubes operates and the other does not. Since the two halves of the circuit are symmetric, either tube may be closed; i.e., the circuit has two stable positions of equilibrium. In fact, if the left tube is closed, and the right one is open, then on the plate of the left tube (O_1) there will be a high voltage, and on the plate of the right tube (O_0) a low voltage (because of the voltage drop on the plate resistance R_a from the current through the tube. These voltages are transferred through the divider resistors R_1

and R_2 to the grids of the opposite tube, and consequently there will be a small voltage on the grid of the left tube and a high voltage on the grid of the right tube. With a proper choice of the parameters of the circuit, these grid voltages will keep the tubes in the given state.

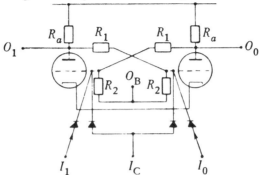

FIG. 5.   The circuit of a flip-flop.

Similarly, if the left tube is open and the right one closed, there will be low voltages on the plate of the left tube and on the grid of the right tube and high voltages on the plate of the right tube and on the grid of the left tube.

The flipping of a flip-flop from one state to the other may be brought about by negative pulses placed on the grids of the tubes through diodes. If we place a negative pulse on the grid of the left tube, then the left tube is closed, and its plate voltage will increase. This produces a higher voltage on the grid of the right tube, which opens the right tube. In this manner, the trigger assumes the first position of equilibrium (high voltage on the plate of the left tube). But if a negative pulse is placed on the grid of the right tube, the flip-flop assumes the second stable equilibrium position (a high voltage on the plate of the right tube). If a negative pulse is placed

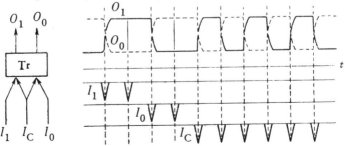

FIG. 6.   The operation of a flip-flop.

simultaneously on the grids of both tubes, then each such pulse will cause the flip-flop to move from one state of equilibrium to the other.

If we consider the circuits by which pulses are placed on the grids of the tubes as inputs of the system and the plate voltages as outputs, we have the diagram in figure 6 for the operation of a flip-flop.

The properties of flip-flops make them convenient for use in the various units of an electronic computing machine. To one equilibrium state of the flip-flop we may assign the code value "0," for example, to high voltage on the right output (O_0)—and to the other the code value "1," high voltage at the left output (O_1). Correspondingly, the inputs may be denoted by I_0, I_1, and I_C (the counting input).

Flip-flops are used in electronic machines for the temporary storage of codes (receiving registers) (figure 7). Initially all the flip-flops are set

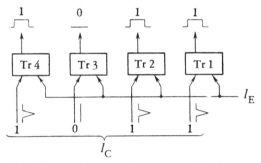

FIG. 7.   Diagram of a receiving register of flip-flops.

in the code position "0" by means of negative pulses (I_E) on the zero inputs of all cells. The code of a number or of an instruction is placed on the unit inputs of the flip-flops in the form of negative pulses. In those positions in which there are code pulses the flip-flops pass to the code position "1" and hold this position until they receive an extinguishing pulse (I_E). Receiving registers are used in the arithmetic units for storing the code of an instruction, for giving the number of a required location of the memory unit, and so forth.

A second realm of application of flip-flops is in addition circuits. Here use is made of the property of a flip-flop that it changes its state of equilibrium every time a negative pulse is applied to the counting input (simultaneously to both inputs). If the flip-flop starts in the code position "0," then the application of a pulse moves it into code position "1." But if the flip-flop starts in code position "1," then the application of a pulse moves it to code position "0." In the absence of a pulse the flip-flop remains in its previous position. The initial position of the flip-flop may be considered as a code for a given digit of the first position of the second number. Here it is easy to see that the behavior of the flip-flop exactly corresponds to the rules of addition of binary numbers for one digit $(0 + 0 = 0; 0 + 1 = 1; 1 + 0 = 1; 1 + 1 = 10$, i.e., "0" in the given position and the carrying of "1" to the next position). In order that the addition circuit may work for several binary digits, it is necessary to guarantee the carry from one digital position to the next. A carry in the original position is caused by the addition of two units, i.e., by the passage of the flip-flop from the code position "1" to the code position "0." In this passage the voltage on the left output of the trigger is changed from high to low. If this voltage is passed through a circuit containing a condenser and a resistor, then in leaving the circuit it causes a negative pulse. Through a delay line this carry pulse may be directed into the counting input of the next position.

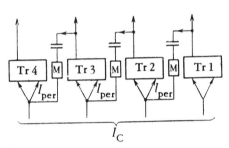

Figure 8 represents the simplest addition circuit with flip-flops. Initially all the flip-flops are set in the code position "0" by a pulse I_0 placed on their zero inputs.

FIG. 8.   Addition circuit with flip-flops.

On reception of the code of the first number, which appears in the form of negative pulses on the counting inputs, the flip-flops assume a position corresponding to the code of the first number. On reception of the code of the second number, there occurs digit-by-digit addition of the binary numbers, and in those positions where the addition has produced two ones, there arise carry pulses that after a time delay t_d are applied to the counting inputs of the flip-flops in the higher positions. These carry pulses may move the flip-flops from the code position "1" to the code position "0." In this case there arises a carry pulse to the next higher position. In the worst case, when in the addition of the codes all the positions are set in the code position "1," and the lowest position passes from code position "1" to code position "0," the carry pulse arises successively in each position after a time delay t_d. In this manner, the total time required for the passage of the carry pulses will be equal to one time delay multiplied by the number of positions. More complicated electronic circuits of flip-flops allow the elimination of such step-by-step carries with consequent shortening of the time required for addition.

For multiplication of numbers an arithmetic unit of flip-flops (figure 9) has two receiving registers for storage of the multiplicand and the multiplier (R_1, R_2) and with them an adder (Add). Multiplication is carried out in the following manner. The code of the multiplier is shifted one place to the right. If in the lowest place the multiplier has the code "1," then in the right output of the register of the multiplier there arises a pulse that is applied to the circuits governing the application of the code in the multiplicand register to the adder (the circuit $+$ N). After this has been done the partial product

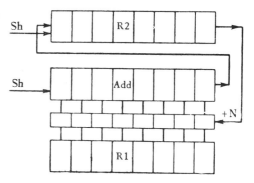

Fig. 9. Multiplication circuit with flip-flops.

in the adder is moved one place to the right and the operations are repeated. In this manner the sum of the partial products is accumulated in the adder. These operations are repeated as many times as there are digit positions in the number codes. In the multiplication of two numbers each of which takes up "n" positions, the product will take up "$2n$" positions. The highest "n" positions of the product are distributed in the adder, and the lowest "n" positions of the product may be entered one after the other, as the shifts to the right successively set free the positions in the register of the multiplier. With the completion of the multiplication, the lowest "n" digits of the product are placed in the multiplier register. The time required for multiplication is roughly equal to the time required for addition multiplied by the number of digit positions in the number code.

A code shift with flip-flops is produced by the circuit illustrated in figure 10. Applying the shift pulse (I_{sh}) to the zero inputs of all the flip-flops places them in code position "0." From these flip-flops which are in the code position "1," carry pulses arise which put the adjacent flip-

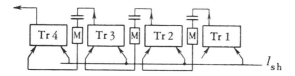

Fig. 10. Circuit for shifting a code with flip-flops.

flops into code position "1" with a time delay t_d. In this way, every application of a carry pulse moves the code one place.

An arithmetic unit with flip-flops which consists of two receiving registers and an adder also enables us to divide one number by another.

Usually an arithmetic unit with flip-flops is constructed so as to serve in a universal way for all the arithmetic and logical operations.

Flip-flops are also used in electronic machines for counting pulses, which is necessary in a number of different control arrangements. The circuit for an electronic counter (figure 11) differs from the circuit for

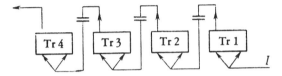

FIG. 11.   Circuit of an electronic computer
with flip-flops.

an elementary adder (figure 8) only in the omission of the delay line in the carry pulse links. A counter of this sort can count up to 2^n pulses (n is the number of places in the counter), after which the position of the counter is repeated. At the cost of some complication in the system it is possible to construct an electronic counter for an arbitrary number of pulses (not equal to 2^n).

For the realization of logical operations and control circuits in electronic computing machines, we make use of coincidence units (the so-called "AND" elements), of inverters, and of divider diode links ("OR" elements).

The AND elements work on the logical principle of "both—and" ("one and also the other"); i.e., at the output of such a unit a signal will occur only in case there are signals at all inputs. Inverters work on the logical principle of "yes—no"; i.e., if there is a signal on an input, then there will be no signal at the output, and conversely, when there is no signal on the input, then there is an output signal. The OR elements obey the logical law "either—or"; i.e., at the output there will be a signal in the case when there is a signal at any one input.

AND elements are widely used for "channeling" electric signals in a machine, i.e., for directing signals to the required circuits. For example, figure 12 illustrates a code bus bar for one of the digits of a number. This code bus bar is joined through an AND element to the inputs and outputs of the locations of the memory unit, to the inputs of two receiving registers of the arithmetic organ and to the output of an adder. Applying a control signal to the output of an AND element of any location of the memory unit, we thereby put the code stored in this location onto the code bus bar. If we simultaneously put a control signal on the input AND elements of the first receiving register, for example, then the code on the bus bar is entered into the first register. Similarly, if we put a control signal on the output AND units of the adder, then the code which is produced in the adder is transferred to the code bus bar. If here we place a control on the AND-circuit inputs of any location of the memory unit, then the codes being transferred by the code bus bars will be received in this location. Of course, before receiving codes in locations of the memory unit or in the receiving registers of the arithmetic unit, it is necessary to clear the codes which were in them previously.

This example does not exhaust all the various applications of AND elements for channelling electric signals in an electronic computing machine. They also are widely applied in the

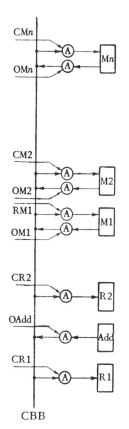

FIG. 12. Channeling of signals by an AND element.

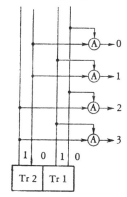

FIG. 13. Circuit of an electronic commutator with four output links.

memory unit, in the arithmetic unit, and in the control unit of the machine.

In addition to solving problems of channelling signals, AND elements perform more complicated functions. For example, when we are having access to the location of the memory unit, there often arises the problem of converting the number of the location (given in binary form) to a control voltage placed on this location. This problem is handled by the electronic commutator, constructed from AND elements. Figure 13 illustrates a circuit for an electronic commutator with four output links. The number of the location is given in the form of a binary code on two flip-flops. All four possible combinations of the state of these locations are given in Table 5.

Table 5.

Code	2nd Trigger		1st Trigger	
	Left Output	Right Output	Left Output	Right Output
"00"	L	H	L	H
"01"	L	H	H	L
"10"	H	L	L	H
"11"	H	L	H	L

L = Low voltage at ouput,   H = High voltage at output

If in an AND element the high voltage is controlling, then to get a signal on the zero-output link it is necessary that the inputs of the AND elements be connected to the right outputs of the first and second flip-flops. In this case on the output of this AND element, there will be a signal only when the flip-flops are found in the code position "00." Similarly, to get a signal on the first output link (the code "01"), the inputs of the corresponding AND element must be connected to the left output of the first flip-flop and to the right output of the second flip-flop. The connections of the AND elements for the second (code "10") and third (code "11") links will also be made on the same principle.

In a number of cases the AND elements together with inverters and OR elements are used in the construction of the arithmetic units. For digit-by-digit addition of numbers with two binary digits, we have the four possible combinations in Table 6.

Table 6.

No.	Value of the Addends		Value of the sum	Transfer to the next Higher Order
	1st	2nd		
1	0	0	0	0
2	0	1	1	0
3	1	0	1	0
4	1	1	0	1

These relations may be realized, for example, by the circuit shown in figure 14. Such circuits are called "semiadders." The carry signal for the higher of the two positions is produced by an AND element (combination 4). To get the signal of the sum (combinations 2 and 3), it is sufficient to have a signal on one of the two outputs with the absence of an output carry signal, which may be done by an AND element, an inverter, and a diode link unifier. For addition of numbers it is necessary to consider not only the digits in a given position but also the carry from the preceding position. The

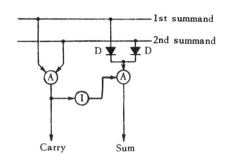

Fig. 14. Circuit for a one-place semiadder.

carry may be taken as repeated addition to the result produced by carrying from the previous position. In this manner, the union in series of two semiadders fully guarantees the addition of one position in two binary numbers.

The circuit of an adder for one position may also be realized directly by considering the possible combinations and taking account of carrying from the preceding lower position.

It is most effective to use adder circuits in AND elements in machines with sequential code distribution. In this case the code of a number is transferred by one of the code bus bars. The digits of the number follow one after another at strictly determined intervals of time. In this case for the addition of numbers, we may use a one-place adder (figure 15).

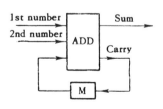

Fig. 15. Circuit for a series adder with AND elements.

The codes of both numbers are placed in advance in the lowest positions on the two basic inputs of the one-place adder. The carry output is run through a delay line to the third input of the adder. The time of the delay is taken equal to the interval between pulses. In this manner, if in the addition of any digit of the numbers there occurs a carry pulse, it is placed in the input of the adder at exactly the same time as the occurrence of the pulses in the next higher position. The time required for addition of two numbers is equal to the time required for the passage of the code of one number.

Multiplication of two numbers in a series code may also be done with a one-place adder, and here it is necessary to put the numbers through the adder a number of times equal to the number of positions occupied by the number code, i.e., the time required for multiplication is "n" times as long as the time for addition.

Memory units. The possibilities of a machine are to a great extent determined by the capacity of its memory unit, i.e., by the number of numbers that can be stored in the machine. For contemporary universal electronic computing machines this capacity is usually 500-4,000 numbers.

For code storage it is possible to use flip-flops. However, the amount of apparatus here turns out to be so large that this form of memory unit is almost never used.

For machines with series operation, widespread application has been found for memory units consisting of electroacoustic mercury tubes (figure 16). An electric signal in the form of a pulse is placed on a quartz

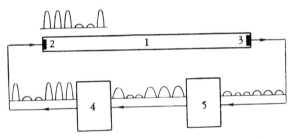

FIG. 16.   Basic circuit for dynamic storage of a code
in an electroacoustic tube:
(1) mercury tube;   (2) transmitting quartz crystal;
(3) receiving quartz crystal;   (4) transmitted form
of the pulse;   (5) received form of the pulse.

crystal at the input of the tube. The quartz crystal has the property of transforming an electrical pulse into a mechanical oscillation, and conversely. In this manner the entering electrical signal is transformed into a mechanical (ultrasonic) vibration, which is propagated along the tube with a specific velocity. When the signal reaches the end of the tube, it falls on a receiving quartz crystal and is transformed again into an electrical pulse. After being amplified and put into its original form, the signal is again directed toward the input of the tube. In this manner, the codes of the numbers introduced in the form of pulses in the mercury tube are circulated indefinitely in the tube. To introduce the numbers into the tube, a code from the machine is placed on the input of the tube, and simultaneously the circuit for the return of pulses from the end of the tube is broken for the same period of time. For the selection of numbers in the corresponding instant of time, when the required code reaches the end of the tube, the output links are opened, thereby transmitting the code to the other units of the machine. The entry and removal of the numbers is accomplished automatically by appropriate electronic circuits. Usually, with the goal of simplifying the apparatus, several numbers are stored in each mercury tube. Thus for access to a number, it is necessary to wait while the required code goes to the end of the tube. The more numbers there are stored in the tube, the greater the time required to find a required number.

Series machines with memory units composed of electroacoustic mercury tubes operate at a rate of 1,000-2,000 operations per second.

For memory units one often applies the principle of magnetic recording of electrical signals, similar to the recording of sound. The record may be made either on a magnetic tape or on a continuously revolving drum covered with a ferromagnetic material (figure 17). Along the generator of the drum there are placed magnetic heads. If at a specific instant of time current pulses are passed through the windings of the magnetic heads, then in the corresponding places on the surface of the drum the signals will be recorded in the form of residual magnetization. With the

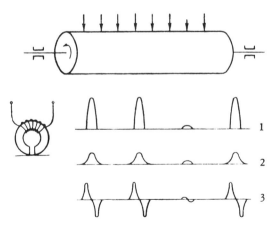

FIG. 17. Basic scheme of a magnetic drum:
(1) current through the coil;   (2) residual magnetization;
(3) emf in the coil in read-out.

rotation of the drum the field resulting from the residual magnetization, passing under the heads, causes in them electric signals, which are amplified and transmitted to the other units of the machine.

A magnetic drum may be used both for a series system and for a parallel system of transmitting codes. However, the drawback of electroacoustic mercury tubes, namely the delay in access to numbers, is even more characteristic of the magnetic drum. Thus memory units with magnetic drums are used for machines of comparatively low speed (of the order of several hundred operations per second). On the other hand, a magnetic drum allows a marked increase in the capacity of the memory unit with only a tolerable increase in the amount of apparatus. Thus the magnetic drum and the magnetic tape are often used in universal machines as complementary (exterior) memory units in addition to fast-acting (operative) memory units.

In high-speed electronic computing machines with parallel operation, cathode-ray tubes are often used for the memory unit (figure 18). If the electron beam is directed at any point of the screen, then at this point there is accumulated an electric charge. The charge will be preserved for a considerable time, so that it is possible to record number codes on the screen. In the process of making a computation, a beam of electrons is again directed to the required point. If the given element has not been charged, it now receives a charge, and through the signal plate and the output amplifier there emerges a code pulse. But if the element is charged, the signal does not emerge. In this way we can determine whether a signal has been recorded at a given point or not. After access to the code we must re-establish the previous state of the given element, which is done automatically by a special circuit. In exactly the same way it is necessary to renew the code recordings periodically, in order to avoid an essential change in the charge by stray electrons and leakage through the dielectric.

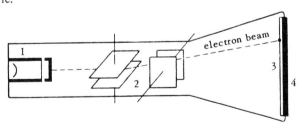

FIG. 18.   Basic scheme of a cathode-ray tube:
(1) source of electrons;   (2) deflection plates;
(3) screen;   (4) signal plate.

Usually there are 1,024 (32 × 32) or 2,048 (32 × 64) points distributed over the screen. The direction of the beam of electrons to the required point is accomplished by appropriate voltages on two pairs of deflecting plates.

In machines with parallel operation, every digit of a binary number requires its own cathode-ray tube and access to the number is made simultaneously for all tubes. The access time, including the entire operation of the element, may be reduced to a few microseconds.

Recently use has been made of memory units with magnetic elements that have rectangular hysteresis loops (figure 19). If we put a positive signal through the coil, then the core is positively magnetized and for a negative signal it will be negatively magnetized.

With the removal of the signal, the core remains magnetized either positively or negatively. Thus, the state of the core characterizes the signal recorded. In the computing process, there passes through the coil a signal of specific polarity, for example, a positive one. If in this case the core was magnetized negatively, then a remagnetization will occur (a change in the magnetic flux),

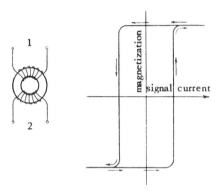

FIG. 19.   Basic scheme of a memory element with a rectangular hysteresis loop:
(1) input coils;   (2) output coils.

and in the output coil there will be induced an electromotive force, which is fed into an amplifier. But if the core was magnetized positively, then a change in its state will not take place, and no signal will arise in the output coil. In this way it is possible to distinguish which signal has been placed on a given element. Of course, after access has been had to the code, it is necessary to restore the original state of the core, which is done by a special circuit.

§4. Prospects for the Development and Use of Electronic Computing Machines

The use of electronic computing machines will inevitably have a great influence on the development of many fields in contemporary science and technology, especially in the physical and mathematical sciences. Thus it is appropriate for us to indicate the basic prospects for further application of computing machines and their significance for mathematics.

Further extension of the areas of application of mathematical machines; 1. Improved machines. At the present time there is continuous and intensive technological progress going on in the production of high-speed computing machines, in further improvement of their construction, and in the use of new physical principles and of combinations of new types. Thus we may expect better technical properties for these machines (speed, capacity of the memory, regularity, and reliability of operation), and also a notable simplification in their construction and use which will guarantee their widespread distribution.

The diversity of the types of the machines will be another factor ensuring their widespread use. Along with powerful machines of enormous capacity there will be the small-gauge machines that are simple to use and are within the purchasing power of any scientific or planning institute or of a factory; in addition to the universal ones, there are simpler special machines, intended for some specific range of problems; besides the purely digital machines other types have been invented, which accept data from certain devices, perform digital calculations on them, and then give out the results again continuously in the form of curves or of values of parameters controlling various units of the machine.

2. Better programming. A second path to new effectiveness in the use of these machines is further improvement in methods of programming. The construction of programs in the usual manner, described in §2, is easy for comparatively simple mathematical problems; in actual problems of any magnitude, it involves very complicated and detailed labor. This work may be lightened to a certain extent by the use of a "library" of standard subroutines, set up permanently for the calculation of basic functions and for performing certain necessary mathematical operations, such as inversion of a matrix or numerical integration. In spite of this, the fitting of subroutines into the basic program, addressing and re-addressing the results, and testing and rearranging the program is a quite complicated and detailed task calling for definite skill. This fact may essentially delay the setting up of new problems for electronic machines.

There are two possibilities for further development in this direction. One of them consists of constructing the program automatically by using the machine itself for this purpose, i.e., by converting the basic formulas and the logical structure of the problem, placed in the machine in coded form, into the desired program through the operation of the machine in accordance with a special "programming program."

The second direction consists of having the machine operate on a certain special universal program, which immediately examines and performs the operations in accordance with a general plan of computation introduced into the machine; this general plan would contain a number of important problems (for example, the solution of a system of equations) and, without setting up the detailed working of the program, would guarantee that the correct results were worked out and assigned to each particular problem.

3. More intellectual tasks. Further progress in the application of computing machines in mathematics is connected with the use of the machines for the performance not only of numerical calculations but also of the standard calculations of analysis.

Basically such a possibility is, in well-known cases, altogether practicable. For example, if we describe a polynomial by its set of coefficients, then such operations as multiplication and division of polynomials consist of arithmetic operations on sequences of coefficients, which are easily programmed on machines. By the use of specific coding in describing a function, it is completely possible to construct a program which gives the derivative of an elementary function (described in the same code), i.e., which allows one to perform the analytic process of differentiation. All these facts ensure the possibility in the future of solving problems by a specific method (for example, of solving a system of differential equations by means of power series), with complete carrying out of all the analytic and numerical calculations. In this manner, computing machines may be used for performing quite subtle and typically intellectual tasks (but only of a standard character), just as the present machines of the everyday world have replaced the physical labor not only of the stevedore but also of the seamstress.

The influence of high-speed machines on numerical and approximative methods. The means and instruments used in any task naturally influence the methods of the work itself. For example, trigonometric formulas computed by using logarithms are unsuitable for use on computing machines, on which only multiplication and division can be carried out directly. The use of a desk machine calls for entirely different computational schemes in approximation methods (for example, nondifference schemes in differential equations).

The fundamental changes in computational instruments and the possibilities that have been opened up by the use of electronic computing machines have naturally brought about a change of attitude not only toward the methods of computational analysis but also, to a great extent, toward the problems of mathematics in general and their applications.

Let us consider a few questions where the changes are most evident.

Mathematical tables and other ways of introducing functions into the computation. First of all, electronic machines made a fundamental change in our powers of computing tables. In place of a single table of functions, we witness an annual output of hundreds of tables, including complete and exact tables for all the basic special functions, not only for one but for several variables. But at the same time an essential change must be made in the structure of the tables. For use in high-speed machines, compact tables are appropriate, containing widely spread basic values and designed for interpolation of a high order.

In many cases, in place of tables, it is convenient to use other methods of introducing functions into the machines, namely polynomials of best approximation over subintervals, expansions in continued fractions, approximating formulas based on numerical calculation of an integral which represents the function, and so forth; all of these may profitably be introduced, in various cases, into the program of computation of a given function.

Special functions and partial analytic solutions. The special functions themselves and the solutions of problems in finite analytic form still retain their significance for qualitative investigation of a problem and for clearing up the character of its singularities, both of which are important for a numerical solution. In certain large-scale problems, the use of such special functions may provide the most economical means of finding the solution numerically. Nevertheless, the construction in many particular cases of an exact or approximate solution, by means of complicated apparatus or of the special functions that were formerly introduced for greater ease of computation, has turned out to be a mistaken policy. For machine calculations it is much simpler and shorter to find the solution by general numerical methods without making use of any of the analytic representations discussed earlier.

Thus the very considerable efforts that have been made to put into complicated analytic form the solutions of various particular problems in technology and mechanics have in many cases turned out to be wasted.

The choice of computational methods. It is incorrect to say that, because of the high productivity of electronic machines, there is no need to develop approximating methods further and that we may always use the most primitive methods. In reality, only for the simplest one-dimensional problems where, independently of the choice of method, the calculation will not run to more than a few thousand steps, can the solution be found on the machine in a few seconds or minutes.

For the systematic solution of newer, more complicated problems the number of steps may well amount to several hundred million, so that a proper choice of methods to decrease this number is quite essential. Consequently, it is a matter of great practical importance to work out effective methods of approximation, especially for multidimensional problems such as interpolation of functions of many variables, computation of multiple integrals, solution of systems of nonlinear algebraic or transcendental equations, solution of three-dimensional integral equations, systems of partial differential equations, and so forth.

At the same time there has been a considerable change in our attitude of mind in estimating the value of approximative methods; they must be judged by the ease with which they can be carried out on the machine or by their universality, that is, by the extent of their applicability to massive problems. Methods lose a great deal in value if they depend on special peculiarities of the problem or on the skill of the person who is directing the computation

The greatest value must be attached to universal methods that apply to a wide range of problems: difference methods, variational methods, the gradient method, iterative methods, linearization, and so forth.

Of course, in choosing a computational method and the manner of carrying it out, one must remember that the method is in fact carried out on the machine, so that in some cases one ought to take into account the peculiarities of construction of the given machine. In particular, one must consider maximal use of the operative memory, minimization of the data introduced from outside, the possibility of introducing intermediate checks, and the convenience of programming the problem.

But one must not think that the machine can carry out only the simplest methods, based on one kind of operation. The wide possibilities in programming and the latest improvements in its methods allow us to carry out very complicated computational programs with many different branches, so that we can change the course of the computation according to the results obtained, which is hard to do even with hand computations. The only essential requirement is that all these possibilities be completely provided for in advance.

Also one must not think that no methods can be carried out which require algebraic operations. As mentioned above, it is also completely possible to carry out some of the operations of analysis.

Significance of the estimates of error. In estimates of error for approximation methods, greater significance must be attached to those of an asymptotic character, since large values of n (for example, the number of equations replacing an integral equation by an algebraic system), small steps in difference methods, and so forth, are fully realizable on high-speed machines. In any comparison of the value of various approximative methods, primary consideration must be given to asymptotic estimates describing the rapidity of convergence of the method.

To increase the usefulness of machine methods, greater attention must be paid to *a posteriori* estimates of the error; that is, estimates made on the basis of the solution already computed. Such estimates may be included in the program and will then help to determine the future course of the computation. For example, if it is seen that the error is unacceptably large, the computation may be automatically repeated with steps decreased by half. In this connection *a posteriori* estimates may turn out to be more convenient and practical than *a priori* ones, which are inevitably too high and considerably more complicated.

The possibility of theoretical analysis of the problem. There is still another possible use for the information obtained in the numerical solution of a problem. In fact, by applying the methods of functional analysis to the approximation obtained, we may judge the existence and uniqueness of the solution, and also establish the range of the solution. Since the investigation of such questions by purely theoretical methods is sometimes extremely complicated and lengthy, and in many cases altogether impracticable, the possibility of making use for this purpose of numerical calculations produced on the machine is undoubtedly of interest.

New problems in numerical methods. The sharp increase in computational possibilities and the accumulation of skill in their use has given rise to an entirely new range of problems in the investigation of numerical methods. Instead of being used in isolated cases as in the past, the solution of systems of linear equations with a large number of unknowns has now become established as a fixed element in the solution of mathematical problems. This fact has given great practical importance to the following question: How important for the accuracy of our determination of the unknowns is the influence of rounding off, not only of the coefficients but also of various processes in the course of the solution? This question has led to a series of interesting investigations.

The possibility of numerical integration on the machine of a system of differential equations over a large interval with small steps has given acute importance to the question of stability of the process of numerical integration. Experimental analysis of this question and subsequent theoretical investigation have produced a considerable change in our estimates of the value of various methods of numerical integration of differential equations.

Questions of stability have primary significance also for the application of difference methods to partial differential equations.

New methods. The possibility of using machines had led to the appearance of completely new types of approximative and numerical methods or on the other hand has made it quite possible and convenient to employ the older methods in cases where up to now they had seemed completely impracticable. A characteristic example is the method of random sampling or, as it is often called, the "Monte-Carlo method." This method consists of finding a probability problem whose solution (probability, mathematical expectation) is identical with the desired quantity. In this probability problem the solution is found experimentally, by random sampling, as the mean value in a series of experiments. For example, to find the area of a figure defined by the inequality $F(x, y) \geqslant 0$ and contained in the square $(0, 1; 0, 1)$, we make as long a sequence as we like of random choices of pairs of numbers (x, y) contained in this square and then determine what fraction of these pairs satisfy the given inequality. Of course, such a method will be very ineffective if the trials are made by hand, but if they are done on a machine, then it is fully practicable. The trials themselves may be carried out by means of a table of random numbers. For certain problems, e.g., for calculating a multiple integral without great exactness, such a method may even be more effective than any other.

A similar method may also be used for the problem of inverting a matrix, if we apply it to samples forming a Markov chain, and also for the solution of partial differential equations, if we have found a stochastic (probabilistic) process connected with it.

The significance of high-speed machines for mathematical analysis, mechanics, and physics. In mathematical analysis great interest and practical importance is attached to investigations of multidimensional problems leading to the integral equations and boundary-value problems of mathematical physics. These investigations and the resulting methods of solution are no longer impracticable but will now be put into effect as a result of the new computing techniques, especially since the solution of such problems is of urgent importance at the present time.

Of course, the value of these newly developed methods must be judged by the ease with which it is possible to put them into practice.

On the other hand the possibility, thanks to machines, of carrying out with sufficient exactness a computation involving a large number of trials has led to an enormous extension in the range of application of "mathematical experiments" for the preliminary investigation of a mathematical problem and to a great increase in their effectiveness. This fact has made it important to work out applications of this Monte-Carlo method not only in general but also for particular problems; for example, the qualitative investigation of differential equations.

It is interesting also to note that the machines may be used in problems of analysis not only in applications but also for purely theoretical questions. Thus machine computation may prove necessary to increase the accuracy of the constants in certain inequalities and estimates in functional analysis; applications of this sort occur not only in analysis but also in the theory of numbers.

Finally, machines may be used for testing the correctness of formulas of mathematical logic, and since many mathematical propositions and proofs can be written by means of the symbols of mathematical logic, it becomes theoretically possible to test on high-speed machines the logical correctness of certain mathematical deductions.

As for mechanics and physics, we must first of all emphasize the vast increase in the application of mathematics in these sciences. Up to the present time the application of mathematics to concrete problems of mathematical physics was restricted by the enormous volume and complicated character of the necessary computations. In the problems arising in actual practice, this volume was usually such that the computation for one problem required several months and in some cases even several years of computational work. Thus, in spite of the fact that general mathematical formulations of many problems were known in mechanics and theoretical physics, and methods of their solution had been worked out in theory, in actual fact mathematical solutions, exact or numerical, had been obtained only for a few idealized and highly simplified cases, such as plane or axially symmetric problems, especially simple boundaries, or an airplane wing of infinite length.

As a result the mathematical solutions were used not so much for finding the necessary numerical values as for a qualitative and tentative investigation of the problem, which in practice had to be supplemented by costly experiments.

On the other hand, the application of new computing methods opens up the possibility of large-scale solutions of problems of mechanics and physics with all their actual complications (space problems, problems with complicated boundary contours, and nonlinear partial differential equations).

Of course, the actual carrying out of this possibility requires further development of the methods of numerical analysis and of machine solution for these problems. However, the practicability of treating such problems in this way has been strikingly demonstrated by successful experience with solution on high-speed machines of systems of partial differential equations in meteorology, in gas dynamics, in the equations of friable materials, and in other questions.

The application of theoretical mathematical analysis to problems of mechanics and physics with a close approximation to the actual physical problems and the increase in rapidity and flexibility resulting from the use of high-speed machines has made it possible in many cases to replace physical experiments by mathematical ones. This possibility will lead to further improvement in the methods of investigating problems in physics and mechanics and will increase the role played in them by theoretical and computational methods.

The significance of electronic machines for technology and industry. The rapidity and effectiveness of numerical solutions of problems of mathematical analysis also allow us to make much greater use in the various branches of technology (structural mechanics, electrical engineering and radiotechnology, the exploitation of water power, and so forth) of theoretical methods and consequently to produce much more accurate and practical results. It is now possible to apply mathematical analysis to many technical problems where it has not been used before.

In addition to the numerical solution of problems of mathematical analysis encountered in technology, a completely different application of mathematical machines to technology has been discovered. It will be possible to apply mathematical machines, for example in technical planning, to the choice of various possibilities for the construction or distribution of various objects. In questions of the organization of an industry many solutions are possible to the problem of distributing the various tasks and determining their proper sequence. The choice of the best, the most productive, and the most economical solution presents great difficulty. Here also one may find applications of machines; if it is possible to program a systematic examination of various solutions that takes account of the features of interest to us, then with the help of the machines we may compare several hundred thousand variants, which would be impossible by usual methods.

In particular, a series of relay-contact circuits allows us to analyze and verify these solutions by the methods of mathematical logic, which may be carried out on high-speed machines. In this way it is possible to select a set of such variants on the basis of any desired criteria and then to choose the best one among this selected set.

Of great promise is the use of machines in the automatic control of industry, if such machines are used in conjuction with servomechanisms and transmission devices. For example, if geometric data concerning a manufactured article are introduced into the machine, together with a specific program for the purpose, it will determine and transmit parameters that will govern the motion of a power press and make necessary changes in the article. Because of its high speed, the same electronic machine may be used for simultaneous control of the work of several presses. It is also easy to see the significance of such machines for automatic guidance of moving objects, for example interplanetary rocket projectiles, since the guidance program can take into account not only the data originally introduced but also the changes in position indicated by various recording devices.

In this way, the construction and analysis of computing machines and the possibilities of their application present a wide field of activity for mathematicians. The use of mathematical machines in the coming years will undoubtedly play a great role in the development of our technology and culture.

Suggested Reading

F. L. Alt, *Electronic digital computers: their use in science and engineering*, Academic Press, New York, 1958.

A. D. Booth and K. H. V. Booth, *Automatic digital calculators*, Academic Press, New York, 1953.

P. von Handel (Editor), *Electronic computers: fundamentals, systems and applications*, Prentice-Hall, Englewood Cliffs, N. J., 1961.

G. N. Lance, *Numerical methods for high-speed computers*, Iliffe and Sons, London, 1960.

D. D. McCracken, *Digital computer programming*, Wiley, New York, 1957.

J. J. Murray, *Mathematical machines*, I, *Digital computers*, II, *Analog devices*, Columbia University Press, New York, 1961.

J. von Neumann, *The computer and the brain*, Yale University Press, New Haven, Conn., 1958.

R. K. Richards, *Arithmetic operations in digital computers*, Van Nostrand, New York, 1955.

G. R. Stibitz and J. A. Larrivee, *Mathematics and computers*, McGraw-Hill, New York, 1957.

M. V. Wilkes, *Automatic digital computers*, Wiley, New York, 1956.

VOLUME THREE

PART 5

CHAPTER XV

THEORY OF FUNCTIONS
OF A REAL VARIABLE

§1. Introduction

At the end of the 18th and the beginning of the 19th century, the differential and integral calculus was essentially worked out. Up to that time (in fact, throughout the 19th century) mathematicians were engaged in constructing its several branches, in discovering more and more new facts, and in developing more and more new domains of application of the differential and integral calculus to various problems of mechanics astronomy, and technology. Now it became possible to survey the results obtained, to study them systematically, and to delve into the meaning of the basic concepts of analysis. And here it became apparent that all was not well with the foundations of analysis.

Already in the 18th century there was no consensus among the greatest mathematicians of that time as to what a function is. This came out in prolonged controversies whether this or that solution of a problem, this or that concrete mathematical result were correct or incorrect. Gradually it became clear that also other basic concepts of analysis had to be made more precise. An inadequate understanding of the meaning of continuity and of the properties of continuous functions led to a number of erroneous statements, for example that a continuous function is always differentiable. Mathematics came to operate with such complicated functions that it became impossible to rely on intuition and guesswork. So there arose a real need to bring order into the fundamental concept of analysis.

The first serious attempt in this direction was made by Lagrange, and then Cauchy followed on the same path. Cauchy sharpened the definitions of limit, continuity, and integral and brought them into common use, as they survive to our days. Approximately at the same time, the Czech mathematician Bolzano made a rigorous study of the basic properties of continuous functions.

Let us consider these properties of continuous functions in more detail. Suppose that a continuous function $f(x)$ is given on some interval $[a, b]$, i.e., for all numbers satisfying the inequalities $a \leqslant x \leqslant b$. Previously it was regarded as obvious that if the function assumes values of opposite signs at the end points of the interval, then it must be zero at some intermediate point. Now this fact received a rigorous foundation. In the same way it was proved rigorously that a continuous foundation given on an interval assumes at certain points its greatest and its least value.

The study of these properties of continuous functions made it necessary to go deeper into the nature of the real numbers. As a result the theory of real numbers appeared; the basic properties of the numerical line were clearly formulated.

Further developments of mathematical analysis necessitated the study of more and more "bad," in particular discontinuous, functions. Discontinuous functions appear, for example, as limits of continuous functions, where it is not known *a priori* whether the limit function is continuous or not, and also in schematizing processes with sudden sharp variations. Here was a new task, namely to generalize the apparatus of analysis to discontinuous functions.

Riemann investigated the problem to what classes of discontinuous functions the concept of integral could be extended. As a result of this work on the foundation of analysis, there arose a new mathematical discipline: the theory of functions of a real variable.

If the classical mathematical analysis operates essentially with "good" (for example, continuous or differentiable) functions, the theory of functions of a real variable investigates considerably wider classes of functions. If in mathematical analysis the definition of some operation (for example integration) is given for continuous functions, then it is characteristic of the theory of functions of a real variable to find out to what classes of functions this definition is applicable, how the definition has to be modified so as to become wider. In particular, only the theory of functions of a real variable could give a satisfactory answer to the question what the length of curve is and for what curves it makes sense to talk of length.

The foundation on which this theory of functions of a real variable is built is the *theory of sets*.

Accordingly, we begin our exposition with an account of the elements of the theory of sets, next we turn to the study of point sets, and we conclude the chapter with an explanation of one of the fundamental concepts of the theory of functions of a real variable, namely the Lebesgue integral.

§2. Sets

People have constantly to deal with various collections of objects. As was already explained in Chapter I, this entailed the development of the concept of *number* and later that of a *set*, which is one of the basic primitive mathematical concepts and does not lend itself to an accurate definition. The following remarks are meant to illustrate what a set is but do not pretend to serve as a definition.

Set is the name for an aggregate, ensemble, or collection of things that are combined under a certain criterion or according to a certain rule. The concept of a set arises by an abstraction. By considering a certain collection of objects as a set, we disregard all the connections and relations between the various objects that make up the set, but we preserve the individual features of the objects. Thus, the set consisting of five coins and the set consisting of five apples are different sets. But the set of five coins arranged in a circle and the set of the same coins arranged one next to the other is one and the same set.

Let us give some examples of sets. We can talk of the grains forming a heap of sand, of the set of all planets of our solar system, of the set of all people that are in a certain house at a given moment, or of the set of all pages of this book. In mathematics we constantly come across various sets such as the set of all roots of a given equation, the set of all natural numbers, the set of all points on a line, etc.

The mathematical discipline that studies general properties of sets, i.e., properties that do not depend on the nature of the constituent objects, is called the *theory of sets*. This discipline began to be developed rigorously at the end of the 19th and the beginning of the 20th century. The founder of the scientific theory of sets is the German mathematician G. Cantor.

Cantor's work on the theory of sets grew from studying questions of convergence of trigonometric series. This is a very common phenomenon: Very often the occupation with concrete mathematical problems leads to the construction of very abstract and general theories. The value of such abstract constructions lies in the fact that they turn out to be connected not only with the concrete problem from which they have sprung but have also applications to a number of other problems. In particular, this is the case in the theory of sets. The ideas and concepts of the theory of sets penetrated literally into all branches of mathematics and changed its face entirely. Therefore it is impossible to form a proper picture of contemporary mathematics without being acquainted with the elements of the theory of sets. For the theory of functions of a real variable the theory of sets is of particularly great significance.

A set is considered as given when one can tell of every object whether it belongs to the set or not. In other words, a set is completely determined by all the objects that belong to it. If a set M consists of the objects $a, b, c, \cdots$ and of no others, then we write

$$M = \{a, b, c, \cdots\}.$$

The objects that form a certain set are usually called its *elements*. The fact that an object m is an element of a set M is written in the form

$$m \in M$$

and is read: "m belongs to M" or "m is an element of M". If an object n does not belong to a set M, then one writes: $n \bar{\in} M$. Every object can only be one element of a given set; in other words, all the elements of one and the same set are distinct from one another.

The elements of a set M can themselves be sets; however, to avoid contradiction it is convenient to postulate that a set M cannot be one of its own elements, $M \bar{\in} M$.

The set that contains no elements is called the *empty set*. For example, the set of all real roots of the equation

$$x^2 + 1 = 0$$

is empty. Henceforth the empty set will be denoted by ϕ.

If for two sets M and N every element x of M is also an element of N, then we say that M enters into N, that M is part of N, that M is a subset of N, or that M is contained in N; this is written in the form

$$M \subseteq N \quad \text{or} \quad N \supseteq M.$$

For example, the set $M = \{1, 2\}$ is part of the set $N = \{1, 2, 3\}$.

Clearly we always have $M \subseteq M$. It is convenient to regard the empty set as part of any set.

Two sets are *equal* if they consist of the same elements. For example, the set of roots of the equation $x^2 - 3x + 2 = 0$ and the set $M = \{1, 2\}$ are equal.

Now we define rules of *operations* on sets.

Union or sum. Suppose that $M, N, P, \cdots$ are sets. The union or sum of these sets is the set X consisting of all elements that belong to at least one of the "summands" $M, N, P, \cdots$

$$X = M + N + P + \cdots.$$

Here, even if an element x belongs to several summands, it occurs in the sum X only once. Clearly

$$M + M = M,$$

and if $M \subseteq N$, then

$$M + N = N.$$

Intersection. The intersection or common part of the sets M, N, P, $\cdots$ is the set Y consisting of all those elements that belong to all the sets M, N, P, $\cdots$.

Clearly $M \cdot M = M$ and if $M \subseteq N$, then $M \cdot N = M$.

If the intersection of the sets M and N is empty, $M \cdot N = \phi$, then we say that these sets are *disjoint*.

As a notation for the operations of sum and intersection of sets, we also use the symbols Σ and Π. Thus,

$$E = \Sigma E_i$$

is the sum of the sets E_i, and

$$F = \Pi E_i$$

their intersection.

We recommend that the reader prove that sum and intersection of sets are connected by the usual distributive law

$$M(N + P) = MN + MP,$$

and also by the law

$$M + NP = (M + N)(M + P).$$

Difference. The difference of two sets M and N is the set Z of all those elements of M that do not belong to N,

$$Z = M - N.$$

If $N \subseteq M$, then the difference $Z = M - N$ is also called the *complement* of N in M.

It is not hard to show that always

$$M(N - P) = MN - MP$$

and

$$(M - N) + MN = M.$$

Thus, the rules for operations on sets differ considerably from the usual rules of arithmetic.

Finite and infinite sets. Sets consisting of a finite number of elements are called finite sets. If the number of elements of a set is unbounded, then the set is called infinite. For example, the set of all natural numbers is infinite.

Let us consider two arbitrary sets M and N and ask whether the number of elements in those sets is the same or not.

If the set M is finite, then the collection of its elements is characterized by a certain natural number, namely, the number of its elements. In this case, in order to compare the numbers of elements of M and N, it is sufficient to count the number of elements in M and the number of elements in N and to compare the numbers so obtained. Also it is natural to reckon that if one of the sets M and N is finite and the other infinite, then the infinite set contains more elements than the finite.

However, if both sets M and N are infinite, then a simple count of the elements yields nothing. Therefore the following problem arises at once: Do all infinite sets have the same number of elements or do there exist infinite sets with larger or smaller numbers of elements? If the latter is true, then how can the numbers of elements in infinite sets be compared? We shall now turn our attention to these problems.

One-to-one correspondences. Again let M and N be two finite sets. How can we find out which of these sets contains more elements without counting the number of elements in each set? To this end let us form *pairs* by combining in a pair one element of M and one element of N. Then, if for some element of M there is no longer an element of N to be paired with it, M has more elements than N. Let us illustrate this argument by an example.

Suppose that in a room there are a certain number of people and a certain number of chairs. In order to find out of which there are more, it is sufficient to ask the people to sit down. If somebody is left without a place, it means that there are more people and if, say, all are placed and all places are taken, then there are as many people as chairs. This method of comparing the number of elements in sets has the advantage over a direct count of the elements because it is applicable without essential modifications not only to finite but also to infinite sets.

Let us consider the set of all natural numbers

$$M = \{1, 2, 3, 4, \cdots\}$$

and the set of all even numbers

$$N = \{2, 4, 6, 8, \cdots\}.$$

Which set contains more elements? At first sight it seems to be the former. However, we can form pairs from elements of these sets, as set out in Table 1:

Table 1.

M	1	2	3	4	$\cdots$
N	2	4	6	8	$\cdots$

No element of M nor of N remains without a partner. True, we could also have formed pairs as in Table 2:

Table 2.

M	1	2	3	4	5	$\cdots$
N	—	2	—	4	—	$\cdots$

Then many elements of M remain without a partner. On the other hand, we could have formed pairs as in Table 3:

Table 3.

M	—	1	—	2	—	3	—	$\cdots$
N	2	4	6	8	10	12	14	$\cdots$

Now many elements of N remain without a partner.

Thus, if the sets A and B are infinite, then distinct methods of forming pairs lead to different results. If there is one method of forming pairs in which every element of A and every element of B is paired off with some element, then we say that a one-to-one correspondence can be set up between A and B. For example, we can establish a one-to-one correspondence between the sets M and N as is clear from Table 1.

If between the sets A and B a one-to-one correspondence can be set up, then we say that they have the *same* number of elements. If for *every* method of pairing there are always some elements of A without a partner, then we say that the set A contains more elements than B or that A has a greater *cardinality* than B.

Thus we have obtained an answer to one of the questions raised earlier: how to compare the number of elements in infinite sets. However, this has by no means brought us nearer to an answer to the other question: Do there exist infinite sets at all having distinct cardinalities? In order to get an answer to this question let us study some simple types of infinite sets.

Countable sets.   If we can set up a one-to-one correspondence between the elements of a set A and the elements of the set of all natural numbers

$$Z = \{1, 2, 3, \cdots\},$$

then we say that the set A is *countable*. In others words, a set is countable if all its elements can be enumerated by means of the natural numbers i.e., written down in the form of a sequence

$$a_1, a_2, \cdots, a_n, \cdots.$$

Table 1 shows that the set of all even numbers is countable (the upper numbers can now be regarded as the suffix of the corresponding lower number).

Countable sets are, so to speak, the very smallest infinite sets: Every infinite set contains a countable subset.

If two nonempty finite sets do not intersect, then their sum contains more elements than either summand. For infinite sets this cannot hold. For example, let E be the set of all even numbers, O the set of all odd numbers, and Z the set of all natural numbers. As Table 4 shows, the sets E and O are countable. However, the set $Z = E + O$ is again countable.

Table 4.

E	2	4	6	8	$\cdots$
O	1	3	5	7	$\cdots$
Z	1	2	3	4	$\cdots$

The violation of the rule "the whole is larger than the parts" in infinite sets shows that the properties of infinite sets differ *qualitatively* from those of finite sets. The transition from the finite to the infinite proceeds in complete agreement with the well-known principle of dialectics, qualitative variation of properties.

Let us show that *the set of all rational numbers is countable*. For this purpose we arrange the rational numbers in Table 5.

Here all the natural numbers are placed in the first row in ascending order, in the second row zero and all the negative numbers in decreasing order, in the third row the positive reduced fractions with denominator 2 in ascending order, in the fourth row the negative reduced fractions with denominator 2 in descending order, etc. It is clear that every rational number occurs once and only once in this table. Let us now enumerate all

Table 5.

(1)	(2)	(3)	(4)	(5)	(6)	
1	2	3	4	5	6	$\cdots$
0	1	-2	-3	-4	-5	$\cdots$
$\dfrac{1}{2}$	$\dfrac{3}{2}$	$\dfrac{5}{2}$	$\dfrac{7}{2}$	$\dfrac{9}{2}$	$\dfrac{11}{2}$	$\cdots$
$-\dfrac{1}{2}$	$-\dfrac{3}{2}$	$-\dfrac{5}{2}$	$-\dfrac{7}{2}$	$-\dfrac{9}{2}$	$-\dfrac{11}{2}$	$\cdots$
$\dfrac{1}{3}$	$\dfrac{2}{3}$	$\dfrac{4}{3}$	$\dfrac{5}{3}$	$\dfrac{7}{3}$	$\dfrac{8}{3}$	$\cdots$
$-\dfrac{1}{3}$	$-\dfrac{2}{3}$	$-\dfrac{4}{3}$	$-\dfrac{5}{3}$	$-\dfrac{7}{3}$	$-\dfrac{8}{3}$	$\cdots$
$\cdots$	$\cdots$	$\cdots$	$\cdots$	$\cdots$	$\cdots$	$\cdots$

the numbers of the table in the order indicated by the arrows. Then all the rational numbers are arranged in a single sequence:

Number of the place occupied by rational number	1	2	3	4	5	6	7	8	9	$\cdots$
Rational number	1,	2,	0,	3,	-1,	$\frac{1}{2}$,	4,	-2,	$\frac{3}{2}$,	$\cdots$

So we have established a one-to-one correspondence between all the rational numbers and all the natural numbers. Therefore the set of all rational numbers is countable.

Sets with the cardinal number of the continuum. If we can set up a one-to-one correspondence between the elements of a set M and the points of the interval $0 \leqslant x \leqslant 1$, then we say that the set M has the *cardinal number of the continuum*. In particular, by this definition the set of points of the segment $0 \leqslant x \leqslant 1$ has itself the cardinal number of the continuum.

From figure 1 it is clear that the set of points of any interval AB has the cardinal number of the continuum. Here the one-to-one correspondence is established geometrically by means of a projection.

It is not hard to show that the sets of points of any open interval $a < x < b$ and of the whole numerical line have the cardinal number of the continuum.

Of considerably greater interest is the following fact: The set of points

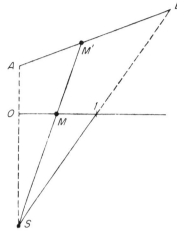

of the square $0 \leqslant x \leqslant 1, 0 \leqslant y \leqslant 1$ has the cardinal number of the continuum. Thus, roughly speaking, there are "as many" points in the square as on the segment.

Fig. 1.

§3. Real Numbers*

The development of the concept of number has been described in detail in Chapter I. Here we shall give the reader a brief account of the theories of the real numbers that have arisen in the 19th century in connection with the foundation of the basic concepts of analysis.

Rational numbers. We assume that the reader is familiar with the main properties of rational numbers. Without going into details we recall these properties. Rational numbers, i.e., numbers of the form m/n, where m and n are integers and $n \neq 0$, form a set of numbers in which two operations (addition and multiplication) are defined. These operations are subject to a number of laws (axioms). In what follows a, b, c, $\cdots$ denote rational numbers

I. *Axioms of addition.*

1. $a + b = b + a$ (commutativity)

2. $a + (b + c) = (a + b) + c$ (associativity)

3. the equation

$$a + x = b$$

has a unique solution (existence of the inverse operation).

From these axioms it follows immediately that the expression $a + b + c$ has a unique meaning, that there exists a rational number 0 (null) for which $a + 0 = a$ and that addition has an inverse operation (subtraction) so that the expression $b - a$ has a meaning.

---

* During the writing of this section, we have had valuable consultations with A. N. Kolmogorov.

Thus, from the algebraic point of view, all the rational numbers form a commutative group under the operation of addition.

II. *Axioms of multiplication.*

1. $ab = ba$ (commutativity)
2. $a(bc) = (ab)c$ (associativity)
3. the equation

$$ay = b,$$

where $a \neq 0$ has a unique solution (existence of the inverse operation).

From these axioms it follows that the expression abc has a meaning, that there exists a rational number 1 for which $a \cdot 1 = a$ and that for rational numbers other than 0 the inverse operation (division) exists. All the rational numbers except 0 form a commutative group under the operation of multiplication.

III. *Axiom of distributivity.*

1. $(a + b) c = ac + bc$.

The axioms I-III together indicate that under the operations of addition and multiplication the rational numbers form a so-called *algebraic field*.

IV. *Axioms of order.*

1. For any two rational numbers a and b one and only one of the following three relations holds: either $a < b$, or $a > b$, or $a = b$.

2. If $a < b$, and $b < c$, then $a < c$.

3. If $a < b$, then $a + c < b + c$ (monotonicity of addition).

4. If $a < b$ and $c > 0$, then $ac < bc$ (monotonicity of multiplication by $c > 0$).

All these axioms together allow us to call the set of rational numbers an *ordered field*.

Apart from the rational numbers there exist also other systems of objects that satisfy these axioms and are therefore ordered fields.

We mention two important properties of rational numbers.

Density: For arbitrary a and b, $a < b$, there is a c such that $a < c < b$.

Countability: The set of all rational numbers is countable (see §2).

On measuring quantities. The insufficiency of the rational numbers alone in mathematics already becomes apparent in dealing with such an important task as that of measuring quantities. As one of the simplest examples, we shall here consider the problem of measuring the length of intervals.

Let us take a line on which a definite direction, an origin (the point 0), and a unit of scale are marked. Then it is clear what a segment OA with its end point at 1/2, 1/3, 2/3, −1/3, etc. is. More generally, with every rational number a we can associate a point a on the line, namely the point with the coordinate $x = a$. In this case the number a determines the length of the directed segment OA. However, not every segment gets a certain (rational) number as the measure of its length in this construction. For example, it was already known to the ancient Greeks that the length of the diagonal of a square with unit side cannot be measured by any rational number. The natural outcome of this situation is the setting up of a one-to-one correspondence between numbers and lengths, i.e., a further extension of the concept of number.

Real numbers. We were led to the conclusion that the rational numbers alone are insufficient to measure quantities and that the concept of number must be extended so that there exists a one-to-one correspondence between numbers and points on a line. With this in mind let us try to find out whether it is not possible to determine the position of an *arbitrary* point on a line by means of the rational points only. A similar construction within the domain of rational numbers will then lead us to the concept of a real number.

Let α be an arbitrary point on the line. Then all the rational points a can be divided into two parts: We put into one part all the points a that are to the left of α, and into the other all those to the right of α. As regards the point α itself (if it happens to be rational), it can be put into either part. Such a division of the rational points is usually called a *cut*. Cuts are taken to be identical if the collection of rational points in the left and the right parts of the cuts coincide (to within a single point). Now it is not difficult to see that distinct points α and β determine different cuts. For since the rational points are everywhere dense on the line, we can find rational points r_1 and r_2 lying strictly between α and β. Then in one of the cuts they come into the right part and in the other into the left part.

Thus every point on the line determines a cut in the domain of rational points, and different cuts correspond to different points. It is very important that a cut can also be defined in another way and without specific reference to the number α. For let us define a cut in the domain of rational points as a division of all rational points into two nonempty disjoint sets A and B

such that $a < b$ for every $a \in A$, $b \in B$. Under this definition we can assign to a cut in a unique fashion that point (boundary) which produces it. In other words, by means of cuts in the domain of rational points, we can determine *every* point on the line. The construction we have just explained was proposed by the German mathematician R. Dedekind and is known under the name of *Dedekind cut*.

A cut is not the only possible method of determining the position of an arbitrary point by means of the rational points. Nearer to the usual practice of measuring is the following method of G. Cantor. Again let α be an arbitrary point on the line. Then we can find two arbitrarily close rational points a and b such that α lies between a and b. The points a and b determine the position of α approximately. Let us imagine this process of approximate determination of α continued indefinitely and in such a way that at each successive step the accuracy is increased more and more. Then we obtain a system of intervals $[a_n, b_n]$ with their ends at rational points such that $[a_{n+1}, b_{n+1}] \subseteq [a_n, b_n]$ and $b_n - a_n \to 0$ $(n \to \infty)$. A system of intervals satisfying these conditions is called a *nest of intervals*. It is clear that such a nest of intervals determines the position of α uniquely.

By means of similar constructions in the domain of rational numbers, we can define the real numbers. Next we define the operations among real numbers and ascertain that they satisfy the same axioms as the operations on rational numbers. Now every point on the line corresponds to a real number and vice versa. On account of this, the set of all real numbers is often called the numerical line.

Principles of continuity. There are essential differences between the set of all rational numbers and the set of all real numbers. In fact, the set of all real numbers has a number of properties that characterise the *continuity* of this set, whereas the set of all rational numbers does not have these properties. They are usually called principles of continuity. We shall enumerate the most important of them.

Dedekind's principle. If the set of all real numbers is divided into two nonempty sets X and Y without elements in common, so that for arbitrary $x \in X$, $y \in Y$ the inequality $x < y$ holds, then there exists a unique number ξ (the boundary) for which $x \leqslant \xi \leqslant y$ for arbitrary $x \in X$, $y \in Y$.

The set of all real numbers x satisfying the inequalities $a \leqslant x \leqslant b$ is called an interval of the numerical line and is denoted by $[a, b]$. A system of intervals $[a_n, b_n]$ is called *nested* if $[a_{n+1}, b_{n+1}] \subseteq [a_n, b_n]$ and $b_n - a_n \to 0$ $(n \to \infty)$.

Cantor's principle. For every nested system of intervals $[a_n, b_n]$ there exists one and only one real number ξ that belongs to each of these intervals.

Weierstrass' principle. Every nondecreasing sequence of real numbers that is bounded above converges.

Let us say that a sequence of real numbers $\{x_n\}$ is a *fundamental* sequence if for every $\epsilon > 0$ we can find a natural number N such that for all $n > N$ and all natural p

$$| x_{n+p} - x_n | < \epsilon.$$

Cauchy's principle. Every fundamental sequence of real numbers converges.

Since we have not given an accurate construction of the real numbers we are not in a position to establish that these principles hold for the set of real numbers. Our next object is to investigate how these principles are interrelated. Let us then assume that one of the principles of continuity holds for the real numbers and examine which of the remaining principles of continuity follows from it.

The over-all result that we shall arrive at is that all the principles of continuity are equivalent

We say that a number b is the (least) upper bound of a set E

$$b = \sup E,$$

if (1) $x \leqslant b$ for every $x \in E$ and (2) there exists no number $b' < b$ with the same property.

Let us show that the following proposition follows from Dedekind's principle: Every nonempty set E of numbers that is bounded above has a least upper bound. We divide all the real numbers into two classes X and Y according to the following criterion: We put $x \in X$ if there exists an $a \in E$ such that $a \geqslant x$, and we put $y \in Y$ if for every $a \in E$ we have $a < y$. It is easy to verify that this is a cut. By Dedekind's principle it has a boundary ξ; this boundary is the least upper bound of E.

We shall now show that Weierstrass' principle follows from Dedekind's. Let $\{x_n\}$ be a nondecreasing sequence of real numbers, bounded above. By what we have just proved it has a least upper bound ξ. By definition of an upper bound $x_n \leqslant \xi$ ($n = 1, 2, \cdots$); for every $\epsilon > 0$ we can find an index n_0 such that $x_{n_0} > \xi - \epsilon$. Since the sequence $\{x_n\}$ is monotonic, this implies that $\xi - \epsilon < x_n \leqslant \xi$ for all $n > n_0$, i.e., the sequence $\{x_n\}$ converges to the limit ξ.

To prove the converse relation between the principles of Dedekind and Weierstrass we note that Weierstrass' principle implies:

Archimedes' principle. No matter what the real numbers $a > 0$ and b are, we can always find a natural number n such that $na > b$.

This principle means that for every real number b the sequence $\{b/n\}$ ($n = 1, 2, \cdots$) converges to zero.

Suppose that Weierstrass' principle holds, but Archimedes' does not hold. The latter means that there exists an $a > 0$ such that the sequence $x_n = na$ is bounded. Moreover, it is increasing. By Weierstrass' principle it has a limit ξ. Hence it follows that the interval $[\xi - a/2, \xi]$ contains some point $x_n = na$ of our sequence. But then $x_{n+1} = (n + 1) a > \xi$, and this contradicts the fact that ξ is the least upper bound of $\{x_n\}$.

Weierstrass' principle implies Dedekind's principle. Let the set of all real numbers be divided into two disjoint sets X and Y such that $x < y$ for all $x \in X$, $y \in Y$. We shall show that this cut has a unique boundary ξ. Let m be an integer and n a natural number. We denote by x_n the largest element of the form $m/2^n \in X$ such that $x_n + 1/2^n \in Y$. Since the set of elements of the form $m/2^n$ is contained in the set of elements of the form $m/2^{n+1}$, we have $x_n \leqslant x_{n+1}$. Moreover, the sequence $\{x_n\}$ is bounded (for example, by the number $x_1 + 1/2$). Hence by Weierstrass' principle it has a limit ξ. We shall show that ξ is the boundary of our cut. For if $x < \xi$ then $x \in X$. And if $y > \xi$, then $y \in Y$, because it follows from Archimedes' principle that we can find a number n such that $1/2^n < y - \xi = a$. But $x_n < \xi, x_n + 1/2^n \in Y$, and then $y = \xi + a > x_n + 1/2^n$, therefore $y \in Y$.

One can also show that Cantor's and Cauchy's principles are equivalent. However, when Cauchy's principle holds it does not follow that Dedekind's principle holds. This statement has to be understood in the following sense: There exists an ordered field for which Cauchy's principle holds, but Dedekind's does not hold. If it is assumed beforehand that Archimedes' principle holds, then all four principles are equivalent.

Uncountability of the continuum. Let us show that the set of all points of the segment $0 \leqslant x \leqslant 1$ is uncountable. We shall give an indirect proof. Suppose that the set of all points of the segment $0 \leqslant x \leqslant 1$ is countable. Then all the points x of this segment can be indexed by means of the natural numbers

$$x_1, x_2, \cdots, x_n, \cdots. \tag{1}$$

In $[0, 1]$ we choose an interval σ_1 so that its length is less than 1 and that it does not contain the point x_1. Such an interval can readily be found. Next, within σ_1 we choose an interval σ_2 so that its length is less than $1/2$ and that σ_2 does not contain the points x_1, x_2. Generally, when an interval σ_{n-1} has already been chosen we choose in it an interval σ_n so that its length is less than $1/n$ and that it does not contain the points x_1, x_2, $\cdots$, x_n. In this way we construct an infinite sequence of intervals

$$\sigma_1, \sigma_2, \cdots, \sigma_n, \cdots,$$

such that each is contained in the preceding one and their lengths tend to zero with increasing n. Then by Cantor's principle there exists a unique point x in the interval $[0, 1]$ that belongs to all the intervals σ_n. Since by our hypothesis all the points of $[0, 1]$ are accounted for in (1), the point x which is common to all σ_n coincides with some point x_m of that sequence. But by our construction σ_m does not contain x_m so that $x \neq x_m$. Thus we have arrived at a contradiction. Therefore the initial hypothesis, that the set of all points of $[0, 1]$ is countable, is false and so this set is uncountable. This is what we set out to prove.

This theorem shows that there exist distinct infinite cardinalities and therefore gives a positive answer to the first question raised.

§4. Point Sets

In the preceding section we have already come across sets whose elements are *points*. In particular, we have considered the set of all points of an arbitrary interval and the set of all points (x, y) of the square $0 \leqslant x \leqslant 1, 0 \leqslant y \leqslant 1$. We shall now turn to a more detailed study of properties of such sets.

A set whose elements are points is called a *point set*. Thus, we can speak of point sets on a line, in a plane, or in an arbitrary space. For simplicity's sake we shall here confine ourselves to the study of points sets *on a line*.

There is a close connection between the real numbers and the points on a line: With every real number we can associate a point on the line and vice versa. Therefore, in speaking of point sets we may include with them sets consisting of real numbers, sets on the numerical line. Conversely, in order to define a point set on a line we shall, as a rule, give the coordinates of all the points of the set.

Point sets (and, in particular, point sets on a line) have a number of special properties that distinguish them from arbitrary sets and make the theory of point sets into a self-contained mathematical discipline. First of all, it makes sense to speak of the *distance* between two points. Furthermore, we can establish a relation of *order* (left, right) between the points on a line; accordingly, one says that a point set on a line is an *ordered* set. Finally, as we have already mentioned earlier, Cantor's principle holds for the line; this property of the line is usually characterized as *completeness* of the line.

We introduce a notation for the simplest sets on a line.

An *interval* $[a, b]$ is the set of points whose coordinates satisfy the inequality $a \leqslant x \leqslant b$.

An open *interval* (a, b) is the set of points whose coordinates satisfy the conditions $a < x < b$.

The *semi-intervals* $[a, b)$ and $(a, b]$ are defined by the conditions $a \leqslant x < b$ and $a < x \leqslant b$, respectively.

Open intervals and semi-intervals can be *improper*. Thus, (∞, ∞) denotes the whole line and, for example, (−∞, b] the set of all points for which $x \leqslant b$.

We begin with an account of the various possibilities for the position of a set *as a whole* on a line.

Bounded and unbounded sets. A set E of points on a line can either consist of points whose distances from the origin of coordinates do not exceed a certain positive number or it has points arbitrarily far from the origin of coordinates. In the first case E is called *bounded*, in the latter *unbounded*. An example of a bounded set is the set of all points of the interval [0, 1], and an example of an unbounded set is the set of all points with integral coordinates

It is easy to see that, when a is a fixed point on the line, a set E is bounded if and only if the distances from a of arbitrary points $x \in E$ do not exceed a certain positive number.

Sets bounded above and below. Let E be a set of points on a line. If there is a point A on the line such that every point $x \in E$ lies to the left of A, then we say that E is *bounded above*. Similarly, if there is a point a on the line such that every $x \in E$ lies to the right of a, then E is called *bounded below*. Thus, the set of all points on the line with positive coordinates is bounded below, and the set of all points with negative coordinates bounded above.

It is clear that the definition of a bounded set is equivalent to the following: A set E of points on a line is called bounded if it is bounded above and below. Notwithstanding that these two definitions are very similar, there is an essential difference between them: The first is based on the fact that a distance is defined between the points on a line, and the second that these points form an ordered set.

We can also say that a set is bounded if it lies entirely in some interval [a, b].

The least upper and greatest lower bound of a set. Suppose that a set E is bounded above. Then there exist points A on the line such that there are no points of E to their right. Using Cantor's principle we can show that among all the points A having these properties there is a leftmost. This point is called the *least upper bound* of E. The *greatest lower bound* of a point set is defined similarly.

If there is a rightmost point in E, then it is obviously the least upper bound of E. However, it can happen that E has no rightmost point.

For example, the set of points with the coordinates

$$\frac{0}{1}, \frac{1}{2}, \frac{2}{3}, \frac{3}{4}, \frac{4}{5}, \cdots$$

is bounded above and has no rightmost point. In this case the least upper bound a does not belong to E, but there are points of E arbitrarily near to a. In the example above $a = 1$.

Distribution of a point set near an arbitrary point on the line. Let E be a point set and x an arbitrary point on the line. We consider the various possibilities for the distribution of the set E near x. The following cases are possible:

1. Neither the point x nor the points sufficiently near to it belong to E.
2. The point x does not belong to E, but there are points of E arbitrarily near it.
3. The point x belongs to E, but all points sufficiently near to it do not belong to E.
4. The point x belongs to E and there are other points of E arbitrarily near it.

In the case 1, x is called *exterior* to E, in the case 3, an *isolated* point of E, and in the cases 2 and 4, a *limit* point of E.

Thus, if $x \bar{\in} E$, then x can be either exterior to E or a limit point, and if $x \in E$, it can be either an isolated point of E or a limit point.

A limit point may or may not belong to E and is characterized by the condition that there are points of E arbitrarily near to it. In other words, a point x is a limit point of E if every open interval δ containing x contains infinitely many points of E. The concept of a limit point is one of the most important in the theory of points sets.

If x and all points sufficiently near to it belong to E, then x is called an *interior* point of E. Every point x that is neither an exterior nor an interior point of E is called a *boundary* point of E.

Let us give some examples to illustrate all these concepts.

Example 1. Let E_1 consist of the points with the coordinates

$$1, \frac{1}{2}, \frac{1}{3}, \cdots, \frac{1}{n}, \cdots.$$

Then every point of this set is an isolated point of it, the point 0 is a limit point of E_1 (and does not belong to it), and all the remaining points on the line are exterior to E_1.

Example 2. Let E_2 consist of all the *rational* points of the interval [0, 1]. This set has no isolated points, every point of the interval [0, 1] is a limit point of E_2, and all the remaining points on the line are exterior to E_2. Clearly among the limit points of E_2 there are some that belong and others that do not belong to the set.

Example 3. Let E_3 consist of *all* points of the interval [0, 1]. As in the preceding example, E_3 has no isolated points and every point of [0, 1] is a limit point. However, in contrast to the preceding example, all the limit points of E_3 belong to the set.

Example 4. Let E_4 consist of all the points on the line with integral coordinates. Every point of E_4 is isolated; E_4 has no limit points.

We also point out that in Example 3 every point of the open interval (0, 1) is an interior point of E_3, and in Example 2 every point of the interval [0, 1] is a boundary point of E_2.

From the preceding examples it is clear that an infinite point set on a line may have isolated points (E_1, E_4) or not (E_2, E_3), that it may have interior points (E_3) or not (E_1, E_2, E_4). As regards limit points, only E_4 in Example 4 does not have any. As the following important theorem shows, this is connected with the fact that E_4 is unbounded.

The Theorem of Bolzano-Weierstrass: *Every bounded infinite point set on a line has at least one limit point.*

Let us prove this theorem. Suppose that E is a bounded infinite point set on a line. Since E is bounded, it lies entirely in some interval $[a, b]$. We divide this interval in half. Since E is infinite, at least one of the intervals so obtained contains infinitely many points of E. We denote that interval by σ_1 (if both halves of $[a, b]$ contain infinitely many points of E, then σ_1 shall denote, say, the left half). Next we divide σ_1 into two equal halves. Since the part of E that lies in σ_1 is infinite, at least one of the intervals so obtained contains infinitely many points of E. We denote it by σ_2. We continue the process of dividing an interval in half indefinitely and each time select that half which contains infinitely many points of E. So we obtain a sequence of intervals $\sigma_1, \sigma_2, \cdots, \sigma_n, \cdots$. This sequence of intervals has the following properties: Every interval σ_{n+1} is contained in the preceding one σ_n; every interval σ_n contains infinitely many points of E; and the lengths of the intervals tend to zero. The first two properties of the sequence follow immediately from its construction, and to prove the last property it is sufficient to note that if the length of $[a, b]$ is l, then the length of σ_n is $l/2^n$. By Cantor's principle there exists a unique point x that belongs to all σ_n. We shall show that this x is a limit point of E.

For this it is sufficient to make sure that if δ is some open interval containing x, then it contains infinitely many points of E. Since every interval σ_n contains x and the lengths of the σ_n tend to zero, for a sufficiently large n the interval σ_n is entirely contained in δ. But by hypothesis σ_n contains infinitely many points of E. Therefore δ too contains infinitely many points of E. Thus, x is in fact a limit point of E and the theorem is proved.

Exercise. Show that if a set E is bounded above and has no rightmost point, then its least upper bound is a limit point of E (and does not belong to E).

Closed and open sets. One of the fundamental tasks of the theory of point sets is the study of properties of various types of points sets. We shall acquaint the reader with this theory for two examples. We shall now study properties of the so-called closed and open sets.

A set is called *closed* if it contains all its limit points. If a set has no limit points, then it is usually also taken to be closed. Apart from its limit points a closed set can also contain isolated points. A set is called *open* if every point of it is interior.

Let us give some examples of closed and open sets. Every interval $[a, b]$ is a closed set, and every open interval (a, b) an open set. The improper semi-intervals $(-\infty, b]$ and $[a, \infty)$ are closed, and the improper intervals $(-\infty, b)$ and (a, ∞) are open. The whole line is at the same time closed and open. It is convenient to regard the empty set also as both closed and open. Every finite point set on the line is closed, since it has no limit points. The set consisting of the points

$$0, 1, \frac{1}{2}, \frac{1}{3}, \frac{1}{4}, \cdots \frac{1}{n}, \cdots$$

is closed; this set has the single limit point $x = 0$ which belongs to the set.

Our task is to examine the structure of an arbitrary closed or open set. For this purpose we require a number of auxiliary facts which we assume without proof.

1. The intersection of any number of closed sets is closed.

2. The union of any number of open sets is open.

3. If a closed set is bounded above, then it contains its least upper bound. Similarly, if a closed set is bounded below, then it contains its greatest lower bound.

Let F be an arbitrary point set on a line. The *complement* of E, denoted by CE, is defined as the set of all points on the line that do not belong to E. Clearly, if x is an exterior point of E, then it is an interior point of CE and vice versa.

4. If a set F is closed, then its complement CF is open and vice versa.

Proposition 4 shows that there is a very close link between closed and open sets: They are complements of each other. Because of this it is sufficient to study either closed or open sets only. A knowledge of the properties of sets of one type enables us at once to read off properties of sets of the other type. For example, every open set is obtained by deleting some closed set from the line.

Let us now proceed to study properties of closed sets. We make one definition. Let F be a closed set. An open interval (a, b) having the property that none of its points belong to F, while a and b belong to F, is called an *adjacent interval* of F. Among the adjacent intervals we also count the improper intervals (a, ∞) or $(-\infty, b)$, provided a or b belong to F but the intervals do not intersect F. We shall show that if a point x does not belong to the closed set F, then it belongs to one of its adjacent intervals.

We denote by F_x that part of F that lies to the right of x. Since x itself does not belong to F, we can represent F_x as an intersection

$$F_x = F \cdot [x, \infty).$$

Both F and $[x, \infty)$ are closed. Therefore, by proposition 1, F_x is closed. If F_x is empty, then the whole semi-interval $[x, \infty)$ does not belong to F. Let us assume then that F_x is not empty. Since this set lies entirely on the semi-interval $[x, \infty)$, it is bounded below. We denote by b its greatest lower bound. By proposition 3, $b \in F_x$ so that $b \in F$. Furthermore, since b is the greatest lower bound of F_x, the semi-interval $[x, b)$ lying to the left of b does not contain points of F_x, consequently not of F either. Thus we have constructed a semi-interval $[x, b)$ containing no points of F, and either $b = \infty$ or b belongs to F. Similarly, we can construct a semi-interval $(a, x]$ not containing points of F with either $a = -\infty$ or $a \in F$. Now it is clear that the open interval (a, b) contains x and is an adjacent interval of F. It is easy to see that if (a_1, b_1) and (a_2, b_2) are two adjacent intervals of F, then they either coincide or are disjoint.

From the preceding it follows that every closed set on the line is obtained by deleting from the line a certain number of open intervals, namely the adjacent intervals of F. Since every open interval contains at least one rational point and all the rational points on the line form a countable set, we see that the number of adjacent intervals cannot be more than countable. Hence we reach a remarkable conclusion. Every closed set on the line is obtained by deleting from the line at the most a countable set of disjoint open intervals.

By proposition 4 it follows at once that every open set on the line is the sum of not more than a countable number of open intervals. By propositions 1 and 2 it is also clear that every set of the structure we have indicated is in fact closed (open).

It will be seen from the following example that closed sets can have a very complicated structure.

Cantor's perfect set. We shall construct a particular closed set that has a number of remarkable properties. First of all we delete from the line the improper intervals $(-\infty, 0)$ and $(1, \infty)$. After this operation we are left with the interval $[0, 1]$. Next we delete from this the open interval $(1/3, 2/3)$ which forms its middle third. From each of the remaining intervals $[0, 1/3]$ and $[2/3, 1]$, we delete its middle third. This process of deleting the middle thirds of the remaining intervals can be continued indefinitely. The point set on the line that remains after all these open intervals have been deleted is called Cantor's perfect set; we shall denote it by the letter P.

Let us investigate some properties of this set: P is closed, because it is formed by deleting from the line a certain set of disjoint open intervals; P is not empty; in any case it contains the end points of all the removed intervals.

A closed set F is called *perfect* if it has no isolated points, i.e., if every point of it is a limit point. We shall show that P is perfect. For if x were an isolated point of P, then it would have to be a common end point of two adjacent intervals of the set. But by our construction the adjacent intervals of P do not have common end points.

The set P does not contain any open interval. For suppose that a certain open interval δ entirely belongs to P. Then it belongs entirely to one of the intervals obtained at the nth step of the construction of P. But this is impossible, because for $n \to \infty$ the length of these intervals tends to zero.

One can show that P has the cardinality of the continuum. From this it follows, in particular, that Cantor's perfect set contains, apart from the end points of the adjacent intervals, other points. For the end points of the adjacent intervals form only a countable set.

Various types of point sets occur constantly in the most diverse branches of mathematics, and a knowledge of their properties is absolutely indispensable in studying many mathematical problems. Of particularly great importance is the theory of point sets in mathematical analysis and topology.

We shall now give a few examples of point sets that appear in the classical parts of analysis. Let $f(x)$ be a continuous function given on the interval $[a, b]$. We fix a number α and consider the set of all points x for which $f(x) \geqslant \alpha$. It is easy to show that this can be an arbitrary closed set on the interval $[a, b]$. Similarly, the set of those points x for which $f(x) > \alpha$ can be any open set $G \subset [a, b]$. If $f_1(x), f_2(x), \cdots, f_n(x), \cdots$ is a sequence of continuous functions given on $[a, b]$, then the set of all points x where this sequence converges cannot be arbitrary and belongs to a certain well-defined type.

The mathematical discipline whose object is to study the structure of point sets is called the *descriptive theory of sets*.

In the development of the descriptive theory of sets Soviet mathematicians have made great contributions, N. N. Luzin and his pupils, P. S. Aleksandrov, M. Ja. Suslin, A. N. Kolmogorov, M. A. Lavrent'ev, P. S. Novikov, L. V. Keldyš, A. A. Ljapunov, and others.

The investigations of N. N. Luzin and his pupils have shown that there are strong ties between descriptive set theory and mathematical logic. Difficulties arising in the study of a number of problems of descriptive set theory (in particular, problems of determining the cardinality of certain sets) are difficulties of a logical nature. On the other hand, the methods of mathematical logic enable us to penetrate more deeply into certain problems of descriptive set theory.

§5. Measure of Sets

The concept of the measure of a set is a far-reaching generalization of the concept of the length of an interval. In the simplest case (the only one we shall consider here) the task is to give a definition of length not only for intervals but also for more complicated point sets on a line.

Let us agree that the unit of measurement is the interval $[0, 1]$. Then the length of an arbitrary interval $[a, b]$ is obviously $b - a$. Similarly, if we have two disjoint intervals $[a_1, b_1]$ and $[a_2, b_2]$, it is natural to interpret the length of the set E consisting of these two intervals as the number $(b_1 - a_1) + (b_2 - a_2)$. However, it is by no means clear what we have to understand by the length of a set on the line of a more complicated nature; for example, what is the length of the Cantor set of §4 of this chapter? Hence the conclusion, the concept of length of a set on the line requires a rigorous mathematical definition.

The problem of defining the lengths of sets or, as we now say, of measuring sets is very important, because it is of vital significance in generalizing the concept of an integral. The concept of measure of a set also has applications to other problems in the theory of functions, in probability, topology, functional analysis, etc.

We shall give an account of the definition of measure of sets which is due to the French mathematician H. Lebesgue and is the foundation of the definition of integral given by him.

Measure of an open and a closed set. We begin with the definition of measure of an arbitrary open or closed set. As we have mentioned in §4, every open set on the line is a finite or countable sum of pairwise disjoint open intervals.

The measure of an open set is defined as the sum of the lengths of its constituent open intervals.

Thus, if

$$G = \sum (a_i, b_i)$$

and the intervals (a_i, b_i) are pairwise disjoint, then the measure of G is equal to $\sum (b_i - a_i)$. Generally, the measure of a set E being denoted by μE, we can write

$$\mu G = \sum (b_i - a_i).$$

In particular, the measure of a single open interval is equal to its length

$$\mu(a, b) = b - a.$$

Every closed set F contained in $[a, b]$ and such that the end points of $[a, b]$ belong to F is obtained from $[a, b]$ by deleting from it a certain open set G. The *measure of the closed set* $F \subseteq [a, b]$, where $a \in F$, $b \in F$ is defined as the difference between the length of $[a, b]$ and the measure of the open set G complementary to F (relative to $[a, b]$).

Thus

$$\mu F = (b - a) - \mu G. \tag{2}$$

It is not difficult to verify that according to this definition the measure of an arbitrary interval is equal to its length

$$\mu[a, b] = b - a,$$

and the measure of a set consisting of a finite number of points is zero.

General definition of measure. In order to give a definition of measure of sets of more general nature than open and closed sets, we have to make use of an auxiliary concept. Let E be a set lying on the interval $[a, b]$. We consider all possible *coverings* of E, i.e., all open sets $V(E)$ containing

E. The measure of each of these sets $V(E)$ is already defined. The aggregate of measures of all sets $V(E)$ is a certain set of positive numbers. This set is bounded below (for example by 0) and therefore has a greatest lower bound which we denote by $\mu_e E$. The number $\mu_e E$ is called the *outer measure* of E.

Let $\mu_e E$ be the outer measure of a set E and $\mu_e CE$ the outer measure of its complement relative to $[a, b]$.

If the relation

$$\mu_e E + \mu_e CE = b - a \tag{3}$$

holds, then the set E is called *measurable* and the number $\mu_e E$ its *measure:* $\mu E = \mu_e E$; if the relation (3) does not hold, then we say that E is not measurable; a nonmeasurable set has no measure.

We note that always

$$\mu_e E + \mu_e CE \geqslant b - a. \tag{4}$$

Let us give a few clarifications. The length of the simplest sets (for example, open or closed intervals) has a number of remarkable properties. We mention the most important of these.

1. If the sets E_1 and E are measurable and $E_1 \subseteq E$, then

$$\mu E_1 \leqslant \mu E;$$

i.e., the measure of a part of E does not exceed the measure of the whole set E.

2. If E_1 and E_2 are measurable, then the set $E = E_1 + E_2$ is measurable and

$$\mu(E_1 + E_2) \leqslant \mu E_1 + \mu E_2;$$

i.e., the measure of a sum does not exceed the sum of the measures of the summands.

3. If sets $E_i\,(i = 1, 2, \cdots)$ are measurable and pairwise disjoint, $E_i E_j = \phi$ $(i \neq j)$, then their sum $E = \sum E_i$ is measurable and

$$\mu\Big(\sum E_i\Big) = \sum \mu E_i\,;$$

i.e., the measure of a finite or countable sum of pairwise disjoint sets is equal to the sum of the measures of the summands.

This property of the measure is called its full additivity.

4. The measure of a set E does not change if it is displaced as a rigid body.

It is desirable that the fundamental properties of length are preserved for the more general concept of measure of sets. But it can be proved quite rigorously that this turns out to be *impossible* if a measure is to be ascribed to an arbitrary point set on the line. Consequently, in the sense of this definition there are sets that have a measure or are measurable and others that have no measure or are nonmeasurable. Besides, the class of measurable sets is so wide that this circumstance does not lead to any essential disadvantages. In fact, the construction of an example of a nonmeasurable set is rather difficult.

We shall now present some examples of measurable sets.

Example 1. *The measure of Cantor's perfect set P* (see §4). In constructing the set P from the interval $[0, 1]$, we have first thrown out an adjacent interval of length $1/3$, then two adjacent intervals of length $1/9$, then four adjacent intervals of length $1/27$, etc. Generally, at the nth step we have thrown out 2^{n-1} adjacent intervals of length $1/3^n$. Thus, the sum of the lengths of the intervals removed is equal to

$$S = \frac{1}{3} + \frac{2}{9} + \frac{4}{27} + \cdots + \frac{2^{n-1}}{3^n} + \cdots .$$

The terms of this series form a geometric progression with the first term $1/3$ and the common ratio $2/3$. Therefore the sum S of the series is

$$\frac{\dfrac{1}{3}}{1 - \dfrac{2}{3}} = 1.$$

Thus, the sum of the length of the intervals adjacent to the Cantor set is 1. In other words, the measure of the open set G complementary to P is 1. Therefore, the set itself has the measure

$$\mu P = 1 - \mu G = 1 - 1 = 0.$$

This example shows that a set may have the cardinality of the continuum and yet have measure zero.

Example 2. *Measure of the set R of all rational points of the interval* $[0, 1]$. First of all we shall show that $\mu_e R = 0$. In §2 we had found that R is countable. We arrange the points of R in a sequence

$$r_1, r_2, \cdots, r_n, \cdots .$$

Next, for given $\epsilon > 0$ we enclose the point r_n by an open interval δ_n of length $\epsilon/2^n$. The sum $\delta - \Sigma \delta_n$ is an open set covering R. The open intervals δ_n may intersect so that

$$\mu(\delta) = \mu\left(\sum \delta_n\right) < \sum \mu\delta_n = \sum \frac{\epsilon}{2^n} = \epsilon.$$

Since ϵ can be chosen arbitrarily small, we have $\mu_e R = 0$.

Further, by (2)

$$\mu_e R + \mu_e CR \geqslant 1,$$

i.e., $\mu_e CR \geqslant 1$. But since CR is contained in $[0, 1]$, we have $\mu_e CR \leqslant 1$. Hence

$$\mu_e R + \mu_e CR = 1,$$

and*

$$\mu R = 0, \quad \mu CR = 1. \tag{5}$$

This example shows that a set may be everywhere dense on an interval and yet have measure zero.

Sets of measure zero play no role in many problems of the theory of functions and can be neglected. For example, a function $f(x)$ is Riemann integrable if and only if it is bounded and the set of its points of discontinuity has measure zero. We could add a considerable number of such examples.

Measurable functions. We now proceed to one of the most brilliant applications of the concept of measure of sets, namely, a description of that class of functions with which mathematical analysis and the theory of functions actually operate. The precise statement of the problem is as follows. If a sequence $\{f_n(x)\}$ of functions given on a certain set E converges at every point of E except at the points of a set N of measure zero, then we shall say that the sequence $\{f_n(x)\}$ converges *almost everywhere*.

What functions can be obtained from continuous functions by repeated application of the operation of forming the limit of an almost everywhere convergent sequence of functions and of algebraic operations?

To answer this question we require some new concepts.

Let $f(x)$ be a function defined on a set E and α an arbitrary real number. We denote by

$$E[f(x) > \alpha]$$

the set of all points of E for which $f(x) > \alpha$. For example, if $f(x)$ is defined on $[0, 1]$ and $f(x) = x$ on this interval, then set $E[f(x) > \alpha]$ is equal to $[0, 1]$ for $\alpha < 0$, to $(\alpha, 1]$ for $0 \leqslant \alpha < 1$, and empty for $\alpha \geqslant 1$.

---

* The same argument shows that every countable point set on the line has measure zero.

A function $f(x)$ defined on a set E is called *measurable* if E itself is measurable and the set $E[f(x) > \alpha]$ is measurable for every real number α.

One can show that every continuous function given on an interval is measurable. However, among the measurable functions there are also many discontinuous functions, for example the Dirichlet function which is equal to 1 at the irrational points of [0, 1] and 0 elsewhere.

We mention without proof that measurable functions have the following properties.

1. If $f(x)$ and $\phi(x)$ are measurable functions defined on one and the same set E, then the functions

$$f + \phi, f - \phi, f \cdot \phi \quad \text{and} \quad \frac{f}{\phi}$$

are also measurable (the latter if $\phi \neq 0$).

This property shows that algebraic operations on measurable functions lead again to measurable functions.

2. If a sequence $\{f_n(x)\}$ of measurable functions defined on a set E converges almost everywhere to a function $f(x)$, then this function is also measurable.

Thus, the operation of forming the limit of an almost everywhere convergent sequence of measurable functions again leads to measurable functions.

These properties of measurable functions were established by Lebesgue. A deeper study of measurable functions was carried out by the Soviet mathematicians D. F. Egorov and N. N. Luzin. In particular, N. N. Luzin has proved that every measurable function on an interval can be made continuous by changing its values on a certain set of arbitrarily small measure.

This classical result of N. N. Luzin and the properties of measurable functions listed enable us to prove that measurable functions form that class of function of which we talked at the beginning of this subsection. Measurable functions are also of great importance in the theory of integration, namely, the concept of an integral can be generalized in such a way that every bounded measurable function turns out to be integrable. A detailed account of this will be given in the next section.

§6. The Lebesgue Integral

We shall now proceed to the central theme of this chapter, the definition of the Lebesgue integral and an account of its properties.

To understand the underlying principle of this integral, let us consider the following example. Suppose that there is a large collection of coins of various denominations and we have to add up the total sum of money involved. This can be done in two ways. We can arrange the coins in a row and add the value of each new coin to the total value of the preceding ones. However, we can also proceed differently: We put the coins in heaps such that the coins in each heap are of equal value; then we count the number of coins in each heap, multiply this number by the value of the corresponding coin, and finally add up the numbers so obtained. The first method of counting money corresponds to the process of Riemann integration, and the second to the process of Lebesgue integration.

Going over from coins to functions, we can say that for the computation of the Riemann integral the domain of definition of the function (the axis of abscissas; figure 2a) is divided into small parts, while for the computa-

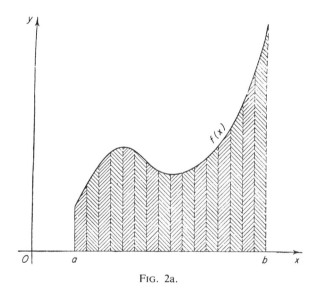

Fig. 2a.

tion of the Lebesgue integral it is the domain of values of the function (the axis of ordinates; figure 2b) that is so divided. The latter principle was applied in practice long before Lebesgue for the computation of integrals of functions of oscillating character; however, Lebesgue was the first to develop it in all generality and to give it a rigorous foundation by means of the theory of measure.

Let us examine how the measure of sets and Lebesgue integral are connected. Let E be an arbitrary measurable set on an interval $[a, b]$.

We construct a function $\phi(x)$ which is equal to 1 when x belongs to E and zero when x does not belong to E. In other words, we set

$$\phi(x) = \begin{cases} 1 & x \in E. \\ 0 & x \bar{\in} E. \end{cases}$$

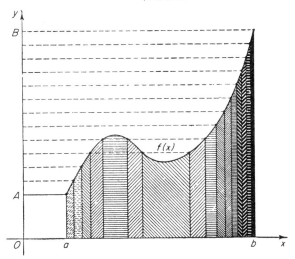

FIG. 2b.

The function $\phi(x)$ is usually called the characteristic function of E. We consider the integral

$$I = \int_{a}^{b} \phi(x)\, dx.$$

We are well accustomed to regarding the integral as equal to the area of the figure D bounded by the axis of abscissas, the lines $x = a$, $x = b$, and the curve $y = \phi(x)$ (see Chapter II). Since in our case the "height" of D is different from zero and is equal to 1 for the points $x \in E$ and these points only, the area (by the formula that area is length times width) must be equal numerically to the length (measure) of E. Thus, I must be equal to the measure of E

$$I = \mu E. \tag{6}$$

The Lebesgue integral of the function $\phi(x)$ is defined just so.

The reader should realize clearly that the equation (6) is the *definition* of the integral $\int_{a}^{b} \phi(x)\, dx$ as a Lebesgue integral. It can happen that the integral I does not exist in the sense in which it was understood in Chapter II, i.e., as a limit of integral sums. But if this is the case, then the integral I exists as a Lebesgue integral and is equal to μE.

As an example let us calculate the integral of the Dirichlet function $\Phi(x)$ equal to 0 at the rational points and to 1 at the irrational points of $[0, 1]$. Since the measure of the set of irrational points of $[0, 1]$ is 1, by (5), the Lebesgue integral

$$\int_0^1 \Phi(x)\, dx$$

is equal to 1. It is easy to verify that the Riemann integral of this function does not exist.

An auxiliary proposition. Suppose now that $f(x)$ is an arbitrary bounded measurable function defined on $[a, b]$. We shall show that every such function can be represented with arbitrarily prescribed accuracy as a linear combination of characteristic functions of sets. In order to see this we divide up the interval of the ordinate axis between the greatest lower bound A and the least upper bound B of the values of the function by points $y_0 = A, y_1, \cdots, y_n = B$ into intervals of length less than ϵ, where ϵ is an arbitrary fixed positive number. Next, if at the point $x \in [a, b]$,

$$y_i \leqslant f(x) < y_{i+1} \quad (i = 0, 1, \cdots, n - 1),$$

then we set at that point

$$\phi(x) = y_i,$$

and if at x

$$f(x) = y_n = B,$$

then we set

$$\phi(x) = y_n.$$

The construction of the function $\phi(x)$ is shown in figure 3.

By the construction of $\phi(x)$ we have for every point of $[a, b]$

$$|f(x) - \phi(x)| < \epsilon.$$

Moreover, since the function $\phi(x)$ assumes only a finite number of values $y_0, y_1, \cdots, y_n$, it can be written in the form

$$\phi(x) = y_0 \cdot \phi_0(x) + y_1 \cdot \phi_1(x) + \cdots + y_n \cdot \phi_n(x), \tag{7}$$

where $\phi_i(x)$ is the characteristic function of the set of points for which $\phi(x) = y_i$, i.e., $y_i \leqslant f(x) < y_{i+1}$ (at every point $x \in [a, b]$ only one summand on the right-hand side of (7) is different from zero)! Thus, our proposition is proved.

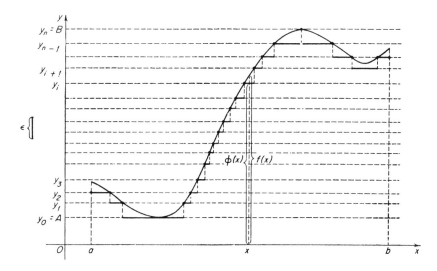

FIG. 3.

Definition of the Lebesgue integral.   We now proceed to the definition of the Lebesgue integral of an arbitrary measurable function. Since $\phi(x)$ differs by little from $f(x)$, we can take, as an approximation of the value of the integral of $f(x)$, that of $\phi(x)$. But when we bear in mind that the functions $\phi_i(x)$ are characteristic functions of sets and formally use the ordinary rules of computation for integrals, we find

$$\int_a^b \phi(x)\, dx = \int_a^b \{\, y_0\phi_0(x) + y_1\phi_1(x) + \cdots + y_n\phi_n(x)\,\}\, dx$$

$$= y_0 \int_a^b \phi_0(x)\, dx + y_1 \int_a^b \phi_1(x)\, dx + \cdots + y_n \int_a^b \phi_n(x)\, dx$$

$$= y_0\mu e_0 + y_1\mu e_1 + \cdots + y_n\mu e_n \,,$$

where μe_i is the measure of the set e_i of those x for which

$$y_i \leqslant f(x) < y_{i+1} \,.$$

Thus, an approximate value of the Lebesgue integral of $f(x)$ is the "Lebesgue integral sum"

$$S = y_0\mu e_0 + y_1\mu e_1 + \cdots + y_n\mu e \,.$$

In accordance with this the Lebesgue integral is defined as the limit of the Lebesgue integral sum S when

$$\max | y_{i+1} - y_i | \to 0,$$

which corresponds to uniform convergence of $\phi(x)$ to $f(x)$.

It can be shown that the Lebesgue integral sums have a limit for every bounded measurable function, i.e., every bounded measurable function is Lebesgue integrable. The Lebesgue integral can also be extended to certain classes of unbounded measurable functions, but we shall not deal with this here.

Properties of the Lebesgue integral. The Lebesgue integral has all the desirable properties of the ordinary integral, namely, the integral of a sum is equal to the sum of the integrals, a positive factor can be taken before the integral sign, etc. However, the Lebesgue integral has one remarkable property that the ordinary integral does not have: If the functions $f_n(x)$ are measurable and uniformly bounded:

$$| f_n(x)| < K$$

for every n and every x in $[a, b]$ and if the sequence $\{f_n(x)\}$ converges almost everywhere to $f(x)$, then

$$\int_a^b f_n(x)\, dx \to \int_a^b f(x)\, dx.$$

In other words, the Lebesgue integral permits unrestricted passage to the limit. In fact, this property of the Lebesgue integral makes it a very convenient and often an indispensable tool in many investigations. In particular, the Lebesgue integral is absolutely necessary in the theory of trigonometric series, the theory of function spaces (see Chapter XIX), and other branches of mathematics.

Let us give an example. Let $f(x)$ be a periodic function with period 2π and

$$\frac{a_0}{2} + \sum_{n=1}^{\infty} (a_n \cos nx + b_n \sin nx)$$

its Fourier series. If, for example, $f(x)$ is continuous, then it is easy to show that

$$\frac{1}{\pi} \int_0^{2\pi} f^2(x)\, dx = \frac{a_0^2}{2} + \sum_{n=1}^{\infty} (a_n^2 + b_n^2). \tag{8}$$

This identity is known as Parseval's equality. Now we raise the question: For what class of periodic functions is Parseval's equality (8) valid? This is the answer: Parseval's equality (8) is valid if and only if $f(x)$ is measurable on $[0, 2\pi]$ and $f^2(x)$ is Lebesgue integrable on the same interval.

Suggested Reading

F. Hausdorff, *Set theory*, 2nd ed., Chelsea, New York, 1962.

J. M. Hyslop, *Real variable*, Interscience, New York, 1960.

E. Kamke, *Theory of sets*, Dover, New York, 1950.

E. Landau, *Foundations of analysis: the arithmetic of whole*, *rational*, *irrational and complex numbers*, Chelsea, New York, 1951.

E. J. McShane and A. T. Betts, *Real analysis*, Van Nostrand, New York, 1959.

CHAPTER XVI

LINEAR ALGEBRA

§1. The Scope of Linear Algebra and Its Apparatus

Linear functions and matrices. Among the functions of a single variable, by far the simplest is the so-called *linear function* $l(x) = ax + b$. Its graph is, of course, the simplest of curves, namely the straight line.

All the same, the linear function is one of the most important. This is due to the fact that every "smooth" curve on a small segment is like a straight line, and the less curved the segment is, the nearer it comes to a straight line. In the language of the theory of the functions, this means that every "smooth" (continuously differentiable) function is, for a small change of the independent variable, close to a linear function. The linear function can be characterized by the fact that its increment is proportional to the increment of the independent variable. Indeed: $\Delta l(x) = l(x_0 + \Delta x) - l(x_0) = a(x_0 + \Delta x) + b - (ax_0 + b) = a \Delta x$. Conversely, if $\Delta l(x) = a \Delta x$, then $l(x) - l(x_0) = a(x - x_0)$ and $l(x) = ax + l(x_0) - ax_0 = ax + b$, where $b = l(x_0) - ax_0$. But from the differential calculus, we know that in the increment of an arbitrary differentiable function we can single out in a natural way the principal part, the so-called differential of the function, which is proportional to the increment of the independent variable, and that the increment of the function differs from its differential by an infinitesimal of higher order than the increment of the independent variable. Thus, a differentiable function is, for an infinitely small change of the independent variable, really close to a linear function to within an infinitesimal of higher order.

The situation is similar with functions of several variables.

A *linear function of several variables* is a function of the form $a_1 x_1 + a_2 x_2 + \cdots + a_n x_n + b$. If $b = 0$, the linear function is said to be *homogeneous.* A linear function of several variables is characterized by the following two properties:

1. The increment of a linear function, computed under the assumption that only one of the independent variables receives some increment while the values of the remaining variables are unchanged, is proportional to the increment of this independent variable.

2. The increment of a linear function, computed under the assumption that all the independent variables obtain increments, is equal to the algebraic sum of the increments obtained by changing each variable separately.

The linear function of several variables plays the same role among all the functions of these variables as the linear function of one variable among all the functions of one variable. For every "smooth" function (i.e., a function having continuous partial derivatives with respect to all variables) is close to some linear function for small changes of the independent variables. In fact, the increment of such a function $w = f(x_1, x_2, \cdots, x_n)$ is equal, to within infinitesimals of higher order, to the total differential $(\partial f/\partial x_1)\, dx_1 + \cdots + (\partial f/\partial x_n)\, dx_n$, which is a linear homogeneous function of the increments $dx_1, \cdots, dx_n$ of the independent variables. Hence it follows that the function w itself, which is equal to the sum of its initial value and its increment, can be expressed in terms of its independent variables for small changes of these in the form of a linear inhomogeneous function to within infinitesimals of higher orders.

Problems whose solution requires the investigation of functions of several variables arise in connection with the study of the dependence of one quantity on several factors. A problem is called *linear* if the dependence under consideration turns out to be linear. By the properties of linear functions that we have indicated earlier, a linear problem can be characterized by the following properties.

1. *The property of proportionality.* The result of the action of each separate factor is proportional to its value.

2. *The property of independence.* The total result of an action is equal to the sum of the results of the actions of the separate factors.

The fact that every "smooth" function can be replaced in a first approximation by a linear one, for small changes of the variables, is a reflection of a general principle, namely that every problem on the change of some quantity under the action of several factors can be regarded in a first approximation, for small actions, as a linear problem, i.e., as having the properties of independence and proportionality. It often turns out that this attitude gives an adequate result for practical purposes (the classical theory of elasticity, the theory of small oscillations, etc.)

The physical quantities to be studied are often characterized by certain numbers (a force by the three projections on the coordinate axes, the tension at a given point of an elastic body by the six components of the so-called stress tensor, etc.). Hence there arises the necessity of considering simultaneously several functions of several variables, and, in a first approximation, of several linear functions.

A linear function of one variable is so simple in its properties that it does not require any special study. Things are different with linear functions of several variables, where the presence of many variables introduces some special features. The situation is still more complicated when we go from a single function of several variables x_1, x_2, $\cdots$, x_n to a set of several functions y_1, y_2, $\cdots$, y_m of the same variables. As a "first approximation" there appears here a set of linear functions:

$$
\begin{aligned}
y_1 &= a_{11}x_1 + \cdots + a_{1n}x_n + b_1, \\
y_2 &= a_{21}x_1 + \cdots + a_{2n}x_n + b_2, \\
&\cdots\cdots\cdots\cdots\cdots\cdots\cdots\cdots\cdots\cdots\cdots\cdots\cdots \\
y_m &= a_{m1}x_1 + \cdots + a_{mn}x_n + b_m.
\end{aligned}
$$

A set of linear functions is already a rather complicated mathematical object, and the study of its full of interesting and nontrivial results.

The study of linear functions and their systems also constitutes the initial object of that branch of algebra that is called linear algebra.

Historically, the first task of linear algebra is that of solving a system of linear equations:

$$
\begin{aligned}
a_{11}x_1 + \cdots + a_{1n}x_n &= b_1, \\
a_{21}x_1 + \cdots + a_{2n}x_n &= b_2, \\
&\cdots\cdots\cdots\cdots\cdots\cdots\cdots\cdots\cdots\cdots \\
a_{m1}x_1 + \cdots + a_{mn}x_n &= b_m.
\end{aligned}
$$

The simplest case of this problem is treated in a school course on elementary algebra. The problem of finding methods for the simplest possible and least laborious numerical solution of systems for large n still attracts the close attention of many researchers, because the numerical solution of such systems enters as an important constituent part into many calculations and investigations.

Linear homogeneous functions are also known as *linear forms*. A given system of linear forms

$$
\begin{aligned}
y_1 &= a_{11}x_1 + \cdots + a_{1n}x_n, \\
&\cdots\cdots\cdots\cdots\cdots\cdots\cdots\cdots\cdots\cdots \\
y_m &= a_{m1}x_1 + \cdots + a_{mn}x_n
\end{aligned}
$$

is completely described by its system of coefficients, since the properties of such a system of forms depend only on the numerical values of the coefficients and the names of the variables are inessential.

For example, the system of forms

$$3x_1 + x_2 - x_3, \qquad 3t_1 + t_2 - t_3,$$
$$2x_1 + x_2 + 3x_3, \quad \text{and} \quad 2t_1 + t_2 + 3t_3,$$
$$x_1 - x_2 - x_3 \qquad t_1 - t_2 - t_3$$

obviously have identical properties and need not be regarded as essentially distinct.

The set of coefficients of a system of linear forms can be given in a natural way in the form of a rectangular array

$$\begin{bmatrix} a_{11} \cdots a_{1n} \\ \cdots\cdots\cdots \\ a_{m1} \cdots a_{mn} \end{bmatrix}.$$

Such arrays bear the name of *matrices*. The numbers a_{ij} are called the elements of the matrix. The need of considering matrices arises necessarily from the very scope of linear algebra.

Important special cases of matrices are the matrices that consist of a single column, which are simply called columns, those that consist of a single row, called rows, and finally the square matrices, i.e., those in which the number of rows is equal to the number of columns. The number of rows (or columns) of a square matrix is called its *order*. The "matrix" (a) consisting of a single number is identified with that number.

In connection with the simplest operations on a set of linear forms, it is natural to define operations on matrices.

Suppose that two systems of linear forms are given,

$$y_1 = a_{11}x_1 + \cdots + a_{1n}x_n,$$
$$\cdots\cdots\cdots\cdots\cdots\cdots\cdots\cdots\cdots$$
$$y_m = a_{m1}x_1 + \cdots + a_{mn}x_n,$$

and

$$z_1 = b_{11}x_1 + \cdots + b_{1n}x_n,$$
$$\cdots\cdots\cdots\cdots\cdots\cdots\cdots\cdots\cdots$$
$$z_m = b_{m1}x_1 + \cdots + b_{mn}x_n.$$

Let us add these forms,

$$y_1 + z_1 = (a_{11} + b_{11})\,x_1 + \cdots + (a_{1n} + b_{1n})\,x_n,$$
$$\cdots\cdots\cdots\cdots\cdots\cdots\cdots\cdots\cdots\cdots\cdots\cdots\cdots\cdots\cdots\cdots$$
$$y_m + z_m = (a_{m1} + b_{m1})\,x_1 + \cdots + (a_{mn} + b_{mn})\,x_n.$$

It is natural to say that the matrix of the system of forms so obtained

$$\begin{bmatrix} a_{11} + b_{11} \cdots a_{1n} + b_{1n} \\ \hline \\ a_{m1} + b_{m1} \cdots a_{mn} + b_{mn} \end{bmatrix}$$

is the sum of the matrices

$$\begin{bmatrix} a_{11} \cdots a_{1n} \\ \cdots \\ a_{m1} \cdots a_{mn} \end{bmatrix} \quad \text{and} \quad \begin{bmatrix} b_{11} \cdots b_{1n} \\ \cdots \\ b_{m1} \cdots b_{mn} \end{bmatrix}.$$

Similarly, the product of the matrix

$$\begin{bmatrix} a_{11} \cdots a_{1n} \\ \cdots \\ a_{m1} \cdots a_{mn} \end{bmatrix}$$

by the number c is defined as the matrix of the coefficients in the system of forms cy_1, cy_2, $\cdots$, cy_m, where y_1, y_2, $\cdots, y_m$ are the forms whose coefficients constitute the matrix

$$\begin{bmatrix} a_{11} \cdots a_{1n} \\ \cdots \\ a_{m1} \cdots a_{mn} \end{bmatrix}.$$

From this definition it is clear that

$$c \begin{bmatrix} a_{11} \cdots a_{1n} \\ \cdots \\ a_{m1} \cdots a_{mn} \end{bmatrix} = \begin{bmatrix} ca_{11} \cdots ca_{1n} \\ \cdots \\ ca_{m1} \cdots ca_{mn} \end{bmatrix}.$$

Finally, the operation of multiplication of a matrix by a matrix is defined as follows. Suppose that

$$\begin{aligned} z_1 &= a_{11}y_1 + \cdots + a_{1m}y_m, \\ &\cdots \cdots \cdots \cdots \cdots \cdots \cdots \cdots \\ z_k &= a_{k1}y_1 + \cdots + a_{km}y_m \end{aligned} \tag{1}$$

and

$$\begin{aligned} y_1 &= b_{11}x_1 + \cdots + b_{1n}x_n, \\ &\cdots \cdots \cdots \cdots \cdots \cdots \cdots \cdots \\ y_m &= b_{m1}x_1 + \cdots + b_{mn}x_n. \end{aligned}$$

When we substitute in (1) the expressions of y_1, y_2, $\cdots, y_m$ in terms of

x_1, x_2, $\cdots$, x_n, we find that z_1, z_2, $\cdots$, z_k can also be expressed in terms of x_1, x_2, $\cdots$, x_n by linear forms

$$z_1 = c_{11}x_1 + \cdots + c_{1n}x_n,$$
$$\dotfill$$
$$z_k = c_{k1}x_1 + \cdots + c_{kn}x_n.$$

The matrix of coefficients

$$\begin{bmatrix} c_{11} \cdots c_{1n} \\ \cdots\cdots\cdots \\ c_{k1} \cdots c_{kn} \end{bmatrix}$$

is called the product of the matrices

$$\begin{bmatrix} a_{11} \cdots a_{1m} \\ \cdots\cdots\cdots \\ a_{k1} \cdots a_{km} \end{bmatrix} \quad \text{and} \quad \begin{bmatrix} b_{11} \cdots b_{1n} \\ \cdots\cdots\cdots \\ b_{m1} \cdots b_{mn} \end{bmatrix}$$

and is denoted by

$$\begin{bmatrix} a_{11} \cdots a_{1n} \\ \cdots\cdots\cdots \\ a_{k1} \cdots a_{km} \end{bmatrix} \begin{bmatrix} b_{11} \cdots b_{1n} \\ \cdots\cdots\cdots \\ b_{m1} \cdots b_{mn} \end{bmatrix}.$$

It is easy to calculate how the elements of the product of two matrices are expressed in terms of the elements of its factors. The element c_{ij} is the coefficient of x_j in the expression for z_i in terms of x_1, x_2, $\cdots$, x_n.

But $z_i = a_{i1}y_1 + \cdots + a_{im}y_m$ and

$$y_1 = \cdots + b_{1j}x_j + \cdots,$$
$$\dotfill$$
$$y_m = \cdots + b_{mj}x_j + \cdots.$$

Therefore,

$$z_i = \cdots + (a_{i1}b_{1j} + \cdots + a_{im}b_{mj})\, x_j + \cdots,$$

hence

$$c_{ij} = a_{i1}b_{1j} + \cdots + a_{im}b_{mj}.$$

Thus, the element in the ith row and the jth column of the product of two matrices is equal to the sum of the products of the elements of the ith row of the first factor by the corresponding elements of the jth column of the second factor. For example:

$$\begin{pmatrix} 2 & 1 & 3 \\ 3 & 1 & 1 \end{pmatrix} \begin{pmatrix} 3 & 1 \\ 1 & 2 \\ 2 & 4 \end{pmatrix} = \begin{pmatrix} 2\cdot3 + 1\cdot1 + 3\cdot2 & 2\cdot1 + 1\cdot2 + 3\cdot4 \\ 3\cdot3 + 1\cdot1 + 1\cdot2 & 3\cdot1 + 1\cdot2 + 1\cdot4 \end{pmatrix} = \begin{pmatrix} 13 & 16 \\ 12 & 9 \end{pmatrix}.$$

Although a matrix is, so to speak, a "composite" object and many elements enter into its formation, it is useful and convenient to denote it by a single letter and to preserve the usual notation for operations of addition and multiplication. We shall use the capital letters of the Latin alphabet to denote matrices. The application of such a concise notation brings simplicity and lucidity into the theory of matrices by embracing in short formulas, that remind us of the formulas of ordinary algebra, complicated relations connecting a set of numbers, namely the elements of the matrix that occur in these formulas. Thus, for example, the set of linear forms

$$a_{11}x_1 + \cdots + a_{1n}x_n,$$
$$\cdots\cdots\cdots\cdots\cdots\cdots\cdots\cdots\cdots$$
$$a_{m1}x_1 + \cdots + a_{mn}x_n$$

appears in matrix notation as AX, where A is the coefficient matrix and X the "column" formed by the variables $x_1, x_2, \cdots, x_n$. The system of linear equations

$$a_{11}x_1 + \cdots + a_{1n}x_n = b_1,$$
$$\cdots\cdots\cdots\cdots\cdots\cdots\cdots\cdots\cdots\cdots$$
$$a_{m1}x_1 + \cdots + a_{mn}x_n = b_m$$

is written as

$$AX = B,$$

where A is the coefficient matrix, X the column of the unknowns, and B the column of the absolute terms.

The fundamental operations on matrices, namely addition and multiplication, are, of course, not always defined. The operation of addition makes sense for matrices of *equal structure*, i.e., having the same number of rows and of columns. As the result of addition, we obtain a matrix of the same structure. The operation of multiplication makes sense if the number of columns of the first matrix is equal to the number of rows of the second. As the result we obtain a matrix in which the number of rows is equal to the number of rows of the first factor and the number of columns is equal to the number of columns of the second factor.

The operations on square matrices are subject to most of the laws for operations on number, but some of the laws turn out to be violated.

Let us enumerate the fundamental properties for operations on matrices:

1. $A + B = B + A$ (commutative law for addition).
2. $(A + B) + C = A + (B + C)$ (associative law for addition).
3. $c(A + B) = cA + cB$ (distributive laws for multiplication by a number. Here c, c_1, c_2 are
3'. $(c_1 + c_2)A = c_1A + c_2A$ numbers and not matrices).

4. $(c_1 c_2)\, A = c_1(c_2 A)$ (associative law for multiplication by a number).

5. There exists a "null" matrix

$$O = \begin{pmatrix} 0 \cdots 0 \\ \cdots \cdots \\ 0 \cdots 0 \end{pmatrix}$$

such that $A + O = A$ for every matrix A.

6. $c \cdot O = 0 \cdot A = O$; conversely, if $cA = O$, then $c = 0$ or $A = O$ (here c is a number).

7. For every matrix A there exists an opposite matrix $-A$, i.e., such that $A + (-A) = O$.

8. $(A + B) \cdot C = AC + BC$ (distributive laws for addition and
8'. $C(A + B) = CA + CB$ multiplication of matrices).

9. $(AB)\, C = A(BC)$ (associative law for multiplication).

10. $(cA)\, B = A(cB) = c(AB)$.

These properties hold not only for square matrices but also for arbitrary rectangular matrices with the sole proviso that the operations that occur in each of the numbered formulas must be defined. For square matrices of equal order this proviso is automatically fulfilled.

All these properties of the operations are similar to the properties of operations on numbers.

We shall now point out two peculiarities of the operations on matrices. First, the commutative law for the multiplication of matrices, even square ones, need not hold; i.e., AB is not always equal to BA. For example:

$$\begin{pmatrix} 3 & -2 \\ -1 & 4 \end{pmatrix}\begin{pmatrix} 1 & 2 \\ 3 & 2 \end{pmatrix} = \begin{pmatrix} -3 & 2 \\ 11 & 6 \end{pmatrix};$$

$$\begin{pmatrix} 1 & 2 \\ 3 & 2 \end{pmatrix}\begin{pmatrix} 3 & -2 \\ -1 & 4 \end{pmatrix} = \begin{pmatrix} 1 & 6 \\ 7 & 2 \end{pmatrix}.$$

Second, the product of two numbers is, of course, equal to zero if and only if one of the factors is equal to zero. This theorem is well known to be fundamental in the theory of algebraic equations. But under multiplication of matrices it turns out to be false. For the product of two matrices may be equal to the null matrix, although neither factor is equal to the null matrix. For example:

$$\begin{pmatrix} 1 & 1 \\ 1 & 1 \end{pmatrix}\begin{pmatrix} 1 & 1 \\ -1 & -1 \end{pmatrix} = \begin{pmatrix} 0 & 0 \\ 0 & 0 \end{pmatrix}.$$

Let us mention yet another property of the multilplication of matrices. The matrix $\bar{A}$ is called the *transpose* of A if in every row of $\bar{A}$ there stand the elements of the corresponding column of A in the same order. For example, for the matrix

$$A = \begin{bmatrix} 1 & 2 \\ 3 & 4 \\ 5 & 6 \end{bmatrix}$$

the transpose is the matrix

$$\bar{A} = \begin{bmatrix} 1 & 3 & 5 \\ 2 & 4 & 6 \end{bmatrix}.$$

The operation of multiplication is connected with that of transposition by the formula

$$\overline{AB} = \bar{B}\bar{A} \,,$$

which is easily verified on the basis of the multiplication rule for matrices.

The theory of matrices forms an indispensable part of linear algebra in that it plays the role of an apparatus for stating and solving its problems.

Geometric analogies in linear algebra. Apart from the earlier described source for the emergence of the ideas and problems of linear algebra, there are also the needs of mathematical analysis and geometry, in particular analytic geometry, that lead to the development of linear algebra and, in turn, enrich it by important ideas and analogies. It is well known that the analytic geometry of the plane, and, in an even greater measure, of the space, as far as the theory of straight lines and planes is concerned, makes use of the apparatus of linear algebra in its simplest form. For a straight line in the plane is given by a linear equation in two variables that links the two coordinates of an arbitrary point of the line. A plane in space is given by a linear equation in three variables (the coordinates of an arbitrary point of this plane), a line in space by two linear equations.

A special simplicity and clarity is, of course, brought into analytical geometry and consequently into the theory of the simplest systems of linear equations by the use of a concept of a vector. Now a similar simplicity and clarity is brought into linear algebra, in particular into the general theory of systems of linear equations, by the use of the concept of a vector in a generalized sense. The way to this generalization is the following. A vector (in space) is given by three numbers, namely its three projections on the coordinate axes. Every triplet of real numbers in turn can be represented geometrically in the form of a vector (in space).

For vectors the operations of addition ("by the parallelogram rule") and multiplication by a number are defined. These operations are defined in accordance with similar operations on forces, velocities, accelerations, and other physical quantities that can be represented by means of vectors.

If vectors are given by their coordinates (i.e., their projections on the coordinate axes), then the operations of addition and multiplication by a number performed on vectors correspond to the analogous operations on the rows (or columns) of their coordinates.

Thus, it is convenient to interpret a row or column of three elements geometrically as a vector in three-dimensional space, and then the basic operations on "rows" (or "columns") are interpreted by the corresponding operations on vectors in space, so that the algebra of rows (or columns) of three elements formally does not differ at all from the algebra of the vectors of three-dimensional space. This circumstance makes it natural to introduce a geometric terminology into linear algebra.

A column (or row) of n numbers

$$\begin{pmatrix} x_1 \\ x_2 \\ \vdots \\ x_n \end{pmatrix}$$

is regarded as a "vector", i.e., as an element of some "n-dimensional vector space." The sum of the vectors

$$\begin{pmatrix} x_1 \\ x_2 \\ \vdots \\ x_n \end{pmatrix} \quad \text{and} \quad \begin{pmatrix} y_1 \\ y_2 \\ \vdots \\ y_n \end{pmatrix}$$

is taken to be the vector

$$\begin{pmatrix} x_1 + y_1 \\ x_2 + y_2 \\ \dots\dots \\ x_n + y_n \end{pmatrix} ;$$

the product of the vector

$$\begin{pmatrix} x_1 \\ x_2 \\ \vdots \\ x_n \end{pmatrix}$$

by the number c is taken to be the vector

$$\begin{pmatrix} cx_1 \\ cx_2 \\ \vdots \\ cx_n \end{pmatrix}.$$

The set of all vectors (columns) forms, by definition, the n-dimensional arithmetical vector space.

Together with the n-dimensional arithmetical vector space we can introduce the concept of an n-dimensional point space, by associating with each column of n real numbers a geometrical image, namely a point. Then the n-dimensional vector space is defined in the following way.

With every pair of points A and B we associate the vector $\overrightarrow{AB}$ leading from A to B by taking as its coordinates (its projections on the coordinate axes), by definition, the difference of the corresponding coordinates of the points B and A. Two vectors are taken to be equal if their corresponding coordinates are equal, just as in three-dimensional space we regard vectors as equal if one of them is obtained from the other by a parallel shift.

Between the vectors of an n-dimensional vector space and the points of an n-dimensional point space, there exists a natural one-to-one correspondence.

The point

$$\begin{pmatrix} 0 \\ 0 \\ \vdots \\ 0 \end{pmatrix}$$

is taken as the "origin of coordinates," and to every point there corresponds the vector that joins the origin to that point. Then we associate with every vector the point that is the end point of this vector, assuming that its beginning coincides with the origin. The introduction of the point space creates new analogies that enable us to "see" better in n-dimensional space.

However, in the further generalizations (§2) a rigorous definition of a point space becomes rather more complicated, and we shall, therefore, not make use of this concept. The reader who wishes to use the analogies arising from the investigation of a point space should visualize the elements of a vector space as vectors emanating from the origin of coordinates.

The introduction of a geometric terminology enables us to use in linear algebra analogies based on the geometric intuition which originates in the study of the geometry of three-dimensional space.

Of course, these analogies must be used with a certain care, bearing in mind that every intuitive-geometric argument can be checked in a strictly logical way applying only precise definitions of "geometric" concepts and rigorous proofs of theorems.

A characteristic feature of the elements of an n-dimensional vector space is the existence of the operations of addition and multiplication by a number, with properties reminiscent of the operations on numbers. Namely, as we have already mentioned in the account of the properties of operations on matrices, for the operation of addition the commutative and associative laws are satisfied, the distributive laws (for multiplication by a number) hold, the operation of addition has a unique inverse, and the product of a vector by a number gives the null vector if and only if either the vector is the null vector or the number is zero.

However, not only these columns (and rows) possess the features referred to. Such features also belong to the set of matrices of equal structure and to physical vector quantities: forces, velocities, accelerations, etc. They also belong to some mathematical objects of an altogether different nature, for example: the set of all polynomials in one variable, the set of all continuous functions on a given interval $[a, b]$, the set of all solutions of a linear homogeneous differential equation, etc.

This circumstance motivates a further generalization of a vector space, namely the introduction of general linear spaces. The elements of such generalized spaces may be arbitrary mathematical or physical objects for which the operations of addition and multiplication by a number are defined in a natural fashion. Such a very general and abstract approach to the concept of a linear space does not bring any complications into the theory, as we have seen earlier: Every linear space (of course, n-dimensional; the meaning of this will be clarified in the next section) does not differ in its structure and its properties from the arithmetical linear space, but the field of applicability is considerably extended by this generalization and it becomes possible to apply the methods of linear algebra to a very wide range of problems of theoretical science.

§2. Linear Spaces

Definition of a linear space.   We now proceed to a rigorous definition of a linear space.

A *linear space* is a collection of objects of an arbitrary nature for which the concepts of a sum and of a product by a number make sense and which satisfy the following postulates:

1. $(X + Y) + Z = X + (Y + Z)$.

2. There exists a "null" element $\mathbf{0}$ such that $X + \mathbf{0} = X$ for every X.

3. For every element X there exists an opposite $-X$ such that $X + (-X) = \mathbf{0}$.

4. $X + Y = Y + X$.

5. $1 \cdot X = X$.

6. $c_1(c_2 X) = c_1 c_2 X$.

7. $(c_1 + c_2) X = c_1 X + c_2 X$.

8. $c(X + Y) = cX + cY$.

Here X, Y, Z are elements of the linear space; 1, c_1, c_2, c are numbers.

These postulates (which are also called the *axioms of a linear space*) are very natural and constitute a formal account of those properties of the operations of addition and multiplication by a number that are necessarily linked with the concept of these operations in whatever generalized sense they are to be understood. Operations having one physical meaning or another are, in fact, treated as addition and multiplication by a number in all cases when these operations satisfy the postulates 1–8.

We mention some consequences of these axioms:

a. The null element $\mathbf{0}$ of the space is unique, i.e., there exists only one element satisfying axiom 2.

b. The opposite element of a given element X is unique.

c. "Subtraction" has a meaning; i.e., when a sum and one of the summands is given, the other summand is always defined, in fact, uniquely: If $X + Z = Y$, then $Z = Y + (-X)$.

d. $0 \cdot X = c \cdot \mathbf{0} = \mathbf{0}$.

e. If $cX = \mathbf{0}$, then either $c = 0$, or $X = \mathbf{0}$.

f. $-X = (-1) X$.

The proof of these consequences are very simple and will be omitted. In what follows the elements of a linear space will be called vectors.

Linear dependence and independence of vectors. We now proceed to the important concept of linear dependence and independence of vectors.

The vector $c_1 X_1 + c_2 X_2 + \cdots + c_m X_m$ with arbitrary numerical values of the coefficients $c_1, c_2, \cdots, c_m$ is called a linear combination of the vectors $X_1, X_2, \cdots, X_m$. If among the vectors $X_1, X_2, \cdots, X_m$ there is at least one that is a linear combination of the remaining ones, then the vectors $X_1, X_2, \cdots, X_m$ are called linearly dependent. But if none of the vectors $X_1, X_2, \cdots, X_m$ is a linear combination of the remaining ones, then the vectors $X_1, X_2, \cdots, X_m$ are called linearly independent.

It is easy to see that for linear independence of the vectors $X_1, X_2, \cdots, X_m$ it is necessary and sufficient that the relation $c_1 X_1 + c_2 X_2 + \cdots + c_m X_m = 0$ should hold for $c_1 = c_2 = \cdots = c_m = 0$ only.

For vectors of the ordinary three-dimensional space the concepts of linear dependence and independence have a simple geometrical meaning.

Suppose two vectors X_1 and X_2 are given. Their linear dependence means that one of the vectors is a "linear combination" of the other, i.e., that they differ simply by a numerical factor. This means that both vectors belong to a common straight line, i.e., that they have equal or opposite direction.

Conversely, if two vectors are contained in one straight line, then they are linearly dependent. Consequently linear independence of two vectors X_1 and X_2 means that these vectors cannot be placed on one straight line; their directions are essentially distinct.

Let us now investigate what linear dependence and independence of three vectors means. Suppose that the vectors X_1, X_2 and X_3 are linearly dependent and, for the sake of definiteness, that the vector X_3 is a linear combination of the vectors X_1 and X_2. Then X_3 obviously lies in a plane containing the vectors X_1 and X_2; i.e., all three vectors X_1, X_2, X_3 belong to one plane. It is easy to see that if the vectors X_1, X_2, X_3 lie in one plane, then they are linearly dependent. For if the vectors X_1 and X_2 do not lie on one line, then X_3 can be decomposed with respect to X_1 and X_2, i.e., represented as a linear combination of X_1 and X_2. But if X_1 and X_2 lie on one line, then already X_1 and X_2 are linearly dependent.

Thus, linear dependence of three vectors X_1, X_2, X_3 is equivalent to the fact that they lie in one plane. Therefore X_1, X_2, X_3 are linearly independent if and only if they do not belong to one plane.

Four vectors in three-dimensional space are always linearly dependent. For if the vectors X_1, X_2, X_3 are linearly dependent, then the vectors X_1, X_2, X_3, X_4 are also linearly dependent for any X_4. But if X_1, X_2, X_3 are linearly independent, then they do not lie in one plane and every vector X_4 can be decomposed with respect to X_1, X_2, X_3, i.e., represented as a linear combination of them.

The preceding arguments can be generalized in the following way.

In three-dimensional space the vectors $X_1, X_2, \cdots, X_k$ $(k \geqslant 3)$ are linearly dependent if and only if they belong to a space (straight line or plane) of a dimension less than k.

In what follows we shall see after a rigorous definition of subspace and dimension that also in the general case linear dependence of the vectors $X_1, X_2, \cdots, X_k$ is equivalent to the fact that they belong to a space whose dimension is less than k; i.e., the "geometrical" meaning of linear dependence remains the same as for vectors in three-dimensional space.

The following theorem plays a fundamental role in the theory of linear spaces. If the vectors $X_1, X_2, \cdots, X_m$ are linear combinations of the vectors $Y_1, Y_2, \cdots, Y_k$ and $m > k$, then $X_1, X_2, \cdots, X_m$ are linearly dependent (theorem on the linear dependence of linear combinations).

For $k = 1$ the theorem is obvious. For $k > 1$ it is easily proved by the method of mathematical induction with respect to k.

Basis and dimension of a space. In three-dimensional space any three vectors X_1, X_2, X_3 that do not lie in one plane (i.e., that are linearly independent) form a *basis* of the space, which means that every vector of the space can be decomposed with respect to X_1, X_2, X_3, i.e., represented as a linear combination of them.

General linear vector spaces can be divided into two types.

It can happen that a space contains an arbitrarily large number of linearly independent vectors. Such spaces are called infinite-dimensional and their study leads to a branch of linear algebra that is the topic of a special mathematical discipline, functional analysis (see Chapter XIX).

A linear space is called finite-dimensional if there exists a finite bound for the number of linearly independent vectors, i.e., a number n such that there exist in the space n linearly independent vectors, but that any vectors more than n in number are linearly dependent. The number n is called the *dimension* of the space.

Thus, the space of vectors of the ordinary geometrical three-dimensional space is three-dimensional also in the sense of the general definition we have given. For in the three-dimensional geometric space, there exist many triplets of linearly independent vectors, but any four vectors are linearly dependent.

The space of n-term columns is n-dimensional in the sense of our definition. For there are n linearly independent vectors in the space, for example

$$\boldsymbol{e}_1 = \begin{pmatrix} 1 \\ 0 \\ \vdots \\ 0 \end{pmatrix}, \quad \boldsymbol{e}_2 = \begin{pmatrix} 0 \\ 1 \\ \vdots \\ 0 \end{pmatrix}, \cdots, \quad \boldsymbol{e}_n = \begin{pmatrix} 0 \\ 0 \\ \vdots \\ 1 \end{pmatrix} \tag{2}$$

but every vector

$$\begin{pmatrix} x_1 \\ x_2 \\ \vdots \\ x_n \end{pmatrix}$$

of the space is a linear combination of them, namely: $x_1 e_1 + x_2 e_2 + \cdots + x_n e_n$. Therefore by the theorem of linear dependence of linear combinations any vectors more than n in number are linearly dependent.

Polynomials in one variable form a linear space. For there is a natural definition of the operations of addition and of multiplication by a number for polynomials, and they satisfy the axioms 1–8. However, this space is infinite-dimensional, since the vectors $1, x, \cdots, x^N$ are linearly independent for any N. But the set of polynomials whose degree does not exceed a given number N form a finite-dimensional space whose dimension is $N + 1$. For the vectors $1, x, \cdots, x^N$ are linearly independent and their number is $N + 1$. Now every polynomial whose degree does not exceed N is a linear combination of $1, x, \cdots, x^N$ so that by the theorem on linear independence any polynomials of degree $\leqslant N$, if they are more than $N + 1$ in number, are linearly dependent.

We now introduce the important concept of a *basis* for an n-dimensional space. A basis is defined as a set of linearly independent vectors of the space such that every vector of the space is a linear combination of vectors of this set. Thus, in the space of columns a basis is, for example, the set of vectors (2). In the space of polynomials of degree $\leqslant N$ the "vectors" $1, x, \cdots, x^N$ can be taken as a basis. In the three-dimensional geometrical space any triplet of linearly independent vectors plays the role of a basis.

In an n-dimensional linear space, every set of n linearly independent vectors (and the existence of at least one such set is part of the definition of an n-dimensional space) form a basis of the space. For let $e_1, e_2, \cdots, e_n$ be linearly independent vectors of an n-dimensional linear space and X an arbitrary vector of the space. Then the vectors $X, e_1, \cdots, e_n$ are linearly dependent (since their number is more than n), i.e., there are numbers $c, c_1, c_2, \cdots, c_n$, not all equal to zero, such that $cX + c_1 e_1 + \cdots + c_n e_n = 0$. Here $c \neq 0$, because if we had $c = 0$, then the vectors $e_1, e_2, \cdots, e_n$ would be linearly dependent. Therefore $X = -(c_1/c) e_1 - \cdots - (c_n/c) e_n$; i.e., every vector of the space is a linear combination of the vectors $e_1, e_2, \cdots, e_n$.

Every basis of an n-dimensional linear space consists of exactly n vectors. For the vectors of a basis are linearly independent and therefore their number cannot be larger than n. On the other hand, let $e_1, e_2, \cdots, e_k$ be an arbitrary basis of an n-dimensional space. We have already established that $k \leqslant n$. But every vector of the space, by definition of a basis, is a linear combination of the vectors $e_1, e_2, \cdots, e_k$ and by the theorem on linear dependence of linear combinations any vectors, more than k in number, are linearly dependent, from which it follows that the dimension n of the space is not larger than the number k of vectors of a basis. Thus, $k = n$, and this is what we had to prove.

We now introduce *coordinates* of a vector with respect to a given basis $e_1, e_2, \cdots, e_n$. As we have shown earlier, every vector X is a linear combination of the vectors of the basis. This representation is unique. For suppose that the vector X is expressed in terms of the basis $e_1, e_2, \cdots, e_n$ in two ways:

$$X = x_1 e_1 + x_2 e_2 + \cdots + x_n e_n,$$

$$X = x_1' e_1 + x_2' e_2 + \cdots + x_n' e_n.$$

Then $(x_1 - x_1') e_1 + (x_2 - x_2') e_2 + \cdots + (x_n - x_n') e_n = 0$, and from this it follows by the linear independence of $e_1, e_2, \cdots, e_n$ that $x = x_1', \cdots, x_n = x_n'$.

The coefficients $x_1, x_2, \cdots, x_n$ in the decomposition of an arbitrary vector X in terms of the vectors of a basis are called the *coordinates* of X in this basis. In this way every vector, *once a basis of the space is chosen*, can in a natural manner be associated with the row (or column) of its coordinates and vice versa: every row (or column) of n numbers can be regarded as the set of coordinates of a certain vector.

The operations of addition of vectors and multiplication of a vector by a number correspond to the similar operations on the rows (or columns) of their coordinates.

Therefore every n-dimensional linear space, irrespective of the nature of its elements (they may be functions, matrices, any physical quantities whatsoever, etc.), does not differ at all from the space of rows (or columns) with respect to these operations. Thus, as we have already mentioned, the generalized axiomatic approach to the concept of a linear space does not lead to any complications in comparison with the treatment of the space as a space of rows, but it extends the domain of applicability of this concept considerably.

An identity of properties of two sets of objects in relation to a given system of operations (or arbitrary other relations between their elements) is called in mathematics an isomorphism. An exact definition of isomorphism of algebraic systems will be given in Chapter XX. Using this term we can say that all n-dimensional linear spaces, irrespective of the nature of their elements, are isomorphic to one another and isomorphic to a single model, namely the space of rows.

Subspaces. A set of vectors of an n-dimensional linear space R_n, satisfying the condition that every linear combination of arbitrary vectors of the set under consideration also belongs to it, is called a *subspace* of the space. Obviously a subspace of the space R_n is itself a linear space and has, therefore, bases and a dimension. It is also obvious that the dimension of the subspace does not exceed the dimension of the whole space and can be equal to it if and only if the subspace coincides with the whole space.

Examples of subspaces of the three-dimensional vector space are the planes and lines we have studied, to within a translation, more accurately, the sets of all vectors that lie in a plane or on a line.

Very frequently we have to investigate the subspaces "spanned" by a system of vectors. These subspaces are defined as follows. Suppose that a system of linearly independent or dependent vectors X_1, X_2, $\cdots$, X_m of the space R_n is given. Then the set of all linear combinations of these vectors $\{c_1 X_1 + c_2 X_2 + \cdots + c_m X_m\}$ forms a subspace of R_n which is called the subspace spanned by the vectors X_1, X_2, $\cdots$, X_m.

The dimension of this subspace is called the *rank* of the system of vectors X_1, X_2, $\cdots$, X_m. It is easy to see that the rank of a system of vectors is equal to the maximal number of linearly independent vectors contained in the system.

The "set" consisting only of the null vector formally satisfies the conditions imposed on a subspace. The dimension of this subspace is taken to be zero.

If two subspaces of a space R_n are given, then we can form from them in a natural manner two other subspaces, their vector sum (or union) and their intersection.

The *vector sum* of two subspaces P and Q is defined as the set of all sums of vectors belonging to the subspaces P and Q. The vector sum can also be regarded as the subspace spanned by the union of the bases of the subspaces P and Q.

The *intersection* of two subspaces is defined as the set of all vectors that belong to both subspaces. For example, the vector sum of two planes (i.e., two-dimensional vector subspaces) in the ordinary three-dimensional space is the whole space (provided only that the planes do not coincide) and their intersection is a straight line (under the same proviso).

The dimensions p and q of the two given subspaces, the dimension t of their vector sum, and the dimension s of their intersection satisfy the following interesting relation:

$$p + q = t + s.$$

We omit the proof of this statement.

From this relation we can make certain deductions concerning the intersection of subspaces in special cases. For example, two noncoincident planes (i.e., two-dimensional subspaces) in a space of four dimensions intersect in general only in a point (the dimension of their intersection is zero) and two planes intersect in a straight line only if their vector sum is three-dimensional, i.e., if both planes belong to some three-dimensional subspace. For in this case $t + s = 2 + 2 = 4$, from which it follows that $s = 1$ only when $t = 3$.

Complex linear spaces. In the description of the space of rows and of the general linear space, we have not specified what sort of numbers we are dealing with in the definition of the operation of multiplication of a vector by a number. Since we have started out from a generalization of the ordinary vectors, i.e., the directed segments in the geometrical three-dimensional space, we have had in mind arbitrary real numbers. The so-constructed linear spaces, which are called real linear spaces, generalize in the most natural way the three-dimensional space of ordinary vectors. However, in many problems of contemporary mathematics it turns out to be useful to consider a *complex linear space.* By this we mean a collection of objects for which the operations of addition and of multiplication by an arbitrary complex number are defined so that these operations satisfy all the axioms 1–8. As an example of a complex space, we can take the space of rows whose elements are arbitrary complex numbers.

Formally the theory of complex spaces does not differ essentially from the theory of real spaces.

However, even a two-dimensional complex space does not have an intuitive geometric interpretation. The fact is that an n-dimensional complex space can also be regarded as a real one, in view of the fact that since the operation of multiplication by an arbitrary complex number is defined for it, the operation of multiplication by a real number is defined just as well. But the dimension of a complex n-dimensional space regarded as a real one is equal to $2n$, i.e., twice as much. For if $e_1, e_2, \cdots, e_n$ is a basis of the complex space, then we can take as a basis of the same space, regarded as a real one, for example the vectors

$$e_1, ie_1, e_2, ie_2, \cdots, e_n, ie_n, \quad \text{where} \quad i = \sqrt{-1}.$$

Therefore a two-dimensional complex space can be interpreted as a real one, but four-dimensional.

Furthermore, the theory of linear spaces does not undergo any changes formally if as the collection of numbers by which the "vectors" of the space may be multiplied we take an arbitrary set of numbers, other than that of all real or all complex numbers, provided only that the results of the basic arithmetical operations (addition, subtraction, multiplication, and division) performed on numbers of the set again belong to the set. A set of numbers satisfying these postulates is called a number field. (This concept will be studied in more detail in Chapter XX.) As an example of a number field, we can take the field of rational numbers.

In some parts of algebra that are close to the theory of numbers, the theory of linear spaces over an arbitrary field is successfully applied.

The *n*-dimensional Euclidean space. Some important concepts of the ordinary vector space have not yet been generalized in the preceding account, in particular, the concept of the length of a vector and the angle between vectors. It is well known that in analytic geometry problems relating to the intersection of lines and planes, parallelism, and many others make no use of these concepts. The properties of a space whose description does not require the concepts of the length of a vector and of angle can be characterized as the properties that remain unchanged under arbitrary affine transformations [see Chapter III, §11]. For this reason linear spaces in which the concept of the length of a vector is not defined are called *affine spaces*.

However, many problems of mathematics require generalizations of the concepts of the length of a vector and of an angle to *n*-dimensional spaces. These generalizations proceed by means of an analogy with the theory of vectors in a plane or in space.

Let us consider, first of all, the real space of rows. The length of the vector $X = (x_1, x_2, \cdots, x_n)$ is defined to be the number

$$|X| = \sqrt{x_1^2 + x_2^2 + \cdots + x_n^2}.$$

This is quite natural, since for $n = 2$ and $n = 3$ the length of a vector is computed precisely by this formula in terms of its coordinates with respect to Cartesian coordinate axes.

The concept of the angle between vectors is introduced in a natural way by the following considerations. In a plane and in space the angle between the vectors X and Y is the angle at the vertex A in the triangle with the sides $AB = |X|$, $AC = |Y|$ and $BC = |X - Y|$.

In an *n*-dimensional space it is natural to take this as the definition of the angle between vectors, i.e., to proceed as if we could "draw" a pair of vectors in an *n*-dimensional space and "place" them in a plane preserving their lengths and the angle between them. However, such a definition would lack rigor; the existence of a triangle ABC with the lengths of the vectors $|X|$, $|Y|$, and $|X - Y|$ is needed in the proof.

Disregarding this inaccuracy, we introduce a formula for the computation of an angle. By a well-known formula of trigonometry we have

$$BC^2 = AB^2 + AC^2 - 2AB \cdot AC \cos \phi;$$

hence,

$$\cos \phi = \frac{|X|^2 + |Y|^2 - |X - Y|^2}{2|X| \cdot |Y|}$$

$$= \frac{x_1^2 + \cdots + x_n^2 + y_1^2 + \cdots + y_n^2 - (x_1 - y_1)^2 - \cdots - (x_n - y_n)^2}{2|X| \cdot |Y|}$$

$$= \frac{x_1 y_1 + \cdots + x_n y_n}{|X| \cdot |Y|}.$$

If we retain the term "scalar product," as in three-dimensional space, for the product of the lengths of vectors by the cosine of the angle between them, we find that the scalar product of the vectors is computed by the formula

$$X \cdot Y = x_1 y_1 + \cdots + x_n y_n ,$$

which for $n = 2$ and $n = 3$ coincides with the well-known formulas for the scalar product of ordinary vectors.

Strictly speaking, the expression $x_1 y_1 + \cdots + x_n y_n$ should be taken as the definition of the *scalar product* (because there is a lack of rigor in the definition of the scalar product by means of the angle) and then the angle between vectors can be defined by the formula

$$\cos \phi = \frac{X \cdot Y}{|X| \cdot |Y|} . \tag{3}$$

This is what we shall do.

To justify this definition of an angle we have to show that the absolute value of the right-hand side of formula (3) does not exceed 1, i.e., that $(X \cdot Y)^2 \leqslant |X|^2 \cdot |Y|^2$.

In expanded form this inequality becomes

$$(x_1 y_1 + \cdots + x_n y_n)^2 < (x_1^2 + \cdots + x_n^2)(y_1^2 + \cdots + y_n^2).$$

It is known as the Cauchy-Bunjakovskiĭ inequality and can be proved directly, by a fairly tedious computation. We shall prove it by the following indirect argument.

First of all we mention that the scalar multiplication of vectors has the following properties:

1′. $X \cdot X = |X|^2 > 0$ for $X \neq 0$.

2′. $X \cdot Y = Y \cdot X$.

3′. $(cX) \cdot Y = c(X \cdot Y)$.

4′. $(X_1 + X_2) \cdot Y = X_1 \cdot Y + X_2 \cdot Y$.

That these properties hold follows immediately from the expression of the scalar product in terms of the coordinates.

We now introduce the vector $Y + tX$, where t is an arbitrary real number. We have $|Y + tX|^2 \geqslant 0$, because the square of the length of a vector cannot be negative. But by the properties of the scalar product $|Y + tX|^2 = |Y|^2 + 2tX \cdot Y + t^2 |X|^2$. Moreover, it is known that a quadratic trinomial is nonnegative for all values of the real variable t if and only if its roots are imaginary or equal, i.e., if its discriminant is negative or zero. But the discriminant of the trinomial $|Y|^2 + 2tX \cdot Y$ $+ t^2 |X|^2$ is equal to $4(X \cdot Y)^2 - 4|X|^2|Y|^2$ so that $(X \cdot Y)^2$ $- |X|^2|Y|^2 \leqslant 0$, and this is equivalent to the Cauchy-Bunjakovskiĭ inequality.

From this inequality it follows that

$$\frac{|X \cdot Y|}{|X||Y|} \leqslant 1$$

and therefore the definition of an angle by the formula (3) is justified.

Furthermore, it is easy to deduce the inequalities

$$|X| - |Y| \leqslant |X \pm Y| \leqslant |X| + |Y|,$$

which imply, in particular, the existence of a triangle with the sides $|X|$, $|Y|$ and $|X - Y|$, so that the nonrigorous definition of an angle given previously, which was based on geometric intuition, now also becomes valid.

Axiomatic definition of an *n*-dimensional Euclidean space.   In the preceding section we have introduced the concepts of the length of a vector, of angle, and of the scalar product in the space of rows. In the general axiomatic definition of an *n*-dimensional real linear space, these concepts are also defined axiomatically, and the concept of a scalar product comes first.

Scalar multiplication of vectors of a linear real space is the name for an operation which associates with every pair of vectors X and Y a real number, their so-called scalar product $X \cdot Y$, where this operation must satisfy the following postulates (axioms):

1′. $X \cdot X > 0$ for $X \neq 0$, $0 \cdot 0 = 0$.

2′. $X \cdot Y = Y \cdot X$.

3′. $(cX) \cdot Y = c(X \cdot Y)$.

4′. $(X_1 + X_2) \cdot Y = X_1 \cdot Y + X_2 \cdot Y$.

Furthermore, by the length of a vector we mean the number $\sqrt{X \cdot X}$, by the cosine of the angle between the vectors X and Y the number $X \cdot Y/|X| \cdot |Y|$. To justify this latter inequality, it is necessary to establish the Cauchy-Bunjakovskiĭ inequality $(X \cdot Y)^2 \leqslant |X|^2|Y|^2$. But this can be done exactly as we have done it in the preceding section. In our proof we have made use only of the properties 1′, 2′, 3′ and 4′ of the scalar product, the specific nature of the space of rows playing no role in the proof. A real linear space in which a scalar multiplication satisfying the axioms 1′–4′ has been introduced is called a *Euclidean space*.

In various concrete linear spaces that are studied in mathematics, scalar products are introduced by various methods whose choice is dictated by the nature of the problem. For example, in the spaces whose elements are the functions of one variable $X(t)$ defined on a given interval $a \leqslant t \leqslant b$, the scalar product of two elements $X(t)$ and $Y(t)$ is often taken to be the number $\int_a^b X(t) \, Y(t) \, dt$ or $\int_a^b X(t) \, Y(t) \, p(t) \, dt$, where $p(t)$ is some positive function. It is easy to see that all the axioms 1′- 4′ are satisfied for either of these definitions.

Orthogonality; orthonormal bases. Two vectors of a Euclidean space are called *orthogonal* (or perpendicular) if their scalar product is zero. It is easy to see that pairwise orthogonal nonzero vectors are always linearly independent. For suppose that $X_1 , X_2 , \cdots, X_m$ are pairwise orthogonal nonzero vectors and that $c_1 X_1 + c_2 X_2 + \cdots + c_m X_m = \mathbf{0}$.

By the property of the scalar product $X_1(c_1 X_1 + c_2 X_2 + \cdots + c_m X_m)$ $= c_1 \mid X_1 \mid^2 = 0$; hence $c_1 = 0$. In the same way we can show that $c_2 = \cdots = c_m = 0$. Therefore $X_1 , \cdots, X_m$ are linearly independent.

From what we have proved, it follows that in an n-dimensional space there cannot be more than n pairwise orthogonal nonzero vectors and that every set of n pairwise orthogonal vectors forms a basis of the space. If, moreover, the lengths of all the n pairwise orthogonal vectors are 1, then the basis they form is called *orthonormal.*

It is not difficult to show, but we shall omit the proof, that a Euclidean space has orthonormal bases, in fact infinitely many. Moreover, if in a space R some subspace P is chosen, then an orthonormal basis of the subspace can be extended to an orthonormal basis of the whole space by adjoining certain vectors.

It is often convenient to define vectors in a Euclidean space by their coordinates in an arbitrary orthonormal basis, because in this case we obtain a particularly simple expression for the scalar product. For if a vector X has the coordinates $(x_1 , x_2 , \cdots, x_n)$ in the orthonormal basis $e_1 , e_2 , \cdots, e_n$ and the vector Y the coordinates $(y_1 , y_2 , \cdots, y_n)$, i.e.,

$$X = x_1 e_1 + x_2 e_2 + \cdots + x_n e_n \quad \text{and} \quad Y = y_1 e_1 + y_2 e_2 + \cdots + y_n e_n ,$$

then by the property of the scalar product

$$\begin{aligned}
X \cdot Y &= x_1 y_1 e_1 e_1 + x_1 y_2 e_1 e_2 + \cdots + x_1 y_n e_1 e_n \\
&\quad + x_2 y_1 e_2 e_1 + x_2 y_2 e_2 e_2 + \cdots + x_2 y_n e_2 e_n \\
&\quad \cdots\cdots\cdots\cdots\cdots\cdots\cdots\cdots\cdots\cdots\cdots\cdots\cdots \\
&\quad + x_n y_1 e_n e_1 + x_n y_2 e_n e_2 + \cdots + x_n y_n e_n e_n \\
&= x_1 y_1 + x_2 y_2 + \cdots + x_n y_n ,
\end{aligned}$$

since $e_i e_k = 0$ for $i \neq k$ and $e_i e_i = |e_i|^2 = 1$ for every $i = 1, 2, \cdots, n$. In particular, $X \cdot X = x^2_1 + x^2_2 + \cdots + x^2_n$.

Thus, the length of a vector and the scalar product are expressed in terms of the coordinates of an orthonormal basis by the same formulas as in the space of rows.

The transition from one of the models of a Euclidean space, namely the space of rows, to the general axiomatically defined Euclidean space does not introduce any complications, but extends the domain of applicability of the theory.

Now let us deal with the problem of orthogonal projection of vectors on a subspace. Let R_n be an n-dimensional Euclidean space and P_m an m-dimensional subspace of it. Further, let $e_1, e_2, \cdots, e_m, f_1, \cdots, f_{n-m}$ be an orthonormal basis of R_n including an orthonormal basis of the subspace P_m. The subspace Q_{n-m} spanned by the vectors $f_1, f_2, \cdots, f_{n-m}$ is called the *orthogonal complement* of the subspace P_m. Its dimension is $n - m$. The orthogonal complement Q_{n-m} can be characterized as the subspace consisting of all vectors that are orthogonal to every vector of the subspace P_m.

Every vector Z belonging to R_n can be expressed uniquely as a sum of vectors X and Y of which one belongs to P_m, the other to Q_{n-m}. This is clear, because the vector Z can be expressed uniquely in the form

$$Z = x_1 e_1 + \cdots + x_m e_m + y_1 f_1 + \cdots + y_{n-m} f_{n-m},$$

so that $X = x_1 e_1 + \cdots + x_m e_m$, $Y = y_1 f_1 + \cdots + y_{n-m} f_{n-m}$.

The vector X is called the *orthogonal projection* of Z onto P_m.

Unitary spaces.   The concepts of the length of a vector and the scalar product of vectors can also be defined in a complex space. As before, the concept of the scalar multiplication is put first, and this is defined as follows. With every pair X and Y of vectors of a complex space we associate a complex (not necessarily real) number, their so-called scalar product $X \cdot Y$. The operation of scalar multiplication must satisfy the following axioms:

1″. $X \cdot X$ is real and positive for $X \neq 0$, $0 \cdot 0 = 0$.

2″. $Y \cdot X = (X \cdot Y)'$. Here the prime denotes transition to the conjugate complex number.

3″. $(cX) \cdot Y = c(X \cdot Y)$ for an arbitrary complex c.

4″. $(X_1 + X_2) \cdot Y = X_1 \cdot Y + X_2 \cdot Y$ (distributive law).

In the space of rows with complex elements the scalar product of the vectors $(x_1, \cdots, x_n)$ and $(y_1, \cdots, y_n)$ can be taken to be the number $x_1 y_1' + \cdots + x_n y_n'$. It is easy to verify that all the axioms $1''$-$4''$ are satisfied for this definition.

The length of a vector is defined as the number $\sqrt{X \cdot X}$. The concept of angle between vectors is not defined.

A complex linear space with a scalar product satisfying the axioms $1''$-$4''$ is called a *unitary space*.

§3. Systems of Linear Equations

Systems of two equations with two unknowns and of three equations with three unknowns. A system of two linear equations with two unknowns appears in the following general form

$$a_1 x + b_1 y = c_1,$$
$$a_2 x + b_2 y = c_2.$$

To solve this system we multiply the first equation by b_2, the second by $-b_1$ and add. We obtain

$$(a_1 b_2 - a_2 b_1)\, x = c_1 b_2 - c_2 b_1.$$

Similarly, by multiplying the first equation by $-a_2$, the second by a_1, and adding, we obtain

$$(a_1 b_2 - a_2 b_1)\, y = a_1 c_2 - a_2 c_1.$$

From these equations it is easy to determine x and y, if only the expression $a_1 b_2 - a_2 b_1$ formed from the coefficients of the unknown x and y is different from zero. This expression is called the *determinant* of the matrix

$$\begin{bmatrix} a_1\, b_1 \\ a_2\, b_2 \end{bmatrix}$$

formed from the coefficients of the system. The determinant is denoted:

$$\begin{vmatrix} a_1\, b_1 \\ a_2\, b_2 \end{vmatrix}.$$

From the definition it follows that the determinant is computed by the scheme

which requires no further explanation.

Let us return to the solution of the system. The expressions $c_1 b_2 - c_2 b_1$ and $a_1 c_2 - a_2 c_1$ also appear as determinants in accordance with our definition, namely,

$$\begin{vmatrix} c_1 & b_1 \\ c_2 & b_2 \end{vmatrix} \quad \text{and} \quad \begin{vmatrix} a_1 & c_1 \\ a_2 & c_2 \end{vmatrix}.$$

Thus, if the determinant

$$\begin{vmatrix} a_1 & b_1 \\ a_2 & b_2 \end{vmatrix}$$

is different from zero, then we obtain the following formulas for the solution of the system:

$$x = \frac{\begin{vmatrix} c_1 & b_1 \\ c_2 & b_2 \end{vmatrix}}{\begin{vmatrix} a_1 & b_1 \\ a_2 & b_2 \end{vmatrix}}; \quad y = \frac{\begin{vmatrix} a_1 & c_1 \\ a_2 & c_2 \end{vmatrix}}{\begin{vmatrix} a_1 & b_1 \\ a_2 & b_2 \end{vmatrix}}. \tag{4}$$

Strictly speaking, these arguments are not complete. The operations on the equations that we have carried out to deduce the formulas for the solution of the system make sense only under the assumption that x and y are in fact numbers that form a solution of the system. The logical substance of our argument is the following: If the determinant of the coefficients of the system is not zero and the solution of the system exists, then it can be computed by the formulas (4). Therefore it is still necessary to verify that the values of the unknown that we have found do in fact satisfy both equations of the system. This can be done without any difficulty.

Thus, if the determinant of the matrix of the coefficients of the system is different from zero, then the system has a unique solution given by the formulas (4).

For a system of three equations with three unknowns

$$a_1 x + b_1 y + c_1 z = d_1,$$
$$a_2 x + b_2 y + c_2 z = d_2,$$
$$a_3 x + b_3 y + c_3 z = d_3,$$

it is easy to carry out similar arguments and computations; for this purpose it is sufficient to add up the equations after multiplying them by factors such that after addition two of the unknowns disappear. To make the unknown y and z disappear, we have to take for these factors $b_2 c_3 - b_3 c_2$, $b_3 c_1 - b_1 c_3$ and $b_1 c_2 - b_2 c_1$, as is easy to verify by computation.

We obtain the result that if the expression

$$\Delta = a_1b_2c_3 - a_1b_3c_2 + a_2b_3c_1 - a_2b_1c_3 + a_3b_1c_2 - a_3b_2c_1$$

is different from zero, then the system has a unique solution obtained by the formulas

$$x = \frac{\Delta_1}{\Delta}, \quad y = \frac{\Delta_2}{\Delta}, \quad z = \frac{\Delta_3}{\Delta},$$

where Δ_1, Δ_2, Δ_3 are the expressions obtained from Δ by replacing the coefficients of the corresponding unknown by the absolute terms.

The expression Δ is called the determinant of the matrix

$$\begin{bmatrix} a_1\, b_1\, c_1 \\ a_2\, b_2\, c_2 \\ a_3\, b_3\, c_3 \end{bmatrix}$$

and is denoted by

$$\begin{vmatrix} a_1\, b_1\, c_1 \\ a_2\, b_2\, c_2 \\ a_3\, b_3\, c_3 \end{vmatrix}.$$

For the computation of a determinant the following scheme is useful:

In the first of these schemes, the lines (a diagonal and two triangles) connect the positions of the elements whose product occurs in the composition of the determinant with a plus sign; and in the second scheme, they connect the terms occuring in the determinant with a minus sign.

For systems of two equations with two unknowns and of three equations with three unknowns, we have obtained entirely similar results. In both cases the system has a unique solution, provided the determinant of the matrix of the coefficients is different from zero. The formulas for the solution are also similar: In the denominator of each of the unknowns stands the determinant of the matrix of coefficients, and in the numerators the determinants of the matrices that arise from the matrix of coefficients by replacing the coefficients of the unknown to be computed by the absolute terms.

An immediate generalization of these results to systems of n equations in n unknowns for arbitrary n is somewhat difficult. It becomes comparatively easy by an indirect method: First we generalize the concept of a determinant to square matrices of arbitrary order, and having studied the properties of determinants we apply their theory to the investigation of systems of equations.

Determinants of the *n*th order.   When we consider the explicit expression for the determinants

$$\begin{vmatrix} a_1\, b_1 \\ a_2\, b_2 \end{vmatrix} = a_1 b_2 - a_2 b_1$$

and

$$\begin{vmatrix} a_1\, b_1\, c_1 \\ a_2\, b_2\, c_2 \\ a_3\, b_3\, c_3 \end{vmatrix} = a_1 b_2 c_3 - a_1 b_3 c_2 + a_2 b_3 c_1 - a_2 b_1 c_3 + a_3 b_1 c_2 - a_3 b_2 c_1 \,,$$

we notice that in every term there occurs as a factor exactly one element from each row and one from each column of the determinant, and that all possible products of this form occur in the determinant with a plus or a minus sign. This property is at the bottom of the generalization of the concept of the determinant to square matrices of arbitrary order. In fact, the determinant of a square matrix of order *n* or, briefly, a *determinant of the nth order* is defined as the algebraic sum of all possible products of the elements of the matrix, precisely one from each row and one from each column; these products are given plus or minus signs by a certain well-defined rule. This rule is somewhat complicated to explain, and we shall not dwell on its formulation. It is sufficient to mention that it is arranged in such a way that the following important basic properties of a determinant are secured:

1. When two rows are interchanged, the determinant changes its sign.

For determinants of order 2 and 3, this property is easy to verify by an immediate computation. In the general case it is proved on the basis of the rule for the signs that we have not formulated here.

Determinants have quite a number of other remarkable properties that enable us to apply determinants successfully in diverse theoretical and numerical calculations, notwithstanding the fact that determinants are extraordinarily cumbersome: A determinant of order *n* contains, as is easy to see, *n*! terms, each term consists of *n* factors, and the factors are provided with their signs according to a complicated rule.

We now proceed to enumerate the basic properties of determinants but omit their detailed proofs. The first of these properties has been formulated.

2. A determinant does not change when its matrix is transposed, i.e.,

when the rows are replaced by the columns, preserving their order. The proof is based on a detailed study of the rule of the distribution of signs in the terms of the determinant. This property enables us to transfer every statement concerning the rows of the determinant to a statement on columns.

3. A determinant is a linear function of the elements of each row (or column). In detail,

$$
\begin{vmatrix}
a_{11} & \cdots & a_{1n} \\
\cdots\cdots\cdots\cdots \\
a_{i1} & \cdots & a_{in} \\
\cdots\cdots\cdots\cdots \\
a_{n1} & \cdots & a_{nn}
\end{vmatrix}
= a_{i1}A_{i1} + a_{i2}A_{i2} + \cdots + a_{in}A_{in} , \tag{5}
$$

where A_{i1}, A_{i2}, $\cdots$, A_{in} are expressions that do not depend on the elements of the ith row.

This property follows evidently from the fact that every term contains one and only one factor from each row, in particular the ith row.

The equation (5) is called the expansion of the determinant with respect to the elements of the ith row, and the coefficients A_{i1}, A_{i2}, $\cdots$, A_{in} are called the algebraic complements of the elements a_{i1}, a_{i2}, $\cdots$, a_{in} of the determinant.

4. The algebraic complement A_{ij} of the element a_{ij} is equal, apart from the sign, to the so-called minor Δ_{ij} of the determinant, i.e., the determinant of order $(n-1)$ that arises from the given one by crossing out the ith row and jth column. To obtain the algebraic complement the minor must be taken with the sign $(-1)^{i+j}$. The properties 3 and 4 reduce the computation of a determinant of order n to the computation of n determinants of order $n-1$.

The fundamental properties that we have enumerated have a number of interesting consequences. We now mention some of these.

5. A determinant with two equal rows is zero.

For if a determinant has two equal rows, then the determinant does not change when they are interchanged, because the rows are identical; but on the other hand, by our first property it should change its sign. Therefore it is equal to zero.

6. The sum of the products of the elements of any row by the algebraic complements of another row is zero.

For such a sum is the result of expanding a determinant with two equal rows with respect to one of them.

7. A common factor of the elements of any row can be taken before the determinant sign.

This follows from property 3.

8. A determinant with two proportional rows is zero.

It is sufficient to take out the factor of proportionality and then we have a determinant with two equal rows.

9. A determinant does not change if we add to the elements of any one row numbers that are proportional to the elements of another row.

For by property 3 the modified determinant is equal to the sum of the original determinant and a determinant with two proportional rows, which is zero.

The last property gives us a good method for computing determinants. Without changing the value of a determinant, we can transform its matrix by applying this rule so that in one row (or column) all the elements except one become zero. Then by expanding the determinant with respect to the elements of this row (column) we reduce the computation of a determinant of order n to that of a *single* determinant of order $n - 1$, namely, the algebraic complement of the only nonzero element of the row in question.

For example, suppose we have to compute the determinant

$$\Delta = \begin{vmatrix} 1 & 1 & -1 & 2 \\ 2 & -1 & 1 & 1 \\ -1 & 2 & 0 & 1 \\ 1 & 1 & -2 & 1 \end{vmatrix}.$$

We add to the second column the first, multiplied by -1, to the third, the first, and to the fourth, the first, multiplied by -2 and obtain

$$\Delta = \begin{vmatrix} 1 & 0 & 0 & 0 \\ 2 & -3 & 3 & -3 \\ -1 & 3 & -1 & 3 \\ 1 & 0 & -1 & -1 \end{vmatrix}.$$

Expanding Δ by the elements of the first row, we obtain

$$\Delta = 1 \cdot (-1)^{1+1} \begin{vmatrix} -3 & 3 & -3 \\ 3 & -1 & 3 \\ 0 & -1 & -1 \end{vmatrix}.$$

Finally, we add to the first row the second and expand with respect to the elements of the first column; we obtain

$$\Delta = \begin{vmatrix} 0 & 2 & 0 \\ 3 & -1 & 3 \\ 0 & -1 & -1 \end{vmatrix} = 3 \cdot (-1)^{1+2} \begin{vmatrix} 2 & 0 \\ -1 & -1 \end{vmatrix} = -3 \cdot (-2) = 6.$$

The determinant of a matrix A is denoted by $|A|$.

In conclusion, we mention a further very important property of determinants.

The determinant of the product of two square matrices is equal to the product of the determinants of the factors, i.e., in short notation, $|AB| = |A||B|$.

This property enables us in particular to multiply determinants of equal order by the rule for multiplication of matrices.

Systems of n linear equations in n unknowns. With the apparatus of determinants, it is now easy to generalize the results obtained earlier for systems of two equations with two unknowns and of three equations with three unknowns to systems of n equations with n unknowns, under the assumption that the determinant of the coefficient matrix is different from zero.

Let

$$a_{11}x_1 + \cdots + a_{1j}x_j + \cdots + a_{1n}x_n = b_1 ,$$
$$\dotfill$$
$$a_{n1}x_1 + \cdots + a_{nj}x_j + \cdots + a_{nn}x_n = b_n$$

be such a system. We denote by $\varDelta$ the determinant of the coefficient matrix of the system. By assumption it is different from zero. Furthermore, we denote by A_{ij} the algebraic complement of the element a_{ij}. We multiply the first equation by A_{1j}, the second by A_{2j}, $\cdots$, the nth by A_{nj} and add. We obtain

$$\varDelta x_j = b_1 A_{1j} + \cdots + b_n A_{nj} .$$

For the coefficients of all the unknowns except x_j vanish, because they appear as sums of the products of the algebraic complements of the elements of the jth column with the elements of other columns (property 6, applied to columns); but the coefficients of the unknown x_j is equal to the sum of the products of the elements of the jth column with their algebraic complements, i.e., is equal to $\varDelta$.

Thus,

$$x_j = \frac{b_1 A_{1j} + \cdots + b_n A_{nj}}{\varDelta} \quad \text{for all} \quad j = 1, 2, \cdots, n. \tag{6}$$

As we have said, these arguments are valid only if we understand by $x_1 , x_2 , \cdots, x_n$ a solution of the system, the existence of which must be assumed in the first instance.

Hence the result of the argument is the following.

If a solution of the system exists, then it is given by the formulas (6) and is therefore unique.

To complete the exposition, it is necessary to prove the existence of a solution, and this can be done by substituting the values we have found for the unknowns into all the equations of the original system. It is easy to verify by using the same property of the determinant (but this time applied to the rows) that these values in fact satisfy all the equations.

Thus the following theorem is true: If the determinant of the coefficient matrix of a system of n equations with n unknowns is different from zero, then the system has a unique solution given by the formulas (6).

These formulas can be transformed by remarking that the sum $b_1A_{1j} + \cdots + b_nA_{nj}$ can be written in the form of a determinant, namely:

$$\Delta_j = b_1A_{1j} + \cdots + b_nA_{nj} = \begin{vmatrix} a_{11} \cdots b_1 \cdots a_{1n} \\ \cdots\cdots\cdots\cdots\cdots \\ a_{n1} \cdots b_n \cdots a_{nn} \end{vmatrix}$$

(the absolute terms occur in the jth column).

Hence the results we have stated for systems of equations with two and three unknowns have been completely generalized to a system of n equations, and even the formulas for the solution are formally exactly the same.

We mention one corollary of this theorem: If it is known of a system of equations that it has no solution at all or that the solution is not unique, then the determinant of the coefficient matrix is equal to zero.

This corollary is particularly often applied to homogeneous systems, i.e., those in which the absolute terms b_1, b_2, $\cdots$, b_n are all zero. Homogeneous systems always have the obvious "trivial" solution $x_1 = x_2 = \cdots = x_n = 0$.

If a homogeneous system has, apart from the trivial one, also a nontrivial solution, then its determinant is zero.

This statement opens up the possibility of using the theory of determinants in other branches of mathematics and its applications.

Let us consider, for example, a problem in analytical geometry: to find the equation of the plane passing through three given points (x_1, y_1, z_1), (x_2, y_2, z_2) and (x_3, y_3, z_3) that do not lie on one line.

From elementary geometry it is known that such a plane exists. Suppose that its equation is of the form $Ax + By + Cz + D = 0$. Then

$$Ax_1 + By_1 + Cz_1 + D = 0,$$
$$Ax_2 + By_2 + Cz_2 + D = 0,$$
$$Ax_3 + By_3 + Cz_3 + D = 0.$$

Let x, y, z be the coordinates of an arbitrary point in that plane. Then we also have

$$Ax + By + Cz + D = 0.$$

We regard these four equations as a system of linear homogeneous equations for the coefficients A, B, C, D of the required plane. This system has a nontrivial solution, because the required plane exists. Therefore the determinant of the system is zero; i.e.,

$$\begin{vmatrix} x_1 & y_1 & z_1 & 1 \\ x_2 & y_2 & z_2 & 1 \\ x_3 & y_3 & z_3 & 1 \\ x & y & z & 1 \end{vmatrix} = 0. \tag{7}$$

Now this is the equation of the required plane. For it is an equation of the first degree in x, y, z, a fact which follows from the linearity of the determinant with respect to the elements of the last row.

By making use of the fact that the given points do not lie on one line, it is easy to verify that not all the coefficients of this equation are zero. Consequently the equation (7) is indeed the equation of a plane. This plane passes through the given points, because their coordinates obviously satisfy the equation.

Matrix notation for a system of n equations in n unknowns. A system of n linear equations in n unknowns

$$a_{11}x_1 + \cdots + a_{1n}x_n = b_1 \,,$$
$$\dotfill$$
$$a_{n1}x_1 + \cdots + a_{nn}x_n = b_n$$

can be written in matrix notation in the form of a single equation

$$AX = B.$$

Here A denotes the coefficient matrix, X the column formed by the unknowns, and B the column of the absolute terms.

The solution of the system (if the determinant of the matrix A is different from zero) can be written explicitly as follows [see formula (6)]:

$$x_1 = \frac{A_{11}}{\Delta} b_1 + \frac{A_{21}}{\Delta} b_2 + \cdots + \frac{A_{n1}}{\Delta} b_n \,,$$
$$x_2 = \frac{A_{12}}{\Delta} b_1 + \frac{A_{22}}{\Delta} b_2 + \cdots + \frac{A_{n2}}{\Delta} b_n \,,$$
$$\dotfill$$
$$x_n = \frac{A_{1n}}{\Delta} b_1 + \frac{A_{2n}}{\Delta} b_2 + \cdots + \frac{A_{nn}}{\Delta} b_n$$

or in matrix form

$$
X =
\begin{bmatrix}
\dfrac{A_{11}}{\varDelta} & \dfrac{A_{21}}{\varDelta} & \cdots & \dfrac{A_{n1}}{\varDelta} \\[2ex]
\dfrac{A_{12}}{\varDelta} & \dfrac{A_{22}}{\varDelta} & \cdots & \dfrac{A_{n2}}{\varDelta} \\[1ex]
\cdots\cdots\cdots\cdots\cdots \\[1ex]
\dfrac{A_{1n}}{\varDelta} & \dfrac{A_{2n}}{\varDelta} & \cdots & \dfrac{A_{nn}}{\varDelta}
\end{bmatrix}
B.
$$

The matrix standing as first factor on the right-hand side of the equation is called the *inverse* to the matrix A and is denoted by A^{-1}. Using this notation we obtain the solution of the system $AX = B$ in the following simple and natural form, which recalls the formula for the solution of a single equation in one unknown:

$$X = A^{-1}B.$$

We can easily give another proof of the result obtained, in terms of the algebra of matrices.

For this purpose we must first of all mention the special role of the matrix

$$
E =
\begin{pmatrix}
1 & 0 & 0 & \cdots & 0 \\
0 & 1 & 0 & \cdots & 0 \\
& & \cdots\cdots\cdots \\
0 & 0 & 0 & \cdots & 1
\end{pmatrix},
$$

the so-called *unit matrix*.

The unit matrix plays among square matrices the same role as the number 1 among numbers. In fact, for every matrix A the following equations hold: $AE = A$ and $EA = A$. This is easy to verify by the rule for the multiplication of matrices.

The matrix A^{-1} defined previously, the inverse to A, plays in relation to it a similar role to that played by the inverse of a given number:

$$AA^{-1} = A^{-1}A = E.$$

The validity of these equations can be verified by the rule for the multiplication of matrices and by the properties 3 and 6 of a determinant.

Knowing these properties of the unit and the inverse matrix we can obtain the solution of the system $AX = B$ in the following way.

Suppose that $AX = B$. Then $A^{-1}(AX) = A^{-1}B$. But $A^{-1}(AX) = (A^{-1}A)X = EX = X$ and therefore $X = A^{-1}B$.

Suppose now that $X = A^{-1}B$, then $AX = AA^{-1}B = EB = B$.

Thus the "equation" $AX = B$ has the unique solution $X = A^{-1}B$, provided only that A^{-1} exists.

We have established the existence of the inverse matrix A^{-1} for A under the assumption that the determinant of A is different from zero. This condition is not only sufficient but also necessary for the existence of the inverse matrix. For suppose that the matrix A has an inverse A^{-1} such that $AA^{-1} = E$. Then by the property of the determinant of the product of two matrices

$$|A| |A^{-1}| = |E| = 1,$$

and from this it follows that the determinant of A is not zero.

A matrix whose determinant is different from zero is called *nondegenerate* or *nonsingular*. We have therefore established that an inverse matrix always exists for nondegenerate matrices and only for them.

The introduction of the concept of an inverse matrix turns out to be useful not only in the theory of systems of linear equations but also in many other problems of linear algebra.

In conclusion, we mention that the formulas we have derived for the solution of linear systems are an irreplaceable tool in theoretical considerations but are not convenient for the numerical solution of systems. As we have already mentioned, various methods and computational schemes have been worked out for the numerical solution of systems, and in view of the great importance of this problem for practical investigations, the work of simplifying the numerical solution of systems (especially with large numbers of unknowns) is intensively pursued even at present.

The general case of systems of linear equations. We now turn to the investigation of systems of linear equations in the most general case when it is not assumed that the number of equations is equal to the number of unknowns. In such a general setting it cannot be expected, naturally, that a solution of the system always exists or that, in case it exists, it turns out to be unique. It is natural to assume that if the number of equations is less than the number of unknowns the system has infinitely many solutions. For example, two equations of the first degree in three unknowns are satisfied by the coordinates of every point on the straight line that is the intersection of the planes defined by the equations. However it can happen in this case that the system has no solution at all, namely when the planes are parallel. And if the number of equations is greater than the number of unknowns, then as a rule the system has no solution. However, in this case it is possible that the system has solutions, even infinitely many.

In order to investigate the existence and the character of the set of solutions of a system in this general setting, we turn to a "geometrical" interpretation of the system.

We interpret the system of equations

$$\begin{aligned}
a_{11}x_1 + a_{12}x_2 + \cdots + a_{1n}x_n &= b_1\,, \\
a_{21}x_1 + a_{22}x_2 + \cdots + a_{2n}x_n &= b_2\,, \\
&\cdots\cdots\cdots\cdots\cdots\cdots\cdots\cdots\cdots\cdots\cdots\cdots\cdots \\
a_{m1}x_1 + a_{m2}x_2 + \cdots + a_{mn}x_n &= b_m
\end{aligned} \tag{8}$$

in the m-dimensional space of columns in the form

$$x_1 A_1 + x_2 A_2 + \cdots + x_n A_n = B.$$

Here A_1, A_2, $\cdots$, A_n denote the columns of the coefficients of the corresponding unknowns and B the column of the absolute terms.

In this interpretation the problem of the existence of a solution of the system turns into the problem of whether the given vector B is a linear combination of the vectors A_1, A_2, $\cdots$, A_n.

The answer to this problem is almost obvious. For the vector B to be a linear combination of the vectors A_1, A_2, $\cdots$, A_n it is necessary and sufficient that B should be contained in the subspace spanned by A_1, A_2, $\cdots$, A_n or, in other words, that the subspaces spanned by the systems of vectors A_1, A_2, $\cdots$, A_n and A_1, A_2, $\cdots$, A_n, B should coincide.

Since the first of these subspaces is contained in the second, they coincide if and only if their dimensions are equal. We recall that the dimension of the subspace spanned by a given system of vectors is called the rank of this system. Thus, a necessary and sufficient condition for the existence of a solution of the system $x_1 A_1 + x_2 A_2 + \cdots + x_n A_n = B$ is the equality of the ranks of the vector systems A_1, A_2, $\cdots$, A_n and A_1, A_2, $\cdots$, A_n, B.

It can be proved, but we shall not do it here, that the rank of a system of vectors is equal to the rank of the matrix formed from the coordinates of these vectors. Here we understand by the *rank of a matrix* the largest order of a nonzero determinant that can be formed from the given matrix by omitting part of its rows and columns.

Since the coordinates of the vectors A_1, A_2, $\cdots$, A_n (in the natural basis for the space of columns) are the coefficients of the system and the coordinates of the vector B its absolute terms, we obtain the following final formulation of the condition for the existence of a solution of a system.

For the existence of at least one solution of the system of linear equations

$$\begin{aligned}
a_{11}x_1 + \cdots + a_{1n}x_n &= b_1\,, \\
&\cdots\cdots\cdots\cdots\cdots\cdots\cdots\cdots\cdots\cdots \\
a_{m1}x_1 + \cdots + a_{mn}x_n &= b_m
\end{aligned}$$

it is necessary and sufficient that the rank of the matrix formed from the coefficients of the system should be equal to the rank of the matrix formed from the coefficients and the absolute terms.

Now let us investigate the character of the set of solutions if they exist. Let $x_1^0, x_2^0, \cdots, x_n^0$ be any solution of the system (8). We set $x_1 = x_1^0 + y_1$, $x_2 = x_2^0 + y_2, \cdots, x_n = x_n^0 + y_n$. Then in view of the fact that x_1^0, $x_2^0, \cdots, x_n^0$ form a solution of the system (8), the new unknowns $y_1, y_2, \cdots, y_n$ must satisfy the homogeneous system

$$
\begin{aligned}
a_{11}y_1 + \cdots + a_{1n}y_n &= 0, \\
&\cdots\cdots\cdots\cdots\cdots\cdots\cdots \\
a_{m1}y_1 + \cdots + a_{mn}y_n &= 0
\end{aligned}
\tag{9}
$$

with the same coefficient matrix. Conversely, if we add to the original solution $x_1^0, x_2^0, \cdots, x_n^0$ of the system (8) an arbitrary solution of the homogeneous system (9), then we obtain another solution of the system (8).

Thus, in order to obtain the general solution of the system (8), it is only necessary to take an arbitrary particular solution of it and to add it to the general solution of the homogeneous system (9).

In this way the problem of the character of the set of solutions of the system (8) is reduced to the same problem for the homogeneous system (9). We shall consider this problem in the next section.

Homogeneous systems. We shall interpret the homogeneous system of linear equations

$$
\begin{aligned}
a_{11}y_1 + a_{12}y_2 + \cdots + a_{1n}y_n &= 0, \\
a_{21}y_1 + a_{22}y_2 + \cdots + a_{2n}y_n &= 0, \\
&\cdots\cdots\cdots\cdots\cdots\cdots\cdots\cdots\cdots\cdots\cdots \\
a_{m1}y_1 + a_{m2}y_2 + \cdots + a_{mn}y_n &= 0
\end{aligned}
$$

in the n-dimensional Euclidean space. (Here we assume that the coefficients of the system are real. For systems with complex coefficients, we can give a similar interpretation in unitary space and obtain similar results.)

Let $A_1', A_2', \cdots, A_m', Y$ be the vectors of a Euclidean space whose coordinates in an orthonormal basis are, respectively,

$$(a_{11}, a_{12}, \cdots, a_{1n}), (a_{21}, a_{22}, \cdots, a_{2n}), \cdots, (a_{m1}, a_{m2}, \cdots, a_{mn}),$$
$$(y_1, y_2, \cdots, y_n).$$

Then the system assumes the form

$$A_1'Y = 0, A_2'Y = 0, \cdots, A_m'Y = 0;$$

i.e., every solution of the system determines a vector orthogonal to all the vectors formed by the coefficients of the various equations.

Therefore, the set of solutions forms the subspace that is the orthogonal complement to the subspace spanned by the vectors A_1', A_2', $\cdots$, A_m'. The dimension of the latter subspace is equal to the rank r of the matrix formed from the coefficients of the system. The dimension of the orthogonal complement, i.e., of the "solution space," is then equal to $n - r$.

Now every subspace has a basis, i.e., a system of linearly independent vectors equal in number to the dimension of the subspace and such that their linear combinations fill the whole subspace. Therefore, among the solutions of a homogeneous system there exist $n - r$ linearly independent solutions such that all the solutions of the system are linear combinations of them. Here n denotes the number of unknowns and r the rank of the coefficient matrix.

Thus, the structure of the solutions of a homogeneous system and, consequently, also of an inhomogeneous system is completely clarified. In particular, a homogeneous system has the unique trivial solution $x_1 = x_2 = \cdots = x_n = 0$ if and only if the rank of the coefficient matrix is equal to the number of unknowns. By what we have said at the end of the preceding paragraph, the same condition is also the condition for uniqueness of the solution for systems of inhomogeneous equations (providing the consistency condition is satisfied).

Our investigation of these systems shows clearly how the introduction of generalized geometrical concepts leads to simplicity and lucidity in a complicated algebraic problem.

§4. Linear Transformations

Definition and examples. In many mathematical investigations it becomes necessary to change the variables, i.e., to go over from one system of variables x_1, x_2, $\cdots$, x_n to another y_1, y_2, $\cdots$, y_n, connected with the first by means of a functional dependence:

$$y_1 = \phi_1(x_1, x_2, \cdots, x_n),$$
$$y_2 = \phi_2(x_1, x_2, \cdots, x_n),$$
$$\cdots\cdots\cdots\cdots\cdots\cdots\cdots\cdots$$
$$y_n = \phi_n(x_1, x_2, \cdots, x_n).$$

For example, if the variables are the coordinates of a point in a plane or in space, then the transition from one system of coordinates to another system gives rise to a transformation of coordinates that is defined by the expressions for the original coordinates in terms of the new ones or vice versa.

Moreover, a transformation of variables arises in studying the changes due to a transition from one position or configuration to another for objects whose position or configuration is described by the values of the variables. As a typical example of this kind of tranformation, we can take the change of coordinates of the points of some body under deformations.

An abstractly given transformation of a system of n variables is usually interpreted as a transformation (deformation) of an n-dimensional space, i.e., as an association between each vector of the space (or part of it) with coordinates x_1, x_2, $\cdots$, x_n and a corresponding vector with coordinates y_1, y_2, $\cdots$, y_n.

As we have said previously, every "smooth" function (having continuous partial derivatives) of several variables is close to a linear function for small changes of these variables. Therefore every "smooth" transformation (i.e., one for which the functions ϕ_1, ϕ_2, $\cdots$, ϕ_n in its analytical expression have continuous partial derivatives) is close to a linear transformation in a small part of the space:

$$
\begin{aligned}
y_1 &= a_{11}x_1 + a_{12}x_2 + \cdots + a_{1n}x_n + b_1, \\
y_2 &= a_{21}x_1 + a_{22}x_2 + \cdots + a_{2n}x_n + b_2, \\
&\cdots\cdots\cdots\cdots\cdots\cdots\cdots\cdots\cdots\cdots\cdots\cdots\cdots\cdots\cdots \\
y_n &= a_{n1}x_1 + a_{n2}x_2 + \cdots + a_{nn}x_n + b_n.
\end{aligned}
\tag{10}
$$

This circumstance alone makes the study of the properties of linear transformations one of the most important problems of mathematics. For example, from the theory of n linear equations in n unknowns, we know that a necessary and sufficient condition for the system of equations (10) with respect to x_1, x_2, $\cdots$, x_n to have a unique solution, i.e., for the corresponding linear transformation to be invertible, is the nonvanishing of the determinant of the coefficients. This circumstance is the foundation of an important theorem of analysis: For a transformation

$$
\begin{aligned}
y_1 &= \phi_1(x_1, x_2, \cdots, x_n), \\
y_2 &= \phi_2(x_1, x_2, \cdots, x_n), \\
&\cdots\cdots\cdots\cdots\cdots\cdots\cdots\cdots\cdots \\
y_n &= \phi_n(x_1, x_2, \cdots, x_n),
\end{aligned}
$$

which is smooth in the neighborhood of a given point, to have a smooth inverse transformation it is necessary and sufficient that at the given point the determinant

$$\begin{vmatrix} \dfrac{\partial \phi_1}{\partial x_1} \cdots \dfrac{\partial \phi_1}{\partial x_n} \\[2mm] \dfrac{\partial \phi_2}{\partial x_1} \cdots \dfrac{\partial \phi_2}{\partial x_n} \\[2mm] \cdots\cdots\cdots\cdots \\[2mm] \dfrac{\partial \phi_n}{\partial x_1} \cdots \dfrac{\partial \phi_n}{\partial x_n} \end{vmatrix}$$

should be different from zero.

The study of the general linear transformation (10) essentially reduces to the study of the homogeneous transformation with the same coefficients

$$\begin{aligned} y_1 &= a_{11}x_1 + \cdots + a_{1n}x_n \,, \\ y_2 &= a_{21}x_1 + \cdots + a_{2n}x_n \,, \\ &\cdots\cdots\cdots\cdots\cdots\cdots\cdots\cdots \\ y_n &= a_{n1}x_1 + \cdots + a_{nn}x_n \,, \end{aligned} \tag{11}$$

and in what follows, when speaking of linear tranformations, we shall always have homogeneous transformations in mind.

Linear transformations of an n-dimensional space can also be defined by their intrinsic properties, apart from the formulas (11) that connect the coordinates of corresponding points. Such a coordinate-free definition of the concept of a linear transformation is useful in that it does not depend on the choice of a basis. This definition is as follows.

A linear transformation of an n-dimensional linear space is a function $Y = A(X)$ where X and Y are vectors. This function satisfies the postulate of linearity

$$A(c_1 X_1 + c_2 X_2) = c_1 A(X_1) + c_2 A(X_2). \tag{12}$$

In what follows, when speaking of a linear transformation of a space, we shall understand it in the sense of this definition.

This definition is equivalent to the preceding one in terms of coordinates. For the function $Y = A(X)$, which to the vector X with the coordinates $x_1, x_2, \cdots, x_n$ associates the vector Y with the coordinates $y_1, y_2, \cdots, y_n$ in such a way that the coordinates $y_1, y_2, \cdots, y_n$ are expressed in terms of the coordinates $x_1, x_2, \cdots, x_n$ in the form of linear homogeneous functions, obviously satisfies the postulate (12). Conversely, if the function $Y = A(X)$ satisfies the postulate (12) and if $e_1, e_2, \cdots, e_n$ is an arbitrary basis of the space, then

$$A(x_1 e_1 + x_2 e_2 + \cdots + x_n e_n) = x_1 A(e_1) + x_2 A(e_2) + \cdots + x_n A(e_n).$$

We denote the coordinates (in the same basis) of the vector $A(e_j)$ by $a_{1j}, \cdots, a_{nj}; j = 1, \cdots, n$. Then the coordinates of the vector $Y = A(X)$ are

$$y_1 = a_{11}x_1 + a_{12}x_2 + \cdots + a_{1n}x_n ,$$
$$y_2 = a_{21}x_1 + a_{22}x_2 + \cdots + a_{2n}x_n ,$$
$$\dotfill$$
$$y_n = a_{n1}x_1 + a_{n2}x_2 + \cdots + a_{nn}x_n .$$

Thus, to every linear transformation of a linear space there corresponds a certain square matrix with respect to a given basis. This transformation can be written in matrix language in the form $Y = AX$. Here X is the column of the coordinates of the original vector, Y the column of the coordinates of the transformed vector, and A the coefficient matrix of the transformation. The columns of the matrix A are formed by the coordinates of those vectors into which the vectors of the basis are transformed. In accordance with the matrix notation, we shall subsequently often write a linear transformation itself in the form $Y = AX$, omitting the parentheses.

From the formula

$$A(x_1e_1 + x_2e_2 + \cdots + x_ne_n) = x_1A(e_1) + x_2A(e_2) + \cdots + x_nA(e_n)$$

it follows that the whole space is mapped under a linear transformation into the subspace spanned by the vectors $A(e_1), \cdots, A(e_n)$. The dimension of this subspace is equal to the rank of the system of vectors $A(e_1)$, $A(e_2), \cdots, A(e_n)$, or, what is the same, to the rank of the matrix formed by their coordinates, i.e., the rank of the matrix A associated with the transformation. This subspace coincides with the whole space if and only if the rank of the matrix A is equal to n, i.e., if the determinant of A is different from zero. In this case the linear transformation is called *nonsingular* or *nondegenerate*.

From the theory of systems of linear equations, we know that non-degenerate transformations are uniquely invertible and that the coordinates of the original vector are expressed in terms of the coordinates of the transformed vector by the formula $\mathbf{X} = \mathbf{A}^{-1}\mathbf{Y}$.

A transformation whose matrix has the determinant zero is called *singular* or *degenerate*. A degenerate transformation is not invertible. This follows from the theory of linear transformations or more intuitively from the fact that it transforms the whole space into part of it.

As a first example of a nondegenerate transformation, we take the identity transformation that maps every vector into itself. The matrix of the identity transformation in any basis is the unit matrix E. A nonsingular transformation is also given by a similarity that consists in multiplying all the vectors of the space by one and the same number. The matrix of a similarity transformation does not depend on the choice of a basis and has the form aE, where a is the similarity factor.

An important special case of nondegenerate transformation are the orthogonal transformations. The concept of an *orthogonal transformation* has a meaning when applied to a Euclidean space and is defined as a linear transformation preserving the lengths of vectors. An orthogonal transformation is a generalization to n-dimensional space of a rotation of the space around the fixed origin of coordinates or a rotation combined with a reflection in an arbitrary plane passing through the origin.

It is easy to see that under an orthogonal transformation not only the lengths of vectors are preserved but also scalar products and that, consequently, orthogonal transformations carry an orthogonal basis of the space into a system of pairwise orthogonal unit vectors which in turn is then necessarily also a basis.

The matrix connected with an orthogonal transformation with respect to an orthonormal basis has the following specific properties.

First, the sum of the squares of the elements of each column is 1, since these sums are the squares of the lengths of the vectors into which the vectors of the given basis are transformed. Second, the sums of the products of corresponding elements taken from distinct columns are zero, since these sums are the scalar products of the vectors into which the vectors of the basis are transformed.

In matrix notation both these properties can be written by the single formula

$$\tilde{P}P = E.$$

Here P is the matrix of the orthogonal transformation (with respect to an orthonormal basis), and $\tilde{P}$ is its transposed matrix, i.e., the matrix whose rows are the columns of P in the same order.

For the diagonal elements of the matrix $\tilde{P}P$ are by the rule for the multiplication of matrices equal to the sum of the squares of the elements of the corresponding column of P, and the nondiagonal elements are equal to the sum of the products of the corresponding elements taken from distinct columns of P.

As an example for a degenerate transformation, we can take the orthogonal projection of all vectors of a Euclidean space onto some subspace (see §2). For in this transformation the whole space is mapped onto part of it.

Transformation of coordinates. We now consider the problem of the transformation of coordinates in n-dimensional space, i.e., the problem how the coordinates of vectors are changed on transition from one basis to another.

Let the original basis be e_1, e_2, $\cdots$, e_n and let f_1, f_2, $\cdots$, f_n be an arbitrary other basis of the space. Suppose further that

$$C = \begin{bmatrix} c_{11}, & \cdots, & c_{1n} \\ \cdots\cdots\cdots\cdots \\ c_{n1}, & \cdots, & c_{nn} \end{bmatrix}$$

is the matrix whose columns are the coordinates of the vectors of the new basis f_1, f_2, $\cdots$, f_n with respect to the original one. The matrix C is obviously nondegenerate, because the vectors f_1, f_2, $\cdots$, f_n are linearly independent. It is called the *matrix of the coordinate transformation*.

We denote by x_1, x_2, $\cdots$, x_n the coordinates of a certain vector X with respect to the basis e_1, e_2, $\cdots$, e_n and by x'_1, x'_2, $...$, x'_n the coordinates of the same vector with respect to the basis f_1, f_2, $\cdots$, f_n. Then $X = x'_1 f_1 + x'_2 f_2 + \cdots + x'_n f_n$ and therefore the coordinates of the vector X with respect to the original basis form the column

$$\begin{bmatrix} x_1 \\ x_2 \\ \cdot \\ x_n \end{bmatrix} = \begin{bmatrix} c_{11}x'_1 + c_{12}x'_2 + \cdots + c_{1n}x'_n \\ c_{21}x'_1 + c_{22}x'_2 + \cdots + c_{2n}x'_n \\ \cdots\cdots\cdots\cdots\cdots\cdots\cdots\cdots \\ c_{n1}x'_1 + c_{n2}x'_2 + \cdots + c_{nn}x'_n \end{bmatrix} = \begin{bmatrix} c_{11} \cdots c_{1n} \\ c_{21} \cdots c_{2n} \\ \cdots\cdots\cdots \\ c_{n1} \cdots c_{nn} \end{bmatrix} \begin{bmatrix} x'_1 \\ x'_2 \\ \cdot \\ x'_n \end{bmatrix}.$$

Thus, the original coordinates are expressed linearly and homogeneously in terms of the transformation with the matrix C.

The formulas that express the connection between the coordinates with respect to the original and the transformed basis coincide formally with the formulas that link the coordinates of corresponding vectors in a nondegenerate linear transformation of the space. This circumstance enables us to interpret an abstractly given linear homogeneous transformation of variables with a nondegenerate matrix either as a transformation of coordinates or as a transformation of the space. In each concrete case the choice of one of these two interpretations is determined by the context of the problem under consideration.

Let us now deal with the question how the matrix of a linear transformation of the space is changed under a coordinate transformation.

Suppose that the given linear transformation has the matrix A in the basis e_1, e_2, $\cdots$, e_n so that the column Y of the coordinates of the transformed vector is linked with the column X of the original one by the formula

$$Y = AX.$$

Suppose now that a transformation of coordinates with the matrix C is made; X', Y' denote, respectively, the columns of the coordinates of the original and the transformed vectors with respect to the new basis. Then $X = CX'$, $Y = CY'$ and hence

$$Y' = C^{-1}Y = C^{-1}AX = C^{-1}ACX'.$$

Thus, the matrix of our transformation with respect to the new basis is $C^{-1}AC$.

Two matrices A and B connected by the relation $B = C^{-1}AC$, where C is a nonsingular matrix, are called *similar*. One and the same linear transformation corresponds with respect to various bases to a class of pairwise similar matrices.

Eigenvectors and eigenvalues of a linear transformation. A very important class of linear transformations consists of the transformations that come about in the following way.

Let e_1, e_2, $\cdots$, e_n be arbitrary linearly independent vectors of the space. Suppose that under the transformation they are multiplied by certain numbers λ_1, λ_2, $\cdots$, λ_n. If the vectors e_1, e_2, $\cdots$, e_n are taken as a basis of the space, then the transformation can be described by

$$\begin{bmatrix} \lambda_1 & 0 & \cdots & 0 \\ 0 & \lambda_2 & \cdots & 0 \\ & & \cdots\cdots\cdots\cdots & \\ 0 & 0 & \cdots & \lambda_n \end{bmatrix}.$$

The transformations of this class have a simple and intuitive geometrical meaning (of course, only for real spaces and for $n = 2$ or $n = 3$). Namely, if all the numbers λ_i are positive, then the transformation that we describe consists in a stretching (or compressing) of the space in the directions of the vectors e_1, e_2, $\cdots$, e_n with coefficients $\lambda_1, \lambda_2, \cdots, \lambda_n$. If some of the λ_i are negative, then the deformation of the space is accompanied by a change of direction of some of the vectors e_1, e_2, $\cdots$, e_n into the opposite. Finally, if for example $\lambda_1 = 0$, then a projection of the space parallel to e_1 takes place onto the subspace spanned by e_2, $\cdots$, e_n with a subsequent deformation in these directions.

The class of transformations we have considered is important, because in spite of its simplicity it is very general. In fact, it can be established that every linear transformation satisfying certain not very severe restrictions belongs to our class; i.e., we can find for it a basis in which it is described by a diagonal matrix.

The restrictions to be imposed on the transformation become particularly clear if we consider linear transformation of a complex space. In what follows this will be assumed.

We introduce the following definition.

A nonzero vector X which under a linear transformation A of the space goes into a collinear vector λX is called an *eigenvector of the transformation*. In other words, a nonzero vector X is an eigenvector of the transformation A if and only if $AX = \lambda X$. The number λ is called an *eigenvalue* of the transformation A.

It is obvious that if in some basis a transformation has a diagonal matrix, then this basis consists of eigenvectors and the diagonal elements are eigenvalues. Conversely, if there exists in the space a basis consisting of eigenvectors of the transformation A, then in this basis the matrix of the transformation A is diagonal and consists of the eigenvalues corresponding to the vectors of the basis.

We now proceed to study the properties of eigenvectors and eigenvalues. With this aim we write the definition of an eigenvector in coordinate notation. Let A be the matrix corresponding to the transformation A with respect to a certain basis and X the column of coordinates of the vector X in the same basis. The equation $AX = \lambda X$ in coordinate notation is written as $AX = \lambda X$ or

$$(A - \lambda E) X = 0.$$

In expanded form this equation turns into the system

$$
\begin{aligned}
(a_{11} - \lambda) x_1 + a_{12}x_2 + \cdots + a_{1n}x_n &= 0, \\
a_{21}x_1 + (a_{22} - \lambda) x_2 + \cdots + a_{2n}x_n &= 0, \\
\cdots\cdots\cdots\cdots\cdots\cdots\cdots\cdots\cdots\cdots\cdots\cdots\cdots\cdots\cdots\cdots \\
a_{n1}x_1 + a_{n2}x_2 + \cdots + (a_{nn} - \lambda) x_n &= 0.
\end{aligned}
$$

We can regard this system of equations as a system of linear and homogeneous equations for $x_1, x_2, \cdots, x_n$. We are interested in the case when this system has a nontrivial solution, because the coordinates of an eigenvector must not all be equal to zero. Now we know that a necessary and sufficient condition for the existence of a nontrivial solution of a system of linear homogeneous equations is that the rank of the coefficient matrix should be less than the number of unknowns, and this is equivalent to the vanishing of the determinant of the system

$$
\begin{vmatrix}
a_{11} - \lambda & & \cdots & a_{1n} \\
a_{21} & a_{22} - \lambda & \cdots & a_{2n} \\
\cdots\cdots\cdots\cdots\cdots\cdots\cdots\cdots\cdots\cdots \\
a_{n1} & a_{n2} & \cdots & a_{nn} - \lambda
\end{vmatrix} = 0.
$$

Thus, all eigenvalues of the transformation A are roots of the polynomial $| A - \lambda E |$ and, conversely, every root of this polynomial is an eigenvalue of the transformation since to every root there corresponds at least one eigenvector. The polynomial $| A - \lambda E |$ is called the *characteristic polynomial* of the matrix A. The equation $| A - \lambda E | = 0$ is called the *characteristic* or *secular equation* and its roots *characteristic numbers* of the matrix.*

By the fundamental theorem of higher algebra (Chapter IV), every polynomial has at least one root; therefore every linear transformation has at least one eigenvalue and hence at least one eigenvector. But, of course, it is quite possible that even in the case when the transformation can be expressed by a real matrix it turns out that all or some of its eigenvalues are complex. Consequently, the theorem on the existence of (real) eigenvalues and eigenvectors for an arbitrary linear transformation is not true in a real space. For example, the transformation of the plane that consists in a rotation around the origin of coordinates by any angle other than 180° changes the directions of all the vectors of the plane so that there are no eigenvectors for this transformation.

The roots of the characteristic polynomial of a matrix A are the eigenvalues of the transformation A; therefore matrices that correspond to one and the same transformation in distinct bases have identical sets of roots of the characteristic polynomial. This leads to the plausible asssertion that the characteristic polynomial of a linear transformation also depends on the transformation only and not on the choice of a basis. This can be verified by the following elegant calculation, which is based on the properties of operations on matrices and determinants.

We know that if a matrix A corresponds to a transformation A in some basis, then in any other basis the transformation A has a similar matrix $C^{-1}AC$, where C is some nonsingular matrix. But

$$| C^{-1} AC - \lambda E | = | C^{-1}AC - C^{-1}\lambda EC | = | C^{-1}(A - \lambda E) C |$$
$$= | C^{-1} | | C | | A - \lambda E | = | C^{-1}C | | A - \lambda E | = | A - \lambda E |.$$

Thus, matrices corresponding to one and the same transformation A in distinct bases have in fact one and the same characteristic polynomial, which can therefore be called the polynomial of the transformation.

We shall now make the assumption that all the eigenvalues of the transformation A are distinct. Let us prove that the eigenvectors, one for each eigenvalue, are linearly independent. For if we suppose that some

---

* The name "secular equation" has arisen in celestial mechanics in connection with the problem of the so-called secular disturbances in the motions of the planets.

of them, say $e_1, \cdots, e_k$, are linearly independent and the remaining ones, among them e_{k+1}, are linear combinations of these, then

$$e_{k+1} = c_1 e_1 + c_2 e_2 + \cdots + c_k e_k. \tag{13}$$

When we apply the linear transformation to both sides of this equation, we obtain

$$A e_{k+1} = c_1 A e_1 + c_2 A e_2 + \cdots + c_k A e_k,$$

from which it follows by the definition of an eigenvector that

$$\lambda_{k+1} e_{k+1} = c_1 \lambda_1 e_1 + c_2 \lambda_2 e_2 + \cdots + c_k \lambda_k e_k.$$

When we multiply equation (13) by λ_{k+1} and subtract from it the equation just obtained, we have

$$c_1(\lambda_{k+1} - \lambda_1)\, e_1 + c_2(\lambda_{k+1} - \lambda_2)\, e_2 + \cdots + c_k(\lambda_{k+1} - \lambda_k)\, e_k = 0,$$

hence it follows by the linear independence of $e_1, e_2, \cdots, e_k$ that

$$c_1(\lambda_{k+1} - \lambda_1) = c_2(\lambda_{k+1} - \lambda_2) = \cdots = c_k(\lambda_{k+1} - \lambda_k) = 0.$$

But we had assumed that all the eigenvalues are distinct and the vectors are chosen one for each eigenvalue. Therefore $\lambda_{k+1} - \lambda_1 \neq 0$, $\lambda_{k+1} - \lambda_2 \neq 0, \cdots, \lambda_{k+1} - \lambda_k \neq 0$ and the equation (13) is impossible, since the coefficients $c_1, c_2, \cdots, c_k$ cannot all be zero.

Now it is clear that if all the eigenvalues of a linear transformation are distinct, then there exists a basis in which the matrix of the transformation has diagonal form. For we can choose as such a basis a system of eigenvectors, one for each eigenvalue. As we have shown, they are linearly independent and their number is equal to the dimension of the space; i.e., they do in fact form a basis.

The theorem we have proved can be stated in terms of the theory of matrices as follows. If all the eigenvalues of a matrix are distinct, then the matrix is similar to the diagonal matrix whose diagonal elements are these eigenvalues.

The problem of transforming the matrix of a linear transformation to its simplest form is considerably more complicated if there are equal ones among the roots of the characteristic polynomial. We shall confine ourselves to a short account of the final result.

A "canonical box" of order m is defined as a matrix of the form

$$I_{m, \lambda_i} = \begin{bmatrix} \lambda_i & 1 & & & \\ & \lambda_i & 1 & & \\ & & \ddots & \ddots & \\ & & & \ddots & 1 \\ & & & & \lambda_i \end{bmatrix}.$$

All the unnamed elements are equal to zero.

A canonical Jordan matrix is defined as a matrix in which there are "canonical boxes" along the main diagonal and all the remaining elements are zero:

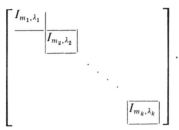

The numbers λ_i in the distinct "boxes" are not necessarily pairwise distinct. Every matrix can be reduced to a canonical Jordan matrix similar to it. The proof of this theorem is rather complicated. We ought to mention that this theorem plays an important role in many applications of algebra to other problems of mathematics, in particular in the theory of systems of linear differential equations.

A matrix can be reduced to diagonal form if and only if the orders m_i of all boxes are equal to 1.

§5. Quadratic Forms

Definition and simplest properties.   A *quadratic form* is a homogeneous polynomial of degree 2 in several variables.

A quadratic form in n variables x_1, x_2, $\cdots$, x_n consists of terms of two types: squares of the variables and products of two variables, both with certain coefficients. A quadratic form can be written in the following quadratic scheme:

$$
\begin{aligned}
f(x_1, x_2, \cdots, x_n) = \quad & a_{11}x_1^2 \; + a_{12}x_1x_2 + \cdots + a_{1n}x_1x_n \\
+ \; & a_{21}x_2x_1 + \quad a_{22}x_2^2 \; + \cdots + a_{2n}x_2x_n \\
& \cdots\cdots\cdots\cdots\cdots\cdots\cdots\cdots\cdots\cdots\cdots\cdots\cdots\cdots\cdots \\
+ \; & a_{n1}x_nx_1 + a_{n2}x_nx_2 + \cdots + a_{nn}x_n^2.
\end{aligned}
$$

Pairs of similar terms $a_{12}x_1x_2$ and $a_{21}x_2x_1$, etc., are written with equal coefficients so that each of them gets half the coefficient of the corresponding product of the variables. Thus, every quadratic form is uniquely connected with its coefficient matrix, which is symmetric.

A quadratic form can conveniently be represented in the following matrix notation. We denote by X the column of the variables x_1, x_2, $\cdots$, x_n by $\bar{X}$ the row $(x_1, x_2, \cdots, x_n)$, i.e., the transposed matrix of X. Then

$$
\begin{aligned}
f(x_1, x_2, \cdots, x_n) &= x_1(a_{11}x_1 + a_{12}x_2 + \cdots + a_{1n}x_n) \\
&\quad + x_2(a_{21}x_1 + a_{22}x_2 + \cdots + a_{2n}x_n) + \cdots \\
&\quad + x_n(a_{n1}x_1 + a_{n2}x_2 + \cdots + a_{nn}x_n)
\end{aligned}
$$

$$
= (x_1, x_2, \cdots, x_n)
\begin{bmatrix}
a_{11}x_1 + a_{12}x_2 + \cdots + a_{1n}x_n \\
a_{21}x_1 + a_{22}x_2 + \cdots + a_{2n}x_n \\
\cdots\cdots\cdots\cdots\cdots\cdots\cdots\cdots\cdots \\
a_{n1}x_1 + a_{n2}x_2 + \cdots + a_{nn}x_n
\end{bmatrix}
$$

$$
= (x_1, x_2, \cdots, x_n)
\begin{bmatrix}
a_{11}a_{12} \cdots a_{1n} \\
a_{21}a_{22} \cdots a_{2n} \\
\cdots\cdots\cdots\cdots \\
a_{n1}a_{n2} \cdots a_{nn}
\end{bmatrix}
\begin{bmatrix}
x_1 \\
x_2 \\
\cdot \\
\cdot \\
x_n
\end{bmatrix}
= \bar{X}AX.
$$

Quadratic forms occur in many branches of mathematics and its applications.

In the theory of numbers and in crystallography, we consider quadratic forms under the proviso that the variables $x_1, x_2, \cdots, x_n$ assume only integral values. In analytical geometry a quadratic form arises in setting up the equation of a curve (or surface) of the second order. In mechanics and physics a quadratic form appears in the expression for the kinetic energy of a system in terms of the components of the generalized velocities, etc. Furthermore, a study of quadratic forms is necessary even in analysis for the investigation of functions of several variables in connection with problems where it is important to clarify how a given function in the neighborhood of a given point deviates from its approximating linear function. An example of problems of this type is the investigation of maxima and minima of a function.

Let us consider, for example, the problem of finding the maxima and minima of a function of two variables $w = f(x, y)$ having continuous partial derivatives of the third order. A necessary condition for the point (x_0, y_0) to give a maximum or minimum of the function w is that the partial derivatives of the first order should vanish at the point (x_0, y_0). Let us assume that this condition is satisfied. We give to the variables x and y small increments h and k and consider the corresponding increment of the function $\Delta w = f(x_0 + h, y_0 + k) - f(x_0, y_0)$. By Taylor's formula this increment is equal, to within terms of higher order of smallness, to the quadratic form $\frac{1}{2}(rh^2 + 2shk + tk^2)$, where r, s and t are the values of the second derivatives $\partial^2 w/\partial x^2$, $\partial^2 w/\partial x \partial y$, $\partial^2 w/\partial y^2$, computed at the point (x_0, y_0). If this quadratic form is positive for all values of h and k (except

$h = k = 0$), then the function w has a minimum at the point (x_0, y_0); if it is negative, then a maximum. Finally, if the form assumes positive as well as negative values, then there is neither a maximum nor a minimum. Functions with a larger number of variables can be investigated in a similar way.

The study of quadratic forms essentially consists in the investigation of the problem of the equivalence of forms under one set or another of linear transformations of the variables. Two quadratic forms are called *equivalent* if one of them can be carried into the other by one of the transformations of the given set. Closely connected with the problem of equivalence is the problem of *reduction* of a form, i.e., its transformation into a certain form, as simple as possible.

In various problems connected with quadratic forms, we consider various sets of admissible transformations of the variables.

In problems of analysis arbitrary nonsingular transformations of the variables are admitted; for the purposes of analytical geometry, the greatest interest lies in orthogonal transformations, i.e., those that correspond to the transition from one system of variable Cartesian coordinates to another. Finally, in the theory of numbers and in crystallography, we consider linear transformations with integer coefficients and with a determinant equal to 1.

Here we shall consider two of these problems: the problem of the reduction of a quadratic form to the simplest possible form by means of arbitrary nonsingular transformations and the same problem for orthogonal transformations. Above all, we shall find out how the matrix of the quadratic form is changed under a linear transformation of the variables.

Suppose that $f(x_1, x_2, \cdots, x_n) = \bar{X}AX$, where A is the symmetric matrix of the coefficients of the form and X is the column of the variables.

We take a linear transformation of the variables, writing it briefly $X = CX'$. Here C denotes the coefficient matrix of this transformation and X' the column of the new variables. Then $\bar{X} = \bar{X}'\bar{C}$ and consequently $\bar{X}AX = \bar{X}'(\bar{C}AC)X'$ so that the matrix of the transformed quadratic form is $\bar{C}AC$.

The matrix $\bar{C}AC$ is automatically symmetric, as is easy to verify. Thus, the problem of reducing a quadratic form to the simplest possible form is equivalent to the task of reducing a symmetric matrix to the simplest form by multiplying it on the left and on the right by transposed matrices.

Transformation of a quadratic form to the canonical form by successive completion of squares. We shall show that every (real) quadratic form can be reduced to a sum of squares of new variables with certain coefficients by a real nonsingular linear transformation.

To prove this we shall show first of all that if the form is not identically zero, then we can make the coefficient of the square of the first variable different from zero by the application of a nonsingular transformation of the variables.

For suppose that

$$
\begin{aligned}
f(x_1, x_2, \cdots, x_n) = \quad & a_{11}x_1^2 + a_{12}x_1x_2 + \cdots + a_{1n}x_1x_n \\
+ \; & a_{21}x_2x_1 + \; a_{22}x_2^2 + \cdots + a_{2n}x_2x_n \\
& \cdots\cdots\cdots\cdots\cdots\cdots\cdots\cdots\cdots\cdots\cdots\cdots\cdots\cdots\cdots \\
+ \; & a_{n1}x_nx_1 + a_{n2}x_nx_2 + \cdots + \; a_{nn}x_n^2 .
\end{aligned}
$$

If $a_{11} \neq 0$, then no transformation is required. If $a_{11} = 0$, but some one of the diagonal coefficients $a_{kk} \neq 0$, then we set $x_1 = x_k', x_k = x_1'$ equating the remaining original variables to the corresponding new ones. This nonsingular transformation achieves our object. Finally, if all the diagonal coefficients are equal to zero, then at least one of the nondiagonal coefficients is different from zero, for example a_{12}.

By taking the nonsingular transformation

$$
\begin{aligned}
x_1 &= x_1', \\
x_2 &= x_1' + x_2'
\end{aligned}
$$

and equating the remaining original variables to the new ones, we achieve our aim.

Thus, without loss of generality we can assume that $a_{11} \neq 0$.

Let us now separate out the square of a linear function such that all the terms containing x_1 occur in this square.

This can easily be done. For

$$
\begin{aligned}
f(x_1, x_2, \cdots, x_n) = \quad & a_{11}x_1^2 + a_{12}x_1x_2 + \cdots + a_{1n}x_1x_n \\
+ \; & a_{21}x_2x_1 + \; a_{22}x_2^2 + \cdots + a_{2n}x_2x_n \\
& \cdots\cdots\cdots\cdots\cdots\cdots\cdots\cdots\cdots\cdots\cdots\cdots\cdots\cdots\cdots \\
+ \; & a_{n1}x_nx_1 + a_{n2}x_nx_2 + \cdots + \; a_{nn}x_n^2 \\
= \; a_{11} & \left[x_1 + \frac{a_{12}}{a_{11}}x_2 + \cdots + \frac{a_{1n}}{a_{11}}x_n \right]^2 - a_{11}\left[\frac{a_{12}}{a_{11}}x_2 + \cdots + \frac{a_{1n}}{a_{11}}x_n \right]^2 \\
+ \; & a_{22}x_2^2 + \cdots + a_{2n}x_2x_n \\
& \cdots\cdots\cdots\cdots\cdots\cdots\cdots\cdots\cdots\cdots\cdots\cdots\cdots \\
+ & a_{n2}x_nx_2 + \cdots + a_{nn}x_n^2.
\end{aligned}
$$

By removing the parentheses in the second summand and collecting similar terms, we obtain

$$f(x_1, x_2, \cdots, x_n) = a_{11}\left[x_1 + \frac{a_{12}}{a_{11}} x_2 + \cdots + \frac{a_{1n}}{a_{11}} x_n\right]^2 + f_1(x_2, \cdots, x_n),$$

where f_1 is a form in $n - 1$ variables.

The transformation

$$x_1 + \frac{a_{12}}{a_{11}} x_2 + \cdots + \frac{a_{1n}}{a_{11}} x_n = x_1',$$
$$x_2 = x_2',$$
$$\cdots\cdots\cdots\cdots\cdots$$
$$x_n = x_n',$$

is obviously nonsingular. By making this transformation we reduce our form to

$$a_{11}x_1'^2 + f_1(x_2', \cdots, x_n').$$

Continuing the process in a similar manner we reduce the form to the required "canonical" form

$$\alpha_1 z_1^2 + \alpha_2 z_2^2 + \cdots + \alpha_n z_n^2.$$

Here $z_1, z_2, \cdots, z_n$ are the new variables introduced at the last step.

The law of inertia of quadratic forms. In the reduction of a quadratic form to canonical form there always is a considerable arbitrariness in the choice of the transformation of variables that brings about this reduction. This arbitrariness comes in, among other things, from the fact that we can precede the previous method of successive separation of squares by an arbitrary nonsingular transformation of the variables.

However, notwithstanding this arbitrariness, as a result of the reduction of the given form we obtain an almost unique canonical quadratic form independent of the choice of the reducing transformation. Namely the number of squares of the new variables that occur with positive, negative, and zero coefficients is always one and the same, irrespective of the method of reduction. This theorem is known as the *law of inertia* of quadratic forms. We shall not prove it here.

The law of interia of a quadratic form solves the problem of equivalence of a real quadratic form under all nonsingular transformations. For two forms are equivalent if and only if their reduction to canonical form leads to forms with the same number of squares with positive, negative, and zero coefficients.

Of special interest for the applications are the quadratic forms that under reduction to canonical form turn into a sum of squares of the new variables with all the coefficients *positive*. Such forms are called *positive definite*.

Positive-definite quadratic forms are characterized by the property that their values for real values of the variables not all equal to zero are always positive.

Orthogonal transformation of quadratic forms to canonical form. Of special interest among all possible methods of reducing a quadratic form to canonical form are the orthogonal transformations, i.e., those that can be obtained by a linear transformation of the variables with an orthogonal matrix. Such transformations are of interest, for example, in analytical geometry, in the problem of reducing the general equation of a curve or surface of the second order to canonical form.

In order to convince ourselves of the possibility of such a transformation, it is convenient to regard the quadratic form as a function of vectors in a Euclidean space by considering the variables x_1, x_2, $\cdots$, x_n as the coordinates of a variable vector in an orthonormal basis. Then an orthogonal transformation of the variables can be interpreted as a transition from one orthonormal basis to another.

With the quadratic form

$$f(x_1, x_2, \cdots, x_n) = a_{11}x_1^2 + \cdots + a_{1n}x_1x_n$$
$$\cdots\cdots\cdots\cdots\cdots\cdots\cdots\cdots\cdots$$
$$+ a_{n1}x_nx_1 + \cdots + a_{nn}x_n^2$$

we connect the linear transformation A that has with respect to the chosen basis the matrix

$$A = \begin{bmatrix} a_{11} \cdots a_{1n} \\ \cdots\cdots\cdots \\ a_{n1} \cdots a_{nn} \end{bmatrix}.$$

Then our quadratic form can be regarded as the scalar product $AX \cdot X$ (where X is the vector with the coordinates x_1, x_2, $\cdots$, x_n), and its coefficients a_{ij} are the scalar product $Ae_i \cdot e_j$, where e_1, e_2, $\cdots$, e_n is the chosen orthonormal basis.

It is easy to see that in consequence of the symmetry of the matrix A the following equation holds for arbitrary vectors X and Y

$$AX \cdot Y = X \cdot AY.$$

We shall show first of all that the transformation A has at least one real eigenvalue and eigenvector corresponding to it.

For this purpose we consider the values of the form $AX \cdot X$ under the assumption that the vector X ranges over the unit sphere, i.e., the set of all unit vectors. Under these conditions the form $AX \cdot X$ will have a maximum. Let us show that this maximum $AX \cdot X$ is an eigenvalue of the tranformation A and the vector X_0 for which this maximum is assumed is a corresponding eigenvector; i.e., $AX_0 = \lambda_1 X_0$.

The proof of this statement is by an indirect method, namely, by establishing that the vector AX_0 is orthogonal to all vectors orthogonal to X_0.

We note that for an arbitrary vector Z we have the inequality $AZ \cdot Z \leqslant \lambda_1 \mid Z \mid^2$. This is obvious from the fact that $X = Z/\mid Z \mid$ is a unit vector and λ_1 the maximum of the values of the form $AX \cdot X$ on the unit sphere. We consider $Z = X_0 + \epsilon Y$, where ϵ is a real number and Y an arbitrary vector orthogonal to the vector X_0. Then

$$AZ \cdot Z = (AX_0 + \epsilon A Y) \cdot (X_0 + \epsilon Y) = AX_0 \cdot X_0 + 2\epsilon AX_0 \cdot Y + \epsilon^2 A Y \cdot Y$$
$$= \lambda_1 + 2\epsilon AX_0 \cdot Y + \epsilon^2 A Y \cdot Y.$$

Moreover,

$$\mid Z \mid^2 = (X_0 + \epsilon Y) \cdot (X_0 + \epsilon Y) = \mid X_0 \mid^2 + \epsilon^2 \mid Y \mid^2 = 1 + \epsilon^2 \mid Y \mid^2,$$

because

$$X_0 \cdot Y = 0, \quad \mid X_0 \mid^2 = 1.$$

Therefore,

$$\lambda_1 + 2\epsilon AX_0 \cdot Y + \epsilon^2 A Y \cdot Y \leqslant \lambda_1 + \epsilon^2 \lambda_1 \mid Y \mid^2,$$

and from this we obtain on dividing by ϵ^2

$$\frac{2}{\epsilon} AX_0 \cdot Y < \lambda_1 \mid Y \mid^2 - A Y \cdot Y. \tag{14}$$

This last inequality must be satisfied for arbitrary real ϵ of sufficiently small absolute value.

But it can only be satisfied under the condition $AX_0 \cdot Y = 0$, because if $AX_0 \cdot Y > 0$, then the inequality (14) is impossible for sufficiently small positive ϵ, and if $AX_0 \cdot Y < 0$, then it is impossible for negative ϵ of sufficiently small absolute value. Thus, $AX_0 \cdot Y = 0$, i.e., AX_0 is in fact orthogonal to every vector orthogonal to X_0. Therefore AX_0 and X_0 are collinear; i.e., $AX_0 = \lambda' X_0$, where λ' is a real number. The fact that $\lambda' = \lambda_1$ is easy to verify, for

$$\lambda_1 = AX_0 \cdot X_0 = \lambda' X_0 \cdot X_0 = \lambda'.$$

Now it is easy to show that every quadratic form can in fact be reduced to canonical form by an orthogonal transformation.

Let $e_1, e_2, \cdots, e_n$ be the original orthonormal basis of the space and $f_1, f_2, \cdots, f_n$ a new orthonormal basis in which the first vector f_1 is an eigenvector X_0 of the transformation A. Let $x_1, x_2, \cdots, x_n$ be the coordinates of the vector X in the original basis and $x'_1, x'_2, \cdots, x'_n$ its coordinates in the new basis. Then

$$\begin{bmatrix} x_1 \\ x_2 \\ \vdots \\ x_n \end{bmatrix} = P \begin{bmatrix} x'_1 \\ x'_2 \\ \vdots \\ x'_n \end{bmatrix}$$

where P is an orthogonal matrix.

Let us carry out the transition to the new variables in the quadratic form $AX \cdot X$. In the new variables the quadratic form has the coefficients $a'_{ij} = Af_i \cdot f_j$. Therefore

$$a'_{11} = Af_1 \cdot f_1 = \lambda_1 f_1 \cdot f_1 = \lambda_1,$$

$$a'_{1j} = a'_{j1} = Af_1 \cdot f_j = \lambda_1 f_1 \cdot f_j = 0 \quad \text{for} \quad j \neq 1;$$

i.e., the form is now

$$\lambda_1 x'^2_1 + \phi(x'_2, \cdots, x'_n).$$

Thus, by means of an orthogonal transformation we have succeeded in separating out one square of a new variable.

By repeating the same arguments with the new form $\phi(x'_2, \cdots, x'_n)$, etc., we eventually arrive at the conclusion that the form turns out to be reduced to canonical form by a chain of orthogonal transformations. But it is obvious that a chain of orthogonal transformations is equivalent to a single orthogonal transformation. This concludes the proof of the theorem.

§6. Functions of Matrices and Some of Their Applications

Functions of matrices. The applications of linear algebra to other branches of mathematics are very numerous and diverse. It is not an exaggeration to say that the ideas and results of linear algebra are used in a large part of contemporary mathematics and theoretical physics in one form or another, particularly in the form of matrix calculus.

We shall deal briefly with one of the methods of applying the matrix calculus to the theory of ordinary differential equations. Here functions of matrices play an important role.

First of all we define the powers of a square matrix A. We set $A^0 = E$, $A^1 = A$, $A^2 = AA$, $A^3 = A^2A$, $A^4 = A^3A$, etc. By the associative law it is easy to show that $A^m A^n = A^{m+n}$ for arbitrary natural numbers m and n. The operations of addition and of multiplication by a number have been defined for matrices. This enables us to define in a natural way the meaning of a polynomial (in one variable) of a matrix. For if $\phi(x) = a_0 x^n + a_1 x^{n-1} + \cdots + a_n$, then we set (by definition) $\phi(A) = a_0 A^n + a_1 A^{n+1} + \cdots + a_n E$. Thus, the concept of the simplest function of a matrix argument, namely the polynomial, is defined.

By means of a limit process it is easy to generalize the concept of a function of a matrix argument to a considerably wider class of functions than polynomials in one variable. Without treating this problem in all its generality we shall restrict ourselves here to the examination of analytic functions.

First of all we introduce the concept of the limit of a sequence of matrices. A sequence of matrices

$$
A_1 = \begin{bmatrix} a_{11}^{(1)} \cdots a_{1n}^{(1)} \\ \cdots\cdots\cdots\cdots \\ a_{n1}^{(1)} \cdots a_{nn}^{(1)} \end{bmatrix}, \quad
A_2 = \begin{bmatrix} a_{11}^{(2)} \cdots a_{1n}^{(2)} \\ \cdots\cdots\cdots\cdots \\ a_{n1}^{(2)} \cdots a_{nn}^{(2)} \end{bmatrix}, \cdots
$$

is said to *converge* to the matrix

$$
A = \begin{bmatrix} a_{11} \cdots a_{1n} \\ \cdots\cdots\cdots \\ a_{n1} \cdots a_{nn} \end{bmatrix}
$$

(or to have the matrix A as its limit) if $\lim_{k \to \infty} a_{ij}^{(k)} = a_{ij}$ for all i, j. Further, the sum of a series $A_1 + A_2 + \cdots + A_k + \cdots$ is defined as the limit of the sums of its segments $\lim_{k \to \infty} (A_1 + A_2 + \cdots + A_k)$ if this limit exists.

Let $f(z)$ be an analytic function regular in a neighborhood of $z = 0$. Then $f(z)$ can be expanded in a power series

$$
f(z) = a_0 + a_1 z + a_2 z^2 + \cdots + a_k z^k + \cdots.
$$

For every square matrix A it is natural to set

$$
f(A) = a_0 E + a_1 A + a_2 A^2 + \cdots + a_k A^k + \cdots.
$$

Now it turns out that this series converges for all matrices A whose eigenvalues lie within the circle of convergence of the power series $a_0 + a_1 z + \cdots + a_k z^k + \cdots$.

Of interest for the applications are the elementary functions of matrices.

For example, the geometric series $E + A + A^2 + \cdots + A^k + \cdots$ is convergent for matrices whose eigenvalues are of absolute value less than 1, and the sum of the series is the matrix $(E - A)^{-1}$, in complete accordance with the formula

$$1 + x + \cdots + x^k + \cdots = \frac{1}{1 - x}.$$

The representation of $(E - A)^{-1}$ in the form of an infinite series gives an effective method for the approximate solution of systems of linear equations whose coefficient matrix is close to the unit matrix.

For when we write such a system in the form

$$(E - A)\, X = B,$$

we obtain

$$X = (E - A)^{-1}\, B = B + AB + A^2B + \cdots \tag{15}$$

and this gives a convenient formula for the solution of the system, if only the series (15) converges sufficiently fast.

It is useful to consider the binomial series

$$(E + A)^m = E + \frac{m}{1}\, A + \frac{m(m - 1)}{2!}\, A^2 + \cdots$$

which can be applied (provided the eigenvalues of A are less than 1 in modulus) not only for natural exponents m but also for fractional and negative exponents.

Particularly important for the applications is the exponential matrix function

$$e^A = E + A + \frac{A^2}{2!} + \frac{A^3}{3!} + \cdots.$$

The series defining the exponential function converges for every matrix A. The exponential matrix function has properties reminding us of properties of the ordinary exponential function. For example, if A and B commute under multiplication, i.e., $AB = BA$, then $e^{A+B} = e^A \cdot e^B$. However, for noncommuting A and B the formula ceases to be true.

Application to the theory of systems of ordinary linear differential equations. In the theory of systems of ordinary linear differential equations it is appropriate to consider matrices whose elements are functions of an independent variable:

$$U(t) = \begin{bmatrix} a_{11}(t) \cdots a_{1n}(t) \\ \cdots\cdots\cdots\cdots\cdots \\ a_{m1}(t) \cdots a_{mn}(t) \end{bmatrix}.$$

For such matrices the concept of the derivative with respect to the argument t is defined in a natural way, namely:

$$\frac{dU(t)}{dt} = \begin{bmatrix} a'_{11}(t) \cdots a'_{1n}(t) \\ \cdots\cdots\cdots\cdots\cdots \\ a'_{m1}(t) \cdots a'_{mn}(t) \end{bmatrix}.$$

It is not difficult to verify that some elementary formulas of differentiation are valid for matrices. For example,

$$\frac{d(U + V)}{dt} = \frac{dU}{dt} + \frac{dV}{dt},$$

$$\frac{d(cU)}{dt} = c\,\frac{dU}{dt},$$

$$\frac{d(UV)}{dt} = \frac{dU}{dt}\,V + U\,\frac{dV}{dt}.$$

(The multiplication must be carried out strictly in the order as given in the formula!)

A system of ordinary linear homogeneous differential equations

$$\frac{dy_1}{dt} = a_{11}(t)\,y_1 + a_{12}(t)\,y_2 + \cdots + a_{1n}(t)\,y_n,$$

$$\frac{dy_2}{dt} = a_{21}(t)\,y_1 + a_{22}(t)\,y_2 + \cdots + a_{2n}(t)\,y_n$$

$$\cdots\cdots\cdots\cdots\cdots\cdots\cdots\cdots\cdots\cdots\cdots\cdots\cdots\cdots\cdots$$

$$\frac{dy_n}{dt} = a_{n1}(t)\,y_1 + a_{n2}(t)\,y_2 + \cdots + a_{nn}(t)\,y_n$$

can be written in this notation in the form

$$\frac{dY}{dt} = A(t)\,Y,$$

where

$$Y = \begin{bmatrix} y_1 \\ \vdots \\ y_n \end{bmatrix}, \qquad A(t) = \begin{bmatrix} a_{11}(t) \cdots a_{1n}(t) \\ \cdots\cdots\cdots\cdots\cdots \\ a_{n1}(t) \cdots a_{nn}(t) \end{bmatrix},$$

i.e, in a form similar to a single linear homogeneous differential equation.

If the coefficients of the system are constants, i.e., if the matrix A is constant, then the solution of the system also looks outwardly like the solution of the equation $y' = ay$. For in this case $Y = e^{At}C$, where C is a column of arbitrary constants.

The solution in this form is very convenient for computations. The fact is that for an arbitrary analytic function $f(z)$ we have the equation

$$f(B^{-1}LB) = B^{-1}f(L)\,B.$$

Since every matrix can be reduced to the canonical Jordan form (see §4), the computation of a function of an arbitrary matrix reduces to the computation of a function of a canonical matrix, which is easy to carry out. Therefore, if $A = B^{-1}LB$, where L is a canonical matrix, then

$$Y = e^{At}C = B^{-1}e^{Lt}BC = B^{-1}e^{Lt}C',$$

where $C' = BC$ is a column of arbitrary constants.

From this formula it is easy to derive an explicit expression for all the components of the required column Y.

The Soviet mathematician I. A. Lappo-Danilevskiĭ has successfully developed the apparatus of the theory of matrix functions and was the first to apply it to the investigation of systems of differential equations, including those with variable coefficients. His results count among the most brilliant achievements of mathematics in the last fifty years.

Suggested Reading

G. Birkhoff and S. MacLane, *A survey of modern algebra*, Macmillan, New York, 1941.

V. N. Faddeeva, *Computational methods of linear algebra*. Translated by C. D. Benster, Dover, New York, 1959.

F. R. Gantmacher, *Applications of the theory of matrices*. Translated by J. L. Brenner, Interscience, New York, 1959.

——, *The theory of matrices*, Vols. 1, 2. Translated by K. A. Hirsch, Chelsea New York, 1959.

[This book and the one above are translations from the Russian original, published in 1953, but with various additions, some of which were supplied by the author. The two volumes contain a great deal of material not available in other texts.]

I. M. Gel'fand, *Lectures on linear algebra*, Interscience, New York, 1961.

F. Klein, *Elementary mathematics from an advanced standpoint*. Vol. I. *Arithmetic, algebra, analysis*. Translated by E. R. Hedrick and C. A. Noble, Dover, New York, 1953.

O. Schreier and E. Sperner, *Introduction to modern algebra and matrix theory*, Chelsea, New York, 1951.

V. I. Smirnov, *Linear algebra and group theory*. Translated and revised by R. A. Silverman, McGraw-Hill, New York, 1961.

H. W. Turnbull and A. C. Aitken, *An introduction to the theory of canonical matrices*, Dover, New York, 1961.

B. L. van der Waerden, *Modern algebra*, Vol. I. Translated from the second revised German edition by Fred Blum. With revisions and additions by the author. Frederick Ungar, New York, 1949.

CHAPTER XVII

NON-EUCLIDEAN
GEOMETRY

Ever since N. I. Lobačevskiǐ first demonstrated the possibility of a non-Euclidean geometry and put forward a new notion concerning the relationship of geometry to the material reality, the scope of geometry, its methods and applications, have been enlarged exceedingly. Nowadays mathematicians study several "spaces": apart from the Euclidean space they deal with Lobačevskiǐ space, projective space, various n-dimensional and even infinite-dimensional spaces, Riemannian, topological, and other spaces; the number of such spaces is boundless and each of them has its own properties, its own "geometry." In physics we use the concept of the so-called phase and configuration "spaces"; the theory of relativity utilizes the notion of a curved space and other results of abstract geometrical theories.

How and whence have these mathematical abstractions arisen? What real basis, what real value and application do they have? What is their relation to reality? How are they defined and how are they studied in mathematics? What is the significance of the general ideas of contemporary geometry in mathematics?

These questions will be answered in the present chapter. We shall not give an account of the theory of abstract mathematical spaces as such; that would require a far longer explanation and far more attention to the specific mathematical apparatus. Our task is to throw light on the essence of the new ideas in geometry, i.e., to answer the questions raised here, and this can be done without complicated proofs and formulas.

The history of our problem goes right back to Euclid's "Elements" and the axiom or, as one also says, the postulate on parallel lines.

§1. History of Euclid's Postulate

In his "Elements" Euclid formulates the fundamental premises of geometry in the form of so-called postulates and axioms. Among these there is Postulate V (in other copies of the "Elements" Axiom XI) which is now usually stated as follows: "Through a point not lying on a given line not more than one line parallel to the given line can be drawn." We recall that a line is called parallel to a given line if both lie in one plane and do not intersect; we think here of the infinite lines, so to speak, not of their finite segments.

It is easy to prove that we can always draw at least one parallel to a given line a through a point A not lying on it.

For let us drop a perpendicular b from A to a and draw through A a line c perpendicular to b (figure 1). The figure so obtained is completely

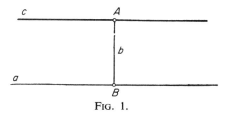

FIG. 1.

symmetric with respect to b, since the angles formed by b with a and c at both its ends are equal. Therefore, when we turn the plane about b, we bring the half lines of a and c into coincidence. Hence it is clear that if a and c were to intersect on one side of AB, then they would have to intersect also on the other side. It would then come out that the lines a and c have two points in common; but this is impossible, because by a fundamental property of the line only one line can be drawn through two points (so that lines having two common points must necessarily coincide).

Thus, from the basic properties of the line and of the motion of a figure (inasmuch as a turn about the line AB is a rotation of a semiplane with this line as axis) it follows that at least one parallel to a given line can always be drawn through a given point. Now Euclid's postulate supplements this result by the statement that this parallel must be the only one and no other can exist.

Among the other postulates (axioms) of geometry this one occupies a somewhat special place. Euclid's own formulation of it is rather complicated, but even in its usual form cited here it contains a certain difficulty. This difficulty is already inherent in the very concept of parallel lines, here we deal with the whole line. But how are we to convince ourselves that two given lines are parallel? For this purpose we ought to produce them on both sides "to infinity" and convince ourselves that they do not intersect anywhere *on their whole infinite extent*. This notion clearly has its difficulties. And all this was apparently the reason why the parallel postulate occupied a somewhat special position even for Euclid: In his "Elements" this postulate is used beginning with the 29th proposition only, whereas in the first 28 propositions he dispenses with it. In view of the complicated nature of the postulate the wish to do without it is quite natural, and already in antiquity attempts were made to change the definition of parallel lines, to change the formulation of the postulate itself or, better still, to deduce it as a theorem from other axioms and basic concepts of geometry.

Thus, the theory of parallel lines, founded on the Fifth Postulate, became an object of comment and criticism in the works of many geometers, from the days of antiquity. In the course of these investigations it was the principal aim to get rid of the Fifth Postulate altogether, by deducing it as a theorem from other basic propositions of geometry.

This task attracted many geometers: the Greek Proclus (5th century A. D.) who wrote a commentary to Euclid, the Persian Nasir ed Din et Tusi* (13th century), the Englishman Wallis (1616-1703), the Italian Saccheri (1667-1733), the German philosopher and mathematician Lambert (1728-1777), the Frenchman Legendre (1752-1833), and many others; in the span of more than two thousand years since Euclid's "Elements" appeared they all outdid themselves in subtlety and geometrical ingenuity trying to prove the Fifth Postulate.

However, the result of these attempts invariably remained negative. Every time it became clear that the author of one proof or another had in fact relied on some proposition, no matter how obvious, that did not at all follow with logical necessity from the other premises of geometry. In other words, each time what they had done amounted to replacing the Fifth Postulate by some other statement from which in fact this postulate followed, but which itself required a proof.†

---

* Translator's note: Nasiraddin

† Of such statements, equivalent to the Fifth Postulate, quite a large number were set up. Here are some examples: (1) a line parallel to a given line has a constant distance from it (Proclus); (2) there exist similar (but not equal) triangles, i.e., triangles whose angles are equal but whose sides are unequal (Wallis); (3) there exists at least one rectangle, i.e., a quadrangle whose angles are all right angles (Saccheri); (4) a line perpendicular to one arm of an acute angle also intersects the other arm (Legendre); (5) the sum of the angles of a triangle is equal to two right angles (Legendre); (6) there exist triangles of arbitrarily large area (Gauss). This list could now be continued indefinitely.

Saccheri and Lambert penetrated deeper into the problem than the others. Saccheri was the first to attempt a proof of the Fifth Postulate by a *reductio ad absurdum*; i.e., he took as a starting point the opposite assertion and by developing its consequences hoped to come to a contradiction. When he arrived, in these inferences, at results that appeared entirely unimaginable, he thought that he had solved the problem. But he was mistaken, because a contradiction to intuitive ideas does not yet indicate a logical contradiction. After all, the problem was to prove the Euclidean postulate on the basis of other propositions of geometry and not to convince oneself once more of its intuitive truth. Intuitively, the postulate itself is convincing enough. But, let us repeat, intuitive conclusiveness and logical necessity are two different things.

Lambert proved to be a deeper thinker than Saccheri and his predecessors. Starting out on the same path he did not find a logical contradiction and did not make the mistakes of the others; he did not claim to have proved the Fifth Postulate. But after him, at the beginning of the 19th century, Legendre once more "proves" the Fifth Postulate by falling into the old mistake: Again he replaces the postulate by other assertions which require proof themselves.

Thus, at the beginning of the 19th century the problem of proving the Fifth Postulate remained as unsolved as it was in Euclid's time. The efforts remained in vain and the problem, it seemed, would not yield. Here, indeed, was a deep enigma of geometry; a problem whose solubility was not doubted by the best geometers did not yield a solution in two thousand years.

The theory of parallel lines became one of the central problems of geometry in the 19th century. It attracted many geometers: Gauss, Lagrange, d'Alembert, Legendre, Wachter, Schweikart, Taurinus, Farkas Bólyai, and others.

However, a proof of the postulate did not come forth. What, then, was the matter: Was it due to lack of ability or was the problem perhaps inaccurately stated? This question began to cross the minds of some geometers that surpassed the others in depth of thought. Gauss, the most famous German mathematician, occupied himself with the problem from 1792 onward and the correct statement of the question gradually dawned upon him. Finally he decided to abandon the Fifth Postulate and from 1813 on he developed a sequence of theorems that are consequences of the opposite assertion. A little later the German mathematician Schweikart, who was then a professor of Law at Kharkov, and Taurinus followed on the same path. But none of these arrived at a final answer to the problem. Gauss carefully concealed his investigations, Schweikart confined himself to private correspondence with Gauss, and only Taurinus went into print with the elements of a new geometry based on the negation of the Fifth Postulate. However, he himself ruled out the possibility of such a geometry. Thus, none of these solved the problem and the question whether the statement was altogether correctly put remained without an answer. The answer was first given by N. I. Lobačevskiĭ, a young professor at the University of Kazan: On February 23, 1826 he read a paper on the theory of parallels at a meeting of the physical-mathematical faculty and in 1829 he published its contents in the Journal of the University of Kazan.

§2. The Solution of Lobačevskiĭ

1. The ideas of Lobačevskiĭ. The essence of Lobačevskiĭ's solution of the problem of the Fifth Postulate was expressed by him in his work "New Elements of Geometry" (1835) in the following words:

"It is well known that in geometry the theory of parallels has so far remained incomplete. The futile efforts from Euclid's time on throughout two thousand years have compelled me to suspect that the concepts themselves do not contain the truth which we have wished to prove, but that it can only be verified like other physical laws by experiments, such as astronomical observations. Convinced, at last, of the truth of my conjecture and regarding the difficult problem as completely solved, I put down my arguments in 1826."

Let us examine what Lobačevskiĭ had in mind in this statement in which his new idea is so to speak focused, which not only gave the solution of the problem of the Fifth Postulate, but revolutionized our whole conception of geometry and of other branches of mathematics as well.

Already in 1815 I. N. Lobačevskiĭ had begun to work on the theory of parallels, trying at first, like the other geometers, to prove the Fifth Postulate. In 1823 he realized clearly that all the proofs, "no matter of what kind, can only be regarded as clarifications, but do not deserve to be called mathematical proofs in the full sense."* He was aware, then, that "the concepts themselves do not contain the truth which we have wished to prove"; i.e., in other words, the Fifth Postulate cannot be deduced from the fundamental propositions and concepts of geometry. How did he convince himself that this deduction is impossible?

He did this by going farther along the path on which Saccheri and Lambert had taken the first steps. As a hypothesis he took the statement contradicting the Euclidean postulate, namely: "through a point not

---

* So wrote Lobačevskiĭ in 1823 in his course of geometry, which was not published during his lifetime. This course "Geometry" was first edited in 1910.

lying on a given line not one, but at least two lines parallel to the given one can be drawn." Let us take this statement conditionally as an axiom and, adjoining it to the other propositions of geometry, let us develop further consequences of it. Then, if this assertion is incompatible with the other propositions of geometry, we shall arrive at a contradiction and so the Fifth Postulate will be proved indirectly: The opposite proposition leads to a contradiction. However, in view of the fact that no such contradiction is detected, we come to two conclusions which Lobačevskiĭ also reached.

The first conclusion is that the Fifth Postulate is not provable. The second conclusion is that on the basis of the opposite axiom just formulated we can develop a chain of consequences, i.e., theorems that do not contain a contradiction. These consequences form in their own right a certain logically possible noncontradictory theory that can be regarded as a new non-Euclidean geometry. Lobačevskiĭ cautiously called it "imaginary," because he could not yet find a real explanation for it. But its logical possibility was clear to him. By expressing and defending this strong conviction Lobačevskiĭ displayed the true grandeur of a genius who defends his convictions without wavering and does not hide them from public opinion for fear of misunderstanding and criticism.

Thus, the first two conclusions reached by Lobačevskiĭ consisted in the statements that the Fifth Postulate is not provable and that it is possible to develop on the basis of a contrary axiom a new geometry that is logically just as rich and perfect as the Euclidean, notwithstanding the fact that its results are at variance with the intuitive picture of space. Lobačevskiĭ in fact developed this new geometry, which now bears his name. A general result of enormous importance was involved here: *More than one geometry is logically conceivable.* The significance of this result in its full extent will be discussed later; in it is really contained not a small part of the solution of the problems concerning abstract mathematical spaces that were raised at the beginning of this chapter.

But let us return to Lobačevskiĭ's statement quoted previously. He says that geometrical truth like other physical laws can only be verified by experiments. This means, first, that we must interpret truth as a correspondence between abstract concepts and reality. This correspondence can only be established by experiment so that the verification of one result or another requires experimental investigations and that mere logical inference is insufficient for the purpose. Although Euclidean geometry reflects the real properties of space very accurately, it cannot be certain that further investigations might not reveal that Euclidean geometry is only approximately true as a theory of the properties of real space. Geometry as a science of real space (and not as a logical system) would then have to be changed and made more precise in accordance with the new experimental data.

This brilliant idea of Lobačevskiĭ has been completely vindicated in a new branch of physics, the theory of relativity.

Lobačevskiĭ himself undertook computations on the basis of astronomical observations with the object of verifying the accuracy of Euclidean geometry. These computations corroborated at the time its truth within the limits of the available accuracy. The situation has now changed, although it must be emphasized at once that Lobačevskiĭ geometry, too, did not prove to be more accurate in its application to space, whose properties turned out to be different and more complicated. But even before this the Lobačevskiĭ geometry had become well based and applicable in another connection, of which we shall speak in more detail later.

It must be emphasized that Lobačevskiĭ did not at all regard his geometry as a mere logical scheme constructed on arbitrarily chosen premises. The important task he saw not in the logical analysis of the foundations of geometry but in the investigation of its relationship to reality. Since an experiment cannot give an absolutely accurate solution of the problem of truth of Euclid's postulate, it makes sense to investigate those logical possibilities that are represented by the most fundamental premises of geometry. This mathematical investigation helps to mark a path on which the physical study of the properties of real space must proceed. Furthermore, Euclid's geometry is a limiting case of Lobačevskiĭ's geometry, so that the latter includes wider possibilities. From this point of view the restriction to Euclid's postulate would have been a hindrance to the development of the theory. The theory must go beyond the known frontiers so as to search for ways of disclosing new facts and laws. A deeper understanding of the links between mathematics and reality enables us to select from the diverse logical possibilities precisely those that have the best chance of being useful in the study of nature. If geometry after Lobačevskiĭ had not developed the mathematical doctrine of the possible properties of space, contemporary physics would not possess the mathematical tools that made it possible to formulate and develop the theory of relativity.

Thus, we can summarize Lobačevskiĭ's solution of the problem of the Fifth Postulate.

1. The postulate is not provable.

2. By adding the opposite axiom to the basic propositions of geometry a logically perfect and comprehensive geometry, different from the Euclidean, can be developed.

3. The truth of the results of one logically conceivable geometry or another in its application to real space can only be verified experimentally. A logically conceivable geometry must be elaborated not only as an arbitrary logical scheme, but also as a theory indicating possible ways and methods of developing physical theories.

The solution is altogether different from what the geometers wished to obtain when they tried to prove the Euclidean postulate. It went so much against the established ideas that it did not meet with much understanding among mathematicians. For them it was too new and radical. Lobačevskiĭ, as it were, cut the Gordian knot of the theory of parallels instead of trying to disentangle it, as others had expected to do.

2. Other geometers and philosophers. Almost simultaneously with Lobačevskiĭ the Hungarian geometer János Bólyai (1802-1860) also discovered the impossibility of proving the Fifth Postulate and the possibility of a non-Euclidean geometry; he published his results as an appendix to the geometric treatise of his father Farkas Bólyai of 1832. Previously the father had sent his son's paper for an opinion to Gauss and had received an encouraging reply, in which Gauss mentioned that he too had reached the same results long ago. However, Gauss refrained from publishing anything. In one of his letters he explains that he was afraid of being misunderstood.

In science it always happens that the time is ripe for some results and that they are obtained almost simultaneously and independently by several scholars. The integral and differential calculus was developed simultaneously by Newton and Leibnitz; Darwin's ideas were independently reached at the same time by Wallace; the elements of the theory of relativity were found simultaneously by Einstein and also by Poincaré, and of such examples there are many more. They show once more that science grows inevitably by the solution of problems for which it is ripe, and not by accidental discoveries and guesses. So it was also with the discovery of the possibility of non-Euclidean geometry, which was made simultaneously by several geometers: Lobačevskiĭ, Bólyai, Gauss, Schweikart, and Taurinus.

However, as it also happens constantly in science, not all the scholars who arrive at a new result play an equal role in its establishment and not all of them have an equal share in the service performed. Priority is of importance here, but also clarity and depth of results, and coherence and sound arguments in their derivation. Neither Schweikart nor Taurinus were convinced of the equal status of the new geometry, and this was a decisive feature of the case, all the more since partial results had already been obtained by Saccheri and Lambert. Gauss, although he apparently had this conviction, was not resolute enough to risk coming out with it into the open.

Bólyai did not display any indecision, but he did not develop the new ideas as far and as deeply as Lobačevskiĭ. For Lobačevskiĭ was the first to express the new ideas openly, orally in 1826 and in print in 1829, and continued to develop and propagate them in a number of papers culminating in the "Pangeometry" of 1855, which he dictated as a blind old man in his declining years, retaining his strength of mind and his confidence in his work. And so the new geometry bears his name.

§3. Lobačevskiĭ Geometry

1. Some striking results. Thus, Lobačevskiĭ took as his starting point the statement contradictory to the Fifth Postulate: in a given plane at least two straight lines can be drawn through a point that do not intersect a given line. From this he derived a number of far-reaching consequences that formed the new geometry. This geometry was, therefore, constructed as a conceivable theory, as a collection of theorems that can be proved logically proceeding from the postulate, in conjuction with other* basic assumptions of Euclidean or, as Lobačevskiĭ used to say, "customary" geometry.

Among his deductions Lobačevskiĭ obtained all the results analogous to those of the "customary" elementary geometry, i.e., right up to non-Euclidean trigonometry and the solution of triangles, to the calculation of areas and volumes. We cannot here follow this chain of Lobačevskiĭ's deductions not because they are too complicated but merely for lack of space. After all, even a school course in "customary" geometry is rather long, and Lobačevskiĭ's deductions are neither simpler nor shorter than these "customary" deductions. Therefore we shall mention here only some striking results of Lobačevskiĭ, and refer the reader who is interested in a deeper study of non-Euclidean geometry to the special literature. Later on we shall explain a simple interpretation of non-Euclidean geometry in the actual world.

Let us begin with the theory of parallel lines. Suppose that a line a and a point A not on it are given. We drop the perpendicular AB from A to a. By the fundamental assumption there exist at least two lines passing through A and not intersecting our line a. Then every line in the angle between these two lines also does not intersect a. It is true that in figure 2 the lines b and b', if produced far enough, would actually intersect a,

---

* These so-called "remaining" propositions of geometry will also be formulated accurately in §5.

against Lobačevskiĭ's assumption. But there is nothing surprising in this. For of course Lobačevskiĭ did not argue from figures as we can draw them in the ordinary plane; he developed logical consequences from his

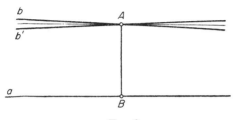

FIG. 2.

assumption, which contradicts what we are accustomed to see in diagrams. Figures play only an auxiliary role here; in them the facts of non-Euclidean geometry cannot be expressed accurately, because in a figure we draw ordinary lines in an ordinary plane, entirely Euclidean within the limits of accuracy of the figure.

This contradiction between logical possibility and visual representation was an important obstacle to the understanding of Lobačevskiĭ geometry. But if we are concerned with geometry as a logical theory, then we must look for logical rigor of the reasoning and not for agreement with customary figures.

2. The angle of parallelism. Let us turn again to our line a and point A. Through A we draw a half line x that does not intersect a (for example perpendicular to AB) and rotate it around A so that the angle ϕ between AB and x decreases, but without bringing it to an intersection with a. Then the half line x reaches a limiting position corresponding to the least value of ϕ. This limiting half line c also does not intersect a.

For if it did intersect a in some point X (figure 3), then we could take

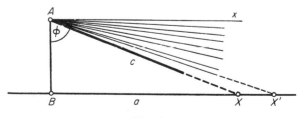

FIG. 3.

a point X' to the right and obtain a half line AX' intersecting a, but forming a larger angle with AB. But this is impossible, since by the construction of c every half line x that forms a larger angle with AB does not intersect a.

Therefore c does not intersect a and is, moreover, the extreme one of all the half lines passing through A and not intersecting a.

By symmetry, it is obvious that on the other side we can also draw a half line c' not intersecting a and also extreme among all such half lines. If c and c' were continuations of one another, then together they would form a single line $c + c'$. This line would then be the unique parallel to a through the given point A, so that under the slightest rotation either c or c' would intersect a. So once it has been assumed that the parallel is not unique, but that there are at least two, the half lines c and c' cannot be continuations of one another.

Thus, we have proved the first theorem of Lobačevskiĭ geometry:

Through a point A not lying on a given line a two half lines c and c' can be drawn such that they do not intersect a, but that every half line in the angle between them intersects a.

If the half lines c and c' are produced, then we obtain (figure 4) two lines not intersecting a with the additional property that every line passing through A in the angle α between these lines does not intersect a, but every line in the angle β intersects a. Lobačevskiĭ called these lines c, c' *parallels* to a: c parallel on the right, c' parallel on the left. Half of the angle β is called the *angle of parallelism* by Lobačevskiĭ; it is less than a right angle, because β is less than two right angles.

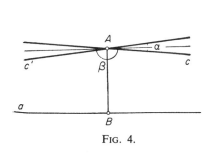

FIG. 4.

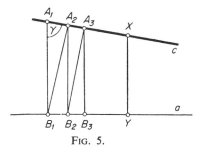

FIG. 5.

3. Convergence of parallel lines; the equidistant curve. Let us now investigate how the distance from a of a point X on c changes when X is shifted along c (figure 5). In Euclidean geometry the distance between parallel lines is constant. But here we can convince ourselves that when X

moves to the right, its distance from a (i.e., the length of the perpendicular XY) decreases.

We drop the perpendicular A_1B_1 from a point A_1 to a. From B_1 we drop the perpendicular B_1A_2 to c (A_2 lies to the right of A_1, since γ is an acute angle). Finally we drop the perpendicular A_2B_2 from A_2 to a. Let us show that A_2B_2 is less than A_1B_1.

The theorem that the perpendicular is shorter than a slant line is valid in Lobačevskiĭ geometry, because its proof (which can be found in every school book on geometry) does not depend on the concept of parallel lines nor on deductions connected with them. Now since the perpendicular is shorter than a slant line, B_1A_2 as a perpendicular to c is shorter than A_1B_1, and similarly A_2B_2 as a perpendicular to a is shorter than B_1A_2. Therefore A_2B_2 is shorter than A_1B_1.

When we now drop the perpendicular B_2A_3 to c from B_2 and repeat these arguments, we see that A_3B_3 is shorter than A_2B_2. Continuing this construction we obtain a sequence of shorter and shorter perpendiculars; i.e., the distances of A_1, A_2, $\cdots$ from a decrease. Furthermore, by supplementing our simple argument we could prove that, generally, if a point X'' on c lies to the right of X', then the perpendicular $X''Y''$ is shorter than $X'Y'$. We shall not dwell on this point. The preceding arguments, we trust, make the substance of the matter sufficiently clear and a rigorous proof is not one of our tasks.

But it is remarkable that, as can be proved, the distance XY not only decreases when X moves on c to the right, but actually tends to zero as X tends to infinity. That is, *the parallel lines a and c converge asymptotically*! Moreover, it can be proved that in the opposite direction the distance between them not only increases but tends to infinity.

In Euclidean geometry a line parallel to a given line has a constant distance from it. In Lobačevskiĭ geometry, in general, such pairs of lines do not exist, since a line always diverges to infinity from a given line either on one side or on both sides. So the line that has a constant distance from a given line can never be straight but is a curve called an *equidistant*.

These conclusions of Lobačevskiĭ geometry are indeed remarkable and are not at all compatible with the customary visual representation. But as we have already said, such a discrepancy cannot be an argument against Lobačevskiĭ geometry as an abstract theory, logically developed from the premises assumed.

4. The magnitude of the angle of parallelism. We shall now study the angle of parallelism, i.e., the angle γ that the line c parallel to a given line a forms with the perpendicular CA (figure 6). Let us show that this

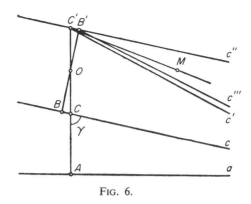

FIG. 6.

angle is smaller, the further C is from a. For this purpose we begin by proving the following. If two lines b and b' form equal angles α, α' with a secant BB', then they have a common perpendicular (figure 7).

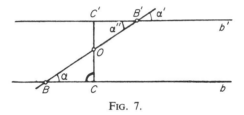

FIG. 7.

For the proof we draw through the midpoint O of BB' the line CC' perpendicular to B. We obtain two triangles OBC and $OB'C'$. Their sides OB and OB' are equal by construction. The angles at the common vertex O are equal as vertically opposite. The angle α'' is equal to α' since they are also vertically opposite. But α' is equal to α by assumption. Therefore α is equal to α''. Thus, in our triangles OBC and $OB'C'$ the sides OB and OB' and their adjacent angles are equal. But then, by a well-known theorem, the triangles are equal, in particular their angles at C and C'. But the angle at C is a right angle, since the line CC' is by construction perpendicular to b. Therefore the angle at C' is also a right angle; i.e., CC' is also perpendicular to b'. Thus, the segment CC' is a common perpendicular to both b and b'. This proves the existence of a common perpendicular.

Now let us prove that the angle of parallelism decreases with increasing distance from the line. That is, if the point C' lies further from a than C, then, as in figure 6, the parallel c' passing through C' forms with the perpendicular $C'A$ a smaller angle than the parallel c passing through C.

For the proof we draw through C' a line c'' under the same angle to $C'A$ as the parallel c. Then the lines c and c'' form equal angles with CC'. Therefore, as we have just shown, they have a common perpendicular BB'. Then we can draw through B' a line c''' parallel to c and forming with the perpendicular an angle less than a right angle, since we know already that a parallel forms with the perpendicular an angle less than a right angle. Now we choose an arbitrary point M in the angle between c'' and c''' and draw the line $C'M$. It lies in the angle between c'' and c''' and cannot intersect c'. *A fortiori*, it cannot intersect c. But it forms with AC' a smaller angle than c'' does, i.e., smaller than γ. Then, *a fortiori*, the parallel c' forms an even smaller angle, because it is the extreme one of all the lines passing through C' and not intersecting a. Therefore c' forms with $C'A$ an angle less than c does and this means that the angle of parallelism decreases on transition to a farther point C'; this is what we set out to prove.

We have shown, then, that the angle of parallelism decreases for increasing distance of C from a. Even more can be shown: *If the point C recedes to infinity, then this angle tends to zero.* That is, for a sufficiently large distance from the line a a parallel to it forms with the perpendicular to it an arbitrarily small angle.* In other words, if at a point very far from a the line perpendicular to a is tilted by a very small angle, the "tilted" line will no longer intersect a. This fact of Lobačevskiĭ geometry, too, makes an amazing impression. But further on we shall obtain other no less amazing results.

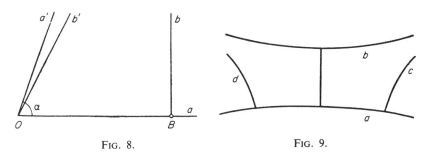

FIG. 8.   FIG. 9.

For example, let us take the acute angle α formed by two half lines a and a'. We erect the perpendicular b to a at a point sufficiently far from

---

* If h is the length of the perpendicular and γ is the angle of parallelism, then as Lobačevskiĭ proved $\tan \gamma/2 = e^{-h/k}$, where k is a constant depending on the unit of length and e is the base of natural logarithms. Obviously $e^{-h/k}$, and with it γ, tends to zero for $h \to \infty$.

the vertex O of α so that the angle of parallelism corresponding to the chosen distance OB (figure 8) is less than α. Once the angle α is larger than the angle of parallelism, the line b' through O parallel to b forms a smaller angle with a. But it does not intersect b. Therefore a', *a fortiori*, does not intersect b. Thus we have shown that the perpendicular to one arm of an acute angle erected sufficiently far from the vertex does not intersect the other arm.

5. More striking theorems. We have drawn all the preceding conclusions with a twofold aim. First of all, and this is the main point, we wanted to show by some simple examples how theorems of Lobačevskiĭ geometry can in fact be obtained from the premises assumed. This is a very simple instance of the way in which mathematicians reach conclusions in abstract geometry, of how conclusions can be reached at all that are not connected with the usual visual representation. Second, we wanted to show what peculiar results are obtained in Lobačevskiĭ geometry. Let us give a few more examples.

Two lines in a Lobačevskiĭ plane either intersect or they are parallel in the sense of Lobačevskiĭ, and then they converge asymptotically on the one side and on the other they diverge infinitely, or else they have a common perpendicular and diverge infinitely on both sides of it.

If the lines a, b have a common perpendicular (figure 9), then two perpendiculars c, d can be drawn to a that are parallel (in the sense of Lobačevskiĭ) to b and the whole line b lies in the strip between c and d.

The limit of a circle of infinitely increasing radius is not a line but a certain curve, a so-called *limiting circle*. It is not always possible to draw a circle through three points not on one line, but either a circle or a limiting circle or an equidistant (i.e., a line formed by the points that are equidistant from a certain line) can be drawn through the three points.

The sum of the angles of a triangle is always less than two right angles. If a triangle is increased so that all three heights grow without bound, then its three angles tend to zero.

There are no triangles of arbitrarily large area.

Two triangles are equal when their angles are equal.

The length l of the circumference of a circle is not proportional to the radius r but grows more rapidly (essentially by an exponential law). In fact, the following formula holds

$$l = \pi k \left(e^{r/k} - e^{-r/k} \right), \tag{1}$$

where k is a constant depending on the unit of length. Since

$$e^{r/k} = 1 + \frac{r}{k} + \frac{1}{2}\left(\frac{r}{k}\right)^2 + \cdots, \quad e^{-r/k} = 1 - \frac{r}{k} + \frac{1}{2}\left(\frac{r}{k}\right)^2 - \cdots,$$

we obtain from (1):

$$l = 2\pi r \left(1 + \frac{1}{6}\frac{r^2}{k^2} + \cdots \right).$$ (2)

Only for small ratios r/k is it true with sufficient accuracy that $l = 2\pi r$.

All these conclusions are logical sequences of the premises assumed: "the axioms of Lobačevskiĭ" in conjunction with the basic propositions of the "customary" geometry.

6. Lobačevskiĭ's geometry compared with Euclid's. An extremely important property of Lobačevskiĭ geometry consists in the fact that for sufficiently small domains it differs but little from Euclid's geometry; the smaller the domain, the less the difference. Thus, for sufficiently small triangles the connection between sides and angles is expressed with sufficient accuracy by the formulas of ordinary trigonometry, and the more accurately, the smaller the triangle.

The formula (2) shows that for small radii the length of a circle is proportional to the radius, with a good accuracy. Similarly the sum of the angles of a triangle differs by little from two right angles, etc.

In the formula for the length of the circumference of a circle, there occurs a constant k depending on the unit of length. If the radius is small in comparison with k, i.e., if r/k is small, then, as is clear from formula (2), the length l is nearly $2\pi r$. Generally, the smaller the ratio of the dimensions of a figure to this constant, the more accurately the properties of the figure approach the properties of the corresponding figure in Euclidean geometry.*

A measure for the deviation of the properties of a figure in Lobačevskiĭ geometry from the properties of a figure of Euclidean geometry is the

---

* For example, if a, b, c are the sides and the hypotenuse of a right-angled triangle then instead of Pythagoras' theorem the following relation holds

$$2(e^{c/k} + e^{-c/k}) = (e^{a/k} + e^{-a/k})(e^{b/k} + e^{-b/k}).$$

Expanding in series we obtain

$$c^2 + \frac{c^4}{12k^2} + \cdots = a^2 + b^2 + \frac{a^4 + 6a^2b^2 + b^4}{12k^2} + \cdots,$$

so that for large l we have the theorem of Pythagoras $c^2 = a^2 + b^2$. Furthermore, from Lobačevskiĭ's formula for the angle of parallelism γ (see the previous footnote), $\tan(\gamma/2) = e^{-h/k}$. If h/k is small, i.e., if the parallels are close together, then $\tan \gamma/2 = e^{-h/k} \approx 1$ and $\gamma \approx 90°$. Thus, for small distances parallels in a Lobačevskiĭ plane differ little from Euclidean parallels.

ratio r/k if r measures the dimensions of the figure (radius of a circle, sides of a triangle, etc.).

This has an important consequence.

Suppose we have to do with the actual space of the external world and measure distances in kilometers. Let us assume that the constant k is very large, say 10^{12}.

Then, for example, by the formula (2), for a circle with a radius of even 100 km the ratio of its length to the radius differs from 2π by less than 10^{-9}. Of the same order are the deviations from other ratios of Euclidean geometry. Within the limits of 1 kilometer they would even be of the order $1/k$, i.e., 10^{-12}, and within the limits of a meter of the order 10^{-15}; i.e., they would be altogether negligible. Such deviations from Euclidean geometry could not be observed, because the dimensions of an atom are a hundred times larger (they are of the order of 10^{-13} km). On the other hand, on the astronomical scale the ratio r/k could turn out to be not too small.

Therefore Lobačevskiĭ also assumed that, although on the ordinary scale Euclid's geometry is true with great accuracy, the deviation from it could be noted by astronomical observations. As we have already mentioned, this assumption has been justified, but the insignificant deviations from Euclidean geometry that have now been observed on the astronomical scale turn out to be even more complicated.

Finally, the arguments given have another important consequence. It is this: Since the deviation from Euclidean geometry becomes smaller for increasing values of the constant k, in the limit when k grows without bound, Lobačevskiĭ geometry goes over into Euclid's geometry. That is, *Euclid's geometry is just a limiting case of Lobačevskiĭ geometry*. Therefore, if this limiting case is added to Lobačevskiĭ geometry, then it comprises also Euclid's geometry and so it turns out, in this sense, to be a more general theory. In view of this situation Lobačevskiĭ called his theory "pangeometry," i.e., universal geometry. Such a relationship of theories constantly appears in the development of mathematics and the natural sciences: A new theory includes the old one as a limiting case, in accordance with the advance of our knowledge from more special to more general deductions.

However, all the reasonings and deductions we have made would remain, as it were, a hardly intelligible game of the mind if we could not establish a comparatively simple real meaning of Lobačevskiĭ geometry within the system of the usual concepts of Euclidean geometry. The solution of this problem was not finally reached by Lobačevskiĭ himself; it fell to the lot of his successors and was found almost forty years after his first paper had appeared. This solution is described in the next section.

§4. The Real Meaning of Lobačevskiĭ Geometry

1. Beltrami's interpretation on the pseudosphere. An intuitive interpretation of Lobačevskiĭ geometry was first given in 1868, when the Italian geometer Beltrami noted that the intrinsic geometry of a certain surface, namely the pseudosphere, coincided with the geometry on part of the Lobačevskiĭ plane. We recall that by the intrinsic geometry of a surface one understands the collection of properties of figures on it that can be determined by measuring lengths only on the surface itself. In figure 10 on the left we have drawn the so-called *tractrix*. This is the curve

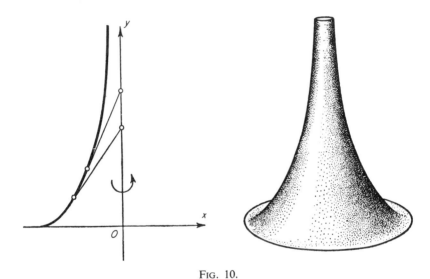

FIG. 10.

with the property that the length of the segment of its tangent from the point of contact to the intersection with the Oy-axis is constant for all points of the curve. The Oy-axis is its asymptote. By rotating the tractrix around its asymptote, we obtain a surface, which is called a *pseudosphere* and is illustrated in figure 10 on the right.

The interpretation of Lobačevskiĭ geometry by Beltrami comes to this, that all geometrical relations on part of the Lobačevskiĭ plane coincide with the geometrical relations on a suitable part of the pseudosphere provided the following convention is adopted. The role of straight line segments is taken over by shortest lines on the surface, the so-called geodesics. The distance between points is defined as the length of the shortest line joining them on the surface. Figures are called equal if their points can be put in correspondence with each other in such a way that intrinsic distances between corresponding points are equal. A motion of figures on the pseudosphere that preserves their dimensions from the point of view of the intrinsic geometry, although accompanied by bending, represents a motion in the Lobačevskiĭ plane. Lengths, angles, and areas are measured on the surface as usual and correspond to lengths, angles, and areas in Lobačevskiĭ geometry.

Beltrami's interpretation shows that, given these conditions, to every statement of Lobačevskiĭ geometry referring to part of the plane there corresponds an immediate fact of the intrinsic geometry of the pseudosphere. Lobačevskiĭ geometry consequently has a perfectly real meaning: It is nothing but an abstract account of the geometry on the pseudosphere.

We ought to mention that, thirty years before Beltrami's discovery, the intrinsic geometry of the pseudosphere had already been investigated by F. Minding, who had in fact established the properties that show that it coincides with Lobačevskiĭ geometry. However, neither he nor anyone else noted this, until Lobačevskiĭ's ideas had been sufficiently propagated. Beltrami had only to compare the results of Lobačevskiĭ and Minding to become aware of the connection between them.

Beltrami's discovery at once changed the attitude of mathematicians to Lobačevskiĭ geometry; from being "fictitious" it became real.*

2. Klein's interpretation in the circle and the sphere. However, as we have emphasized, not the whole Lobačevskiĭ plane is realized on the pseudosphere, but only part of it.†

---

* The history of the gradual establishment of the real meaning of Lobačevskiĭ geometry was in fact even more complicated. First of all, Lobačevskiĭ himself had the means of proving its noncontradictory character by a so-called analytical model, but he did not succeed in carrying the proof right through. This was done much later. Second, the German mathematician Riemann came forward in 1854 with a theory (see §10) in which Beltrami's results are already contained, although Riemann did not express them clearly; his paper was not understood and was not published until after his death in 1868 in the same year as the appearance of Beltrami's paper. The whole history of Lobačevskiĭ geometry from the attempts at proving Euclid's postulate to the complete clarification of the significance of non-Euclidean geometry is highly instructive in that it shows what struggles and roundabout ways are often needed to discover a truth which afterwards turns out to be simple and intelligible.

† The pseudosphere has everywhere the same negative Gaussian curvature. All surfaces of constant negative curvature have (at least in small parts) the same intrinsic geometry and can therefore serve as models of Lobačevskiĭ geometry. However, as Hilbert proved in 1901, none of these surfaces can be extended infinitely in all directions without singularities, so that they cannot serve as a model of the *whole* Lobačevskiĭ plane. On the other hand, the young Dutch mathematician Kuiper showed in 1955 that there exist smooth surfaces that represent in the sense of their intrinsic geometry the whole Lobačevskiĭ plane, but such surfaces although smooth cannot be bent continuously, they do not have a definite curvature.

Let us also note the following. When Lobačevskiĭ geometry is represented on a

Therefore the task of giving an actual interpretation of Lobačevskiĭ geometry on the whole plane, and all the more for his geometry in space, still remained unsolved. This was done later in 1870 by the German mathematician Klein. Let us explain what his solution was.

In an ordinary Euclidean plane we take a circle and consider only the interior of the circle; i.e., we exclude from our investigation its circumference and the domain outside. We agree to call this interior of the circle a "plane," since it will turn out to play the role of a Lobačevskiĭ plane. The chords of our circle will be called "lines" and in accordance with the agreement we have just made the end points of chords, as lying on the circumference, are excluded. Finally, a "motion" shall be any transformation of the circle that carries it into itself and carries lines into lines, i.e., does not distort its chords. The simplest example of such a transformation is a rotation of the circle around its center, but it turns out that there are far more of them. What these transformations are will be stated in the following paragraphs.

Now if we introduce these conventions of nomenclature, then the facts of the ordinary geometry within our circle are transformed into theorems of Lobačevskiĭ geometry. And conversely, every theorem of Lobačevskiĭ geometry is interpreted as a fact of the ordinary geometry within the circle.

For example, by Lobačevskiĭ's axiom at least two lines can be drawn through a point not lying on a given line that do not intersect the line. Let us translate this axiom into the language of ordinary geometry by our conventions, i.e., replace lines by chords. Then we obtain the statement: At least two chords can be drawn through a point inside the circle not lying on a given chord that do not intersect the chord. The truth of this statement is obvious from figure 11. Therefore Lobačevskiĭ's axiom is satisfied here.

We recall further that in Lobačevskiĭ geometry among the lines passing through the given point and not intersecting the given line there are two extreme ones, namely the ones that are called by Lobačevskiĭ parallels to the given line. This means that among the chords passing through the given point A and not intersecting the given chord BC there are two extreme chords. And indeed, these extreme chords are the ones that pass through B or C, respectively. They do not have common points with BC, because we exclude points on the circumference. Thus, this theorem of Lobačevskiĭ is satisfied here.

For a further translation of Lobačevskiĭ's theorems into the language

---

surface of negative constant curvature K, the constant k that figures in the formulas of the preceding section assumes a simple meaning: $k^2 = -1/K$.

of ordinary geometry within the circle, it is necessary to explain how segments and angles are to be measured in the circle in such a way that these measurements correspond to Lobačevskiĭ geometry. Of course, the measuring cannot be the same as the usual one, because in the ordinary sense a chord has finite length and the line that the chord represents is infinite. This could perhaps be regarded even as a contradiction, but we shall see that no contradiction exists.

Let us recall first of all that in ordinary geometry lengths of segments are measured as follows. Some segment AB is chosen whose length is taken to be the unit and the length of any other segment XY is determined by comparing it with AB. For this purpose the segment AB is laid off along XY. If there remains a part of XY less than AB, then AB is divided into, say, 10 equal parts (equal in the sense that each is obtained from the other by a translation); these parts are laid off on the remaining portion of XY; if necessary, AB is then divided into 100 parts, etc. As a result the length of XY is expressed as a decimal fraction, which may be infinite. Consequently, lengths are measured by means of a motion of the whole or part of the segment chosen as unit; i.e., measurement is based on motion. And once motions are defined (in our case we have defined them as transformations of the circle that carry lines into lines), then it is known what segments are to be regarded as equal and how length has to be measured. The term that defines motion already contains, though in an implicit form, the rule for measuring lengths. Angles are measured in just the same way, by laying off the angle taken to be the unit. Thus, the rule for measuring angles is also contained in the definition of motion.

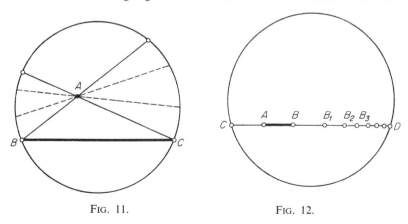

FIG. 11.   FIG. 12.

The corresponding rules for measuring lengths and angles in Lobačevskiĭ geometry are fairly simple, although essentially different from the ordinary rules. We shall not derive them here, because this is not a point of principal significance in our arguments.*

By the rule for measuring lengths it comes out that a chord has infinite length. And this is so because if by a transformation that we have taken to be a motion the segment AB is carried into BB_1, then into B_1B_2, etc., these segments B_kB_{k+1} become shorter and shorter in the usual sense (although equal in the sense of our model of Lobačevskiĭ geometry; figure 12). The points B_1, B_2, $\cdots$, B_k, $\cdots$ accumulate at the end point of the chord. But for us the chord has no end point; the end point is excluded by agreement, and in this sense it is "at infinity." In the sense of Lobačevskiĭ geometry the points B_1, B_2, $\cdots$ do not accumulate anywhere, they extend to infinity. By means of transformations that we have taken to be motions, laying off equal segments one after another, we cannot reach the circumference of the circle from within.

In order to understand clearly how segments are laid off in the model, let us consider the transformation that plays the role of a translation along a line.

Suppose that rectangular coordinates are introduced in the plane with the center of the circle as origin. To fix our ideas let us assume that our circle has unit radius so that its circumference is represented by the equation $x^2 + y^2 = 1$ and the points of the interior satisfy the inequality $x^2 + y^2 < 1$.

We consider the transformation given by the formulas

$$x' = \frac{x + a}{1 + ax}, \quad y' = \frac{y\sqrt{1 - a^2}}{1 + ax}, \tag{3}$$

where x', y' are the coordinates of the point into which the point with the original coordinates x, y is carried by the transformation, and where a is an arbitrarily given number of absolute value less than 1.

When we find x and y from (3) we obtain, as is easy to verify, for the inverse expressions of x and y in terms of x', y'

$$x = \frac{x' - a}{1 - ax'}, \quad y = \frac{y'\sqrt{1 - a^2}}{1 - ax'}. \tag{4}$$

The transformation (3) satisfies the two conditions for a "motion" in our model: (1) it carries the circle into itself; (2) it carries lines into lines.

---

* The rule for measuring length turns out to be the following. Let the segment AB lie on the chord CD (figure 12). Measuring segments in the usual way we form the so-called cross ratio $CB/CA : DB/DA$. Its logarithm is taken to be the length of AB.

To prove the first property we have to convince ourselves, strictly speaking, that the inequality or equality $x^2 + y^2 \leqslant 1$ implies the corresponding relation $x'^2 + y'^2 \leqslant 1$ and vice versa. Let us show, for example, that when $x^2 + y^2 = 1$, then necessarily $x'^2 + y'^2 = 1$, i.e., that the points on the circumference of the given circle remain on it.

We compute $x'^2 + y'^2$, using (3) and taking into account that $x^2 + y^2 = 1$, i.e., $y^2 = 1 - x^2$,

$$\begin{aligned}
x'^2 + y'^2 &= \frac{(x + a)^2 + y^2(1 - a^2)}{(1 + ax)^2} = \frac{(x + a)^2 + (1 - x^2)(1 - a^2)}{(1 + ax)^2} \\
&= \frac{x^2 + 2ax + a^2 + 1 - x^2 - a^2 + a^2x^2}{(1 + ax)^2} = \frac{1 + 2ax + a^2x^2}{1 + 2ax + a^2x^2} \\
&= 1.
\end{aligned}$$

Therefore, when $x^2 + y^2 = 1$, then also $x'^2 + y'^2 = 1$. The remaining cases are verified similarly.

The second property of (3) is established very simply. For we know that every line is represented by a linear equation and, conversely, every linear equation represents a line. Suppose that the given line is

$$Ax + By + C = 0. \tag{5}$$

After the transformation (4) we obtain

$$A\frac{x' - a}{1 - ax'} + B\frac{y'\sqrt{1 - a^2}}{1 - ax'} + C = 0,$$

or, by reducing to the common denominator,

$$(A - aC)\,x' + B\sqrt{1 - a^2}\,y' + (C - aA) = 0.$$

This equation is linear and consequently represents a line. This is the line into which (5) is carried by the transformation. Let us also note that the transformation (3) carries the Ox-axis into itself, causing only a displacement of the points along it. This is clear, because on this axis $y = 0$ and by (3) we then also have $y' = 0$. On the Ox-axis the transformation is given by the single formula

$$x' = \frac{x + a}{1 + ax} \quad (|a| < 1). \tag{3'}$$

On this line the segment $x_1 x_2$ goes over into $x_1' x_2'$ by the formula (3′) and by our convention these arguments are to be regarded as equal. This is the manner in which the "laying off a segment" proceeds.

For the center O of the circle $x = 0$ so that $x' = a$; i.e., under (3′) the center goes over into the point A with the coordinate $x = a$.

Since a can be arbitrary subject only to $|a| < 1$, the center can be carried into any point on the diameter along the Ox-axis.

Under the same transformation, the point that was at A before is carried into the point A_1 with coordinate

$$x_1 = \frac{a + a}{1 + a^2} = \frac{2a}{1 + a^2}.$$

Thus, the segment OA goes over under (3) into AA_1 and so it is "laid off" on the "line" representing the diameter of the circle.

By repeating the same transformation, we can again lay off the same segment arbitrarily often. The point A_n with the coordinate x_n goes over into the point A_{n+1} with the coordinate

$$x_{n+1} = \frac{x_n + a}{1 + ax_n}.$$

So we obtain points A, A_1, A_2, $\cdots$ with the coordinates

$$x_0 = a, \quad x_1 = \frac{2a}{1 + a^2}, \quad x_2 = \frac{x_1 + a}{1 + ax_1} = \frac{3a + a^3}{1 + 3a^2}, \cdots.$$

Since all the segments $A_n A_{n+1}$ are obtained from OA by transformations expressing a motion, they are all "equal" to one another, equal in the sense of Lobačevskiǐ geometry as it is represented in the model. It is easy to show that the points A_n converge to the end point of the diameter. In the sense of the model they recede to infinity.

Since the Ox-axis can be given any direction, the same shift transformations are possible along any diameter. By combining them with rotations around the center of the circle and reflections in a diameter, we obtain all "motions" as they are to be understood in the model; they consist of shifts, rotations, and reflections. These transformations will be studied in more detail in the next section, where it will be proved rigorously that Lobačevskiǐ geometry is really satisfied in our model and that, in particular, the transformations we have defined as motions satisfy all the conditions (axioms) to which motions are subject in geometry.

Let us repeat once more what sort of a model of Lobačevskiǐ geometry Klein proposed. In a plane we have taken the interior of a circle; a point is regarded as a point, a line as a chord (with the end points excluded), a motion is taken to be a transformation carrying the circle into itself and chords into chords; the situation of points (a point lies on a line; a point lies between two others) is to be understood in the usual sense. The rule for measuring lengths and angles (and also areas) already follows from the way in which motion is defined, equality of segments and angles (and of arbitrary figures) is also defined and the same definition is applicable to the operation of laying off one segment along another.

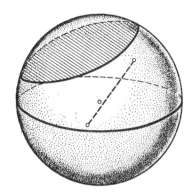

FIG. 13.

Under all these conditions to every theorem of Lobačevskiĭ geometry in the plane there corresponds a fact of Euclidean geometry within the circle, and vice versa: every such fact can be reinterpreted in the form of a theorem of Lobačevskiĭ geometry.

A model of Lobačevskiĭ geometry in space can be constructed similarly. For the space we take the interior of some sphere (figure 13), a line is interpreted as a chord, and a plane as a circle with its circumference on the sphere, but the surface of the sphere itself, and hence the end points of chords and the circumferences of these circles, are excluded; finally, a motion is defined as a transformation of the sphere into itself that carries chords into chords.

When this model of Lobačevskiĭ geometry was given, it was established incidentally that his geometry has a simple real meaning. Lobačevskiĭ geometry is valid, because it can be understood as a specific account of geometry in a circle or a sphere. At the same time its noncontradictory character was proved: Its results cannot lead to a contradiction, since every one can be translated into the language of ordinary Euclidean geometry within a circle (or a sphere, if we are concerned with Lobačevskiĭ geometry in space).*

3. Other interpretations. After Klein another model of Lobačevskiĭ geometry was given by the French mathematician Poincaré, who used it to derive important results in the theory of functions of a complex

---

* Mathematicians usually say that Lobačevskiĭ geometry can be represented within Euclid's geometry, and so it is just as noncontradictory as Euclid's geometry.

variable.* Thus, in his hands Lobačevskiĭ geometry led to the solution of difficult problems from an entirely different branch of mathematics. Lobačevskiĭ geometry has found a number of other applications in mathematics and theoretical physics; for example, in 1913 the physicist Varičak applied it in the theory of relativity.

Lobačevskiĭ geometry is growing successfully; the theory of geometrical constructions, the general theory of curves and surfaces, the theory of convex bodies, and other subjects are being developed in it.

§5. The Axioms of Geometry; Their Verification in the Present Case

1. Precise formulation of the axioms. In order to give a strict mathematical proof that Klein's model really provides an interpretation of Lobačevskiĭ geometry, we have first of all to state accurately what it is that has to be proved. To verify Lobačevskiĭ's theorems one after the other would be absurd; there are too many of them, in fact infinitely many, because one can prove more and more new theorems. However, it will be sufficient to show that in Klein's model the fundamental propositions of Lobačevskiĭ geometry are satisfied, since from these the remaining ones can be deduced. But in that case these fundamental propositions must be formulated precisely.

Thus, the problem of proving that Lobačevskiĭ geometry is noncontradictory reduces to the problem of stating its fundamental propositions, i.e., its axioms, accurately and completely. And since the assumptions of Lobačevskiĭ geometry differ from those of Euclid's geometry by the axiom of parallelism only, our task comes to a precise and complete formulation of the axioms of Euclidean geometry. In Euclid such a formulation does not yet exist; in particular, a definition of the properties of motion or superposition of figures is altogether absent, although of course, he makes use of them. The task of making Euclid's axioms accurate and complete came to the fore precisely in connection with the development of Lobačevskiĭ geometry; and also with the earlier mentioned general trend at the end of the last century toward making the foundations of mathematics more rigorous.

As a result of the investigations of a number of geometers, the problem of formulating the axioms of geometry was finally solved.

Generally speaking, the axioms can be chosen in various ways, taking

---

* Poincaré's model amounts to this: The Lobačevskiĭ plane is again taken to be the interior of a circle, but lines are interpreted as arcs of a circle perpendicular to the circumference of the given circle; a motion is defined as an arbitrary conformal transformation of the circle into itself. (The connection with conformal transformation also yields the connection with the theory of functions of a complex variable.)

various concepts as starting points. Here we shall give an account of the axioms of geometry in a plane which is based on the concepts of point, straight line, motion, and such concepts as: The point X lies *on* the line a; the point B lies *between* the points A and C; a *motion* carries the point X into the point Y. (In our case other concepts can be defined in terms of these; for example, a segment is defined as the set of all points that lie between two given ones.)

The axioms fall into five groups.

I. *Axioms of incidence.*

1. One and only one straight line passes through any two points.

2. On every straight line there are at least two points.

3. There exist at least three points not lying on one straight line.

II. *Axioms of order.*

1. Of any three points on a straight line, just one lies between the other two.

2. If A, B are two points of a straight line, then there is at least one point C on the line such that B lies between A and C.

3. A straight line divides the plane into two half planes (i.e., it splits all the points of the plane not lying on the line into two classes such that points of one class can be joined by segments without intersecting the line, and points of distinct classes cannot).

III. *Axioms of motion.*

(A motion is to be understood as a transformation not of an individual figure, but of the whole plane.)

1. A motion carries straight lines into straight lines.

2. Two motions carried out one after the other are equivalent to a certain single motion.

3. Let A, A' and a, a' be two points and half lines going out from them, and α, α' half planes bounded by the lines a and a' produced; then there exists a unique motion that carries A into A', a into a' and α into α'. (Speaking intuitively, A is carried in A' by a translation, then the half line a is carried by a rotation into a', and finally the half plane α either coincides with α' or else it has to be subjected to a "revolution" around a as axis.)

IV. *Axiom of continuity.*

1. Let $X_1, X_2, X_3, \cdots$ be points situated on a straight line such that each succeeding one lies to the right of the preceding one, but that there *is* a point A lying to the right of them all.* Then there exists a point B that also lies to the right of all the points $X_1, X_2, \cdots$, but such that a point X_n is arbitrarily near to it (i.e., no matter what point C is taken to the left of B, there is a point X_n on the segment CB).

V. *Axiom of parallelism (Euclid).*

1. Only one straight line can pass through a given point that does not intersect a given straight line.

These axioms, then, are sufficient to construct Euclidean geometry in the plane. All the axioms of a school course of plane geometry can in fact be derived from them, though their derivation is very tedious.

The axioms of Lobačevskiĭ geometry differ only in the axiom of parallelism.

V'. *Axiom of parallelism (Lobačevskiĭ).*

1. At least two straight lines pass through a point not lying on a given straight line that do not intersect the line.

It may appear somewhat strange that in the list of axioms there is, for example, this one: "On every straight line there are at least two points." Surely in our idea of a line there are even infinitely many points on it. No wonder that neither to Euclid nor to any one of the mathematicians up to the end of the last century did it occur that such an axiom had to be stated: it was assumed tacitly. But now the situation has changed. When we give a new interpretation of geometry, we may understand by a straight line not the usual line, but something else: a geodesic on a surface, a chord of a circle, or what have you. Therefore the need clearly arises for stating accurately and exhaustively *everything* we have to postulate of those objects that will be described as straight lines. The same applies to all the other concepts and axioms.

So the appearance of various interpretations of geometry, as we have already said, was one of the important stimuli towards an accurate account of its fundamental statements. This is also the historical order: The precise formulation of the axioms came after the models of Beltrami, Klein, and Poincaré.

---

* "Right" can be replaced by "left."

2. Verification of the axioms for Klein's model. We shall now show that in Klein's model all the enumerated axioms are satisfied except the Euclidean axiom of parallelism. As we have mentioned in the preceding section (figure 11), it is clear that here not this axiom but Lobačevskiĭ's axiom holds. It remains to verify the axioms I–IV.

The plane in the model is the interior of a circle (whose radius will be taken to be 1). Points play the role of points, chords the role of straight lines; the concepts "a point lies on a straight line" and "a point lies between two others" are understood in the usual sense. Hence it is obvious that the axioms of incidence, order, and continuity are satisfied. For example, the third axiom of order simply means that a chord divides the circle into two parts.

It remains to verify the axioms of motion. A motion is defined as a transformation that carries the circle into itself and straight lines into straight lines. From this definition it is obvious that these transformations fulfill the first two axioms of motion: The first, because straight lines are just chords and consequently preservation of chords means preservation of straight lines; the second, because if two transformations carrying the circle into itself and the chords into chords are carried out, then the resulting transformation also carries the circle into itself and chords into chords; i.e., it is one of those regarded as "motions."

Thus, only the third axiom of motion remains and the verification of it is the only point of difficulty here.

First of all we note that this axiom contains two statements.

Let A, A' be two points, a, a' two half lines going out from them, α, α' two half planes bounded by the lines a, a'.

The first claim is that there *exists* a motion carrying A into A', a into a', and α into α'.

The second claim is that there is *only one* such motion.

We could perhaps refer to the fact that both these statements have already been proved in Chapter III, §14, but we prefer to prove them here without getting involved, as was the case in Chapter III, with other more general problems.

Let us show that the first statement is true in our model (i.e., in the appropriate interpretation of the terms "half line," "half plane," "motion").

To begin with let us assume that the point A' is at the center of the circle. We take coordinate axes so that the origin is at the center of the circle and the Ox-axis passes through the point A (figure 14). In the preceding section we have investigated the transformation

$$ x' = \frac{x + a}{1 + ax}, \quad y' = \frac{y\sqrt{1 - a^2}}{1 + ax}. \tag{6} $$

We have proved there that it is a "motion" (i.e., that it carries the given circle into itself and straight lines into straight lines).

Let x_0 be the abscissa of A, its ordinate being $y_0 = 0$. Therefore, when we choose $a = -x_0$, then, by (6), A goes over into the point with the coordinates $(0, 0)$, i.e., into A'.

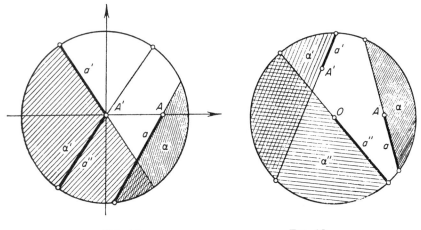

FIG. 14.   FIG. 15.

Now since straight lines go over into straight lines, the "half line" (i.e., segment of a chord) a assumes some position a'' (figure 14). By a rotation around the center we can now carry a'' into a'. The "half plane" α is one of the segments bounded by the "straight line" (chord) a. If after our motion it coincides with a', then the transformation is complete; if not, then by a revolution (reflection in the diameter a') we carry it into the semicircle α'.

Thus, by combining the "shifting" process (6) with a rotation and, if necessary, a revolution, we have carried A, a, α into A', a', α'. But the result of all these "motions" is again a "motion"; this "motion" carries A, a, α into A', a', α'; i.e., the existence of the required motion is now proved.

So far we have restricted ourselves to the special case when the point A' is at the center. Let us now assume that it has some other position. Then by what we have already proved we can carry it into the center by a certain "motion" which we denote by D_1. Then the "half line" a' goes over into some "half line" a'' passing through the center, and the "half plane" α' into a certain "half plane" (semicircle) α'' (figure 15). As we have already proved, by means of a certain "motion" D_2 we can also carry A into the center, the "half line" a into a'', the "half plane" α

into α''. Finally by the inverse "motion" of D_1 we carry A' into its former place and at the same time a'', α'' return to their original positions a', α'*

Thus, as a result of combining the "motion" D_2 and the "motion" inverse to D_1 we carry A, a, α into A', a', α'. But a combination of "motions" is again a "motion"; and so we have ascertained that there

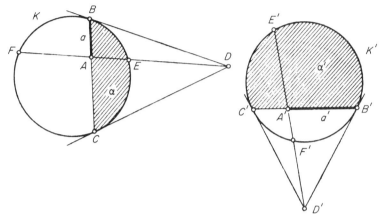

Fig. 16.

exists a motion carrying A, a, α into A', a', α' whatever the position of A and A' in the circle. So the first statement included in the third axiom of motion is proved in its full extent.

Let us now show that the second statement is also satisfied in our model. In the sense of this statement, we have to prove the following.

Let A, A' be two points inside the circle, a, a' segments of chords emanating from them, and α, α' parts of the circle bounded by these chords. For the sake of clarity we illustrate these points, chords, and parts of the circle in two different drawings (figure 16) although, of course, they lie in the one circle in question. It is claimed that the "motion" carrying A into A', a into a', α into α', respectively, is unique, i.e., is completely determined by these data.

For the proof we shall consider a transformation not only of the given circle, but of the whole plane.† A "motion," by definition, carries straight lines into straight lines. A transformation having these properties is

---

* The "motion" inverse to D_1 is expressed by the formulas (4) (§4) if D_1 is expressed by (3).

† It can be proved that a transformation of a circle into itself carrying straight lines into straight lines can be extended uniquely with preservation of this property to the whole projective plane, i.e., the plane to which points at infinity have been adjoined.

called *projective*. Consequently, we can say that for us a "motion" means a projective transformation carrying the given circle into itself. (In figure 16 we have illustrated this so that the circle K goes over into the circle K'. We only have to imagine that by a parallel shift the circle K' is super-imposed on K.)

Projective transformations have been studied in Chapter III, §13, and we shall make use here of the following important theorem that was proved there: A projective transformation is completely determined by the images of four points no three of which are collinear.

Let us return to the "motion" in question. It carries the chord segment a into a' and therefore carries B into B'. Since it carries chords into chords, it also carries C into C'.

Furthermore, since a "motion" generally carries straight lines into straight lines and the given circle into itself but in our illustration the circle K into K', it carries the tangents at B and C into the tangents at B' and C'. Therefore the point D of intersection of the first tangents goes over into the point D' of intersection of the second tangents* (figure 16).

Now since A goes over into A' and straight lines into straight lines, the line AD goes over into $A'D'$. But AD intersects our circle in two points E, F, and $A'D'$ intersects it in E', F'. (In the drawing these points lie on the circumferences of K and K'.) Since the circle goes over into itself, the points E, F go over into E', F'. Let E lie on the arc bounding the part α and E' on that bounding α'. Then, since α by assumption goes over into α', E goes over precisely into E' and F into F'.

Thus we have found that under the "motion" in question the points B, C, E, F on the circumference go over into B', C', E', F'. But it is obvious that of the points B, C, E, F and B', C', E', F' no three can be collinear. Therefore, by the theorem quoted earlier, the projective transformation carrying B, C, E, F into B', C', E', F' is unique. But a "motion" is a projective transformation. Consequently our "motion" carrying A, a, α into A', a', α' is unique, and this is what we had to prove.

Thus we have shown that in the model in question all the axioms of Euclidean geometry are in fact satisfied, except the axiom of parallelism; or in other words, that all the axioms of Lobačevskiĭ geometry are satisfied in the model. The model therefore is a realization of Lobačevskiĭ geometry. This geometry is, as it were, reduced to Euclid's geometry inside the circle, presented in a special manner with the agreed inter-pretation of the terms "straight line" and "motion" in the model.

---

* If, for example, the tangents at B and C are parallel, then the point D is "at infinity."

Incidentally, this enables us to develop the Lobačevskiĭ geometry on the given concrete model, a method that turns out to be more convenient in many problems.

From the point of view of a logical analysis of the foundations of geometry, the proof we have given shows, first, that the Lobačevskiĭ geometry is noncontradictory and, second, that the parallel postulate cannot possibly be deduced from the remaining axioms enumerated previously.

§6. Separation of Independent Geometric Theories from Euclidean Geometry

1. Projective geometry. A fundamental development of geometry parallel with the creation of Lobačevskiĭ geometry came about in yet another way. Within the wealth of all the geometric properties of space, separate groups of properties, distinguished by a peculiar interrelatedness and stability, were singled out and subjected to an independent study. These investigations, with their separate methods, gave rise to new chapters of geometry, i.e., to the science of spatial forms, just as for example anatomy or physiology form distinct chapters in the science of the human organism.

Initially, geometry was not divided up at all. It studied mainly the metrical properties of space connected with the measurement of the dimensions of figures. Circumstances not connected with the measurement but the qualitative character of the natural location of figures were considered in passing only, although it was noted long ago that a part of these properties is distinguished by a peculiar stability, in that they are

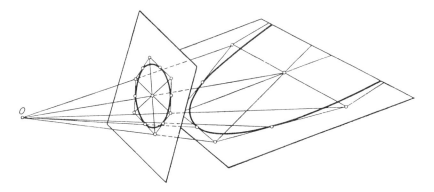

FIG. 17.

preserved under rather substantial distortion of the form and displacement of the location of figures.

Let us consider, for example, the projection of a figure from one plane into another (figure 17). The lengths of segments are changed in the process and so are the angles, the outlines of objects are visibly distorted. However, for example, the property of a number of points of lying on one straight line is preserved and so is the property of a straight line of being a tangent to a given curve, etc.

Projections and projective transformations have already been treated in Chapter III, where we mentioned their obvious connection with perspective, i.e., the drawing of spatial figures on a plane. The study of properties of perspective goes back in antiquity right to Euclid, to the work of the ancient architects; artists concerned themselves with perspective: Dürer, Leonardo da Vinci, and the engineer and mathematician Desargues (17th century). Finally, at the beginning of the 19th century Poncelet was the first to separate out and study systematically the geometrical properties that are preserved under arbitrary projective transformations of the plane (or of space) and so to create an independent science, namely *projective geometry.**

It might seem that there are only a few, very primitive properties that are preserved under arbitrary projective transformations, but this is by no means so. For example, we do not notice immediately that the theorem stating that the points of intersection of opposite sides (produced) of a hexagon inscribed in a circle lie on a straight line also holds for an ellipse, parabola, and hyperbola. The theorem only speaks of projective properties, and these curves can be obtained from the circle by projection. It is even less obvious that the theorem to the effect that the diagonals of a circumscribed hexagon meet in a point is a peculiar analogue of the theorem just mentioned; the deep connection between them is revealed only in projective geometry. Also it is not obvious that under a projection, irrespective of the distortion of distances, for any four points A, B, C, D (figure 18) lying on a straight line the cross ratio $AC/CB : AD/DB$ remains unaltered

$$\frac{AC}{CB} : \frac{AD}{DB} = \frac{A'C'}{C'B'} : \frac{A'D'}{D'B'}.$$

This implies that many relations are maintained in a perspective. For example, by using this fact it is easy to determine the distance of the

---

* Poncelet, a French military engineer, carried out his geometrical investigations during his captivity in Russia after 1812. His "Traité des propriétés projectives des figures" appeared in 1822.

telegraph poles A, B, C from the point D (figure 18) on a photograph of the road leading into the distance, when their spacing is known.

With projective geometry and its application to aerial photography, we have dealt in Chapter III. It stands to reason that its laws are used in architecture, in the construction of panoramas, in decorating, etc.

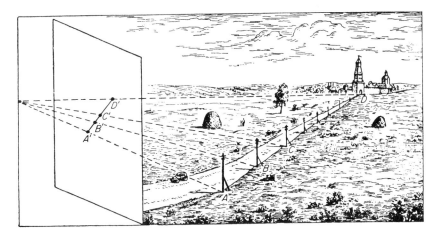

Fig. 18.

The separation of projective geometry played an important role in the development of geometry itself.

2. Affine geometry. As another example of an independent geometry we can take *affine geometry*. Here one studies the properties of figures that are not changed by arbitrary transformations in which the Cartesian coordinates of the original (x, y, z) and the new (x', y', z') position of each point are connected by linear equations:

$$x' = a_1x + b_1y + c_1z + d_1,$$
$$y' = a_2x + b_2y + c_2z + d_2,$$
$$z' = a_3x + b_3y + c_3z + d_3$$

(where it is assumed that the determinant

$$\begin{vmatrix} a_1 & b_1 & c_1 \\ a_2 & b_2 & c_2 \\ a_3 & b_3 & c_3 \end{vmatrix}$$

is different from zero).

It turns out that every affine transformation reduces to a motion, possibly a reflection, in a plane and then to a contraction or extension of space in three mutually perpendicular directions.

Quite a number of properties of figures are preserved under each of these transformations. Straight lines remain straight lines (in fact all "projective" properties are preserved); moreover, parallel lines remain parallel; the ratio of volumes is preserved, also the ratio of areas of figures that lie in parallel planes or in one and the same plane, the ratio of lengths of segments that lie on one straight line or on parallel lines, etc. Many well-known theorems belong essentially to affine geometry. Examples are the statements that the medians of a triangle are concurrent, that the diagonals of a parallelogram bisect each other, that the midpoints of parallel chords of an ellipse lie on a straight line, etc.

The whole theory of curves (and surfaces) of the second order is closely connected with affine geometry. The very division of these curves into ellipses, parabolas, hyperbolas is, in fact, based on affine properties of the figures: Under affine transformations an ellipse is transformed precisely into an ellipse and never into a parabola or a hyperbola; similarly a parabola can be transformed into any other parabola, but not into an ellipse, etc.

The importance of the separation and detailed investigation of general affine properties of figures is emphasized by the fact that incomparably more complicated transformations turn out to be essentially linear, i.e., affine in the infinitely small, and the application of the methods of the differential calculus is linked exactly with the consideration of infinitely small regions of space.

3. Klein's Erlanger Programm.   In 1872 Klein, in a lecture at the University of Erlangen which is now known as the "Erlanger Programm," in summing up the results of the developments of projective, affine, and other "geometries" gave a clear formulation of the general principle of their formation: We can consider an arbitrary group of single-valued transformations of space and investigate the properties of figures that are preserved under the transformations of this group.*

From this point of view the properties of space are stratified, as it were, with respect to their depth and stability. The ordinary Euclidean

---

* The word "group" is used here not merely in the sense of a collection. When we speak of a group of transformations (see Chapter XX) we have in mind a set of transformations which must contain the identity transformation (leaving all points in place), which contains together with every nonidentical transformation the one inverse to it (restoring all points to their previous place), and which contains together with any two transformations of the set also the transformation that is equivalent to the two carried out in succession.

geometry was created by disregarding all properties of real bodies other than the geometrical; here, in the special branches of geometry, we perform yet another abstraction within geometry, by disregarding all geometrical properties except the ones that interest us in the given branch of geometry.

In accordance with this principle of Klein, we can construct many geometries. For example, we can consider the transformations that preserve the angle between arbitrary lines (conformal transformations of space), and when studying properties of figures preserved under such transformations we talk of the corresponding conformal geometry. We can consider transformations of not necessarily the whole space. Thus, by considering the points and chords of a circle under all its transformations into itself that carry chords into chords and by singling out the properties that are preserved under such transformations, we obtain a geometry which as we have shown in §§4 and 5 coincides with Lobačevskiĭ geometry.

4. Abstract spaces. The further development of the theories thus separated, even from the theoretical point of view (to say nothing of their factual content), did not stop at what we have said here.

If we are interested, for example, only in affine properties of figures we can, by abstracting from all other properties, imagine a space and geometrical figures in it that have *only* properties of interest to us and, as it were, no other properties at all. In this "space," figures do not have any properties except affine ones. It is natural to try and give also an axiomatic account of the geometry of such an abstract space, i.e., to assume that we are dealing with some abstract objects: "points," "straight lines," and "planes" whose properties are expressed in certain axioms (there are, of course, fewer of these properties than in the case of Euclidean geometry) so that consequences derived from these axioms correspond to affine properties of figures of the ordinary space.

This can indeed be done; and such a collection of abstract "points," "straight lines" and "planes" with the system of their properties is called an *affine space*.

In exactly the same way we can imagine an abstract system of objects having only that range of properties that correspond to projective properties of figures of Euclidean space. (This time the difference of the axiom system from the axioms of ordinary geometry turns out to be even more substantial.)

5. Abstract topological space. When we go deeper into the nature of geometric forms, we may note that in quite a number of problems we are concerned with properties, even deeper than projective ones, which are so firmly connected with the given figure that they are preserved under arbitrary distortions, provided only that these distortions do not cause the figure to break up or parts of it to become "pasted together." To make the notion of such a continuous distortion more precise than the intuitive description does, we refer to the definition of a continuous function known to us from analysis and say that we are dealing with an arbitrary transformation of all the points of the figure into a new position under which the Cartesian coordinates of the points in the new position are expressed as continuous functions of the old coordinates, while the old coordinates in turn can be expressed as continuous functions of the new ones.

Properties of figures that are preserved under arbitrary transformations of this kind are called topological and the science which investigates them is *topology* (see Chapter XVIII).

Topological properties connected with figures are in the simplest cases distinguished by their exceptionally intuitive character. For example, it is almost obvious that every line in a plane that can be obtained by a continuous deformation of the circumference of a circle divides the plane into two parts, the interior and exterior, no matter how much the contour winds; therefore

FIG. 19.

the property of a circumference of dividing the plane is topological. Also it is visibly obvious, no doubt, that the torus surface (figure 19) cannot possibly be turned into a sphere by a continuous transformation, so that the property of an arbitrary surface of admitting a continuous transformation into, say, a torus surface is a topological property that distinguishes it from many other surfaces.

Arguments in connection with continuity are of an intuitive character and often clarify the essence of the matter so well that it is very tempting to try to turn them into rigorous proofs and, even more, to extend such methods to other, incomparably more complicated problems.

For example, take the argument that establishes the truth of the fundamental theorem of algebra to the effect that every equation

$$z^n + a_1 z^{n-1} + a_2 z^{n-2} + \cdots + a_{n-1} z + a_n = 0 \tag{7}$$

has at least one real or complex root.

Let z be a point of one complex plane and $w = f(z)$ the corresponding point of another complex plane w, where $f(z)$ denotes the left-hand side of (7). For very large absolute values of z, the function $f(z)$ differs relatively little from z^n; but the function z^n is very simple. In particular, it is easy to verify that if z by a continuous motion describes in the complex plane a circumference with center at the origin, then the point $w_1 = z^n$ goes precisely n times around a similar circumference of radius $|z|^n$ in the w-plane.* So the point $w = f(z)$ describes n loops forming some contour Γ comparatively near to the line described by the point $w_1 = z^n$ (figure 20).

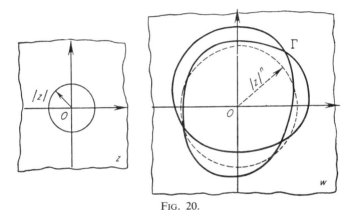

FIG. 20.

Now when the circumference described by z is continuously contracted to one point, then the n times looped-contour Γ described by $w = f(z)$ is continuously deformed and also contracts to a point. But it is rather obvious that it cannot contract to a point without passing through the origin O which this contour surrounds initially. Hence it goes through O at least once, and for such a z we have $w = f(z) = 0$. This z is, then, a root of the equation (7). Strictly speaking it is also clear that in a certain sense there must be exactly n roots, since each of the n loops of the contour Γ on contraction passes through O.

Our argument requires, of course, a rigorous establishment of just those topological properties of the contour and its deformations that we have used here.

* For if $z = \rho(\cos \phi + i \sin \phi)$ then [see Chapter IV, §3] we know that
$$z^n = \rho^n(\cos \phi n + i \sin \phi n);$$
therefore, when the argument ϕ of z changes from 0 to 2π, then the argument of z^n changes from 0 to $2\pi n$, i.e., the radius vector leading to z^n makes n complete revolutions under this motion.

We could give many examples of the use of topological properties in various branches of mathematics often very far removed from geometry.

With the study of topological properties before us, we can again imagine an abstract set of objects having only properties of this kind (see §7, Chapter XX). Such a set is called an *abstract topological space*.

This point of view is already incomparably wider than the study of topological properties of geometric figures only. Topological spaces can be extremely varied; for example, all the points of a torus surface with their specific properties of adjacency to one another form one topological space, all the points of the plane another, the whole Euclidean space a third; all the points of the various many-sheeted Riemann surfaces of which we have talked in Chapter IX, §5, in connection with the theory of functions of a complex variable, form other distinct topological spaces. But it is most remarkable that often the concept of neighborhood and adjacency can clearly be established between objects that do not fall at all under our notion of geometric points. For example, for all possible positions of a hinged mechanism we can clearly indicate what a "neighboring" position means, or that one position is "adjacent" to an infinite range of others among which there are positions arbitrarily near to the given one.

We see that the concept of a topological space is extremely general. We shall return to this point in §8.

The object of this section was not only to give the reader an idea of the various geometries, but also to show that certain concrete problems lead to the isolation and investigation of separate groups of geometric properties; that these investigations entail the creation of the idea of abstract geometric objects having *only* these properties, i.e., that the isolation of these properties in their pure form leads us to the idea of the corresponding abstract space.

Other developments, leading to the construction of a different kind of abstract spaces, will be discussed in the following section.

§7. Many-Dimensional Spaces

1. The geometry of *n*-dimensional space. An important step in the development of new geometric ideas was the creation of the geometry of many-dimensional spaces to which we have already referred in the preceding chapter. One of the moving forces was the tendency to use geometric arguments for the solution of problems in algebra and analysis. The geometric approach to the solution of analytical problems is based on the method of coordinates. Let us give a simple example.

We want to know how many integral solutions the inequality $x^2 + y^2 < N$

has. By regarding x and y as Cartesian coordinates in a plane, we see that the problem reduces to the following: How many points with integer coordinates are contained in a circle of radius $\sqrt{N}$. The points with integer coordinates are the vertices of squares with sides of unit length covering the plane (figure 21). The number of such points inside the circle is approximately equal to the number of squares lying in the circle, i.e., to the area of a circle of radius $\sqrt{N}$. Thus, the number of solutions of the inequality we are interested in is approximately equal to πN. Furthermore, it is not difficult to prove that the relative error occurring here tends to zero for $N \to \infty$. A more accurate study of this error is a

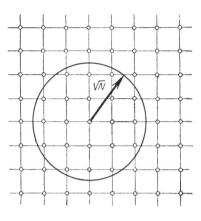

FIG. 21.

very difficult problem in the theory of numbers and has become the object of deep investigations in comparatively recent years.

In our example it was sufficient to translate the problem into geometric language in order to obtain at once a result which is by no means obvious from the point of view of "pure algebra." The corresponding problem for an inequality with three unknowns can be solved in exactly the same way. However, when there are more than three unknowns, this method is not applicable because our space is three-dimensional, i.e., the position of a point in it is determined by a triple of coordinates. To preserve the convenient geometrical analogy in similar cases we introduce the idea of an abstract "n-dimensional space" whose points are determined by n coordinates $x_1, x_2, \cdots, x_n$. The fundamental concepts of geometry are then generalized in such a way that the geometric arguments turn out to be applicable to the solution of problems with n variables; this makes it much easier to obtain results. The possibility of such a generalization is based on the uniformity of the algebraic laws thanks to which many problems can be solved simultaneously for an arbitrary number of variables. This enables us to apply geometric arguments that are valid for three dimensions to an arbitrary number of dimensions.

2. Coordinates in n-dimensional geometry. Rudiments of the concepts of a four-dimensional space can already be found in Lagrange who, in his papers on mechanics, formally regarded the time as a "fourth coordinate" beside the three spatial coordinates. But the first systematic account of the elements of many-dimensional geometry was given in 1844 by the German mathematician Grassmann and independently by the Englishman Cayley. They proceeded by way of a formal analogy with the ordinary analytical geometry. A general outline of this analogy in an up-to-date exposition is as follows.

A point in an n-dimensional space is determined by n coordinates

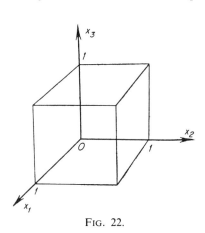

FIG. 22.

$x_1, x_2, \cdots, x_n$. A figure in n-dimensional space is a geometric locus or set of points satisfying certain conditions. For example, an n-dimensional "cube" is defined as the geometric locus of points whose coordinates are subject to the inequalities $a \leqslant x_i \leqslant b$ ($i = 1, 2, \cdots, n$). The analogy with the ordinary cube is here completely evident: When $n = 3$, i.e., when the space is three-dimensional, our inequalities in fact define the cube whose edges are parallel to the coordinate axes and of length $b - a$ (figure 22 illustrates the case $a = 0$, $b = 1$).

The distance between two points can be defined as the square root of the sum of the squares of the differences of the coordinates

$$d = \sqrt{(x_1' - x_1'')^2 + (x_2' - x_2'')^2 + \cdots + (x_n' - x_n'')^2}.$$

This is a direct generalization of the well-known formulas for the distance in a plane or in three-dimensional space, i.e., for $n = 2$ or 3.

It is now possible to define equality of figures in n-dimensional space. Two figures are regarded as equal if a correspondence can be established between their points under which the distances between pairs of corresponding points are equal. A transformation preserving distances can be called a generalized motion.* Then we can say by analogy with the usual Euclidean geometry that the objects of study in n-dimensional geometry are the properties of figures that are preserved under generalized motions.

---

* The generalization consists not only in the transition to n variables but also in the fact that a reflection in a plane is incorporated among motions, because it also does not alter distances between points.

This definition of the content of n-dimensional geometry was set up in the 1870's and provided a precise foundation for its development. Since then n-dimensional geometry has been the object of numerous investigations in all directions similar to those of Euclidean geometry (elementary geometry, general theory of curves, etc.).

The concept of distance between points also enables us to transfer to n-dimensional space other concepts of geometry such as segment, sphere, length, angle, volume, etc. For example, an n-dimensional sphere is defined as the set of points whose distance from a given point is not more than a given R. Therefore a sphere is given analytically by an inequality

$$(x_1 - a_1)^2 + \cdots + (x_n - a_n)^2 \leqslant R^2,$$

where $a_1, \cdots, a_n$ are the coordinates of its center. The surface of the sphere is given by the equation

$$(x_1 - a_1)^2 + \cdots + (x_n - a_n)^2 = R^2.$$

The segment AB can be defined as the set of points X such that the sum of the distances from X to A and B is equal to the distance from A to B. (The length of a segment is the distance between its end points.)

3. Hyperplanes. Let us dwell in some detail on planes of various dimensions.

In three-dimensional space there are the one-dimensional "planes" (namely, the straight lines), and the ordinary (two-dimensional) planes. In n-dimensional space for $n > 3$ we also have to take many-dimensional planes into account, of dimensions 3 to $n - 1$.

In three-dimensional space a plane is, of course, given by one linear equation, and a straight line by two such equations.

By a direct generalization we come to the following definition: A *k-dimensional plane* (usually called a *hyperplane*) in an n-dimensional space is the geometric locus of points whose coordinates satisfy a system of $n - k$ linear equations

$$
\begin{aligned}
a_{11} x_1 + a_{12} x_2 + \cdots + a_{1n} x_n + b_1 &= 0, \\
a_{21} x_1 + a_{22} x_2 + \cdots + a_{2n} x_n + b_2 &= 0, \\
&\cdots\cdots\cdots\cdots\cdots\cdots\cdots\cdots\cdots\cdots\cdots\cdots\cdots \\
a_{n-k,1} x_1 + a_{n-k,2} x_2 + \cdots + a_{n-k,n} x_n + b_n &= 0,
\end{aligned}
\qquad (8)
$$

provided the equations are compatible and independent (i.e., none of them is a consequence of the others). Each of these equations represents an $(n - 1)$-dimensional hyperplane and together they determine the common points of $n - k$ such hyperplanes. The fact that the equations (8) are compatible means that there are some points that satisfy them, i.e., that the $n - k$ given $(n - 1)$-dimensional hyperplanes intersect. The fact that none of the equations is a consequence of the others means that none of them can be omitted. Otherwise the system would reduce to a smaller number of equations and would define a hyperplane of a larger number of dimensions. Thus, to speak geometrically, a k-dimensional hyperplane is determined as the intersection of $n - k$ $(n - 1)$-dimensional hyperplanes represented by independent equations. In particular, when $k = 1$, we have $n - 1$ equations which determine a "one-dimensional hyperplane," i.e., a straight line. Thus, this definition of a k-dimensional hyperplane is a natural formal generalization of well-known results of analytical geometry. The advantage of this generalization becomes apparent in the fact that conclusions concerning systems of linear equations receive a geometric interpretation which makes them more lucid. This geometric approach to problems of linear algebra was also discussed in Chapter XVI.

An important property of a k-dimensional hyperplane is the fact that it can be regarded as a k-dimensional space. For example, a three-dimensional hyperplane is itself an ordinary three-dimensional space. This enables us to transfer to a space of higher dimension many con-clusions obtained for spaces of lower dimension, by means of the usual argument from n to $n + 1$.

If the equations (8) are compatible and independent, then it is proved in algebra that we can choose k of the n variables x_i at will and then the remaining $n - k$ variables can be expressed in terms of them.* For example:

$$x_{k+1} = c_{11} x_1 + c_{12} x_2 + \cdots + c_{1k} x_k + d_1 ,$$
$$x_{k+2} = c_{21} x_1 + c_{22} x_2 + \cdots + c_{2k} x_k + d_2 ,$$
$$\ldots\ldots\ldots\ldots\ldots\ldots\ldots\ldots\ldots\ldots\ldots\ldots\ldots\ldots\ldots\ldots\ldots\ldots\ldots\ldots\ldots\ldots$$
$$x_n \ \ = c_{n-k,1} x_1 + c_{n-k,2} x_2 + \cdots + c_{n-k,k} x_k + d_k .$$

Here arbitrary values can be given to x_1, x_2, $\cdots$, x_k, and the remaining x_i are determined by them. This means that the position of a point in a k-dimensional hyperplane is determined by k coordinates that can assume arbitrary values. It is in this sense that the hyperplane has k dimensions.

---

* These k variables cannot, in general, be chosen arbitrarily from the x_i. For example, in the system $x_1 + x_2 + x_3 = 0$, $x_1 - x_2 - x_3 = 0$ the value of x_1 is uniquely deter-mined: $x_1 = 0$, and obviously neither x_2 nor x_3 can be expressed in terms of it. All that is stated, however, is that the necessary k of the x_i can always be found.

From the definition of the hyperplanes of various dimensions, we can derive in a purely algebraic way the following fundamental theorems.

1. One and only one k-dimensional hyperplane passes through $k + 1$ points that do not lie in a $(k - 1)$-dimensional hyperplane.

The complete analogy with known facts of elementary geometry is obvious here. The proof of this theorem is based on the theory of systems of linear equations and is somewhat complicated, so that we shall not write it out.

2. If an l-dimensional and a k-dimensional hyperplane in an n-dimensional space have at least one point in common and $l + k \geqslant n$, then they intersect in a hyperplane of dimension not less than $l + k - n$.

Hence it follows as a special case that two two-dimensional planes in a three-dimensional space, if they do not coincide and are not parallel, intersect on a straight line ($n = 3$, $l = 2$, $l + k - n = 1$). But in a four-dimensional space two two-dimensional planes may well have a single point in common. For example, the planes given by the system of equations:

$$\left.\begin{matrix} x_1 = 0 \\ x_2 = 0 \end{matrix}\right\} , \qquad \left.\begin{matrix} x_3 = 0 \\ x_4 = 0 \end{matrix}\right\} ,$$

obviously intersect only in the point with the coordinates $x_1 = 0$, $x_2 = 0$, $x_3 = 0$, $x_4 = 0$.

The proof of the theorem is extremely simple: An l-dimensional hyperplane is given by $n - l$ equations, a k-dimensional one by $n - k$; the coordinates of the points of intersection must satisfy simultaneously all these $(n - l) + (n - k) = n - (l + k - n)$ equations. If none of the equations is a consequence of the others, then by the very definition of a hyperplane we have as intersection an $(l + k - n)$-dimensional hyperplane; otherwise we have a hyperplane with a larger number of dimensions.

To these two theorems we can add another two.

3. In each k-dimensional hyperplane there are at least $k + 1$ points that do not lie in a hyperplane of smaller dimension. In an n-dimensional space there are at least $n + 1$ points that do not lie in any hyperplane.

4. If a straight line has two points in common with a hyperplane (of an arbitrary number of dimensions), then it lies entirely in that hyperplane. Generally, if an l-dimensional hyperplane has $l + 1$ points in common with a k-dimensional hyperplane that do not lie in an $(l - 1)$-dimensional hyperplane, then it lies entirely in this k-dimensional hyperplane.

Note that n-dimensional geometry can be built up starting from axioms that generalize the axioms listed in §5. In this approach the four theorems mentioned here assume the role of axioms of incidence. This shows, by the way, that the concept of axiom is relative: One and the same statement can emerge as a theorem in one buildup of a theory, and as an axiom in another.

4. Various examples of an *n*-dimensional space. We have now obtained a general idea of the mathematical concept of a many-dimensional space. In order to clarify the actual physical meaning of this concept, let us turn again to the problem of graphical illustration. Suppose, for example, we wish to illustrate the dependence of the pressure of a gas on its volume. We take coordinate axes in a plane and plot on one axis the volume v, on the other the pressure p. The dependence of the pressure on the volume under the given conditions is then illustrated by a certain curve (by the well-known Boyle-Mariotte law this would be a hyperbola for an ideal gas with a fixed temperature). But when we have a more complicated physical system, whose state is given not by two data (like volume and pressure in the case of the gas) but by say five, then the graphical illustration of its behavior leads to the notion of a five-dimensional space.

Suppose, for example, that we are concerned with an alloy of three metals or a mixture of three gases. The state of the mixture is determined by four data: the temperature T, the pressure p, and the percentage contents c_1, c_2 of two gases (the percentage content of the third gas is then determined by the fact that the sum total of the percentage contents is 100% so that $c_3 = 100 - c_1 - c_2$). The state of such a mixture is, therefore, determined by four data. A graphical illustration of it either requires a combination of several diagrams, or else we have to represent the state in the form of a point of a four-dimensional space with four coordinates T, p, c_1, c_2. Such a representation is, in fact, used in chemistry; the application of the methods of many-dimensional geometry to chemistry was developed by the American scientist Gibbs and the school of Soviet physicochemists of Academician Kurnakov. Here the introduction of a many-dimensional space is dictated by the endeavor to preserve the useful geometrical analogies and arguments arising from the simple device of graphical illustration.

Let us give an example from the realm of geometry. A sphere is given by four data: the three coordinates of its center and its radius.* The special geometry of spheres which was built up about a century ago by several mathematicians can, therefore, be regarded as a four-dimensional geometry.

---

* Translator's note: Therefore a sphere can be represented as a point in four-dimensional space.

From all we have said the real basis for the introduction of the concept of a many-dimensional space will be clear. If some figure or the state of some system, etc., is given by n data, then this figure, this state, etc., can be conceived as a point of some n-dimensional space. The advantage of this representation is approximately the same as that of ordinary graphs: It consists in the possibility of applying well-known geometric analogies and methods to the study of the phenomena in question.

There is, therefore, no mysticism in the mathematical concept of a many-dimensional space. It is not more than a certain abstract concept developed by the mathematicians for the purpose of describing in geometric language those things that do not admit a simple geometric illustration in the usual sense. This abstract concept has an entirely real basis, it reflects actuality and was created by the demands of science, not by idle play of the imagination. It reflects the fact that there exist such things as a sphere or a mixture of three gases that are characterized by several data so that the collection of all these things is many-dimensional. The number of variables in a given case is just the number of these data; just as a point moving in space changes its three coordinates, so a sphere moving, expanding, and contracting changes its four "coordinates," i.e., the four quantities that determine it.

In the subsequent sections, we shall dwell upon many-dimensional geometry. It is important here to understand that this is a method of mathematical description of real things and phenomena. The idea that there exists some sort of four-dimensional space in which our real space is embedded, an idea that has been used by certain litterateurs and spiritualists, has no relation to the mathematical concept of a four-dimensional space. If one can speak here of a relationship to science at all, it is perhaps possible only in the sense of an imaginative distortion of scientific concepts.

5. Polyhedra in n-dimensional space. As we have already said, the geometry of a many-dimensional space was built up at first by way of a formal generalization of the usual analytic geometry to an arbitrary number of variables. However, such an approach to the matter could not satisfy the mathematicians completely. As a matter of fact, the purpose was not so much a generalization of the geometric concepts as of the geometric method of investigation itself. It was, therefore, important to give a purely geometric exposition of n-dimensional geometry, independently of the analytical apparatus. This was first done by the Swiss mathematician Schlaefli in 1852, one of whose articles deals with the problem of regular polyhedra of a many-dimensional space. True, Schlaefli's article was not appreciated by his contemporaries, because to

understand it one has to rise, to a certain extent, to an abstract view
of geometry. Only subsequent developments of mathematics have brought
complete clarity into this problem, by an exhaustive elucidation of the
mutual relationship of the analytic and geometric approach. Since we
cannot go deeper into this problem, we confine ourselves to examples of
a geometric exposition of n-dimensional geometry.

Let us consider the geometric definition of an n-dimensional cube.
When we move a segment in a plane perpendicular to itself by a distance
equal to its length, we sweep out a square, i.e., a two-dimensional cube
(figure 23a). Similarly, when we move a square in the direction perpen-

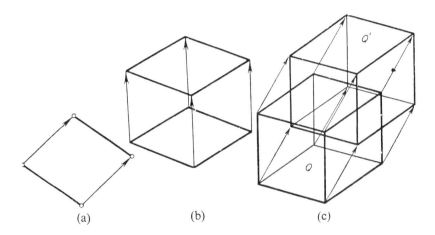

(a)   (b)   (c)

Fig. 23.

dicular to its plane by a distance equal to its side, we sweep out a three-
dimensional cube (figure 23b). In order to obtain a four-dimensional cube,
we use the same construction: We take a three-dimensional hyperplane
in a four-dimensional space and in it a three-dimensional cube and we
move it in the direction perpendicular to this three-dimensional hyperplane
by a distance equal to its edge (by definition a straight line is perpendicular
to a k-dimensional hyperplane if it is perpendicular to every straight line
lying in that hyperplane). This construction is symbolically represented in
figure 23c. Two three-dimensional cubes Q and Q' are drawn here,
namely the given cube in its initial and its final position. The lines joining
the vertices of these cubes illustrate the segments traced out by the vertices
in the movement of the cube. We see that a four-dimensional cube has
16 vertices in all: 8 for the cube Q and 8 for Q'. Further, it has 32 edges:
12 edges of the moving three-dimensional cube in its initial position Q,

12 edges in the final position Q', and 8 "lateral" edges. It has 8 three-dimensional faces which are themselves cubes. In the motion of the three-dimensional cube each of its faces sweeps out a three-dimensional cube so that we obtain 6 cubes as the lateral faces of the four-dimensional cube, and in addition there are two faces: "front" and "back" or the initial and final position of the moving cube. Finally, a four-dimensional cube also has two-dimensional square faces, 24 in number: 6 each in the cubes Q and Q', and another 12 squares that are swept out by the edges of Q in its motion.

So a four-dimensional cube has 8 three-dimensional faces, 24 two-dimensional faces, 32 one-dimensional faces (edges), and finally 16 vertices; every face is a "cube" of the appropriate number of dimensions: a three-dimensional cube, a square, a segment, and a vertex (which we can regard as a zero-dimensional cube).

Similarly, by shifting a four-dimensional cube "into the fifth dimension" we obtain a five-dimensional cube and so, by repeating the construction, we can build up a cube of an arbitrary number of dimensions. All the faces of an n-dimensional cube are themselves cubes of a smaller number of dimensions: $(n - 1)$-dimensional, $(n - 2)$-dimensional, etc., finally one-dimensional, i.e., edges. For the inquisitive reade. it is not a difficult task to find out how many faces of each number of dimensions an n-dimensional cube has. It is easy to see that the number of $(n - 1)$-dimensional faces is $2n$ and that there are 2^n vertices. But how many edges are there, for example?

Let us look at another polyhedron in an n-dimensional space. In the plane the simplest polygon is a triangle; it has the least possible number of vertices. In order to obtain a polyhedron with the least number of vertices it is sufficient to take a point not in the plane of a triangle and join it by segments to each point of the triangle. The segments so obtained fill a three-sided pyramid, i.e., a tetrahedron (figure 24). In order to obtain

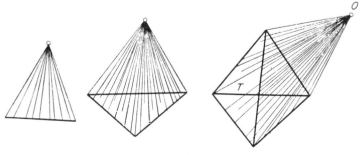

F$_{IG}$. 24.

the simplest polyhedron in a four-dimensional space we argue as follows. We take an arbitrary three-dimensional hyperplane and in it a certain tetrahedron T. Next we take a point not in the given three-dimensional hyperplane and join it by segments to all the points of the tetrahedron T. On the right of figure 24 we have illustrated this construction symbolically. Each of the segments joining the point O to a point of the tetrahedron T has no other points in common with the tetrahedron, because otherwise it would lie entirely in the three-dimensional space containing T. All these segments, as it were, "go into the fourth dimension." They form the simplest four-dimensional polyhedron, the so-called *four-dimensional simplex*. Its three-dimensional faces are tetrahedra: one at the base and 4 lateral faces resting on the two-dimensional faces of the basis; altogether 5 faces. Its two-dimensional faces are triangles; there are 10 of them: 4 in the basis and 6 lateral. Finally, it has 10 edges and 5 vertices.

By repeating the same construction for an arbitrary number n of dimensions we obtain the simplest n-dimensional polyhedron, the so-called *n-dimensional simplex*. As is clear from the construction, it has $n + 1$ vertices. One can see that all its faces are also simplexes of a smaller number of dimensions: $(n - 1)$-dimensional, $(n - 2)$-dimensional, etc.*

It is also easy to generalize the concepts of a prism and a pyramid.

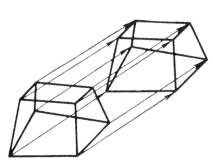

FIG. 25.

If we give a parallel shift to a plane polygon into the third dimension, then it sweeps out a prism. Similarly, by shifting a three-dimensional polyhedron into the fourth dimension, we obtain a four-dimensional prism (illustrated symbolically in figure 25). A four-dimensional cube is, of course, a special case of a prism.

A pyramid is constructed as follows. We take a polygon Q and a point O not in the plane of the polygon. Each point of Q is joined

---

* Any m vertices of a simplex determine the $(m - 1)$-dimensional simplex "spanned" by them: an $(m - 1)$-dimensional face of the given n-dimensional simplex. The number of $(m - 1)$-dimensional faces of an n-dimensional simplex is therefore equal to the number of combinations of m of its vertices from $n + 1$, i.e.,

$$C_m^{n+1} = \frac{(n + 1)!}{m! \, (n - m + 1)!}.$$

by a segment to O and these segments fill a pyramid with the base Q (figure 26). Similarly, if a three-dimensional polyhedron Q is given in a four-dimensional space and a point O not in the same three-dimensional plane, then the segments joining the points of Q to O form a four-dimensional pyramid with the base Q. A four-dimensional simplex is nothing but a pyramid with a tetrahedron as base.

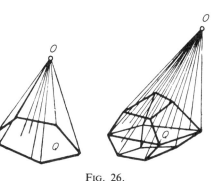

FIG. 26.

In exactly the same way, by starting from an $(n-1)$-dimensional polyhedron Q, we can define an n-dimensional prism and an n-dimensional pyramid.

Generally, an n-dimensional polyhedron is a part of an n-dimensional space bounded by a finite number of portions of $(n-1)$-dimensional hyperplanes; a k-dimensional polyhedron is a part of a k-dimensional hyperplane bounded by a finite number of portions of $(k-1)$-dimensional hyperplanes. The faces of a polyhedron are themselves polyhedra of a smaller number of dimensions.

The theory of n-dimensional polyhedra is a generalization of the theory of ordinary three-dimensional polyhedra and is full of concrete results. In a number of cases theorems on three-dimensional polyhedra generalize to an arbitrary number of dimensions without special difficulties, but there are also problems whose solution for n-dimensional polyhedra runs into enormous difficulties. We might mention here the deep investigations of G. F. Voronoǐ (1868-1908) which arose, by the way, in connection with problems of the theory of numbers; they were continued by Soviet geometers. One of the problems, the so-called "Voronoǐ problem," is still not completely solved.*

As an example for the essential difference that can hold between spaces of different dimensions, we can take the regular polyhedra. In the plane a regular polyhedron can have an arbitrary number of sides. In other words, there are infinitely many distinct forms of regular "two-dimensional

---

* It concerns the search for those convex polyhedra by which the space can be filled by joining them one to another parallel and along whole faces. In the case of three-dimensional space this problem was raised and solved by Fedorov in connection with the needs of crystallography; Voronoǐ and his successors made some progress with the same problem for the n-dimensional space, but the final solution is known only for the spaces of two, three, and four dimensions.

polyhedra." There are altogether five forms of three-dimensional regular polyhedra: the tetrahedron, the cube, the octahedron, the dodecahedron, and the icosahedron. In four-dimensional space there are six forms of regular polyhedra, but in any space of a larger number of dimensions there are only three. They are: (1) the analogue to the tetrahedron, the regular n-dimensional simplex, i.e., the simplex whose edges are all equal; (2) the n-dimensional cube; (3) the analogue to the octahedron which is constructed as follows: The centers of the faces of the cube become the vertices of this polyhedron so that it is spanned by them, as it were. In the case of a three-dimensional space this construction is carried out in figure 27. So we see that as far as regular polyhedra are concerned, spaces of two, three and four dimensions occupy a special position.

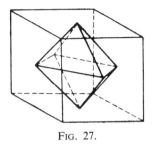

FIG. 27.

6. Calculation of volumes. Now let us discuss the problem of the volume of a body in n-dimensional space. The volume of an n-dimensional body is defined similarly to the way it is done in ordinary geometry. Volume is a numerical characteristic attached to a figure, and of the volume it is postulated that equal bodies have equal volumes, i.e., that the volume does not change when the figure moves as a rigid body, and that in case one body is composed of two others, its volume is equal to the sum of their volumes. As unit of volume one takes the cube with edges of unit length. It is then shown that the volume of a cube with edge a is a^n. This is done exactly as in the plane and in three-dimensional space, by filling up the cube with layers of cubes (figures 28). Since the cubes are packed in n directions, this gives us a^n.

In order to define the volume of an arbitrary n-dimensional body, we

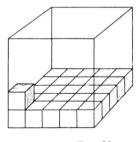

FIG. 28.

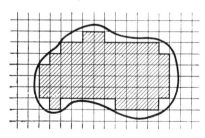

FIG. 29.

replace it by an approximate body composed of very many n-dimensional cubes similarly to the way in which in figure 29 a plane figure is replaced by a figure of squares. The volume of the body is defined as the limit of such step-shaped bodies when the cubes that make it up decrease in size *ad infinitum*.

The k-dimensional volume of a k-dimensional figure lying in some k-dimensional hyperplane is defined in exactly the same way. From the definition of the volume it is easy to deduce an important property of it: Under a similarity magnification of a body, i.e., when all its linear dimensions are increased λ fold, the k-dimensional volume is increased λ^k fold.

If a body is divided into parallel layers, then its volume is the sum of the volumes of these layers

$$V = \Sigma V_i .$$

The volume of each layer can be represented approximately as the product of its height Δh_i with the $(n-1)$-dimensional volume ("area") of the corresponding section S_i. As a result, the total volume of the body is represented approximately by the sum

$$V \approx \Sigma S_i \, \Delta h_i .$$

Passing to the limit for all $\Delta h_i \to 0$ we obtain a representation of the volume in the form of an integral

$$V = \int_0^H S(h) \, dh, \tag{9}$$

where H is the length of the body in the direction perpendicular to the section.

All this is completely analogous to the calculation of volumes in three-dimensional space. For example, for a prism all the sections are equal and, therefore, their "area" does not depend on h. Hence for the prism $V = SH$; i.e., the volume of a prism is equal to the product of the "area" of the base and the height. Let us determine the volume of an n-dimensional pyramid. Suppose that a pyramid is given with height H and area of the base S. We cut it by a plane parallel to the base at a distance h from the vertex. Then a pyramid of height h is cut off. We denote the area of its base by $s(h)$. This smaller pyramid is, obviously, similar to the original one: All its dimensions are smaller in the ratio of h to H, i.e., they are multiplied by h/H. Therefore the $(n-1)$-dimensional volume (i.e., the "area") of its base is

$$s(h) = \left(\frac{h}{H}\right)^{n-1} S,$$

because when the linear dimensions of an $(n-1)$-dimensional figure are changed by a factor λ, then the volume is multiplied by λ^{n-1}.

By formula (9) the volume of the whole pyramid is equal to

$$V = \int_0^H s(h)\, dh,$$

hence

$$V = \int_0^H S\left(\frac{h}{H}\right)^{n-1} dh = \frac{S}{H^{n-1}} \int_0^H h^{n-1}\, dh = \frac{S}{H^{n-1}} \cdot \frac{H^n}{n} = \frac{1}{n} SH;$$

i.e., the volume of an n-dimensional pyramid is equal to $1/n$th of the product of the "area" [$(n-1)$-dimensional volume] of the base and the height. For n equal to 2 or 3 we obtain as special cases the well-known results: The area (two-dimensional volume) of a triangle is equal to half the product of the base and the height, and the volume of a three-dimensional pyramid is one third of the product of the area of the base and the height.

A sphere can be represented approximately as composed of very narrow pyramids with a common vertex at the center of the sphere. The heights of these pyramids are equal to the radius R and the areas of their bases σ_i cover the whole surface S of the sphere approximately. Since the volume of each pyramid is equal to $1/n\, R\sigma_i$, we obtain by adding up these volumes that the volume of the sphere is

$$V \approx \frac{1}{n} R \sum \sigma_i \approx \frac{1}{n} RS.$$

In the limit this gives us the exact formula: $V = 1/n\, RS$; i.e., the volume of a sphere is equal to $1/n$th of the product of its radius and its surface. For n equal to 2 or 3 this relation is widely known.*

We mention one important property of a sphere which can be proved, generally speaking, for an n-dimensional space in exactly the same way as for a three-dimensional space: Among all bodies of a given volume the sphere alone has the least surface area.

7. "Higher" n-dimensional geometry. So far we have confined ourselves to the elementary geometry of the n-dimensional space; but we can also develop the "higher" geometry, for example the general theory

---

* The calculation of the volume of a sphere can also be effected by applying formula (9); a section of an n-dimensional sphere is an $(n-1)$-dimensional sphere, and therefore the volume of an n-dimensional sphere can be calculated by going from $n-1$ to n.

of curves and surfaces. In n-dimensional space surfaces may have various numbers of dimensions: one-dimensional "surfaces," i.e., curves, two-dimensional surfaces, three-dimensional, $\cdots$, and finally $(n-1)$-dimensional surfaces. A curve can be defined as the geometric locus of points whose coordinates depend continuously on a variable or parameter t

$$x_1 = x_1(t), \quad x_2 = x_2(t), \cdots, \quad x_n = x_n(t).$$

A curve is so to speak the trajectory of the motion of a point in n-dimensional space with varying t. If our space serves as an illustration of the state of some physical system such as we have discussed in subsection 4, then a curve illustrates a continuous sequence of states or the course of the change of state in dependence on the parameter t (for example the time). This generalizes the usual graphical illustration of the process of change of state by means of curves.

With every point of a curve in n-dimensional space, we connect not only a tangent ("one-dimensional tangential plane") but also tangent hyperplanes of all dimensions from 2 to $n-1$. The rate of rotation of each of these hyperplanes with respect to the rate of increase of the arc length of the curve gives the corresponding curvature. Thus, a curve has $n-1$ tangent hyperplanes, from the one-dimensional to the $(n-1)$-dimensional, and accordingly $n-1$ curvatures. The differential geometry in an n-dimensional space turns out to be far more complicated than in three-dimensional space.

So far we have only talked of the n-dimensional geometry which is an immediate generalization of the ordinary Euclidean geometry. But we know already that, apart from Euclidean geometry, there exist also Lobačevskiĭ geometry, projective geometry and others. These geometries are just as easily generalized to an arbitrary number of dimensions.

§8. Generalization of the Scope of Geometry

1. The space of colors. When we spoke in the preceding section of the real meaning of n-dimensional space, we came up against the problem of generalizing the scope of geometry, the problem of the general concept of space in mathematics. But before giving the corresponding general definition, let us consider a number of examples.

Experience shows, and this was already mentioned by M. V. Lomonosov, that the normal human vision is three-colored, i.e., every chromatic perception, of a color C, is a combination of three fundamental perceptions: red R, green G and blue B, with specific intensities.* When we

---

* We are concerned here with the perception of color, not of light. Perception of color is also an objective phenomenon—a reaction to light. One and the same per-

denote these intensities in certain units by x, y, z, we can write down that $C = xR + yG + zB$. Just as a point can be shifted in space up and down, right and left, back and forth, so a perception of color, of a color C, can be changed continuously in three directions by changing its constituent parts red, green, and blue. By analogy we can say, therefore, that the set of all possible colors is the "three-dimensional color space." The intensities x, y, z play the role of coordinates of a point, of a color C. (An important difference from the ordinary coordinates consists in the fact that the intensities cannot be negative. When $x = y = z = 0$, we obtain a perfectly black color corresponding to complete absence of light.)

A continuous change of color can be represented as a line in the "color space"; the colors of the rainbow, for example, from such a line; a color line is also formed by a number of perceptions produced on an object of homogeneous coloration by a continuous change of the brightness of illumination. In this case only the intensity of the perception changes, its "coloredness" remains unchanged.

Further, when two colors are given, say red R and white W, then by mixing them in varying proportions* we obtain a continuous sequence of colors from R to W which we can call the segment RW. The conception that a rose color lies between red and white has a clear meaning.

In this way there arises the concept of the simplest geometric figures and relations in the "color space." A "point" is a color, the "segment" AB is the set obtained by mixing the colors A and B; the statement that "the point D lies on the segment AB" means that D is a mixture of A and B. The mixture of three colors gives a piece of a plane—a "color triangle." All this can also be described analytically by using the color coordinates x, y, z, and the formulas giving color lines and planes are entirely analogous to the formulas of ordinary analytic geometry.†

In the color space the relations of Euclidean geometry concerning the disposition of points and segments are satisfied. The system of these relations forms an affine geometry, and we can say that the set of all

---

ception may be produced by different light waves. For example, a green color may be obtained not only from spectrally pure green light, but also from a mixture of red and blue. On the other hand, people suffering from "color blindness" (Daltonism) have only two fundamental perceptions; cases of "complete color blindness," when there is only one fundamental perception of color, are extremely rare.

* Such a mixture can be obtained by mixing in varying proportions very fine colored powders provided the illumination remains unchanged.

† For example, if the colors C_0 and C_1 are determined by the intensities, namely the coordinates x_0, y_0, z_0 and x_1, y_1, z_1, then a color C between C_0 and C_1 has the coordinates $x = (1 - t)x_0 + tx_1$, $y = (1 - t)y_0 + ty_1$, $z = (1 - t)z_0 + tz_1$, where t is the portion of C_1 and $1 - t$ the portion of C_0 in the mixture that makes up C.

possible color perceptions realizes an affine geometry. (This is not quite accurate, because, as we have already said, the color coordinates x, y, z cannot be negative. Therefore the color space corresponds only to that part of the space where in the given coordinate system all the coordinates of points are positive or zero.)

Further, we have a natural idea of the degree of distinctness of colors. For example, it is clear that pale pink is nearer to white than deep pink, and crimson nearer to red than to blue, etc. Thus, we have a qualitative concept of distance between colors as the degree of their distinctness. This qualitative concept can be made into a quantitative measure. However, to define the distance between colors as in Euclidean geometry by the formula $r = \sqrt{(x_0 - x_1)^2 + (y_0 - y_1)^2 + (z_0 - z_1)^2}$ turns out to be unnatural. A distance so defined does not correspond to real perception; with this definition it would happen in a number of cases that two colors that differ from a given one in varying degree would have one and the same distance from it. The definition of distance must reflect the real relations between color perceptions.

Guided by this principle we introduce a peculiar measure of distance in the space of colors. This is done as follows. When a color is altered continuously, a human being does not perceive this change at once, but only when it reaches a certain extent exceeding the so-called threshold of distinction. In this connection it is assumed that all colors that are exactly on the threshold of distinction from a given one are equidistant from it. We are then led automatically to the idea that the distance between any two colors must be measured by the smallest number of thresholds of distinction that can be laid between them. The length of a color line is measured by the number of such thresholds covering it. The distance between two colors is defined as the length of the shortest line joining them. This is similar to the fact that distances between two points in a plane are measured by the length of the shortest line joining them.

Thus, measurement of length and distance in the color space proceeds in very small, as it were infinitely small, steps.

As a result, a certain peculiar non-Euclidean geometry is defined in the color space. This geometry has a perfectly real meaning: It describes in geometrical language properties of the set of all possible colors, i.e., properties of the reaction of the eye to a light stimulus.

The concept of color space arose about a century ago. Many physicists have studied the geometry of this space; for example, we may mention Helmholtz and Maxwell. These investigations continue; they have not only theoretical but also practical value. They give an accurate mathematical foundation for the solution of problems on the difference of color signals, on dyes in the textile industry, and others.

2. Phase spaces in physics and chemistry.   Now let us turn to another example already mentioned in the preceding section.

Suppose that we study some physicochemical system such as a mixture of gases, an alloy, etc. Suppose that the state of this system is determined by n values (as the state of a gaseous mixture is determined by pressure, temperature, and the concentrations of its constituent components). Then one says that the system has n degrees of freedom, meaning that its state can be changed in n independent directions under a change of each of the values that determine this state. These values that determine the state of the system play as it were the role of its coordinates. Therefore the set of all its states can be regarded as an n-dimensional space, the so-called phase space of the system.

Continuous changes of state, i.e., processes occurring in the system, are presented by lines in this space. Separate domains of states, distinguished by one feature or another, are domains of the phase space. The states bordering two such domains form a surface in this space.

In physical chemistry it is particularly important to study the form and the mutual contiguity of those domains of the phase space of a system

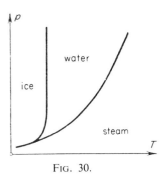

FIG. 30.

that correspond to qualitatively distinct states. The surfaces dividing these domains correspond to such qualitative transitions as melting, evaporation, precipitation of a sediment, etc. A state of a system with two degrees of freedom is illustrated by a point in a plane. As an example we can take a homogeneous substance whose state is determined by the pressure p and temperature T; they are the coordinate points describing the state. Then the question reduces to studying the lines of division between domains corresponding to qualitatively distinct states. In the case of water, for example, these domains are ice, liquid water, and steam (figure 30). Their division lines correspond to melting (freezing), evaporation (condensation), sublimation of ice (precipitation of ice crystals from steam).

For an investigation of systems with many degrees of freedom, the methods of many-dimensional geometry are required.

The concept of phase space applies not only to physicochemical but also to mechanical systems, and generally it can be applied to any system if its possible states form a certain continuous collection. In the kinetic theory of gases one considers, for example, the phase spaces of systems of material particles, the molecules of the gas. The state of motion of one particle at each moment is determined by its position and velocity, which gives altogether six values: three coordinates and three velocity components (with respect to the three coordinate axes). The state of N particles is given by $6N$ values and since there are very many molecules, $6N$ is an enormous number. This does not disturb the physicists in the least, who speak of a $6N$-dimensional phase space of a system of molecules.

A point in this space describes the state of the whole mass of molecules with coordinates and velocities. A motion of a point describes a change of state. This abstract presentation turns out to be very useful in many deep theoretical developments. In a word, the concept of phase space has a secure place in the arsenal of the exact natural sciences and is applicable in diverse problems.

3. The generalization of geometry.   The examples we have given enable us to reach a conclusion on how the scope of geometry is to be generalized.

Suppose that we wish to study some continuous collection of objects, events, or states of one kind or another, for example the set of all possible colors or of the states of a group of molecules. The relations holding in such a collection may happen to be similar to the ordinary spatial relations; for example "distance" between colors or the "mutual position" of domains of a phase space. In that case, by abstracting from the qualitative peculiarities of the objects in question and by taking into account only the aforementioned relations between them, we can regard the given collection as a space of its own kind. The "points" of this "space" are the objects, events or states themselves. A "figure" in this space is an arbitrary aggregate of its points, as for example the "line" of rainbow colors or the "domain" of steam in the "space" of states of water. The "geometry" of such a space is determined by those spacelike relations that hold between the given objects, phenomena or states. Thus, the "geometry" of the color space is determined by the laws of color mixing and the distances between colors.

The real significance of this point of view is that it makes it possible to use the concepts and methods of abstract geometry for the investigation of diverse phenomena. The realm of applicability of geometric concepts and methods is extended immensely in this way. As a result of the generalization of the concept of space the term "space" assumes two meanings in science: On the one hand it is the ordinary real space (the universal form of existence of matter), on the other hand it is the "abstract space," a collection of homogeneous objects (events, states, etc.) in which spacelike relationships hold.

It is worth noting that the ordinary space as we visualize it in a somewhat simplified manner can also be regarded as a collection of homogeneous states. Namely, as the collection of all possible positions of an infinitely small body, a "material point." This remark does not pretend to give a definition of space but aims at making the connection between the two concepts of space clearer. The concept of an abstract space will be further expounded in the next subsection, and the relation of abstract geometry to the ordinary real space will be treated in the last section of this chapter.

4. Generalized spaces in mathematics.   The widest application of the concept of an abstract space occurs in mathematics itself. In geometry one considers the "space" of certain figures as, for example, the "space of spheres" of which we have spoken, "the space of straight lines," and so forth.

This method turns out to be particularly fruitful in the theory of polyhedra. For example, in §5 of Chapter VII a theorem was mentioned on the existence of a convex polyhedron with a given development. The proof of this theorem is based on the discussion of two "spaces": the "space of polyhedra" and the "space of developments." The set of convex polyhedra having a given number of vertices is regarded as a space of its own whose points represent polyhedra; similarly, the set of admissible developments is also treated as a certain space whose points represent developments. The process of fitting together polyhedra from developments establishes a correspondence between polyhedra and developments, i.e., a correspondence between the points of the "space of polyhedra" and the "space of developments." The problem consists in showing that to every development there corresponds a polyhedron; i.e., to every point of one space there corresponds a point of the other. And precisely this can be proved by means of an application of topology.

A whole series of other theorems on polyhedra can be proved similarly, and this "method of abstract spaces" turns out in a number of cases (like the theorem on the existence of a polyhedron with a given development) to be the simplest of the known methods of proving such theorems. Unfortunately, however, the method itself is still rather complicated, and we cannot give a more accurate account of it here.

Extensive applications of the generalized concept of space also occur in analysis, algebra, and number theory. This stems from the usual representation of functions by means of curves. The values of one variable x are usually plotted as points on a line. Similarly, the values of two variables are plotted as points in a plane, the values of n variables as points in an n-dimensional space; we represent the set of values of the variables x_1, x_2, $\cdots$, x_n by the point with the coordinates x_1, x_2, $\cdots$, x_n. We speak of the "domain of variation of the variables" or of the "domain of values" of a function $f(x_1, x_2, \cdots, x_n)$ of these variables; we speak of points, lines, or surfaces of discontinuity of the function, etc. This geometric language is in constant use and is not only a mode of expression; the geometric representation makes many facts of analysis "intuitive" by analogy with the ordinary space and permits the use of geometric methods of proof, generalized to n-dimensional space.

The same takes place in algebra, when equations with n unknown or algebraic functions of n variables are under discussion. In the preceding section it was mentioned that a linear equation with n unknowns determines a hyperplane in n-dimensional space, that m such equations determine m hyperplanes and every solution of them represents a point that is common to all these hyperplanes. The hyperplanes need not intersect at all, or intersect in a single point or in a whole straight line in a two-dimensional or, generally, a k-dimensional hyperplane. All in all, the problem of solubility of systems of linear equations is expressed as a problem on the intersection of hyperplanes. This geometric approach has a number of advantages. Quite generally, "linear algebra," which comprises the study of linear equations and linear transformations is usually set forth to a large extent in geometrical form, as it was done in Chapter XVI.

5. Infinite-dimensional space; definition of a "space." In all our examples we were concerned with a continuous collection of objects of one sort or another being treated as a space of its own particular kind. These objects were colors, states of one system or another, figures, or aggregates of values of variables. In all cases one object was given by a finite number of data so that the corresponding space had a finite number of dimensions, namely the number of these data.

However, at the beginning of the present century mathematicians began to discuss also "infinite-dimensional spaces," namely collections of objects each of which cannot be given by a finite number of data. This is so, above all, in a "functional space."

The idea of treating the collection of functions of one type or another as a space of its own is one of the basic ideas of a new branch of analysis, namely functional analysis, and turns out to be extremely fruitful in the solution of many problems. The reader will find an account of it in Chapter XIX, which is devoted specially to functional analysis.

One can discuss the spaces of continuous functions of one or several variables. One also regards as "spaces" various classes of discontinuous functions, for which one defines a distance between functions by one method or another, depending on the character of the problems awaiting solution. In a word, the number of possible "function spaces" is unlimited, and in fact many such spaces are used in mathematics.

In just the same way one can discuss the "space of curves," the "space of convex bodies," the "space of possible motions of a mechanical system," etc. For example, in §5 of Chapter VII the theorem was mentioned that on every closed convex surface there exist at least three closed geodesics and that any two points can be joined by an infinite number of geodesics. For the proofs of these theorems one uses the space of curves on the surface: in the first the space of closed curves, in the second the space of curves joining two given points. We introduce in the set of all possible curves joining two given points a kind of distance and so turn this set into a space. The proof of the theorem is based on an application of certain deep results of topology to this space.

Let us now formulate a general conclusion.

By a "space" we understand in mathematics quite generally an arbitrary collection of homogeneous objects (events, states, functions, figures, values of variables, etc.) between which there are relationships similar to the usual spatial relations (continuity, distance, etc.). Moreover, in regarding a given collection of objects as a space we abstract from all properties of these objects except those that are determined by these spacelike relationships in question. These relations determine what we can call the structure or the "geometry" of the space. The objects themselves play the role of "points" of such a space; "figures" are sets of its "points."

The scope of the geometry of a given abstract space consists in those properties of the space and the figures in it that are determined by the spacelike relationships taken into account. For example, in discussing the space of continuous functions the properties of an individual function on its own are completely ignored. The function here plays the role of a point and consequently "has no parts," has no structure at all in this sense, no properties unconnected with other points; more accurately, all this is neglected. In a function space, properties of functions are determined only by their relations to one another, by their distance and by other relations that can be derived from distance.

To the variety of possible sets of objects and diverse relations between them, there corresponds the unlimited variety of spaces studied in mathematics. Spaces can be classified with respect to the types of those spacelike relations that underlie their definition. Without aiming at a full account of all the various types of abstract spaces, let us mention, above all, two very important types: topological and metric spaces.

6. Topological spaces. A *topological space* (see Chapter XVIII) is any collection of points (an arbitrary set of elements), in which a relation of neighborhood of one point to a set of points is defined and, consequently, a relation of neighborhood or adherence of two sets (figures) to one another. This is a generalization of the intuitive intelligible relation of neighborhood or adherence of figures in the ordinary space.

Already Lobačevskiĭ with remarkable foresight pointed out that of all the relations of figures, the most fundamental is the relation of neighborhood. "Neighborhood forms a distinctive appurtenance of bodies and gives them the same *geometric* when we retain in them this property and do not take into consideration all others whether they be essential or accidental."* For example, every point on the circumference is adherent to the set of all interior points of a circle; two parts of a connected body are adherent to one another. As the subsequent development of topology has shown, it is precisely the property of neighborhood that underlies all other topological properties.

The concept of neighborhood expresses the notion of a point being infinitely near to a set. Therefore every collection of objects in which there is a natural concept of continuity, of being infinitely near, turns out to be a topological space.

The concept of a topological space is extremely general and the study of such spaces, abstract topology, represents the most general mathematical study of continuity.

A rigorous mathematical definition of a general topological space can be given in the following way.

A set R of arbitrary elements "points" is called a general topological space if for every set M contained in it neighborhood points are defined such that the following conditions, namely the axioms of the space, are satisfied:

1. Every point of M is counted among its neighborhood points. (It is perfectly natural to assume that a set is adherent to each of its points.)

2. If a set M_1 contains a set M_2, then the neighborhood points of M_1 must contain all the neighborhood points of M_2. (To put it briefly, but less accurately, the larger set does not have fewer neighborhood points.)

Usually other axioms are added to these, so that various types of topological spaces are thereby defined.

With the help of the concept of neighborhood, it is easy to define a number of very important topological concepts. These are at the same time the most fundamental and general concepts of geometry and their definitions are intuitively altogether clear. Let us give some examples.

---

* N. I. Lobačevskiĭ, "Collected works," Vol. II, Gostehizdat, 1949, page 168.

1. *Adherent.* We shall say that sets M_1 and M_2 are adherent to one another if one of them contains at least one neighborhood point of the other. (In this sense, for example, the circumference of a circle is adherent to the interior.)

2. Continuity or, as the mathematicians say, *connectedness* of a figure. A figure, i.e., a set of points M, is connected if it cannot be split into parts that are not adherent to one another. (For example, a segment is connected, but a segment without its midpoint is disconnected.)

3. *Boundary.* The boundary of a set M in a space R is the set of all points that are adherent both to M and to its complement $R - M$, i.e., to the remaining part of the space R. (This is, obviously, a perfectly natural concept of boundary.)

4. *Interior point.* A point of a set M is called interior if it does not lie on its boundary, i.e., if it is not adherent to the complement $R - M$.

5. *Continuous mapping or transformation.* A transformation of a set M is called continuous if it does not disrupt neighborhoods. (One could hardly give a more natural definition of a continuous transformation.)

Other important definitions could be added to this list, such as for example a definition of the concept of convergence of a sequence of figures to a given figure or the concept of the number of dimensions of a space.

We see that the most fundamental geometric concepts can be defined in terms of neighborhoods. The significance of topology, in particular, lies in the fact that it gives rigorous general definitions for these concepts, thereby providing a basis for the strict application of arguments connected with the intuitive conception of continuity.

Topology is the study of those properties of spaces, of figures in them, and of their transformations that are defined by the relation of neighborhood.

The generality and fundamental nature of this relation makes topology into a very general geometrical theory that penetrates the diverse branches of mathematics, wherever continuity only is under discussion. But precisely because of its generality topology in its most abstract parts goes beyond the framework of geometry proper. All the same, at its basis lies a generalization of the properties of real space and the most fruitful and powerful of its results are connected with the application of methods that spring from intuitive geometric ideas. An example is the method of approximating general figures by polyhedra which was developed by P. S. Aleksandrov and was extended by him, though in an abstract form, to extremely general types of topological spaces.

Nowadays every specialist, no matter what his subject of study, investigates when he discovers that there is a natural way of introducing into it the concept of nearness or adherence, and immediately has at his disposal the ready-made, widely ramified apparatus of topology, which enables him to draw conclusions that are far from trivial even in their application to his special field.

7. Metric spaces. A *metric space* is a set of arbitrary elements, to be called points, between which a distance is defined; i.e., with each pair of points X, Y a number $r(X, Y)$ is associated so that the following conditions, namely the axioms of a metric space, are satisfied:

1. $r(X, Y) = 0$ if and only if the points X, Y coincide.

2. For any three points X, Y, Z

$$r(X, Y) + r(Y, Z) \geqslant r(Z, X).$$

This condition is called the "triangle inequality," since it is quite analogous to the well-known property of the ordinary distance between points A_1, A_2, A_3 of Euclidean space (figure 31):

$$r(A_1, A_2) + r(A_2, A_3) \geqslant r(A_3, A_1).$$

As examples of metric spaces we may take

1. the Euclidean space of an arbitrary number n of dimensions,

2. the Lobačevskiĭ space,

3. any surface in its intrinsic metric (Chapter VII, §4),

4. the space C of continuous functions with distance defined by the formula $r(f_1, f_2) = \max |f_1(x) - f_2(x)|$,

5. the Hilbert space to be described in Chapter XIX, which is an "infinite-dimensional Euclidean" space.

The Hilbert space is the most important of the spaces used in functional analysis; it is closely connected with the theory of Fourier series and more generally with the expansion of functions in series by orthogonal functions (the coordinates x_1, x_2, x_3, $\cdots$ are then the coefficients of such series). This space also plays an important role in mathematical physics and has acquired much significance in quantum mechanics. It turns out that the set of all

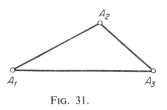

FIG. 31.

possible states (not only stationary) of an atomic system, for example a hydrogen atom, can be regarded from an abstract point of view as a Hilbert space.

The number of examples of metric spaces that are actually discussed in mathematics could be increased considerably; in the next section we shall become acquainted with one important class of metric spaces, the so-called Riemannian spaces, but the examples we have given so far will be sufficient to show how widely the general concept of a metric space extends.

In a metric space it is always possible to define all topological concepts and, moreover, to introduce other "metric" concepts; for example, the concept of the length of a curve. Length is defined in any metric space precisely in the same way as usual, and the basic properties of length are then preserved. Indeed, by the length of a curve we understand the limit of the sum of the distances between points X_1, X_2, $\cdots$, X_n that are arranged in a sequence on the curve, subject to the condition that the points come to lie closer and closer on the curve.

8. Advantages of the geometric method. Many types of spaces besides the general topological or the metric spaces are discussed in mathematics. In fact, we have already become acquainted in §6 with a whole class of such spaces. These are the spaces in which some group of transformations is given (for example, the projective or affine spaces). In such spaces we can define "equality" of figures. Figures are "equal" if they can be carried into one another by a transformation of the given group.

We shall not go deeper into the definitions of possible types of spaces; they are rather diverse and the reader can turn to the special literature on the various branches of present-day geometry.

But what is the sense of extending the range of geometric concepts so much? For what purpose, for example, does one have to introduce the concept of the space of continuous functions? Is it not sufficient to solve problems of analysis by the usual means without resorting to such abstract spaces?

The general answer to these questions consists briefly in this: that by introducing one space or another into the discussion we open the way to applications of geometric concepts and methods, which are extremely numerous.

A characteristic feature of geometric concepts and methods is that they are based, all things considered, on intuitive ideas and preserve their advantages even though in an abstract form. What the analyst achieves by long calculations, the geometer can occasionally grasp at once. An elementary example of this can be seen in a graph that gives a completely clear picture of one dependence or another between quantities. The geometrical method can be characterized as an all-embracing synthetic method, in contrast to the analytical method. Of course, in abstract geometrical theories the immediate intuitiveness fades away, but the intuitive arguments by analogy remain and so does the synthetic character of the geometrical method.

The reader is already familiar with the application of geometric pictures in analysis, with a geometric representation of complex numbers and functions of a complex variable, with geometric arguments in a proof of the fundamental theorem of algebra, and with other applications of geometric concepts and methods. Everywhere he may observe what we have described here in general terms. We recall the examples, in the beginning of §7, and also here §8, under subsection 5, of theorems that can be proved by applying many-dimensional geometry. Let us give one further example of a problem in analysis that can be solved by an application of the concept of a function space.

It is proved in topology that if we take in the ordinary plane any domain that has the form of a distorted circle and then deform it in a continuous manner as we please, but so that in the end it becomes embedded inside its original contour, then at least one point of the domain comes to lie after the transformation where it was before. This is a purely topological fact.

Now let us consider a problem altogether remote from geometry: A function $y(x)$ is to be found that satisfies the differential equation,*

$$y' = f(x, y) \tag{10}$$

and assumes for $x = 0$ the value $y = 0$.

Obviously, instead of this equation we can look for a solution of the equation

$$y = \int_0^x f[t, y(t)]\, dt. \tag{11}$$

Naturally the problem arises: Does there exist, in general, a function $y(x)$ satisfying this condition?

Let us look at the problem in another way. We represent every continuous function $y(x)$ by a point of some abstract space. The result of computing the integral

$$\int_0^x f[t, y(t)]\, dt = z(x)$$

is again a continuous function of x, i.e., a "point" of our abstract space. By taking various "points" y, i.e., various functions $y(x)$, we obtain, generally speaking, various points z. In this way, the set of points of

---

* Here, $f(x, y)$ is assumed to be a continuous function of the variables x and y.

our space is again mapped into points of the same space. The problem of finding a solution of the equation (11) has been reduced to the question: Can a "point" of our space be found such that after this transformation it coincides with its previous "place"?

A natural problem in the theory of differential equations has become a problem concerning a property of an abstract function space. The analogy with the aforementioned theorem tells us that we are evidently dealing here with a topological property of the corresponding space.

In this way we obtain, by means of the requisite topological investigations, what are probably the shortest proofs of many theorems on the existence of solutions of differential equations; in particular we can make it clear that the equation (10) does in fact have a solution for every continuous function $f(x, y)$.

§9. Riemannian Geometry

1. History of Riemannian geometry. The ideas explained previously that every continuous collection of homogeneous phenomena can be treated as a space of its own was first expressed by Riemann in his lecture "On the hypotheses that underlie geometry," given at the University of Göttingen in 1854. This was a sort of test lecture, somewhat like a report or a dissertation that a lecturer or professor had to make to the faculty before taking up his post. In his lecture Riemann set out in general lines, without calculations or mathematical proofs, the original idea of a vast geometric theory that is now called Riemannian geometry. It is said that nobody in the audience understood it except the aged Gauss. Riemann provided the formal apparatus of his theory in another paper, with an application to the problem of heat conduction, so that the abstract Riemannian geometry was born in close connection with mathematical physics. In the development of geometry, Riemann's ideas came next after Lobačevskiĭ's decisive step. However, Riemann's papers were not at once duly appreciated. His lectures and papers on heat conduction were published only posthumously in 1868. It is worth while mentioning that in 1868 the first interpretation of Lobačevskiĭ's geometry also appeared, by Beltrami, and in 1870 the second one, by Klein. In 1872 Klein expounded his general view of the various geometries: Euclidean, Lobačevskiĭ, projective, affine, etc., as the study of properties of figures that remain unchanged under the transformations of one group or another. In the same year many-dimensional geometry was finally consolidated in mathematics. Thus, the seventies of the 19th century were that critical period in the history of geometry when the new geometrical ideas, accumulated in the course of the preceding fifty years, were finally understood by a wide circle of mathematicians and assumed a secure place in the science.

Riemann's work was then continued and at the end of the 19th century Riemannian geometry had reached a considerable development and had found applications in mechanics and physics. When, in 1915, Einstein in his general theory of relativity applied Riemannian geometry to the theory of universal gravitation, this event drew particular attention to Riemannian geometry and resulted in its brisk development and in various generalizations.

2. The basic ideas of Riemannian geometry. Riemann's ideas, which had such a brilliant success, are really rather simple if one sets aside the mathematical details and concentrates on the basic essentials. Such an intrinsic simplicity is a feature of all great ideas. Was not Lobačevskiĭ's idea simple: to regard the consequences of the negation of the Fifth Postulate as a possible geometry? Was not the idea of evolution of organisms simple, or the idea of the atomic structure of matter? All of these are simple and at the same time very complicated, because new ideas must, first of all, work their way over a wide field and must not be pressed into a rigid framework, and second, their foundation, development, and application is a many-sided task, requiring an immense amount of labor and ingenuity, and impossible without the specialized apparatus of science. For Riemannian geometry this apparatus consists in its formulas; they are complicated and therefore accessible to a specialist only. But we shall not deal with complicated formulas and turn now to the essence of Riemann's ideas. As we have already said, Riemann began by considering an arbitrary continuous collection of phenomena as a space of its own. In this space the coordinates of points are quantities that determine the corresponding phenomenon among others, as for example the intensities x, y, z that determine the color $C = xR + yG + zB$. If there are n such values, say x_1, x_2, $\cdots$, x_n, then we speak of an n-dimensional space. In this space we may consider lines and introduce a measurement of their length in small (infinitely small) steps, similar to the measurement of the length of a curve in ordinary space.

In order to measure lengths in infinitely small steps, it is sufficient to give a rule that determines the distance of any given point from another infinitely near to it. This rule of determining (measuring) distance is called a *metric*. The simplest case is when this rule happens to be the same as in Euclidean space. Such a space is Euclidean in the infinitely small. In other words, the geometrical relations of Euclidean geometry are satisfied in it, but only in infinitely small domains; it is more accurate to say that they are satisfied in any sufficiently small domain, though not exactly, but with an accuracy that is the greater, the smaller the domain. A space in which distance is measured by such a rule is called *Riemannian*; and the geometry of such spaces is also called *Riemannian*. A Riemannian space is, therefore, a space that is Euclidean "in the infinitely small."

The simplest example of a Riemannian space is an arbitrary smooth surface in its intrinsic geometry. The intrinsic geometry of a surface is a Riemannian geometry of two dimensions. For in the neighborhood of each of its points a smooth surface differs only a little from its tangent plane, and this difference is the smaller, the smaller the domain of the surface that we consider. Therefore the geometry in a small domain of the surface also differs little from the geometry in a plane; the smaller the domain, the smaller this difference. However, in large domains the geometry of a curved surface turns out to be different from the Euclidean, as was explained in §4 of Chapter VII and is easy to see in the examples of the sphere or pseudosphere. Riemannian geometry is nothing but a natural generalization of the intrinsic geometry of a surface with two dimensions to an arbitrary number *n*. Like a surface, considered only from the point of view of its intrinsic geometry, a three-dimensional Riemannian space, although Euclidean in small domains, may differ from the Euclidean in large domains. For example, the length of a circle may not be proportional to the radius; it will be proportional to the radius with a good approximation for small circumferences only. The sum of the angles of a triangle may not be two right angles; here the role of rectilinear segments in the construction of a triangle is played by the lines of shortest distance, i.e., the lines having the smallest length among all the lines joining the given points.

One can speculate that the real space is Euclidean only in domains that are small in comparison with the astronomical scale. The smaller a domain is, the more accurately Euclidean geometry holds, but we can imagine (and, in fact, it turns out to be so) that on a very large scale the geometry differs somewhat from the Euclidean. This idea, as we know, was already put forward by Lobačevskiĭ. Riemann generalized it so that it applied to an arbitrary geometry and not only to Lobačevskiĭ geometry, which now appears as a special case of Riemannian geometry.

From what we have said it is clear that Riemannian geometry has grown by a synthesis and generalization of three ideas that contributed to the successful development of geometry. First came the idea of the possibility of a geometry other than the Euclidean, second was the concept of the intrinsic geometry of a surface, and third the concept of a space of an arbitrary number of dimensions.

3. Measurement of distance. In order to make it clear how a

Ricmannian space is defined mathematically, we recall first of all the rule for measuring distances in a Euclidean space.

If rectangular coordinates x, y are introduced in a plane, then by Pythagoras' theorem the distance between two points whose coordinates differ by Δx and Δy is expressed by the formula

$$s = \sqrt{\Delta x^2 + \Delta y^2}.$$

Similarly in a three-dimensional space

$$s = \sqrt{\Delta x^2 + \Delta y^2 + \Delta z^2}.$$

In a n-dimensional Euclidean space the distance is defined by the general formula

$$s = \sqrt{\Delta x_1^2 + \cdots + \Delta x_n^2}.$$

Hence it is easy to conclude how the rule for measuring distance in a Riemannian space ought to be given. The rule must coincide with the Euclidean, but only for an infinitely small domain in the neighborhood of each point. This leads to the following statement of the rule.

A Riemannian n-dimensional space is characterized by the fact that in the neighborhood of each of its points A coordinates x_1, x_2, $\cdots$, x_n can be introduced such that the distance from A of an infinitely near point X is expressed by the formula

$$ds = \sqrt{dx_1^2 + \cdots + dx_n^2}, \tag{12}$$

where $dx_1, \cdots, dx_n$ are the infinitely small differences of the coordinates of A and X. This can also be expressed more accurately in another way: The distance from A to an arbitrarily near point X is expressed by the same formula as in Euclidean geometry, but only with a certain accuracy which is the greater the nearer the point X is to A, i.e.,

$$s(AX) = \sqrt{\Delta x_1^2 + \cdots + \Delta x_n^2} + \epsilon,$$

where ϵ is a small quantity in comparison with the first term and is smaller, the smaller the coordinate differences $\Delta x_1, \cdots, \Delta x_n$ are.*

---

* Usually the precise meaning of the formula (12) is expressed as follows. Suppose that a curve starts out from A so that the coordinates of the point X on it are given as functions $x_1(t), x_2(t), \cdots, x_n(t)$ of some variable t. Then the differential ds of the arc length of this curve at A is expressed by (12).

Now this is the exact mathematical definition of a Riemannian metric and a Riemannian space. The difference of Riemannian metric, i.e., the rule for measuring distances, from Euclidean consists in that this rule holds only in the neighborhood of each given point. Moreover, the coordinates in which it is expressed so simply have to be taken differently for different points.* The difference between the general Riemannian metric and the Euclidean will be further specified later on.

The fact that a Riemannian space coincides with a Euclidean in the infinitely small enables us to define in it the fundamental geometric quantities similarly to the way this was done for the intrinsic geometry of a surface by approximating an infinitely small portion of the surface by a plane (Chapter VII, §4). For example, an infinitely small volume is expressed just as in Euclidean space. The volume of a finite domain is obtained by summing infinitely small volumes, i.e., by integrating the differential of the volume. The length of a curve is determined by summing infinitely small distances between infinitely near points on it, i.e., by integrating the differential of the length ds along the curve. And this is a rigorous analytic expression for the fact that the length is determined by laying off a small (infinitely small) measuring rod along the curve. The angle between curves at a common point is defined exactly as in a space. Further, in an n-dimensional Riemannian space we can define surfaces of various numbers of dimensions from 2 to $n - 1$. Moreover, it is easy to prove that each such surface in its turn represents a Riemannian space of the corresponding number of dimensions, just as a surface in the ordinary Euclidean space turns out to be a two-dimensional Riemannian space.

It has also been proved that a Riemannian space can always be represented as a surface in a Euclidean space of a sufficiently large number of dimensions, namely: for every n-dimensional Riemannian space one can find in an $n(n + 1)/2$-dimensional Euclidean space an n-dimensional surface which from the point of view of its intrinsic geometry does not differ from this Riemannian space (at least in a given limited part of it).

4. The fundamental quadratic form.   In order to obtain the actual analytical expression for various geometric quantities in a Riemannian geometry, we have to define, first of all, a general expression for the rule of measuring lengths in a Riemannian space independent of the specific coordinates at each point. True, the formula (12) holds at every point A for a special choice of coordinates at that point, so that on transition

---

* If in the whole space coordinates could be introduced so that for any pair of neighboring points this rule for measuring distance would hold, then the space would be Euclidean.

from one point to another the coordinates themselves must be changed, and this is of course inconvenient. But this can easily be avoided, for we can prove the following.

Suppose that in some domain of a Riemannian space coordinates $y_1, y_2, \cdots, y_n$ are introduced arbitrarily. Then the "infinitely small distance" or, as one says, the "element of length" from the point A with the coordinates $y_1, y_2, \cdots, y_n$ to the point X with the coordinates $y_1 + dy_1, y_2 + dy_2, \cdots, y_n + dy_n$ is expressed by the formula

$$ds = \sqrt{\sum_{i,k=1}^{n} g_{ik}\, dy_i\, dy_k}, \quad \text{or} \quad ds^2 = \sum g_{ik}\, dy_i\, dy_k, \tag{13}$$

where the coefficients g_{ik} are functions of the coordinates $y_1, y_2, \cdots, y_n$ of A.

The expression on the right of the last formula is called a quadratic form* in the differentials of the coordinates $dy_1, \cdots, dy_n$. In expanded form it can be written as follows:

$$\sum g_{ik}\, dy_i\, dy_k = g_{11}\, dy_1^2 + g_{12}\, dy_1\, dy_2 + g_{21}\, dy_2\, dy_1 + g_{22}\, dy_2^2 + \cdots.$$

Since $dy_1\, dy_2 = dy_2\, dy_1$, it is convenient to take the second and third term as equal: $g_{12} = g_{21}$ and generally $g_{ik} = g_{ki}$; this is possible, because only their sum $(g_{ik} + g_{ki})dy_i\, dy_k$ is important.

The quadratic form is positive definite, since obviously $ds^2 > 0$, except when all the differentials dy_i are equal to zero.

The converse also holds. Namely, if in an n-dimensional space, where coordinates $y_1, y_2, \cdots, y_n$ have been introduced, the element of length is given by the formula (13) with the condition that the quadratic form is positive definite (i.e., always greater than zero except when all the $dy_i = 0$), then the space is Riemannian. In other words, in the neighborhood of each point A one can introduce new special coordinates $x_1, x_2, \cdots, x_n$ so that in the new coordinates the element of length at this point is expressed in the simple form (12)

$$ds^2 = dx_1^2 + dx_2^2 + \cdots + dx_n^2.$$

Thus, Riemannian metric (i.e., a definition of length that is Euclidean in the infinitely small) can be given by any positive definite quadratic form (13) with coefficients g_{ik} that are functions of the coordinates y_i. This is the general method of giving a Riemannian metric.

---

* A quadratic form of several quantities is an algebraic expression that is a homogeneous polynomial of degree 2 in these quantities.

A curve in a Riemannian space is given by the fact that all n coordinates of a point vary in dependence on a single parameter t which ranges over a certain interval

$$y_1 = y_1(t), y_2 = y_2(t), \cdots, y_n = y_n(t) \qquad (a \leqslant t \leqslant b). \tag{14}$$

The length of the curve is expressed by the integral

$$s = \int ds = \int \sqrt{\sum g_{ik} \, dy_i \, dy_k}.$$

In the case of the curve given by the equations (14) we have

$$dy_1 = y_1' \, dt, \cdots, dy_n = y_n' \, dt,$$

therefore

$$s = \int_a^b \sqrt{\sum g_{ik} y_i' y_k'} \, dt. \tag{15}$$

Since the g_{ik} are known functions of the coordinates $y_1, \cdots, y_n$ and the latter depend in a known manner on t in accordance with the formulas (14), the function of t under the integral sign in (15) is completely determined for the given curve. Consequently its integral has a definite value and so the curve has a definite length.

The length of the shortest curve joining two given points A, B is taken to be the distance between these points. This curve itself, called a geodesic, plays the role of an analogue to the rectilinear segment AB. One can show that in a small domain any two points are joined by a unique shortest line. The problem of finding the geodesic (shortest) lines is that of minimizing the integral (15). This is a problem in the calculus of variations which was discussed in Chapter VIII. A standard application of the methods of the calculus of variations permits us to derive a differential equation that determines the geodesic lines and to establish their general properties for every Riemannian space.

Let us prove the principal statement made previously, namely, that a Riemannian metric is given in arbitrary coordinates by the general formula (13).

Suppose that in some domain of a Riemannian space certain coordinates $y_1, y_2, \cdots, y_n$ are introduced. We take an arbitrary point A in this domain and assume that $x_1, x_2, \cdots, x_n$ are the special coordinates in which the element of length at A is expressed by the formula (12) or, what is the same,

$$ds^2 = dx_1^2 + dx_2^2 + \cdots + dx_n^2. \tag{16}$$

The coordinates x_i are expressed in terms of the y_j $(i, j = 1, \cdots, n)$ by certain formulas,

$$
\begin{aligned}
x_1 &= f_1(y_1, y_2, \cdots, y_n), \\
x_2 &= f_2(y_1, y_2, \cdots, y_n), \\
&\cdots\cdots\cdots\cdots\cdots\cdots \\
x_n &= f_n(y_1, y_2, \cdots, y_n).
\end{aligned}
$$

Then

$$
dx_1 = \frac{\partial f_1}{\partial y_1}\, dy_1 + \frac{\partial f_2}{\partial y_2}\, dy_2 + \cdots + \frac{\partial f_n}{\partial y_n}\, dy_n
$$

and similarly for $dx_2, \cdots, dx_n$. We substitute these expressions in (16). When we then square the right-hand side and combine the terms with dy_1^2, $dy_1\, dy_2$, dy_2^2, etc., we obtain an expression of the form

$$
ds^2 = g_{11}\, dy_1^2 + 2g_{12}\, dy_1\, dy_2 + g_{22}\, dy_2^2 + \cdots + g_{nn}\, dy_n^2,
$$

where the coefficients $g_{11}, g_{12}, \cdots, g_{nn}$ are expressed in terms of the partial derivatives $\partial f_i/\partial y_j$ (the form of these expressions is of no interest to us). But this is simply formula (13) written in expanded form, and so our statement is proved.

Let us now show that, conversely, the formula (13) defines a Riemannian metric, i.e., that at each point by a special choice of the coordinates x_i it can be transformed into the simple form (16). Suppose that

$$
ds^2 = \Sigma\, g_{ik}\, dy_i\, dy_k,
$$

where the g_{ik} are functions of the coordinates $y_1, \cdots, y_n$ and the quadratic form on the right-hand side is positive definite. Then the coefficients g_{ik} are given numbers and the variables on which the form depends are $dy_1, \cdots, dy_n$. From algebra it is known that every positive-definite quadratic form (with arbitrary numerical coefficients) can be reduced to a sum of squares by a linear transformation of the variables (see Chapter XVI),* i.e., that there exists a transformation

$$
\begin{aligned}
dy_1 &= a_{11}dx_1 + \cdots + a_{1n}dx_n, \\
&\cdots\cdots\cdots\cdots\cdots\cdots\cdots\cdots \\
dy_n &= a_{n1}dx_1 + \cdots + a_{nn}dx_n,
\end{aligned}
\tag{17}
$$

---

* It does not matter that in our case the variables of the form are differentials; we can regard them simply as certain independent variables.

such that when these expressions are substituted in (13), we obtain

$$ds^2 = dx_1^2 + \cdots + dx_n^2.$$

If we make the change of the coordinates $y_1, \cdots, y_n$ to $x_1, \cdots, x_n$ by

$$y_1 = a_{11}x_1 + \cdots + a_{1n}x_n,$$
$$\dots\dots\dots\dots\dots\dots\dots\dots$$
$$y_n = a_{n1}x_1 + \cdots + a_{nn}x_n,$$

then the differentials dy_j are expressed in terms of the differentials dx_i precisely by the formulas (17). Consequently this change of coordinates solves our problem: in the coordinates $x_1, \cdots, x_n$ at the point we have chosen the square of the differential ds^2 is expressed in the simple "Euclidean" form (16). So we have proved that the general formula (13) does in fact give a Riemannian metric.

5. The curvature tensor. A Euclidean space is the simplest special case of a Riemannian space.* It is an important task of Riemannian geometry to give an analytical expression for the difference of a general Riemannian space from a Euclidean by defining a measure, so to speak, for the non-Euclideanness of a Riemannian space. This measure is the so-called curvature of the space.

We must emphasize right away that the concept of curvature of a space is not at all connected with the idea that the space is situated in some higher enveloping space in which it is somehow curved. Curvature is defined within the given space and expresses its difference from a Euclidean space in the sense of its intrinsic geometric properties. This must be clearly understood in order to avoid linking the concept of a curved space with something extraneous. When it is said that our real space has curvature, this only means that its geometric properties differ from the properties of a Euclidean space. But it does not mean at all that our space lies within some higher space in which it is somehow curved. Such an idea has no relation whatsoever with an application of Riemannian geometry to the real space and belongs in the realm of speculative phantasy.

---

* In a Euclidean space the element of length in rectangular coordinates is expressed by the formula (16): $ds^2 = \sum dx_i^2$. If we go over to other coordinates, then by what we have deduced under subsection 4 earlier ds^2 is expressed by some quadratic form (13). Consequently the same general formula (13) for the element of length holds in *arbitrary* coordinates in a Euclidean space. The Euclidean space, however, differs from any other by the fact that in it coordinates can be introduced (and these will be rectangular coordinates) such that the formula (16) holds everywhere with one and the same coordinate system and not only near one point or another, as is the case in a general Riemannian space.

The concept of curvature of a Riemannian space generalizes to n dimensions that of the Gaussian curvature of a surface. As was explained in §4 of Chapter VII, the Gaussian curvature is a measure of the deviation of the intrinsic geometry of the surface from the geometry in a plane and can be treated purely from an internal-geometric point of view. It is nothing other than the curvature of that two-dimensional Riemannian space that represents the given surface.

Let us recall, for example, two formulas of the intrinsic geometry in which the Gaussian curvature occurs. Suppose that there is a small triangle on the surface near a certain point O whose sides are geodesic lines; let its angles be α, β, γ and its area σ. The quantity $\alpha + \beta + \gamma - \pi$ expresses the difference of the sum of its angles from the sum of the angles of a triangle in the plane.

When the triangle is contracted towards the point O, then the ratio of $\alpha + \beta + \gamma - \pi$ to its area σ tends to the Gaussian curvature K at O. In other words, for a small triangle

$$\frac{\alpha + \beta + \gamma - \pi}{\sigma} = K + \epsilon,$$

where $\epsilon \to 0$ as the triangle is contracted to O. This shows exactly that the Gaussian curvature K is a measure for the difference of the sum of the angles of a triangle in a plane and the sum of the angles of a triangle on the surface.

Now let us consider a small circle on the surface with its center at O (i.e., the geometric locus of points equidistant from O in the sense of the distance on the surface). If r is the radius of the circle and l its length, then

$$l = 2\pi r - \frac{\pi}{3} K r^3 + \epsilon,$$

where K is again the Gaussian curvature at O and ϵ denotes a quantity that is small compared with r^3.

Here the Gaussian curvature emerges as a measure of the deviation of the length of a small circle from the value $2\pi r$ to which it is equal in Euclidean geometry.

Now the curvature of a Riemannian space plays a similar role. It can be defined, for example, in the following manner. In the given Riemannian space we construct a smooth surface F formed from geodesic lines passing through a given point O. The Gaussian curvature of this surface is taken to be the curvature of the space at O in the direction of the surface F. Generally speaking, this curvature will differ not only at different points

O, but also for various geodesic surfaces *G* passing through one and the same point *O*. The curvature of a space at a given point is, therefore, not characterized by a single number. Already Riemann introduced a general rule connecting the curvatures of the various surfaces *F* at one and the same point. Owing to these connections, the curvature at a point is completely characterized by a certain system of numbers, the so-called *curvature tensor*.

However, we cannot dwell here on an explanation of this situation, since it would require extensive mathematical apparatus. The only important thing is to grasp that the curvature is a measure of the non-Euclideanness of a Riemannian space; it is defined intrinsically as a measure of the deviation of its metric from the metric of Euclidean space. It determines, for example, the difference of the sum of the angles of a triangle from π or the difference of the length of a circle from $2\pi r$. At different points it has, in general, distinct values and at one and the same point it is given not by one number but by a certain system of numbers.

A Riemannian space need not be homogeneous in its properties, and in that case free mobility of figures without altering the distances between their points is impossible. So there arises the question in what Riemannian spaces free motion for figures is possible with the same number of degrees of freedom as in a Euclidean space. These are the most homogeneous Riemannian spaces.

It turns out that a Euclidean space is homogeneous without curvature (a space of zero curvature). Another type of homogeneous space is the Lobačevskiĭ space, so that Lobačevskiĭ geometry, just like Euclid's geometry, is a special case of the general Riemannian geometry.

Generally, a Riemannian space in which free motion of figures is possible is a space of constant curvature: In it the curvature has one and the same value at all points and for all geodesic surfaces. (Instead of the "curvature tensor" which changes from point to point it is given this time by a single number common to all points.) A space of zero curvature is Euclidean; a space of negative curvature is a Lobačevskiĭ space; a space of positive curvature has the same geometry as an *n*-dimensional sphere in an $(n + 1)$-dimensional Euclidean space.

6. Applications of Riemannian geometry. Riemannian geometry did not have long to wait for applications. Riemann himself, as we have already said, applied its formal apparatus to the solution of a problem of heat conduction, but this was merely an application of its formulas, not of the idea of an abstract space with a Euclidean measure of distance in infinitely small domains. Such an application was made to the color space, where the distance between neighboring colors can be expressed by using a Riemannian metric; the color space has been treated as a special three-dimensional Riemannian space.

Another important application of Riemannian geometry emerged in mechanics. In order to understand its essence let us consider, to begin with, the motion of a point on a surface. We imagine a material point, for example a particle that can move freely on a certain smooth surface without leaving the surface. The point moves, as it were, in the surface itself. We can introduce arbitrary coordinates x_1, x_2 on the surface; then the motion of the point is completely determined by the dependence of these coordinates on the time, and its velocity on the velocities of the change of coordinates, i.e., on their derivatives with respect to the time $\dot{x}_1$, $\dot{x}_2$. So we see that the point moves, as it were, in two-dimensional space; but this space is not Euclidean and has its own geometry, the intrinsic geometry of the surface. The laws of motion can be transformed so as to contain only the coordinates x_1, x_2 of a point on the surface and their first and second derivatives.

If a force acts on the point, then its component perpendicular to the surface is annihilated by the resistance (by the reaction of the surface) and there remain only the components tangential to the surface; so the force acts only along the surface. In this manner the forces acting on a point can also be regarded as acting in the surface itself. The intrinsic geometry of the surface is a special case of a Riemannian geometry. Therefore the motion of the point on the surface is a motion in a two-dimensional Riemannian space. The laws of this motion have the same character as the usual laws of motion, with the difference only that the intrinsic geometry of the surface is taken into account. This becomes perfectly clear from the following fact, mentioned before, in §4 of Chapter VII: A point moving on the surface under inertia and without friction moves on a geodesic line with constant velocity. Since the geodesic lines play the role of straight lines on the surface, this fact is analogous to the law of inertia: it is the same law of inertia, but for the motion on a surface or, abstractly, in a two-dimensional Riemannian space.

Of course, so far there is no advantage visible in this abstract presentation, because we are concerned only with the motion on an ordinary surface.

The benefit of the abstract point of view makes itself felt at once when we go over to mechanical systems whose state is given by more than two quantities. Then a representation of the motion as the motion of a point on a surface becomes impossible. We have encountered this circumstance in §7 when we talked of how graphical methods fail in an abstract presentation of a many-dimensional space.

Suppose, then, that there is a mechanical system whose configuration, i.e., the distribution of whose parts, is given by n quantities $x_1, x_2, \cdots, x_n$.

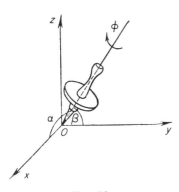

If we are concerned with a system of several material points, then their distribution is determined by giving all their coordinates, three for each point. As another example we can take a gyroscope (a wheel spinning on an axis that itself can turn around a stationary point). The rotation of the gyroscope around the axis is given by the angle of deflection and the inclination of its axis by the two angles it forms with two given directions. Altogether there are three quantities that determine the position of such a gyroscope (figure 32).

FIG. 32.

Every configuration (every position of the parts of the system) can be thought of as a "point" in the space of all possible configurations. This is the so-called configuration space of the system.* The number of its dimensions is equal to the number of quantities $x_1, x_2, \cdots, x_n$ that determine the configuration. These quantities serve as coordinates of a "point" in the configuration space. For a system of say three material points, we obtain three coordinates each for three points, or nine coordinates altogether. For the case of a gyroscope, we have three coordinates, namely the three angles, so that the configuration space of a gyroscope is three-dimensional.

The motion of the system is represented as the motion of a point in a configuration space. The velocity of the motion is determined by the velocities of change of the coordinates $x_1, x_2, \cdots, x_n$.

Such spaces will be discussed again in Chapter XVIII in connection with their topological structure. Here we wish only to emphasize that we can introduce in a configuration space a special rule of measuring distances which is closely connected with the mechanical properties of the system. Indeed, if the kinetic energy of the system is expressed by the formula

$$T = \frac{1}{2} \sum_{i,k=1}^{n} a_{ik}\dot{x}_i\dot{x}_k ,$$

---

* This must not be confused with the "phase space" mentioned in §8 under subsection 2. In the phase space a point determines not only the position but also the velocity of motion of the points of the system at every moment.

where the $\dot{x}_i$ are the velocities of change of the corresponding coordinates, then the square of an infinitely small distance is given by the formula

$$ds^2 = \sum_{i,k=1}^{n} a_{ik}\,dx_i\,dx_k \,.$$

Thus, the configuration space becomes a Riemannian space. Moreover, not only is the motion of the system represented as the motion of a point in the configuration space but the very equations that describe the motion of the system coincide with the equations of motion of this point; in a word, the mechanics of the system are represented as the mechanics of a point in the configuration space. In particular, the motion of the system under inertia, i.e., without the action of forces (like the free rotation of a gyroscope) becomes a uniform motion of a point on a geodesic line in that space.

This representation is expedient in a number of cases and is used, along with certain generalizations and modifications, in theoretical mechanics.

Thus, Riemannian geometry has its applications as a method of abstract-geometrical description of physical phenomena. This description is not at all arbitrary and is not an idle play of the mathematical mind; it reflects the real mechanism of the phenomena in question but reflects it in an abstract form. But this is the nature of every mathematical description of physical phenomena. This is also the nature of every application of abstract geometry; the difference consists only in that more powerful, more delicate abstractions are applied, but the essence remains the same.

The most brilliant application of Riemannian geometry came in the theory of relativity. Of this we shall speak in the following section, where we will be concerned with the important and difficult problem of the relationship between abstract geometry and properties of the real space.

7. Generalizations. In the last thirty years the geometry of various non-Euclidean spaces has been the subject of remarkable developments and generalizations in several directions. New theories have arisen in which Riemannian geometry is included as a special case. The first of these was the so-called Finsler geometry, the idea of which goes back right to Riemann;* then came a general theory of spaces of the eminent French geometer E. Cartan, which combines Riemannian geometry with Klein's Erlanger Programm, and other theories. We cannot possibly give

---

* Finsler was the German mathematician who in 1916 initiated a detailed treatment of the geometry mentioned here.

an account of these new directions of geometry and shall mention only that they are worked out essentially by means of a special adaptation of their analytical apparatus. A group of Soviet geometers has been taking part in the development of these new directions; we could perhaps name here the new "polymetric" geometry created by P. K. Raševskiĭ and the investigations of V. V. Vagner which extend from the most general problems of the theory of curved spaces to the applications of non-Euclidean geometry in mechanics.

§10. Abstract Geometry and the Real Space

1. Difficulties of visualizing our actual space as non-Euclidean. In the course of the preceding account of the development of geometrical ideas beginning with Lobačevskiĭ, we have gone deep into abstract spaces and quite far away from the original object of geometry, that real space in which all phenomena take place. We shall now return to this space in the usual sense and shall set ourselves the task of explaining what the development of abstract geometry has contributed to our knowledge of its properties.

We know that geometry has grown from experiment, from an experimental investigation of spatial forms and relations of bodies: from the measurement of lots of land, of volumes of containers, etc. So in origin it is a physical theory such as, say, mechanics. The axioms of Euclidean geometry were conclusions clearly formulated on the basis of protracted experiments; they express laws of nature, and they can be called laws of geometry, just as the fundamental laws of mechanics are now often called axioms of mechanics.* But it is wrong to assert that these laws are absolutely exact and never require modification or generalization in connection with new experimental data; the real properties of space may differ more or less from what Euclidean geometry states.

We have already brought forward these arguments and now they must appear, we would think, quite obvious. But this was not so a hundred years ago, when Lobačevskiĭ's ideas failed to achieve general recognition. Before Lobačevskiĭ and Gauss, it had not entered into anybody's head that the Euclidean geometry could turn out to be not entirely accurate,

---

* An abstract conception of axioms that dissociates the axioms from their original content has arisen during the last fifty years; this changes nothing in the fact that the axioms of Euclidean geometry express laws of nature. In speaking of axioms and not of laws of geometry or mechanics we wish to place the logical deductive construction of these sciences in the forefront, but they do not lose their experimental foundation on that account. Any statement of a theorem is called an axiom when it is taken as the basis for the deductive construction of the theory, and other statements of the theory (theorems) are deduced from the basic ones (the axioms) by logical reasoning.

that the real properties of space could be somewhat different. Lobačevskiĭ developed his geometry as a theory of possible properties of the real space. Later Riemann and some other scientists also raised the problem of possible properties of space, of possible laws of measuring lengths that could be discovered by more accurate measurements. Quite generally, even abstract geometry in some of its parts can be regarded as a theory of possible properties of space. All this remained, however, in the realm of hypothesis until in 1915 Einstein in his general theory of relativity corroborated the ideas of Lobačevskiĭ and Riemann. This theory claims that the geometry of the real space in fact differs somewhat from Euclidean geometry, and this was discovered on just that astronomical scale that Lobačevskiĭ had anticipated.

In what we have just said about space at least three difficulties are involved. The problem of the relation of abstract geometry to physical geometry, i.e., to the properties of real space, reduces in the ultimate analysis to a clarification of these difficulties.

The first difficulty consists in visualizing at all how and in what sense the properties of real space can possibly differ from the statements of Euclidean geometry. We are so accustomed to it that we cannot easily imagine anything else, and explanations are obviously necessary here.

The second difficulty lies in the very expression "properties of the real space." Space by itself is conceived of as empty and homogeneous. It would seem that in the concept of space itself the idea of its homogeneity is already included. How then can the empty space, i.e., the "emptiness," have any properties? We speak of "properties of space" without thinking of these problems, but they are worth thinking about, as is apparent from the difficulty just mentioned.

The third difficulty lies in the concept of the truth of one geometry or another. The question may appear very simple: That geometry is true which corresponds to reality. This is so, of course. But on the other hand we have seen, for example, that the geometry inside a circle can be regarded as a Lobačevskiĭ geometry, because every geometrical fact inside the circle can be presented as a theorem of Lobačevskiĭ geometry. Consequently it turns out that the same geometrical facts can be presented both as theorems of Euclid's geometry and as theorems of Lobačevskiĭ's geometry. Hence both geometries correspond to reality. So which of them is true and in what sense, and why do we assume all the same that, in fact, Euclidean geometry is satisfied in the circle and that Lobačevskiĭ geometry is only illustrated or interpreted in it?

Clearly, in these problems lies a considerable difficulty which has baffled at times even some eminent mathematicians.

Our explanation must begin with the second of the difficulties mentioned, because an understanding of what these "properties of space" are will lead us to a solution of the other difficulties.

2. Space and matter. The subject matter of geometry, namely the "properties of space," is made up of the properties of real bodies, their material relationship and forms. In real space a "place," a "point," a "direction," etc., is determined by material bodies. "Here" and "there," "hither" and "thither" have a meaning only in connection with one material object or another. "Here" can mean "on the earth," "in this room" or something else of the sort; in a word, "here" always denotes a place determined by material criteria of one kind or another. Similarly, for example, a straight line does not exist by itself, but only as a taut thread or the edge of a ruler or a ray of light. A straight line, "a line like this," is altogether an abstraction which reflects the common properties of these material lines, just as, say, the "house in itself" is an abstraction reflecting the common properties of houses; the "house in itself" does not exist outside of or independently of the various real houses.

This objective character of the properties of space is expressed by the well-known statement of dialectic materialism: *Space is the form of existence of matter.* The form of an object is determined by the connections and relations of its parts. The structure of space is the common regularity of a number of relations of material bodies and events. There are the spatial relationships, the spatial order of objects, their mutual positions, distances, etc. But as every form is inseparable from its content, except in abstractions and in certain contexts, so is space inseparable from matter. The idea of a space "in itself," of a space without matter is an abstraction that must not be abused. Real spatial relationships and forms: "here," "between," "inside," "straight," "sphere," etc., these are always relations and forms of material bodies. Geometry, however, considers them abstractly. This abstraction is necessary, because otherwise it would not be possible to perceive generality in the diverse concrete relations of objects. But this abstraction must not be made absolute by substituting abstract concepts for the objective reality itself.

In an absolutely empty space, void of all traces of matter, nothing would distinguish a place, a direction, consequently there are no places, no directions, so that the absolutely empty space reduces to nothing. Even in the abstract idea of an empty space we imply tacitly that various places and directions are distinguishable in it. In other words, in the abstract idea of space we retain the properties of distinctness of places, directions, distances which exist in real space precisely because this space is inseparably connected with material bodies.

Thus, space is the form of existence of matter; "properties of space" are, therefore, properties of matter, properties of certain relations of material bodies, their mutual positions, dimensions, etc.

Furthermore, if the theorems of geometry are to have a physical meaning, we must know what we shall understand by "straight line," "distance," and other geometrical concepts in them. In §4 we have seen that one and the same geometrical theory admits different interpretations.

Consequently, for comparing geometry with experiment it is necessary to define as accurately as possible the physical meaning of geometrical concepts, because geometry describes the properties of real space only on condition that the corresponding physical meaning is attributed to its concepts. Without this physical meaning the theorems of geometry are of an abstract mathematical, formal character. Herein lies the solution to the second difficulty specified earlier. This difficulty arises because, instead of the real space, which is inseparably connected with matter, one wants to think of a "pure" space, a space "as such," which is, however, nothing more than an abstraction.

3. Intuition and understanding. Now it is easy to understand how the other two difficulties are resolved.

First of all, how can one imagine that the real space in its properties is other than Euclidean? Suppose we wish to verify some statement of Euclidean geometry, for example, that the sum of the angles of a triangle is 180° or that the length of a circle is equal to $2\pi R$. To verify the first we have to determine what physically defined triangles are to be considered and how their angles are to be defined. Suppose a side of a triangle is a light ray in empty space. In that case there is nothing inconceivable in the fact that very accurate experiments may show that sums of the angles of a triangle are different from 180°. In the same way one can imagine that measurement of a radius and circumference on one and the same scale leads to results that do not satisfy the relation $l = 2\pi R$ accurately. In fact, this is so on the surface of the earth, where the length of a circle is not proportional to the radius but grows slowly and reaches a maximum when the radius is made equal to half a meridian. But it may be objected that on the surface of the earth the role of straight lines is played by arcs of great circles so that the radius is understood here in another sense and our result, therefore, does not contradict Euclidean geometry. However, according to the theory of relativity, near a body of large mass the ratio of the length of a circle to the "genuine" radius is all the same somewhat different from 2π and, in fact, the following approximate formula holds for the ratio of the length of the equator of a homogeneous sphere-shaped body and its radius:

$$\frac{\text{length of circumference}}{\text{radius}} = 2\pi \left(1 - \frac{kM}{3Rc^2}\right),$$

where M is the mass of the body (in metric tons), R the radius of the body (in kilometers), $c = 300,000$ km/sec the velocity of light and k the constant of gravitation which for this choice of units of measurement is equal to $66.6 \cdot 10^{-18}$.

So we see that the ratio of the length of a circle to its radius is not equal to 2π, but a little less. Computations show that on the surface, of the sun this ratio differs from 2π by approximately .000004, and on the surface of the companion of Sirius, whose average density is 50,000 times that of water, the deviation reaches .00014.

It may be objected, of course, that all this is nevertheless impossible to imagine, that in an intuitive picture space is always Euclidean. This objection need not disturb us, first of all, because the task of science does not consist in giving intuitive pictures of the phenomena but in arriving at an understanding of them. An intuitive picture is restricted and conditioned by the customary forms transmitted by our sense organs. Therefore we are not in a position to have an intuitive picture of ultraviolet rays, of the propagation of radio waves, of the motion of an electron in an atom, or of many other phenomena, except by substituting models in place of them. But this does not mean at all that these phenomena are incomprehensible for us. On the contrary, the successes of radio technology, for example, show clearly that we have mastered radio waves completely and therefore understand them quite well. Second, the solution of the problem of what we can and what we cannot imagine depends on habit and training of the imagination. Can we imagine the antipodes, where from our point of view people walk with their heads hanging downwards? Nowadays we are able to imagine this, but there was a time when the "unimaginability" of the antipodes served as an argument against the spherical shape of the earth.

4. Geometry and truth.   Now let us turn to the last difficulty, i.e., to the problem as to which geometry can be regarded as true. When we posed this problem, we indicated that the geometrical facts inside a circle can be interpreted as theorems of Euclid's geometry and of Lobačevskiĭ's geometry. Therefore, both these geometries correspond to reality; i.e., both are true. And after all, there is nothing astonishing here. One and the same phenomenon can always be described by various methods; one and the same quantity can be measured in various units; one and the same curve can be given by various equations depending on the choice of the coordinate system. Similarly a given isolated collection of geometrical relations (in the example in question, the relations inside a circle) can be described by various methods. But we raise the problem not of some isolated collection of geometrical facts but of the spatial relationships in their entire generality. Space is the universal form of the existence of matter and, consequently, when the problem of the properties of space is raised, no domain of facts can be separated artificially.

When the problem is posed in this way, then geometrical quantities or geometrical facts cannot be considered isolated from other phenomena with which they are necessarily connected. For example, the length of a segment is determined by laying off a rigid rule so that the measure of length is necessarily connected with the motion of rigid bodies, i.e., with mechanics. Geometry is inseparable from mechanics. However, the measurement of length inside a circle in the interpretation of Lobačevskiĭ geometry proceeds quite differently, as we have explained in §4; a chord here becomes infinitely long. It is clear that the so-defined measure does not correspond to the original idea of measurement which has grown on the basis of mechanical transportation of the real bodies to be compared. In general, figures that are equal in the sense of Euclidean geometry are, by the very origin of Euclidean geometry, figures that can be superimposed on one another by means of a mechanical motion. In the interpretation of Lobačevskiĭ geometry equality is defined differently, the role of motions is taken here by other transformations. Therefore, when the geometrical facts inside a circle are taken in conjunction with their necessary link with mechanics, then we must admit that it is in fact the Euclidean geometry which holds inside a circle (with great accuracy).

Euclidean geometry is the one in which the role of motion is played by the ordinary mechanical motion of rigid bodies. It was precisely for this reason that Euclidean geometry, and not any other geometry, was the first to be discovered. But the development of physics has now led to the conclusion that the laws of Newtonian mechanics and with them the laws of Euclidean geometry are only approximations to more accurate and general laws. In this change of the laws of geometry, in the transition from Euclidean to Riemannian geometry which was accomplished in the theory of relativity, mechanics was not the only branch of physics to play a role: Of equal, if not greater, importance were the theories of electromagnetic phenomena and optics. Geometry as the science of the properties of space is connected with physics, depends on it, and can be separated from it only in abstraction and only in certain contexts.

The dependence of geometry on physics, or in other words the dependence of the properties of space on matter, was clearly indicated by Lobačevskiĭ, who foresaw the possibility of a change of the laws of geometry on transition to a new domain of physical phenomena. In contrast to this materialistic point of view, the famous mathematician Poincaré stated rather recently that the choice of one geometry or another is dictated solely by considerations of simplicity or "economy of thought" in the phrase of the well-known idealist Mach. On this basis Poincaré predicted further that science would more readily give up the law of rectilinear propagation of light than the Euclidean geometry, because it is the "simplest." However, Poincaré died three years before the general theory of relativity was finally set up, in which just the opposite is done: Euclidean geometry is abandoned, but the law of propagation of light is preserved, though in a generalized form; light is propagated on a geodesic line.

So we have reached the following conclusions. One and the same isolated collection of facts can, generally speaking, be described in a variety of ways and all these descriptions are true provided they reflect reality. However, it is wrong to consider the geometrical facts in their entire generality severed from other phenomena. Only so can the properties of space be established, because it is the universal form of existence of matter. But by taking geometry in conjunction with physics we must necessarily adapt them to one another, and then we see the essential difference between the various "geometries" which, when dissociated from physics, can only be distinguished by their greater or lesser simplicity. Euclidean geometry appeared not because it was simpler than the others, but because it corresponded to mechanics. In fact, in connection with the development of physics in the theory of relativity we now go over to a more complicated geometry, namely Riemannian.

To sum up, in reference to the properties of real space that geometry is true which reflects the properties of spatial relationships in their entire generality with sufficient accuracy and which consequently corresponds not only to the purely geometrical facts but also to mechanics and to the whole of physics.

5. The space-time of relativity.   What little we have said above on the theory of relativity does not touch on its main contents. In the understanding of the problem of space, it went substantially further than Lobačevskiĭ and Riemann had thought of.

The most essential and basic proposition of the theory of relativity is: Space is completely inseparable from time and that together they form a single form of existence of matter, the four-dimensional manifold of space-time. An event in the world is characterized by its place and time and consequently by four coordinates: three spatial and the fourth temporal, the time of the event. The events form in this sense a four-dimensional collection. The theory of relativity is concerned above all with this four-dimensional collection from the point of view of its structure, disregarding the properties of the individual phenomena. It is not fundamentally a theory of fast motions, nor of cosmology, nor a new theory of space or time, but just a theory of space-time as a single form of existence of matter.

Of course, in Newtonian mechanics also we can combine space and time in a single four-dimensional manifold. As we have had occasion to recall, the idea itself of a many-dimensional space was born in Lagrange's work, when in considering the motion of a material point he added to the spatial coordinates x, y, z the time coordinate t. The motion of a point is then represented as a line in the four-dimensional space with the coordinates x, y, z, t; under a motion of the point all four coordinates change: the position (x, y, z) and the time t. However, this unification of space and time has a purely formal character. No internal necessary connection between space and time is set up here. Of course, in the law of motion of each given body there is its dependence on spatial position and on time. But this concerns only each given motion, no universal internal connection between space and time was established before the theory of relativity, neither in mechanics nor in physics generally. The spatial relationships, the spatial order of objects and phenomena were always carefully distinguished from their relationships and order in time. The temporal sequence of events, the duration of time intervals were regarded as absolute, as definite irrespective of what happened. In short, the concept of absolute time ruled supreme.

Einstein's greatest discovery, which not only laid the corner stone of the theory of relativity but revolutionized the whole physical and philosophical understanding of the problem of space and time, was the discovery that absolute time does not really exist. Shortly after Einstein developed his theory in 1905,* Minkowski showed that its essence consists not as much in the rejection of absolute time as in the institution of a mutual link of space and time, in virtue of which there exists a single absolute form of existence of matter: space-time. The separation of space (the spatial coordinates) from time (from the time coordinate t) is to a certain degree relative, depending on the material system (the "system of reference") in relation to which the spatial and temporal order of the phenomena is determined. Events that are simultaneous with reference to one system need not be so with reference to another system.

The definition of the order of phenomena cannot be, of course, completely dependent on the system of reference. The order of events that are connected by direct interaction, it stands to reason, remains one and the same with reference to all systems, so that an action always precedes its result. But for events that are not connected by interaction the order

---

* The theory that Einstein developed in 1905 is called the special theory of relativity in contradiction to Einstein's "general" theory of 1915.

of time turns out to be relative. Since the spatial order (in its pure form) refers to simultaneous events and simultaneity is relative, the separation of purely spatial relationships from the general aggregate of space-time relationships turns out to be relative, depending on the system of reference. Space in the abstract sense is, as it were, a "section" of the four-dimensional manifold of space-time that is laid through simultaneous events (in reference to a given system).

It cannot be our task to explain the foundations of the theory of relativity and so we shall try only to characterize in a few words its basic features in the form in which it is most natural to consider them in connection with the ideas of abstract geometry. This interpretation, incidentally, is quite different from that which comes from Einstein himself.

The world, the universe, can be regarded as a set of diverse events. By an event we understand here not an arbitrary phenomenon extending in space and lasting in time but as it were an instantaneous, pointlike phenomenon such as a momentary flash of a point source of light. To use geometrical language, events are points in the four-dimensional manifold of the universe.

Space-time is the form of existence of matter, the form of this world manifold. The structure of space-time, its "geometry," is nothing but a certain general world structure, i.e., in accordance with our analysis, the "geometry" of the set of events. This structure is determined by certain universal material connections and relations of events.

First, as we have explained previously for the spatial relationships, we must, in fact, be dealing here with material relationships and connections. This is also true for the relationships of phenomena in time. Spatial and temporal relationships "as such," in a pure form, are only abstractions.

Second, the relationships of events that determine the structure of space-time must have universal character in accordance with the universal character of space-time.

Such a universal material relation of events is their cause-effect connection. Every event acts in one way or another, directly on certain other events and, in turn, experiences the action of other events. This relation of the action of some events on others determines the structure of space-time.

Thus, the theory of relativity allows us to make the following definition. Space-time is the set of all events, irrespective of their concrete properties and relations except for the general relation of action of some events on others. This relation too must be understood here in the general sense, irrespective of its various concrete forms.

In the special theory of relativity, space-time is regarded as maximally

homogeneous. This means that the manifold of events admits transformations that do not disturb the relations of action between events, and the
group of these transformations is in a certain sense as large as possible.
Suppose, for example, that there are two pairs of events A, B and A', B'
and that neither A acts on B nor B on A and similarly for A' and B'.
Then there exists a transformation between the events under which A
corresponds to A' and B to B' and such that for any pairs of events the
relation of action (or inaction) is not infringed; i.e., if X acts on Y, then
the corresponding event X' also acts on Y' and if X does not act on Y,
then X' does not act on Y'.

In accordance with these explanations, it turns out that from the point
of view of the special theory of relativity, space-time is a four-dimensional
space of its own kind whose geometry is determined by a certain group
of transformations. These transformations are nothing other than the
famous Lorentz transformations. The laws of geometry and physics do
not change under these transformations. This outlook corresponds to the
view of geometry put forward by Klein in his Erlanger Programm of
which we have spoken in §6.

6. Gravitation and curvature. The general theory of relativity goes
further and abandons the idea of homogeneity of space-time. It assumes
that space-time is homogeneous only to a certain approximation in
sufficiently small domains but is on the whole inhomogeneous. The
inhomogeneity of space-time is determined according to Einstein's
theory by the distribution and motion of matter. In its turn the structure
of space-time determines the laws of motion of bodies, and this appears
in the phenomenon of universal gravitation. The general theory of
relativity is, properly speaking, a theory of gravitation that explains the
gravitational link of the structure of space-time with the motion of matter.

The idea of space-time as homogeneous only to a certain approximation
in small domains is similar to the idea of Riemannian space which is
Euclidean only "in the infinitely small." The mathematical space-time of
the general theory of relativity is treated as a kind of Riemann space,
though in a substantially altered sense.

In fact, in a four-dimensional Riemannian space we can introduce
coordinates in the neighborhood of every point such that the square of
the line elements is expressed by the formula $ds^2 = dx_1^2 + dx_2^2 + dx_3^2 + dx_4^2$.

In space-time we can introduce coordinates x, y, z, t in the neighborhood
of every event such that the line element is represented by the formula
$ds^2 = dx^2 + dy^2 + dz^2 - c^2dt^2$, where c is the velocity of light, which
for a suitable choice of the units of measurement can be taken to be 1.
Here x, y, z are the spatial coordinates and t is the time. The minus

sign for dt^2 gives a formal expression to the essential radical difference of the time coordinate from the spatial ones, of time from space.

In the theory of gravitation the concept of curvature of space-time plays a very important role. The fundamental equations of the theory given by Einstein at once connect the quantities that characterize the curvature of space-time with the quantities that characterize the distribution and motion of matter. These equations are at the same time the equations of the gravitational field and thus, as Einstein has proved in collaboration with V. A. Fok, the laws of motion of bodies in a gravitational field can be derived from them.

The structure of space-time according to the general theory of relativity is complicated, and space cannot be separated from time even to the extent permitted by the special theory of relativity. However, with a certain approximation and under certain assumptions this can be done. Space turns out to be Euclidean with a sufficient accuracy in domains that are small in comparison with the cosmic scale, but in large domains the deviation from Euclidean geometry becomes apparent. This deviation depends on the distribution and motion of masses of matter and reaches appreciable, though still very small, values near a star of large mass, or in general when the magnitudes involved are on a cosmic scale. In a number of hypotheses on the structure of the universe as a whole it is assumed that on the average the distribution of mass is approximately uniform. In one of these hypothetical theories proposed by the Soviet physicist and mathematician Fridman the geometry of space on the whole coincides with Lobačevskiĭ geometry.

In the theory of relativity abstract geometry finds an application not only as a mathematical apparatus; the very ideas of an abstract space provide the means for a deeper formulation of the foundations of this theory. Possibilities contemplated in abstract geometry are discovered in reality, and theoretical thinking celebrates here its most brilliant triumph. Abstract geometry, which itself has grown from an experimental study of the spatial relationships and forms of bodies, now faces, as a well-developed mathematical method, the study of real space. Such is the general path of science: From what is immediately given by experiment it rises to theoretical generalizations and abstractions, and then turns again to the experiment as the instrument for deeper understanding of the essence of the phenomena; by thus giving explanations of known phenomena and predictions of new ones, it guides the practical activities of its investigators and in return finds herein its own justification and the source of its future development.

Suggested Reading

H. S. M. Coxeter, *Introduction to geometry*, Wiley, New York, 1961.

L. P. Eisenhart, *An introduction to differential geometry with use of the tensor calculus*, Princeton University Press, Princeton, N. J., 1940.

——, *Riemannian geometry*, Princeton University Press, Princeton, N. J., 1949.

W. C. Graustein, *Elementary differential geometry*, Macmillan, New York, 1935.

D. Hilbert, *Foundations of geometry*, Open Court, LaSalle, Ill., 1959.

A. V. Pogorelov, *Differential geometry*, Noordhoff, Groningen, 1959.

G. Y. Rainich, *Mathematics of relativity*, Wiley, New York, 1950.

D. J. Struik, *Lectures on classical differential geometry*, Addison-Wesley, Cambridge, Mass., 1950.

J. L. Synge, *Relativity: the special theory*, Interscience, New York, 1956.

PART 6

XVIII

TOPOLOGY

§1. The Object of Topology

"*Adjacency* is the distinguishing appurtenance of bodies and permits us to call them *geometric*, when we retain in them this property and abstract from all others, whether they be essential or accidental." With these words I. N. Lobačevskiĭ begins the first chapter of his work "New Elements of Geometry".*

Explaining by a diagram (figure 1) the words just quoted Lobačevskiĭ continues: "Two bodies *A, B* that touch each other form a single geometric body *C* ⋯ . Conversely, every body *C* can be split by an arbitrary section *S* into two parts *A, B*."

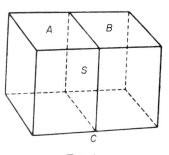

FIG. 1.

These concepts of adjacency, neighborhood, infinite proximity, and also the concept of a dissection of a body, which in a certain sense is dual to them, these are the concepts that Lobačevskiĭ places at the foundation of the whole structure of geometry and they are also in essence the fundamental, primordial concepts of topology in the full extent in which we now understand this discipline. Therefore contemporary commentators on the great geometer are right in saying† that "Lobačevskiĭ makes the first attempt in the history of the mathematical sciences to start, in the construction of

---

* I. N. Lobačevskiĭ, "Collected works," Vol. II, Gostehizdat, 1949, page 168.

† Comments to "New Elements of Geometry," loc. cit., page 465.

geometry, from topological properties of bodies .... . The concepts of
surface, line, point are defined by Lobačevskiĭ in terms of dissections and
adjacencies of bodies." Some idea of the diversity of concrete geometric
content reflected in the concepts of adjacency and dissection of bodies,
as Lobačevskiĭ imagined them, can be obtained from the following
diagrams (figure 2) taken from his work.

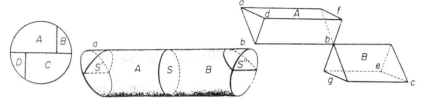

FIG. 2.

Every transformation of a geometric figure in which the relations of
adjacency of various parts of the figure are not destroyed is called
continuous; if the adjacencies are not only not destroyed, but also no
new ones arise, then the transformation is called *topological*. Therefore,
under a topological transformation of an arbitrary figure the parts of
this figure that are in contact remain in contact, and the parts that are
not in contact cannot come into contact; to put it briefly, in a topological
transformation neither breaks nor fusions can arise. In particular, two
distinct points cannot be united into a single point (in that case a new
contact would arise: figure 3). Therefore a topological transformation of

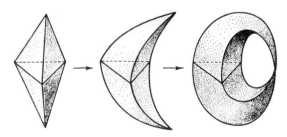

FIG. 3.

any geometrical figure, considered as the set of points forming it, is not
only a continuous, but also a one-to-one transformation: Any two distinct
points of the figure are transformed into two distinct points. Thus,
topological transformations are single-valued and continuous both ways.

Intuitively a topological transformation of an arbitrary geometric figure (a curve, a surface, etc.) can be represented in the following way.

Let us imagine that our figure is made of some flexible and stretchable material, for example of rubber. Then it can be subjected to all possible continuous deformations under which it will be extended in some of its

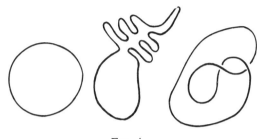

FIG. 4.

parts and contracted in others, and altogether will change its size and shape in every way. For example, when a closed rubber ring is given in the form of a circle, then we can stretch it into the shape of an extremely elongated ellipse, we can give it the form of a regular or an irregular polygon and even of very complicated closed curves, some of which are illustrated in figure 4. But we cannot, by a topological transformation, turn a circle into a figure of eight (for this would require the fusion of two distinct points of the circle; figure 5) or into an interval (for this

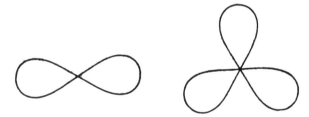

FIG. 5.

would require the fusion of one semicircle with another or else a break of the circle at an arbitrary point). The circle is the simplest closed curve, since it forms only one loop in contrast to the figure of eight, which forms two loops, or the trefoil curve (figure 5), which forms three loops. The property of a circle of being a simple closed curve is a property preserved under an arbitrary topological transformation or, as one says, it is a topological property.

If we take a spherical surface, which we can imagine in the form of a thin rubber sheet, then we can again make exceedingly great changes in its shape by means of a topological transformation (figure 6). But we

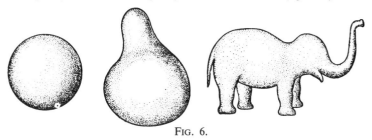

FIG. 6.

cannot, by a topological transformation, turn our spherical surface into a square or a ring-shaped surface (the surface of a steering wheel or a life belt), which is called a torus (figure 7). For the surface of a sphere

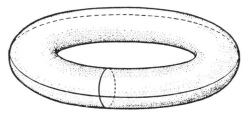

FIG. 7.

has the following two properties which are both preserved under an arbitrary topological transformation. First, our surface is closed: There are no edges on it (but the square has edges); second, every closed curve on a spherical surface is in Lobačevskiĭ's expression a dissection of it; if we make a cut along a given closed curve traced out on our rubber sheet, then the surface splits into two disconnected parts. The torus does not have this property: If a torus is cut along a meridian (figure 7), then it is not split into parts but is turned into a surface having the form of

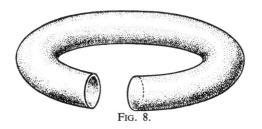

FIG. 8.

a bent tube (figure 8) which we can then easily turn (straighten) into a cylinder by a topological transformation. Thus, in contrast to the sphere, not every closed curve on the torus is a dissection. Therefore the spherical surface cannot be turned into a torus by a topological transformation. We say that the sphere and the torus are topologically distinct surfaces or surfaces that belong to distinct topological types or, finally, that these surfaces are not homeomorphic to each other. Conversely, a sphere and an ellipsoid and quite generally any bounded convex surface belong to one and the same topological type; i.e., they are *homeomorphic*. This means that they can be carried into one another by a topological transformation.

§2. Surfaces

As mentioned earlier, every property of a geometrical figure that is preserved under an arbitrary topological transformation of it is called a topological property. Topology studies topological properties of figures; furthermore, it studies topological transformations and also arbitrary continuous transformations of geometrical figures.

We have just given some examples of topological properties. Such properties are: the property of a curve or a surface of being closed, the property of a closed curve of being simple (i.e., of forming only one loop), the property of a surface that every closed curve lying on it is a dissection of the surface (the spherical surface has this property, but the ring-shaped one has not), etc.

The largest number of closed curves that can be drawn on a given surface in such a way that these curves do not form dissections, i.e., that the surface does not split into parts when cuts are made along all these curves, is called the *order of connectivity* of the surface. This number gives us some important information on the topological layout of the surface. We have seen that for a spherical surface it is equal to zero (every closed curve on this surface is a dissection). On the torus we can find two closed curves that taken together do not form sections: One

FIG. 9.

of them can be taken as an arbitrary meridian and the other as a parallel of the torus (figure 7). However, it is impossible to draw on the torus three closed curves that taken together do not form a dissection of it; the order of connectivity of the torus is 2. The order of connectivity of the pretzel surface (figure 9) is 4, etc. Quite generally, let us take a spherical surface and cut $2p$ spherical holes in it (in figure 10, the case $p = 3$ is

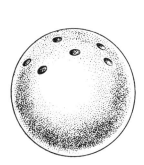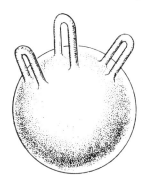

FIG. 10.

illustrated). We divide these holes into p pairs and attach to each pair of holes (at the edges) a cylindrical tube (a "handle"). We obtain a sphere with p "handles" or as it is called, a normal surface of genus p. The order of connectivity of this surface is $2p$.

All these surfaces, in Lobačevskiĭ's expression, are "dissections" of space: Each of them divides the space into two domains, an interior and an exterior, and they are the common boundary of these two domains. This fact is connected with another, namely that every one of our surfaces has two sides: an interior and an exterior (one side can be painted in one color, and the other in another).

However, apart from these there also exist the so-called one-sided surfaces on which there are not two distinct sides. The simplest of these

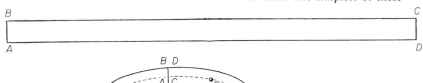

FIG. 11.

is the well-known "Möbius band," which is obtained when we take a rectangular strip of paper $ABCD$ and paste together the two opposite short sides AB and CD so that the vertex A coincides with C, and B with D. The surface so obtained is illustrated in figure 11; it is called the Möbius band or strip. It is easy to verify that there are not two sides on it that could be painted in different colors: When we go along the middle line of the surface beginning our path at the point E, say, then we arrive again at the point E by proceeding on the surface, but on the other side of it, although we have not crossed an edge on our path. Incidentally, the edge of the Möbius surface consists of a single closed line.

The problem now arises: Do there exist closed one-sided surfaces, i.e., one-sided surfaces that do not have edges? It turns out that they exist, but that such surfaces, no matter how we arrange them in three-dimensional space, always have self-intersections. A typical example of a closed one-sided surface is illustrated in figure 12; this is the so-called "one-sided

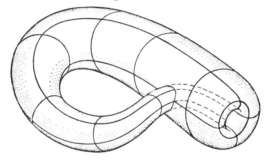

FIG. 12.

torus" or the Klein bottle. If without worrying about self-intersections we imagine two copies of a Möbius strip pasted together along their edges, (the edge of a Möbius strip, as was mentioned earlier, consists of a single contour), then we obtain a Klein bottle.

Now we can formulate the fundamental theorem of the topology of surfaces as applied to two-sided surfaces: Every closed two-sided surface is homeomorphic to some *normal surface* of genus p, i.e., to a "sphere with p handles"; two closed two-sided surfaces are homeomorphic if and only if they are of one and the same *genus p* (the same order of connectivity $2p$), i.e., when they are homeomorphic to a sphere with one and the same number of handles p.

For one-sided surfaces there also exist "normal forms," similar to the normal forms of two-dimensional surfaces of genus p, but they are complicated to describe. For this purpose we have to take a sphere, cut p circular holes in it, and attach to each of them a Möbius surface by pasting the edge of this surface to the edge of the corresponding hole. The complication that arises in an attempt to imagine such a pasting comes from the fact that there is no physical realization of it: Self-intersections of the surface arise at once in such a pasting, and they are unavoidable in every realization of a one-sided closed surface in the form of a spatial model.

One must not think that closed one-sided surfaces belong to the domain of mathematical curiosities, unconnected with serious problems of science. To see how wrong such an opinion is we need only recall that one of the fundamental achievements of geometric thinking was the creation of the so-called projective geometry, the elements of which occur nowadays in geometry courses of universities and teacher training colleges. Practical sources of projective geometry lie in the theory of perspective which dates back to the Renaissance (Leonardo da Vinci) in connection with the needs of architecture, pictorial art, and technical projection. The 16th and 17th centuries saw the discovery of the first theorems of projective geometry. Thus, arising in connection with quite a definite practical requirement, projective geometry became in its full development one of the most significant generalizations of geometry, as far as theoretical ideas are concerned. In particular, it was in the framework of projective geometry that Lobačevskiĭ's non-Euclidean geometry was for the first time completely understood.*

The transition from the ordinary plane, as it is studied in elementary geometry, to the projective plane consists in a completion of the plane by new abstract elements, the so-called improper or "infinitely distant" points. Only after such a completion does the operation of projecting one plane onto another (for example, the projection onto a screen by means of a projection lantern) become a one-to-one transformation of the one plane onto the other. The completion of the plane by improper points, which in coordinate geometry corresponds to a transition from the ordinary Cartesian coordinates to homogeneous coordinates, proceeds in the following way. Every straight line is completed by a single improper point ("at infinity"), and two straight lines have the same improper point if and only if they are parallel. A straight line completed by a single point at infinity becomes a closed line and the set of all points at infinity of all possible straight lines forms by definition an improper line or a line at infinity.

---

* See, for example, Chapter XVII, §6, or the book by P. S. Aleksandrov "What is non-Euclidean geometry," Moscow, 1951.

Since parallel lines have a common point at infinity, in the representation of the whole process of completion of the plane by improper points it is sufficient to consider the lines passing through an arbitrary point of the plane, for example through the origin of coordinates O (figure 13). The improper points of these lines already exhaust all the improper points of the whole projective plane (since every line has the same improper point as the line through O parallel to it). Therefore we obtain a "model" of the projective plane by regarding it as a circle of "infinitely large" radius with center at O and assuming that every pair of diametrically opposite points A, A' of the circumference of this circle are united into the single "infinitely distant" point of the line AA'.

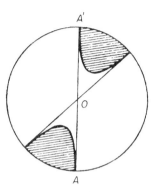

Fig. 13.

The circumference of our circle then becomes the line at infinity, but we must strictly keep in mind that any two diametrically opposite points of this circumference must be thought of as identified with one another. Hence it is at once clear that the projective plane is a closed surface, that there are no edges on it.

If we take a curve of the second order in a projective plane, which we can illustrate in the form of a hyperbola (see figure 13), then it is obvious that this hyperbola in the projective plane is a closed curve (only cut into two branches by the line at infinity). Bearing in mind that diametrically opposite points of the circumference of our fundamental circle are identified with one another, we can see without difficulty that the interior of our hyperbola which is shaded in figure 13 is homeomorphic to the interior of an ordinary circle, and the complement, the unshaded part of the projective plane, is homeomorphic to a Möbius band. Thus, from the point of view of topology the projective plane is the result of pasting together a circle (in our case the interior of a hyperbola) with a Möbius band along their edge. Hence it follows that the projective plane, i.e., the basic object of study of plane projective geometry, is a closed one-sided surface.

The example of the projective plane, apart from its great intrinsic geometrical value, is also interesting because it throws into relief one peculiarity of contemporary geometric thinking as it is moulded on the basis of Lobačevskiĭ's discovery. Geometric thinking has always been abstract, by the very character of the concept of a geometrical figure.

Now it rises to a new degree of abstraction, which in our case becomes apparent in the completion of the ordinary plane by new abstract elements, namely the improper points. Of course, even these abstract elements form a real entity (every "improper point" is nothing but an abstraction of a pencil of parallel lines), but they are introduced into our discussion as distinct geometric elements which we can only indirectly imagine as the result of the "fusion" (which does not exist physically) of diametrically opposite points of the circumference of some circle. Similar abstract constructions are of very great value in the whole of contemporary topology, especially on transition from planes to manifolds of three or more dimensions.

§3. Manifolds

Let us consider the following simple apparatus, sometimes called a compound plane pendulum (figure 14). It consists of two rods OA and AB, hinged together at A; the point O remains immovable, the rod OA turns freely in a fixed plane around O, and the rod AB turns freely in the same plane around A. Every possible position of our system is completely determined by the magnitude of the angles ϕ and

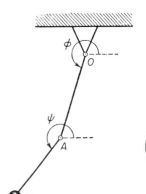

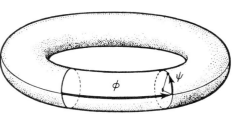

FIG. 14.   FIG. 15.

ψ that the rods OA and AB form with an arbitrary fixed direction in the plane, for example with the positive direction of the abscissa axis. We can regard these two angles, which change from O to 2π, as "geographical coordinates" of a point on a torus, counting from the "equator" of the torus and one of its "meridians," respectively, (figure 15).

Thus, we can say that the manifold of all possible states of our mechanical system is a manifold of two dimensions, namely a torus. When we replace each of the two angles ϕ, ψ by a corresponding point on the circumference of a circle on which an initial point and a direction are given (figure 16), then we can also say that every possible

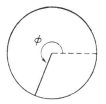 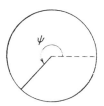

FIG. 16.

state of our mechanical system is completely characterized by giving one point on each of two circles (one of these is taken as the latitude ϕ and the other as the longitude ψ). In other words, just as in analytic geometry we identify a point of the plane with a pair of numbers, namely its coordinates, so in our case we can identify a point of the torus (and hence an arbitrary position of our pendulum) with the pair of its geographic coordinates, i.e., with a pair of points one of which lies on one circle and the other on another. The essence of the situation is expressed by saying that the manifold of all possible states of our compound plane pendulum, i.e., the torus, is the topological product of two circles.

Now we modify our apparatus in the following way. Suppose that, as before, it consists of two rods OA and AB and that OA can turn freely in a definite plane around the point O but that AB is now hinged to OA by a spherical hinge at A, so that for a given position of this point it can freely turn around it in space, keeping parallel to an arbitrary original ray through A. Now the position of our system is given by three parameters of which the first is the previous angle ϕ formed by the rod OA with the positive direction of the abscissa axis, and the other two determine the direction of the rod AB in space. The latter direction can be determined, for example, by giving the point B' on the unit sphere with the centre at the origin of coordinates O at which the radius OB' parallel to AB intersects the sphere, or by giving on the sphere the two geographical coordinates of B'. Thus, the manifold of all positions of our new hinged system is a certain three-dimensional manifold, and the reader will easily realize that it can be treated as the topological product of a circle and a sphere. This manifold is closed, i.e., it has no edges, therefore it cannot be realized in the form of a figure lying in three-dimensional space. If nevertheless we wish to get a somewhat intuitive idea of it, we can consider the part of space lying between two concentric spheres. Each ray emanating from the common center of these spheres pierces them in two points. If we regard each pair of such points as identified (fused into a single point), then we obtain a three-dimensional manifold which is the topological product of a sphere with a circle.

We can make our hinged apparatus even more complicated if we not only connect the rods OA and AB by a spherical hinge at A, but also assume that the rod OA can turn freely in space around the point O. The set of possible positions of the system so obtained is then a four-dimensional closed manifold, namely the product of two spheres.

Thus we have seen that even the simplest mechanical (kinematical) considerations lead us to topological manifolds of three and more dimensions. Of even greater value in the practical, more detailed discussion of mechanical problems are certain manifolds (in general, many-dimensional), the so-called *phase spaces* of dynamical systems. Here we take into account not only the configurations that the given mechanical system can have, but also the velocities with which its various constituent points move. Let us confine ourselves to one of the simplest examples. Suppose we have a point that can move freely on a circle with an arbitrary velocity. Every state of this system is determined by two data: the position of the point on the circle and the velocity at the given instant. The manifold of states (the phase space) of this mechanical system is, of course, an infinite cylinder (a product of a circle with a straight line).

The number of dimensions of the phase space increases as we increase the number of degrees of freedom of the given system. Many of the dynamical characteristics of a mechanical system can be expressed in terms of the topological properties of its phase space. For example, to every periodic motion of the given system there corresponds a closed curve in its phase space.

The study of the phase spaces of dynamical systems occurring in various problems of mechanics, physics, and astronomy (celestial mechanics, cosmogony) drew the attention of mathematicians to the topology of many-dimensional manifolds. It was precisely in connection with these problems that the famous French mathematician Poincaré in the 1890's inaugurated the systematic construction of the topology of manifolds, by applying the so-called combinatorial method, which up to the present day is one of the fundamental methods of topology.

§4. The Combinatorial Method

Historically the first theorem in topology is the theorem or formula of Euler (which was apparently known even to Descartes). It consists in the following. Let us take the surface of an arbitrary convex polyhedron. We denote by α_0 the number of its vertices, by α_1 the number of its edges, and by α_2 the number of its faces; then the following relation is known as *Euler's formula*

$$\alpha_0 - \alpha_1 + \alpha_2 = 2. \tag{1}$$

This geometrical theorem belongs to topology, because our formula obviously remains true when we subject the convex polyhedron in question to an arbitrary topological transformation. Under such a transformation the edges will, in general, cease to be rectilinear, the faces cease to be plane, the surface of the polyhedron goes over into a curved surface, but the relation (1) between the number of vertices and the numbers of edges and faces, now curved, remains valid. The most important case is when all the faces are triangles and then we have a so-called *triangulation* (a division of our surface into triangles, rectilinear or curvilinear). It is easy to reduce the general case of arbitrary polygonal faces to this case: It is sufficient to divide these faces into triangles (for example by drawing diagonals from an arbitrary vertex of the given face). Thus, we can restrict our attention to the case of a triangulation. The combinatorial method in the topology of surfaces consists in replacing the study of such a surface by the study of one of its triangulations, and of course we are only interested in properties of the triangulation that are independent of the accidental choice of one triangulation or another and so, being common to all triangulations of the given surface, express some property of the surface itself.

Euler's formula leads us to one of such properties, and we shall now consider it in more detail. The left-hand side of Euler's formula, i.e., the expression $\alpha_0 - \alpha_1 + \alpha_2$, where α_0 is the number of vertices, α_1 the number of edges, and α_2 the number of triangles of the given triangulation, is called the *Euler characteristic* of this triangulation. Euler's theorem states that for all triangulations of a surface homeomorphic to a sphere the Euler characteristic is equal to two. Now it turns out that for every surface (and not only for a surface homeomorphic to a sphere) all triangulations of the surface have one and the same Euler characteristic.

It is easy to figure out the value of the Euler characteristic for various surfaces. First of all, for the cylindrical surface it is equal to zero. For when we remove from an arbitrary triangulation of the sphere two nonadjacent triangles but preserve the boundaries of these triangles, then we obviously obtain a triangulation of a surface homeomorphic to the curved surface of a cylinder. Here the number of vertices and of edges remains as before, but the number of triangles is decreased by two, therefore the Euler characteristic of the triangulation so obtained is zero. Now let us take the surface obtained from a triangulation of a sphere after removal of $2p$ triangles of this triangulation that are pairwise not

adjacent (i.e., do not have any common vertices nor common sides).*
Here the Euler characteristic is decreased by $2p$ units. It is easy to see
that the Euler characteristic does not change when cylindrical tubes are
attached to each pair of holes made in the surface of the sphere. This
comes from the fact that the characteristic of the tube to be pasted in
is, as we have seen, zero and on the rim of the tube the number of vertices
is equal to the number of edges. Thus, a closed two-sided surface of
genus p has the Euler characteristic $2 - 2p$ (a fact that was first proved
by the French admiral de Jonquières).

We now give an important property of triangulations which satisfies
the so-called condition of topological invariance (i.e., every triangulation
of the given surface has the property if at least one of them does). This
is the property of *orientability*. Before we formulate it, let us observe
that every triangle can be oriented, i.e., that a definite direction of traversing its boundary can be furnished. Each of the two possible orientations
of a triangle is given by a definite order of the sequence of its vertices.†
Now let us suppose that on an arbitrary surface we are given two triangles
which have a common side and no other common points (figure 17).
Two orientations of these triangles are called compatible if they generate
opposite directions on the common side of the triangle. (In the plane or
on any other two-sided surface this means that the two triangles, when
they are regarded as lying on one side of the surface, are traversed in the
same direction, i.e., either both counterclockwise or both clockwise.) A
triangulation of a given closed surface is called orientable if the orientations
of all the triangles occurring in it can be so chosen that any two triangles,
adjoining in a common side, turn out to be compatibly oriented. Then
the following fact holds: Every triangulation of a two-sided surface is
orientable. every triangulation of a one-sided surface is nonorientable.
Therefore two-sided surfaces are also called orientable, and one-sided
nonorientable. Choosing an arbitrary triangulation of a Möbius band the
reader can easily convince himself of its nonorientability. In order to
obtain the simplest triangulation of the projective plane, we have only
to draw in it any three straight lines that do not pass through one and
the same point (figure 18). They divide the projective plane into four
triangles one of which lies in the finite part of the plane, while each of the

---

* In order to do this we have only to make the triangulation in question sufficiently
"fine." This can always be achieved by a suitable subdivision of an arbitrary triangulation.

† Moreover, it is easy to see that two orders of the vertices determine one and the
same orientation (one and the same direction of circuit) if and only if they go over
into one another by an "even" permutation. Thus (ABC), (BCA), (CAB) determine
one orientation of the triangle, and (BAC), (ACB), (CBA) the other. (About even and
odd permutations see, e.g., Chapter XX, §3.)

other three is cut by the line at infinity into two parts. In figure 18 one of these triangles, extending to infinity, is shaded. From the same figure it is clear that when we attempt to give all four triangles compatible orientations we are inevitably doomed to failure. In particular, in the

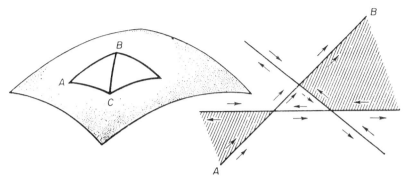

FIG. 17.   FIG. 18.

choice of orientations of our four triangles as it is made in figure 18, we obtain for the algebraic sum of its boundaries not zero, as it should be for compatible orientations, but the line AB taken twice.

Euler's characteristic and the property of orientability or nonorientability of closed surfaces give us, so to speak, a complete system of topological invariants of closed surfaces. The meaning of this statement is that two surfaces are homeomorphic if and only if, first, their triangulations have one and the same Euler characteristic, and second, both are orientable or not orientable.

The combinatorial method is applicable not only to the study of surfaces (two-dimensional manifolds) but also to manifolds of an arbitrary number of dimensions. But in the case of three-dimensional manifolds, for example, the role of the ordinary triangulations is now taken by decompositions into tetrahedra. They are called three-dimensional triangulations or simplicial divisions of the manifold. The Euler characteristic of a three-dimensional triangulation is defined as the number $\alpha_0 - \alpha_1 + \alpha_2 - \alpha_3$, where α_i, $i = 0, 1, 2, 3$, is the number of i-dimensional elements of this triangulation (i.e., α_0 the number of vertices, α_1 the number of edges, α_2 the number of two-dimensional boundaries, α_3 the number of tetrahedra). For a number of dimensions $n > 3$ the manifolds are divided into n-dimensional simplexes, i.e., the simplest convex n-dimensional polyhedra analogous to triangles ($n = 2$) and tetrahedra ($n = 3$). The simplexes into which an n-dimensional manifold is divided and their boundaries form an n-dimensional triangulation of this manifold. As before we can speak of an Euler characteristic, interpreting it as the sum $\Sigma_{i=0} (-1)^i \, \alpha_i$, where α_i is the number of i-dimensional elements occurring in this triangulation ($i = 0, 1, 2, \cdots, n$), and as before the Euler characteristic has one and the same value for all triangulations of a given n-dimensional manifold (and for all manifolds homeomorphic to it); i.e., it is a topological invariant. But in the present state of our knowledge, we cannot dream of a complete system of invariants even for three-dimensional manifolds (in the sense in which it is given for surfaces by the Euler characteristic and orientability).

The value of the combinatorial method in contemporary topology is very great. It opens the door to an application of certain algebraic devices in the solution of topological problems. The attentive reader will have noticed the possibility of such an algebraic approach when we talked above of the algebraic sum of the boundaries of the oriented triangle in a triangulation of the projective plane. For if a triangle is oriented, i.e., if a direction of traversing it is defined, then it is natural to take as its boundary the collection of its sides each with a definite direction, namely the one that continues the existing circuit of the triangles.

Now let us consider all the triangles T_i^2 , $i = 1, 2, \cdots, \alpha_2$, that occur in a given triangulation of a surface. To each of them we can give two orientations; let us denote the triangle T_i^2 with one of the two possible orientations by t_i^2 , and the same triangle with the other (opposite) orientation by $- t_i^2$. In exactly the same way we can orient each of the one-dimensional elements (edges) T_k^1 ($k = 1, 2, \cdots, \alpha_1$) that occur in the given triangulation, i.e., provide it with one of the two possible directions. We denote the segment T_k^1 with one of these orientations by t_k^1 , and with the other by $- t_k^1$. Now if the sides of the triangle T_i^2 are T_1^1 , T_2^1 , T_3^1 , then the boundary of the oriented triangle t_i^2 is the set of the same sides, but taken with a definite direction, i.e., the boundary consists of the directed segments $\epsilon_1 t_1^1$, $\epsilon_2 t_2^1$, $\epsilon_3 t_3^1$; here $\epsilon_i = 1$ if this direction for the edge T_i^1 coincides with its appropriate direction t_i^1 , and $\epsilon_i = -1$ in the opposite case. The boundary of t_i^2 is denoted by $\varDelta t_i^2$. As we have seen, $\varDelta t_i^2 = \Sigma \, \epsilon_k t_k^1$ and this sum can be imagined as extending over all edges of our triangulation when we consider the coefficients ϵ_k for segments not in the boundary of t_i^2 as being equal to zero.

It now becomes natural to consider more general sums of the form $x^1 = \Sigma \, a_k t_k^1$ extended over all edges of the given triangulation.* The geometric meaning of such sums is very simple: Every summand of the sum is a certain segment that occurs in our triangulation, taken with a

---

* The coefficients a_k are assumed to be integers.

definite direction and a definite coefficient (a definite "multiplicity"). The whole algebraic sum so described expresses a path composed of segments in which every segment is assumed to occur as often as its coefficient indicates. For example, if we begin by running around the polygon $ABCDEF$ (figure 19) in the direction of the arrow, then go along AA' to the

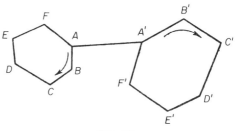

FIG. 19.

polygon $A'B'C'D'E'F'$ and traverse it in the indicated direction, and then return along $A'A$ and go around $ABCDEF$ again in the same direction as before, we obtain a sum in which the segments AB, BC, CD, DE, EF, FA, occur with the coefficient 2, the segments $A'B', B'C', C'D', D'E', E'F', F'A'$, with the coefficient 1, and the segment AA' does not occur at all (it has the coefficient zero because it is traversed twice in opposite directions).

Sums of the form $x^1 = \sum a_k t_k^1$ are called *one-dimensional chains* of the given triangulation. From the algebraic point of view they represent linear forms (homogeneous polynomials of the first degree); they can be added and subtracted and also multiplied by an arbitrary integer according to the usual rules of algebra. Of particular importance among the one-dimensional chains are the so-called one-dimensional cycles. Geometrically they correspond to closed paths (and in figure 19 we have just dealt with such a path).

For a purely algebraic definition of a cycle, let us make the convention that of the two vertices of a segment AB the end point B occurs in the boundary of $\overrightarrow{AB}$ with the plus sign (with the coefficient $+1$) and the initial point A with the minus sign (with the coefficient -1). Then the *boundary* of the segment $\overrightarrow{AB}$ can be written in the form $\Delta(\overrightarrow{AB}) = B - A$.

When we accept this convention, we observe immediately that the sum of the boundaries of the segments $\overrightarrow{AB}, \overrightarrow{BC}, \overrightarrow{CD}, \cdots, \overrightarrow{FA}$, which form a closed path (in the usual sense of the word), is zero. This leads us naturally to the general definition of a one-dimensional *cycle* as a one-dimensional chain $z^1 = \sum_k a_k t_k^1$ for which the sum of the boundaries of the terms, i.e., the sum $\sum_k a_k \Delta t_k^1$, is zero. It is easy to verify that the sum of two cycles is a cycle. When we multiply a cycle as an algebraic expression by an arbitrary integer, we obtain again a cycle. This enables us to speak of linear combinations of cycles $z_1^1, z_2^1, \cdots, z_s^1$, i.e., of cycles of the form $z = \sum_{\nu=1}^s c_\nu z_\nu^1$, where c_ν are integers.

In analogy to the concept of a one-dimensional chain of a given triangulation, we can also speak of two-dimensional chains of this triangulation, i.e., of expressions of the form $x^2 = \sum_i a_i t_i^2$, where the t_i^2 are oriented triangles of the given triangulation. Since the boundary of each oriented triangle is a one-dimensional cycle, the chain $\sum_i a_i \Delta t_i^2$ is also a cycle. This cycle is taken to be the boundary Δx^2 of the chain $x^2 = \sum_i a_i t_i^2$.

The concept of a *boundary of a chain* now enables us to formulate the concept of homology: A one-dimensional cycle z^1 of a given triangulation is called *homologous to zero* in this triangulation if it is the boundary of some two-dimensional chain of this triangulation. In every triangulation of a closed convex surface and more generally of any surface homeomorphic to a sphere, every one-dimensional cycle is homologous to zero; geometrically this is perfectly clear: Every closed polygon on a convex surface is the boundary of some piece of the surface. But this is not so on the torus: A meridian of the torus as well as its equator are not boundaries of any piece of the surface. If we take an arbitrary triangulation of the torus, then we can find in it cycles similar to the meridian or the equator of the torus and these cycles are not homologous to zero.

We observe an entirely new phenomenon in the triangulation of the projective plane we have constructed. If we regard a straight line, for example AB (see figure 18), as a cycle of this triangulation, then this cycle is not homologous to zero in it. However, the same line taken with the coefficient 2 does turn out to be homologous to zero. Thus, the introduction of coefficients other than $+1$ in the definition of chains, which appears in the first instance formal and unnecessary, enables us to discern important geometric properties of surfaces and more generally of manifolds. In the given case this is the so-called property of torsion which consists in the existence of cycles that in the given manifold are not homologous to zero (do not bound any piece of the surface) but become homologous to zero when they are provided with certain integer coefficients.

In connection with what we have said, let us finally introduce the extremely important concept of *homological independence* of cycles. The cycles $z_1, \cdots, z_s$ are called homologically independent in the given triangulation if no linear combination $\sum c_i z_i$ of them in which at least one coefficient c_i is different from zero is homologous to zero in this triangulation. As examples of homologically independent cycles on the torus, we can take an arbitrary meridian and equator considered as cycles of some triangulation of the torus.

The fundamental concepts of the whole of combinatorial topology (the concepts of boundary, cycle, homology) were defined by us for one-dimensional formations, but they can be extended verbatim to an arbitrary number of dimensions. For example, a two-dimensional chain $z^2 = \sum a_i t_i^2$ is called a cycle if its boundary $\Delta z^2 = \sum a_i \Delta t_i^2$ is equal to zero. A three-dimensional chain is an expression of the form $x^3 = \sum a_h t_h^3$, where the t_h^3 are oriented three-dimensional simplexes (tetrahedra).

As in the case of the triangle, the orientation of the three-dimensional simplex (tetrahedron) is given by a definite order of its vertices, where two orders of the vertices that can be carried into one another by an even permutation determine one and the same orientation. The boundary of a three-dimensional oriented simplex $t^3 = (ABCD)$ is the two-dimensional chain (cycle) $\Delta t^3 = (BCD) - (ACD) + (ABD) - (ABC)$ (figure 20).

The boundary of a three-dimensional chain is defined as the sum of the boundaries of its simplexes taken with the same coefficient with which these simplexes occur in the given chain. The reader can easily verify that the boundary of an arbitrary three-dimensional chain is a two-dimensional cycle (it is sufficient to prove this for the boundary of a single three-dimensional simplex). We say that a two-dimensional cycle is homologous to zero in a given manifold if it is the boundary of some three-dimensional chain of this manifold, and so on. Observe that from

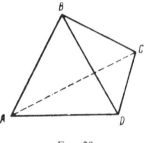

Fig. 20.

the definition of orientable and nonorientable triangulations* given previously it follows easily that in every orientable triangulation there occur cycles (two-dimensional in the case of surfaces) different from zero, and in nonorientable triangulations there are no such cycles; this result can also be generalized immediately to an arbitrary number of dimensions.

The concepts introduced enable us to define the order of the one-dimensional, two-dimensional, etc., connectivity of given manifolds of an arbitrary number of dimensions. The maximal number of homologically independent one-dimensional, two-dimensional, etc., cycles occurring in an arbitrary triangulation of a given manifold does not depend on the choice of the triangulation of this manifold and is called its *order of connectivity*, or its Betti number (of the corresponding dimension).

The one-dimensional Betti number of a closed orientable surface of genus p is equal to $2p$ (i.e., to the order of connectivity of the surface,

---

* This definition, given earlier for triangulations of surfaces, can be extended to the case of triangulations of manifolds of an arbitrary number of dimensions.

as defined in §2). The one-dimensional Betti number of the projective plane is equal to zero. (Here every cycle not extending to infinity bounds part of the plane, i.e., is homologous to zero, but a cycle extending to infinity, for example a projective line, turns out to be homologous to zero if it is taken twice.) The two-dimensional Betti number of every non-orientable surface is equal to zero (there is not a single two-dimensional cycle different from zero on such a surface).

The two-dimensional Betti number of every orientable surface is equal to one. For if we orient all the triangles of an arbitrary triangulation of an orientable surface in a suitable manner, then we obtain a cycle (the so-called fundamental cycle of the surface). It is not difficult to observe that every two-dimensional cycle is obtained from a fundamental cycle by multiplying it by an arbitrary integer. These results can be generalized immediately to n-dimensional manifolds. We remark that the zero-dimensional Betti number of a connected manifold (i.e., one that does not split into parts) is taken to be one.

The Betti numbers of the various dimensions are connected with the Euler characteristic of the manifold by a remarkable formula that was proved by Poincaré and generalizes Euler's theorem. This formula, which is known as the *Euler-Poincaré formula*, has the following simple form:

$$\sum_{r=0}^{n} (-1)^r \, \alpha_r = \sum_{r=0}^{n} (-1)^r \, p_r \, .$$

Here we have on the left-hand side the Euler characteristic of an arbitrary triangulation of the given manifold, and the numbers p_r on the right are the Betti numbers of the various dimensions r of this manifold. In particular, for orientable surfaces we have, as we have just seen, $p_0 = p_2 = 1, p_1 = 2p$, where p is the genus of the surface. This gives us Euler's theorem for orientable surfaces

$$\alpha_0 - \alpha_1 + \alpha_2 = 2 - 2p.$$

§5. Vector Fields

Let us consider the simplest differential equation

$$\frac{dy}{dx} = F(x, y), \tag{2}$$

in a given plane domain G. Its geometrical meaning is that at every point (x, y) of G a direction is defined whose slope is equal to $F(x, y)$, where $F(x, y)$ is a certain continuous function of the point (x, y). We say that in G a continuous field of directions is given; we can easily turn it into a continuous vector field by taking, for example, a vector of unit length in each of the given directions. The task of integrating the differential equation (2) consists in splitting, if this is possible, the given plane domain into pairwise nonintersecting curves (the "integral curves" of the equation) such that at each point of the domain the direction given in it is the direction of the tangent to the unique integral curve passing through this point.

Let us consider, for example, the equation

$$\frac{dy}{dx} = \frac{y}{x}.$$

At every point $M(x, y)$ of the plane the direction corresponding to it is obviously that of the ray $\overrightarrow{OM}$ (where O is the origin of coordinates). The integral curves are the straight lines passing through O. Through every point of the plane other than O there passes a unique integral curve. As regards the origin of coordinates, this is a singular point of the given differential equation (a so-called "node") through which all the integral curves pass.

If we take the differential equation

$$\frac{dy}{dx} = -\frac{x}{y},$$

then we see that it associates with every point $M(x, y)$ other than O the direction that is perpendicular to $\overrightarrow{OM}$. In this case the integral curves are circles with their center at O which is again a singular point of our differential equation, but a singular point of an altogether different type. It is not a "node", but a so-called "center." There are also other types of singular points (see Chapter V, §6), some of which are illustrated in figures 21 and 22. The differential equation $dy/dx = y/x$ has no closed integral curves. In contrast, the differential equation $dy/dx = -x/y$ has only closed integral curves. There are also possible integral curves that wind in the form of a spiral around a singular point, which in this case is called a focus.

Of extreme importance in various applications is the case of a so-called limit cycle, namely a closed integral curve, around which other integral curves wind spirally. Many other cases of the mutual arrangement of the integral curves are possible, and also their position with respect to the singular points. All problems concerning the forms and positions of the integral curves of a differential equation, and also the number, character, and mutual arrangement of its singular points, belong to the qualitative theory of differential equations. As the name implies, the qualitative theory of differential equations leaves aside the direct integration of the differential equation "in finite form," as well as methods of approximation or of numerical integration. The basic object of the qualitative theory is in essence the topology of the field of directions and the system of integral curves of the given differential equation.

FIG. 21.   FIG. 22.

The qualitative approach to differential equations, including such problems as the existence of closed integral curves, in particular all problems connected with the existence, number, and occurrence of limit cycles, was dictated in the first instance by problems of mechanics, physics, and technology. These problems arose first in connection with the investigations of Poincaré in celestial mechanics and cosmogony which, as we have mentioned above, were also the cause for the topological researches of the French geometer. The circle of topological problems in the theory of differential equations has occupied one of the central places in the outstanding investigations by Soviet scholars in the theory of oscillations and radiotechnology; we think here of the school of L. I. Mandel'stam, or of the school of A. A. Andronov that has developed from it, which constitutes one of the most important centers of research in the qualitative and essentially topological theory of differential equations. Another center of research in the qualitative theory of differential equations by essentially topological methods is the school of V. V. Stepanov and V. V. Nemyckiĭ at the University of Moscow. In papers of mathematicians in Leningrad, Sverdlovsk and Kazan on problems of the qualitative theory of differential equations topological method also play a prominent role.

The theory of differential equations leads to the study of vector fields not only in the plane but also in many-dimensional manifolds; even the simplest systems of several differential equations are interpreted geometrically as fields of directions in many-dimensional Euclidean spaces. The introduction of so-called first integrals of an equation means a selection among the set of all integral curves of those that lie in a certain manifold defined by the given first integral. Every dynamical system (in the classical meaning of this word) gives rise, generally speaking, to the many-dimensional manifold of its possible states (see §3) and to a system of differential equations whose integral curves, filling the given phase space, represent possible motions of the given system. Each individual one of these motions is determined by some set of initial conditions. Therefore, a fundamental object of study in this case is the field of directions and the system of trajectories on the given manifold. Numerous applications, especially in recent years, have made it understandable that the qualitative theory of differential equations should be developed in its widest aspects, with a consequent development of topology also as the basis of this theory. Precisely these mechanical, physical, and even astronomical topics have caused the rapid growth of contemporary topology which forms such a significant part of the general development of mathematics in the present half century. The reader who wishes to acquaint himself with the topological problems in the theory of differential equations and its concrete physical and technological aspects can consult the well-known book by A. A. Andronov and S. E. Haĭkin "Theory of oscillations," an English-language edition of which was published by the Princeton University Press, Princeton, N.J., 1949.

As an example of a concrete problem in the theory of vector fields on many-dimensional manifolds let us consider the problem of the algebraic number of singular points of such a vector field.

Suppose that a smooth manifold is given. For simplicity we shall imagine a smooth closed surface. Let us assume that at every point of this manifold a tangent vector to it is given that depends continuously on the point both in length and in direction. The singular points of such a vector field are those points of the manifold at which the vector associated with them is the null vector, i.e., where there is no definite direction. We shall assume that each of these points is isolated. In the case of a closed manifold, this means that there is only a finite number of singular points. (Otherwise part of these points would be condensed near a certain limit point, which by the continuity of the field would also be a singular point but not isolated.)

We can define the index of an isolated singular point, a concept that in a certain sense is similar to the concept of multiplicity of a root of an algebraic equation. In order to define the index we surround the given singular point by an arbitrary closed curve C that "isolates it" (in the plane simply a circle), i.e., a curve that does not pass through any singular point and contains in its interior only the one singular point in question. At all points of C the direction of the field is uniquely determined. For simplicity we shall assume that a neighborhood of our singular point including C is flat (in the general case we can transform the neighborhood that interests us together with the field given on it into a plane). When we go around the curve in the positive direction, the angle that the direction of the field forms with an arbitrary fixed direction returns to its initial value increased as a result of our going around the closed curve by a certain summand of the form $2k\pi$, where k is a certain well-defined integer. This number is called the *index of the singular point* or the *winding number* of the field along C. Note that it does not depend on the special choice of the closed curve that isolates the given singular point. In figures 21 and 22, we have illustrated singular points with the indices -2 and $+3$, respectively. In a similar but more complicated manner, we can also define the index for a singular point of a field of vectors (directions) defined on an n-dimensional manifold for $n > 2$. Now the following remarkable theorem was proved in 1926 by the German mathematician Hopf: If on a given manifold a continuous vector field is defined having only a finite number of singular points, then the sum of the indices or, as one says, the algebraic number of these singular points, does not depend on the field and is always equal to the Euler characteristic of the manifold.

From Hopf's theorem it follows that vector fields without singularities are possible only on manifolds whose Euler characteristic is equal to zero; it turns out that on such manifolds a vector field without singularities can always be constructed. Thus, among all the closed surfaces only on the torus and the so-called one-sided torus (the Klein bottle) can vector fields without singularities be constructed.

Closely connected with the theory of vector fields is the theory of continuous mappings of manifolds into themselves and particularly the results concerning the existence of fixed points for such mappings. A point x is called a fixed point of a given mapping f if its image under this mapping coincides with the point itself, i.e., if

$$f(x) = x.$$

In order to explain the character of this connection, let us consider the simplest case, namely the case of a continuous mapping f of a circle K into itself. By joining each point x of K to its image $f(x)$, we obtain a vector $\boldsymbol{u}_x = \overrightarrow{x, f(x)}$. This vector is the null vector if and only if $f(x) = x$, i.e, if x is a fixed point of the given mapping. Let us show that such a point actually exists. For this purpose we assume the contrary and determine the winding number of our vector field along the circumference C of the circle K.

Under a continuous transformation of our field, its winding number along C can obviously change only continuously. But since it is an integer, it must remain constant. Hence it follows that the winding number of the field along C is equal to 1. Indeed, since every point of K is mapped into the same circle, the vector $\boldsymbol{u}_x$ of a point x on the circumference C (which according to our assumption is not the null vector) points into the interior of the circle and therefore forms an acute angle with the radius Ox, which we regard as a vector pointing to the center O.

Now let us subject the directions of all the vectors $\boldsymbol{u}_x$ for the points x lying on C to a continuous transformation. This transformation consists of turning each of these vectors through such an acute angle that it comes to point in the direction of the center O. As we have just said, the winding number of the field along C does not change in this process. But as a result of this transformation, our original field goes over on C into the field of radial vectors, which obviously has the winding number one. Thus, our initial field also had the winding number one along the circumference C.

In virtue of the continuity of the original vector field its winding numbers along two circumferences with one and the same centre O and radii of slightly different length have one and the same value.* Therefore the winding number of the field along all circles with center O lying within K has one and the same value, namely one. But since by assumption the vector $\boldsymbol{u}_x$ is defined and different from zero for all points of the circle and among them for its center O, the winding number of the field along a circumference of sufficiently small radius with center at O is certainly equal to zero. So we have arrived at a contradiction and have proved that under a continuous mapping of a circle into itself there is always at least one fixed point. This theorem is a special case of a very important theorem of Brouwer which states that under every continuous mapping of an n-dimensional sphere into itself there is always at least one fixed point.

In recent years the problem of the existence of fixed points under mappings of one type or another has been studied in detail and forms an essential part of the topology of manifolds.

---

* In virtue of the assumption that the field in question, which is everywhere defined and different from zero, has no fixed points under f, we can also talk of its winding number along an arbitrary curve within K.

§6. The Development of Topology

The topology of closed surfaces is the only domain of topology that was more or less worked out at the end of the last century. The construction of this theory was connected with the development of the theory of functions of a complex variable in the course of the 19th century. The latter, which forms one of the most significant phenomena in the history of mathematics during the past century, has been built up by several distinct methods. One of the most fruitful in the sense of understanding the essence of the phenomena to be studied was the geometric method of Riemann. Riemann's method, which showed very convincingly that it is impossible in the general theory of functions of a complex variable to restrict ourselves to single-valued functions only, led to the construction of the so-called Riemann surfaces. In the simplest case of algebraic functions of a complex variable, these surfaces always turn out to be closed orientable surfaces. The investigation of their topological properties is in a certain sense equivalent to the investigation of the given algebraic function. Further development of Riemann's idea is due to Poincaré, Klein, and their followers and led to the discovery of unexpected and deep connections between the theory of functions, the topology of closed surfaces, and non-Euclidean geometry, namely the theory of the group of motions in a Lobačevskiĭ plane.* Thus, topology from the first showed itself to be organically related to a whole group of problems of fundamental importance connected with very diverse domains of mathematics.

In the further development of this circle of problems, it turned out that the topology of surfaces alone was insufficient, that the solution of problems in n-dimensional topology was necessary. The first of these was the problem of the topological invariance of the number of dimensions of a space. This problem consists in proving the impossibility of a topological mapping of an n-dimensional Euclidean space into an m-dimensional for $n \neq m$. This difficult problem was solved in 1911 by Brouwer.† In connection with its solution, new topological methods were discovered that led to a rapid construction of the beginnings of the theory of continuous mappings of many-dimensional manifolds and the theory of vector fields on them. All these investigations were found to involve the first fundamental concepts of the so-called set-theoretical topology

---

* In connection with this see the book by A. I. Markuševič "Theory of analytic functions," Gostehizdat, 1950.

† Actually, for the development of the theory of functions of a complex variable it was necessary to solve an even more difficult problem, namely to prove that the topological image of an n-dimensional domain lying in an n-dimensional space is always again a domain. This problem was also solved by Brouwer.

that arise on the basis of the general theory of sets constructed by Cantor in the last quarter of the 19th century.

In set-theoretical topology the very object of investigation, i.e., the class of geometric figures under consideration, is extremely wide and comprises if not altogether all sets in Euclidean spaces, then at least all the closed sets. Scholars of many countries collaborated in the rapid development of the new set-theoretical direction of topology, but above all we must mention the Polish topological school.

An essentially new direction of development of set-theoretical topology was taken in papers of Soviet topologists; in particular the outstanding Soviet mathematician P. S. Uryson (1898–1924), who met an untimely death, developed the general theory of dimension which laid the foundation of a classification of very general point sets by a fundamental criterion, namely the number of dimensions. This classification turned out to be extremely fruitful and involved entirely new points of view in the study of the most general geometric forms.* Uryson's ideas, as developed in his theory of dimension, were a stepping stone for the remarkable work of L. A. Ljusternik (jointly with L. G. Šnirel'man) on the variational calculus.

These papers contain, apart from other results, an exhaustive positive solution of a famous problem of Poincaré on the existence of three closed geodesic lines without multiple points on every surface homeomorphic to a sphere.

On the other hand, P. S. Aleksandrov, on the basis of the theory of dimension, transferred the algebraic methods of combinatorial topology to the realm of set theory and this led in turn to new directions of topological investigations in which mathematicians of the Soviet Union, including the younger generation, hold a leading place.†

As regards the proper combinatorial topology, after the papers of Poincaré and Brouwer, approximately in 1915, there begins a group of

---

* The inductive definition of dimension of sets proposed by Uryson can be regarded as the fullest development of Lobačevskiĭ's idea of dissection as the fundamental geometric operation. In a rough approximation it amounts to this. A set is zero-dimensional if it can be represented in the form of a sum of arbitrarily small parts no two of which are in contact with one another. A set is n-dimensional if it can be "dissected" by $(n-1)$-dimensional subsets into arbitrarily small parts no two of which are in contact and if this cannot be achieved with sets of dimensions smaller than $n-1$. (A precise definition of contact as it is understood in contemporary topology will be given in §7.)

† Here we must refer to the so-called homological theory of dimension of P. S. Aleksandrov, to the related remarkable constructions of L. N. Pontrjagin and to further developments of the homological theory of dimension in papers by M. S. Bokšteĭn, V. G. Boltjanskiĭ and particularly K. A. Sitnikov. The duality law of L. S. Pontrjagin will be discussed later.

investigations by American topologists, Veblen, Birkhoff, Alexander, and Lefschetz. They achieved very remarkable results. For example, Alexander proved the topological invariance of the Betti numbers and also a very important duality theorem that served as a starting point for the subsequent investigations of L. S. Pontrjagin; Lefschetz gave a certain formula for the algebraic number of fixed points under arbitrary continuous mappings of manifolds and so laid the foundation of the general algebraic theory of continuous mappings that was further developed by Hopf; to Birkhoff our science owes an essential advance in the theory of dynamical systems in its purely topological aspect, its metrical aspect, etc.. Further very deep developments of the topology of manifolds and their continuous mappings were obtained in papers by Hopf, who proved along with many other results the existence of an infinite number of continuous mappings of a three-dimensional sphere onto a two-dimensional one, which are essentially different from one another in the sense that no two of these mappings can be carried into one another by a continuous change. So Hopf became the founder of a new direction, the so-called homotopic topology. Recently a powerful new impetus has come to homotopic topology, as to the whole of combinatorial topology, from the work of the new French topological school (Leray, Serre, and others).

As we have mentioned earlier, the fundamental investigations of Uryson were the beginning of the activities of Soviet mathematicians in the domain of topology. These investigations were concerned with set-theoretical topology, but already at the end of the 1920's Soviet topology comprised also combinatorial topology in the range of its interests. This came about in a rather original manner, namely by the application of combinatorial methods to the study of closed sets, i.e., to objects of a very general nature. On this foundation there arose one of the most remarkable geometric discoveries of the present century, the statement and proof by L. S. Pontrjagin of his general law of duality, which establishes deep and in a certain sense exhaustive connections between the topological structure of a given closed set in an n-dimensional Euclidean space and the parts of the space complementary to it. In connection with his duality law, Pontrjagin constructed a general theory of characters of commutative groups and this led him to further investigations in the domain of the general topological and the classical continuous Lie groups, a domain that has been completely transformed by the work of Pontrjagin. Subsequently Pontrjagin and his pupils made a number of notable investigations on the topology of manifolds and their continuous mappings (Z. G. Boltjanskiĭ, M. M. Postnikov, and others). In these investigations a new method was applied, the so-called ∇-homology (cohomology) introduced into combinatorial topology by A. N. Kolmogorov and, independently, by Alexander. This method, which now occupies the first place in the whole of homotopic topology, has made it possible to continue Pontrjagin's duality theory in the most diverse directions, and this has led to the duality theorems of A. N. Kolmogorov (and Alexander), P. S. Aleksandrov, and K. A. Sitnikov, which belong to the most significant results of contemporary topology. The same method has also found important applications in very recent papers by L. A. Ljusternik on the calculus of variations.

§7. Metric and Topological Spaces

At the beginning of our account, we have talked of adjacency (of different parts of a given figure) as of a fundamental topological concept, and we have defined continuous transformations as those that preserve this relation. However, we have not given a rigorous definition of this fundamental concept; to do this in sufficient generality we have to use concepts of set theory. This will be our task in the present section; we shall finally solve it by introducing the concept of a topological space.

The theory of sets made it possible to give the concept of a geometrical figure a breadth and generality that were inaccessible in the so-called "classical" mathematics. The object of a geometrical, in particular a topological, investigation now becomes an arbitrary point set, i.e., an arbitrary set whose elements are points of an n-dimensional Euclidean space. Between points of an n-dimensional space a *distance* is defined: namely, the distance between the points $A = (x_1, x_2, \cdots, x_n)$ and $B = (y_1, y_2, \cdots, y_n)$ is by definition equal to the nonnegative number

$$\rho(A, B) = \sqrt{(x_1 - y_1)^2 + (x_2 - y_2)^2 + \cdots + (x_n - y_n)^2}.$$

The concept of distance permits us to define adjacency first between a set and a point, and then between two sets. We say that a point A is an adherent point of the set M if M contains points whose distance from A is less than any preassigned positive number. Obviously every point of the given set is an adherent point of it, but there may be points that do not belong to the given set and are adherent to it. Let us take, for example, the open interval $(0, 1)$ on the numerical line, i.e., the set of all points lying between 0 and 1; the points 0 and 1 themselves do not belong to this interval, but are adherent to it, since in the interval $(0, 1)$ there are points arbitrarily near to zero and points arbitrarily near to one. A set is called *closed* if it contains all its adherent points. For example the closed interval $[0, 1]$ of the numerical line, i.e., the set of all points x satisfying the inequality $0 \leqslant x \leqslant 1$, is closed. Closed sets in a plane and all the more in a space of three or more dimensions can have an extremely complicated structure; indeed, they form the main study object of the set theoretical topology of an n-dimensional space.

Next we say that two sets P and Q adjoin one another if at least one of them contains adherent points of the other. From the preceding it follows that two closed sets can adjoin only when they have at least one point in common; but, for example, the intervals [0, 1] and (1, 2), which do not have common points, adjoin because the point 1 which belongs to [0, 1] is at the same time an adherent point of (1, 2). Now we can say that a set R is divided ("dissected") by a set S lying in it, or that S is a "section" of $R - S$ consisting of all the points of R that do not belong to S can be represented as the sum of two nonadjoining sets.

Thus, Lobačevskiĭ's ideas on adjacency and dissection of sets receive in contemporary topology a rigorous and highly general expression. We have already seen how Uryson's definition of dimension of an arbitrary set (see the remark in §6) is founded on these ideas; the statement of this definition now becomes completely rigorous. The same applies to the definition of a continuous mapping or transformation; a mapping f of a set X onto a set Y is called continuous if adjacency is preserved under this mapping, i.e., if the fact that a certain point A of X is an adherent point of an arbitrary subset P of Y implies that that image $f(A)$ of A is an adherent point of the image $f(P)$ of P. Finally a one-to-one mapping of a set X onto a set Y is called *topological* if it is continuous and if its inverse mapping of Y onto X is also continuous. These definitions give an accurate basis to all that has been said in the first sections of the present account.

However, set theoretical topology is not restricted to the degree of generality that is achieved by considering as geometrical figures all the point sets. It is natural to introduce the concept of distance not only between points of an arbitrary Euclidean space but also between other objects that do not appear to refer at all to geometry.

Let us consider, for example, the set of all continuous functions defined, say, on the interval [0, 1]. We can define the distance $\rho(f, g)$ between the two functions f and g as the maximum of the expression $|f(x) - g(x)|$, when x ranges over the whole segment [0, 1]. This "distance" has all the basic properties of distance between two points in space: $\rho(f, g)$ between the two functions f and g is equal to zero if and only if the functions coincide, i.e., if $f(x) = g(x)$ for every point x; further, the distance is obviously symmetrical, i.e., $\rho(f, g) = \rho(g, f)$; finally, it satisfies the so-called triangle axiom: For any three functions f_1, f_2, f_3 we have $\rho(f_1, f_2) + \rho(f_2, f_3) \geqslant \rho(f_1, f_3)$. It is customary to say that the so-defined distance turns our set of functions into a metric space (usually denoted by C). By a metric space we understand more generally, a set of arbitrary objects that are to be called *points* of the metric space if between any two points there is defined a *distance*, a nonnegative number satisfying the "axioms of distance" just stated.

Now when an arbitrary metric space is given, we can talk of adherent points of its subsets and consequently of adjacency of its subsets to one another and of topological concepts in general (closed sets, continuous mappings, and further concepts to be introduced on the basis of these simplest ones). This course opens up an extensive and extremely fruitful field of application for topological and general geometrical ideas to ranges of mathematical objects where it would appear completely impossible to talk of any kind of geometry. Let us give an illustrative example.

We take again the differential equation (2)

$$\frac{dy}{dx} = F(x, y).$$

If $y = \phi(x)$ for $0 \leqslant x \leqslant 1$ is a solution of this equation that assumes at $x = 0$ the value $y = 0$, say, then the function $\phi(x)$ obviously satisfies the integral equation

$$\phi(x) = \int_0^x F[x, \phi(x)] \, dx. \tag{3}$$

Now we consider the integral $G(f) = \int_0^x F[x, f(x)] \, dx$, where $0 \leqslant x \leqslant 1$ and $f(x)$ is an arbitrary continuous function defined on the interval $[0, 1]$. This integral is a certain continuous function $g(x)$ also defined on $[0, 1]$. So the expression $G(f) = \int_0^x F[x, f(x)] \, dx$ associates with every function f a function $g = G(f)$; in other words, we have a mapping G, easily seen to be continuous, of the metric space C into itself. How can we characterize here a function $\phi(x)$ (there may be several of them) that is a solution of the equation (2) or the equation (3) equivalent to it? Obviously under our mapping it goes over into itself; i.e., it is a fixed point of our mapping G. Now it turns out that such a fixed point of the mapping G actually exists, as follows from a very general theorem on fixed points of continuous mappings of metric spaces that was proved in 1926 by P. S. Aleksandrov and V. V. Nemyckiĭ. Nowadays the study of various metric spaces whose points are functions of one kind or another (such spaces are called functional spaces) is a constantly used tool of analysis, and the study of functional spaces by methods which are partly topological, but mainly algebraic, in a wide sense of the word, forms the content of functional analysis (see Chapter XIX).

Functional analysis, as was mentioned in the introductory chapter, occupies an extremely prominent place in the contemporary mathematical scene in view of the variety of its connections with all sorts of other parts of mathematics and its value in natural science, above all in theoretical physics. The investigation of topological properties of functional spaces is closely connected with the calculus of variations and the theory of partial differential equations (investigations by Ljusternik, Morse, Leray, Schauder, Krasnosel'skiĭ, and others). Problems on the existence of fixed points under continuous mappings of functional spaces play an important role in these investigations.

The topology of functional and general metric spaces is not the last word in generality in contemporary topological theories. The fact of the matter is that in metric spaces the fundamental topological concept of adjacency is introduced on the basis of distance between points, which in turn is not a topological concept. The problem therefore arises of a direct, axiomatic definition of adjacency. Thus we are led to the concept of a topological space, the most general concept of present-day topology.

A *topological space* is a set of objects of an arbitrary nature (which are called points of the space) in which for every subset its *adherent* points are given in one way or another. Furthermore, a few natural conditions are supposed to be satisfied, the so-called axioms of a topological space (for example, every point of the given set is an adherent point of it, an adherent point of the sum of two sets is an adherent point of at least one of the summands, etc.). Profound work is going on at present in the theory of topological spaces; in its development the Soviet mathematicians P. S. Uryson, P. S. Aleksandrov, A. N. Tihonov, and others have taken a leading part. Of the latest results in the theory of topological spaces we must mention one of fundamental value: The young mathematician Ju. M. Smirnov has found necessary and sufficient conditions for the metrizability of a topological space, i.e., conditions under which a distance between the points of the space can be defined such that the "topology" which the space carries can be regarded as generating this concept of distance; in other words, such that the adherent points of all possible sets in the metric space obtained are the same as those defined initially in the given topological space.

Suggested Reading

P. S. Aleksandrov, *Combinatorial topology*, 3 vols., Graylock Press, Albany, N. Y., vol. 1, 1956; vol. 2, 1957; vol. 3, 1960.

S. S. Cairns, *Introductory topology*, Ronald Press, New York, 1961.

D. W. Hall and G. L. Spencer, *Elementary topology*, Wiley, New York, 1955.

P. J. Hilton and S. Wiley, *Homology theory: an introduction to algebraic topology*, Cambridge University Press, New York, 1960.

J. G. Hocking and G. S. Young, *Topology*, Addison-Wesley, Reading, Mass., 1961.

J. L. Kelley, *General topology*, Van Nostrand, New York, 1955.

S. Lefschetz, *Introduction to topology*, Princeton University Press, Princeton, N. J., 1949.

M. H. A. Newman, *Elements of the topology of plane sets of points*, 2nd ed., Cambridge University Press, New York, 1951.

L. S. Pontrjagin, *Foundations of combinatorial topology*, Graylock Press, Albany, N. Y., 1952.

CHAPTER XIX

FUNCTIONAL ANALYSIS

The rise and spread of functional analysis in the 20th century had two main causes. On the one hand it became desirable to interpret from a uniform point of view the copious factual material accumulated in the course of the 19th century in various, often hardly connected, branches of mathematics. The fundamental concepts of functional analysis were formed and crystalized under various aspects and for various reasons. Many of the fundamental concepts of functional analysis emerged in a natural fashion in the process of development of the calculus of variations, in problems on oscillations (in the transition from the oscillations of systems with a finite number of degrees of freedom to oscillations of continuous media), in the theory of integral equations, in the theory of differential equations both ordinary and partial (in boundary problems, problems on eigenvalues, etc.) in the development of the theory of functions of a real variable, in operator calculus, in the discussion of problems in the theory of approximation of functions, and others. Functional analysis permitted an understanding of many results in these domains from a single point of view and often promoted the derivation of new ones. In recent decades the preparatory concepts and apparatus were then used in a new branch of theoretical physics—in quantum mechanics.

On the other hand, the investigation of mathematical problems connected with quantum mechanics became a crucial feature in the further development of functional analysis itself: It created, and still creates at the present time, fundamental branches of this development.

Functional analysis has not yet reached its completion by far. On the contrary, undoubtedly in its further development the questions and requirements of contemporary physics will have the same significance for it as classical mechanics had for the rise and development of the differential and integral calculus in the 18th century.

It is impossible here to include in this chapter all, or even only all the fundamental, problems of functional analysis. Many important branches exceed the limitations of this book. Nevertheless, by confining ourselves to certain selected problems, we wish to acquaint the reader with some fundamental concepts of functional analysis and to illustrate as far as possible the connections of which we have spoken here. These problems were analyzed mainly at the beginning of the 20th century on the basis of the classical papers of Hilbert, who was one of the founders of functional analysis. Since then functional analysis has developed very vigorously and has been widely applied in almost all branches of mathematics; in partial differential equations, in the theory of probability, in quantum mechanics, in the quantum theory of fields, etc. Unfortunately these further developments of functional analysis cannot be included in our account. In order to describe them we would have to write a separate large book, and therefore, we restrict ourselves to one of the oldest problems, namely the theory of eigenfunctions.

§1. *n*-Dimensional Space

In what follows we shall make use of the fundamental concepts of n-dimensional space. Although these concepts have been introduced in the chapters on linear algebra and on abstract spaces, we do not think it superfluous to repeat them in the form in which they will occur here. For scanning through this section it is sufficient that the reader should have a knowledge of the foundations of analytic geometry.

We know that in analytic geometry of three-dimensional space a point is given by a triplet of numbers (f_1, f_2, f_3), which are its coordinates. The distance of this point from the origin of coordinates is equal to $\sqrt{f_1^2 + f_2^2 + f_3^2}$. If we regard the point as the end of a vector leading to it from the origin of coordinates, then the length of the vector is also equal to $\sqrt{f_1^2 + f_2^2 + f_3^2}$. The cosine of the angle between nonzero vectors leading from the origin of coordinates to two distinct points $A(f_1, f_2, f_3)$ and $B(g_1, g_2, g_3)$ is defined by the formula

$$\cos\phi = \frac{f_1 g_1 + f_2 g_2 + f_3 g_3}{\sqrt{f_1^2 + f_2^2 + f_3^2}\sqrt{g_1^2 + g_2^2 + g_3^2}}.$$

From trigonometry we know that $|\cos\phi| \leqslant 1$. Therefore we have the inequality

$$\frac{|f_1 g_1 + f_2 g_2 + f_3 g_3|}{\sqrt{f_1^2 + f_2^2 + f_3^2}\sqrt{g_1^2 + g_2^2 + g_3^2}} \leqslant 1,$$

and hence always

$$(f_1 g_1 + f_2 g_2 + f_3 g_3)^2 \leqslant (f_1^2 + f_2^2 + f_3^2)(g_1^2 + g_2^2 + g_3^2). \qquad (1)$$

This last inequality has an algebraic character and is true for any arbitrary six numbers (f_1, f_2, f_3) and (g_1, g_2, g_3), since any six numbers can be the coordinates of two points of space. All the same, the inequality (1) was obtained from purely geometric considerations and is closely connected with geometry, and this enables us to give it an easily visualized meaning.

In the analytic formulation of a number of geometric relations, it often turns out that the corresponding facts remain true when the triplet of numbers is replaced by n numbers. For example, our inequality (1) can be generalized to $2n$ numbers $(f_1, f_2, \cdots, f_n)$ and $(g_1, g_2, \cdots, g_n)$. This means that for any arbitrary $2n$ numbers $(f_1, f_2, \cdots, f_n)$ and $(g_1, g_2, \cdots, g_n)$ an inequality analogous to (1) is true, namely:

$$(f_1 g_1 + f_2 g_2 + \cdots + f_n g_n)^2 \leqslant (f_1^2 + f_2^2 + \cdots + f_n^2)(g_1^2 + g_2^2 + \cdots + g_n^2). \qquad (1')$$

This inequality, of which (1) is a special case, can be proved purely analytically.* In a similar way many other relations between triplets of numbers derived in analytic geometry can be generalized to n numbers. This connection of geometry with relations between numbers (numerical relations) for which the cited inequality is an example becomes particularly lucid when the concept of an n-dimensional space is introduced. This concept was introduced in Chapter XVI. We repeat it here briefly.

A collection of n numbers $(f_1, f_2, \cdots, f_n)$ is called a point or *vector* of n-dimensional space (we shall more often use the latter name). The vector $(f_1, f_2, \cdots, f_n)$ will from now on be abbreviated by the single letter f.

Just as in three-dimensional space on addition of vectors their components are added, so we define the sum of the vectors

$$f = \{f_1, f_2, \cdots, f_n\} \quad \text{and} \quad g = \{g_1, g_2, \cdots, g_n\}$$

as the vector $\{f_1 + g_1, f_2 + g_2, \cdots, f_n + g_n\}$ and we denote it by $f + g$.

The product of the vector $f = \{f_1, f_2, \cdots, f_n\}$ by the number λ is the vector $\lambda f = \{\lambda f_1, \lambda f_2, \cdots, \lambda f_n\}$.

The length of the vector $f = \{f_1, f_2, \cdots, f_n\}$, like the length of a vector in three-dimensional space, is defined as $\sqrt{f_1^2 + f_2^2 + \cdots + f_n^2}$.

---

* See Chapter XVI.

The angle ϕ between the two vectors $f = \{f_1, f_2, \cdots, f_n\}$ and $g = \{g_1, g_2, \cdots, g_n\}$ in n-dimensional space is given by its cosine in exactly the same way as the angle between vectors in three-dimensional space. For it is defined by the formula*

$$\cos \phi = \frac{f_1 g_1 + f_2 g_2 + \cdots + f_n g_n}{\sqrt{f_1^2 + f_2^2 + \cdots + f_n^2} \; \sqrt{g_1^2 + g_2^2 + \cdots + g_n^2}} . \tag{2}$$

The scalar product of two vectors is the name for the product of their lengths by the cosine of the angle between them. Thus, if $f = \{f_1, f_2, \cdots, f_n\}$ and $g = \{g_1, g_2, \cdots, g_n\}$, then since the lengths of the vectors are $\sqrt{f_1^2 + f_2^2 + \cdots + f_n^2}$ and $\sqrt{g_1^2 + g_2^2 + \cdots + g_n^2}$, respectively, their scalar product, which is denoted by (f, g), is given by the formula

$$(f, g) = f_1 g_1 + f_2 g_2 + \cdots + f_n g_n . \tag{3}$$

In particular, the condition of orthogonality (perpendicularity) of two vectors is the equation $\cos \phi = 0$; i.e., $(f, g) = 0$.

By means of the formula (3) the reader can verify that the scalar product in n-dimensional space has the following properties:

1. $(f, g) = (g, f)$.
2. $(\lambda f, g) = \lambda(f, g)$.
3. $(f, g_1 + g_2) = (f, g_1) + (f, g_2)$.
4. $(f, f) \geqslant 0$, and the equality sign holds for $f = 0$ only, i.e., when $f_1 = f_2 = \cdots = f_n = 0$.

The scalar product of a vector f with itself (f, f) is equal to the square of the length of f.

The scalar product is a very convenient tool in studying n-dimensional spaces. We shall not study here the geometry of an n-dimensional space but shall restrict ourselves to a single example.

As our example we choose the theorem of Pythagoras in n-dimensional space: The square of the hypotenuse is equal to the sum of the squares of the sides. For this purpose we give a proof of this theorem in the plane which is easily transferred to the case of an n-dimensional space.

Let f and g be two perpendicular vectors in a plane. We consider the right-angled triangle constructed on f and g (figure 1). The hypotenuse of this triangle is equal in length to the vector $f + g$. Let us write down in vector form the theorem of Pythagoras in our notation. Since the square of the length of a vector is equal to the scalar product of the vector with

---

* The fact that $|\cos \phi| \leqslant 1$ follows from the inequality (1′).

itself, Pythagoras' theorem can be written in the language of scalar products as follows:

$$(f + g, f + g) = (f, f) + (g, g).$$

The proof immediately follows from the properties of the scalar product. In fact,

$$(f + g, f + g) = (f, f) + (f, g) + (g, f) + (g, g),$$

and the two middle summands are equal to zero owing to the orthogonality of f and g.

In this proof we have only used the definition of the length of a vector, the perpendicularity of vectors, and the properties of the scalar product. Therefore nothing changes in the proof when we assume that f and g are two orthogonal vectors of an n-dimensional space. And so Pythagoras' theorem is proved for a right-angled triangle in n-dimensional space.

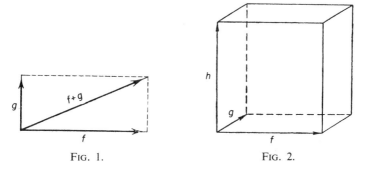

FIG. 1.   FIG. 2.

If three pairwise orthogonal vectors f, g and h are given in n-dimensional space, then their sum $f + g + h$ is the diagonal of the right-angled parallelepiped constructed from these vectors (figure 2) and we have the equation

$$(f + g + h, f + g + h) = (f, f) + (g, g) + (h, h),$$

which signifies that the square of the length of the diagonal of a parallelepiped is equal to the sum of the squares of the lengths of its edges. The proof of this statement, which is entirely analogous to the one given earlier for Pythagoras' theorem, is left to the reader. Similarly, if in an n-dimensional space there are k pairwise orthogonal vectors $f^1, f^2, \cdots, f^k$ then the equation

$$(f^1 + f^2 + \cdots + f^k, f^1 + f^2 + \cdots + f^k) \\ = (f^1, f^1) + (f^2, f^2) + \cdots + (f^k, f^k), \tag{4}$$

which is just as easy to prove, signifies that the square of the length of the diagonal of a "k-dimensional parallelelipiped" in n-dimensional space is also equal to the sum of the squares of the lengths of its edges.

§2. Hilbert Space (Infinite-Dimensional Space)

Connection with n-dimensional space. The introduction of the concept of n-dimensional space turned out to be useful in the study of a number of problems of mathematics and physics. In its turn this concept gave the impetus to a further development of the concept of space and to its application in various domains of mathematics. An important role in the development of linear algebra and of the geometry of n-dimensional spaces was played by problems of small oscillations of elastic systems.

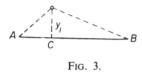

FIG. 3.

Let us consider the following classical example of such a problem (figure 3). Let AB be a flexible string spanned between the points A and B. Let us assume that a weight is attached at a certain point C to the string. If it is moved from its position of equilibrium, it begins to oscillate with a certain frequency ω, which can be computed when we know the tension of the string, the mass m and the position of the weight. The state of the system at every instant is then given by a single number, namely the displacement y_1 of the mass m from the position of equilibrium of the string.

Now let us place n weights on the string AB at the points C_1, C_2, $\cdots$, C_n. The string itself is taken to be weightless. This means that its mass is so small that compared with the masses of the weights it can be neglected. The state of such a system is given by n numbers y_1, y_2, $\cdots$, y_n equal to the displacements of the weights from the position of equilibrium. The collection of numbers y_1, y_2, $\cdots$, y_n can be regarded (and this turns out to be useful in many respects) as a vector $(y_1, y_2, \cdots, y_n)$ of an n-dimensional space.

The investigation of the small oscillations that take place under these circumstances turns out to be closely connected with fundamental facts of the geometry of n-dimensional spaces. We can show, for example, that the determination of the frequency of the oscillations of such a system can be reduced to the task of finding the axes of a certain ellipsoid in n-dimensional space.

Now let us consider the problem of the small oscillations of a string spanned between the points A and B. Here we have in mind an idealized string, i.e., an elastic thread having a finite mass distributed continuously along the thread. In particular, by a homogeneous string we understand one whose density is constant.

Since the mass is distributed continuously along the string, the position of the string can no longer be given by a finite set of numbers y_1, y_2, $\cdots$, y_n, and instead the displacement $y(x)$ of every point x of the string has to be given. Thus, the state of the string at each instant is given by a certain function $y(x)$.

The state of a thread with n weights attached at the points with the abscissas x_1, x_2, $\cdots$, x_n, is represented graphically by a broken line with n members (figure 4), so that when the number of weights is increased, then the number of segments of the broken line

FIG. 4.

increases correspondingly. When the number of weights grows without bound and the distance between adjacent weights tends to zero, we obtain in the limit a continuous distribution of mass along the thread, i.e., an idealized string. The broken line that describes the position of the thread with weights then goes over into a curve describing the position of the string (figure 5).

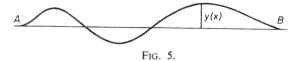

FIG. 5.

So we see that there exists a close connection between the oscillations of a thread with weights and the oscillations of a string. In the first problem the position of the system was given by a point or vector of an n-dimensional space. Therefore it is natural to regard the function $f(x)$ that describes the position of the oscillating string in the second case as a vector or a point of a certain infinite-dimensional space. A whole series of similar problems leads to the same idea of considering a space whose points (vectors) are functions $f(x)$ given on a certain interval.*

---

* As another such problem let us consider the electrical oscillations set up in a series of connected electrical circuits (figure 6).

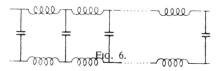

FIG. 6.

This example of oscillation of a string, to which we shall return again in §4, suggests to us how we shall have to introduce the fundamental concepts in an infinite-dimensional space.

Hilbert space. Here we shall discuss one of the most widespread concepts of an infinite-dimensional space of the greatest importance for the applications, namely the concept of the Hilbert space.

A vector of an n-dimensional space is defined as a collection of n numbers f_i, where i ranges from 1 to n. Similarly a vector of an infinite-dimensional space is defined as a function $f(x)$, where x ranges from a to b.

Addition of vectors and multiplication of a vector by a number is defined as addition of the functions and multiplication of the function by a number.

The length of a vector f in an n-dimensional space is defined by the formula

$$\sqrt{\sum_{i=1}^{n} f_i^2}$$

Since for functions the role of the sum is taken by the integral, the length of the vector $f(x)$ of a Hilbert space is given by the formula

$$\sqrt{\int_a^b f^2(x)\, dx}. \tag{5}$$

The distance between the points f and g in an n-dimensional space is defined as the length of the vector $f - g$, i.e., as

$$\sqrt{\sum_{i=1}^{n} (f_i - g_i)^2}.$$

Similarly the "distance" between the elements $f(t)$ and $g(t)$ in a functional space is equal to

$$\sqrt{\int_a^b [f(t) - g(t)]^2\, dt}.$$

The state of such a series can be expressed by the set of n numbers $u_1, u_2, \cdots, u_n$, where u_i is the voltage on the condensor of the ith circuit of the chain. The collection of the n numbers $(u_1, \cdots, u_n)$ is a vector of an n-dimensional space.

Now let us imagine a two-wire line, i.e., a line consisting of two conductors having finite capacity and inductance, distributed along the line. The electric state of the line is expressed by a certain function $u(x)$, which gives the distribution of the voltage along the line. This function is a vector of the infinite-dimensional space of functions given on the interval (a, b).

The expression $\int_a^b [f(t) - g(t)]^2 \, dt$ is called the mean-square deviation of the functions $f(t)$ and $g(t)$. Thus, the mean-square deviation of two elements of Hilbert space is taken to be a measure of their distance.

Let us now proceed to the definition of the angle between vectors. In an n-dimensional space the angle ϕ between the vectors $f = \{f_i\}$ and $g = \{g_i\}$ is defined by the formula

$$\cos \phi = \frac{\sum_{i=1}^n f_i g_i}{\sqrt{\sum_{i=1}^n f_i^2} \, \sqrt{\sum_{i=1}^n g_i^2}}.$$

In an infinite-dimensional space the sums are replaced by the corresponding integrals and the angle ϕ between the two vectors f and g of Hilbert space is defined by the analogous formula

$$\cos \phi = \frac{\int_a^b f(t) \, g(t) \, dt}{\sqrt{\int_a^b f^2(t) \, dt} \, \sqrt{\int_a^b g^2(t) \, dt}}. \tag{6}$$

This expression can be regarded as the cosine of a certain angle ϕ, provided the fraction on the right-hand side is an absolute value less than one, i.e., if

$$\left| \int_a^b f(t) \, g(t) \, dt \right| < \sqrt{\int_a^b f^2(t) \, dt} \, \sqrt{\int_a^b g^2(t) \, dt}. \tag{7}$$

This inequality in fact holds for two arbitrary functions $f(t)$ and $g(t)$. It plays an important role in analysis and is known as the Cauchy-Bunjakovskiĭ inequality. Let us prove it.

Let $f(x)$ and $g(x)$ be two functions, not identically equal to zero, given on the interval (a, b). We choose arbitrary numbers λ and μ and form the expresson

$$\int_a^b [\lambda f(x) - \mu g(x)]^2 \, dx.$$

Since the function $[\lambda f(x) - \mu g(x)]^2$ under the integral sign is nonnegative, we have the following inequality

$$\int_a^b [\lambda f(x) - \mu g(x)]^2 \, dx \geqslant 0;$$

i.e.,

$$2\lambda\mu \int_a^b f(x) \, g(x) \, dx \leqslant \lambda^2 \int_a^b f^2(x) \, dx + \mu^2 \int_a^b g^2(x) \, dx.$$

For brevity we introduce the notation

$$\left| \int_a^b f(x) \, g(x) \, dx \right| = C, \quad \int_a^b f^2(x) \, dx = A, \quad \int_a^b g^2(x) \, dx = B. \tag{8}$$

In this notation the inequality can be rewritten as follows:*

$$2\lambda\mu C \leqslant \lambda^2 A + \mu^2 B. \tag{9}$$

This inequality is valid for arbitrary values of λ and μ; in particular we may set

$$\lambda = \sqrt{\frac{C}{A}}, \mu = \sqrt{\frac{C}{B}}. \tag{10}$$

Substituting these values of λ and μ in (9), we obtain

$$\frac{C}{\sqrt{AB}} \leqslant 1.$$

When we replace A, B and C by their expressions in (8), we finally obtain the Cauchy-Bunjakovskiĭ inequality.

In geometry the scalar product of vectors is defined as the product of their lengths by the cosine of the angle between them. The lengths of the vectors f and g in our case are equal to

$$\sqrt{\int_a^b f^2(x)\, dx} \quad \text{and} \quad \sqrt{\int_a^b g^2(x)\, dx},$$

and the cosine of the angle between them is defined by the formula

$$\cos\phi = \frac{\int_a^b f(x)\, g(x)\, dx}{\sqrt{\int_a^b f^2(x)\, dx}\, \sqrt{\int_a^b g^2(x)\, dx}}.$$

When we multiply out these expressions, we arrive at the following formula for the scalar product of two vectors of Hilbert space:

$$(f, g) = \int_a^b f(x)\, g(x)\, dx. \tag{11}$$

From this formula it is clear that the scalar product of the vector f with itself its the square of its length.

If the scalar product of the nonzero vectors f and g is equal to zero, it means that $\cos\phi = 0$, i.e., that the angle ϕ ascribed to them by our definition is 90°. Therefore functions f and g for which

$$(f, g) = \int_a^b f(x)\, g(x)\, dx = 0,$$

are called orthogonal.

Pythagoras' theorem (see §1) holds in Hilbert space as in an n-dimen-

---

* For C we have to take the modulus of the integral because of the arbitrary sign of λ or μ.

sional space. Let $f_1(x), f_2(x), \cdots, f_N(x)$ be N pairwise orthogonal functions and

$$f(x) = f_1(x) + f_2(x) + \cdots + f_N(x).$$

Then the square of the length of f is equal to the sum of the squares of the lengths of $f_1, f_2, \cdots, f_N$.

Since the lengths of vectors in Hilbert space are given by means of integrals, Pythagoras' theorem in this case is expressed by the formula

$$\int_a^b f^2(x)\,dx = \int_a^b f_1^2(x)\,dx + \int_a^b f_2^2(x)\,dx + \cdots + \int_a^b f_N^2(x)\,dx. \qquad (12)$$

The proof of this theorem does not differ in any respect from the one given previously (§1) for the same theorem in n-dimensional space.

So far we have not made precise what functions are to be regarded as vectors in Hilbert space. For such functions we have to take all those for which $\int_a^b f^2(x)\,dx$ has a meaning. It might appear natural to confine ourselves to continuous functions for which $\int_a^b f^2(x)\,dx$ always exists. However, the theory of Hilbert space becomes more complete and natural if $\int_a^b f^2(x)\,dx$ is interpreted in a generalized sense, namely as a Lebesgue integral (see Chapter XV).

This extension of the concept of integrals (and correspondingly of the class of functions to be discussed) is necessary for functional analysis in the same way as a strict theory of the real numbers is necessary for the foundation of the differential and integral calculus. Thus, the generalization of the ordinary concept of an integral that was created at the beginning of the 20th century in connection with the development of the theory of functions of a real variable turned out to be quite essential for functional analysis and the branches of mathematics connected with it.

§3. Expansion by Orthogonal Systems of Functions

Definition and examples of orthogonal systems of functions. If in a plane two arbitrary mutually perpendicular vectors e_1 and e_2 of unit length are chosen (figure 7), then every vector of the same plane can be decomposed in the directions of these two vectors, i.e., can be represented in the form

$$f = a_1 e_1 + a_2 e_2,$$

where a_1 and a_2 are the numbers equal to the projections of the vector f in the direction of the axis of e_1 and e_2. Since

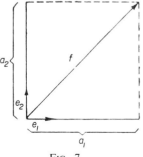

Fig. 7.

the projection of f on an axis is equal to the product of the length of f by the cosine of the angle between f and the axis, we can write, remembering the definition of the scalar product,

$$a_1 = (f, e_1),$$
$$a_2 = (f, e_2).$$

Similarly if in a three-dimensional space any three mutually perpendicular vectors e_1, e_2, e_3 of unit length are chosen, then every vector f in this space can be written in the form

$$f = a_1 e_1 + a_2 e_2 + a_3 e_3,$$

where

$$a_k = (f, e_k) \ (k = 1, 2, 3).$$

In Hilbert space we can also consider systems of pairwise orthogonal vectors of the space, i.e., functions $\phi_1(x), \phi_2(x), \cdots, \phi_n(x), \cdots$.

Such systems of functions are called orthogonal and play an important role in analysis. They occur in very diverse problems of mathematical physics, integral equations, approximate computations, the theory of functions of a real variable, etc. The ordering and unification of the concepts relating to such systems formed one of the motivations that led at the beginning of the 20th century to the creation of the general concept of a Hilbert space.

Let us give a precise definition. A system of functions

$$\phi_1(x), \phi_2(x), \cdots, \phi_n(x), \cdots$$

is called *orthogonal* if any two functions of the system are orthogonal, i.e., if

$$\int_a^b \phi_i(x) \phi_k(x) \, dx = 0 \quad \text{for} \quad i \neq k. \tag{13}$$

In three-dimensional space we required that the vectors of the system should be of unit length. Recalling the definition of length of a vector we see that in the case of Hilbert space this requirement can be written as follows:

$$\int_a^b \phi_k^2(x) \, dx = 1. \tag{14}$$

A system of functions satisfying the conditions (13) and (14) is called orthonormal.

Let us give examples of such systems of functions.

1. On the interval $(-\pi, \pi)$ we consider the sequence of functions

$$1, \cos x, \sin x, \cos 2x, \sin 2x, \cdots, \cos nx, \sin nx, \cdots .$$

Any two functions of this sequence are orthogonal to each other. This can be verified by the simple computation of the corresponding integrals. The square of the length of a vector in Hilbert space is the integral of the square of the function. Thus, the squares of the lengths of the vectors of the sequence

$$1, \cos x, \sin x, \cos 2x, \sin 2x, \cdots, \cos nx, \sin nx, \cdots$$

are the integrals

$$\int_{-\pi}^{\pi} dx = 2\pi, \ \int_{-\pi}^{\pi} \cos^2 nx \, dx = \pi, \ \int_{-\pi}^{\pi} \sin^2 nx \, dx = \pi,$$

i.e., the vectors of our sequence are orthogonal, but not normalized. The length of the first vector of the sequence is equal to $\sqrt{2\pi}$, and all the others are of length $\sqrt{\pi}$. When we divide every vector by its length, we obtain the orthonormal system of trigonometric functions

$$\frac{1}{\sqrt{2\pi}}, \frac{\cos x}{\sqrt{\pi}}, \frac{\sin x}{\sqrt{\pi}}, \frac{\cos 2x}{\sqrt{\pi}}, \frac{\sin 2x}{\sqrt{\pi}}, \cdots, \frac{\cos nx}{\sqrt{\pi}}, \frac{\sin nx}{\sqrt{\pi}}, \cdots .$$

This system is historically one of the first and most important examples of orthogonal systems. It appeared in the works of Euler, D. Bernoulli, and d'Alembert in connection with problems on the oscillations of strings. The study of it plays an essential role in the development of the whole of analysis.*

The appearance of the orthogonal system of trigonometrical functions in connection with problems on oscillations of strings is not accidental. Every problem on small oscillations of a medium leads to a certain system of orthogonal functions that describe the so-called characteristic oscillations of the given system (see §4). For example, in connection with problems on the oscillations of a sphere there appear the so-called spherical functions, in connection with problems on the oscillations of a circular membrane or a cylinder there appear the so-called cylinder functions, etc.

2. We can give an example of an orthogonal system of functions in

---

* See Chapter XII, §1.

which every function is a polynomial. Such an example is the sequence of Legendre polynomials

$$P_n(x) = \frac{1}{2^n n!} \frac{d^n(x^2 - 1)^n}{dx^n},$$

i.e., $P_n(x)$ is (apart from a constant factor) the nth derivative of $(x^2 - 1)^n$. Let us write down the first few polynomials of this sequence:

$$P_0(x) = 1;$$
$$P_1(x) = x;$$
$$P_2(x) = \tfrac{1}{2}(3x^2 - 1);$$
$$P_3(x) = \tfrac{1}{2}(5x^3 - 3x).$$

Obviously $P_n(x)$ is a polynomial of degree n. We leave it to the reader to convince himself that these polynomials are an orthogonal sequence on the interval $(-1, 1)$.

The general theory of orthogonal polynomials (the so-called orthogonal polynomials with weights) was developed in the second half of the 19th century by the famous Russian mathematician P. L. Čebyšev.

Expansion by orthogonal systems of functions. Just as in three-dimensional space every vector can be represented in the form of a linear combination of three pairwise orthogonal vectors e_1, e_2, e_3 of unit length

$$f = a_1 e_1 + a_2 e_2 + a_3 e_3,$$

so in a functional space there arises the problem of the decomposition of an arbitrary function f in a series with respect to an orthonormal system of functions, i.e., of the representation of f in the form

$$f(x) = a_1 \phi_1(x) + a_2 \phi_2(x) + \cdots + a_n \phi_n(x) + \cdots. \tag{15}$$

Here the convergence of the series (15) to the function f has to be understood in the sense of the distance between elements in Hilbert space. This means that the mean-square deviation of the partial sum of the series

$$S_n(t) = \sum_{k=1}^{n} a_k \phi_k(t)$$

from the function $f(t)$ tends to zero for $n \to \infty$; i.e.,

$$\lim_{n \to \infty} \int_a^b [f(t) - S_n(t)]^2 \, dt = 0. \tag{16}$$

This convergence is usually called "convergence in the mean."

Expansions in various systems of orthogonal functions often occur in analysis and are an important method for the solution of problems of mathematical physics. For example, if the orthogonal system is the system of trigonometric functions on the interval $(-\pi, \pi)$

$$1, \cos x, \sin x, \cos 2x, \sin 2x, \cdots, \cos nx, \sin nx, \cdots,$$

then this expansion is the classical expansion of a function in a trigonometric series*

$$f(x) = a_0 + a_1 \cos x + b_1 \sin x + a_2 \cos 2x + b_2 \sin 2x + \cdots.$$

Let us assume that an expansion (15) is possible for every function f of a Hilbert space and let us find its coefficients a_n. For this purpose we multiply both sides of the equation scalarly by one and the same function ϕ_m of our system. We obtain the equation

$$(f, \phi_m) = a_1 (\phi_1, \phi_m) + a_2(\phi_2, \phi_m) + \cdots + a_m(\phi_m, \phi_m)$$
$$+ a_{m+1}(\phi_{m+1}, \phi_m) + \cdots,$$

in virtue of the fact that $(\phi_m, \phi_n) = 0$ for $m \neq n$ and $(\phi_m, \phi_m) = 1$, this determines the value of the coefficient a_m

$$a_m = (f, \phi_m) \ (m = 1, 2, \cdots).$$

We see that, as in ordinary three-dimensional space (see the beginning of this section), the coefficients a_m are equal to the projections of the vector f in the direction of the vectors ϕ_k.

Recalling the definition of the scalar product we see that the coefficients of the expansion of $f(x)$ by the normal orthogonal system of functions $\phi_1(x), \phi_2(x), \cdots, \phi_n(x), \cdots$

$$f(x) = a_1\phi_1(x) + a_2\phi_2(x) + \cdots + a_n\phi_n(x) + \cdots \tag{17}$$

are determined by the formulas

$$a_m = \int_a^b f(t) \phi_m(t) \, dt. \tag{18}$$

As an example let us consider the normal orthogonal trigonometric system of functions mentioned previously:

$$\frac{1}{\sqrt{2\pi}}, \frac{\cos x}{\sqrt{\pi}}, \frac{\sin x}{\sqrt{\pi}}, \frac{\cos 2x}{\sqrt{\pi}}, \frac{\sin 2x}{\sqrt{\pi}}, \cdots.$$

---

* Such a decomposition often occurs in various problems of physics in the decomposition of an oscillation into its harmonic constituents. See Chapter VI, §5.

Then

$$f(x) = \frac{a_0}{2} + \sum_{n=1}^{\infty} (a_n \cos nx + b_n \sin nx),$$

where

$$a_0 = \frac{1}{\pi} \int_{-\pi}^{\pi} f(x) \, dx, \quad a_n = \frac{1}{\pi} \int_{-\pi}^{\pi} f(x) \cos nx \, dx,$$

$$b_n = \frac{1}{\pi} \int_{-\pi}^{\pi} f(x) \sin nx \, dx.$$

So we have obtained the formula for the computation of the coefficients of the expansion of a function in trigonometric series, assuming of course that this expansion is possible.*

We have established the form of the coefficients of the expansion (18) of the function $f(x)$ by an orthogonal system of functions under the assumptions that this expansion holds. However, an infinite orthogonal system of functions $\phi_1, \phi_2, \cdots, \phi_n, \cdots$ may turn out to be insufficient for every function of a Hilbert space to have such an expansion. For such an expansion to be possible, the system of orthogonal functions must satisfy an additional condition, namely the so-called condition of completeness.

An orthogonal system of functions is called *complete* if it is impossible to add to it even one function, not identically equal to zero, that is orthogonal to all the functions of the system.

It is easy to give an example of an incomplete orthogonal system. For this purpose we choose an arbitrary orthogonal system, for example that of the trigonometric functions, and remove one of the functions of the system, for example $\cos x$. The remaining infinite system of functions

$$1, \sin x, \cos 2x, \sin 2x, \cdots, \cos nx, \sin nx, \cdots$$

is orthogonal as before, but of course it is not complete, since the function $\cos x$ which we have excluded is orthogonal to all the functions of the system.

If a system of functions is incomplete, then not every function of a Hilbert space can be expanded by it. For if we attempt to expand by such a system a nonzero function $f_0(x)$ that is orthogonal to all the functions of the system, then by (18) all the coefficients turn out to be zero, whereas the function $f_0(x)$ is not equal to zero.

The following theorem holds: If a complete orthonormal system of

---

* On trigonometric series see also Chapter XII, §7.

functions in a Hilbert space $\phi_1(x)$, $\phi_2(x)$, $\cdots$, $\phi_n(x)$, $\cdots$, is given, then every function $f(x)$ can be expanded in a series by functions of this system*

$$f(x) = a_1\phi_1(x) + a_2\phi_2(x) + \cdots + a_n\phi_n(x) + \cdots.$$

Here the coefficients a_n of the expansion are equal to the projections of the vectors f on the elements of the normal orthogonal system

$$a_n = (f, \phi_n) = \int_a^b f(x)\,\phi_n(x)\,dx.$$

Pythagoras' theorem in Hilbert space, which was established in §2, enables us to find an interesting relation between the coefficients a_k and the function $f(x)$. We denote by $r_n(x)$ the difference between $f(x)$ and the sum of the first n terms of its series; i.e.,

$$r_n(x) = f(x) - [a_1\phi_1(x) + \cdots + a_n\phi_n(x)].$$

The function $r_n(x)$ is orthogonal to $\phi_1(x)$, $\phi_2(x)$, $\cdots$, $\phi_n(x)$. Let us verify for example that it is orthogonal to $\phi_1(x)$, i.e., that $\int_a^b r_n(x)\,\phi_1(x)\,dx = 0$. We have

$$\int_a^b r_n(x)\,\phi_1(x)\,dx = \int_a^b [f(x) - a_1\phi_1(x) - a_2\phi_2(x) - \cdots - a_n\phi_n(x)]\,\phi_1(x)\,dx$$

$$= \int_a^b f(x)\,\phi_1(x)\,dx - a_1 \int_a^b \phi_1^2(x)\,dx.^\dagger$$

Since $a_1 = \int_a^b f(x)\,\phi_1(x)\,dx$, and $\int_a^b \phi_1^2(x)\,dx = 1$, it follows from this that $\int_a^b r_n(x)\phi_1(x)\,dx = 0$.

Thus, in the equation

$$f(x) = a_1\phi_1(x) + a_2\phi_2(x) + \cdots + a_n\phi_n(x) + r_n(x) \qquad (19)$$

the individual terms on the right-hand side are orthogonal to each other. Hence, by Pythagoras' theorem as formulated in §1, the square of the length of $f(x)$ is equal to the sum of the squares of the lengths of the summands of the right-hand side in (19); i.e.,

$$\int_a^b f^2(x)\,dx = \int_a^b [a_1\phi_1(x)]^2\,dx + \cdots + \int_a^b [a_n\phi_n(x)]^2\,dx + \int_a^b r_n^2(x)\,dx.$$

---

* This series is related to its sum in the sense defined in formula (16).

† The remaining integrals are equal to zero, because the functions $\phi_k(x)$ are orthogonal to each other.

Since the system of functions $\phi_1, \phi_2, \cdots, \phi_n$ is normalized [equation (14)], we have

$$\int_a^b f^2(x)\,dx = a_1^2 + a_2^2 + \cdots + a_n^2 + \int_a^b r_n^2(x)\,dx. \qquad (20)$$

The series $\Sigma_{k=1}^\infty a_k \phi_k(x)$ converges in the mean. This means that

$$\int_a^b [f(x) - a_1\phi_1(x) - \cdots - a_n\phi_n(x)]^2\,dx \to 0,$$

i.e., that

$$\int_a^b r_n^2(x)\,dx \to 0.$$

But then we obtain from the formula (20) the equation

$$\sum_{k=1}^\infty a_k^2 = \int_a^b f^2(x)\,dx, * \qquad (21)$$

which states that the integral of the square of a function is equal to the sum of the squares of the coefficients of its expansion by a closed orthogonal system of functions. If the condition (21) holds for an arbitrary function of the Hilbert space, it is called the condition of completeness.

We wish to draw attention to the following important question. Which numbers a_k can be the coefficients of the expansion of a function in Hilbert space? The equation (21) asserts that for this purpose the series $\Sigma_{k=1}^\infty a_k^2$ must converge. Now it turns out that this condition is also sufficient; i.e., a sequence of numbers a_k is the sequence of coefficients of the expansion by an orthogonal system of functions in Hilbert space if and only if the series $\Sigma_{k=1}^\infty a_k^2$ converges.

We remark that this fundamental theorem holds if Hilbert space is interpreted as the collection of all functions with integrable square in the sense of Lebesgue (see §2). If we were to confine ourselves in Hilbert space, for example, to the continuous functions, then the solution of the problem as to which numbers a_k can be the coefficients of an expansion would become unnecessarily complicated.

The arguments given here are only one of the reasons that have led to the use of an integral in a generalized (Lebesgue) sense in the definition of Hilbert space.

---

* Geometrically, this means that the square of the length of a vector in Hilbert space is equal to the sum of the squares of its projections onto a complete system of mutually orthogonal directions.

§4. Integral Equations

In this section the reader will become acquainted with one of the most important and, historically, one of the first branches of functional analysis, namely the theory of integral equations, which has also played an essential role in the subsequent development of functional analysis. Quite apart from internal requirements of mathematics [for example, boundary problems for partial differential equations (Chapter VI)], various problems of physics were of great importance in the development of the theory of integral equations. Side by side with differential equations, the integral equations are, in the 20th century, one of the most important means of the mathematical investigation of various problems of physics. In this section we shall give a certain amount of information concerning the theory of integral equations. The facts we shall explain here are closely connected and have essentially sprung up (directly or indirectly) in connection with the study of small oscillations of elastic systems.

The problem of small oscillations of elastic systems. We return to the problem of small oscillations discussed in §2. Let us find equations that describe such oscillations. For the sake of simplicity we assume that we are dealing with the oscillation of a linear elastic system. As examples of such systems we can take, say, a string of length l (figure 8) or an elastic rod (figure 9). We shall assume that in the position of equilibrium our elastic system is situated along the segment Ol of the x-axis. We apply a unit force at the point x. Under the action of this force all the points of

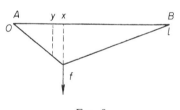

FIG. 8.

the system receive a certain displacement. The displacement arising at the point y (figure 8) is denoted by $k(x, y)$.

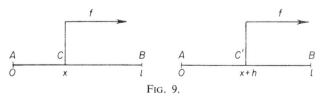

FIG. 9.

The function $k(x, y)$ is a function of two points: the point x at which the force is applied, and the point y at which we measure the displacement. It is called the influence function (Green's function).

From the law of conservation of energy, we can deduce an important property of the Green's function $k(x, y)$, namely the so-called reciprocity law: The displacement arising at the point y under the action of a force applied at the point x is equal to the displacement arising at the point x under the action of the same force applied at the point y. In other words, this means that

$$k(x, y) = k(y, x). \tag{22}$$

Let us find, for example, the Green's function for the longitudinal oscillations of an elastic rod (in figure 8 we have illustrated other transverse displacements). We consider a rod AB of length fixed at the ends (figure 9). At the point C we apply a force f acting in the direction of B. Under the action of this force the rod is deformed and the point C is shifted into the position C'. We denote the magnitude of the shift of C by h. Let us find the value of h. By means of h we can then find the shift at an arbitrary point y. For this purpose we shall make use of Hooke's law, which states that the force is proportional to the relative extension (i.e., to the ratio of the amount of displacement to the length). A similar relation holds for compressions.

Under the action of the force f the part AC of the rod is stretched. We denote the reaction arising here by T_1. At the same time the part CB of the rod is compressed, giving rise to a reaction T_2. By Hooke's law

$$T_1 = \kappa \frac{h}{x}, \quad T_2 = \kappa \frac{h}{l - x},$$

where κ is the coefficient of proportionality that characterizes the elastic properties of the rod. The position of equilibrium of the forces acting at the point C gives us

$$f = \kappa \frac{h}{x} + \kappa \frac{h}{l - x}, \quad \text{i.e.,} \quad f = \frac{\kappa l h}{x(l - x)}.$$

Hence

$$h = \frac{f}{\kappa l} x(l - x).$$

In order to find the displacement arising at a certain point y on the segment AC, i.e., for $y < x$, we note that it follows from Hooke's law that under an extension of the rod the relative extension (i.e., the ratio of the displacement of the point to its distance from the fixed end) does not depend on the position of the point. We denote the displacement of the point y by k.

Then by comparing the relative displacements at the points x and y we obtain

$$\frac{k}{y} = \frac{h}{x};$$

hence

$$k = h\frac{y}{x} = \frac{f}{\kappa_l}y(l - x) \quad \text{for} \quad y < x.$$

Similarly, if the point lies on the segment CB ($y > x$), we obtain

$$k = h\frac{l - y}{l - x} = \frac{f}{\kappa_l}x(l - y).$$

Bearing in mind that the Green's function $k(x, y)$ is the displacement at the point y under the action of a unit force applied at the point x, we see that on the longitudinal oscillations of an elastic rod the Green's function has the form

$$k(x, y) = \begin{cases} \dfrac{1}{\kappa_l}y(l - x) & \text{for} \quad y < x, \\[2mm] \dfrac{1}{\kappa_l}x(l - y) & \text{for} \quad y > x. \end{cases}$$

In a more or less similar way we could have found the Green's function for a string. If the tension of the string is T and the length l, then under the action of a unit force applied at the point x the string assumes the form illustrated in figure 8, and the displacement $k(x, y)$ at the point y is given by the formula

$$k(x, y) = \begin{cases} \dfrac{1}{Tl}x(l - y), & \text{for} \quad x < y, \\[2mm] \dfrac{1}{Tl}y(l - x), & \text{for} \quad x > y, \end{cases}$$

which coincides with the Green's function for the rod which we have derived.

In terms of the Green's function we can express the displacement of the system from its position of equilibrium provided that it is acted upon by a continuously distributed force of density $f(y)$. Since on an interval of length Δy there acts a force $f(y)\,\Delta y$, which we can regard approximately as concentrated at the point y, under the action of this force at the point x there arises a displacement $k(x, y)f(y)\,\Delta y$. The displacement under the action of the whole load is approximately equal to the sum

$$\sum k(x, y)f(y)\,\Delta y.$$

Passing to the limit for $\Delta y \to 0$ we see that the displacement $u(x)$ at the point x under the action of the force $f(y)$ distributed along the system is given by the formula

$$u(x) = \int_a^b k(x, y) f(y)\, dy. \tag{23}$$

Let us assume that our elastic system is not subject to the action of external forces. If it is displaced from its position of equilibrium, it then begins to move. These motions are called the free oscillations of the system.

Now let us write down in terms of the Green's function $k(x, y)$ the equation that the free oscillations of the elastic system in question have to obey. For this purpose we denote by $u(x, t)$ the displacement from the position of equilibrium at the point x and the instant of time t. Then the acceleration of x at the time t is equal to $\partial^2 u(x, t)/\partial t^2$.

If ρ is the linear density of the field, i.e., $\rho\, dy$ the mass of the element of length dy, then we obtain by a fundamental law of mechanics the equation of motion by replacing in (23) the force $f(y)\, dy$ by the product of the mass and the acceleration $[\partial^2 u(y, t)/\partial t^2]\, \rho\, dy$ taken with the opposite sign.

Thus, the equation of the free oscillations has the form

$$u(x, t) = -\int_a^b k(x, y) \frac{\partial^2 u(y, t)}{\partial t^2}\, \rho\, dy.$$

An important role in the theory of oscillations is played by the so-called harmonic oscillations of the elastic system, i.e., the motions for which

$$u(x, t) = u(x) \sin \omega t.$$

They are characterized by the fact that every fixed point performs harmonic oscillations (moves according to a sinusoidal law) with a certain frequency ω, and that this frequency is one and the same for all the points x.

Later on we shall see that every free oscillation is composed of harmonic oscillations.

We set

$$u(x, t) = u(x) \sin \omega t$$

in the equation of the free oscillations and cancel $\sin \omega t$. Then we obtain the following equation to determine the function $u(x)$

$$u(x) = \rho \omega^2 \int_a^b k(x, y) u(y)\, dy. \tag{24}$$

Such an equation is called a homogeneous integral equation for the function $u(x)$.

Obviously the equation (24) has for every ω the uninteresting solution $u(x) \equiv 0$, which corresponds to the state of rest. Those values of ω for which there exist other solutions of the equation (24), different from zero, are called the eigenfrequencies of the system.

Since nonzero solutions do not exist for every value of ω, the system can perform free oscillations only with definite frequencies. The smallest of these is called the fundamental tone of the system, and the remaining ones are overtones.

Now it turns out that for every system there exists an infinite sequence of eigenfrequencies, the so-called frequency spectrum

$$\omega_1, \omega_2, \cdots, \omega_n, \cdots$$

The nonzero solution $u_n(x)$ of the equation (24) corresponding to the the eigenfrequency ω_n gives us the form of the corresponding characteristic oscillation.

For example, if the elastic system is a string stretched between the points O and l and fastened at these points, then the possible frequencies of the characteristic oscillations of the system are equal to

$$a\,\frac{\pi}{l},\; 2a\,\frac{\pi}{l},\; 3a\,\frac{\pi}{l},\; \cdots,\, na\,\frac{\pi}{l},\; \cdots,$$

where a is a coefficient depending on the density and the tension of the

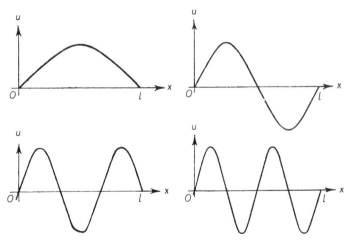

Fig. 10.

string, namely, $a = \sqrt{T/\rho}$. The fundamental tone is here $\omega_1 = a(\pi/l)$, and the overtones are $\omega_2 = 2\omega_1$, $\omega_3 = 3\omega_1$, $\cdots$, $\omega_n = n\omega_1$. The form of the corresponding harmonic oscillations is given by the equation

$$u_n(x) = \sin \frac{n\pi}{l} x$$

and are illustrated for $n = 1, 2, 3, 4$ in figure 10.

So far we have discussed free oscillations of elastic systems. Now if an exterior harmonic force acts on the elastic system during the motion, then, in determining the harmonic oscillations under the action of this force, we arrive at the function $u(x)$ at the so-called inhomogeneous integral equation

$$u(x) = \rho\omega^2 \int_a^b k(x, y)\, u(y)\, dy + h(x). \tag{25}$$

Properties of integral equations. Previously we have become acquainted with examples of integral equations

$$f(x) = \lambda \int_a^b k(x, y) f(y)\, dy \tag{26}$$

and

$$f(x) = \lambda \int_a^b k(x, y) f(y)\, dy + h(x), \tag{27}$$

the first of which was obtained in the solution of the problem on the free oscillations of an elastic system, and the second in the discussion of forced oscillations, i.e., oscillations under the action of external forces.

The unknown function in these equations is $f(x)$. The given function $k(x, y)$ is called the *kernel* of the integral equation. The equation (27) is called an *inhomogeneous linear integral equation*, and the equation (26) is *homogeneous*. It is obtained from the inhomogeneous one by setting $h(x) = 0$.

It is clear that the homogeneous equation always has the zero solution, i.e., the solution $f(x) = 0$. A close connection exists between the solutions, of the inhomogeneous and the homogeneous integral equations. By way of example we mention the following theorem: If the homogeneous integral equation has only the zero solution, then the corresponding inhomogeneous equation is soluble for every function $h(x)$.

If for a certain value λ a homogeneous equation has the solution $f(x)$, not identically equal to zero, then this value λ is called an *eigenvalue* and the corresponding solution $f(x)$ an *eigenfunction*. We have seen earlier that when an integral equation describes the free oscillations of an elastic system, then the eigenvalues are closely connected with the frequencies of the oscillations of the system (namely $\lambda = \rho\omega^2$). The eigenfunctions then give the form of the corresponding harmonic oscillations.

In the problems on oscillations it followed from the law of conservation of energy that

$$k(x, y) = k(y, x). \tag{28}$$

A kernel satisfying the condition (28) is called *symmetric.*

The eigenfunctions and eigenvalues of an equation with a symmetric kernel have a number of important properties. One can prove that such an equation always has a sequence of real eigenvalues

$$\lambda_1, \lambda_2, \cdots, \lambda_n, \cdots.$$

To every eigenvalue there correspond one or several eigenfunctions. Here eigenfunctions corresponding to distinct eigenvalues are always orthogonal to each other.*

Thus, for every integral equation with a symmetric kernel the system of eigenfunctions is an orthogonal system of functions. There arises the question of when this system is complete, i.e., when can every function of the Hilbert space be expanded in a series by a system of eigenfunctions of the integral equation. In particular, if the equation

$$\int_a^b k(x, y) f(y)\, dy = 0 \tag{29}$$

is satisfied for $f(y) \equiv 0$ only, then the system of eigenfunctions of the integral equation

$$\lambda \int_a^b k(x, y) f(y)\, dy = f(x)$$

is a complete orthogonal system.†

Thus, every function $f(x)$ with integrable square can in this case be expanded in a series by eigenfunctions. By discussing various types of integral equations, we obtain a general and powerful method of proving

---

* The latter statement will be proved in the next section.

† In the case when $k(x, y)$ is the Green's function of an elastic system, the equation (29) assumes a simple physical meaning. In fact [see formula (23)] we have seen that under the action of a force $f(y)$ distributed along the system the displacement of the system from the position of equilibrium is expressed by the formula $u(x) = \int_l^a k(x, y) f(y)\, dy$. Thus, the condition (29) signifies that every nonzero force takes the system out of its position of equilibrium.

that various important orthogonal systems are closed, i.e., that the functions are expandable in series by orthogonal functions. By this method we can prove the completeness of the system of trigonometric functions, of cylinder functions, spherical functions, and many other important systems of functions.

The fact that an arbitrary function can be expanded in a series by eigenfunctions means in the case of oscillations that every oscillation can be decomposed into a sum of harmonic oscillations. Such a decomposition yields a method that is widely applicable in solving problems on oscillations in various domains of mechanics and physics (oscillations of elastic bodies, acoustic oscillations, electromagnetic waves, etc.).

The development of the theory of linear integral equations gave the impetus to the creation of the general theory of linear operators of which the theory of linear integral equations forms an organic part. In the last few decades the general methods of the theory of linear operators have vigorously contributed to the further development of the theory of integral equations.

§5. Linear Operators and Further Developments of Functional Analysis

In the preceding section we have seen that problems on the oscillations of an elastic system lead to the search for the eigenvalues and eigenfunctions of integral equations. Let us note that these problems can also be reduced to the investigation of the eigenvalues and eigenfunctions of linear differential equations.* Many other physical problems also lead to the task of computing the eigenvalues and eigenfunctions of linear differential or integral equations.

Let us give one more example. In modern radio technology the so-called wave guides are widely used for the transmission of electromagnetic oscillations of high frequencies, i.e., hollow metallic tubes in which electromagnetic waves are propagated. It is known that in a wave guide only electromagnetic oscillations of not too large a wave length can be propagated. The search for the critical wave length amounts to a problem on the eigenvalues of a certain differential equation.

Problems on eigenvalues occur, moreover, in linear algebra, in the theory of ordinary differential equations, in questions of stability, etc.

So it became necessary to discuss all these related problems from one single point of view. This common point of view is the general theory of linear operators. Many problems on eigenfunctions and eigenvalues in various concrete cases came to be fully understood only in the light of

---

* See Chapter VI, §5.

the general theory of operators. Thus, in this and a number of other directions the general theory of operators turned out to be a very fruitful research tool in those domains of mathematics in which it is applicable.

In the subsequent development of the theory of operators, quantum mechanics played a very important role, since it makes extensive use of the methods of the theory of operators. The fundamental mathematical apparatus of quantum mechanics is the theory of the so-called self-adjoint operators. The formulation of mathematical problems arising in quantum mechanics was and still is a powerful stimulus for the further development of functional analysis.

The operator point of view on differential and integral equations turned out to be extremely useful also for the development of practical methods for approximate solutions of such equations.

Fundamental concepts of the theory of operators. Let us now proceed to an explanation of the fundamental definitions and facts in the theory of operators.

In analysis we have come across the concept of a function. In its simplest form this was a relation that associates with every number x (the value of the independent variable) a number y (the value of the function). In the further development of analysis it became necessary to consider relations of a more general type.

Such more general relations are discussed, for example, in the calculus of variations (Chapter VIII), where we associated with every function a number. If with every function a certain number is associated, then we say that we are given a functional. As an example of a functional we can take the association between an arbitrary function $y = f(x)$ ($a \leqslant x \leqslant b$) and the arc length of the curve represented by it. We obtain another example of a functional if we associate with every function $y = f(x)$ ($a \leqslant x \leqslant b$) its definite integral $\int_a^b f(x)\, dx$.

If we regard $f(x)$ as a point of an infinite-dimensional space, then a functional is simply a function of the points of the infinite-dimensional space. From this point of view the problems of the calculus of variations concern the search for maxima and minima of functions of the points of an infinite-dimensional space.

In order to define what we mean by a continuous functional it is necessary to define first what we mean by proximity of two points of an infinite-dimensional space. In §2 we gave the distance between two functions $f(x)$ and $g(x)$ (points of an infinite-dimensional space) as

$$\sqrt{\int_a^b [f(x) - g(x)]^2\, dx}.$$

This method of assigning a distance in infinite-dimensional space is often used, but of course it is not the only possible one. In other problems other methods of giving the distance between functions may turn out to be better. We may point, for example, to the problem of the theory of approximation of functions (see Chapter XII, §3), where the distance between functions, which characterizes the measure of proximity of the two functions $f(x)$ and $g(x)$, is given, for example, by the formula

$$\max |f(x) - g(x)|.$$

Other methods of giving a distance between functions are used in the investigation of functionals in the calculus of variations. Distinct methods of giving the distance between functions lead us to distinct infinite-dimensional spaces.

Thus, various infinite-dimensional (functional) spaces differ from each other by their set of functions and by the definition of distance between them. For example, if we take the set of all functions with integrable square and define distance as

$$\sqrt{\int_a^b [f(x) - g(x)]^2 \, dx},$$

then we arrive at the Hilbert space that was introduced in §2; but if we take the set of all continuous functions and define distance as $\max |f(x) - g(x)|$, then we obtain the so-called space (C).

In the discussion of integral equations we come across expressions of the form

$$g(x) = \int_a^b k(x, y) f(y) \, dy.$$

For a given kernel $k(x, y)$ this equation indicates a rule by which every function $f(x)$ is set in correspondence with another function $g(x)$.

This kind of a correspondence that relates with one function f another function g is called an *operator*.

We shall say that we are given a linear operator A in a Hilbert space if we have a rule by which we associate with every function f another function g. The correspondence need not be given for all the functions of the Hilbert space. In that case the set of those functions f for which there exists the function $g = Af$ is called the *domain of definition* of the operator A (similar to the domain of definition of a function in ordinary analysis). The correspondence itself is usually denoted as follows:

$$g = Af. \tag{30}$$

The linearity of the operator means that the sum of the functions f_1 and f_2 is associated with the sum of Af_1 and Af_2, and the product of f and a number λ with the function λAf; i.e.,

$$A(f_1 + f_2) = Af_1 + Af_2 \tag{31}$$

and

$$A(\lambda f) = \lambda Af. \tag{32}$$

Occasionally continuity is also postulated for linear operators; i.e., it is required that the convergence of a sequence of functions f_n to a function f should imply that the sequence Af_n should converge to Af.

Let us give examples of linear operators.

1. Let us associate with every function $f(x)$ the function $g(x) = \int_a^x f(t)\, dt$, i.e., the indefinite integral of f. The linearity of this operator follows from the ordinary properties of the integral, i.e., from the fact that the integral of the sum is equal to the sum of the integrals and that a constant factor can be taken out of the integral sign.

2. Let us associate with every differentiable function $f(x)$ its derivative $f'(x)$. This operator is usually denoted by the letter D; i.e.,

$$f'(x) = D\, f(x).$$

Observe that this operator is not defined for all the functions of the Hilbert space but only for those that have a derivative belonging to the Hilbert space. These functions form, as we have said previously, the domain of definition of this operator.

3. The examples 1 and 2 were examples of linear operators in an infinite-dimensional space. But examples of linear operators in finite-dimensional spaces have occurred in other chapters of this book. Thus, in Chapter III affine transformations were investigated. If an affine transformation of a plane of space leaves the origin of coordinates fixed, then it is an example of a linear operator in a two-dimensional, or three-dimensional, space. The linear transformations of an n-dimensional space introduced in Chapter XVI now appear as linear operators in n-dimensional space.

4. In the integral equations, we have already met a very important and widely applicable class of linear operators in a functional space, namely the so-called integral operators. Let us choose a certain definite function $k(x, y)$. Then the formula

$$g(x) = \int_a^b k(x, y)\, f(y)\, dy$$

associates with every function f a certain function g. Symbolically we can write this transformation as follows:

$$g = Af.$$

The operator A in this case is called an integral operator. We could mention many other important examples of integral operators.

In §4 we spoke of the inhomogeneous integral equation

$$f(x) = \lambda \int_a^b k(x, y) f(y) \, dy + h(x).$$

In the notation of the theory of operators this equation can be rewritten as follows

$$f = \lambda Af + h, \tag{33}$$

where λ is a given number, h a given function (a vector of an infinite-dimensional space), and f the required function. In the same notation the homogeneous equation can be written as follows:

$$f = \lambda Af. \tag{34}$$

The classical theorems on integral equations, such as, for example, the theorem formulated in §4 on the connection between the solvability of the inhomogeneous and the corresponding homogeneous integral equation, are not true for every operator equation. However, one can indicate certain general conditions to be imposed on the operator A under which these theorems are true.

These conditions are stated in topological terms and express that the operator A should carry the unit sphere (i.e., the set of vectors whose length does not exceed 1) into a compact set.

Eigenvalues and eigenvectors of operators. The problem of eigenvalues and eigenfunctions of an integral equation to which we were led by problems on oscillations can be formulated as follows: to find the values λ for which there exists a nonzero function f satisfying the equation

$$f(x) = \lambda \int_a^b k(x, y) f(y) \, dy.$$

As before, this equation can be written as follows:

$$f = \lambda Af$$

or

$$Af = \frac{1}{\lambda} f. \tag{35}$$

Now we shall understand by A an arbitrary linear operator. Then a vector f satisfying the equation (35) is called an eigenvector of the operator A, and the number $1/\lambda$ the corresponding eigenvalue.

Since the vector $(1/\lambda)f$ coincides in direction with the vector f (differs from f only by a numerical factor), the problem of finding eigenvectors can also be stated as the problem of finding nonzero vectors f that do not change direction under the transformation A.

This way of looking at the eigenvalues enables us to unify the problem of eigenvalues of integral equations (if A is an integral operator), differential equations (if A is a differential operator), and the problem of eigenvalues in linear algebra (if A is a linear transformation in finite-dimensional space; see Chapter VI and Chapter XVI). In the case of three-dimensional space this problem arises in the search for the so-called principal axes of an ellipsoid.

In the case of integral equations a number of important properties of the eigenfunctions and eigenvalues (for example the reality of the eigenvalues, the orthogonality of the eigenfunctions, etc.) are consequences of the symmetry of the kernel, i.e., of the equation $k(x, y) = k(y, x)$.

For an arbitrary linear operator A in a Hilbert space the analogue of of this property is the so-called self-adjointness of the operator.

The condition for an operator A to be self-adjoint in the general case is that for any two elements f_1 and f_2 the equation

$$(Af_1, f_2) = (f_1, Af_2)$$

holds, where (Af_1, f_2) denotes the scalar product of the vector Af_1 and the vector f_2.

In problems of mechanics the condition of self-adjointness of an operator is usually a consequence of the law of conservation of energy. Therefore it is satisfied for operators connected with, say, oscillations for which there is no loss (dissipation) of energy.

The majority of operators that occur in quantum mechanics are also self-adjoint.

Let us verify that an integral operator with a symmetric kernel $k(x, y)$ is self-adjoint. In fact, in this case Af_1 is the function $\int_a^b k(x, y) f_1(y)\, dy$. Therefore the scalar product (Af_1, f_2), which is equal to the integral of the product of this function with f_2, is given by the formula

$$(Af_1, f_2) = \int_a^b \int_a^b k(x, y) f_1(y) f_2(x)\, dy\, dx.$$

Similarly

$$(f_1, Af_2) = \int_a^b \int_a^b k(x, y) f_2(y) f_1(x)\, dy\, dx.$$

The equation $(Af_1, f_2) = (f_1, Af_2)$ is an immediate consequence of the symmetry of the kernel $k(x, y)$.

Arbitrary self-adjoint operators have a number of important properties that are useful in the applications of these operators to the solution of a variety of problems. Indeed, the eigenvalues of a self-adjoint linear operator are always real and the eigenfunctions corresponding to distinct eigenvalues are orthogonal to each other.

Let us prove, for example, the last statement. Let λ_1 and λ_2 be two distinct eigenvalues of the operator A, and f_1 and f_2 eigenvectors corresponding to them. This means that

$$
\begin{aligned}
Af_1 &= \lambda_1 f_1, \\
Af_2 &= \lambda_2 f_2.
\end{aligned}
\tag{36}
$$

We form the scalar product of the first equation (36) by f_2, and of the second by f_1. Then we have.

$$
\begin{aligned}
(Af_1, f_2) &= \lambda_1(f_1, f_2), \\
(Af_2, f_1) &= \lambda_2(f_2, f_1).
\end{aligned}
\tag{37}
$$

Since the operator A is self-adjoint, we have $(Af_1, f_2) = (Af_2, f_1)$. When we subtract the second equation (37) from the first, we obtain

$$0 = (\lambda_1 - \lambda_2)(f_1, f_2).$$

Since $\lambda_1 \neq \lambda_2$, we have $(f_1, f_2) = 0$, i.e., the eigenvectors f_1 and f_2 are orthogonal.

The investigation of self-adjoint operators has brought clarity into many concrete problems and questions connected with the theory of eigenvalues. Let us dwell in more detail on one of them, namely on the problem of the expansion by eigenfunctions in the case of a continuous spectrum.

In order to explain what a continuous spectrum means, let us turn again to the classical example of the oscillation of a string. Earlier we have shown that for a string of length l the characteristic frequencies of oscillations can assume the sequence of values.

$$a\frac{\pi}{l}, 2a\frac{\pi}{l}, \cdots, na\frac{\pi}{l}, \cdots.$$

Let us plot the points of this sequence on the numerical axis $O\lambda$. When we increase the length of the string l, the distance between any two adjacent points of the sequence will decrease, and they will fill the numerical axis more densely. In the limit, when $l \to \infty$, i.e., for an infinite string, the the eigenfrequencies fill the whole numerical semiaxis $\lambda \geqslant 0$. In this case we say that the system has a continuous spectrum.

We have already said that for a string of length l the expansion in a series by eigenfunctions is an expansion in a series by sines and cosines of $n(\pi/l)x$; i.e., in a trigonometric series

$$f(x) = \frac{a_0}{2} + \sum a_n \cos n \frac{\pi}{l} x + b_n \sin n \frac{\pi}{l} x.$$

For the case of an infinite string we can again show that a more or less arbitrary function can be expanded by sines and cosines. However, since the eigenfrequencies are now distributed continuously along the numerical line, this is not an expansion in a series, but in a so-called Fourier integral

$$f(x) = \int_{-\infty}^{+\infty} [A(\lambda) \cos \lambda x + B(\lambda) \sin \lambda x] \, d\lambda.$$

The expansion in a Fourier integral was already well known and widely used in the 19th century in the solutions of various problems of mathematical physics.

However, in more general cases with a continuous spectrum* many problems referring to an expansion of functions by eigenfunctions were not properly clarified. Only the creation of the general theory of self-adjoint operators brought the necessary clarity to these problems.

Let us mention still another set of classical problems that have been solved on the basis of the general theory of operators. The discussion of oscillations involving dissipation (scattering) of energy belongs to such problems.

In this case we can again look for free oscillations of the system in the form $u(x) \, \phi(t)$. However, in contrast to the case of oscillations without dissipation of energy, the function $\phi(t)$ is not simply $\cos \omega t$, but has the form $e^{-kt} \cos \omega t$, where $k > 0$. Thus, the corresponding solution has the form $u(x)e^{-kt} \cos \omega t$. In this case every point x again performs oscillations (with frequency ω), however the oscillations are damped because for $t \to \infty$ the amplitude of these oscillations containing the factor e^{-kt} tends to zero.

It is convenient to write the characteristic oscillations of the system in the complex form $u(x)e^{-i\lambda t}$, where in the absence of friction the number λ is real and in the presence of friction λ is complex.

---

* As examples we can take the oscillations of an inhomogeneous elastic medium and also many problems of quantum mechanics.

The problem of the oscillations of a system with dissipation of energy again leads to a problem on eigenvalues, but this time not for self-adjoint operators. A characteristic feature here is the presence of complex eigenvalues indicative of the damping of the free oscillations.

Using a method of the theory of operators in conjunction with methods of the theory of analytic functions M. V. Keldyš investigated this class of problems in 1950–1951 and proved for it the completeness of the system of eigenfunctions.

Connection of functional analysis with other branches of mathematics and quantum mechanics. We have already mentioned that the creation of quantum mechanics gave a decisive impetus to the development of functional analysis. Just as the rise of the differential and integral calculus in the 18th century was dictated by the requirements of mechanics and classical physics, so the development of functional analysis was, and still is, the result of the vigorous influence of contemporary physics, principally of quantum mechanics. The fundamental mathematical apparatus of quantum mechanics consists of the branches of mathematics relating essentially to functional analysis. We can only briefly indicate the connections existing here, because an explanation of the foundations of quantum mechanics exceeds the framework of this book.

In quantum mechanics the state of the system is given in its mathematical description by a vector of Hilbert space. Such quantities as energy, impulse, and moment of momentum are investigated by means of self-adjoint operators. For example, the possible energy levels of an electron in an atom are computed as eigenvalues of the energy operator. The differences of these eigenvalues give the frequencies of the emitted quantum of light and thus define the structure of the radiation spectrum of the given substance. The corresponding states of the electron are here described as eigenfunctions of the energy operator.

The solution of problems of quantum mechanics often requires the computation of eigenvalues of various (usually differential) operators. In some complicated cases the precise solution of these problems turns out to be practically impossible. For an approximate solution of these problems the so-called perturbation theory is widely used, which enables us to find from the known eigenvalues and functions of a certain self-adjoint operator A the eigenvalues of an operator A_1 slightly different from it. We mention that the perturbation theory has not yet received a full mathematical foundation, which is an interesting and important mathematical problem.

Independently of the approximate determination of eigenvalues, we can often say a good deal about a given problem by means of qualitative investigation. This investigation proceeds in problems of quantum mechanics on the basis of the symmetries existing in the given case. As examples of such symmetries we can take the properties of symmetry of crystals, spherical symmetry in an atom, symmetry with respect to rotation, and others. Since the symmetries form a group (see Chapter XX), the group methods (the so-called representation theory of groups) enables us to answer a number of problems without computation. As examples we may mention the classification of atomic spectra, nuclear transformations, and other problems. Thus, quantum mechanics makes extensive use of the mathematical apparatus of the theory of self-adjoint operators. At the same time the continued contemporary development of quantum mechanics leads to a further development of the theory of operators by placing new problems before this theory.

The influence of quantum mechanics and also the internal mathematical developments of functional analysis have had the effect that in recent years algebraic problems and methods have played a significant role in functional analysis. This intensification of algebraic tendencies in contemporary analysis can well be compared with the growth of the value of algebraic methods in contemporary theoretical physics in comparison with the methods of physics of the 19th century.

In conclusion, we wish to emphasize once more that functional analysis is one of the rapidly developing branches of contemporary mathematics. Its connections and applications in contemporary physics, differential equations, approximate computations, and its use of general methods developed in algebra, topology, the theory of functions of a real variable, etc., make functional analysis one of the focal points of contemporary mathematics.

Suggested Reading

N. Dunford and J. T. Schwartz, *Linear operators*. I. *General theory*, Interscience, New York, 1958.

I. M. Gel'fand and Z. Ja. Šapiro, *Representations of the rotation group of three-dimensional space and their applications*, Amer. Math. Soc. Translations Series 2, vol. 2, 1956, 207-316.

A. N. Kolmogorov and S. V. Fomin, *Elements of the theory of functions and functional analysis*. Vol. 1, *Metric and normed spaces*. Vol. 2, *Measure, Lebesgue integrals and Hilbert space*. Graylock, New York, 1957/1961.

L. D. Landau and E. M. Lifšic, *Course of theoretical physics*. Vol. 3, *Quantum mechanics*, Pergamon, New York, 1958/1960.

F. Riesz and B. Sz.-Nagy, *Functional analysis*, Frederick Ungar, New York, 1955.

A. E. Taylor, *Introduction to functional analysis*, Wiley, New York, 1958.

XX

GROUPS AND OTHER ALGEBRAIC SYSTEMS

§1. Introduction

In Chapter IV, which deals with the algebra of polynomials, we have already talked of the main lines of development of algebra, its place among other mathematical disciplines, and of the changes in the views on the very subject-matter of algebra. The aim of the present chapter is to give the reader an idea of those new algebraic theories that have sprung up in the last century, but have only been fully developed in the present one and have made a deep impact on the contemporary mathematical research.

Contemporary, as well as classical, algebra is the study of operations, of rules of computation. But it is not restricted to the study of properties of operations on numbers, since it strives to study the properties of operations on elements of a far more general nature. This tendency is dictated by practical requirements. For example, in mechanics we add up forces, velocities, or rotations. In linear algebra (see Chapter XVI), whose ideas and methods have wide application in practical calculations, the domains of operations are matrices, linear transformations, or vectors of an n-dimensional space.

The theory of groups plays a particularly prominent role in contemporary algebra, and a large part of this chapter is devoted to it. Among other algebraic theories, we shall dwell on the theory of hypercomplex systems, which is a necessary and important stage in the historical process of the development of the concept of number. Of course, these two theories do not exhaust by any means the content of contemporary algebra, but they illustrate rather well its ideas and methods.

The theory of groups has arisen from the necessity of finding an apparatus for investigating such important regularities of the real world as, for example, symmetry.

A knowledge of the symmetry properties of geometric bodies or other mathematical or physical objects sometimes gives us a key to the clarification of their structure. However, although the concept of symmetry is altogether intuitive, an accurate and general description of what symmetry is, and in particular a quantitative account of the properties of symmetry, requires use of the apparatus of the theory of groups.

The theory of groups arose rather long ago, at the end of the 18th and the beginning of the 19th century. Originally it was developed only as an auxiliary apparatus in problems on the solution by radicals of equations of higher degree. This was due to the fact that precisely in this problem it was first observed that properties of equivalence, of symmetry of the roots of the equation, are fundamental for the solution of the whole problem. In the course of the 19th and 20th centuries the important role of the laws of symmetry appeared in many other branches of science; geometry, crystallography, physics, and chemistry. This led to a wide propagation of the methods and results of the theory of groups. Since every domain of application presented its own peculiar problems to the theory of groups, the growing number of these domains also exerted the opposite effect, in giving rise to new branches of the theory of groups, and the result of all this is that the contemporary theory of groups, which is a single entity in its essential concepts, actually splits into a number of more or less independent disciplines: the general theory of groups, the theory of finite groups, the theory of continuous groups, of discrete groups of transformations, the theory of representations and characters of groups, and so forth. In their gradual evolution, the methods and concepts of the theory of groups turned out to be important not only for the investigation of the laws of symmetry but also for the solution of many other problems.

In our time the concept of a group has become one of the most important general concepts of modern mathematics, and the theory of groups has assumed a conspicuous place among the mathematical disciplines. Outstanding contributors to the development of the theory of groups and its applications were E. S. Fedorov, O. Ju. Šmidt, and L. S. Pontrjagin. The researches of Soviet mathematicians in the realm of group theory occupy a leading place in the present-day development of this theory.

§2. Symmetry and Transformations

The simplest forms of symmetry. We begin with an account of the simplest forms of symmetry with which the reader is familiar from everyday life. One of these is the mirror symmetry of geometric bodies or the symmetry with respect to a plane.

A point A in space is called symmetrical to a point B with respect to a plane α (figure 1) if the plane intersects the segment AB perpendicularly at its midpoint. We also say that B is the mirror image of A in the plane α. A geometric body is called symmetric with respect to a plane if the plane divides the body into two parts each of which is the mirror image of the other in the plane. The plane itself is then called a plane of symmetry of the body. Mirror symmetry is often encountered in nature. For example, the form of the human body, or of the body of birds or animals, usually has a plane of symmetry.

Symmetry with respect to a line is defined in a similar way. We say that the points A, B lie symmetrically with respect to a line if the line intersects the segment AB at its midpoint and is perpendicular to AB (figure 2).

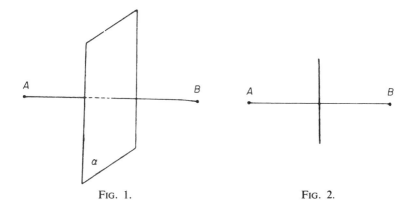

FIG. 1.   FIG. 2.

A geometric body is said to be symmetrical with respect to a line or to have this line as an axis of symmetry of order 2 if for every point of the body the symmetrical point also belongs to the body.

A body having an axis of symmetry of order 2 comes into coincidence with itself when the body is rotated around this axis by a half rotation, i.e., by an angle of 180°.

The concept of an axis of symmetry can be generalized in a natural way. A line is called an axis of symmetry of order n for a given body if the body comes into coincidence with itself on rotation around the axis by an angle $1/n\,360°$. For example, a regular pyramid whose base is a regular n-gon has the line joining the vertex of the pyramid to the center of the base (figure 3) as an axis of symmetry of order n.

A line is called an axis of rotation of a body if the body comes into coincidence with itself on rotation around the axis by an arbitrary angle. For example, the axis of a cylinder or a cone, or any diameter of a sphere, is an axis of rotation. An axis of rotation is also an axis of symmetry of every order.

Finally, an important type of symmetry is symmetry with respect to a point or central symmetry. Points A and B are called symmetrical with respect to a center O if the segment joining A and B is bisected at O. A body is called symmetrical with respect to a center O if all its points fall into pairs of points symmetrical with respect to O. Examples of centrally symmetric bodies are the sphere and the cube, whose centers are their center of symmetry (figure 4).

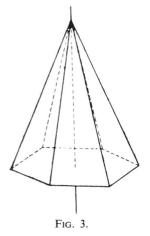

Fig. 3.

A knowledge of all the planes, axes, and centers of symmetry of a body gives a fairly complete idea of its symmetry properties.

But the concept of symmetry has a meaning not only when applied to geometric figures. For example, the statement that in the polynomial $x_1^3 + x_2^3 + x_3^3 + x_4^3$ the variables x_1, x_2, x_3, x_4 occur symmetrically has a perfectly clear meaning; also that in the polynomial $x_1^3 + x_2^2 + x_3^2 + x_4^3$ the variables x_1 and x_4, x_2 and x_3 occur symmetrically, whereas for example, the variables x_1 and x_2 play different roles. The number of such examples could easily be increased. This prompts us to raise the important question: What is symmetry in general and how can we take account mathematically of the relation of symmetry?

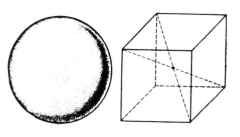

Fig. 4.

Now it turns out that a precise answer to this question is connected with the concept of transformation, which has already occurred many times in this book, right from the very first chapters. In order to be in a position to give a general definition of symmetry comprising such heterogeneous cases as the symmetry of spatial bodies and the symmetry of polynomials, it is necessary to formulate the concept of transformation in a very general way.

Transformations. Let M denote a finite or infinite collection of completely arbitrary objects. For example, M may be the set of numbers 1, 2, $\cdots$, n, the set of independent variables x_1, x_2, x_3, x_4, or the set of all points of a plane. If with every element of M a well-defined element of the same set is associated, then we say that a transformation of M is given. Every transformation of a finite set M can be given by means of a table consisting of two rows: In the upper row we write the names of the elements of M in an arbitrary order and below each of them we write the name of the element corresponding to it. For example, the table

$$\begin{pmatrix} 1 & 2 & 3 & 4 \\ 2 & 3 & 2 & 1 \end{pmatrix}$$

denotes the transformation of the set of numbers 1, 2, 3, 4 in which the numbers 1, 2, 3, 4 go over, respectively, into the numbers 2, 3, 2, 1. When we set out in the upper row the numbers 1, 2, 3, 4 in the order 3, 4, 1, 2, then we can write the same transformation also in form of the table

$$\begin{pmatrix} 3 & 4 & 1 & 2 \\ 2 & 1 & 2 & 3 \end{pmatrix} .$$

If the set M is infinite, but its elements can be counted (enumerated), then the transformation can be given in a similar way be setting out the elements in a single row (for example, if M is the set of all natural numbers 1, 2, 3, ...).

In studying transformations it is necessary to introduce a comprehensive notation for them. We shall denote transformations simply by letters A, B, etc., and if some transformation of the set M is denoted by the letter A, then we denote by mA, where m is an arbitrary element of M, the image of the element m, i.e., that element into which M goes over, under the transformation A. Suppose, for example, that

$$A = \begin{pmatrix} 1 & 2 & 3 & 4 \\ 2 & 3 & 2 & 1 \end{pmatrix} ; \text{ then } 1A = 2, \ 2A = 3, \ 3A = 2, \ 4A = 1.$$

Let us indicate some transformations that play an important role in geometry.

We draw an arbitrary line a in space and associate with every point P of space the point Q obtained by rotating the point P around the axis a by a fixed angle ϕ (figure 5). In this way we have defined a transformation of the set of all points of space, the so-called rotation of space by the angle ϕ around the axis a.

Observe that the word "rotation" in mechanics denotes a certain process as a result of which the points of the body assume a new position. Here we have used the term "rotation" in the sense of a transformation of space. We abstract from the actual process of motion and consider only its final result, namely the correspondence between the initial and the final position of the points.

Another important transformation of space is the parallel shift of all the points in a given direction by a given distance. From figure 6, in which

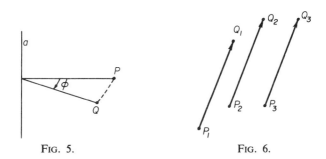

FIG. 5.   FIG. 6.

we have indicated for arbitrary points P_1, P_2, P_3 the corresponding points Q_1, Q_2, Q_3, it is clear that when we know the corresponding point of only one point of space in a parallel shift, then we can find the corresponding points for all other points of space.

Earlier we have defined the concepts of a plane of symmetry and of an axis and center of symmetry of a figure in space. To each of these concepts there corresponds a definite transformation of space: a reflection in the plane, a rotation around the line, and a reflection with respect to the center. For example, a reflection in a plane is the transformation in which every point of space is associated with its symmetrical point with respect to the plane. A rotation around the line and a reflection with respect to the center are similarly defined.

So far we have talked of transformations of space. The corresponding transformations of a plane: rotation of the plane around a point by a given angle, a parallel shift of the plane in itself in a given direction, and a reflection with respect to a line lying in the plane, all these are similarly defined and are even more intuitive than the corresponding transformations of space.

One-to-one transformations.   In discussing all possible transformations of one and the same set, we must first of all observe the fundamental difference between one-to-one transformations of the set onto itself and transformations that are not one-to-one. A transformation A of a set M

is called a one-to-one transformation of the set onto itself if not only to every element of M there corresponds a definite unique element of M (this is part of the definition of transformation) but if also for every element y of M there exists one and only one element x that goes over into y. In other words, a transformation A is one-to-one if the "equation" $xA = y$ has one and only one "solution" x in M for every y in M.

All the transformations of space considered here, reflections, rotations and translations, are one-to-one, since in these cases not only is there for every point X a point into which X goes over but there is also a unique point that goes over into X.

It is easy to give examples to the contrary; thus, the transformation of the set of numbers 1, 2, 3, 4, given by the table

$$\begin{pmatrix} 1 & 2 & 3 & 4 \\ 2 & 1 & 2 & 3 \end{pmatrix}$$

is not one-to-one, since in it no number goes over into 4. The transformation of the set of all natural numbers 1, 2, 3, $\cdots$, given by the table

$$\begin{pmatrix} 1 & 2 & 3 & 4 & 5 & 6 & \cdots \\ 1 & 1 & 2 & 2 & 3 & 3 & \cdots \end{pmatrix}$$

is also not one-to-one. Although here for every number n there is the number $2n$ that goes over into it, the number $2n$ is not the only one having this property, since $2n - 1$ also goes over into n. For transformations given by tables it is very easy to establish a criterion under with the transformation is one-to-one. For this it is obviously necessary and sufficient that the lower line of the table should contain every element of the set once and once only. Occasionally in mathematics one discusses transformations that are not one-to-one. For example, the great importance of the operation of projecting a space onto a plane is well known. This transformation is not one-to-one, because in it every point is the projection not of one but of a whole series of points of space. But in the majority of cases it is convenient to deal only with one-to-one transformations; these transformations, in particular, play a fundamental role when physical processes are considered under which the elements of the system in question are not merged with one another, not annihilated and not created.

Henceforth in talking of transformations we shall tacitly assume that they are one-to-one; they are also often called permutations, especially when we are dealing with transformations of a finite set.

For every (one-to-one) transformation A of a set M onto itself, we can easily define an inverse transformation A^{-1}. If A carries an arbitrary element x of M into y, then the transformation carrying y into x is called the inverse transformation to A and is denoted by A^{-1}. For example, if

$$A = \begin{pmatrix} 1 & 2 & 3 & 4 \\ 2 & 3 & 4 & 1 \end{pmatrix},$$

then

$$A^{-1} = \begin{pmatrix} 2 & 3 & 4 & 1 \\ 1 & 2 & 3 & 4 \end{pmatrix} = \begin{pmatrix} 1 & 2 & 3 & 4 \\ 4 & 1 & 2 & 3 \end{pmatrix};$$

if A is a rotation of space around an axis by an angle ϕ, then A^{-1} is the rotation around the same axis by the angle ϕ in the opposite direction, etc.

Occasionally it happens that the inverse transformation coincides with the given one. In particular, reflections with respect to a plane or a point in space have this property. So has the permutation

$$A = \begin{pmatrix} 2 & 1 & 4 & 3 \\ 1 & 2 & 3 & 4 \end{pmatrix}, \quad \text{since} \quad A^{-1} = \begin{pmatrix} 1 & 2 & 3 & 4 \\ 2 & 1 & 4 & 3 \end{pmatrix} = \begin{pmatrix} 2 & 1 & 4 & 3 \\ 1 & 2 & 3 & 4 \end{pmatrix}.$$

Note that we cannot speak of an inverse transformation for those that are not one-to-one, because an individual element may be such that no elements or several elements go over into it.

The general definition of symmetry.   In mathematics and its applications it is very rarely necessary to consider all transformations of a given set. The fact is that the sets themselves are rarely thought of as merely the collections of their elements completely disconnected from one another. This is natural, because the sets that are discussed in mathematics are abstract images of real collections, whose elements always stand in an infinite variety of interrelations with each other, and of connections with what is going on beyond the limits of the set in question. But in mathematics it is convenient to abstract from the major part of these connections and to preserve and take into account the most essential one. This compels us in the first instance to consider only such transformations of sets as do not destroy the relevant connections of one kind or another between their elements. These are often called admissible transformations or *automorphisms* with respect to the relevant connections between the elements of the set. For example, for points of space the concept of distance between two points is important. The presence of this concept forges a link between points which consists in the fact that any two points stand at a definite distance from one another. Transformations that do not destroy these connections are the same as those under which the distance between points remains unchanged. These transformations are called "motions" of space.

With the help of the concept of automorphism it is not difficult to give a general definition of symmetry. Suppose that a certain set M is given, in which definite connections between the elements are to be taken into account, and that P is a certain part of M. We say that P is symmetrical or invariant with respect to the admissible transformation A of M if A carries every element of P again into an element of P. Therefore, a symmetry of P is characterized by the collection of admissible transformations of the containing set M that transform P into itself. The concept of symmetry of a body in space falls entirely under this definition. The role of the set M is played by the whole space, the role of admissible transformations by the "motions," the role of P by the given body. The symmetry of P is therefore characterized by the collection of motions under which P coincides with itself.

The reflections, parallel shifts, and rotations of space around a given line that we have discussed are special cases of motions, because distances between points obviously remain unchanged under these transformations. A more detailed investigation shows that every motion of a plane is either a parallel shift or a rotation around a center or a reflection in a line or a combination of a reflection in a line with a parallel shift along that line. Similarly, every motion of space is either a parallel shift or a rotation around an axis or a spiral motion, i.e., a rotation around an axis combined with the shift along this axis, or a reflection in a plane combined with, possibly, a shift along the plane of reflection or a rotation around an axis perpendicular to this plane.

Parallel shifts, rotations, and spiral motions of space are called proper motions or motions of the first kind. The remaining "motions" (including reflections) are known as improper motions or motions of the second kind. In a plane, motions of the first kind are parallel shifts and rotations, whereas reflections in a line and reflections combined with a rotation or a translation are motions of the second kind.

It is easy to imagine how transformations that are motions of the first kind can be obtained as a result of a continuous motion of space or of a plane in itself. Motions of the second kind cannot be obtained in this way, because this is prevented by the mirror reflection that occurs in their formation.

One often says that the plane is symmetrical in all its parts or that all points of the plane are equivalent. In the strict language of transformations this statement means that every point of the plane can be superimposed on any other point by means of a suitable "motion."

The cases of symmetry of bodies or figures discussed previously are also comprised under the general definition of symmetry. For example, a body that is symmetrical with respect to a plane α comes into coincidence with itself on reflection in the plane α; a body that is symmetrical with respect to a center O comes into coincidence with itself under reflection in O. Therefore, the degree of symmetry of a body or of a spatial figure can be completely characterized by the collection of all motions of space of the first and second kind that bring the body or the figure into coincidence with itself. The greater and more diverse this collection of motions, the higher is the degree of symmetry of the body or figure. If, in particular, this collection contains no motions except the identity transformation, then the body can be called unsymmetrical.

The degree of symmetry of a square in a plane is characterized by the collection of motions of the plane that bring the square into coincidence with itself. But if the square coincides with itself, then the point of

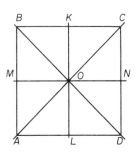

FIG. 7.

intersection of its diagonals must also coincide with itself. Therefore the required motions leave the center of the square invariant, and so they are either rotations around the center or reflections in lines passing through the center. From figure 7 we can easily read that the square *ABCD* is symmetrical with respect to the rotations around its center O by angles that are multiples of 90° and also with respect to reflections in the diagonals *AC*, *BD* and the lines *KL*, *MN*. These eight motions characterize the symmetry of the square.

The collection of symmetries of a rectangle reduces to a rotation around the center by 180° and a reflection in the lines that join the midpoints of opposite sides; and the set of symmetries of a parallelogram (figure 8) consists only of the rotations around the center by angles that are multiples of 180°, i.e., of reflections in the center and the identity transformation.

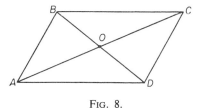

FIG. 8.

Previously we have given an algebraic example of symmetry; we mentioned that the concept of symmetry of a polynomial in several variables also has a meaning.

Let us discuss how the symmetry of a polynomial can be characterized.

We shall say that the permutation of the variables

$$A = \begin{pmatrix} x_1, x_2, \cdots, x_n \\ x_{i_1}, x_{i_2}, \cdots, x_{i_n} \end{pmatrix}$$

or briefly

$$A = \begin{pmatrix} 1, 2, \cdots, n \\ i_1, i_2, \cdots, i_n \end{pmatrix}$$

has been made in the polynomial $F(x_1, x_2, \cdots, x_n)$ if everywhere in the polynomial the letter x_1 has been replaced by x_{i_1}, x_2 by x_{i_2}, etc. The polynomial so obtained will be denoted by FA. Thus, if $F = x_1^2 - 2x_2 + x_3 - x_4$,

$$A = \begin{pmatrix} 1 & 2 & 3 & 4 \\ 3 & 1 & 4 & 2 \end{pmatrix}, \quad \text{then} \quad FA = x_3^2 - 2x_1 + x_4 - x_2.$$

The symmetry of the given polynomial is characterized by the collection of those permutations of the variables that, when carried out on the polynomial, leave it unchanged. For example, the symmetry of the polynomial $x_1^3 + 2x_2 + x_3^3 + 2x_4$ is characterized by the four permutations:

$$\begin{pmatrix} 1 & 2 & 3 & 4 \\ 1 & 2 & 3 & 4 \end{pmatrix}, \begin{pmatrix} 1 & 2 & 3 & 4 \\ 3 & 2 & 1 & 4 \end{pmatrix}, \begin{pmatrix} 1 & 2 & 3 & 4 \\ 1 & 4 & 3 & 2 \end{pmatrix}, \begin{pmatrix} 1 & 2 & 3 & 4 \\ 3 & 4 & 1 & 2 \end{pmatrix},$$

and the symmetry of the polynomial $x_1^3 + 2x_2 + x_3^3 + x_4$ is characterized by the two permutations:

$$\begin{pmatrix} 1 & 2 & 3 & 4 \\ 1 & 2 & 3 & 4 \end{pmatrix} \quad \text{and} \quad \begin{pmatrix} 1 & 2 & 3 & 4 \\ 3 & 2 & 1 & 4 \end{pmatrix}.$$

§3. Groups of Transformations

Multiplication of transformations. In studying properties of transformations it is easy to observe that certain transformations can be constructed from others. For example, a spiral motion is composed of a rotation around the axis and a shift along the axis. This process of forming new transformations from given ones is called multiplication of transformations. When we apply to an arbitrary element x of a set M some transformation A and then apply the transformation B to the new element xA, we obtain the element $(xA)B$. The transformation that carries x immediately into $(xA)B$ is called the product of A and B and is denoted by AB. Therefore, by definition, we have

$$x(AB) = (xA)B.$$

Example:

$$\begin{pmatrix} 1 & 2 & 3 & 4 \\ 2 & 3 & 4 & 1 \end{pmatrix} \begin{pmatrix} 1 & 2 & 3 & 4 \\ 3 & 4 & 1 & 2 \end{pmatrix} = \begin{pmatrix} 1 & 2 & 3 & 4 \\ 4 & 1 & 2 & 3 \end{pmatrix}.$$

Since the first permutation carries 1 into 2 and the second, 2 into 4, therefore the resulting permutation must carry 1 into 4, and so forth. Here are a few more examples:

$$\begin{pmatrix} 1 & 2 & 3 & 4 \\ 3 & 1 & 4 & 2 \end{pmatrix} \begin{pmatrix} 1 & 2 & 3 & 4 \\ 1 & 3 & 2 & 4 \end{pmatrix}^{-1} = \begin{pmatrix} 1 & 2 & 3 & 4 \\ 3 & 1 & 4 & 2 \end{pmatrix} \begin{pmatrix} 1 & 2 & 3 & 4 \\ 1 & 3 & 2 & 4 \end{pmatrix} = \begin{pmatrix} 1 & 2 & 3 & 4 \\ 2 & 1 & 4 & 3 \end{pmatrix};$$

$$\begin{pmatrix} 1 & 2 & 3 & 4 \\ 2 & 1 & 4 & 3 \end{pmatrix} \begin{pmatrix} 1 & 2 & 3 & 4 \\ 3 & 1 & 2 & 4 \end{pmatrix} = \begin{pmatrix} 1 & 2 & 3 & 4 \\ 1 & 3 & 4 & 2 \end{pmatrix}; \begin{pmatrix} 1 & 2 & 3 & 4 \\ 3 & 1 & 2 & 4 \end{pmatrix} \begin{pmatrix} 1 & 2 & 3 & 4 \\ 2 & 1 & 4 & 3 \end{pmatrix} = \begin{pmatrix} 1 & 2 & 3 & 4 \\ 4 & 2 & 1 & 3 \end{pmatrix}.$$

The last two examples show that the multiplication of transformations is, as we say, a noncommutative operation: Its result depends on the order of the factors. This is also easily verified for the multiplication of motions of a plane. Suppose, for example, that A is a rotation of the plane by 90° around the origin O, and B a parallel shift by a unit length along the x-axis.

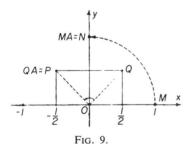

FIG. 9.

Let us find the image of O under the transformations AB and BA. By definition we have (figure 9)

$$O(AB) = (OA) B = OB = M,$$
$$O(BA) = (OB) A = MA = N,$$

i.e., $AB \neq BA$.

For a closer understanding of the geometric nature of the transformation BA, let us consider the point P. We have

$$P(BA) = (PB) A = QA = P,$$

i.e., the point P remains unchanged under the transformation BA. Starting out from this it is easy to show that BA is simply a rotation of the plane by 90° around P. Similarly

$$Q(AB) = (QA) B = PB = Q,$$

and AB is the rotation of the plane by 90° around Q.

The multiplication of motions of the plane or of space generally follows rather complicated rules. However, in two important cases the rules of multiplication are very simple. First, when we multiply rotations of a plane around one and the same point or rotations of space around one and the same line by the angles ϕ and ψ, then the resulting transformation is the corresponding rotation by the angle $\phi + \psi$. Second, when we multiply parallel shifts characterized by the vectors $\overrightarrow{MN}$ and $\overrightarrow{NP}$, then the product is also a parallel shift characterized by the vector $\overrightarrow{MP}$, i.e., the sum of the original vectors.

The very term "multiplication" of transformations points to a certain analogy between the multiplication of numbers and the multiplication of transformations. However, this analogy is incomplete. For example, for the multiplication of numbers we have the commutative law. But we have already seen that in the multiplication of transformations this law may be violated. The second fundamental law of arithmetic, namely the associative law, is completely preserved for transformations. In fact, for arbitrary transformations A, B, C of a set M we have the equation $A(BC) = (AB)\,C$.

For if m is an arbitrary element of M, then

$$m[A(BC)] = (mA)(BC) = [(mA)\,B]\,C = [m(AB)]\,C = m[(AB)\,C].$$

The associative law enables us, instead of speaking of the two products $A(BC)$ and (AB) of the transformations A, B, C, to speak only of the single product $A(BC) = (AB)\,C = ABC$. The same law shows that the product of four or more transformations does not depend on the distribution of parentheses.

Furthermore, among transformations there is the one that plays the role of the number 1, this is the identity or unit transformation E, which leaves every element of M unchanged. Clearly, $AE = EA = A$, whatever the transformation A.

We mention the following important fact: The product of one-to-one transformations is also one-to-one. For in order to find the element x of M that is carried by AB into a given element a, it is sufficient to find the element x_1 that is carried by B into a and then to find the element x_2 that is carried by A into x_1. Since $x_2(AB) = (x_2A)\,B = x_1A = a$, then x_2 is the required element x.

The product of a transformation A and the inverse transformation A^{-1} is the unit transformation; i.e.,

$$AA^{-1} = A^{-1}A = E.$$

This follows immediately from the definition of the inverse transformation.

The example discussed previously of the multiplication of a parallel shift of a plane and a rotation shows that properties of a product of transformations are not always easily discerned starting from properties of the factors. However, the product of the transformations of the form

$C = B^{-1}AB$ is an important exception: The properties of C are here very simply connected with the properties of A and B. For if an element m of M is carried by A into n, then the element mB, which is "shifted" by means of B, is carried by C into the "shifted" element nB.

Proof: $(mB)B^{-1}AB = mAB = nB$.

The transformation $B^{-1}AB$ is said to be obtained from A by transforming it by B or to be conjugate to A by means of B.

Let us transform, for example, a rotation P_0 of a plane around the point O by means of a translation V. By the preceding rule, in order to find the pairs of initial and final positions of points for the transformed motions $C = V^{-1}P_0V$, we have to shift by means of V the corresponding pairs of points for the transformation P_0.

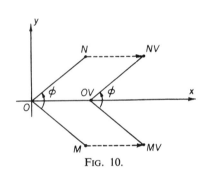

FIG. 10.

Since the point O in the rotation P_0 remains unchanged (figure 10), the point OV will remain unchanged under the transformation C. Furthermore, if a point M is carried by P into N, then the shifted point MV will be carried by C into the point NV. From figure 10 it is then clear that the transformation C is a rotation around the point OV by the same angle ϕ as the rotation P.

Similarly, it can be shown that if a translation of the plane characterized by the vector $\overrightarrow{MN}$ is transformed by means of a rotation P_0 by the angle ϕ, then we obtain again a translation of the plane, characterized by a different vector.

The previous rule for finding the transformation $B^{-1}AB$ can be formulated in a very elegant way, when the transformations are given by tables. Suppose that

$$A = \begin{pmatrix} 1 & 2 & \cdots & n \\ a_1 & a_2 & \cdots & a_n \end{pmatrix}, \quad B = \begin{pmatrix} 1 & 2 & \cdots & n \\ b_1 & b_2 & \cdots & b_n \end{pmatrix},$$

then

$$B^{-1}AB = \begin{pmatrix} b_1 & b_2 & \cdots & b_n \\ 1 & 2 & \cdots & n \end{pmatrix} \begin{pmatrix} 1 & 2 & \cdots & n \\ a_1 & a_2 & \cdots & a_n \end{pmatrix} \begin{pmatrix} 1 & 2 & \cdots & n \\ b_1 & b_2 & \cdots & b_n \end{pmatrix} = \begin{pmatrix} b_1 & b_2 & \cdots & b_n \\ b_{a_1} & b_{a_2} & \cdots & b_{a_n} \end{pmatrix} ;$$

i.e., in order to transform a permutation A by means of permutation B, we have to subject all the elements of the upper and of the lower row of A to the transformation specified by B. For example, if

$$A = \begin{pmatrix} 1 & 2 & 3 & 4 & 5 \\ 3 & 5 & 4 & 1 & 2 \end{pmatrix}, \quad B = \begin{pmatrix} 1 & 2 & 3 & 4 & 5 \\ 2 & 5 & 1 & 3 & 4 \end{pmatrix},$$

then

$$B^{-1}AB = \begin{pmatrix} 1B & 2B & 3B & 4B & 5B \\ 3B & 5B & 4B & 1B & 2B \end{pmatrix} = \begin{pmatrix} 2 & 5 & 1 & 3 & 4 \\ 1 & 4 & 3 & 2 & 5 \end{pmatrix} = \begin{pmatrix} 1 & 2 & 3 & 4 & 5 \\ 3 & 1 & 2 & 5 & 4 \end{pmatrix}.$$

Note that although in general the product of two transformations depends on the order of the factors, in individual cases the products AB and BA may be one and the same. Then the transformations A and B are called permutable or commuting. If $AB = BA$, then

$$B^{-1}AB = B^{-1}BA = A.$$

Thus, the transformation of a permutation by means of another one commuting with it does not change the given permutation.

Groups of transformations. The set of transformations that characterizes the symmetry of a certain figure cannot be arbitrary, it must necessarily have the following properties:

1. The product of two transformations belonging to the set also belongs to the set.

2. The identity transformation belongs to the set.

3. If a transformation belongs to the set, then the inverse transformation also belongs to the set.

These properties turn out to be very important for the study of transformations; in view of this, every set of one-to-one transformations of a set that has these three properties is called a *group of transformations* of M, independently of the fact whether this set characterizes the symmetry of a certain figure or not.

From the point of view of algebra, the properties 1–3 are very important, since they enable us, starting from certain transformations A, B, C, $\cdots$, belonging to a given set, to form various new transformations of the form $ABAC$, $A^{-1}BCB^{-1}$ and so forth, and the properties 1–3 guarantee that all the transformations so obtained do not carry us beyond the limits of the given set of transformations.

The number of transformations that form a group is called the *order of the group*; it may be finite or infinite. Accordingly, groups are divided into finite and infinite. Earlier we discussed the group of symmetry of a square in a plane. This group turned out to consist altogether of eight transformations. On the other hand, the infinite set of points A_i of the plane, illustrated in figure 11, is transformed into itself by the following motions of the plane: translations along the axis

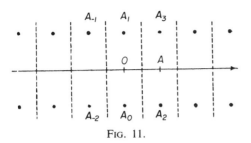

FIG. 11.

OA in either of the two directions by distances that are multiples of OA; reflections in the dotted lines; reflection in the axis OA. Hence it is clear that the group of symmetries of this figure is infinite.

The collection of transformations that preserve a certain object, i.e., characterize its symmetry, is always a group. This method of giving groups in the form of symmetry is one of most significance. Very important groups can be obtained by this principle. Of first importance among these are the groups of motions of a plane and of space. The symmetry groups of the regular polyhedra are also of great interest. It is known that in space there exist altogether five types of regular polyhedra (with 4, 6, 8, 12 and 20 faces). When we take an arbitrary regular polyhedron and consider all the motions of space that bring the given polyhedron into coincidence with itself, we obtain a group, namely the symmetry group of the polyhedron. If instead of all the motions we consider only the motions of the first kind that carry the polyhedron into coincidence with itself, then we obtain again a group that is part of the full group of symmetries of the polyhedron. This group is called the group of rotations of the polyhedron. Since in a superposition of the polyhedron with itself, its center is also superimposed on itself, all motions that occur in the group of symmetries of the polyhedron leave the center of the polyhedron unchanged and can therefore only be either rotations around axes passing through the center or reflections in planes passing through the center or, finally, reflections in such planes combined with rotations around axes passing through the center and perpendicular to these planes.

With the help of these remarks it is easy to find all the groups of symmetry and the groups of rotations of the regular polyhedra. In Table 1 we have given the order of the symmetry groups and the rotation groups of the regular polyhedron. All these groups are finite.

Table 1.

Number of faces	4	6	8	12	20
Order of the symmetry group	24	48	48	120	120
Order of the rotation group	12	24	24	60	60

Permutation groups. Of all the transformation groups, historically, the first to be studied in mathematics were the permutation groups of polynomials in the variables x_1, x_2, $\cdots$, x_n. The investigation of these groups is closely connected with the problem of solving equations of higher degrees by radicals. Obviously, the collection of all permutations of the variables that do not change the values of one or several polynomials in these variables is a group. Polynomials that are unchanged under all permutations of the variables are called symmetric polynomials. For example, $x_1 + x_2 + ... + x_n$ is a symmetric polynomial. Accordingly the set of all permutations of a given set of variables is called the *symmetric* group of the permutations of this set.

The number of the variables to be permuted is called the degree of the symmetric group. Instead of the permutations of the variables x_1, ..., x_n, we can simply consider the permutations of the numbers 1, 2, ..., n. Since every permutation of these numbers can be written in the form

$$\begin{pmatrix} 1 \ 2 \ ... \ n \\ a_1 a_2 \ ... \ a_n \end{pmatrix},$$

where a_1, a_2, $\cdots$, a_n are the numbers 1, 2, $\cdots$, n written in some order, the number of all permutations of n elements; i.e., the order of the symmetric group, is equal to $n! = 1 \cdot 2 \cdot 3 \cdots n$. This order increases very rapidly with n and the group of permutations of 10 variables is already of the order 3,628,800.

Let us consider the polynomial

$$F(x_1, \cdots, x_n)$$
$$= (x_2 - x_1)(x_3 - x_1) \cdots (x_n - x_1)(x_3 - x_2) \cdots (x_n - x_2) \cdots (x_n - x_{n-1}). \quad (1)$$

It is clear that every permutation of the variables either leaves the value of the polynomial F unchanged or changes its sign only. Permutations of the first kind are called even. Permutations that change the sign of F are called odd. The set of even permutations forms the symmetry group of the polynomial (1). It is called the alternating group.

The product of two even permutations is even, because even permutations form a group. The product of two odd permutations is an even permutation.

For if A and B are odd permutations, then

$$FAB = (FA)B = (-F)B = -(-F) = F.$$

In the same way it can be shown that the product of an even and an odd permutation is an odd permutation and that the permutation inverse to an even or an odd permutation is a permutation of the same parity.

An example of an odd permutation is

$$S = \begin{pmatrix} 1, 2, 3, \cdots, n \\ 2, 1, 3, \cdots, n \end{pmatrix},$$

which interchanges the elements 1 and 2.

Decomposition of permutations into cycles. In studying permutation groups it is very helpful to represent permutations in the form of products of so-called cycles. By definition, the symbol $(m_1, m_2, \cdots, m_k)$ denotes the permutation that carries m_1 into m_2, m_2 into m_3, $\cdots$, m_{k-1} into m_k, and m_k again into m_1 and leaves all the remaining elements of the set in question unchanged. For example, if we consider permutations of the numbers 1, 2, 3, 4, 5, then

$$(1, 2, 3, 4, 5) = \begin{pmatrix} 1 & 2 & 3 & 4 & 5 \\ 2 & 3 & 4 & 5 & 1 \end{pmatrix}, \quad (3, 5) = \begin{pmatrix} 1 & 2 & 3 & 4 & 5 \\ 1 & 2 & 5 & 4 & 3 \end{pmatrix}.$$

A permutation of the form $(m_1, m_2, \cdots, m_k)$ is called *cyclic* or a *cycle* of length k, and $m_1, m_2, \cdots, m_k$ are called the elements of the cycle. The unit permutation can be written in the form of cycles $(1) = (2) = \cdots$ of length 1. Cycles of length 2 are called *transpositions*. When we permute the elements of a cycle in cyclic order, we obtain the same permutation, for example $(1, 2, 3) = (2, 3, 1) = (3, 1, 2), (5, 6) = (6, 5)$.

It is easy to verify that cycles without common elements, for example $(2, 3)$ and $(1, 4, 5)$, are permutable, so that in multiplying such cycles we need not take the order of the factors in the product into account.

The significance of cycles in the general theory is based on the following theorem: Every permutation can be represented in the form of a product of cycles without common elements, and this representation is unique to within the order of the factors.

The proof of the theorem is immediately clear from the method of such a representation. Suppose that we wish to decompose the permutation.

$$A = \begin{pmatrix} 1 & 2 & 3 & 4 & 5 & 6 \\ 4 & 5 & 6 & 3 & 2 & 1 \end{pmatrix}.$$

We see that A carries 1 into 4, 4 into 3, 3 into 6, and 6 into 1. As a result we have a first factor $(1, 4, 3, 6)$. Of the remaining numbers we consider the 2 and note that A carries 2 into 5, 5 into 2. Therefore the second factor is $(2, 5)$. Since all numbers are now accounted for, we have

$$\begin{pmatrix} 1 & 2 & 3 & 4 & 5 & 6 \\ 4 & 5 & 6 & 3 & 2 & 1 \end{pmatrix} = (1, 4, 3, 6)\,(2, 5). \tag{2}$$

It is also possible to decompose permutations into cycles with common elements, but this is not unique. For example,

$$\begin{aligned} (a_1, a_2, \cdots, a_n) &= (a_1, a_2)(a_1, a_3) \cdots (a_1, a_n) \\ &= (a_2, a_3)(a_2, a_4) \cdots (a_2, a_n)(a_2, a_1). \end{aligned} \tag{3}$$

Let us show that every cycle of length 2 is an odd permutation. We have already seen this for the cycle $(1, 2)$. But every cycle (i, j) is $S^{-1}(1, 2)\,S$, where S is an arbitrary permutation

$$\begin{pmatrix} 1, 2, \cdots \\ i, j, \cdots \end{pmatrix}$$

carrying 1 into i and 2 into j. The permutation $S^{-1}(1, 2)\,S$ is an odd permutation, because $(1, 2)$ is odd, and S and S^{-1} are simultaneously even or odd.

According to (3) a cycle of length $m + 1$ can be represented as a product of m odd permutations. Therefore a cycle of length $m + 1$ is an odd permutation, when $m + 1$ is even, and even, when $m + 1$ is odd. This enables us to compute rapidly the parity of permutations whose decomposition into cycles is known. Specifically, the permutation

$$\begin{pmatrix} 1 & 2 & 3 & 4 & 5 & 6 \\ 4 & 5 & 6 & 3 & 2 & 1 \end{pmatrix}$$

is even since by (2) it is the product of two odd permutations.

Subgroups. A part of a group that is itself a group is called a *subgroup* of the given group. Thus, the alternating group of permutations of the variables x_1, x_2, $\cdots$, x_n is a subgroup of the symmetric group. The set of proper motions of a plane is a group, which is a subgroup of the group of all proper and improper motions of the plane.

From the formal point of view the unit (identity) transformation forms a subgroup by itself. Equally, every group can be regarded as a subgroup of itself. But almost always groups contain many other subgroups apart from these trivial ones. A knowledge of all the subgroups of a given group gives a fairly complete idea of the internal structure of the group.

One of the most extensively used methods of forming subgroups is that of giving so-called generators of the subgroup.

Let A_1, A_2, $\cdots$, A_m be arbitrary transformations belonging to a group G. The set H of all transformations that can be obtained by multiplying the given permutations and their inverses among each other arbitrarily often is a group. For the unit transformation belongs to this set, since it can be represented in the form $A_1 A_1^{-1}$. Next, if the transformations B and C can be represented as such products, then by multiplying these products we obtain the required representation for BC. Finally, if B is expressed as such a product, e.g., $B = A_1^{-1} A_2 A_1 A_1 A_2^{-1}$, then B^{-1} can also be represented in the required product form since $B^{-1} = A_2 A_1^{-1} A_1^{-1} A_2^{-1} A_1$.

The group H obviously is a subgroup of G and is called the subgroup generated by the transformations A_1, $\cdots$, A_m, and these transformations A_1, $\cdots$, A_m are called the generators of H. It can happen that H coincides with G, and in this case A_1, $\cdots$, A_m are called the generators of the whole group G. It is easy to verify in examples that one and the same subgroup may be generated by several distinct systems of generators.

A subgroup generated by a single transformation A is called cyclic. Its elements are transformations

$$E, A, AA, AAA, \cdots, A^{-1}, A^{-1}A^{-1}, A^{-1}A^{-1}A^{-1}, \cdots,$$

which are naturally called the powers of A. In fact:

$$E = A^0, A = A^1, AA = A^2, \cdots, A^{-1}A^{-1} = A^{-2},$$
$$A^{-1}A^{-1}A^{-1} = A^{-3}, \cdots.$$

It is easy to show, as in the ordinary arithmetic, that

$$A^m A^n = A^{m+n} \quad \text{and} \quad (A^m)^n = A^{mn}. \tag{4}$$

A transformation is called *periodic* if some positive power of it is the identity transformation. The smallest positive exponent of the power to which a periodic transformation must be raised in order to obtain the identity is called the *order of the transformation*. We also say that a transformation that is not periodic is of infinite order.

Let us consider some examples. Let A be a rotation of the plane around the point O by $360°/n$, where n is a given positive integer greater than 1. Then A^2 is the rotation by the angle $2(360°/n)$, A^3 the rotation by $3(360°/n)$, A^{n-1} the rotation by $(n-1)(360°/n)$, and A^n the rotation by $360°$, i.e., the identity transformation. This shows that the rotation by $360°/n$ is a periodic transformation of order n.

Let A be a shift of the plane along a certain line. Then A^2, A^3, $\cdots$ are also shifts along this line by twice the distance, three times, and so forth. Therefore, no positive power of A is the identity transformation, and the order of A is infinite.

The elements of the cyclic group generated by A are

$$\cdots, A^{-2}, A^{-1}, E, A, A^2, \cdots. \tag{5}$$

If A is a transformation of infinite order, then all the transformations in the sequence (5) are distinct, and the group is infinite. For otherwise we would have an equation of the form $A^k = A^l \, (k < l)$, hence $A^{l-k} = E$ $(l - k > 0)$, and this contradicts the fact that A is not periodic.

Now let us assume that A is a periodic transformation of order m. Then

$$A^m = E, A^{m+1} = A, A^{m+2} = A^2, \cdots, A^{m-1} = A^{-1}, A^{m-2} = A^{-2}, \cdots;$$

i.e., the sequence (5) consists precisely of the transformations E, A, A^2, $\cdots$, A^{m-1} repeated periodically. They are distinct from one another, since if we had $A^k = A^l \, (0 \leqslant k < l < m)$, then we would have $A^{l-k} = E$ $(0 < l - k < m)$, in contradiction to the choice of m. Consequently, the cyclic group generated by a transformation of order m contains precisely m distinct transformations.

A group in which all elements commute with one another is called commutative or *Abelian*, in honor of the Norwegian mathematician Abel who discovered the great importance of these groups for the theory of equations

The formulas (4) show that the powers of one and the same transformation always commute with one another: $A^m A^n = A^n A^m = A^{m+n}$. Therefore, cyclic groups are always Abelian.

In the arithmetic of numbers, apart from the multiplication, great importance also attaches to the operation of division. In the theory of groups, as a consequence of the fact that multiplication need not be commutative, we have to speak of two divisions: on the right and on the left. The solution of the equation $Ax = B$, where A, B are given transformations, is naturally called the right quotient, and the solution of the equation $yA = B$ is the left quotient on division of B by A. When we multiply both sides of the first equation by A^{-1} on the left and both sides of the second by A^{-1} on the right, we obtain: $x = A^{-1}B$, $y = BA^{-1}$. Thus, we can regard $A^{-1}B$ or BA^{-1} as the "quotient" of the transformations B and A.

In numerous examples we have seen that in groups, in general, $AB = BA$. The "quotient" $(AB)(BA)^{-1}$ or $(BA)^{-1}(AB)$ can be taken to be a "measure" of the noncommutativity of the permutations A and B. The second of these expressions, namely $(BA)^{-1}(AB) = A^{-1}B^{-1}AB$, is called the *commutator* of A and B and denoted by (A, B). From the formula

$$(A, B) = A^{-1}B^{-1}AB$$

it follows that the commutator can be represented as the quotient on "division" of the conjugate transformation $B^{-1}AB$ by A.

For example, if A is a translation of the plane, then a conjugate transformation is also a translation, and the quotient of two translations obviously is a translation. Therefore the commutator of a translation and an arbitrary motion of the plane is a translation. Now let A be a rotation around a certain point O by an angle ϕ, and B a rotation or translation. Then the conjugate transformation is again a rotation by the angle ϕ, but around another point O'. Therefore the commutator (A, B) in this case is the product of a rotation around O by the negative angle ϕ and the rotation around O' by the positive angle ϕ. From figure 12 it is clear that the resulting transformation is a translation by the distance $2 \cdot \overline{OO'}$ $\sin \phi/2$ in the direction at an angle $\pi/2 - \phi/2$ with the segment $\overline{OO'}$.

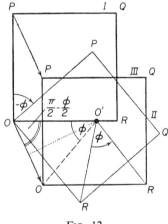

FIG. 12.

Thus, we have arrived at the interesting fact that for a plane the commutator of any two motions of the first kind is a parallel shift or the identity transformation. Since $(A, B) = E$ signifies that $AB = BA$, every noncommutative group of motions of the first kind in the plane contains parallel shifts.

The subgroup generated by the commutators of all elements of a group G is called the *commutator subgroup* or *derived group* of G. Recalling the relevant definitions, we can say that the derived group of G consists precisely of those elements that can be represented in the form of products of commutators. Since in a plane the commutator of any two motions of the first kind is a parallel shift, and products of parallel shifts are again parallel shifts, we can say that the derived group of the group of motions of the first kind in a plane consists only of parallel shifts.

The derived group of an Abelian group consists only of the identity transformation, because from $AB = BA$ it follows that $(A, B) = E$.

Let G be the symmetric group of all permutations of the numbers $1, 2, \cdots, n$. Let us show that the commutator of any two permutations A, B is always an even permutation. Indeed, the permutation AB, BA and consequently also $(BA)^{-1}$, always have the same parity; but then the commutator $(A, B) = (BA)^{-1}(AB)$, as the product of permutations of equal parity, is an even permutation.

We have seen that the derived group of the symmetric group consists of even permutations only. It is easy to show that in fact it coincides with the whole alternating group.

The derived group of a group G is often denoted by G'; the derived group of the derived group of G is called the second derived group of G and is denoted by G''. Repeating this process we can define the derived group of arbitrary order of a group G.

If among the derived groups of a group G at least one (hence all subsequent ones) consists of the identity transformation only, then the group G is called *solvable*. This name has arisen in the theory of equations, where *solvability* of a group corresponds to solvability of an equation by radicals. The group of motions of the first kind in a plane is solvable, because its second derived group is the identity. The symmetric groups of degree 2, 3, and 4 are solvable, because their first, second, and third derived groups, respectively, are the identity. In contrast, the symmetric groups of degree 5 and higher are not solvable, since it can be shown that their second derived group coincides with the first and is different from the identity.

§4. Fedorov Groups (Crystallographic Groups)

The symmetry groups of finite plane figures. As we have already seen, the symmetry of a figure or a body is characterized by the group of motions of the plane or space that bring the figure into coincidence with itself.

The symmetry groups of finite plane figures are the easiest of these groups to find.* For suppose that a finite plane figure is given and that

---

* Finiteness is to be understood in the sense that the whole figure lies in a bounded part of the plane, for example, within a certain circle.

this figure is brought into coincidence with itself by a certain motion A. Then the center of gravity O of the figure must also be brought into coincidence with itself by A; i.e., A is either a rotation around O or a reflection in a line passing through O. Thus, the symmetry group of an arbitrary finite plane figure can consist only of rotations around its center of gravity and of reflections in lines passing through this center.

Let us discuss a number of different cases that can arise in studying symmetry groups of a finite plane figure.

1. The symmetry group K_1 consists only of the unit (identity) transformation. This is the symmetry group of an arbitrary unsymmetric figure (figure 13).

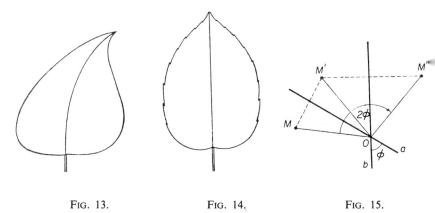

FIG. 13.   FIG. 14.   FIG. 15.

2. The symmetry group K_2 consists of the unit transformation and a reflection in a single line (figure 14).

Observe that if a group K contains reflections in two lines a, b passing through O and forming an angle ϕ between them, then the product of these reflections is a rotation around O by the angle 2ϕ (figure 15). Hence it is clear that the group K_2 is the only symmetry group not containing rotations.

3. The symmetry group K_3 consists of rotations only, and among them there are no rotations by arbitrarily small angles. In that case there is among the rotations of K_3 a rotation by a smallest positive angle. Let this angle be $\alpha°$. We shall show that every other rotation contained in the group is a multiple of $\alpha°$. We denote the number of degrees in such a rotation by β and find the integer h, for which $h\alpha° \leqslant \beta° < (h + 1)\alpha°$, so that $0 \leqslant \beta° - h\alpha° < \alpha°$. The group K_3, which contains rotations by $\alpha°$

and $\beta°$, also contains a rotation by $\beta° - h\alpha°$. But $0 \leqslant \beta° - h\alpha° < \alpha°$, and the group does not contain positive rotations by less than $\alpha°$. Therefore $\beta° - h\alpha° = 0$; i.e., $\beta° = h\alpha°$. In particular, since the group K_3 contains the rotation by 360°, we have $n\alpha° = 360°$, for a certain *integer n*, so that $\alpha° = 360°/n$.

Thus, the group K_3 consists of the rotations by 0°, 360°/n, 2(360°/n), $\cdots$, $(n-1)(360°/n)$. By giving to *n* the values 2, 3, 4, $\cdots$, we obtain all types of groups K_3. In figure 16 we have illustrated figures whose symmetry groups consist only of rotations around O by angles that are multiples of 360°/n, for $n = 19$, $n = 3$.

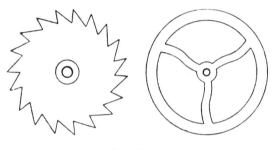

4. The symmetry group K_4 consists only of rotations, but contains arbi-

FIG. 16.

trarily small rotations. Then a rotation of arbitrary angle α can be made up with any degree of accuracy from rotations belonging to the group K_4. Of course, we are interested here only in closed figures, i.e., such that include their boundary points (see Chapter XVII, §9). It is easy to establish that for closed figures the group K_4 contains rotations by any angle ϕ. This is the case of directed circular symmetry (illustrated by a circumference, a circular

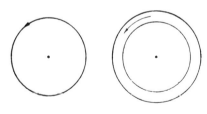

FIG. 17.

annulus, and so forth) provided with a definite sense of direction (figure 17). Here not only the figure must come into coincidence under all admissible transformations but also its directional sense, and this excludes reflections in a line.

It now remains to discuss the mixed cases when the symmetry group K contains both rotations and reflections. Without going into the proof, which would be quite simple, we only state the result: Apart from the groups K_1 through K_4 there only exist groups of the following two types.

5. The symmetry group K_5 consists of *n* reflections in lines passing

through O and dividing the plane into $2n$ equal angles, and of rotations by angles that are multiples of $360°/n$. For example, regular n-gons (figure 18) have such a symmetry group.

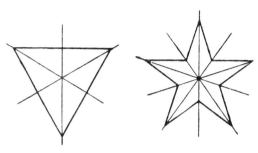

FIG. 18.

6. The symmetry group K_6 consists of all rotations around O and of reflections in all lines passing through the center O. This is the case of complete circular symmetry, which can be illustrated by the symmetry of an unorientated circumference or unorientated annulus.

Symmetry groups of infinite plane figures. The task of finding all possible symmetry groups of infinite plane figures is more complicated. Of course, in practice we are never given a whole infinite plane. However, often a piece of a plane happens to be covered with figures so fine that compared with them the piece appears to be infinitely large. For example, the smoothly polished plane surface of a piece of steel is covered with figures of microscopic dimensions. The regularity of these figures is an indication of the internal homogeneity of the structure of the metal.

Other examples are patterns on wallpaper or tapestries with repeating figures. The art of making such patterns, the art of ornamentation, has been widespread among most nations from antiquity to the present day. In figure 19 we have a specimen of an Egyptian ceiling pattern which dates from the middle of the second millenium B.C.

FIG. 19.

In discussing the symmetry groups of finite figures we were compelled to distinguish between the cases 1, 2, 3, and 5, when the symmetry group does not contain rotations by arbitrarily small angles, and the cases 4 and 6 when the group contains such rotations. In studying the symmetry groups of infinite figures, especially in the three-dimensional case, this division into discrete groups and groups with arbitrarily small transformations becomes even more important. Therefore we shall begin with a more careful discrimination between these cases.

A group of motions of a plane is called *discrete* if every point of the plane can be enclosed in a circle such that every motion of the group either leaves the point unchanged or carries it outside the chosen circle.

In the same way we can find all discrete groups of motions of a plane. All these groups are symmetry groups of plane figures. It is natural here to distinguish three types of discrete symmetry groups:

I. There exists a point in the plane that remains fixed under all symmetry transformations. This type contains the groups K_1, K_2, K_3 and K_5 of our previous list.

II. There are no fixed points in the plane, but there exists a line that is carried into itself under all transformations of the group. This line is called an axis of the group. Symmetry groups of this type occur in ornaments that are set out in the form of an infinite strip (border). Of such groups there exist altogether seven:

1. The symmetry group L_1 consisting only of translations by distances that are multiples of a certain segment a.

2. The group L_2, which is obtained from L_1 by adjoining the rotation by 180° around one of the points on the axis of the group.

3. The group L_3, which is obtained from L_1 by adjoining the reflection in a line perpendicular to the axis of the group.

4. The group L_4, which is obtained from L_1 by adjoining the reflection in the axis.

5. The group L_5, which is obtained from L_1 by adjoining a translation by $a/2$ combined with a reflection in the axis.

6. The group L_6, which is obtained from L_4 by adjoining the reflection in a certain line perpendicular to the axis of the group.

7. The group L_7, which is obtained from L_5 by adjoining the reflection in some line perpendicular to the axis of the group.

Table 2 gives examples of "borders" corresponding to each of the groups L_1 through L_7.

Table 2.

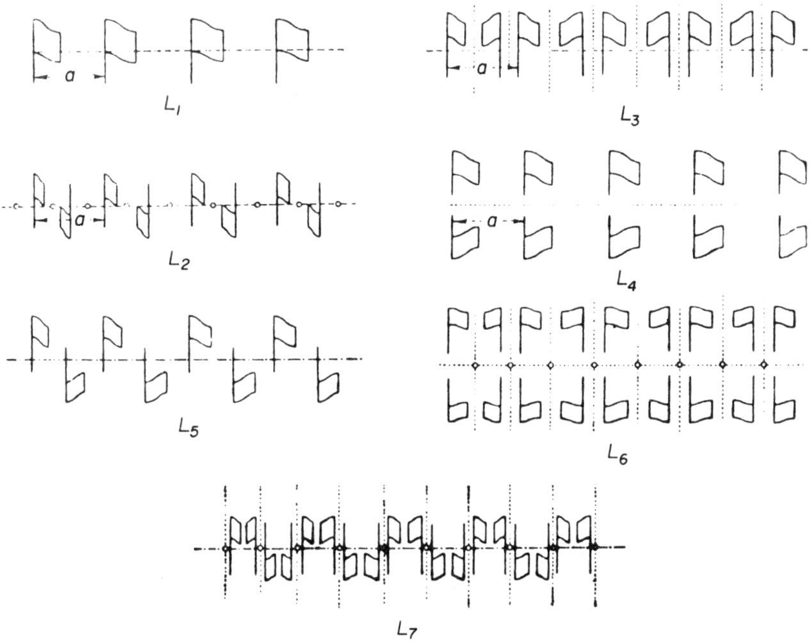

III. There exists neither a point nor a line in the plane that is carried into itself under all the transformations of the group. Groups of this type are called *plane Fedorov groups*. They are the symmetry groups of infinite plane ornaments. There are altogether 17 of them: five consist of motions of the first kind only, and twelve of motions of the first and second kind.

In Table 3 we have given examples of ornaments corresponding to each of the seventeen plane Fedorov groups; every group consists of precisely those motions that carry an arbitrary flag drawn in the diagram into any other flag of the same diagram.

It is interesting to note that the masters of the art of ornamentation have in practice discovered ornaments with all possible symmetry groups; it fell to the theory of groups to prove that other forms do not exist.

Crystallographic groups. In 1890 the eminent Russian crystallographer and geometer E. S. Fedorov solved by group-theoretical methods one of the fundamental problems of crystallography: to classify the regular systems of points in space. This was the first example of a direct application of the theory of groups to the solution of an important problem in natural science and made a substantial impact on the development of the theory of groups.

Table 3.

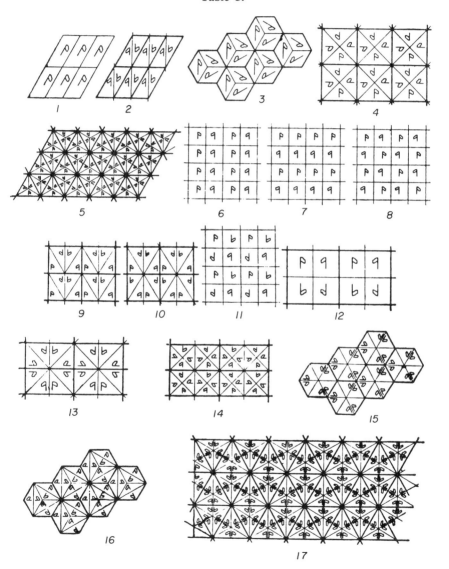

A crystal has the peculiarity that the atoms of which it is composed form in a certain sense a regular system in space. Let us consider the motions of space that carry the points of the system again into points of the system. These motions form a group whose properties enable us to formulate more accurately the concept of a regular point system.

A system of points in space is called a *regular spatial point system* if

1. Every point of the system can be carried into every other point by a motion which brings the system into coincidence with itself;

2. No sphere of finite radius contains infinitely many points of the system;

3. There exists a positive number r such that every sphere of radius r contains at least one of the points of the system.

The problem of studying the structure of crystals turns out to be closely connected with the classification of regular spatial point systems, which in turn is connected with the classification of discrete groups of motion in space. Just as in the case of a plane, a group of motions H of space is called *discrete* if we can describe around every point A of space a sphere of positive radius r with center at A such that every motion occurring in H either leaves the point A fixed or else carries it outside the sphere. It can be shown that the set of motions of space that bring a given regular spatial system of points into coincidence with itself is necessarily a discrete group and that all the points of the system can be obtained from any given point of the system by subjecting it to all the transformations of the group. Conversely, if a certain discrete group H is known, then by taking arbitrary point A in space and shifting it by means of all possible motions occuring in H we obtain a system of points that has the properties 1 and 2. By means of simple additional conditions we can single out from the discrete groups those that for suitably chosen points A give us, in fact, regular spatial point systems, i.e., systems of points with all three properties (1), (2), (3). Such discrete groups are called *Fedorov* or *crystallographic* groups. From what has been shown it is clear that the task of finding the Fedorov groups is the first and most important step in the study of regular spatial point systems. Now for the purposes of natural science it has proved necessary to consider not merely groups consisting of proper motions only, but also those that contain proper and improper motions, i.e., including reflections. The number of Fedorov groups formed from proper motions only is significantly smaller than that of Fedorov groups composed of proper and improper motions, and only in the latter more general case does the variety of regular spatial point systems obtained exhaust the whole diversity of structure of crystals occurring in nature.

It is interesting to note that, in contrast to the plane case we have treated previously, only the theory of groups enable us to analyze this exceptionally large number of possibilities.

The complexity of the space problem compared with the plane one is clear from Table 4.

Table 4. Number of Spatial Fedorov Groups

Groups containing only motions of the first kind	65
Groups containing also motions of the second kind . . .	165
Total	230

Even today a detailed derivation and enumeration of all Fedorov groups in space requires several dozen pages of text. We shall therefore restrict ourselves to reporting these quantitative results and refer the interested reader to the special literature.*

The modern developments of crystallography have made it necessary to introduce a further extension of the concept of symmetry. Such new possibilities in methods are outlined in the book on crystallography by Academician A. V. Šubnikov "Symmetry and anti-symmetry of finite figures," Akad. Nauk SSSR, Moscow, 1951.

§5. Galois Groups

The results explained on the preceding pages give a certain idea of the role played by the theory of groups in the solution of the problem of classification of crystals. However, this problem was not the motivation for the creation of the theory of groups. Approximately a hundred years earlier Lagrange noticed a connection between the symmetry properties of the roots of an algebraic equation and the possibility of solving the equation by radicals. This connection was the object of deep investigations by the famous mathematicians Abel and Galois in the first thirty years of the 19th century; and so they arrived at a solution of the celebrated problem of conditions for the solvability of algebraic equations by radicals. This solution was based entirely on a subtle investigation of properties of permutation groups and was, in fact, the beginning of the theory of groups.

The study of connections between properties of algebraic equations and

---

* A detailed account of the plane discrete groups of motion of the first kind is contained in the book by D. Hilbert and S. E. Cohn-Vossen, *Geometry and the imagination*, Chelsea, New York, 1952. A derivation of the crystallographic groups in space can be found in the fundamental article by E. S. Fedorov, *Symmetry of regular systems of figures* (Collected works, Akad. Nauk SSSR, Moscow, 1949).

properties of groups nowadays forms the object of an extensive theory, which is known as Galois theory.

An idea of the history of the problem and of the significance of Galois theory was given in Chapter IV. However, since Galois theory has played such a decisive role in the development of group theory, we shall repeat here the basic facts of the theory, but in a form more convenient for the purpose of throwing light on the theory of groups itself. The proofs of these facts require many auxiliary concepts and will be omitted.

The group of an algebraic equation.   Consider an equation of degree n

$$x^n + a_1 x^{n-1} + \cdots + a_n = 0, \tag{6}$$

whose coefficients are assumed to have given values; for example, certain complex numbers. The set of all quantities that can be obtained from the coefficients of the equation by means of a finite number of the operations of addition, subtraction, multiplication, and division is called the *ground field* or domain of rationality of the equation.

For example, if the equation has rational coefficients, then the domain of rationality consists of all rational numbers; if the equation has the form $x^2 + \sqrt{2}\,x + 1 = 0$, then the domain of rationality consists of all numbers of the form $a + b\sqrt{2}$, where a, b are rational numbers.

We shall now denote the roots of this equation by $\xi_1, \cdots, \xi_n$. The set of quantities that can be obtained by means of a finite number of the operations of addition, subtraction, multiplication, and division starting out from the roots $\xi_1, \cdots, \xi_n$, is called the *splitting field* of the equation. For example, the splitting field of the equation $x^2 + 1 = 0$ is the set of complex numbers $a + bi$ with rational a, b; and the splitting field of the above equation $x^2 + \sqrt{2}\,x + 1 = 0$ is the set of numbers of the form $a + bi + c\sqrt{2} + di\sqrt{2}$, where a, b, c, d are rational numbers.

By Viète's formulas the coefficients of the equation are obtained from its roots by means of the operation of addition and multiplication, therefore the splitting field of an equation always contains its ground field. Sometimes these fields coincide.

A one-to-one mapping A of the splitting field onto itself is called an automorphism of the splitting field with respect to the ground field, if for every pair of elements of the splitting field their sum goes over into the sum of their images, and their product into the product, and every element of the ground field goes over into itself. These properties can be described by the formulas

$$(a + b)A = aA + bA, \quad (ab)A = aA \cdot bA, \quad \alpha A = \alpha$$
$$(a, b \in K, \quad \alpha \in P), \tag{7}$$

where aA is the image of a; that is, Aa is the element into which a goes over under the mapping A; P is the ground field; and K is the splitting field.

By the general principle, explained in §2, the set of all automorphisms of the splitting field relative to the ground field is a group. This group is called the *Galois group* of the given equation.

To form a more concrete idea of the Galois group, let us note first of all that the automorphisms of the Galois group carry a root of the given equation into another root. For if x is a root of the equation (6), then operating on both sides of the equation with the automorphism A and using the properties (7) we obtain

$$(xA)_n + a_1A(xA)^{n-1} + \cdots + a_nA = 0 \cdot A;$$

since $0 \cdot A = 0$, $a_iA = a_i$, we thus have

$$(xA)^n + a_1(xA)^{n-1} + \cdots + a_n = 0,$$

as required. Consequently, every automorphism A effects a definite permutation of the set of roots of the equation. On the other hand, when we know this permutation, we also know the automorphism, because all the elements of the splitting field are obtained from the roots by means of arithmetical operations only. This shows that instead of the automorphism group we can also consider the group of permutations of the roots of the equation corresponding to it. Hence it follows, in particular, that all Galois groups are finite.

To find the Galois group of a given equation is usually a complicated problem, and only in special cases is the task comparatively easy. Let us consider, for example, the equation (6) with literal coefficients $a_1, \cdots, a_n$. The ground field of this equation is formed by the rational fractions of the coefficients, i.e., the fractions whose numerators and denominators are polynomials in $a_1, \cdots, a_n$. The splitting field is formed by the rational fractions of the roots of the equation $\xi_1, \cdots, \xi_n$, which are connected with the coefficients by the formulas

$$\begin{aligned}
-a_1 &= \xi_1 + \xi_2 + \cdots + \xi_n, \\
a_2 &= \xi_1\xi_2 + \xi_1\xi_3 + \cdots + \xi_{n-1}\xi_n, \\
&\cdots\cdots\cdots\cdots\cdots\cdots\cdots\cdots\cdots\cdots\cdots\cdots\cdots\cdots\cdots\cdots\cdots\cdots\cdots \\
(-1)^n a_n &= \xi_1\xi_2 \cdots \xi_n.
\end{aligned} \tag{8}$$

Since the equation (6) is "general," we can regard its roots as independent variables. Then every permutation of these roots gives rise to an automorphism of the splitting field. The formulas (8) show that under every such automorphism the coefficients go over into themselves and, together with them, all rational fractions formed from them also go over into themselves. Thus, the Galois group of the general equation of degree n is essentially the symmetric group of all permutations of n letters.

We can also indicate equations with numerical coefficients that have the symmetric group for their Galois group. For example, it has been shown that the Galois group of the equation

$$1 - \frac{n}{1} x + \frac{n(n-1)}{1 \cdot 2} \cdot \frac{1}{1 \cdot 2} x^2 - \frac{n(n-1)(n-2)}{1 \cdot 2 \cdot 3} \cdot \frac{1}{1 \cdot 2 \cdot 3} x^3$$
$$+ \cdots + (-1)^n \frac{1}{1 \cdot 2 \cdots n} x^n = 0 \qquad (9)$$

for arbitrary n is the symmetric group of permutations of degree n.

General methods are known for constructing equations with any preassigned group as Galois group, but under the condition that the coefficients can be taken to be arbitrary. However, if a construction is required for equations that are required to have rational coefficients, then this is known at present only for individual types of groups. Remarkable progress in this direction has been made by the Soviet mathematician I. R. Šafarevič, who has found methods of constructing equations with rational coefficients having an arbitrary preassigned solvable group as Galois group. In general, however, this problem is still unsolved.

Solvability of equations by radicals.   The Galois group of an equation characterizes, as is clear from the definition, the intrinsic symmetry of the roots of the equation. All the most fundamental problems concerning the possibility of reducing the solution of a given equation to that of equations of lower degree and also many other problems can be formulated as problems on the structure of the Galois group; and the Galois group of every equation of degree n is a certain group of permutations of degree n, i.e., an entirely finite object, in which all relationships, at least theoretically, can be found by means of trial and error.

The study of the Galois group is a valuable method of solving problems related to algebraic equations of higher degree. For example, it can be shown that an equation is solvable by radicals if and only if its Galois group is solvable (for the definition of a solvable group see §3). We have already mentioned that the symmetric groups of degree 2, 3, and 4 are solvable. This is in complete accord with the well-known fact that equations of degree 2, 3, and 4 are solvable by radicals. The Galois groups of the "general" equations of degree 5, 6, and so forth, are the symmetric groups of the same degrees. But these groups are not solvable. Hence it follows that the general equations of degree higher than 4 cannot be solved by radicals.

Among the equations that are not solvable by radicals there are also the equations (9) for $n > 4$, because their Galois group is the symmetric group.

§6. Fundamental Concepts of the General Theory of Groups

In the 19th century the theory of groups arose primarily as the theory of transformation groups. However, in the course of time it became more and more clear that the most significant of the results obtained depend only on the fact that transformations can be multiplied and that this operation has a number of characteristic properties. On the other hand, objects were found having nothing to do with transformations, on which a certain operation can be carried out (for the time being we shall call it multiplication) having the same properties as in transformation groups and to which the main theorems of the theory of transformation groups were applicable. As a result, the concept of a group was applied at the end of the last century not only to systems of transformations, but also to systems of arbitrary elements.

General definition of a group. The following definition of a group is generally accepted nowadays: Suppose that with every pair of elements a, b, taken in a definite order, of an arbitrary set G another well-defined element c of the same set is associated. Then we say that an operation is given on the set G. It is customary to introduce special names for operations: addition, multiplication, composition. The element of G that corresponds to the pair a, b is then called the sum, product, and compositum of the elements a, b, and is denoted by $a + b$, ab, $a * b$, respectively. The names "addition" or "multiplication" are used even in cases when the operation in question has nothing to do with the ordinary operations of addition and multiplication of numbers.

A set G together with an operation $*$ defined on it is called a *group* with respect to this operation if the following group axioms are satisfied:

1. For any three elements x, y, z of G

$$x * (y * z) = (x * y) * z \text{ (associative law)}.$$

2. Among the elements of G there exists an element e such that for every x of G

$$x * e = e * x = x.$$

3. For every element a of G there exists an element a^{-1} of G such that

$$a * a^{-1} = a^{-1} * a = e.$$

The element e, described in axiom 2, is called the neutral element of the group, and the element a^{-1} whose existence is postulated by axiom 3 is called the inverse of a. If the group operation is called addition or multiplication, then the neutral element is called the zero or the unit element, respectively, and the group axioms assume the form

(1) $x + (y + z) = (x + y) + z$,   (1) $x(yz) = (xy)z$,
(2) $x + 0 = 0 + x = x$,   (2) $xe = ex = x$,
(3) $x + (-x) = (-x) + x = 0$,   (3) $xx^{-1} = x^{-1} x = e$.

In the preceding sections we have discussed many examples of groups. The elements of these groups were transformations, and the group operation was multiplication of transformations. The set of numbers $0, \pm 1, \pm 2, \cdots$ also forms a group under the operation of addition, because the sum of integers is again an integer and addition of integers is associative; the neutral element is the integer 0 and for every number a of our set there is the opposite number $-a$. Another example of a group is the set of all real numbers (except 0) under multiplication. For the product of any two real numbers different from zero is a real number different from zero; the operation of multiplication of real numbers is associative; the neutral element is the number 1; and every nonzero real number a has the inverse $a^{-1} = 1/a$. The number of similar examples could be increased indefinitely.

Although the group operation may be called by different names, let us agree henceforth to call it almost always multiplication. The concepts of a subgroup, of powers of an element of a group, of a cyclic group, of the order of an element of a group are defined exactly as for transformation groups and we shall not repeat this here (see §3). We only mention that an element a of a group is called conjugate to an element b if there is an element x in G such that $b = x^{-1}ax$. Since $a = a^{-1}aa$, every element of a group is conjugate to itself. Furthermore, from $b = x^{-1}ax$ it obviously follows that $xbx^{-1} = a$ or $a = (x^{-1})^{-1}bx^{-1}$. i.e., if a is conjugate to b, then b is conjugate to a. Finally, if $b = x^{-1}ax$ and $c = y^{-1}by$, then

$$c = y^{-1}x^{-1}axy = (xy)^{-1}a(xy).$$

Therefore two elements conjugate to a third are conjugate to each other. These properties show that all elements of a group split into disjoint classes of conjugate elements. Also, if the group is commutative, i.e., $xy = yx$ for every x and y, then conjugate elements coincide and every class of conjugate elements consists of one element only.

Isomorphisms. Two aspects can be distinguished in the concept of a group. In order to give a group we have to: (1) indicate what objects are its elements and (2) indicate the law of multiplication of the elements. Accordingly, the study of group properties can be carried out from distinct points of view. We can study connections between individual properties of elements of the group and of sets of them and their properties in relation to the group operation. This point of view is often adopted in studying individual concrete groups; for example, the group of motions of space or a plane. However, we can also study those group properties that are entirely expressed in terms of properties of the group operation. This point of view is characteristic for the *abstract* or *general* theory of groups. It can be expressed more clearly by means of the concept of isomorphism.

Two groups are called *isomorphic* if the elements of one of them can be associated with the elements of the other in such a way that the product of arbitrary elements of the first group is associated with the product of the corresponding elements of the second group. A one-to-one correspondence between elements of two groups that has this property is called an *isomorphism.*

It is easy to see that elements of two groups that correspond to each other under an isomorphism have identical properties with respect to the group operation. Thus, under an isomorphism the neutral element, inverse elements, elements of a given order n, subgroups of one group go over, respectively, into the neutral element, inverse elements, elements of the same order, subgroups of the second group. We can therefore say that the abstract theory of groups studies only those properties of a group that are preserved under isomorphic mappings. For example, from the point of view of the abstract theory of groups the group of all permutations of four elements and the group of proper and improper motions of space that carry a fixed regular tetrahedron into itself have identical properties, because they are isomorphic. In fact, the motions in question carry the vertices of the tetrahedron again into its vertices. The number of these motions is 24. By associating with every motion the permutation of the vertices that it produces, we obtain a one-to-one correspondence between the elements of the two groups which is the required isomorphism.

A remarkable example of an isomorphic mapping is given by the theory of logarithms. By associating with every positive real number its logarithm, we obtain a one-to-one mapping of the set of positive real numbers onto the set of all real numbers. The relation $\log (xy) = \log x + \log y$ shows that this correspondence is an isomorphic mapping of the group of positive real numbers under multiplication with the group of all real numbers under addition. The practical importance of this isomorphism is well known.

Examples of nonisomorphic groups are finite groups of distinct orders. As we have already mentioned, an abstract group is determined by the law of multiplication of its elements, independent of their nature, so that distinct but isomorphic concretely given groups can be regarded as models of one and the same abstract group.

An abstract group can be given by various methods of which the most natural, at least for finite groups, is by means of the "multiplication table."

For a group of order n whose elements are written down in an arbitrary order, such a multiplication table consists of a square divided into n rows and n columns. In the cell at the intersection of the ith row and the jth column we write down the element that is the product of the element with the number i and that with the number j. This multiplication table for finite groups is sometimes called its Cayley square.

However, in practice it is almost never convenient to give a group by means of the multiplication table, because it is very clumsy.

There are other methods of giving an abstract group. One of them, namely by means of generating elements and defining relations, we have already come across. However, most frequently an abstract group is defined by giving a concrete group isomorphic to it, in particular, a transformation group.

Naturally the problem arises whether every abstract group can be regarded as a transformation group. The following theorem gives us the answer: Every group G is isomorphic to some transformation group of the set of its elements.

For let g be a fixed element of G. We denote by A_g the transformation of the set of elements of G under which to every element x of G there corresponds the element xg. The transformation A_g is one-to-one, because the equation

$$xA_g = xg = a$$

has for every given a the unique solution $x = ag^{-1}$. On the other hand, the product of group elements gh is associated with the product of the corresponding transformations $A_g A_h$, because

$$xA_{gh} = x(gh) = (xg)h = (xA_g)A_h = x(A_g A_h).$$

To the neutral element e of G there corresponds the identity, and to the inverse element g^{-1} the inverse transformation. Therefore the set Γ of all transformations corresponding to the elements of G is a transformation group isomorphic to G. It is easy to verify that if the number of elements of G is greater than 2, then the set Γ does not exhaust all the transformations of G and is only a subgroup of the "symmetric" groups of all transformations of that set.

Normal subgroups and factor-groups. Let P and Q be arbitrary collections of elements of some group G. The product of P by Q, symbolically PQ, is the name for the set of those elements of G that can be represented in the form of a product of some element of P by some element of Q. In particular, the product gP, where g is an element of G, is the set of products of g by every element of the set P.

A subgroup H of G is called a *normal* or *invariant subgroup* of G if $gH = Hg$ for every g of G. The sets of the form gH and Hg, where H is an arbitrary subgroup, are called respectively the *right and left cosets* of G with respect to H, containing the element g. Thus we can say that normal subgroups are entirely characterized by the property that for them the left and right cosets corresponding to one and the same element coincide.

If H is a normal subgroup, then the product of two cosets is again a coset, as it is easy to see, in fact: $aH \cdot bH = ab \cdot H$. The subgroup H by itself is a coset corresponding to the unit element or to any of its elements h, since $hH = H$. Multiplication of cosets is associative

$$(aH \cdot bH)cH = (ab \cdot c)H = (a \cdot bc)H = aH(bH \cdot cH).$$

The subgroup H plays the role of the neutral element in this multiplication: $H \cdot aH = eH \cdot aH = (ea)H = aH$, similarly $aH \cdot H = aH$. The coset $a^{-1}H$ is the inverse of aH, since $aH \cdot a^{-1}H = aa^{-1}H = H$. Therefore, by regarding every coset with respect to a normal subgroup as an element of a new set, we see that this set is a group under the operation of multiplication of cosets. This group is called the *factor group* of G with respect to the normal subgroup H and is denoted by G/H.

It is easy to show that for finite groups every coset with respect to an arbitrary subgroup contains as many distinct elements as the subgroup H contains and that distinct cosets have no elements in common. Hence it follows that the number of cosets of a finite group G with respect to its subgroup H is equal to the order of G divided by the order of H; and this implies the important theorem of Lagrange which states that the order of every subgroup of a finite group is a divisor of the order of a group.

From the definition of a normal subgroup it is clear that in Abelian groups every subgroup is normal. The other extreme case consists of the so-called *simple groups* in which no subgroup other than the unit subgroup and the group itself is normal. Apart from Abelian and simple groups, the solvable groups defined in §3 are also very important. It can be shown that solvable groups have a finite chain of normal subgroups $G, G_1, G_2, \cdots, G_k$, the first of which coincides with the given group G, while every subsequent one is contained in its predecessor, the last group being the unit element, and all the factor groups $G/G_1, G_1/G_2, \cdots, G_{k-1}/G_k$ being Abelian.

Homomorphisms. The concept of a factor group is very closely connected with the concept of a homomorphic mapping, which is fundamental for the whole theory of groups.

A single-valued mapping of the set of elements of a group G onto the set of elements of a group H is called a *homomorphism* or *homomorphic mapping* if the product of any two elements of the first group is mapped onto the product of the corresponding elements of the second.

If for every element x of G we denote the corresponding element of H by x', then a homomorphic mapping can be characterized by the property

$$(x_1 x_2)' = x_1' x_2' .$$

From the definitions of a homomorphism and an isomorphism it is clear that an isomorphic mapping is necessarily one-to-one, whereas a homomorphic mapping is single-valued only in one direction: To every element of G there corresponds a unique element of H, but distinct elements of G may have one and the same image in H. In a certain sense we can say that under an isomorphic mapping the group H is an accurate copy of G, but under a homomorphic mapping on transition from G to H distinct elements of G may coalesce; several elements, as it were, can be "merged" into a single element of H. However, this "coarse" nature of a homomorphic mapping is not a deficiency; on the contrary, it is a great advantage, because it enables us to use homomorphic mappings as a powerful tool in the investigation of group properties.

Homomorphic mappings make their appearance in many situations connected with transformations. For example, let us consider the symmetry group of the regular tetrahedron (figure 20). This group is isomorphic to the symmetric group of permutations of four elements, because there exists one and only one motion (of the first or second kind) that carries the vertices A_1, A_2, A_3, A_4 into any other given arrangement.

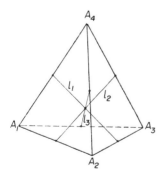

FIG. 20.

Now let us consider the lines l_1, l_2, l_3 that join the midpoints of opposite edges. Every motion that brings the tetrahedron into coincidence with itself generates a certain permutation of l_1, l_2, l_3 and every permutation of l_1, l_2, l_3 is generated by some symmetry of the tetrahedron. Clearly, the product of transformations of the tetrahedron corresponds to the product of the permutations of the lines l_1, l_2, l_3. From figure 20 it is easy to read off how the homomorphic mapping of the symmetric group of permutations of the four elements A_1, A_2, A_3, A_4 onto the symmetric group of the permutations of the three elements l_1, l_2, l_3 can be realized in a natural way. It is not difficult to find the elements of the "larger" group that are "merged" in this homomorphism.

Let us discuss a few more examples. The set of all permutations of n elements is a noncommutative group for $n > 2$. On the other hand, the numbers $+1$ and -1 also form a group under multiplication. Now let us associate with every even permutation of arbitrary n elements the number $+1$ and with every odd permutation the number -1. This gives us a homomorphic mapping of the symmetric group of permutations of n elements onto the group $\{+1, -1\}$, because according to §3 the product of permutations of equal parity is an even permutation and the product of permutations of distinct parity is an odd permutation.

Another example: If we associate with every real number $x \neq 0$ its absolute value $|x|$, then the resulting mapping of the group of positive and negative real numbers under multiplication (zero excluded) onto the group of the positive real numbers only is a homomorphism under multiplication because $|xy| = |x||y|$.

We have already mentioned that in a plane every motion of the first kind A can be represented in the form of a product of a suitable rotation V_A around a fixed point O and a certain parallel shift D_A. Rotations around the point O form a group. Therefore the correspondence $A \rightarrow V_A$ uniquely maps the group of plane motions of the first kind onto the group of rotations of the plane around the point O. Let us show that this mapping is a homomorphism. From the decompositions $A = V_A D_A$, $B = V_B D_B$ it follows that

$$AB = V_A D_A V_B D_B = (V_A V_B)(V_B^{-1} D_A V_B D_B).$$

The first parenthesis is a rotation around O and the second is the product of the transformed translation $V_B^{-1} D_A V_B$ and the translation D_B and is consequently also a translation. This shows that the product of the motions AB is associated with the product of the corresponding rotation $V_A V_B$, i.e., that the mapping in question is a homomorphism.

Finally let us show that the factor group G/N of an arbitrary group G with respect to the normal subgroup N is a homomorphic image of G.

For by associating with every element g of G the coset gN containing g, we obtain the required homomorphic mapping of G onto G/N, since the product gh corresponds to the coset ghN, which is equal to the product of the cosets gN and hN corresponding to the elements g and h.

Turning now to general properties of homomorphic mappings, let us show that the neutral element goes over under any homomorphism into the neutral element and that inverse elements go over into inverse elements.

For if e is the neutral element of G, and e' its image in H, then it follows from $ee = e$ that $e'e' = e'$ so that, if we denote by ϵ the neutral element of H, we obtain: $e' = e'e'^{-1} = \epsilon$. This proves the first statement. Now let x and y be inverse elements in G and x' and y' their images in H. From $xy = e$ it follows that $x'y' = e' = \epsilon$, i.e., that x and y are inverse elements in H and hence

$$(x^{-1})' = x'^{-1}.$$

The facts we have proved make it easy to find the image of an arbitrary product of elements in G. For example,

$$(ab^{-1}c^{-1}dh^{-1})' = a'(b^{-1})'(c^{-1})'d'(h^{-1})' = a'b'^{-1}c'^{-1}d'h'^{-1}.$$

The following theorem is fundamental for the whole theory of homomorphic mappings.

Under homomorphic mapping of an arbitrary group G onto a group H, the set N of elements of G that are mapped into the neutral element e' of H is a normal subgroup of G; the set of elements of G that are mapped into an arbitrary fixed element of H is a coset of G with respect to N, and the one-to-one correspondence so established between the cosets of G with respect to N and the elements of H is an isomorphism between H and the factor group G/N.

Let us now prove the theorem. Let a, b be arbitrary elements of N. This means that $a' = b' = e'$ where, as before, the prime denotes the images of elements of G in H. But then

$$(ab)' = a'b' = e'e' = e',$$
$$(a^{-1})' = a'^{-1} = e'^{-1} = e';$$

i.e., ab and the inverse elements a^{-1}, b^{-1} belong to N so that N is a group. Furthermore, for an arbitrary element g of G we have

$$(g^{-1}ag)' = g'^{-1}a'g' = g'^{-1}e'g' = g'^{-1}g' = e';$$

i.e., $g^{-1}ag$ lies in N for every g of G and every a of N, and from this it follows obviously that N is a normal subgroup. This proves the first statement of the theorem.

To prove the second statement we choose in G an arbitrary element g and consider the set U of all those elements u of G whose image u' coincides with the image g' of g. Suppose that $u \in gN$, i.e., $u = gn$ where $n \in N$, so that $u' = g'n' = g'e' = g'$. Therefore $gN \subset U$. Conversely, if $u' = g'$, then $(g^{-1}u)' = g'^{-1}u' = g'^{-1}g' = e'$, i.e., $g^{-1}u = n$, where n is an element of N. Hence $u = gn$ and so $U \subset gN$. From $gN \subset U$ and $U \subset gN$ it follows that $U = gN$.

Finally, the third statement of the theorem is obvious: To arbitrary cosets gN, hN of the factor group G/N there correspond in H the elements g', h', and to the product of the cosets, by the formula

$$gN \cdot hN = ghN,$$

there corresponds $(gh)' = g'h'$, as required.

The theorem on homomorphisms shows that every homomorphic image H of a group G is isomorphic to the corresponding factor group G/H. Thus, to within an isomorphism all homomorphic images of a given group G are exhausted by its distinct factor groups.

§7. Continuous Groups

Lie groups; continuous groups of transformations. The progress that was made by means of the theory of groups in the solution of algebraic equations of higher degree induced mathematicians of the middle of the last century to attempt to use the theory of groups in the solution of equations of other forms, in the first instance the solution of differential equations, which play such an important role in the applications of mathematics. This attempt was crowned with success. Although the place occupied by groups in differential equations is entirely different from their place in the theory of algebraic equations, the investigations on the application of the theory of groups to the solution of differential equations led to a substantial extension of the very concept of a group and to the creation of a new theory of the so-called continuous groups and Lie groups which have proved to be extremely important for the development of the most diverse branches of mathematics.

Whereas the groups of algebraic equations consist only of a finite number of transformations, the groups of differential equations constructed in a similar way turn out to be infinite. However, the transformations belonging to a group of a differential equation can be given by means of a finite system of parameters, and by changing the numerical values of these all the transformations of a group can be obtained. Suppose, for example, that all the transformations of the group are determined by the values of the parameters a_1, a_2, $\cdots$, a_r. When we give these parameters the values x_1, x_2, $\cdots$, x_r, we obtain a certain transformation X; by giving to the parameters new values y_1, y_2, $\cdots$, y_r we obtain another transformation Y. By hypothesis the product of these transformations $Z = XY$ also occurs in the group and hence is obtained by certain new values of the parameters

$z_1 , z_2 , \cdots, z_r$. The values z_i depend on $x_1 , x_2 , \cdots, x_r , y_1 , y_2 , \cdots, y_r$, i.e., they are certain functions of them

$$z_1 = \phi_1(x_1 , x_2 , \cdots, x_r; \ y_1 , y_2 , \cdots, y_r),$$
$$z_2 = \phi_2(x_1 , x_2 , \cdots, x_r; \ y_1 , y_2 , \cdots, y_r),$$
$$\cdots\cdots\cdots\cdots\cdots\cdots\cdots\cdots\cdots\cdots\cdots\cdots\cdots\cdots\cdots\cdots$$
$$z_r = \phi_r(x_1 , x_2 , \cdots, x_r; \ y_1 , y_2 , \cdots, y_r).$$

Groups whose elements depend continuously on the values of a finite system of parameters and whose multiplication law can be expressed by means of twice-differentiable functions $\phi_1 , \cdots, \phi_r$ are called Lie groups in honor of the Norwegian mathematician Sophus Lie who first investigated these groups.

In the first half of the 19th century, N. I. Lobačevskiĭ developed a new geometric system which now bears his name. At approximately the same time projective geometry emerged as an independent geometric system; somewhat later the geometry of Riemann was created. As a result one could enumerate in the second half of the 19th century a number of independent geometric systems that investigated from different points of view the "spatial forms of the actual world" (Engels). To comprise all these geometric systems in a single point of view, but preserving their most important qualitative differences, proved possible by means of the theory of groups.

Let us consider a one-to-one transformation of the set of points of an arbitrary geometric space that does not change those basic relations between figures that are studied in this geometry. The collection of these transformations forms a group which is usually called the group of motions or of automorphisms of the given geometry. The group of motions completely characterizes the given geometry, in view of the fact that when the group of motions is known, the corresponding geometry can be regarded as the study of those properties of the collection of points that remain unchanged under the transformations of the group. The method of classifying the various geometric systems by their groups of motions was introduced in the second half of the last century by F. Klein. This method and the various geometric systems have been treated in Chapter XVII. Here let us only mention that the groups of motions of all geometric systems that were actually investigated in the last century turned out to be Lie groups. In view of this the task of studying Lie groups assumed particular importance.

Owing to its many connections with the most diverse domains of mathematics and mechanics the theory of Lie groups has been developed energetically from its foundation right to the present time. It so happens that certain problems that have not yet been solved for finite groups were solved comparatively rapidly for Lie groups. For example, little progress has been made so far in the problem of classifying the finite simple groups (i.e., the finite groups that have no nontrivial normal subgroups), but the corresponding classification of simple Lie groups was obtained by Killing and Cartan already at the end of the last century. By developing the theory of Lie groups the Soviet mathematicians V. V. Morozov, A. I. Mal'cev, and E. B. Dynkin have found a complete solution of the important and long outstanding problem of classifying the simple *subgroups* of Lie groups. In another direction the theory of Lie groups was developed by the Soviet mathematicians I. M. Gel'fand and M. A. Naĭmark who have found the so-called continuous representations of the simple Lie groups by unitary transformations of a Hilbert space; the latter task is of particular interest for analysis and physics.

The study of Lie groups proceeds by means of the peculiar apparatus of the so-called "infinitesimal groups" or Lie algebras. These will be discussed in more detail in §13.

Topological groups. Side by side with a wide extension of the classical theory of Lie groups, in the USSR exceptional advances were achieved in the more general theory of topological or continuous groups. In contrast to the concept of a Lie group, where it is required that the elements of a group be defined by a finite system of parameters and that the multiplication rule be expressible by means of differentiable functions, the concept of a topological group is simpler and wider. A group is called *topological* if apart from the ordinary group operation a concept of proximity is defined for its elements and if the proximity of group elements implies the proximity of their products and of their inverse elements.

Originally the concept of a topological group proved to be necessary to bring order into many of the fundamental concepts of the theory of Lie groups. But later the extreme importance of this concept for other branches of mathematics was recognized. The first papers on the theory of general topological groups fall into the early twenties of our century, but the fundamental results that make it possible to speak of the creation of a new discipline were not found until the end of the twenties and the beginning of the thirties. A considerable part of them was obtained by the Soviet mathematician L. S. Pontrjagin who deservedly is regarded as one of the founders of the modern theory of continuous groups. His book "Topological groups," which was the first in the world literature to contain a comprehensive account of the theory of continuous groups, still remains the basic textbook in this domain even after twenty years.

§8. Fundamental Groups

In all the concrete examples discussed in the preceding sections, groups have usually appeared as transformation groups of one set or another. The only exceptions were the groups of numbers with respect to addition and multiplication. We now wish to analyze an important example in which the group originally arises not as a group of transformations but as a certain algebraic system with one operation.

The fundamental group. Let us consider a certain surface S and on it a moving point M. By making M run on the surface along a continuous curve joining a point A to a point B, we obtain a definite path from A to B. This path may intersect itself any number of times and may even retrace part of itself in individual sections. In order to indicate the path it is not enough to give only the curve on which the point M runs. We also have to indicate the sections that the point traverses more than once and also the direction of its passage. For example, a point may range over one and the same circle a different number of times and in different directions, and all these circular paths are regarded as distinct. Two paths with the same beginning and the same end are called equivalent if one of them can be carried into the other by continuous change. In the plane or on a sphere any two paths joining a point A to a point B are equivalent (figure 21). However, on the surface of the torus, for example, the closed paths U and V (figure 22) that begin and end at the point A are not equivalent to each other.

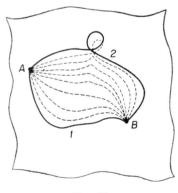

FIG. 21.

If instead of a torus we consider an infinite circular cylinder extending in both directions and take on it the path X (figure 23), then it is easy to figure out that every closed path on the cylinder beginning at A is equivalent to a path of the form X^n ($n = 0, \pm 1, \pm 2, \cdots$), where we have to understand by X^n ($n > 0$) the path X repeated n times; by X^0 the zero path consisting only of the single point A; and by X^{-n} the path X^n traversed in the opposite direction; for example, $Z \sim X^{-1}$, $Y \sim X^2$, $U \sim X^0$ (figure 23). This example shows the significance of the concept of equivalence of paths: Whereas there exists an immense set of distinct closed paths on the cylinder, all these paths reduce, to within equivalence, to the circle X traversed in one or the other direction a sufficient number of times. For $m \neq n$ the paths X^m and X^n are not equivalent.

Turning now to the discussion of an arbitrary surface, let us assume that two paths are given on it, namely a path U leading from a point A to a point B, and a path V leading from B to C. Then, by making a point run first through the path AB and then through BC we obtain a path AC which we naturally call the product of the paths $U = AB$ and $V = BC$

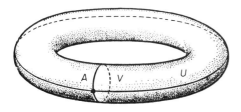

FIG. 22.   FIG. 23.

and denote by UV. If the paths U, V are equivalent to the paths U_1, V_1, respectively, then their products UV and U_1V_1 are also equivalent. The multiplication of paths is associative in the sense that if one of the products $U(VW)$ or $(UV)W$ is defined, then the other is also defined and the two products represent equivalent paths. If the moving point M is made to run through a path $U = AB$ but in the opposite direction, then we obtain the inverse path $U^{-1} = BA$ leading from B to A. The product of the path AB with its inverse path BA is a closed path equivalent to the zero path consisting only of the point A.

According to the definition we cannot multiply any two paths but only those in which the end point of the first coincides with the initial point of the second. This inadequacy disappears when we consider only closed paths starting from one and the same initial point A. Any two such paths can be multiplied and as a result we obtain again a closed path with the initial point A. Furthermore, for every closed path with initial point A its inverse path has the same properties.

Now let us agree to regard equivalent paths as distinct representations of one and the same "path," only drawn in distinct ways on the surface, and nonequivalent paths as representations of essentially distinct "paths." The remarks made previously show that then the set of all closed paths (we now omit the quotation marks) starting out from an arbitrary point A of the surface is a group under the operation of multiplication of paths. The unit (neutral) element of this group is the zero path, and the inverse element of a given path is the same path but traversed in the opposite direction.

The group of paths, in general, depends not only on the form of the surface but also on the choice of the initial point A. However, if the surface does not fall into separate pieces, i.e., if any two of its points can be joined by a continuous path lying on the surface, then the group of paths corresponding to distinct points are isomorphic and in that case we can talk simply of the group of paths of the surface S without indicating A. This group of paths of the surface is also called its *fundamental* group.

If the surface S is a plane or a sphere, then the group of paths consists of the unit element alone, because in the plane and on the sphere every path can be contracted to a point. However, on the surface of an infinite circular cylinder, as we have seen, there are closed paths that do not contract to a single point. Since on the cylinder every closed path starting from A is equivalent to a certain power of the path X (figure 23), and distinct powers of X are not equivalent, the group of paths of the cylinder surface is an infinite cyclic group. It can be shown that the group of paths on the torus (figure 22) consists of the paths of the form $U^m V^n$ ($m, n = 0, \pm 1, \pm 2, \cdots$) with $UV = VU$ and $U^m V^n = U^{m_1} V^{n_1}$ only for $m = m_1$, $n = n_1$ where we recall that in discussing the group of paths equality has to be understood in the sense of equivalence.

The importance of the group of paths is due to the following property. Let us assume that apart from the surface S another surface S_1 is given such that between the points of S and S_1 we can establish a one-to-one continuous correspondence. For example, such a correspondence is possible if the surface S_1 is obtained from S by means of a certain continuous deformation without tearing apart or fusing distinct points of the surface. To every path on the original surface S, there corresponds a path on S_1. Moreover, equivalent paths correspond to equivalent ones, the product of two paths to their product, so that the group of paths on the surface S_1 is isomorphic to the group of paths on S. In other words, the group of paths regarded from the abstract point of view, i.e., to within isomorphism, is an invariant under all possible one-to-one continuous transformations of the surface. If the group of paths of two surfaces are distinct, then the surfaces cannot be carried continuously into each another. For example, the plane cannot be deformed without fusions or tearings into the cylinder surface, because the group of paths of the plane consists of the unit element only and the group of paths of the cylinder is infinite.

Properties of figures that remain unchanged under one-to-one and bicontinuous transformations are studied in the fundamental mathematical discipline of topology, whose basic ideas have been explained in Chapter XVIII. Invariants of bicontinuous transformations are called *topological invariants.* The group of paths is one of the most remarkable examples of topological invariants. It is clear that the group of paths can be defined not only for surfaces but also for arbitrary sets of points, provided only that we can speak of paths in these sets and of their deformations.

Defining relations. In topology methods of computing the group of paths are studied in detail. As a rule it proves convenient to define these groups by a special method that is often applied in the theory of groups for the purpose of defining abstract groups in general and not only for fundamental groups in topology. It consists in the following.

Let G be a group. The elements g_1, g_2, $\cdots$, g_n are called generators of G if every element g can be represented in the form

$$g = g_{i_1}^{\alpha_1} g_{i_2}^{\alpha_2} \cdots g_{i_k}^{\alpha_k},$$

where i_1, i_2, $\cdots$, i_k are some of the numbers 1, 2, $\cdots$, n; indices i that do not stand side by side may be identical; the number of factors k is arbitrary; the exponents α_1, α_2, $\cdots$, α_k are positive or negative integers.

To know the group G it is sufficient to know, apart from the generators, also which products represent one and the same element of the group and which represent distinct elements. Thus, in order to define the group we have to list all equations of the form

$$g_{i_1}^{\alpha_1} g_{i_2}^{\alpha_2} \cdots g_{i_k}^{\alpha_k} = g_{j_1}^{\beta_1} g_{j_2}^{\beta_2} \cdots g_{j_l}^{\beta_l}$$

that hold in G. Since the set of such equations is always infinite, we usually give, instead of describing them all, only such equations as imply all the remaining ones followed by the group axioms. These equations are called defining relations.

It is clear that there are various ways of giving one and the same group by defining relations.

Let us consider, for example, the group H with the generators a, b and the relations

$$a^2 = b^3, \ ab = ba. \tag{10}$$

Setting $c = ab^{-1}$, we have

$$a = bc, \ a^2 = b^2 c^2, \ b^3 = b^2 c^2, \ b = c^2, \ a = c^3.$$

We see that all the elements of the group H can be expressed by the single element c, where

$$a = c^3, \ b = c^2.$$

Since the relations (10) follow immediately from these equations, there are no nontrivial relations for c. Therefore H is an infinite cyclic group with the generating element c.

If we can choose in a group generators that are not connected by any nontrivial relations, then the group is called *free*, and these generators are free generators. For example, if a group has the free generators a, b, then every element of it can be *uniquely* written in the form

$$a^{\alpha_0}b^{\beta_1}a^{\alpha_1}b^{\beta_2}a^{\alpha_2} \cdots b^{\beta_k}a^{\alpha_k},$$

where $k = 0, 1, 2, \cdots, n$ and the exponents $\alpha_0, \beta_1, \alpha_1, \cdots, \beta_k, \alpha_k$ are positive or negative integers except that the "extremes" α_0 and α_k can also assume the value zero. A similar statement holds for free groups with a larger number of generators.

When we write out the generators and defining relations for two groups assuming that the groups have no elements in common, then by combining these relations we obtain a new group, the so-called free product of the given ones.

The theory of free groups, and also the more general theory of free products, has an important place in the theory of groups. From the geometrical point of view the free product of the groups H_1 and H_2 is a group of paths of that figure which can be represented in the form of a sum of two closed figures that are fused at only one point and have H_1 and H_2 as their groups of paths. We know already that the group of paths of a cylinder surface is a free group with one generator. From the remark just made it follows, for example, that the group of paths of the surface illustrated in figure 24 is a free group with two generators.

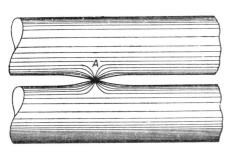

FIG. 24.

Similarly to the way in which the fundamental group of a surface was defined we can also introduce the fundamental group of spatial bodies, finite or infinite.

Knots and groups of knots. As we have already said, from the point of view of topology, two surfaces are regarded as identical if one of them can be carried into the other by a one-to-one and bicontinuous transformation. The problem of a topological classification of all closed surfaces was solved long ago. Every closed surface lying in our ordinary space is topologically equivalent either to a sphere, or to a sphere with a certain number of handles (figure 25). For example, the torus surface, illustrated in figure

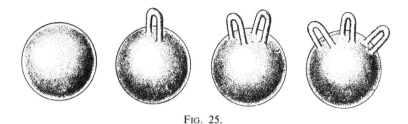

FIG. 25.

22, can be deformed continuously into a sphere with a single handle, the surface of a cube into the surface of a sphere, and so on. In view of this, the study of the fundamental groups of closed surfaces is not very interesting, since closed surfaces are completely classified even without these groups. However, there are very simple problems where so far almost nothing has been achieved without the fundamental groups. Among them is the famous problem of knots.

A knot is a closed curve lying in the ordinary three-dimensional space. As figure 26 shows, its position can be very varied. Two knots are called

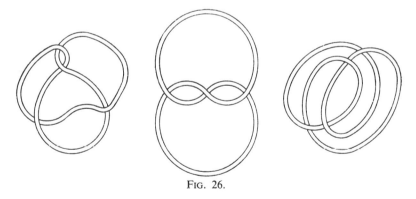

FIG. 26.

equivalent if one of them can be deformed into the other by a continuous process without breaking the curve and without self-penetration. Two problems arise at once: (1) how can we tell whether two knots given by their plane projections are equivalent or not; (2) how can we classify all nonequivalent knots?

Both problems remain as yet unsolved, but the substantial progress that has been made in a partial solution is connected with the theory of groups.

Let us remove from space the points that belong to the given knot and consider the fundamental group of the remaining set of points. This group is called the group of the knot. It is immediately obvious that if knots are equivalent, then their groups are isomorphic. Therefore, if the groups of knots are nonisomorphic, we can conclude that the knots themselves are inequivalent. For example, the group of the knot that can

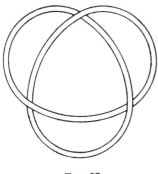

be reduced to a circle is a cyclic group, but the group of the knot that has the form of a trefoil (figure 27) is a more complicated group. The latter group is noncommutative and hence not isomorphic to the group of a circle. We can therefore state that it is impossible to deform the trefoil knot into a circle without breaking it, a fact that is completely obvious but requires a proof by precise mathematical arguments.

FIG. 27.

Unfortunately, in the discussion of the groups of knots there also arise difficult problems that have not so far been solved. The fact is that in topology very simple methods are known of finding generators and defining relations for the group of a knot represented in a given way. But in order to use groups for the comparison of distinct knots we have to be able to tell whether groups given by generators and defining relations are isomorphic or not, and a solution of this problem is not known so far. Indeed, the Soviet mathematician P. S. Novikov has recently proved the remarkable theorem that it is impossible to indicate any single regular process (more accurately, any so-called normal algorithm) by means of which it would always be possible to tell whether two given systems of defining relations for one and the same set of generators define one and the same group or not. This theorem compels us to doubt the existence of any uniform general method to decide the equivalence of knots given by their plane projections.

§9. Representations and Characters of Groups

The general theory of groups has a certain resemblance in its methods to elementary geometry: Both are founded on a definite system of axioms which is the starting point for the construction of the whole content of the theory. But the example of analytic geometry shows to what extent the application of analytic, numerical methods can prove useful in the investigation of geometrical problems.

An application of the tools of analysis and classical algebra to the theory of groups is the so-called theory of group representations. Just as analytic geometry not only gives us methods of solving geometric problems by means of analysis but, conversely, throws a geometric light on many complicated problems of analysis, so to an even higher degree the representation theory not only serves as an auxiliary apparatus for investigating properties of groups, but by forging a link between deep concepts and problems of analysis and the theory of groups enables us to find expressions for group-theoretical facts in terms of numerical relations, and to find a group interpretation for analytic relationships. A large part of the present-day important applications of the theory of groups in physics is connected precisely with the theory of representations.

Representations of groups by matrices. In linear algebra (see Chapter XVI) we have discussed the operation of multiplication for matrices. This operation is associative, but in general noncommutative. The nonsingular square matrices of a given order form a group under multiplication, since the product of two nonsingular matrices is again nonsingular, the role of the neutral element is played by the unit matrix, and for every nonsingular matrix there exists its inverse which is also nonsingular.

Let us assume that a certain group G is given and that with every element g of it there is associated a definite nonsingular matrix of complex numbers A_g of order n such that when elements of the group are multiplied the matrices corresponding to them are also multiplied: $A_{gh} = A_g \cdot A_h$. Then we say that we have a representation of the group G by matrices of degree n. Usually the words "by matrices" are omitted and we simply speak of a representation of degree n of G. A representation of degree n of a given group G is simply a homomorphic mapping of G into the group of nonsingular matrices of degree n. From the general properties of homomorphic mappings it follows that in every representation the neutral element of G goes over into the unit matrix and inverse elements in G go over into inverse matrices.

Matrices of order 1 are simply individual complex numbers. Therefore a representation of degree 1 of G is a relation under which every element of G corresponds to a complex number and the product of the elements of the group corresponds to the product of these complex numbers. For example, the mapping under which we associate with an even permutation the number 1 and with an odd one the number —1 is a representation of degree 1.

By associating with every element of a group G the unit matrix E of degree n we obtain a representation of G which is called the unit representation of degree n. If G is a finite group containing more than one element, then G must necessarily also have infinitely many representations apart from the unit representation of the varying degrees. Methods of finding them will be indicated in the following.

When we know one representation of the group G, we can obtain an infinite set of others. For let $g \to A_g$ be the given representation of G by matrices of degree n. We choose an arbitrary nonsingular matrix P of the same degree n and set $B_g = P^{-1}A_gP$. The correspondence $g \to B_g$ is again a representation of G, since

$$B_{gh} = P^{-1}A_{gh}P = P^{-1}A_gA_hP = P^{-1}A_gPP^{-1}A_hP = B_gB_h .$$

The representations so obtained from a given representation by the choice of various matrices P are called *equivalent* to the given one. In the theory of representations equivalent representations are not regarded as essentially distinct, all representations being usually considered only to within equivalence.

Another method of finding new representations is the direct addition of representations, which consists in the following: Let $g \to A_g$, $g \to B_g$ be arbitrary representations of a group G by matrices of degree m and n, respectively. We consider the mapping

$$g \to \begin{bmatrix} A_g & 0 \\ 0 & B_g \end{bmatrix}.$$

By the rule for multiplication of matrices (see Chapter XVI) we have

$$gh \to \begin{bmatrix} A_{gh} & 0 \\ 0 & B_{gh} \end{bmatrix} = \begin{bmatrix} A_gA_h & 0 \\ 0 & B_gB_h \end{bmatrix} = \begin{bmatrix} A_g & 0 \\ 0 & B_g \end{bmatrix}\begin{bmatrix} A_h & 0 \\ 0 & B_h \end{bmatrix};$$

i.e., our mapping is again a representation of G. It is called the sum of the two given representations and is denoted by $A_g + B_g$. If the summands are rearranged, then we obtain another representation

$$g \to \begin{bmatrix} B_g & 0 \\ 0 & A_g \end{bmatrix},$$

which is, however, equivalent to the given one. Therefore, if we do not distinguish between equivalent representations, the addition of representations is a commutative operation. It is easy to see that under the same condition addition is also an associative operation. Having a certain stock of representations A_g, B_g, C_g, $\cdots$ of a group G we can obtain by means of addition representations of higher and higher degrees: $A_g + B_g + C_g$, $A_g + A_g + A_g + A_g$, and so forth.

For example, the numbers $1, -1, i, -i$ form a group under multiplication. By associating every number of this group with itself, we obtain a representation of degree 1. As a second representation we can take the mapping $1 \to 1, -1 \to -1, i \to -i, -i \to i$. The sum of these representations is the mapping

$$1 \to \begin{bmatrix} 1 & 0 \\ 0 & 1 \end{bmatrix}, -1 \to \begin{bmatrix} -1 & 0 \\ 0 & -1 \end{bmatrix}, i \to \begin{bmatrix} i & 0 \\ 0 & -i \end{bmatrix}, -i \to \begin{bmatrix} -i & 0 \\ 0 & i \end{bmatrix}.$$

Transforming this by means of the matrix

$$P = \begin{bmatrix} 1 & i \\ 1 & -i \end{bmatrix}$$

we obtain the equivalent representation

$$1 \to \begin{bmatrix} 1 & 0 \\ 0 & 1 \end{bmatrix}, -1 \to \begin{bmatrix} -1 & 0 \\ 0 & -1 \end{bmatrix}, i \to \begin{bmatrix} 0 & 1 \\ -1 & 0 \end{bmatrix}, -i \to \begin{bmatrix} 0 & -1 \\ 1 & 0 \end{bmatrix}.$$

It is interesting to note that all the matrices of this representation are real. Suppose now that all the matrices of a certain representation of degree n of a group G have the form

$$g \to A_g = \begin{bmatrix} B_g & C_g \\ 0 & D_g \end{bmatrix},$$

where B_g, D_g are square matrices and the left lower rectangle of A_g is entirely filled with zeros. By multiplying matrices A_g and A_h we obtain

$$A_{gh} = A_g A_h = \begin{bmatrix} B_g B_h & B_g C_h + C_g D_h \\ 0 & D_g D_h \end{bmatrix};$$

i.e., $B_{gh} = B_g B_h$, $D_{gh} = D_g D_h$. This shows that the mappings $g \to B_g$ and $g \to D_g$ are also representations of G, but of smaller degree. Here A_g is called a graduated representation of G and every representation equivalent to it is called *reducible*. A representation that is not equivalent to any graduated representation is called *irreducible*.

If in all the matrices of A_g not only the left lower but also the right upper rectangle C_g is filled with zeros, then A_g is said to *split* into the sum of the representations B_g, D_g. A representation that is equivalent to a sum of irreducible representations is called *completely reducible*.

In the theory of groups it is proved that every representation of a finite group is completely reducible.* Hence it follows that in order to find all the representations of a finite group it is sufficient to know its irreducible representations, because all the others are equivalent to various sums of irreducible ones.

The practical computation of irreducible representations of an arbitrary finite group is, as a rule, a fairly complicated task which is solved in an explicit form only for individual classes of finite groups; for example, for commutative groups, for the symmetric groups, and for some others, though from a theoretical point of view the properties of representations of finite groups have been studied in much detail.

Every finite group has a particular "regular" representation that is constructed as follows. Suppose that $g_1, g_2, \cdots, g_n$ are the elements of the given group G numbered in an arbitrary order and that

$$g_i g_k = g_{i_k} \quad (i, k = 1, 2, \cdots, n).$$

By choosing an arbitrary fixed value for k, we form the matrix of degree n which has a 1 at the i_kth place and zeros in the remaining places ($i = 1, 2, \cdots, n$), and we denote it by R_{g_k}. The correspondence $g_k \to R_{g_k}$ ($k = 1, 2, \cdots, n$) is called the *regular representation* of G. The fact that it is a representation can be shown by simple computations.

It can also be shown that by changing the numbering of the elements of the group we arrive at an equivalent representation and that consequently, to within equivalence, every finite group has only one regular representation.

Let us briefly formulate the fundamental theorems of the theory of representations of finite groups. The number of distinct (inequivalent) irreducible representations of a finite group is finite and is equal to the number of classes of conjugate elements (see §6) of the group. The degree of an irreducible representation is necessarily a divisor of the order of the group, and the regular representation is equivalent to a sum of all inequivalent irreducible representations in which every irreducible summand is repeated as often as its degree indicates.

This implies the following interesting relation between the order of a finite group and the degrees of its irreducible representations.

We denote the number of elements of the group G by n, the number of classes of conjugate elements by k, and the degrees of the irreducible representations of G by $n_1, n_2, \cdots, n_k$, respectively. From the construction of the regular representation it is clear that its degree is n. Furthermore, since the regular representation is equivalent to a sum of n_1 representations

---

* We recall that we are considering representations of groups by matrices whose elements may be arbitrary complex numbers.

equivalent to the first irreducible representation, plus n_2 representations equivalent to the second, and so forth, and since under addition of representations their degrees are added, we must have the following equation

$$n = n_1^2 + n_2^2 + \cdots + n_k^2 . \tag{11}$$

By associating with every element of the group the number 1, we obtain the trivial irreducible representation of degree 1 which every group possesses. If in the formula (11) we take n_1 to be the degree of precisely this unit representation, then we can rewrite (11) in the equivalent form

$$n = 1 + n_2^2 + \cdots + n_k^2 ,$$

where $n_2, \cdots, n_k$ now denote the degrees of the nontrivial irreducible representations.

By using the fact that $n_2, \cdots, n_k$ must be divisors of n, we can occasionally, when k is known, find $n_2, \cdots, n_k$ from the equation (11) only. For example, the symmetric group S_3 of permutations of three elements has three classes of conjugate permutations: (1); (12), (13), (23); (123), (132). For $n = 6$, $k = 3$ the equation (11) admits only one system of solutions: $6 = 1^2 + 1^2 + 2^2$. Therefore S_3 has two distinct representations of degree 1 and one irreducible representation of degree 2.

Another example is finite Abelian groups. Here every element forms an individual class. Therefore $k = n$ and it follows from formula (11) that $n_1 = n_2 = \cdots = n_k = 1$, i.e.; all irreducible representations of these groups are of degree 1 and their number is equal to the order of the group.

The irreducible representations of Abelian groups are also called their characters, whereas for every representation of a non-Abelian group the name "character" is given to the set of the so-called traces (i.e., the sum of the diagonal elements) of the matrices forming the representation. The characters of finite groups have remarkable properties and relationships. The investigation of representations and characters of groups has enriched the theory of groups by interesting general results that have found extensive application in contemporary theoretical physics.

§10. The General Theory of Groups

We have already mentioned that almost throughout the last century the theory of groups was developed primarily as the theory of transformation groups. However, it gradually became clear that the study of groups as such was fundamental and the study of transformation groups can be reduced to that of abstract groups and their subgroups. The transition from the theory of transformation groups to the theory of abstract groups occurred first in the theory of finite groups, but the rapid development of the theory of Lie groups and the penetration of group theory into topology made it necessary to create the general theory of groups in which finite groups are regarded only as a certain special case.

The first textbook on the theory of groups in which this point of view was adopted in its full clarity was the book by O. Ju. Šmidt, which appeared in Kiev in 1916. Šmidt also obtained in the 1920's an important theorem on infinite groups which became the starting point of investigations of a number of other Soviet algebraists. Thanks to the activities of O. Ju. Šmidt and P. S. Aleksandrov, who did a great deal to popularize the ideas of contemporary algebra, a large school of group theory was formed in Moscow which later came under the leadership of their pupil, A. G. Kuroš. He became widely known, in particular, for his proof of the theorem that every subgroup of a free product is itself a free product of subgroups isomorphic to suitable subgroups of the factors and, possibly, a separate free subgroup. Later he published a monograph on the theory of groups which gave the first systematic account of the rich factual material obtained in the general theory of groups. This monograph is still the most complete textbook in the world literature on the general theory of groups and has become internationally famous.

Following the lead of the Moscow school, algebraists in Leningrad and other cities became interested in the general theory of groups and made their own contribution to its development. The researches on the theory of groups that are conducted at present in the Soviet Union comprise all its essential branches, and the results obtained by Soviet mathematicians have repeatedly exerted a decisive influence on the development of the subject.

§11. Hypercomplex Numbers

In solving practical problems by algebraic methods we usually arrive in the simplest cases at one or several equations from which the values of the unknown quantities have to be found. The unknown entities in this context are quantitative characteristics of the objects under investigation; the equations are formed by means of an analysis of the real relationships that hold between the objects.

This is the state of affairs in cases when we are dealing with the simplest quantities, such as mass, volume, or distance, that can be characterized quantitatively by a single number. However, in concrete problems we encounter objects unable to be characterized by a single number. Far from it, in the development of technology all the more important objects are of a much complicated nature and require for their characterization several numbers, even infinitely many. Even such important physical quantities as force, velocity, or acceleration are characterized by directed segments and require three numbers. Also it is well known that the position of a point in space is characterized by three numbers, the position of a plane also by three, the position of a line by four, and the position of a rigid body by six numbers. Therefore, when we wish to solve by algebraic means problems referring to more complicated objects, we obtain equations with a larger number of unknowns and often it turns out to be more tedious to analyze them than to solve the problem directly by making use of its geometric or physical peculiarities. Hence the idea naturally arose of trying to characterize more complicated objects not by systems of ordinary numbers, but by certain more complicated general numbers on which one might perform operations similar to the ordinary arithmetical operations. This statement of the problem was the more natural, since the history of science exhibited not the invariability of the concept of number but its flexibility, the gradual enrichment of the realm of numbers from the natural numbers to the fractional numbers, then to algebraic numbers, to real (rational and irrational) numbers, and finally to complex numbers.

Complex numbers. From Chapter IV the reader is already acquainted with the fundamental properties of complex numbers and their simplest applications. Here we shall only be interested in the foundation of the concept of a complex number. When we begin with a discussion of the ordinary real numbers, we notice that the square root of negative numbers has no meaning, because the square of every real number is positive or zero. One then shows that urgent needs of science compelled the mathematicians to regard expressions of the form $a + b\sqrt{-1}$ also as a special kind of number which became known as imaginary, as opposed to the ordinary real numbers. If it is assumed that these imaginary numbers are subject to the same laws of arithmetical operations as the ordinary numbers, then all square roots of negative numbers can be expressed in terms of the quantity $i = \sqrt{-1}$, and the result of arithmetical operations performed any finite number of times on real or imaginary numbers can always be represented in the form $a + bi$, where a and b are real numbers.

Clearly, this definition of imaginary numbers runs counter, in the highest degree, to common sense: First it was stated that the expressions $\sqrt{-1}$, $\sqrt{-2}$, and so forth, have no meaning, and then it was proposed that these expressions without a meaning should be called imaginary numbers. This circumstance caused many mathematicians of the 17th and 18th century to doubt the validity of the use of complex numbers. However, these doubts were dispelled at the beginning of the 19th century, when a geometrical interpretation was found for the complex numbers by points in a plane. Another purely arithmetical foundation of the theory of complex numbers was given somewhat later by the Hungarian mathematician Bolyai and the Irish mathematician Hamilton. This proceeds as follows.

Instead of the numbers $a + bi$ we shall simply speak of pairs of real numbers (a, b). Two pairs shall be regarded as equal if their first and second terms are equal, i.e., $(a, b) = (c, d)$ if and only if $a = c$ and $b = d$. Addition and multiplication of pairs are defined by the formulas

$$(a, b) + (c, d) = (a + c, b + d); (a, b) \cdot (c, d) = (ac - bd, ad + bc).$$

For example, we have

$$(2, 3) + (1, -2) = (3, 1), \qquad (2, 3)(1, -2) = (8, -1),$$
$$(3, 0) + (2, 0) = (5, 0), \qquad (3, 0)(2, 0) = (6, 0).$$

These examples show, in particular, that the arithmetical operations on pairs with a zero in the second place reduce to the same operations on their first terms, so that such pairs can be simply denoted by their first numbers. If we introduce the notation i for the pair $(0, 1)$ then we have

$$(a, b) = a(1, 0) + b(0, 1) = a + bi,$$
$$i^2 = (0, 1)(0, 1) = (-1, 0) = -1;$$

i.e., we have the usual notation for complex numbers.

Thus, from this point of view complex numbers are pairs of ordinary real numbers and operations on complex numbers are only a special kind of operations on pairs of real numbers.

Hypercomplex numbers.   Various successful applications of the complex numbers induced mathematicians as early as the first decades of the 19th century to turn their attention to the problem whether one could not construct higher complex numbers to be represented by triplets, quadruplets, and so forth, of real numbers, similar to the way in which the complex numbers are constructed in the form of pairs of real numbers. From the middle of the last century onward many distinct special systems of such higher complex numbers or hypercomplex numbers were investigated, and at the end of the last and in the first half of the present century a general theory of hypercomplex numbers was developed that has found a number of important applications in neighboring domains of mathematics and physics.

Hypercomplex number of rank n is the name for a number that can be represented by a collection of n real numbers $(a_1, a_2, \cdots, a_n)$ which for the time being we shall call its coordinates. The hypercomplex numbers $(a_1, a_2, \cdots, a_n)$ and $(b_1, b_2, \cdots, b_n)$ shall be called equal if their corresponding coordinates are equal, i.e., if $a_1 = b_1, a_2 = b_2, \cdots, a_n = b_n$. We define the operation of addition by the natural formula

$$(a_1, a_2, \cdots, a_n) + (b_1, b_2, \cdots, b_n) = (a_1 + b_1, a_2 + b_2, \cdots, a_n + b_n),$$

analogous to the formula of addition for complex numbers.

It is equally natural to introduce the operation of multiplication of a hypercomplex number by a real one: By definition we set

$$a(a_1, a_2, \cdots, a_n) = (aa_1, aa_2, \cdots, aa_n).$$

We now have to define the operation of multiplication of two hypercomplex numbers so that the result of this operation is again a hypercomplex number.

To extend the definition of multiplication of ordinary complex numbers to the general case is tedious. It can be done in various ways, and then we obtain various systems of hypercomplex numbers. Therefore, first of all we have to clarify what such a definition is to achieve. Undoubtedly it is desirable that the operations on hypercomplex numbers we are about to define should resemble in their properties the ordinary operations on real numbers. Now what are these properties of the ordinary operations?

In a careful discussion of the properties of numbers and the operations on them that are used most frequently in algebra, it is easy to observe that they reduce to the following:

1. For any two numbers, their sum is uniquely determined.

2. For any two numbers, their product is uniquely determined.

3. There exists a number zero with the property $a + 0 = a$ for every a.

4. For every number a, there exists the opposite number x satisfying the equation $a + x = 0$.

5. Addition is commutative

$$a + b = b + a.$$

6. Addition has the associative property

$$(a + b) + c = a + (b + c).$$

7. Multiplication is commutative

$$ab = ba.$$

8. Multiplication is associative

$$(ab) \cdot c = a \cdot (bc).$$

9. Multiplication is distributive

$$a(b + c) = ab + ac, \qquad (b + c)\, a = ba + ca.$$

10. For every a and every $b \neq 0$, there exists a unique number x satisfying the equation $bx = a$.

The properties 1 through 10 were selected as a result of a careful analysis; the development of mathematics in the last century proved their great importance. Nowadays every system of quantities satisfying the conditions 1 through 10 is called a *field*. Examples of fields are: the set of all rational numbers, the set of all real numbers, or the set of all complex numbers, because in each of these cases the numbers of the set can be added and multiplied and the result is a number of the same set, and the operations have the properties 1 through 10. Apart from these three very important fields we can determine infinitely many other fields formed from numbers. But beside the fields formed from numbers there is much interest in fields formed from quantities of another nature. For example, already at school we learn to operate with the so-called algebraic fractions, i.e., fractions in which the numerator and denominator are polynomials in certain letters. Algebraic fractions can be added, subtracted, multiplied, and divided, and these operations have the properties 1 through 10. Therefore, algebraic fractions form a system of objects that is a field. We could give many other examples of fields formed from quantities of a more complicated nature. In view of the importance of the properties 1 through 10 that define a field, the original formulation of the problem was to find such an operation of multiplication of hypercomplex numbers that they should form a field. In case of success one would then try to obtain new even more general complex numbers. However, already at the beginning of the last century it was discovered that this is only possible for hypercomplex numbers of rank 2 and that only the ordinary complex numbers could be obtained in this way. This result proved that the complex numbers have a very special position and that it is impossible to obtain an extension of the number system beyond the limits of the complex numbers, provided we insist that all the properties 1 through 10 are fulfilled.

Therefore, in further attempts to construct higher number systems it was necessary to omit one or several of the properties 1 through 10.

Quaternions. Historically, the first hypercomplex system that was discussed in mathematics is the system of quaternions, i.e., "fourfold numbers," which was introduced by the Irish mathematician Hamilton at the middle of the last century. This system satisfies all the requirements 1 through 10, except 7 (commutativity of multiplication).

Quaternions can be described as follows. For the quadruplets $(1, 0, 0, 0)$, $(0, 1, 0, 0)$, $(0, 0, 1, 0)$, $(0, 0, 0, 1)$ we introduce the abbreviations $1, i, j, k$. Then by the equation

$$(a, b, c, d) = a(1, 0, 0, 0) + b(0, 1, 0, 0) + c(0, 0, 1, 0) + d(0, 0, 0, 1)$$

every quaternion can be uniquely represented in the form

$$(a, b, c, d) = a \cdot 1 + b \cdot i + c \cdot j + d \cdot k.$$

The quaternion 1 will play the role of the unit of the system of quantities to be constructed; i.e., we shall assume that $1 \cdot \alpha = \alpha \cdot 1 = \alpha$ for every quaternion α. Further we set by definition: $i^2 = j^2 = k^2 = -1$;

$$ij = -ji = k,$$
$$ik = -ki = -j,$$
$$jk = -kj = i.$$

It is easy to memorize this "multiplication table" by means of figure 28 in which the points i, j, k on the circle represent the corresponding quaternions i, j, k. The product of two adjacent quaternions is equal to the third if the movement from the first sector to the second proceeds clockwise in the figure, and equal to the third with the minus sign if the motion is counterclockwise. Knowing the multiplication table for the quaternions i, j, k, we carry out the multiplication of arbitrary quaternions by using the distributive law 9. In fact:

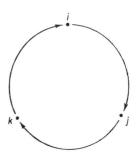

FIG. 28.

$$(a \cdot 1 + b \cdot i + c \cdot j + d \cdot k)(a_1 \cdot 1 + b_1 \cdot i + c_1 \cdot j + d_1 \cdot k)$$
$$= aa_1 \cdot 1 + ab_1 \cdot i + ac_1 \cdot j + ad_1 \cdot k$$
$$+ ba_1 \cdot i + bb_1 \cdot ii + bc_1 \cdot ij + bd_1 \cdot ik$$
$$+ ca_1 \cdot j + cb_1 \cdot ji + cc_1 \cdot jj + cd_1 \cdot jk$$
$$+ da_1 \cdot k + db_1 \cdot ki + dc_1 \cdot kj + dd_1 \cdot kk$$
$$= (aa_1 - bb_1 - cc_1 - dd_1) \cdot 1 + (ab_1 + ba_1 + cd_1 - dc_1) \cdot i$$
$$+ (ac_1 + ca_1 - bd_1 + db_1) \cdot j + (ad_1 + da_1 + bc_1 - cb_1) \cdot k.$$

The factor 1 in the first term of a quaternion is usually omitted and instead of $a \cdot 1$ we write a. The equations $ij = -ji, ik = -ki, jk = -kj$ show that the multiplication of quaternions is not commutative. The multiplicand and the multiplier are not of equal status here. Therefore, in computations with quaternions we have to adhere carefully to the order of the factors. Otherwise the operations with quaternions do not present any difficulties. In particular, the associative law 8 holds for the multiplication of quaternions. It is easily verified for the quaternions of the basis 1, i, j, k by means of the multiplication table; the transition to the general case is obvious.

The number a of the quaternion $a + bi + cj + dk$ is called its real or scalar part, and the sum $bi + cj + dk$ its vector part. The quaternions $a + bi + cj + dk$ and $a - bi - cj - dk$ that differ only in the sign of the vector part are called conjugate. Obviously, the sum of two conjugate quaternions is a real number. Furthermore, on multiplying conjugate quaternions by the previous formula we obtain

$$(a + bi + cj + dk)(a - bi - cj - dk) = a^2 + b^2 + c^2 + d^2; \quad (12)$$

i.e., the product of conjugate quaternions is also a real number.

The sum of the squares of the coefficients $a^2 + b^2 + c^2 + d^2$ of a quaternion $a + bi + cj + dk$ is called its norm. Since the square of every real number is nonnegative, the norm of every quaternion is also nonnegative, and is equal to zero only for a null quaternion.

The formula (12) shows that the product of any quaternion with its conjugate is equal to its norm.

We shall denote by an asterisk the quaternion that is conjugate to a given one. Then a direct multiplication verifies the following formula:

$$(\alpha\beta)^* = \beta^*\alpha^*.$$

This has an interesting consequence: The norm of the product of quaternions is equal to the product of the norms of the factors. For by the preceding we have

norm $(\alpha\beta) = (\alpha\beta)(\alpha\beta)^* = \alpha\beta\beta^*\alpha^* = (\alpha\alpha^*)(\beta\beta^*) =$ norm $\alpha \cdot$ norm β.

The properties of the norm enable us to give a very simple solution to the problem of division of quaternions. Let $\alpha = a + bi + cj + dk$ be an arbitrary nonzero quaternion. Then

$$(a + bi + cj + dk) \frac{1}{a^2 + b^2 + c^2 + d^2} (a - bi - cj - dk)$$

$$= \frac{1}{a^2 + b^2 + c^2 + d^2} (a^2 + b^2 + c^2 + d^2) = 1;$$

i.e., the quaternion

$$\frac{1}{a^2 + b^2 + c^2 + d^2}(a - bi - cj - dk) = \alpha^{-1}$$

is the inverse of the given quaternion α.

Having found the inverse quaternion it is now easy to find the quotient of two quaternions. For suppose that two quaternions α, β are given, the first of them being different from zero. Then the quotients obtained by dividing β by α must be the solutions of the equations

$$\alpha x = \beta, \qquad y\alpha = \beta.$$

Multiplying both sides of the first equation by the inverse quaternion α^{-1} on the left we obtain

$$x = \alpha^{-1}\beta.$$

Multiplying both sides of the second equation by α^{-1} on the right we have

$$y = \beta\alpha^{-1}.$$

Since the products $\alpha^{-1}\beta$ and $\beta\alpha^{-1}$ are in general distinct, we have to distinguish between two divisions for quaternions, on the right and on the left; both are always possible, except of course division by zero.

The algebra of vectors. Although the operations on quaternions are in many respects similar to those on complex numbers, the absence of the commutative law of multiplication makes the properties of quaternions very different from those of numbers. For example, from the algebra of complex numbers it is well known that a quadratic equation has two roots. But if we consider the quadratic equation

$$x^2 + 1 = 0$$

in the domain of quaternions, then we can easily find 6 roots: $\pm i, \pm j, \pm k$, and a more precise analysis shows that there is even an infinite number of other solutions. This circumstance strongly impedes the use of quaternions in mathematics, and notwithstanding the numerous attempts of Hamilton and other mathematicians to introduce quaternions into various branches of mathematics and physics, the role of the quaternions remains to the present day somewhat modest and can in no way be compared with the role of complex numbers.

However, quaternions have given a spur to the development of vector algebra which is an indispensable tool in modern technology and physics. The fact is that in mechanics and physics the concepts of velocity, acceleration, force, and so forth, which require three numbers for their characterization, play an essential role. Earlier we have seen that every quaternion can be regarded as an aggregate of a real number a and the vector part $bi + cj + dk$. Since the vector part of a quaternion is determined by three numbers, the most important physical quantities can be characterized by vector parts of quaternions.

Geometrically the vector part $bi + cj + dk$ of the quaternion $a + bi + cj + dk$ can be taken to represent the vector leading from the origin of a rectangular Cartesian system of coordinates to the point whose projections on the coordinate axes are equal to the numbers b, c, d, respectively. Therefore, every quaternion can be represented geometrically as an aggregate of a number and of a vector in space. Let us see how the operations on quaternions have to be interpreted.

We take two vector quaternions $xi + yj + zk$ and $x_1 i + y_1 j + z_1 k$ whose scalar parts are equal to zero. Geometrically, they are illustrated by vectors from the coordinate origin. The sum of these quaternions is again a vector quaternion $(x + x_1)i + (y + y_1)j + (z + z_1)k$. It is easy to see that the vector representing this sum is the diagonal of the parallelogram constructed on the first two vectors. Thus, the addition of vector quaternions corresponds to the well-known operation of addition of vectors by the parallelogram rule. Similarly, if we multiply a vector quaternion by an arbitrary real number, the representing quaternion vector is also multiplied by that number.

We come to a different situation when we multiply quaternions. Indeed,

$$(xi + yj + zk)(x_1 i + y_1 j + z_1 k)$$
$$= -xx_1 - yy_1 - zz_1 + (yz_1 - y_1 z)i + (zx_1 - z_1 x)j + (xy_1 - x_1 y)k;$$

i.e., on multiplying two vector quaternions we obtain a complete quaternion having a scalar part and a vector part.

The scalar part of the product of vector quaternions taken with the opposite sign is called the scalar product of the vectors representing the given quaternions, and the vector representing the vector part of the product is the vector product of the given quaternions. The scalar product of the vectors α and β is usually denoted by $(\alpha\beta)$ or simply by $\alpha\beta$, and the vector product of the same vectors by $[\alpha\beta]$. Let i, j, k be the vectors corresponding to the quaternions i, j, k, i.e., vectors of unit length lying along the coordinate axes. By definition, if $\alpha = xi + yj + zk$, and $\beta = x_1 i + y_1 j + z_1 k$, then

$$(\alpha\beta) = xx_1 + yy_1 + zz_1 , [\alpha\beta]$$
$$= (yz_1 - y_1 z)i + (zx_1 - z_1 x)j + (xy_1 - x_1 y)k.$$

By means of the latter formulas it is easy to give also a geometrical interpretation of the scalar and the vector product of vectors. As it turns out, the scalar product of two vectors is equal to the product of their lengths and of the cosine of the angle between them, and the vector product of two vectors is the vector whose length is equal to the area of the parallelogram constructed on the given vectors and whose direction is perpendicular to the plane of this parallelogram on that side from which the rotation of the first given vector toward the second looks like the rotation of the x-axis toward the y-axis as seen from the z-axis.

Nowadays in mechanics and physics we do not, as a rule, use operations on quaternions, but instead we consider only the operations on vectors, and these operations are defined in a purely geometrical manner according to the results just stated.

In conclusion, we want to point out one problem in mechanics that can be solved by means of quaternions in a particularly elegant way. Its solution was actually one of the motives for the discovery of quaternions.

Suppose that a rigid body is first rotated by a certain angle ϕ in a given direction around the definite axis OA passing through a given point O, and that it is then rotated by an angle ϕ_1 around another axis OB passing through the same point. The question is: Around what axis and by what angle must the body be rotated in order to bring it from its first position at once to the third? This is the well-known problem of mechanics on the addition of finite rotations. True, it can be solved by means of the ordinary analytic geometry, as was done already by Euler in the 18th century. However, its solution assumes a far more lucid form by means of quaternions.

Let $\xi = xi + yj + zk$ and $\alpha = a + bi + cj + dk$ be two quaternions, the first of which will be regarded as variable and the second as fixed. The expression $\alpha^{-1}\xi\alpha$ is a vector quaternion, as can easily be verified by a computation. Now if the quaternions ξ, $\alpha^{-1}\xi\alpha$ and the vector part of α are represented by the vectors $\vec{\xi}, \vec{\xi_1}, \vec{\alpha}$, then it turns out that the vector $\vec{\xi_1}$ is obtained geometrically from $\vec{\xi}$ by a rotation around the axis passing through the vector $\vec{\alpha}$ by an angle ϕ defined by the formula

$$\cos\frac{\phi}{2} = \frac{a}{\sqrt{a^2 + b^2 + c^2 + d^2}}.$$

Therefore we can say that the quaternion $\alpha = a + bi + cj + dk$ represents the rotation of space by the angle ϕ around the axis $\vec{\alpha} = bi + cj + dk$.

Conversely, knowing an axis of rotation and an angle ϕ we can look for the quaternion that represents this rotation. There is an infinite set of quaternions but they all differ from one another only by a numerical factor.

Now let us consider another rotation by an angle ϕ_1 around a certain axis $\vec{\beta} = b_1 i + c_1 j + d_1 k$. Let this rotation be represented by the quaternion $\beta = a_1 + b_1 i + c_1 j + d_1 k$. Under the action of the first rotation an arbitrary vector $\vec{\xi} = xi + yj + zk$ goes over into the vector $\alpha^{-1}\vec{\xi}\alpha$ and under the action of the second rotation the latter vector goes over into $\beta^{-1}(\alpha^{-1}\vec{\xi}\alpha)\beta$. By the associative law the latter result can be represented in the form

$$\beta^{-1}(\alpha^{-1}\vec{\xi}\,\alpha)\,\beta = (\alpha\beta)^{-1}\vec{\xi}\,\alpha\beta.$$

Since the multiplication of a vector, namely the vector quaternion $\vec{\xi}$ by the quaternion $(\alpha\beta)^{-1}$ on the left and the quaternion $\alpha\beta$ on the right, is equivalent to a rotation of this vector by the corresponding angle around the corresponding axis, we come to the conclusion that the result of two successive rotations characterized by the quaternions α and β is the rotation characterized by the product $\alpha\beta$. In other words, to the addition of the rotations corresponds the multiplication of their representing quaternions.

Apart from geometric and physical applications quaternions have found remarkable applications in the theory of numbers. Of a succession of works in this domain we must mention, in particular, the papers of Ju. V. Linnik.

§12. Associative Algebras

General definition of algebras (hypercomplex systems). We have defined hypercomplex numbers as quantities for the description of which several real numbers are required, in fact for the sake of definiteness we have regarded hypercomplex numbers simply as systems of real numbers. However, this point of view is too narrow, and for theoretical investigations the following more general definition gradually became accepted.

A certain system of quantities S is called an algebra (or a hypercomplex system) over the field P if

a. For every element a of P and every quantity α of the system S, a certain element of the system is defined, which is called the product of a and α is denoted by $a\alpha$;

b. For any two quantities α, β of the system, a certain quantity of the same system is uniquely defined which is called the sum of the first two quantities and is denoted by $\alpha + \beta$;

c. For any two quantities of the system α, β, another quantity of the same system is uniquely defined which is called the product of the first two and is denoted by $\alpha\beta$;

and if these three operations have the following properties:*

1'. $\alpha + \beta = \beta + \alpha$,

2'. $(\alpha + \beta) + \gamma = \alpha + (\beta + \gamma)$,

3'. The system S has a zero quantity θ with the property

$$\alpha + \theta = \alpha,$$

4'. $a(\alpha + \beta) = a\alpha + a\beta$,

5'. $(a + b)\alpha = a\alpha + b\alpha$,

6'. $(ab)\alpha = a(b\alpha)$,

7'. $\theta\alpha = \theta, 1 \cdot \alpha = \alpha$, where 1 is the unit element of the field P,

8'. Among the quantities of S there exist $\alpha_1, \alpha_2, \cdots, \alpha_n$, such that in terms of these every quantity of the system can be uniquely represented in the form $a_1\alpha_1 + a_2\alpha_2 + \cdots + a_n\alpha_n$,

9'. $(a\alpha)\beta = \alpha(a\beta) = a(\alpha\beta)$,

10'. $\alpha(\beta + \gamma) = \alpha\beta + \alpha\gamma$, $(\beta + \gamma)\alpha = \beta\alpha + \gamma\alpha$.

In this definition the elements of the arbitrary field P play the role that so far was played by the real numbers. From the condition 8' it is clear that every hypercomplex quantity is determined by a system of n elements $a_1, a_2, \cdots, a_n$ of P and that it can therefore, depending on the choice of P, be determined by n complex numbers, n rational numbers, n real numbers, and so forth.

The first eight postulates signify that S forms a linear finite-dimensional space (see Chapter XVI, §2) over the field P, which we shall call the ground field of the algebra.

The requirement 9' and 10' can be combined in the form of the equations

$$(a\beta + b\gamma)\alpha = a(\beta\alpha) + b(\gamma\alpha),$$
$$\alpha(a\beta + b\gamma) = a(\alpha\beta) + b(\alpha\gamma),$$

from which it follows that the operation of multiplication is linear with respect to each factor.

Of the two terms "hypercomplex systems" and "algebra" the second has been preferred in recent years, since the elements of very general "hypercomplex systems" may differ in their properties considerably from the ordinary numbers, so that it is inappropriate to call them "hypercomplex

---

* By letters of the Greek alphabet we denote arbitrary quantities of the system S, and by letters of the Latin alphabet elements of the field P.

numbers." The terms "hypercomplex systems," and "hypercomplex numbers" are now applied only to the simplest algebras, for example to the system of the ordinary quaternions.

From the requirements 1' through 10' it is clear that in algebras the commutativity and associativity of multiplication is not assumed, nor the existence of a unit element nor the possibility of "division."

Every algebra S has a basis, i.e., a system of elements α_1, α_2, $\cdots$, α_n in terms of which all the elements of the algebra can be uniquely represented in the form of linear combinations $a_1\alpha_1 + a_2\alpha_2 + \cdots + a_n\alpha_n$ with coefficients from the ground field P. Every algebra can have infinitely many bases, but the number of elements of each basis is one and the same and is called the rank of the algebra.

The system of complex numbers regarded as an algebra over the field of real numbers has a basis of the numbers 1 and i. But the pairs of numbers 2 and $3i$, 1 and $a + bi$ (a, b are real, $b \neq 0$) can also serve as bases.

Let ϵ_1, ϵ_2, $\cdots$, ϵ_n be the basis of an arbitrary algebra over a certain field P. By definition every element of the algebra can be written uniquely in the form

$$\alpha = a_1\epsilon_1 + a_2\epsilon_2 + \cdots + a_n\epsilon_n .$$

If $\beta = b_1\epsilon_1 + \cdots + b_n\epsilon_n$ is any other element of it, then by the properties 1' through 6' we have

$$\alpha + \beta = (a_1 + b_1)\,\epsilon_1 + (a_2 + b_2)\,\epsilon_2 + \cdots + (a_n + b_n)\,\epsilon_n .$$

Similarly, for every a of P we have

$$a\alpha = aa_1\epsilon_1 + aa_2\epsilon_2 + \cdots + aa_n\epsilon_n .$$

Therefore the operation of addition of quantities of the algebras and of their multiplication by elements of the field P are uniquely determined by the given formulas. The operation of multiplication of quantities of the algebra must be specially defined for each algebra; but we need not know how to multiply arbitrary quantities of the algebra, it is sufficient to know the law of multiplication of the basis quantities ϵ_i. Indeed, by the properties 9' and 10'

$$(a_1\epsilon_1 + a_2\epsilon_2 + \cdots + a_n\epsilon_n)(b_1\epsilon_1 + b_2\epsilon_2 + \cdots + b_n\epsilon_n) = \Sigma\, a_i b_j \cdot \epsilon_i\epsilon_j .$$

Each of the products $\epsilon_i\epsilon_j$ is a certain quantity of the algebra and can therefore be expressed in terms of the basis elements

$$\epsilon_i\epsilon_j = c_{ij1}\epsilon_1 + c_{ij2}\epsilon_2 + \cdots + c_{ijn}\epsilon_n .$$

Here c_{ijk} denote elements of the ground field P over which the algebra is constructed. The first index denotes the number of the first factor, the second, the number of the second factor, and the third indicates the number of that element whose coefficient is c_{ijk}. The coefficients c_{ijk} are called the structure constants of the algebra, since a knowledge of these constants completely determines all the operations on the quantities of the algebra.

It is easy to count the number of structure constants of an algebra of rank n. Every constant has three indices i, j, k. Therefore the number of structure constants of an algebra of rank n is equal to the number of triplets formed from the natural numbers $1, 2, \cdots, n$, i.e., to n^3. For example, the system of complex numbers over the field of real numbers has a basis consisting of the numbers $1, i$. In virtue of the equations

$$1 \cdot 1 = 1 \cdot 1 + 0 \cdot i, \qquad i \cdot 1 = 0 \cdot 1 + 1 \cdot i,$$
$$1 \cdot i = 0 \cdot 1 + 1 \cdot i, \qquad i \cdot i = -1 \cdot 1 + 0 \cdot i,$$

the structure constants are equal, respectively, to

$$c_{111} = a, \qquad c_{112} = 0, \qquad c_{211} = 0, \qquad c_{212} = 1,$$
$$c_{121} = 0, \qquad c_{122} = 1, \qquad c_{221} = -1, \qquad c_{222} = 0.$$

Suppose, conversely, that n^3 elements c_{ijk} of an arbitrary field P are given, indexed by triplets of the natural numbers $(i, j, k = 1, 2, \cdots, n)$. Then they can be taken as the structure constants of an algebra over the field P using the equation $\epsilon_i \epsilon_j = \sum_{k=1}^{n} c_{ijk} \epsilon_k$ as the definition of multiplication in the algebra.

Previously we have seen that every algebra has, in general, infinitely many distinct bases. The structure constants depend on the choice of the basis and therefore one and the same algebra can be given by distinct systems of structure constants.

Which algebras should be regarded as distinct and which as equal? In the theory of algebras it is convenient to regard two algebras over one and the same field P as equal if they are isomorphic, i.e., if the quantities of one algebra can be put into one-to-one correspondence with the quantities of the other in such a way that the sum and the product of any two quantities of the first algebra are associated with the sum and the product of the corresponding quantities of the second algebra and that the product of any element of the field P by an element of the first algebra is associated with the product of the same element of P and the corresponding element of the second algebra.

This definition of identity of algebras shows that in the theory of algebras we only study those properties of the quantities and systems of quantities of the algebra that find their expression in the form of certain properties of the three basic operations. To put it briefly, the theory of algebra studies properties of the operations performed on the quantities of the algebra and has nothing to do with the nature of the quantities that form the algebra.

It is easy to show that if two algebras are isomorphic, then quantities that form a basis of one algebra correspond to quantities that form a basis of the other, and that the structure constants computed with respect to corresponding bases are equal. Conversely, if two algebras over one and the same field have equal structure constants in suitable bases, then such algebras are isomorphic.

Among all the algebras, the associative algebras have always played and are still playing a very important part, i.e., the algebras in which the operation of multiplication satisfies the associative law $\alpha(\beta\gamma) = (\alpha\beta)\gamma$. The present section will give an account of the properties of such algebras. Among the nonassociative algebras the most interesting are the Lie algebras, for which the following properties of multiplication are assumed to be satisfied:

$$\alpha\beta = -\beta\alpha, \qquad \alpha(\beta\gamma) + \beta(\gamma\alpha) + \gamma(\alpha\beta) = 0.$$

They are of interest in view of the close connection that exists between Lie algebras and Lie groups, which were discussed in §7.

The algebra of matrices. We have pointed out earlier that in the first period of the development of the theory of hypercomplex systems the main attention was centered on the investigation of various systems which for one reason or another were of particular interest to the investigators. We have already examined some of these systems. The investigation of the algebra of matrices which plays a fundamental role in the general theory of algebras began approximately at the middle of the last century. Let us briefly recall here the definitions of the operations on matrices (see Chapter XVI, §1).

A matrix over a field P is a collection of elements of the field arranged in the form of a rectangular table. Two matrices are called equal if their elements in corresponding places are equal. Here we shall only consider square matrices for which the number of rows is equal to the number of columns. The number of rows or columns of a square matrix is called its order.

To add two matrices of equal order we add their corresponding elements. Multiplication of a matrix by a number is, by definition, multiplication of all the elements of the matrix by that number. The operation of multiplication of a matrix by a matrix is defined in a more complicated fashion: The product of two matrices of order n is the matrix of the same order in which the element in the ith row and the jth column is equal to the sum of the products of the elements of the ith row of the first matrix into the corresponding elements of the jth column of the second. For example:

$$\begin{bmatrix} a & b \\ a_1 & b_1 \end{bmatrix} \begin{bmatrix} x & y \\ x_1 & y_1 \end{bmatrix} = \begin{bmatrix} ax + bx_1 & ay + by_1 \\ a_1x + b_1x_1 & a_1y + b_1y_1 \end{bmatrix}.$$

The motives for the choice of this definition of multiplication of matrices were explained in Chapter XVI.

In virtue of the definitions given, matrices of order n with elements from an arbitrary field P form a system of quantities which can be added, multiplied by elements of P, and multiplied among each other. Straightforward computations show that the properties 1′ through 10′ which define an algebra are satisfied. Furthermore, it is easy to show that the multiplication of matrices satisfies the associative law. Therefore the system of all matrices of a given order n with elements from a given field P form an associative algebra over this field.

The obvious equation

$$\begin{bmatrix} a & b \\ c & d \end{bmatrix} = a \begin{bmatrix} 1 & 0 \\ 0 & 0 \end{bmatrix} + b \begin{bmatrix} 0 & 1 \\ 0 & 0 \end{bmatrix} + c \begin{bmatrix} 0 & 0 \\ 1 & 0 \end{bmatrix} + d \begin{bmatrix} 0 & 0 \\ 0 & 1 \end{bmatrix}$$

shows that the four matrices on the right-hand side form a basis of the algebra of matrices of order 2. More generally, when we denote by ϵ_{ij} the matrix in which there is a 1 in the ith row and jth column and the remaining places are zeros, then we have the equation

$$\begin{bmatrix} a_{11} & \cdots & a_{1n} \\ \vdots & & \vdots \\ a_{n1} & \cdots & a_{nn} \end{bmatrix} = \sum_{i,j} a_{ij}\epsilon_{ij},$$

which shows that the matrices ϵ_{ij} form a basis of the algebra of matrices of order n. Since the number of matrices ϵ_{ij} is equal to n^2, the rank of the algebra of matrices is also equal to n^2. The multiplication table for the basis elements ϵ_{ij} has the form

$$\epsilon_{ij} \cdot \epsilon_{jl} = \epsilon_{il}, \qquad \epsilon_{ij} \cdot \epsilon_{kl} = 0, \qquad j \neq k, \qquad i, j, k, l = 1, 2, \cdots, n.$$

The algebra of matrices contains a unit element, namely the unit matrix.

Representations of associative algebras. Suppose that with every quantity of a certain algebra A over a field P a definite quantity of some other algebra B over the same field P is associated. If the sum and product of any two elements of A are associated with the sum and product of the corresponding elements of B and the product of every element of P by an arbitrary element of A with the product of the same element of P and the corresponding element of B, then we say that the algebra A is homomorphically mapped into the algebra B. A homomorphic mapping of an associative algebra into the algebra of matrices of order n is called a representation of A of degree n. If distinct elements of A correspond to distinct matrices the representation is called faithful or isomorphic. When an algebra A is isomorphically represented by matrices, we may assume that the operations on the quantities of the algebra reduce to the operations on the corresponding matrices. Therefore the task of finding representations of algebras is of considerable interest. Here we shall only consider some of the simplest methods of finding representations; however, these methods play an important role in the general theory.

Let us choose an arbitrary basis ϵ_1, ϵ_2, $\cdots$, ϵ_n in the given associative algebra A, and let α be an arbitrary quantity of A. The products $\epsilon_1\alpha$, $\epsilon_2\alpha$, $\cdots$, $\epsilon_n\alpha$ are again quantities of A and therefore must be expressible linearly in terms of ϵ_1, ϵ_2, $\cdots$, ϵ_n. Suppose that

$$\epsilon_1\alpha = a_{11}\epsilon_1 + a_{12}\epsilon_2 + \cdots + a_{1n}\epsilon_n \,,$$

$$\epsilon_2\alpha = a_{21}\epsilon_1 + a_{22}\epsilon_2 + \cdots + a_{2n}\epsilon_n \,,$$

$$\cdots\cdots\cdots\cdots\cdots\cdots\cdots\cdots\cdots\cdots\cdots\cdots\cdots\cdots\cdots$$

$$\epsilon_n\alpha = a_{n1}\epsilon_1 + a_{n2}\epsilon_2 + \cdots + a_{nn}\epsilon_n \,.$$

As we can see, for a fixed basis we can associate with every element α a definite matrix $|\,a_{ij}\,|$. A very simple calculation shows that this correspondence is a representation of A. This representation is often called the regular representation of A. Its degree is obviously equal to the rank of the algebra.

The complex numbers can be regarded as an algebra of rank 2 over the field of real numbers with the basis 1, i. The equations

$$1 \cdot (a + bi) = a \cdot 1 + b \cdot i,$$

$$i \cdot (a + bi) = -b \cdot 1 + a \cdot i$$

show that in the corresponding regular representation the complex number $a + bi$ corresponds to the matrix

$$\begin{bmatrix} a & b \\ -b & a \end{bmatrix}.$$

The analogous representation of quaternions has the form

$$a + bi + cj + dk \rightarrow \begin{bmatrix} a & b & c & d \\ -b & a & -d & c \\ -c & d & a & -b \\ -d & -c & b & a \end{bmatrix}.$$

These representations of the complex numbers and the quaternions are faithful (i.e., isomorphic to the algebra). Examples show, however, that the regular representation is not always faithful. But if the algebra contains a unit element, then its regular representation is necessarily faithful.

It is easy to show that every associative algebra can be embedded in an algebra with a unit element. The regular representation of the containing algebra is faithful; therefore this representation of the given algebra is also faithful. Thus, every associative algebra has a faithful representation by matrices.

This method of finding representations is insufficient for constructing all the representations of an algebra. A more refined method is connected with the concept of an ideal of an algebra which plays an important role in modern mathematics.

A system I of elements of an algebra is called a right *ideal* if it is a linear subspace of the algebra and if the product of every element of I with an arbitrary element of the algebra again belongs to I. A left ideal is defined similarly (with an interchange of the order of the factors). An ideal that is simultaneously left and right is called two-sided. It is clear that the zero element of an algebra by itself forms a two-sided ideal, the so-called zero ideal of the algebra. Also the whole algebra can be regarded as a two-sided ideal. However, apart from these two trivial ideals, the algebra may contain other ideals, the existence of which is usually connected with interesting properties of the algebra.

Suppose that an associative algebra A contains a right ideal I. Let us choose a basis ϵ_1, ϵ_2, $\cdots$, ϵ_m in this ideal. Since in the general case I forms only part of A, the basis of I will, as a rule, have fewer elements than a basis of A. Let α be an arbitrary element of A. Since I is a right ideal and ϵ_1, ϵ_2, $\cdots$, ϵ_m are contained in I, the products $\epsilon_1\alpha$, ..., $\epsilon_m\alpha$ are also contained in I and hence can be expressed linearly in terms of the basis ϵ_1, $\cdots$, ϵ_m ; i.e.,

$$\epsilon_1\alpha = a_{11}\epsilon_1 + \cdots + a_{1m}\epsilon_m ,$$
$$\cdots\cdots\cdots\cdots\cdots\cdots\cdots\cdots\cdots\cdots\cdots$$
$$\epsilon_m\alpha = a_{m1}\epsilon_1 + \cdots + a_{mm}\epsilon_m .$$

By associating with the element α the matrix $\| a_{ij} \|$ we obtain, as before, a representation of the algebra A. The degree of this representation is equal to the number of elements of the basis of the ideal I and is, therefore, in general smaller than the degree of the regular representation. Obviously, the degree of the representation obtained by means of an ideal will be smallest if the ideal is minimal. Hence one can understand the fundamental role of minimal ideals in the theory of algebras.

The structure of algebras.   By what has been said, every associative algebra A can be isomorphically represented by matrices of a certain order. The aggregate of matrices that correspond in this representation to the quantities of A is itself an algebra, but only part of the algebra of all matrices of the given order. If a certain part of the quantities of an algebra is itself an algebra, then it is called a subalgebra of the given algebra. We can therefore say that every associative algebra is isomorphic to a certain subalgebra of matrices.

Although this result is of interest in principle, since it reduces the problem of finding all algebras to that of finding all possible subalgebras of matrix algebras, it does not give a direct answer to the question of the structure of algebras. The first general answer to this question was given at the end of the last century in the works of F. E. Molin (1861–1941), Professor at the University of Dorpat (Tartu), who taught at the Polytechnic Institute at Tomsk around 1900.

An algebra is called *simple* if it does not contain any two-sided ideals other than the zero ideal and the whole algebra. Molin proved that every simple associative algebra of rank 2 or more over the field of complex numbers is isomorphic to the algebra of all the matrices of a suitable order over this field.

Continuing Molin's fundamental investigations, Wedderburn obtained at the beginning of the 20th century a number of results which give a very complete description of the structure of algebras over an arbitrary field.

An arbitrary system of elements of an algebra A (in particular the algebra A itself or an ideal or a subalgebra of it) is called *nilpotent* if there exists a natural number s such that the s-th power of any element of the system is equal to zero. Every associative algebra has the unique maximal two-sided nilpotent ideal which is called the radical of the algebra. An algebra whose radical is equal to zero is called *semisimple*. It can be shown that every semisimple algebra splits into a sum of a special kind of simple algebras; in virtue of this, the study of semisimple algebras reduces entirely to that of simple ones. Finally, an algebra A is called a *divison algebra* if every equation of the form $ax = b$ ($a \neq 0$) has a solution in A.

The structure of simple algebras over the field of complex numbers is completely described by the theorem of Molin mentioned earlier. But if the ground field P is arbitrary, then the following more general theorem of Wedderburn holds: Every simple algebra of rank 2 or more over a field P is isomorphic to the algebra of all matrices of suitable order with elements from a certain division algebra over the same field P. Thus, Wedderburn's theorem reduces the problem of finding simple algebras over a given field P to that of finding division algebras over P. Over the field of complex numbers there exists only one division algebra, the field of complex numbers itself. Hence it follows by Wedderburn's theorem that all simple algebras over the field of complex numbers are isomorphic to an algebra of matrices over the same field, i.e., Molin's theorem.

Over the field of real numbers there exist only three associative division algebras: the field of real numbers itself, the field of complex numbers, and the algebra of quaternions. The proof of this statement is not very easy, and we shall not dwell on it here. By Wedderburn's theorem this implies that every simple algebra over the field of real numbers is isomorphic to the algebra of matrices of a suitable order either over the field of real numbers or over the field of complex numbers or over the quaternion algebra.

From these examples it is clear how the structure of semisimple algebras is described by the theorems of Molin and Wedderburn. In regard to algebras with a radical, for them the so-called fundamental theorem of Wedderburn is of great importance; according to this theorem, under certain restrictions to be imposed on the ground field, every algebra A with a radical R has a semisimple subalgebra L such that every element of the given algebra A can be uniquely represented in the form of a sum $\lambda + \rho$ ($\lambda \in L$, $\rho \in R$), where the subalgebra L is in a certain sense uniquely determined within A.

The fundamental theorems just formulated give an orderly idea of the possible types of associative algebras and reduce the question of their structure essentially to the analogous problem of the structure of nilpotent algebras. The theory of the latter is at present still in the process of development.

§13. Lie Algebras

In §12 we said that in addition to the theory of associative algebras at the present time the theory of Lie algebras has been worked out in great detail; for these, multiplication is subject to the rules

$$\alpha\beta = -\beta\alpha, \qquad \alpha(\beta\gamma) + \beta(\gamma\alpha) + \gamma(\alpha\beta) = 0.$$

The importance of these algebras can be explained by the fact that they are closely connected with Lie groups (see §7), i.e., with the most important class of continuous groups. As we have seen above, Lie groups play a remarkable role in contemporary geometry. Because of the origin of the theory of Lie groups and Lie algebras, the greatest interest lies in Lie algebras over the field of all real and of all complex numbers.

One of the simplest examples of a Lie algebra is the following. Let us consider the set of all square matrices of a given order n. We introduce an operation of commutation on them; by this we understand the formation of the so-called commutator $AB - BA$ of given matrices A and B, denoted by $[A, B]$.

It is easy to verify that

$$[A, B] = -[B, A],$$
$$[A, [B, C]] + [B, [C, A]] + [C, [A, B]] = 0.$$

Consequently, the set of all square matrices of a given order forms a Lie algebra with respect to the operation of commutation. It is clear that every subalgebra of the Lie algebra formed by matrices, i.e., every set of matrices that is closed with respect to the operations of addition, multiplication by a number of the ground field, and commutation, is in its turn a Lie algebra.

The question whether for every abstractly given Lie algebra there exists a matrix algebra isomorphic to it remained open for a long time. It was solved in the affirmative only in 1935 by *I. D.* Ado, a pupil of the famous algebraist H. G. Čebotarev.

Now let us sketch in general terms, without going into details and without giving rigorous statements, the connections between Lie groups and Lie algebras, restricting ourselves to the case when the Lie group and the Lie algebra are represented by matrices.

Let L be a certain Lie algebra of matrices. With every matrix A belonging to L we associate the matrix $U = e^A = E + A/1! + A^2/2! + \cdots$. Then the collection of all matrices obtained in this way forms a Lie group under the ordinary matrix multiplication. Conversely, for every Lie group we can find a unique Lie algebra (to within isomorphism) such that the group corresponding to it is isomorphic to the given one. For simplicity we have given not an accurate but a simplified formulation of the theorem on the connection between Lie groups and Lie algebras. Actually the relation $U = e^A$ exists only for U sufficiently close to the unit matrix and for A sufficiently close to the null matrix. A rigorous formulation would require the introduction of the rather complicated concepts of a local group and a local isomorphism.

Thus, the transition from the Lie algebra to the corresponding group proceeds by an operation similar to exponentiation and the inverse transition, from the group to the algebra, by an operation similar to taking logarithms.

If L coincides with the algebra of all matrices of order n, then the corresponding Lie group is the group of all nonsingular matrices, because every matrix U close to the unit matrix can be represented in the form $U = e^A$.

A matrix $A = \| a_{ij} \|$ is called skew-symmetric if its elements satisfy the relation $a_{ji} = -a_{ij}$. Skew-symmetric matrices form a Lie algebra, because if A and B are skew-symmetric, then the matrices $AB - BA = [A, B]$ and $\alpha A + \beta B$ are also skew-symmetric.

It is easy to verify that for every skew-symmetric matrix A the expression e^A is an orthogonal matrix and that every orthogonal matrix which is close to the unit matrix can be represented in this exponential form. Therefore the Lie algebra of the group of orthogonal matrices is the algebra of skew-symmetric matrices.

From analytical geometry it is known that every rotation of space around the coordinate origin is given by an orthogonal matrix and that the product of rotations corresponds to the product of the corresponding matrices. In other words, the group of rotations of space around a certain fixed point is isomorphic to the group of orthogonal matrices of order 3. Hence we deduce that the Lie algebra for the group of rotations of space is the algebra of all skew-symmetric matrices of order 3, i.e., the Lie algebra of matrices of the form

$$A = \begin{bmatrix} 0 & -a & -b \\ a & 0 & -c \\ b & c & 0 \end{bmatrix}.$$

Since each of these matrices is completely characterized by the three numbers a, b, c, it can be represented by the vector $\boldsymbol{a}$ having the projections a, b, c on the coordinate axes. Here a linear combination $\alpha A_1 + \beta A_2$ of matrices A_1 and A_2 of the given form obviously is associated with the linear combination of the corresponding vectors $\alpha \boldsymbol{a}_1 + \beta \boldsymbol{a}_2$, and the commutator of the matrices

$$[A_1, A_2] = A_1 A_2 - A_2 A_1$$

$$= \begin{bmatrix} 0 & -a_1 & -b_1 \\ a_1 & 0 & -c_1 \\ b_1 & c_1 & 0 \end{bmatrix} \begin{bmatrix} 0 & -a_2 & -b_2 \\ a_2 & 0 & -c_2 \\ b_2 & c_2 & 0 \end{bmatrix} - \begin{bmatrix} 0 & -a_2 & -b_2 \\ a_2 & 0 & -c_2 \\ b_2 & c_2 & 0 \end{bmatrix} \begin{bmatrix} 0 & -a_1 & -b_1 \\ a_1 & 0 & -c_1 \\ b_1 & c_1 & 0 \end{bmatrix}$$

$$= \begin{bmatrix} 0 & b_2 c_1 - b_1 c_2 & a_1 c_2 - a_2 c_1 \\ b_1 c_2 - b_2 c_1 & 0 & a_2 b_1 - a_1 b_2 \\ a_2 c_1 - a_1 c_2 & a_1 b_2 - a_2 b_1 & 0 \end{bmatrix}$$

is associated with the vector whose components are $b_1 c_2 - b_2 c_1$,

$a_2 c_1 - a_1 c_2$, $a_1 b_2 - a_2 b_1$, i.e., with the vector products of a_1 and a_2. So we have arrived at the remarkable result that the set of ordinary vectors under the operations of addition, multiplication by a scalar, and vector multiplication forms a Lie algebra which corresponds to the group of rotations of space around a fixed point. This shows at once how closely geometric concepts are connected with the group of rotations of space, in other words with the laws of motion of rigid bodies.

At the end of the last and the beginning of the present century, a number of results were obtained for Lie algebras that are similar to the fundamental results on associative algebras, although the proofs and statements are here more complicated. Thus, as a result of the efforts of Lie, Killing, and Cartan the concepts of a radical and of semisimplicity of a Lie algebra were successfully established at the beginning of the 20th century and all simple Lie algebras over the fields of real and complex numbers were found. In the early 1930's the theory of representations of Lie algebras by matrices was constructed, in principle, by Cartan and Weyl and proved to be a remarkable instrument for the solution of many problems. In the last 15 years the development of the theory of Lie algebras has occupied a number of Soviet mathematicians, who have obtained in this domain some significant results. In particular, they made important progress in the theory of representations of Lie algebras and gave definitive solutions to the problems of semisimple subalgebras of Lie algebras, of the structure of algebras with a given radical, and so forth.

§14. Rings

In §11 we have given the general definition of a field as an arbitrary set of elements on which the operations of addition and multiplication satisfying the postulates 1 through 10 are defined. By omitting in this definition the postulate 10, on the existence of a quotient, and the postulates 7 and 8, on commutativity and associativity of multiplication, we obtain a definition of the concept of a ring, one of the most important concepts of contemporary algebra.

Every field and also every algebra considered only with respect to the operations of addition and multiplication is a ring. An even simpler example of a ring is the set of all rational integers with the usual operations of addition and multiplication. Under the same operations the sets of numbers of the form $a + bi$, $a + b\sqrt{2}$, $a + b\sqrt[3]{2} + c\sqrt[3]{4}$ and so forth also form rings, where a, b, c are rational integers. The elements of these rings are numbers and the rings are therefore called number rings. Some important properties and applications of these rings were discussed in Chapters IV and X.

However, there exist important classes of rings whose elements are not numbers. For example, under the usual operations of addition and multiplication the sets of polynomials in given variables x_1, x_2, $\cdots$, x_n with coefficients from any fixed ring or field form rings, also the set of all continuous functions defined on a certain domain, or the set of linear transformations of a linear space or a Hilbert space.

The arithmetic properties of number rings form the subject matter of the profound theory of algebraic numbers, which lies halfway between algebra proper and number theory proper. The investigation of properties of rings of polynomials is the object of the so-called theory of polynomial ideals, which is closely connected with the higher branches of analytical geometry. Finally, rings of functions and transformations play a fundamental role in functional analysis (see Chapter XIX).

On the basis of these and some other concrete theories, the general theory of rings and the theory of topological rings were rapidly developed in the present century.

For reasons of space we shall now give only some individual results relating to the rudiments of the theory of rings.

Ideals. A subset I of elements of a ring K (not necessarily associative) is called an *ideal* if the difference of any two elements of I is again contained in I and if the products ax, xa of an arbitrary element a of I and an arbitrary element x of K are contained in I.

Every ideal of a ring is itself a ring under the operations of addition and multiplication defined in the ring. Such parts are called subrings of the given ring, so that every ideal is at the same time a subring. The converse is not true, as a rule.

The intersection of an arbitrary system of ideals of a ring is again an ideal, in particular the intersection of all the ideals containing an arbitrary fixed element a of the ring is an ideal. This is called the principal ideal generated by the element a and is denoted by (a).

The concept of the ideal generated by two or several elements is defined in the same way. It is easy to show that if an associative commutative ring has a unit element, then the ideal generated by the elements a_1, $\cdots$, a_n is simply the collection of all elements of the ring that admit a representation in the form of a sum $x_1 a_1 + \cdots + x_n a_n$, where x_1, $\cdots$, x_n are arbitrary elements of the ring. In particular, the principal ideal (a) in a commutative associative ring with a unit element is simply the collection of all elements that are multiples of a, i.e., have the form xa.

In the ring of all rational integers every ideal is principal. The ring of polynomials in a single variable with coefficients from an arbitrary field has the same property, and so has the ring of complex numbers of the form $a + bi$, where a and b are rational integers, and also a number of other rings. However, the set of all polynomials in two variables x, y without a free term is an ideal, but not a principal ideal, in the ring of all polynomials in x and y with rational coefficients.

Just as we have done above for a normal subgroup in the theory of groups, so we can construct for every ideal I of a ring K a residue class ring (or factor ring) K/I. This is done as follows. Two elements a, b, of K are called congruent modulo the ideal I, in symbols $a \equiv b(I)$, if their difference $a - b$ is contained in I. It is easy to verify that the congruence relation is symmetric, reflexive, and transitive (see Chapter XV), so that all the elements of K are split into classes of congruent ones (modulo I). If we now consider these classes as elements of a new set, we can introduce for them the concept of a sum and of a product: The "sum" of two classes shall be that class which contains the sum of any two elements that occur in the given classes respectively, and the product is that class which contains the product of these representatives. From the definition of ideals it follows that the sum and product defined in this do not in fact depend on the choice of the representatives and that as a result the set of classes becomes a ring.

The role of the residue class ring in the theory of rings is entirely analogous to the role of the factor group in the theory of groups. In particular, the construction of residue class rings of known rings is a convenient method of forming new rings with various properties. Furthermore, it is easy to show, for example, that an arbitrary commutative ring K is isomorphic to the factor ring of a ring of polynomials with integer rational coefficients in a sufficiently large number of variables.

Arithmetic properties of rings. In number rings and in fields the product of several elements can only be equal to zero if at least one of the factors is equal to zero. In arbitrary rings this need not be true, for example the product of two nonnull matrices can be the null matrix. If in a certain ring $ab = 0$ and $a \neq 0$, $b \neq 0$, then a and b are called *divisors of zero*. If there are no such elements in a ring, then the ring is called a *ring without divisors of zero*.

For the investigation of the laws of divisibility in rings, we usually assume that the ring is commutative and has no divisors of zero. Such rings are often called integral domains. The number rings and polynomial rings mentioned previously are integral domains.

Let K be an integral domain. We say that an element a is divisible in K by the element b if $a = bq$, $q \in K$. From this it follows immediately that a sum of elements divisible by b is divisible by b and that the product of several elements of K is necessarily divisible by b if one of the factors is divisible by b. When we try to introduce in the theory of rings the concept of a prime element similar to that of a prime number, we come across a complication that was already mentioned in Chapter X. Namely, to begin with we have to introduce the concept of associate elements of a ring, calling elements a, b associate if a is divisible by b and b divisible by a. Setting $a = bq_1$, $b = aq_2$, we have $ab = ab \cdot q_1 q_2$; i.e., $q_1 q_2 = e$, where e is the unit element of K. The quotients of associate elements are therefore called divisors of one or units. Every element of the domain is divisible by every unit. In the ring of rational integers the units are ± 1, in the ring of numbers of the form $a + bi$, where a, b are integers, the units are the numbers ± 1, $\pm i$.

Every element of an integral domain K has decompositions of the form $a = a\epsilon \cdot \epsilon^{-1}$, where ϵ is an arbitrary unit. These decompositions are called trivial. If a has no other decompositions, then a is called a prime or indecomposable element of K. In connection with the very important theorem on the unique decomposition of integers into prime factors, it is of interest to find such classes of rings, and among them noncommutative ones, in which a similar theorem holds. For example this theorem holds in principal ideal rings, i.e., in integral domains in which all ideals are principal.

The very concept of ideal arose in connection with the problem of uniqueness of decomposition into prime factors. Approximately at the middle of the last century the German mathematician Kummer, trying to prove the famous proposition of Fermat that the equation $x^n + y^n = z^n$ has no integer solutions for $n \geqslant 3$, had the idea of considering numbers of the form $a_0 + a_1 \zeta + \cdots + a_n \zeta^{n-1}$, where $\zeta = \cos 2\pi/n + i \sin 2\pi/n$ is a solution of the equation $x^n = 1$ and $a_0, \cdots, a_n$ are ordinary integers. The numbers of this type form an integral domain and Kummer at first took it for granted as an obvious proposition that the theorem of unique decomposition into prime factors holds in this domain. On this basis he constructed a proof of Fermat's theorem. However, in checking his arguments he observed that this assumption of the uniqueness of decomposition is not true. Wishing to preserve the uniqueness of decomposition into prime factors, Kummer was compelled to consider decompositions of numbers of the domain into factors that do not occur in the domain itself. These numbers he called ideal. Subsequently, in the construction of the general theory, mathematicians introduced instead of the ideal numbers the sets of elements of the domain that are divisible by one ideal number or another, and they were called ideals. The discovery of the nonuniqueness of the decomposition into prime factors in number rings is one of the most interesting facts found in the last century and has led to the creation of the extensive theory of algebraic numbers.

One of the most striking applications of this theorem to the problem of the decomposition of ordinary integers into a sum of squares was mentioned at the end of Chapter X. The work of mathematicians of an older generation, E I. Zolotarev, G. F. Voronoï, I. M. Vinogradov, and N. G. Čebotarev, has played a significant role in the development of the theory of number rings.

Algebraic varieties. Another source of the theory of ideals lies in algebraic geometry. When one first becomes acquainted with the theory of curves of the second order, one usually learns with astonishment that the single name hyperbola is given to the collection of two disconnected curves, namely the branches of the hyperbola, and also that a pair of straight lines is called a degenerate curve of the second order. This point of terminology is clarified in algebra: If equations of curves are considered in the form $f(x, y) = 0$, where $f(x, y)$ is a polynomial in x, y, then in the first case the left-hand side of this equation is an irreducible polynomial of the second degree, and in the second case a product of two factors of the first degree. A curve whose equation can be represented by means of an irreducible polynomial $f(x, y)$ is called irreducible, and otherwise reducible.

On transition to curves in space the matter becomes more complicated. A space curve can be represented by a system of two equations $f(x, y, z) = 0$, $g(x, y, z) = 0$, where the polynomials f and g are by no means uniquely determined by the curve. What shall we call here an irreducible curve?

The natural answer is given by the theory of ideals. Let $f_1, f_2, \cdots$ be an arbitrary set of polynomials in the variables x, y, z with complex coefficients. The set of points in the (complex) space whose coordinates make all these polynomials vanish is called the algebraic variety defined by the given polynomials. We denote this variety by M and consider all the polynomials in the variables x, y, z that vanish at every point of M. It is easy to see that the set I of all such polynomials is an ideal in the ring of polynomials in x, y, z. Moreover, this ideal has the property that if a power of some polynomial is contained in I, then the polynomial itself is contained in I. Now it turns out that, whereas distinct sets of polynomials may define one and the same algebraic variety, the correspondence between varieties and ideals with the aforementioned additional property is one-to-one.

Thus, in studying properties of varieties, it is natural to discuss not their more or less accidental "equations," but the corresponding ideals. If an ideal I can be represented in the form of the intersection of any two ideals I_1, I_2, then the variety M is the union of the varieties M_1, M_2, corresponding to the ideals I_1, I_2. Hence it is clear that a variety M must naturally be called irreducible when the corresponding ideal I cannot be represented in the form of an intersection of any two containing ideals. To the splitting of a curve into curves of lower orders, to the decomposition of a variety into irreducible ones, there now corresponds the representation of a corresponding ideal in the form of an intersection of indecomposable ones. The problem of uniqueness and possibility of such decompositions is one of the first in the theory of algebraic varieties and the general theory of ideals.

The structure of noncommutative rings. Every algebra is at the same time a ring with respect to the operations of addition and multiplication. Therefore, a considerable number of fundamental concepts and results in the theory of algebras remains valid for arbitrary rings. However, the transfer of more subtle results in the theory of algebras similar, in particular, to the theorems of Molin and Wedderburn (see §11) comes up against great difficulties which have only been partially overcome in the last 10 or 15 years. First of all there is the matter of finding such a definition of the radical of a ring that rings with a zero radical have some resemblance to semisimple algebras and that for all algebras results of the structure theory of algebras should be obtained as special cases from theorems in the theory of rings. There is at present in the theory of rings a number of definitions of a radical that enable us under some restriction or another to construct a satisfactory theory of the structure of semisimple rings. As we have mentioned earlier, the interest in the theory of noncommutative rings is stimulated to a certain extent by the very appreciable value of the theory of rings of operators in functional analysis.

§15. Lattices

As the reader is aware, a set of objects is called partially ordered if for certain pairs of its elements it can be determined which of these objects precedes the other or is subordinate to the other; here it is assumed that: (1) every object is subordinate to itself; (2) if a is subordinate to b and b subordinate to a, then it follows that a and b are identical; (3) if a is subordinate to b and b to c, then it follows that a is subordinate to c. The relation of subordination is usually denoted by the symbol $\leqslant$.

An important example of a partially ordered set is a system of all subsets of an arbitrary set where the relation of subordination means that one subset is part of another.

If the relation of subordination is defined for every pair of elements of a partially ordered set, then the set is called totally (or linearly) ordered. Ordered sets are, for example, the real numbers, where the relation $a \leqslant b$

means that a is not greater than b. By way of contrast, the partially ordered set of all parts of an arbitrary collection containing more than one element is not totally ordered, since subsets without common elements are not comparable with one another.

Suppose that the elements of a partially ordered set M have the property that every pair a, b have a unique nearest common larger element c, i.e., such that $a \leqslant c$, $b \leqslant c$ and that for every d of M satisfying the conditions $a \leqslant d$, $b \leqslant d$ we have $c \leqslant d$. Then M is called an *upper semilattice* and the element c is the "sum" of a and b. It is easy to verify that this "addition" has the following properties:

$$a + b = b + a, \qquad (a + b) + c = a + (b + c), \qquad a + a = a. \quad (13)$$

It is very remarkable that the converse can also be stated. If in a certain set an operation of addition is defined having the properties (13), then, by calling an element a subordinate to an element b if $a + b = b$, we obtain a partially ordered set in which $a + b$ is the unique nearest common larger element for a and b.

Similarly we can define lower semilattices by considering in place of the nearest larger elements the nearest smaller ones, which are here called "products"of the given elements. This operation has the same property as "addition," namely

$$ab = ba, \qquad (ab)c = a(bc), \qquad aa = a. \quad (14)$$

A partially ordered set which is at the same time an upper and a lower semilattice is called a *lattice*. By what has been explained, in every lattice we can define two operations subject to the conditions (13) and (14). However, these operations are connected with one another, since the relation $a \leqslant b$ in a lattice can be written in either of the forms $a + b = b$, $ab = a$. In other words, in lattices the equation $a + b = b$ and $ab = b$ must be equivalent. It turns out that the latter conditions can be written algebraically in the form of equations

$$a + ab = a, \qquad a(a + b) = a, \quad (15)$$

and, by virtue of this, the study of lattices becomes a purely algebraic task of studying systems with two operations subject to the conditions (13), (14), and (15). The significance of the algebraic approach to the study of lattices consists, roughly speaking, in the fact that the peculiarities of one concrete lattice or another in individual cases can be conveniently expressed in the form of algebraic relationships of one kind or another between the elements; also we can take advantage of the rich apparatus of the classical theory of groups and rings.

As we have already mentioned, the set of all subsets of a set is a partially ordered set. It is not difficult to see that it is a lattice and that the lattice sum is here the union and the lattice product is the intersection of the corresponding subsets. If we consider not all but only some of the subsets, then we can obtain a variety of lattices. For example, lattices are the set of all subgroups and also the set of all normal subgroups of an arbitrary group, the set of all subrings and the set of all ideals of an arbitrary ring and so forth. In particular, in the lattices of all normal subgroups of a group and of all ideals of a ring apart from the fundamental identities (13), (14), and (15), the following so-called modular law holds also:

$$a(ab + c) = ab + ac.$$

The theory of lattices with the modular law (modular or Dedekind lattices) is an important chapter in the general theory of lattices.

A considerable number of theorems in the theory of groups and in the theory of rings are statements on the arrangement of subgroups, normal subgroups and ideals; consequently, these theorems can be reformulated as theorems on the lattices of subgroups or ideals. With some restrictions similar theorems hold for general lattices. In this way certain important theorems were transferred from the theory of groups, the theory of rings, and other disciplines to the theory of lattices. On the other hand, the application of the apparatus of the theory of lattices proved useful in finding properties of concrete lattices, for example in the theory of groups and the theory of rings.

The theory of lattices has grown up rather recently, in the twenties and thirties of our century, and has not yet found such important applications as, say, the theory of groups. However, at the present time the theory of lattices is a well-formed mathematical discipline with a rich content and a substantial range of problems.

§16. Other Algebraic Systems

In the preceding sections we have made an attempt to give an idea how the application of algebraic methods to an ever-expanding range of problems has led to an extension of the system of objects that are studied in algebra and to a generalization of the concept of algebraic operations. In this context an important part was played by the development of the axiomatic method which arose in the work of I. N. Lobačevskiĭ on the foundations of geometry and also the development of the general theory of sets.

One of the fundamental results was the gradual clarification of the general concepts of an algebraic operation, of an algebraic system, and the accumulation of the most important facts referring to the definition of algebraic systems. Instead of the concretely defined operations of school algebra, which concern mostly numbers, modern algebra starts from the general concept of an operation. Namely, suppose that a certain system of elements S is given and also a rule that associates with every system $a_1, a_2, \cdots, a_m$ of m elements of S, taken in a definite order, a well-defined element a of the same system. Then we say that on S an m-ary operation is given and that the element a is the result of this operation performed on the elements $a_1, a_2, \cdots, a_m$. A set of elements, together with one or several operations defined on it is called an algebraic system. One of the basic tasks of algebra is the study and classification of algebraic systems. However, in this form the problem has too general a character. In fact, only certain special algebraic systems have proved at present to be really important and capable of interesting theories. For example, of the systems with a single operation only the theory of groups, to which §§1 through 10 of this chapter were devoted, has grown to a deep mathematical science, and among the systems with two or more operations those of the greatest significance are fields, algebras, rings, and lattices. However, the number of algebraic systems that are actually considered for one reason or another increases continually. At the same time certain classical branches of algebra such as, for example, the study of homomorphisms, of free systems and free unions, of direct unions, and recently the study of radicals has been successfully extended to the general theory of algebraic systems. This enables us to speak of this theory as a new branch of algebra.

In discussing the character of algebra as a whole, it is often emphasized that the complete absence or the subordinate role of the concept of continuity is a distinguishing feature of it, so that algebra is regarded as a science with a preference for the discrete. This view undoubtedly reflects one of the important objective peculiarities of algebra. In the real world the discontinuous and the continuous are found in dialectic unity. But in order to know reality, it is sometimes necessary to dissect it into parts and to study these parts separately. Therefore the one-sided attention of algebra to discrete relationships must not be regarded as a deficiency.

From the example of the theory of groups it is clear that individual algebraic disciplines provide not only the tools for technical computation but also the language for the expression of deep laws of nature. However, apart from the direct practical value of a number of branches of algebra for physics, chemistry, crystallography, and other sciences, algebra occupies one of the most important places in mathematics itself. In the words of the well-known Soviet algebraist N. G. Čebotarev, algebra has been the cradle of many new ideas and concepts that arise in mathematics and has fertilized to a remarkable extent the development of branches of mathematics that serve as a direct basis for the physical and technological sciences.

Suggested Reading

W. Burnside, *Theory of groups of finite order*, Dover, New York, 1956.

M. Hall, Jr., *The theory of groups*, Macmillan, New York, 1959.

A. G. Kurosh, *The theory of groups*, 2 vols., Chelsea, New York, 1960.

D. Montgomery and L. Zippin, *Topological transformation groups*, Interscience, New York, 1955.

L. S. Pontryagin, *Topological groups*, Princeton University Press, Princeton, N. J., 1958.

H. Weyl, *Symmetry*, Princeton University Press, Princeton, N. J., 1952.

INDEX

This revised, enlarged index was prepared through the generous efforts of Stanley Gerr.